Human Genetics

Human Genetics

Daniel L. Hartl

Washington University

School of Medicine

St. Louis, Missouri

1817

HARPER & ROW, PUBLISHERS, New York

Cambridge, Philadelphia, San Francisco,
London, Mexico City, São Paulo, Sydney

Sponsoring Editor: Claudia M. Wilson
Project Editor: Bob Greiner
Designer: Robert Sugar
Production Manager: Debi Forrest-Bochner
Photo Researcher: Mira Schachne
Compositor: Black Dot, Inc.
Printer and Binder: R. R. Donnelley & Sons Company
Art Studio: Vantage Art, Inc.

Cover: Computer-generated view from the side of double-helical DNA (B form). The color code is green=carbon, blue=nitrogen, red=oxygen, yellow=phosphorus. (Courtesy of R. Langridge, Computer Graphics Laboratory, University of California, San Francisco. Copyright © Regents, University of California.)

Library of Congress Cataloging in Publication Data
Hartl, Daniel L.
    Human genetics.

    Includes bibliographical references and index.
    1. Human genetics   I. Title.
QH431.H2957 1983   573.2′1   82–15708
ISBN 0–06–042677–2

For three superlative teachers

James F. Crow
Spencer W. Brown
Robert W. Meyer

And for Chris, Betty, and Mister B.

Contents

chapter 3/Mendel's Laws and Dominant Inheritance   64

chapter 4/Mendel's Laws and Recessive Inheritance   93

chapter 5/The Genetic Basis of Sex   125

chapter 6/Abnormalities in Chromosome Number 159

chapter 7/Abnormalities in Chromosome Structure 188

chapter 8/The Molecular Basis of Heredity 215

chapter 12/Viruses and Cancer   380

chapter 13/Immunity and Blood Groups   413

chapter 14/Population Genetics and Evolution   451

chapter 15/The Causes of Evolution   487

chapter 16/Quantitative and Behavior Genetics   521

Preface

This book grows out of my long-standing conviction that human genetics not only is a subject worth knowing for its own sake but is also an ideal theme for teaching virtually all aspects of genetics. The book is intended for courses in human genetics, of course, but it will also prove suitable for courses in basic genetics when the emphasis is to be on human genetics. I even recommend the book for some relatively rigorous courses for nonscience majors because pains have been taken to fill in the appropriate cellular and molecular background—by way of review for science majors and introduction for nonscientists.

Human genetics was at one time a rather specialized field: the human generation time is inconveniently long; pedigree and medical records were difficult to gather and collate; biochemical tests were relatively crude and required unacceptably much tissue mass to carry out; human chromosomes were difficult to study because of their relatively small size, large number, and similarity of appearance; human cells were all but impossible to maintain

for sufficient times in culture to permit their study and manipulation; and few human genes were known, and these few were difficult to localize even as to chromosome let alone position on the chromosome. In those days experimental organisms such as bacteria, yeast, fruit flies, and mice were the material of choice in genetic research. It seemed certain that virtually all important discoveries would be made first in these organisms and only later, if at all, confirmed or denied in humans.

All this has changed. Human genetics is now solidly in the mainstream of genetics, and many cardinal discoveries are made first in humans and later examined in experimental organisms. Pedigree and medical records are now computerized, permitting easy access and sophisticated analysis. Biochemical techniques have been miniaturized to the extent that many can be carried out with single cells or even parts of cells. Special methods and staining procedures have been devised that render human chromosomes not only amenable to study but also favorable for study. Modern cell culture methods permit the long-term maintenance of human cells and also their fusion with cells of other species to allow relatively precise mapping of human genes. Well over a thousand human genes are currently known, and the number continues to grow. And, of course, the sublime new procedures known collectively as recombinant DNA (not to mention the equally important development of monoclonal antibodies) have proven to be so powerful in studies of human genetics that they have accelerated the pace of change in an already fast-moving field.

Indeed, progress in human genetics is now so rapid that important new discoveries and insights will almost certainly be made between the time I put these words to paper and the time they appear in print. Struggle against it as one might, books and research papers in human genetics can quickly be dated. However, in preparing this book, I have tried to emphasize the relatively unchanging core of our understanding of human genetics while at the same time presenting the most up-to-date, state-of-the-art methods and knowledge when these seem sufficiently well established or solid. There may nevertheless be some inaccuracies or outdated conclusions, for which I am sorry.

Along with progress in human genetics on the scientific front have come enormously complex legal, moral, ethical, and social problems. What has been learned from past research and where current research seems headed are woven together in the text as much as possible. This preserves some of the vigor and excitement of today's human genetics. Problems related to wider social or ethical issues are discussed in their appropriate contexts. I have attempted to give a fair and balanced presentation of each of these controversial issues, knowing from experience that, given the facts, students will deliberate the issues and make their own judgements, which is as it should be.

A note on the book's organization and on some special features: Chapter 1 is a brief review and overview of those aspects of cell biology relevant to genetics; its focus is on the process of mitosis, and it gives a brief introduction to DNA and the manner in which genetic information is encoded in this remarkable material. Chapter 2 deals with human chromosomes and with the process of meiosis, an understanding of which is fundamental to virtually all aspects of genetics. Chapters 3 through 5 focus on simple Mendelian inheritance—dominant, recessive, and X linked. The core of Mendelian genetics is found in these chapters, along with such other subjects as somatic cell genetics and normal and abnormal sexual development. Chapters 6 and 7 concern genetics at the chromosomal level—abnormalities in chromosome number and structure.

Chapters 8 through 13 are the modern core of molecular genetics. DNA structure and recombinant DNA are included in Chapter 8. Transcription, RNA processing, and translation are the focus of Chapter 9. Chapter 10 deals with the subject of much current interest and research—gene regulation. Mutation at the molecular level, as well as transposable elements and mutagens, is the topic of Chapter 11. Chapters 12 and 13 deal with the genetics of viruses, cancer genetics, and the genetic basis of blood groups and the immune system. Monoclonal antibodies and their importance are discussed in Chapter 13. Instructors who wish to approach human genetics from the standpoint of molecular biology may wish to begin their course with Chapters 8 through 13.

Chapters 14 through 16 are a review of modern population genetics, including the amount and organization of human genetic diversity (Chapter 14), the principal causes of allele frequency change and the emergent principles of molecular evolution (Chapter 15), and an overview of quantitative genetics, including such important related issues as genetic counseling and the genetics of behavioral differences (Chapter 16).

Each chapter has a complete summary of its contents to serve as material for review and as a sort of study guide. At the end of each chapter is a list of important terms organized by subject so that related terms are found together. Also, approximately 15 problems are provided for students to verify their mastery of the subject. The first of each set pertains to some moral, social, or ethical problem that arises in human genetics, and it is designed to stimulate discussion and to show that many values come into complete or partial conflict where these issues are concerned, so much so that no completely "right" answer exists. Answers to the problems themselves are provided in full near the end of the book. In addition, each chapter is provided with an annotated list of suggestions for further reading for interested students. I have avoided a long bibliography of original research papers and the citation of such papers within the text to keep the size of the book within manageable proportions and to keep the text easy to read and uninterrupted by literature citations. Students at this level seldom care or need to know which investigator performed which studies, but I apologize to the many hundreds of scientists whose work is not overtly acknowledged. When actual data are cited in the text, however, the source is credited in full.

Many people have contributed their time and energy to make this book possible. Includ-

ed among these are the reviewers—Robert J. Huskey (University of Virginia, Charlottesville), Clifford Johnson (University of Florida, Gainesville), Ronald S. Ostrowski (University of North Carolina, Charlotte), Che-Kun James Shen (University of California, Davis), and Glenys Thomson (University of California, Berkeley)—each of whom made many important suggestions for improvement. A special thanks goes to the many physicians and scientists who provided numerous excellent photographs for illustration, particularly Ann P. Craig-Holmes, who contributed more than her share. The manuscript was beautifully prepared by Carla Oliver and Ruth Rafferty, and I wish to thank Ann Honeycutt, Laurel Mapes, Dan Dykhuizen, Nancy Scavarda, David Haymer, Raissa Berg, Jean de Framond, James Jacobson, and Antony Dean for their patience. Laurel Mapes checked the answers to all the problems, and Ann Honeycutt undertook the grueling task of proofreading with her usual competence, grace, and good humor. A special thanks also goes to the Harper & Row editorial and production staff, particularly to Bob Greiner, Claudia Wilson, Robert Sugar, Mira Schachne, and Joanne McManus. They toiled as I did to make the book attractive as well as useful. As always, Dana and Ted were often on my mind as I searched for the right words to make complex things seem clear. Finally, I wish to thank Christine for her patience and understanding while I struggled with the manuscript.

Daniel L. Hartl

Human Genetics

chapter 1
Cells and Cell Division

Genetics is the study of the fundamental units of inheritance called **genes** and of the statistical laws that govern the passage of genes from generation to generation. Genetics as a modern science began with the work of the Austrian monk Gregor Johann Mendel in the years just prior to 1866. Mendel was ahead of his time, however. His work was ignored by the scientific community. Not until 1900—16 years after Mendel's death—was his work rediscovered and its importance finally appreciated.

The science of genetics has been so successful that it has split into several distinct branches. **Molecular genetics** deals with the chemical makeup of the gene. **Cytogenetics** is concerned with the role that genes and their carriers, the chromosomes, play in the life of individual cells in the body. **Physiological genetics** is the study of how genes influence vital chemical transformations in living organisms. **Developmental genetics** is concerned with the manner in which genes control the complex processes of embryonic development. **Behavioral genetics** addresses the important and controversial issue of the extent to which genes influence the behavior of individual organisms. And **population genetics** focuses on the laws of genetics as they affect entire populations of organisms.

Each of these branches of genetics has offshoots that apply particularly to human beings, and the collection of these offshoots is the field of **human genetics.** This book is an introduction to human genetics, so its focus is on the human organism—human genes, human cells, human development, human behavior, human populations. Human genetics is an exciting scientific discipline because it is a lively, challenging field, marked by discoveries of great practical importance, such as the prenatal diagnosis of genetic disease. At the same time, these new discoveries in human genetics are often accompanied by enormously complex legal, moral, ethical, and social problems.

Human genetics is a subject of personal interest to many people, especially those who are affected with some inherited disorder or who have affected relatives. Disorders that can be traced in part to genetic factors are relatively common. Indeed, few families are entirely free of such conditions as high blood pressure, diabetes, schizophrenia, mental retardation, epilepsy, stuttering, deafness, and blindness—conditions that are caused in part by genetic factors. The occurrence of an inherited disorder in a family often evokes feelings of guilt or personal inadequacy on the part of the parents, and it often creates feelings of anxiety in the unaffected relatives because they may be carriers of the harmful genes. Such guilt and anxiety can be counteracted only by an adequate understanding of the principles of human inheritance together with the facts pertaining to the genetic basis of the particular condition in question.

Not only is human genetics of interest to each of us as individuals, but the subject also has wider social implications. In the early years of this century, genetic knowledge was misused. In those years, there developed a politically powerful **eugenics** movement—a pseudoscientific movement that aimed to "improve" the "genetic quality" of human beings. The eugenics movement demonstrated its political clout in the United States with the passage of the Immigration Restriction Act of 1924, which was designed to limit the immigration of ethnic minorities from southern and eastern Europe on the grounds that the "Nordic" types from northern and western Europe were genetically superior. In state legislatures, eugenicists pushed for the passage of laws for the forcible sterilization of "hereditary defectives," "sexual perverts," "drug fiends," "drunkards," "prostitutes," and others. In 1907, Indiana became the first state to adopt such a compulsory sterilization law, but by 1930, the majority of states had adopted such measures. Luckily, these laws were largely ignored by public officials in most of the United States. In California, however, where the eugenics movement was particularly influential, at least 10,000 men and women were involuntarily sterilized.

The eugenics movement was worldwide, with centers of activity in Japan and western Europe as well as in the United States. The lunacy got entirely out of hand in Nazi Germany, where the Eugenic Sterilization Law of 1933 ultimately led to the compulsory sterilization of at least 250,000 men and women deemed "hereditarily defective." Toward the end of the Nazi regime, compulsory sterilization was dispensed with, and millions of "undesirables" were simply killed.

In the United States and elsewhere, reputable geneticists were cowed into silence by the power of the eugenics movement and were rendered ineffectual. Ordinary citizens were largely ignorant of the principles of human genetics, so the eugenicists controlled events. Even today, a surprising amount of eugenic nonsense is found in the public press, and it is hoped that an educated populace will prevent a recurrence of the tragedies that grew out of the eugenics movement.

In short, a knowledge of human genetics serves a social function in allowing people to recognize and thwart those who would misuse genetics for their own ends and prejudices. As mentioned earlier, knowledge of human genetics also satisfies a private interest. Who has never wondered (or, indeed, worried) about his or her own personal hereditary endowment? As it happens, much of what is known about human genetics has come from the study of genetic abnormalities and diseases. As will become evident in later chapters, no family is immune from actual or potential genetic misfortune, and some of the examples discussed may remind you of situations in your own family, among your relatives, or among your friends.

Cells

To a considerable extent, the study of modern genetics involves the study of **cells**—the smallest units of living matter. Cells are living units because they take in nutrients, conduct chemical activities, move, reproduce, and do all the other things that are considered to be part of being alive. Cells are fundamental units in biology because animals, plants, and other forms of life are composed of cells. Cells are particularly important in genetics because inheritance is basically a cellular phenomenon. Sexually mature organisms form specialized reproductive cells called **gametes**. (In animals, the male gamete is the sperm and the female gamete is the egg.) Gametes from a male and female individual then unite in the process of fertilization to form a cell called the **zygote**, which undergoes successive cycles of cellular reproduction to form the embryo and ultimately the newborn organism. This process of gamete formation and fertilization is the basic one in genetics because the gametes carry the hereditary material—the genes—from generation to generation. Figure 1.1 is a photograph of a human egg cell surrounded by sperm.

Cells are of interest in modern genetics for practical reasons, too. In many cases, inherited diseases can be traced to defects in

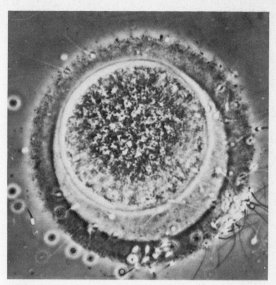

Figure 1.1   A human egg surrounded by human sperm.

certain types of cells in the body. For example, the inherited disorder **sickle cell anemia,** has many complex symptoms, including lack of stamina, susceptibility to infections, recurrent pain in skeletal joints, and often heart and kidney abnormalities. Yet all these symptoms have a single underlying cause—a defect in the oxygen-carrying protein **hemoglobin**, which is a normal constituent of the red blood cells. The seemingly unconnected symptoms of sickle cell anemia all relate to the reduced supply of oxygen to the tissues brought on by the hemoglobin defect. When the cellular basis of a disease is known, it is often possible to compensate for the defect directly rather than to settle for mere treatment of the symptoms.

Before saying more about cells, it is worthwhile to note that most cells are too small to be seen with the naked eye. A convenient measure of size for such small objects is the **micrometer** (or **micron**), abbreviated by the symbol μm (μ is the Greek letter mu). A micrometer is a unit of length equal to 0.001 mm or approximately 0.00004 in. Another convenient unit for small objects is the **ang-strom** unit, abbreviated Å, which is equal to 0.0001 μm. An angstrom unit is extremely small; 1 equals the diameter of a single hydrogen atom. The unaided human eye can resolve objects no smaller than about 80 μm (800,000 Å) in diameter. With an ordinary light microscope, which consists of a series of magnifying and focusing lenses, objects as small as about 0.5 μm (5000 Å) can be resolved. Objects smaller than 5000 Å cannot be resolved in a light microscope because the waves in which visible light travels are too long. Finer resolution requires an **electron microscope,** which uses an arrangement of magnets to focus a beam of electrons. The waves in an electron beam are much shorter than those of visible light, and an electron microscope can resolve objects as small as about 0.001 μm (10 Å).

Cells are found in a variety of shapes and sizes, although many cells in plants and animals are roughly spherical and about 10 μm in diameter. Among the smallest cells are bacteria, which range in diameter from 0.2 to 5 μm and are just at the limit of resolution of the light microscope. The largest known cell is the ostrich egg, which is 75,000 μm (about 3 inches) in diameter. The ostrich egg owes much of its size to its yolk and other stored foods for nourishment of the ostrich embryo. The size of a bacterium is in about the same ratio to the size of an ostrich egg as a thimble is to a domed football stadium. In the human body, the smallest cell is a type of white blood cell that is about the same size as a bacterium. The largest cells are certain nerve cells, which have slim projections that may extend more than 3 feet. The main body of such a nerve cell is not so enormously large, however; it is only 25 times the size of the smallest white blood cell. The human egg, one of the largest cells in the body, is spherical and about 130 μm in diameter. The egg is therefore visible to the naked eye and is roughly the size of the dot over this i. The

human sperm is shaped something like a polliwog, having a head 3 to 5 μm long and 2 to 3 μm wide and a long (30 to 50 μm) threadlike tail. Dimensions as small as these are hard to make vivid because they are so much smaller than those encountered in everyday life. It may help to note that all the sperm that gave rise to all the people that ever lived could be carried in a teaspoon; the eggs, being larger, would require a small bucket.

Human reproduction is accomplished through the production of sperm and eggs, which, recall, are collectively known as **gametes**. The genes in an individual are distributed to the gametes in a highly orderly fashion that will be discussed in Chapter 2. For now it is sufficient to note that a normal human female releases one egg about once a month from one or the other of her ovaries. Rarely will two eggs be ovulated in the same month, very rarely three, and almost never four or more—unless the ovaries have been artificially stimulated by fertility drugs or hormones. A normal male, by contrast, will expel 350 million sperm in a single ejaculate. Thus, any particular sperm is very unlikely to fertilize an egg.

When a sperm penetrates an egg in the process of fertilization (see Figure 1.1), the resulting cell is known as a **zygote**. The single-celled zygote then undergoes a process of cell division called **mitosis**, whereby it produces two daughter cells that are genetically identical. (The process of mitosis will be discussed in detail later in this chapter.) The two daughter cells produced by the first division each undergo mitosis, giving a total of four genetically identical cells. The four cells each undergo mitosis, giving a total of eight genetically identical cells. So on and on the number of cells increases. By means of signals that are still largely mysterious, the cells organize into a recognizable human embryo. Although the cells are genetically identical, they nevertheless undergo changes in form and function—a process called **differentiation**—whereby they become specialized into heart cells, liver cells, brain cells, lung cells, muscle cells, and so on. As the embryo grows, cell divisions continue. At birth, the human infant consists of about 1 trillion (a million million) cells. This would correspond to roughly 40 doublings of the fertilized egg and its products if the cells in the embryo had all divided in unison. They do not divide synchronously, however. Some of the cells lag behind others. Some cease division entirely. Therefore, it is only approximately true to say that about 40 cell doublings occur between fertilization and birth. The growth of a child into an adult is partly due to an increase in the number of cells. A fully grown adult has about 10 trillion cells, each genetically identical with all the others. Some of the cells in an adult are specialized to form gametes, and so the cycle of life continues.

The Architecture of Cells

In addition to wide-ranging sizes and numbers, cells are also found in a variety of shapes (see Figure 1.2 for several examples). Often we picture cells as spheres, but this does an injustice to their variety. The nerve cell has long, thin projections; the muscle cell resembles an earthworm; and the pigment cells (found just beneath the outer layer of skin and containing the pigments responsible for skin color) look very much like a splotch of ink. Some cells can change their shape. The pigment cell can expand and contract, which leads to different intensities of skin color; muscle cells are long and thin when relaxed, short and fat when contracted. The variety of cell shapes in the human body only hints at the many and often bizarre shapes found among other forms of life.

The cell may be thought of as a little package of specialized structures and molecules that carry out the functions of life in a coordinated fashion. The wrapping on the package is a thin film called the **cell membrane.** This membrane is a flexible structure about 0.01 μm thick, so its detailed structure can be seen only through the electron microscope. The cell membrane does not completely isolate the cell from the world outside. It has holes or channels in it that allow water and other small molecules to pass through easily. Large molecules and important charged atoms or molecules known as **ions**, such as sodium ions, cannot easily get through the membrane, however. Energy-requiring "pumps" in the cell membrane move large molecules and ions in and out, although the nature of these pumps is still not understood. For a long time, the cell membrane was thought to be a sandwich of fatty lipid molecules between two sheets of protein molecules. Recent findings, however, suggest that bits of protein may actually float around in a thin sea of lipid.

Most cells manufacture substances that surround or coat the cell membrane. These are usually too porous to hinder the movement of even very large molecules, but they do serve to protect and support the cells. Plant cells have a rigid outer coating called the **cell wall,** which is made up of a fibrous material called **cellulose**. Wood is mostly cellulose, and cotton is more than 90 percent pure cellulose.

Inside, the cell contains not only molecules but also internal structures of various kinds that carry out specialized functions or activities. These intracellular structures are called **organelles**, or little organs. About a dozen different classes of organelles are found in a typical cell (see Figure 1.3 for illustrations and Table 1.1 for a list).

One organelle, the **nucleus**, is of particular importance in genetics because it contains molecules that carry the hereditary information in the cell—the genetic information that makes the cells of a particular species unique.

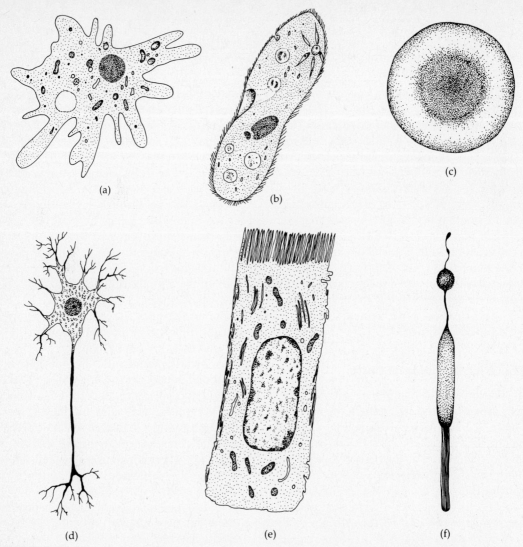

Figure 1.2 Specialized cells of living organisms have a great variety of shapes: *(a)* a single-celled amoeba, *(b)* a paramecium, *(c)* a red blood cell, *(d)* a nerve cell, *(e)* an intestinal cell, and *(f)* a rod cell from the retina of the eye.

The nucleus carries the hereditary information that gives the cell its identity as a human cell, a toad cell, an elm cell, or an apple cell. Genetic information in the nucleus is present in molecules of a substance known as **DNA** (stands for **deoxyribonucleic acid**). The nucleus of non-dividing cells contains strands of a material called **chromatin**, which is dispersed throughout the nucleus and, because it is more dense in some regions than others, gives the nucleus an uneven or granular appearance. The chief chemical constituent of chromatin is DNA. In

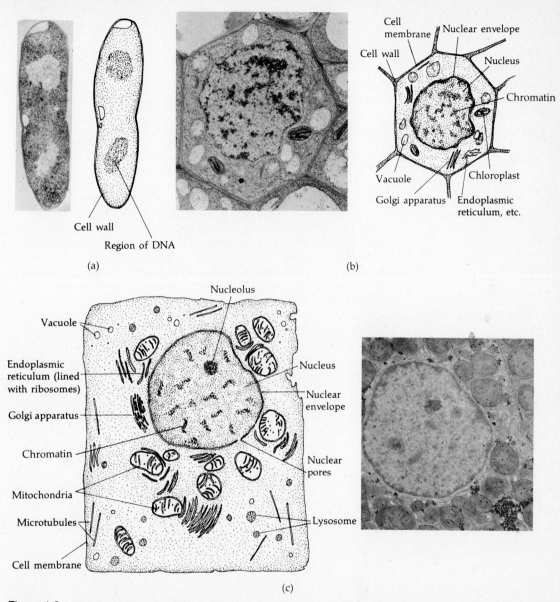

(a)

Cell wall

Region of DNA

Cell
membrane   Nuclear envelope

Cell wall

Nucleus

Chromatin

Vacuole   Chloroplast

Golgi apparatus   Endoplasmic
reticulum, etc.

(b)

Nucleolus

Vacuole

Endoplasmic
reticulum (lined
with ribosomes)

Golgi apparatus

Chromatin

Mitochondria

Microtubules

Cell membrane

Nucleus

Nuclear
envelope

Nuclear
pores

Lysosome

(c)

Figure 1.3 Micrographs and drawings of *(a)* a prokaryotic bacterium, *(b)* a eukaryotic plant cell, and *(c)* a eukaryotic animal cell. Note the absence of a nuclear envelope in the prokaryotic cell. Eukaryotic cells are conveniently considered to consist of two parts—the *nucleus,* which contains the chromosomes (although only chromatin is visible at this stage of the cell's life cycle), and the *cytoplasm,* which includes everything outside the nucleus. The photos are: *(a)* the human intestinal bacterium *Escherichia coli* (20,000X); *(b)* a cell from the stem of a pea seedling, *Pisum sativum* (7000X); and *(c)* a bat liver cell (6000X).

TABLE 1.1 FUNCTIONS OF ORGANELLES AND SOME OTHER STRUCTURES FOUND IN EUKARYOTIC CELLS.

Nucleus

Nuclear envelope	Membranous boundary of nucleus
Nuclear pores	Holelike gaps in nuclear envelope; function uncertain
Nucleolus	Dense spherical body within nucleus; site of synthesis of certain constituents of ribosomes (see below)
Chromatin	Contains genetic material DNA (deoxyribonucleic acid) and other constituents; becomes organized into visible **chromosomes** at certain stages in dividing cells

Cytoplasm

Ribosomes	Tiny spherical particles containing special proteins and RNA (ribonucleic acid); synthesis of all proteins occurs on ribosomes; frequently found in chains of 5 to 10 ribosomes called **polysomes,** actively engaged in protein synthesis
Endoplasmic reticulum (ER)	Network of extensively folded internal membranes; two principal categories are **rough ER** (surface peppered with ribosomes) and **smooth ER** (free of ribosomes)
Golgi apparatus	Collection of smooth membranous structures of characteristic appearance near the nucleus; important in storage and secretion of certain substances
Mitochondria	The "power plants" of eukaryotic cells; site of chemical transformations resulting in the high-energy molecule ATP (adenosine triphosphate), which is a source of energy for other chemical transformations in the cell
Chloroplasts (plants only)	Site of photosynthesis, in which the energy of sunlight is used to combine carbon dioxide and water into the sugar glucose
Lysosomes	Membranous sacs for storage of powerful digestive enzymes
Vacuoles	Membranous sacs serving as storage reservoirs of diverse substances; some contain only water, others dissolved sugars, still others pigments or engulfed food particles
Microtubules and microfilaments	Cablelike aggregates of special proteins; important in cell rigidity, movement, and (particularly microtubules) cell division
Centrioles (animals only)	Bodies with a characteristic arrangement of 27 microtubules; important in chromosome movement during cell division
Cilia and flagella (some cells only)	Short hairlike (cilia) or long whiplike (flagella) projections from the cell surface; important in cell movement; flagellated cells usually have one or two such projections, whereas cilia are numerous
Cell membrane	Membranous outer boundary of cell
Cell wall (plants only)	Rigid boxlike structure enclosing cell membrane; composed primarily of network of tough cellulose molecules

dividing cells, the chromatin strands undergo a little-understood process of coiling and so become visible through the light microscope as discrete entities called **chromosomes.**

The nucleus is the control center of the cell because hereditary instructions in the nucleus are "read" and they trigger the rest of the cell to manufacture certain things or to carry out certain tasks. For this reason, it is customary to consider the cell as consisting of two major regions—the nucleus and the cytoplasm. The **cytoplasm** includes everything in the cell other than the nucleus. The distinction between nucleus and cytoplasm turns out to be useful because most cells have just one nucleus, and this is set off from the cytoplasm by its own membranous **nuclear envelope.** Large molecules can move between the nucleus and the cytoplasm, very likely by means of **nuclear pores** in the nuclear envelope.

Bacteria and blue-green algae have no well-defined nucleus. This creates the major distinction that divides all cells, and indeed all forms of life, into two groups. Organisms represented by cells with a nucleus are called **eukaryotes** (see Figure 1.3b and c); relatively primitive cells such as bacteria and blue-green algae with no nuclear envelope surrounding the hereditary material are called **prokaryotes** (see Figure 1.3a). In accord with their comparatively simple structure, prokaryotic cells lack most of the complex organelles associated with eukaryotic cells. Moreover, cell division in prokaryotes is quite different from that in eukaryotes. Although the genetic material in prokaryotes is DNA, as it is in eukaryotes, the DNA in prokaryotic cells is not organized into chromatin, so prokaryotic cells do not possess true chromosomes. Nevertheless, the DNA in prokaryotic cells is often called a "chromosome," although this terminology is, strictly speaking, incorrect.

Next in prominence to the nucleus in most eukaryotic cells are the **mitochondria**. These organelles range in size from 0.2 to 3 μm and in shape from spheres to sausages. An active cell like a liver cell may contain up to 1000 mitochondria. The covering of the mitochondrion is a double membrane, each layer like the cell membrane, and the inner layer is wrinkled or corrugated, with its numerous folds extending into the interior of the mitochondrion (Figure 1.4a). This folding provides a large surface area inside the mitochondrion to support the chemical processes that occur there—vital processes that generate most of the energy in the cell. (Many of the chemical processes in a cell take place on the surface of membranes.) Thus, the mitochondrion may be thought of as the cell's power plant, converting the energy in the chemical bonds of various substances into a supercharged chemical bond that is stored in a molecule known by the initials **ATP** (stands for **adenosine triphosphate**). In loose terms, the mitochondrion "cascades" an electron down a special sequence of molecules, and at every step the electron gives up a pulse of energy to form a molecule of ATP. Finally the electron has very little usable energy left, and it is incorporated into a molecule of water. The process of releasing energy to ATP in the mitochondrion is known as **cellular respiration.**

The storage battery of the cell is the high-energy bond in ATP. In the same way that some of the energy required to compress a spring is regained when the spring jumps back, some of the energy required to make a chemical bond between atoms is regained when the bond is cleaved. The high-energy bond of ATP is unusual in that it releases more energy than most chemical bonds when broken. All chemical bonds release some energy when broken, but the high-energy bond of ATP provides the energy for most cellular activities.

The ultimate source of energy in the chemical bonds of biological molecules is the sun. Green plants are almost alone in being able to convert the energy in sunlight into chemical bonds usable to living things; green plants are the original source of energy for almost all other forms of life, including humans. The cellular locations of the energy-binding process in eukaryotic plant cells are in organelles called **chloroplasts**, and the process is known as **photosynthesis**. Chloroplasts are present in only certain cells of green plants; they make green plants green. Found in a variety of shapes, often disk-shaped particles 5 to 8 μm in diameter, chloroplasts contain, among other things, the green pigments collectively called **chlorophyll** (see Figure 1.4b). Chlorophyll has an electron that becomes very excited (it gains energy) when struck by sun-

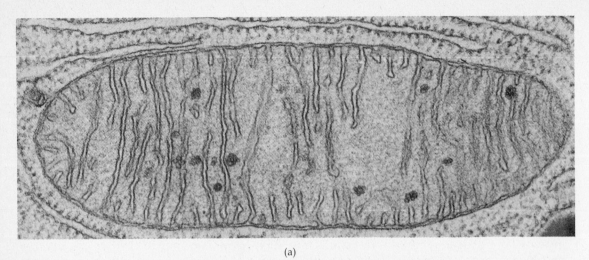

(a)

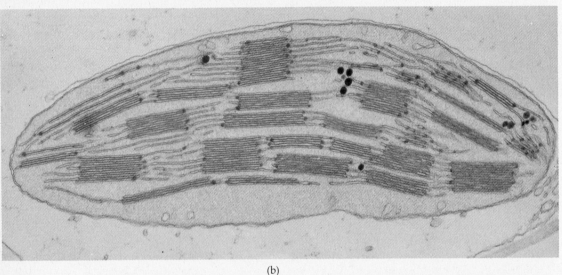

(b)

Figure 1.4   Micrographs of *(a)* a mitochondrion, found in all eukaryotic cells—this one is from a bat pancreatic cell (61,600X); and *(b)* a chloroplast, found only in plant cells—this one is from a mature lettuce leaf (25,550X).

light. The chloroplast can then convert the trapped energy in this excited electron into the chemical bonds of molecules. The chloroplast, therefore, does roughly the opposite of the mitochondrion. Whereas the chloroplast acquires energy by building chemical bonds dur-ing photosynthesis, the mitochondrion releases the energy by breaking them down during respiration.

Mitochondria and chloroplasts contain small ring-shaped molecules of DNA, very similar to the DNA in prokaryotic cells. These organelles also contain ribosomes (particles on which proteins are manufactured) and certain other constituents that are very similar to their prokaryotic counterparts. The contribution of organelle DNA to the cell's heredity is minor, however; the DNA in chromosomes in the cell nucleus is vastly more important in heredity than is the DNA in mitochondria and chloroplasts. Moreover, the bulk of protein synthesis occurs on ribosomes located in the cytoplasm, not on ribosomes in these organelles. Nevertheless, many biologists believe that the ancestors of mitochondria and chloroplasts were once prokaryotic organisms in their own right, not parts of cells, and that the DNA and ribosomes in these organelles today are vestiges of this ancient past. As evolution proceeded, these primitive organisms came to be associated with cells and, along the way, lost their ability to survive independently. The eukaryotic cell, in turn, has become dependent on them for its own survival; the mitochondrion is necessary for respiration and the chloroplast for photosynthesis.

Convoluted throughout the cytoplasm of a eukaryotic cell is the **endoplasmic reticulum,** a maze of membranes that fold back and overlap, providing a large surface area that supports the numerous chemical reactions that occur on them (see Figure 1.3). A second membranous structure in the cell, this one commonly found in cells that produce large amounts of material to be secreted, is the **Golgi apparatus.** This resembles a stack or series of folds of membranes, and it can be distinguished from the endoplasmic reticulum by its regular, seemingly organized folds. The Golgi apparatus evidently serves as a reservoir of materials to be secreted from the cell.

Ribosomes are small granular particles often found in strings of five or six or on the interior surface of parts of the endoplasmic reticulum. These tiny, almost spherical structures are the workers on the cellular assembly lines. Only 0.015 μm in diameter and numbering in the hundreds of thousands in a typical eukaryotic cell, ribosomes assemble the countless protein molecules needed for specific chemical reactions to occur in the cell. (In general, one particular kind of protein is able to aid one particular reaction, although some proteins, rather than accelerating chemical reactions, become parts of intracellular structures.) Back in the nucleus, the cell usually contains one or more dense areas, each called a **nucleolus** (see Figure 1.3). The nucleolus is involved in the production of the **RNA** (stands for **ribonucleic acid**) constituent of ribosomes. (Ribosomes also contain protein constituents.) Even prokaryotic cells contain ribosomes; a typical prokaryotic cell may contain 20,000 to 30,000 of them.

Among the thousands of proteins produced in cells are some that form the **microtubules** and others that form the **microfilaments**. These are long cablelike aggregates of protein that provide structural support for the cell. The cell is evidently able to assemble these supporting structures whenever they are needed and to disassemble them later. A most important function of microtubules is to form the football-shaped network of fibers that arches through the cell during the middle phases of cell division and guides the movement of the chromosomes; this network is called the **spindle**. The microtubules that make up the spindle in animal cells are organized by a pair of organelles that lie just outside the nuclear envelope; these organizers are known as **centrioles** (see Figure 1.3). It is curious that plant cells seem to lack centrioles, although they do have spindles during cell division.

Eukaryotic cells also contain various types of membranous sacs with diverse functions. **Lysosomes** are spherical sacs about the size of mitochondria that appear in animal cells and perhaps in plant cells as well. They contain powerful enzymes that can digest large molecules, even whole organelles. A lysosome may digest a mitochondrion that is no longer functioning, and certain white blood cells use their lysosomes to digest bacteria they have engulfed. If the digestive enzymes inside the lysosomes were released into the surrounding cytoplasm of the cell, the cell would probably be killed. Indeed, cells that are injured or dead are often ultimately destroyed by the release of the enzymes in their own lysosomes.

Many cells also contain globules or blebs of various kinds and sizes called **vacuoles**. These may contain water, pigments, stored food, substances brought into the cell or waste products packaged for transport out of the cell.

Movement is often considered a characteristic of life, and some cells need to move objects or to move about themselves. They sometimes accomplish this by means of small appendages or hairlike structures that can move. The cells in the throat and in the lining of the bronchial tubes leading to the lungs have small hairs called **cilia**, which normally have a back-and-forth brushlike movement that occurs in waves, sweeping out trapped dust, dirt, and small organisms that are inhaled. Another example of a movable cellular appendage is the tail of a sperm, which whips and propels the sperm through its liquid surroundings. Such long whiplike appendages are known as **flagella**.

Figure 1.5 shows a three-dimensional rendition of a "typical" animal cell, and it illustrates the spatial relationships among the cellular components listed in Table 1.1. The cell is obviously quite complex, and something has to coordinate its activities. The coordinator of the eukaryotic cell is the nucleus. More precisely, it is the genetic or hereditary material in the nucleus—the celebrated molecule known as DNA. One key to understanding the cell—and the whole organism—is to understand the nature and behavior of the genetic material in the nucleus.

Cell Division

As noted earlier, the genetic material resides in the chromosomes found only within the nucleus of eukaryotic cells. In this section, we focus on the behavior of chromosomes during cell division. Cell division is a central aspect of genetics because chromosomes are transmitted from one generation to the next by means of cell division.

Actually, two types of cell division occur in humans and most other eukaryotes. Figure 1.6 is an outline of the human life cycle showing the role played by each type of division. One type of division, which occurs only in the ovaries of females and the testes of males, is concerned exclusively with the production of gametes (eggs or sperm). This gamete-producing type of division is known as **meiosis**, and one of its principal features is that it reduces the number of chromosomes from 46 (the characteristic number in most human cells) to 23 (the number in human gametes). Gametes are said to carry a **haploid** complement of chromosomes; the zygote and all subsequent stages of the life cycle are said to carry the **diploid** complement of chromosomes. Details of how the chromosome number is reduced during meiosis will be discussed in the next chapter. For now it is sufficient to note that the fertilization of a 23-chromosome egg by a 23-chromosome sperm produces a 46-chromosome zygote; thus, fertilization is the process that restores the proper chromosome number in each generation.

The other type of cell division in eukaryotes is called **mitosis**. By means of mitosis, a single parent cell gives rise to two genetically identical daughter cells. Although the

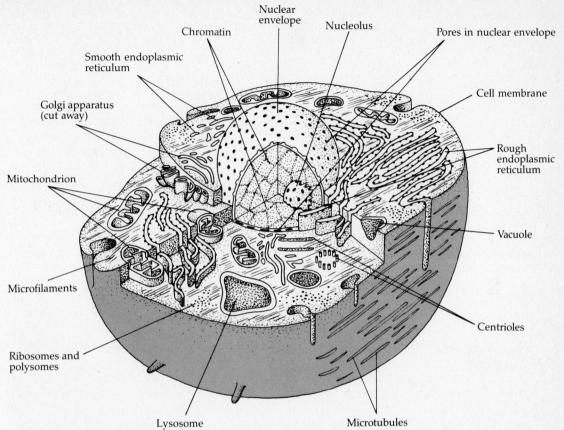

Chromatin

Nuclear envelope

Nucleolus

Pores in nuclear envelope

Smooth endoplasmic reticulum

Cell membrane

Golgi apparatus (cut away)

Rough endoplasmic reticulum

Mitochondrion

Vacuole

Microfilaments

Centrioles

Ribosomes and polysomes

Lysosome

Microtubules

Figure 1.5   Three-dimensional drawing of a "typical" eukaryotic animal cell showing the spatial organization of the various cellular components. Note the extensive membrane structures within the cytoplasm.

term *mitosis* is technically used to denote the division of the nucleus, it will be convenient for our purposes to use the term to include the division of the cytoplasm as well.

Mitosis is the normal manner of cell division through which growth, repair, and replacement of the body's cells and tissues take place. Not all cells in the body of an adult are capable of undergoing mitosis, however. The cells of certain tissues lose their ability to divide in the embryo or child when they become differentiated by acquiring or expressing specializations in structure or function that make them different from other cells. Nerve cells and muscle cells are examples of highly differentiated cells that do not undergo mitosis in the adult. On the other hand, some types of cells in adults do undergo mitosis. Rapid cell division in the body occurs in a number of cell types, including cells in the bone marrow and in the spleen. Cells in the liver and kidney represent an intermediate situation; they are capable of undergoing division, but there are comparatively long time intervals between successive divisions.

The cells in the embryo are very different

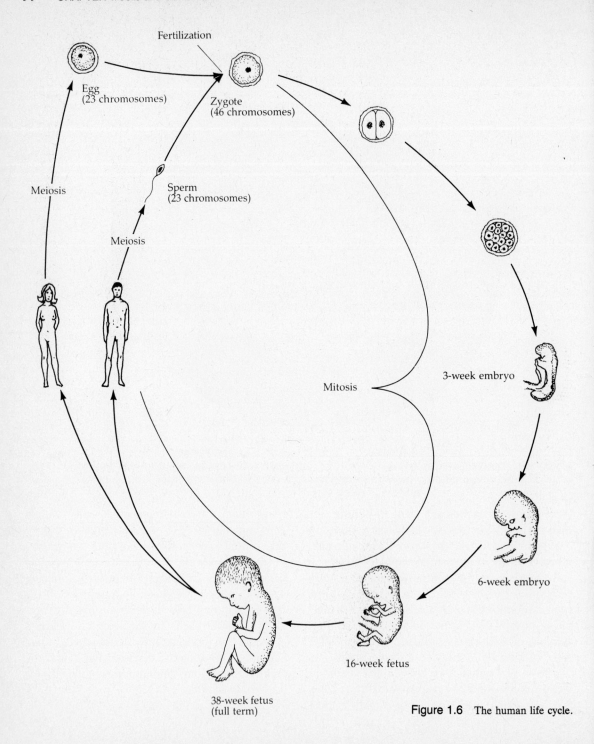

Fertilization

Egg
(23 chromosomes)

Zygote
(46 chromosomes)

Meiosis

Sperm
(23 chromosomes)

Meiosis

Mitosis

3-week embryo

6-week embryo

16-week fetus

38-week fetus
(full term)

Figure 1.6 The human life cycle.

from those in the adult. In the embryo, especially in very young embryos, practically every cell divides again and again very rapidly with only short pauses for growth between divisions. Mitosis is the type of cell division by which the number of cells in the embryo increases. It is a matter of great practical importance that cells undergoing mitosis are highly sensitive to damage by radiation and certain chemicals; for this reason, embryos tend to be much more severely damaged than adults by harmful drugs and radiation along with other forms of environmental stress. This extreme sensitivity of the embryo is one reason why tests of drug safety carried out in animals should include careful examination of the offspring of females who were given the drug when pregnant. The extreme sensitivity of the embryo decreases as the embryo ages, as the number of differentiated cells increases, and as the rate of cell division declines.

Mitosis

In this section, we take a closer look at mitosis, paying particular attention to the behavior of chromosomes during mitosis. Mitosis is a cellular *process*, so the events that occur at any one moment blend imperceptibly into those that occur at the next moment. Nevertheless, for purposes of discussion, mitosis is usually described in terms of landmark stages known as *interphase, prophase, metaphase, anaphase,* and *telophase*. These landmark stages are illustrated by the sketches in Figure 1.7 and in the photographs of nuclei of cells of the desert locust *Schistocerca gregaria* in Figure 1.8. Each stage of mitosis warrants a brief discussion, and repeated reference will be made to these illustrations.

As is apparent in Figure 1.7, mitosis is a cyclical process, with the products of one division being able to recycle into yet another division. It will be convenient for discussion to break into the cycle at **interphase**, as this is the stage of mitosis mainly concerned with cellular growth. During interphase, the cell usually doubles, or nearly doubles, in size; it increases the number or amount of virtually all the cytoplasmic components—mitochondria, ribosomes, endoplasmic reticulum, and so on—but not the nucleus. The nucleus remains single. During this period, however, within the nucleus, the cell makes an exact copy or replica of each chromosome.

Chromosome Replication, as the chromosome-copying process is called, is not apparent in the nuclei of cells in interphase because at this stage the chromosomes are not visible through the light microscope. All that can be seen in the nucleus is one or more nucleoli and the diffuse granular substance called *chromatin* mentioned earlier (see Figures 1.7 and 1.8). In the electron microscope, however, the chromosomes are seen as slender, very thin, coiled rods that meander throughout the nucleus, and the tips of the chromosomes are attached to the inner surface of the nuclear envelope.

During interphase, the cell produces DNA, RNA, protein, lipids, and several other kinds of molecules, and it grows rapidly as the various constituents of the cytoplasm, like mitochondria and ribosomes, are increasing in number. The production of molecules of certain kinds may be restricted to specific places in the cell or to particular times. Almost all DNA synthesis, for example, occurs during the so-called **S period**, which occupies about the middle third of interphase. Very little DNA synthesis occurs outside the nucleus or outside this interval. The synthesis of DNA is part of chromosome replication, during which the cell makes an exact copy of each of its chromosomes. Not every chromosome is replicated at precisely the same time, however; certain

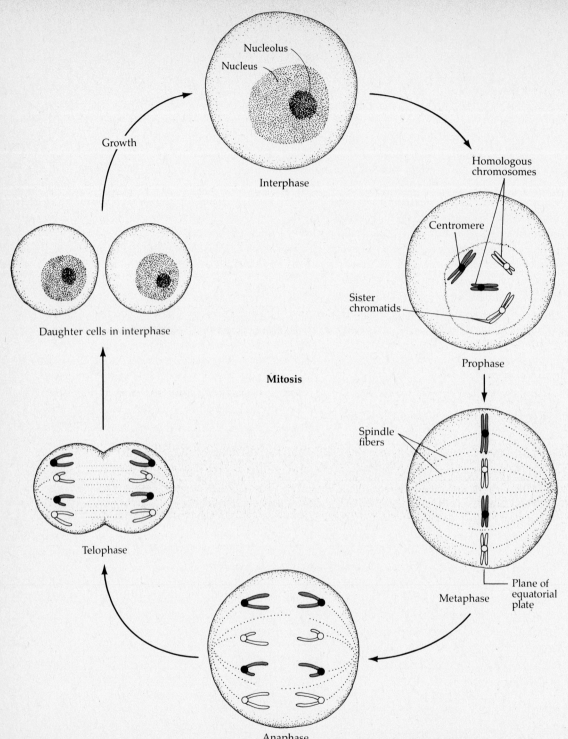

Figure 1.7 Mitosis in a somatic cell of a hypothetical organism that has two pairs of chromosomes.

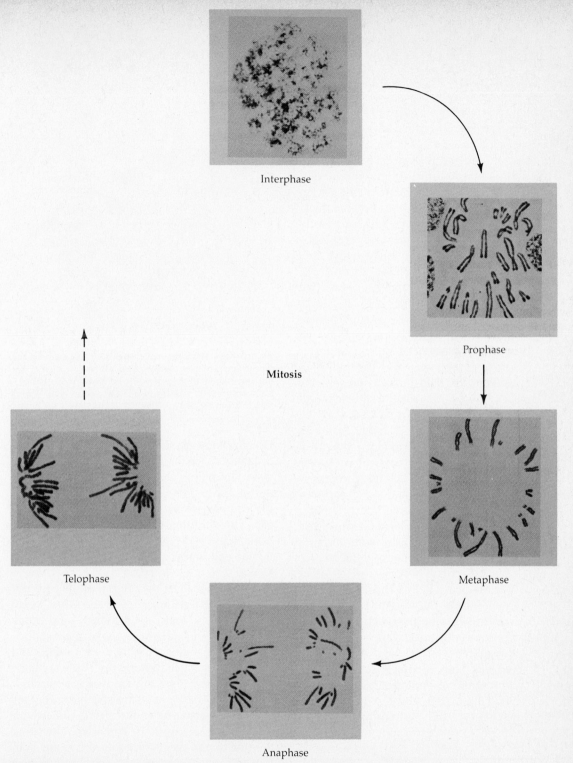

Interphase

Prophase

Mitosis

Telophase

Metaphase

Anaphase

Figure 1.8 Mitosis in cells of the desert locust *Schistocerca gregaria* as viewed through the light microscope (600–850X). The metaphase view is from one pole of the spindle. Note the similarity with the stages illustrated in Figure 1.7.

chromosomes are consistently replicated early in the period of DNA synthesis, while others are consistently replicated late in this period.

The chromosomes that have been replicated develop more coils. Like a string that becomes shorter and thicker by being twisted into coils and twisted further into more coils, each chromosome becomes shorter and thicker by coiling. As the chromosomes become thicker, they provide more surface area for absorbing the dye added to stain the cells; eventually their thickness concentrates enough dye that the chromosomes become visible through the light microscope. The chromosomes are first barely visible as long, slender threads convoluted throughout the nucleus. The first visible appearance of the chromosomes marks the beginning of **prophase**.

During phrophase, the chromosomes continue to shorten and thicken, presumably by further coiling. Gradually the chromosomes become short enough so that they can be recognized as discrete entities rather than as a disorganized jumble of thread (see Figures 1.7 and 1.8). As time goes on, the chromosomes become still shorter and thicker and thus easier to scrutinize. Figure 1.9 shows the chromosomes in a human cell during late prophase. Several features of chromosome structure can readily be observed. First, the chromosomes have replicated; they look double. Each chromosome consists of halves lying close together side by side, and each half of the chromosome is called a **chromatid**. Because the two chromatids in each chromosome are genetically identical, the partner chromatids are often called **sister chromatids.**

A second thing to note is that, although each chromosome has replicated almost everywhere along its length, there is one specific point or region on the chromosome that still appears single, like the crosspoint of an X. This special point is called the **centromere**, and it will become visibly doubled later in mitosis.

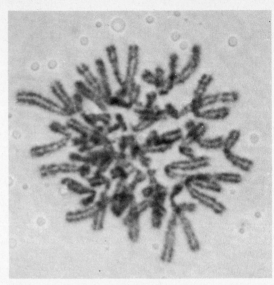

Figure 1.9 Micrograph of chromosomes in a human cell during late prophase of mitosis. As can be seen clearly in a number of chromosomes near the periphery, each chromosome at this stage consists of two chromatids lying side by side and connected to a common centromere.

Because they are still both attached to a single centromere, sister chromatids are technically considered to be parts of a single chromosome during prophase. Only later in mitosis, when the centromere itself finally splits apart, will each sister chromatid be considered a chromosome in its own right. Thus, during prophase, a human cell has 92 chromatids but 46 chromosomes.

Another thing to note is that the chromosomes in a cell do not all look the same. Some are relatively long, some short, and some intermediate in size. Some have their centromere in the middle, others near an end, and still others practically at the tip. But a regularity or pattern can be discerned. Each chromosome has a match, a sort of partner, another chromosome that looks just like it somewhere else in the same nucleus. It may require some hunting to find the chromosomes that match in size and appearance because the chromosomes are in no particular order in the nucleus; their arrangement appears haphazard. The pairs of chromosomes that match in size and appearance are called **homologous chromosomes,** or **homologues** for short. Chromosomes that do not match are called **nonhomologues**. The 46 chromosomes in a cell in a human female are shown in Figure 1.10a. When the individual chromosomes are cut out of the photograph and arranged in pairs by size, as shown in Figure 1.10b, the representation is called a **karyotype** of the individual. Note in the karyotype in Figure 1.10b that there are 23 pairs of homologous chromosomes. Although the homologous pairs of chromosomes in some organisms are sufficiently distinctive to be recognized quite easily (see Figure 1.8 for one example), the homologous pairs of chromosomes in human cells are difficult to identify because many chromosomes are too similar in size and centromere position to be readily

distinguished. As will be discussed in the next chapter, special staining procedures have been devised to reveal which human chromosomes are homologues. It should be pointed out here that in the human male, in contrast to the female, one pair of chromosomes does not match in size. The chromosomes in this nonmatching pair are involved in the determination of sex, and they will be discussed in later chapters. Yet even though the sex chromosomes are of unequal size, they are often considered to be homologous.

Other events occur in prophase in addition to the visible appearance and coiling of the chromosomes. While the shortening and thickening of the chromosomes are progressing, other parts of the cell are preparing for division. Sometime in about the last third of prophase, the nucleoli become smaller and stain less intensely; they fade away and disappear by the next stage in mitosis. The centrioles, ly-

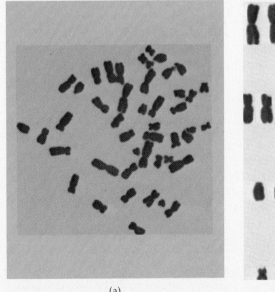
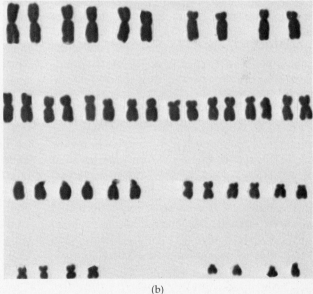

(a)   (b)

Figure 1.10 Metaphase spread *(a)* and karyotype *(b)* of the 46 chromosomes found in the somatic cells of a normal female.

ing just outside the nuclear envelope, separate and the halves slowly move apart, eventually arriving at exactly opposite sides of the nucleus.

Prophase is followed by **metaphase**. The beginning of metaphase is signaled by the breakdown and rapid dissolution of the nuclear envelope. The spindle, which arches between the positions of the centrioles, becomes especially prominent in metaphase (see Figure 1.7), although it actually began to be formed during prophase. The *spindle* is composed of fibers, mainly microtubules, that arch over and through the region occupied by the chromosomes. The spindle has roughly the shape of a football, with the centrioles positioned at its pointed ends; the bulge in the middle engulfs the region occupied by the chromosomes. Some of the spindle fibers become attached to the centromeres of the chromosomes, and a tugging of the fibers on the centromeres begins. Each centromere is pulled or pushed into position approximately in the middle of the spindle. Imagine a plane cutting crosswise through the spindle at an equal distance from each of the centrioles. This imaginary plane is called the **equatorial plate,** or the **metaphase plate,** and the centromeres of the chromosomes are brought to lie on this plane. The metaphase sketch in Figure 1.7 is a view edge on of the equatorial plate because the imaginary plane runs through the centromeres. The metaphase photograph in Figure 1.8 is a view perpendicular to the equatorial plate as seen from one end of the spindle. Although the two views look rather different, they are just different aspects of the same three-dimensional structure. When the centromeres have been aligned on the equatorial plate, the cell is ready for the actual separation of the sister chromatids.

Anaphase commences when the centromeres of the chromosomes become visibly double and begin to separate (see Figures 1.7 and 1.8). At the moment the centromere splits, each chromosome is considered to have divided; each half (formerly called a *chromatid*) is now considered to be a chromosome in its own right. Immediately after the centromeres have split in anaphase, therefore, a human cell contains 92 chromosomes. When a centromere has split, its halves begin to be pulled toward opposite poles of the spindle, toward the pointed ends. This movement apart happens because the spindle fibers that are attached to the centromeres shorten during anaphase, and the centromeres are towed by the spindle fibers. In addition, the whole cell seems to elongate slightly along the direction of the spindle.

Each centromere is part of a chromosome, of course, and as the centromere moves, the arms of the chromosome are pulled passively along. (Very rare cells can sometimes be found that contain an abnormal chromosome with no centromere. This centromereless chromosome does not move like the others; it tends to remain behind as the others are pulled away, and it is often pushed by other forces toward the periphery of the spindle.) Sister chromatids first begin to separate at the centromere, and the tugging of the spindle fiber on the centromere is initially counteracted by the apparent stickiness of the chromatids and sometimes by the winding of the arms of sister chromatids around each other. When the sister chromatids do finally become completely separated, they are rapidly pulled to the opposite poles. The chromosomes assume characteristic shapes as they are pulled along, with the shapes depending on the position of the centromere. Chromosomes with their centromere near the middle are called **metacentric** chromosomes, and they appear V-shaped during anaphase; chromosomes with arms of unequal length are called **submetacentric** chromosomes, and they appear J-shaped at anaphase; and chromosomes with their centromere very near the tip are called **acrocentric** chromosomes, and they appear rod-shaped during anaphase (Figure 1.11).

Type of chromosome	Appearance at metaphase	Appearance at anaphase

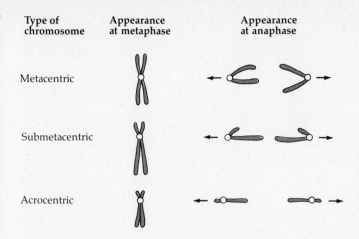

Figure 1.11 Definition of types of chromosomes (metacentric, submetacentric, and acrocentric) in terms of centromere position and appearance at metaphase and anaphase of mitosis.

The chromosomes finally become completely separated into two groups, with one group at each pole of the spindle (see Figures 1.7 and 1.8). In normal human cells, each group contains exactly 46 chromosomes. Barring rare accidents, such as the addition or loss of a chromosome, each group consists of an exact duplicate of the chromosomes that were present in the cell before it began mitosis; that is, the group of chromosomes at one pole is identical with the group at the other pole.

The next stage of mitosis, **telophase**, is marked by the reappearance of a nuclear envelope around each group of chromosomes, the redevelopment of the nucleoli, the breakdown and disappearance of the spindle, and the division of the cytoplasm (see Figure 1.7). In animals, the cell literally pinches itself in two (Figure 1.12). A furrow due to the action of microfilaments develops around the circumference of the cell along the edge of the equatorial plate, which now separates the two groups of chromosomes. This furrow pinches in like a noose tightening around a balloon. Finally the two daughter cells are completely separated. (In plants, a cell wall grows, forming a partition between the daughter cells.) Cytoplasmic division divides the contents of the cytoplasm —mitochondria, ribosomes, and so on—into roughly equal parts. Whereas the division of

the chromosomes is highly exact, the division of the cytoplasmic contents is more irregular. It does not seem to matter whether one daughter cell gets a few more mitochondria or ribosomes than the other. In any case, the process of mitosis is now over. Where before there was one cell, now there are two, and the two daughter cells are genetically identical.

After telophase, each daughter nucleus fades back into an interphaselike appearance, and the mitotic cycle may begin anew. Each telophase nucleus has received one complete diploid set of chromosomes, and, if the cell were to divide again, it would be important to replicate the chromosomes before proceeding into the next prophase.

At this point, it might be helpful to review the major aspects of chromosome behavior during the various stages of mitosis:

INTERPHASE: Chromosomes replicate.
PROPHASE: Chromosomes first become visible through light microscope.
METAPHASE: Chromosomes align with centromeres on equatorial plate.
ANAPHASE: Centromeres split and chromosomes move to opposite poles.
TELOPHASE: Chromosomes resume interphase appearance; cytoplasm divides.

From a genetic point of view, the most impor-

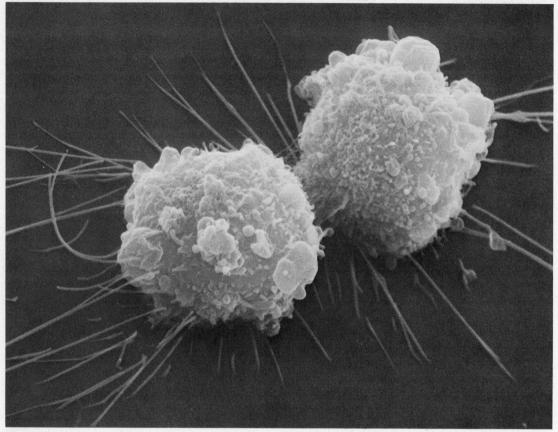

Figure 1.12 A mouse cell late in telophase nearing the completion of cell division (approx. 4000X). The fibers extending out from the cells are "retraction fibers," along which each daughter cell will flatten out after mitosis is completed.

tant feature of mitosis is that the daughter cells receive identical complements of chromosomes.

Capturing Chromosomes in Metaphase

Chromosomes are easiest to examine and count in metaphase (see Figure 1.10). The problem, though, is that metaphase is fairly short, so not many cells are found in metaphase. The lengths of time spent by cells in the different stages of mitosis depend on the type of cell and the organism it comes from, as well as on such other circumstances as temperature and age. Cells in the grasshopper embryo, grown under ideal conditions in the laboratory, complete mitosis in $3\frac{1}{2}$ hours. Interphase takes up 13 percent of this time, prophase 50 percent, metaphase 6 percent, anaphase 4 percent, and telophase 27 percent. Even in these rapidly dividing cells, the chance of actually finding a cell in metaphase is only 6 percent. Grasshopper embryonic cells are especially favorable for chromosome study because their interphase is proportionally much shorter than is interphase in most other animals or plants.

In human tissues, the situation for chromosome study would seem to be very bad. Certain white blood cells cultured in the laboratory divide about every 18 hours; 17 of these hours are spent in interphase, and metaphase takes just a few minutes. Luckily, a chemical trick can be used to capture chromosomes at metaphase. The trick is to treat cells with the drug **colchicine** (or with a less toxic chemical derivative of it). Colchicine is a spindle poison. It binds chemically with the microtubules of the spindle and thereby prevents the formation of the spindle during mitosis. Cells treated with colchicine cannot pass through anaphase. Therefore, a culture of cells treated with colchicine will accumulate cells in metaphase because, no matter where the cell was in the mitotic cycle when the colchicine was added, sooner or later it must reach metaphase. Since a spindle is required to continue into anaphase, cells that lack a spindle simply stop at metaphase. Colchicine is used routinely for chromosome study in virtually all human genetics laboratories.

Sister-Chromatid Exchange

Sister chromatids are, of course, genetically identical because they are replicas of a single chromosome in the parent cell. Nevertheless, in the early 1970s, a technique was developed whereby sister chromatids could be rendered chemically and physically distinct. One important finding from the application of this technique was that sister chromatids frequently undergo an exchange of parts during mitosis. The technique itself is relatively simple. Cells that are undergoing mitosis are treated with a chemical called **BUdR (bromodeoxyuridine),** which is incorporated into DNA in place of one of its normal constituents. After two rounds of chromosome replication in the presence of BUdR, chromosomes are trapped in metaphase by means of colchicine. The chromosomes are then treated with a special fluorescent dye, and one of the sister chromatids is found to fluoresce brightly whereas the other fluoresces dimly. The appearance of the sister chromatids after BUdR treatment is depicted in the ovals at the right side of Figure 1.13. The majority of chromosomes in normal cells appear as in Figure 1.13a; one sister chromatid is uniformly bright fluorescent (unshaded), whereas the other is uniformly dim fluorescent (shaded). However, in many chromosomes, the sister chromatids appear to have undergone an exchange of parts, as illustrated in Figure 1.13b. **Sister-chromatid exchange** in a human cell is shown in Figure 1.14.

The underlying reason for the bright and dim fluorescence is shown in the left part of Figure 1.13. The two intertwined strands inside each chromosome or chromatid represent a single molecule of DNA. Although a single DNA molecule does consist of two such intertwined strands (as will be discussed more fully in Chapter 8), the length of the DNA molecule is relatively much longer than illustrated in Figure 1.13. Nevertheless, when the DNA molecule replicates during chromosome replication, the two strands come apart and a new partner DNA strand is added to each of the original ones. In the presence of BUdR, the newly synthesized strands (shown in black) will contain BUdR instead of the normal constituent. In the first mitotic division after BUdR treatment, therefore, both sister chromatids contain one normal DNA strand (gray) and one BUdR-containing strand (black). After the second round of DNA replication in the presence of BUdR, the DNA in the sister chromatids is different; one sister chromatid contains a DNA molecule with one normal (gray) and one BUdR-containing DNA strand (black), whereas, in the other sister chromatid, both DNA strands contain BUdR.

In the presence of the special fluorescent dye, DNA molecules with one normal and one

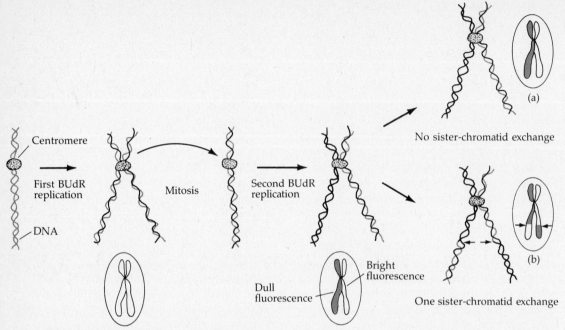

Figure 1.13 Simplified representation of the organization of DNA in chromosomes showing detection of sister-chromatid exchange by means of bright fluorescence of DNA with a single BUdR-containing strand and dull fluorescence of DNA with two BUdR-containing strands.

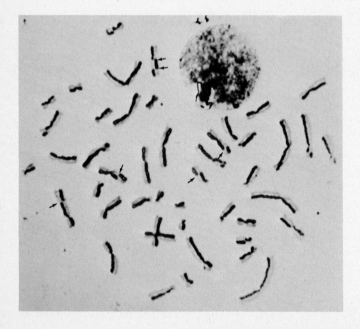

Figure 1.14 Metaphase spread of human chromosomes showing sister-chromatid exchanges (arrows) revealed by BUdR with Giemsa staining, an alternative to the fluorescent procedure discussed in the text. The cell here had been exposed to a chromosome-breaking agent prior to staining.

BUdR-containing strand fluoresce brightly, whereas DNA molecules with two BUdR-containing strands fluoresce dimly. Thus, after two rounds of DNA replication in the presence of BUdR, one sister chromatid will fluoresce brightly and the other will fluoresce dimly.

The presence of sister-chromatid exchange (see Figure 1.13b) indicates a corresponding exchange in the DNA molecules. Although the manner in which sister-chromatid exchange occurs is not understood in detail, it is easiest to visualize it as a sort of breakage and reunion. The process of sister-chromatid exchange is exact, however, which is to say that, whatever the actual process may be, the exchange involves identical positions in the two DNA molecules.

The DNA molecule in a chromosome contains the chromosome's genetic information in a manner that will be described more fully in the following section. Details aside, the most important feature of DNA replication is that the two molecules produced carry exactly the same genetic information. Because of this exactness, the genetic information in a pair of sister chromatids must be identical. Therefore, sister-chromatid exchange normally has no genetic effects because it involves an exchange of parts between two identical DNA molecules.

There are, however, some inherited diseases in which the frequency of sister-chromatid exchanges is much greater than it is in normal individuals. One of these conditions is known as **Bloom syndrome.** (The word *syndrome* is a medical term referring to a collection of symptoms that occur together sufficiently often to warrant recognition as a distinct clinical entity deserving of a special name. Syndromes are often named after the physician who first recognized the symptoms as distinctive.) The symptoms of Bloom syndrome are low birth weight, short stature, and extreme sensitivity to sunlight leading to skin rash, especially on the face (Figure 1.15). The

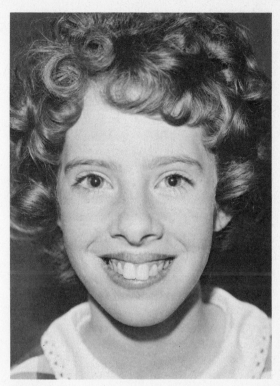

Figure 1.15 A young woman with Bloom syndrome.

relationship between increased sister-chromatid exchange and the other symptoms is not clear for Bloom syndrome, nor is it clear for the other conditions associated with enhanced sister-chromatid exchange. However, the increased frequency of sister-chromatid exchange is important for two reasons. First, sister-chromatid exchange can be used as an aid in the diagnosis of the conditions, even in cells from unborn fetuses. Second, cells from patients with Bloom syndrome frequently exhibit abnormal kinds of chromosome breakage in addition to increased sister-chromatid exchange. Since Bloom syndrome patients (and patients with other conditions associated with increased sister-chromatid exchange) have a predisposition toward leukemia, there is apparently some link between genetic changes in a cell and the occurrence of cancer. Further information regarding the associations between genetics and cancer will be discussed in Chapter 12.

Genetic Information in the Chromosomes

Figure 1.13 depicts the organization of DNA in chromosomes. The figure is greatly oversimplified because an actual molecule of DNA is extremely long—much longer than a chromosome. For example, the length of the longest human chromosome at metaphase is about 10 μm (10^{-3} cm); if the DNA in this chromosome were fully extended, its length would be about 8 cm; that is, the extended length of the DNA is some 8000 times the length of the chromosome itself.

Details of how DNA is packaged in the chromosome are not understood. It is known that the DNA molecule becomes associated with certain types of proteins called **histones**

and that the DNA-histone complex becomes organized by some sort of coiling into a **chromatin fiber** approximately 200 to 300 Å (0.02 to 0.03 μm) in diameter. This fiber then undergoes further coiling to form the chromosome seen at metaphase. Figure 1.16 is an electron micrograph of one of the smallest human chromosomes at metaphase. The complex coils and gyrations of the fundamental chromatin fiber are evident.

The genetic information in a chromosome resides in its DNA. As noted earlier, each DNA molecule consists of two intertwined strands. These strands carry genetic information in the form of a chemical "alphabet," so each strand of DNA may be thought of as a tiny ticker tape with genetic instructions printed on it. The most important unit of genetic information is the gene, which, to use the ticker-tape analogy, corresponds to a segment of the ticker tape. A typical gene is thought to consist of several thousand chemical "letters." Most genes carry the information to direct the manufacture of a particular type of

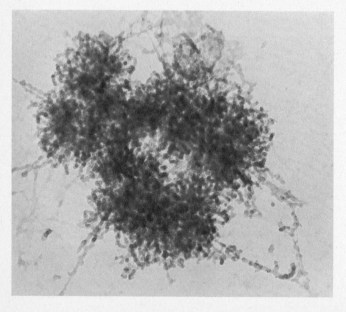

Figure 1.16   Electron micrograph of human E-group chromosome showing extensively coiled and folded fiber.

protein in the cell (such as hemoglobin), and the protein, in turn, carries out some important cellular function. (Exceptional genes that do not code for proteins include, for example, genes that code for the RNA constituents of ribosomes.) Thus, cellular functions are ultimately directed by genes, most of which exert their influence by dictating the makeup of proteins by means of a process to be described in Chapter 9. Each gene has associated with it in the DNA some information about its control, about whether or not its particular protein should be made in a particular cell at a particular time.

The alphabet of DNA consists of only four chemical letters, called **nucleotides**. Nucleotides are the chemical constituents of DNA, and millions of them are chemically linked together, end to end, to form each strand in a molecule of DNA (the ticker tapes). The total number of nucleotides in a human cell is about 6 billion. That is as many letters as would appear in 10,000 books the size of this one. As will be noted later, though, not all the DNA in human cells is thought to carry genetic information. Nevertheless, a fertilized human egg carries sufficient hereditary information to direct its development into an embryo, its growth from newborn infant through childhood, its health as an adult, and its vitality as an old man or woman.

The four nucleotides found in DNA are called

deoxyadenosine phosphate (abbreviated **A**)
thymidine phosphate (abbreviated **T**)
deoxyguanosine phosphate (abbreviated **G**)
deoxycytidine phosphate (abbreviated **C**)

The genetic alphabet is therefore extremely simple because it consists of just four letters— A, T, G, and C. Moreover, each word in a gene consists of exactly three letters. Thus, an actual gene could carry the information CAT ACT TAG GAG, but, as there are no spaces in DNA, the information would read CAT-ACTTAGGAG.

Because a gene corresponds to part of a DNA molecule, and because DNA molecules consist of two intertwined strands, each gene consists of two strands (the ticker tapes). For each gene, however, only one of the two strands carries the genetic information; this information-containing strand is known as the **sense** strand for that gene. The other DNA strand in the gene is called the **antisense** strand. A long DNA molecule will contain many genes, and the same DNA strand that represents "sense" for one gene may represent "antisense" for a different gene down the way. In Figure 1.17, for example, the DNA strands in a molecule that contains several genes are shown as parallel heavy lines. The sense strand for genes 1, 2, and 4 is the top strand (and the antisense strand for these genes is the bottom one), whereas, for gene 3, the situation is the reverse.

On the right of Figure 1.17, corresponding small segments of the DNA strands have been enlarged for closer inspection of their nucleotide sequences. There is a precise relationship between the corresponding nucleotide sequences, although the relationship may not immediately be obvious. However, the relationship between the sense strand and the antisense strand is quite simple: Wherever one strand carries an A, the other carries a T; and wherever one strand carries a G, the other carries a C. It is an amusing coincidence that the genetic information in the sense strand is encoded in the antisense strand in a manner that spies used hundreds of years ago to conceal secret messages, and some important characteristics of DNA and its replication can be understood in terms of this spy analogy without the need for chemical details. The code (actually a type of cipher) is known as

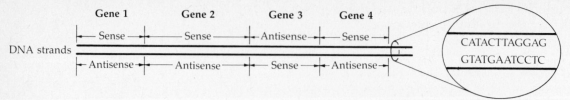

Figure 1.17 Diagram of DNA (the strands are shown uncoiled for ease of representation) with an expanded region at the right showing the nucleotide pairs in the strands (A with T and G with C). Genes correspond to segments of DNA, but the actual genetic information resides in only one strand (the sense strand). As illustrated, a long molecule of DNA contains many genes, and the DNA strand that is the sense strand for one gene can be the antisense strand for another gene along the way.

alphabetic substitution. In alphabetic substitution, each letter of a sense message (the *plaintext*, as code breakers call it) is rendered in the enciphered antisense message (the *ciphertext*) by a different letter of the same alphabet. Thus, for example, if we choose a simple cipher in which plaintext C is replaced with ciphertext G, plaintext A is replaced with T, and T is replaced with A, then the plaintext message CAT would be enciphered as GTA. Recall now that the genetic information in the sense strand is in an alphabet of just four letters—A, T, G, and C. The genetic cipher actually used in DNA is the following:

Replace A in sense strand (plaintext) with T in antisense strand (ciphertext).

Replace T in sense strand (plaintext) with A in antisense strand (ciphertext).

Replace G in sense strand (plaintext) with C in antisense strand (ciphertext).

Replace C in sense strand (plaintext) with G in antisense strand (ciphertext).

Thus, if the sense strand carries CAT, the antisense strand across the way will carry GTA. The longer sequence CATACT-TAGGAG in the sense strand is present in the other strand as GTATGAATCCTC (see Figure 1.17). When the sense and antisense strands are close together so that corresponding nucleotides are paired, as in Figure 1.17, an A in one strand will always be paired with a T in the other strand across the way, and a G in one strand will always be paired with a C in the other strand.

One important feature of the genetic "cipher" is that the encipherment of the ciphertext regenerates the plaintext. Thus, for example, the ciphertext for CAT is GTA, and the cipertext for GTA is CAT; the plaintext message comes back again. This feature of the genetic cipher is important because, when the DNA strands separate during DNA replication, the new strand added to the sense strand will be identical to the old antisense strand, and the new strand added to the antisense strand will be identical to the old sense strand. In this manner, genetic information in the DNA is faithfully replicated generation after generation.

In summary, a gene corresponds to a segment of a DNA molecule. The genetic information in the gene is contained in the sequence of nucleotides (A, T, G, and C) in the sense strand. The antisense strand contains the genetic information in the form of a cipher in which A is replaced with T, T with A, G with C, and C with G. (The underlying reasons for this relationship between the nucleotide sequences of the sense and antisense strands will be discussed in Chapter 8.)

A cell contains as many chromosomes as its nucleus contains molecules of DNA. The

TABLE 1.2 CHROMOSOME NUMBERS IN VARIOUS PLANTS
AND ANIMALS

Organism	Diploid number	Haploid number
Human *(Homo sapiens)*	46	23
Chimpanzee *(Pan troglodytes)*	48	24
Gorilla *(Gorilla gorilla)*	48	24
Dog *(Canis familiaris)*	78	39
Cat *(Felis domesticus)*	38	19
House mouse *(Mus musculus)*	40	20
Carp *(Cyprinus carpio)*	104	52
Fruit fly *(Drosophila melanogaster)*	8	4
Wheat *(Triticum aestivum)*	42	21
Corn *(Zea mays)*	20	10
Tomato *(Lycopersicon esculentum)*	24	12
Tobacco *(Nicotiana tabacum)*	48	24
Garden pea *(Pisum sativum)*	14	7
Pink bread mold *(Neurospora crassa)*	14	7
Yeast *(Saccharomyces cerevisiae)*	34	17

number of chromosomes varies from species to species; some examples are given in Table 1.2. The diploid number of chromosomes is the number found in all cells of the body except the gametes; these nongametic cells are called **somatic cells.** The haploid number of chromosomes is the number found in gametes (sperm or eggs in the case of animals).

Recall that, in somatic cells, chromosomes actually occur in homologous pairs. Each chromosome and its homologue carry genetic information corresponding to the same genes, but the DNA molecules are not necessarily identical in every detail. The DNA in one chromosome may contain errors or "misprints" called **mutations**. The information that may be garbled in one chromosome by a mutation is often supplied in usable form by the homologue, because the odds are that both chromosomes will not have mutations in exactly the same genes. Mutations of the same gene in both homologous chromosomes do occur, though, and the result may be a hereditary

disease in the person who carries these chromosomes. In some cases, moreover, a mutation in only one homologue of a pair may be sufficient to cause a hereditary disease. (One exception to the rule that a chromosome and its homologue carry genetic information for corresponding genes should be mentioned here. The pair of chromosomes that determine sex—the X and Y chromosomes—carry different genes. These chromosomes will be discussed in the next chapter and in detail in Chapter 5.)

When chromosomes are viewed through the light microscope, as in Figure 1.10, none of the underlying chemical details regarding DNA can be seen. Such details lie well below the limit of resolution of the light microscope. Nevertheless, a great deal about genetics can be learned from the study of chromosomes. In the next chapter, we will examine human chromosomes more closely and discuss the important process of cell division—meiosis—in which gametes are formed.

SUMMARY

1. Human genetics is the study of the inherited characteristics of human beings. Knowledge of human genetics is of practical importance because many families have individuals affected with conditions that are, at least in part, due to genetic factors. Knowledge of human genetics is also subject to misinterpretation or misuse, as evidenced by the eugenics movement early in this century.

2. Hereditary factors, called **genes**, are composed of a chemical **deoxyribonucleic acid (DNA)** found in cells. Much of genetics is therefore concerned with cells. Cells are the fundamental units of living things, the smallest units of life.

3. Two fundamentally different types of cells are to be distinguished: **prokaryotic cells,** which are relatively primitive cells lacking a nucleus, and **eukaryotic cells,** which have a nucleus. Examples of prokaryotes are bacteria and blue-green algae. Eukaryotic cells are those of plants, animals, and humans.

4. Eukaryotic cells contain internal structures called **organelles**. The most important of these in genetics is the **nucleus**, which contains the DNA in the form of **chromatin**. The nonnuclear part of a eukaryotic cell is called the **cytoplasm**, and it contains such organelles as the **ribosomes**, which are the sites of protein synthesis.

5. Mitosis is a type of cellular reproduction in which one cell divides to produce two genetically identical daughter cells. For convenience, the process of mitosis is considered to consist of five stages: **interphase** (DNA replicates and chromatin begins to coil into discrete chromosomes), **prophase** (chromosomes first become visible in the light microscope as pairs of **sister chromatids** attached to a common **centromere**), **metaphase** (centromeres of chromosomes align on a cell's equatorial plate) **anaphase** (centromeres split and sister chromatids proceed to opposite poles of the spindle), and **telophase** (chromosomes regain their interphase appearance and the cytoplasm divides).

6. During mitosis, sister chromatids can undergo a precise interchange of parts called **sister-chromatid exchange.** This is a normal process, but in certain inherited disorders such as Bloom syndrome, the level of sister-chromatid exchange is much increased over normal.

7. A **chromosome** consists of a tightly coiled fiber of chromatin, which, in turn, consists of a long, coiled molecule of DNA along with certain other constituents.

8. A molecule of DNA consists of two intertwined strands, each composed of a long sequence of **nucleotides** linked end to end. Four nucleotides are found in DNA; these are denoted by the symbols A, T, G, and C. A gene corresponds to a segment of such a DNA molecule. The genetic information in the gene consists of the sequence of nucleotides along one of the DNA strands (the **sense strand**); the other DNA strand for the gene is called the **antisense strand.** The relationship between the nucleotide sequences in the sense and antisense strands is as follows: Wherever one strand carries an A, the other carries a T; and wherever one strand carries a G, the other carries a C.

9. Most genes exert their influence on cells by providing instructions for the cell to manufacture a particular type of protein, such as hemoglobin. Genes may contain errors called **mutations**, which can lead to the manufacture of defective proteins and thus a genetically caused disorder. However, since homologous chromosomes carry genetic information for corresponding genes, a mutation in a gene in one chromosome is often compensated by a normal gene in the homologue.

WORDS TO KNOW

Cell	**Chromosome**	Metaphase
Eukaryotic	Chromatin	Anaphase
Prokaryotic	Chromatid	Telophase
Gamete	Sister chromatid	Colchicine
Zygote	Homologous	BUdR
Somatic	Centromere	Sister-chromatid exchange
Diploid		Bloom syndrome
Haploid	**Gene**	
Nucleus	Mutation	**DNA** (deoxyribonucleic acid)
Cytoplasm		Nucleotide
Organelle	**Mitosis**	A, T, G, and C
Ribosome		Sense strand
	Interphase	Antisense strand
	Prophase	

PROBLEMS

1. For discussion: A few years ago, a California businessman announced the establishment of a "Nobel sperm bank," in which the sperm of Nobel prize winners is obtained and preserved and made available for artificial insemination of "superior" women. What is your initial reaction to this proposal? If you approve of it, would you also approve the use of frozen sperm from donors who have died? What similarities and differences do you see between this sort of "positive eugenics" and the "negative eugenics" exemplified in the compulsory sterilization of certain individuals?

2. What characteristics of mitochondria and chloroplasts make it a plausible hypothesis that these organelles were once prokaryotic organisms in their own right?

3. Rabbits have 44 chromosomes in somatic cells. How many chromosomes are there in a rabbit sperm?

4. Mosquito eggs and sperm each have three chromosomes. How many are there in somatic cells?

5. A haploid set of human chromosomes, such as is found in the sperm or the egg, contains 23 chromosomes. Diploid somatic cells have 46 chromosomes. How many chromosomes would you expect to find in triploid (three haploid sets) cells? In tetraploid (four haploid sets) cells?

6. How does colchicine interrupt the normal process of mitosis? At which stage of mitosis are cells arrested?

7. If a human cell underwent one abnormal mitosis in which metaphase, anaphase, and telophase were eliminated, how many chromosomes would the cell have?

8. A woman who is pregnant goes to a clinic for genetic counseling. She and her husband already have one child, who is affected with Bloom syndrome. What procedure could the physician use to determine whether the woman's unborn fetus is also affected with the condition?

9. DNA of eukaryotic organisms can be chemically purified, split into its individual nucleotides, and the relative amounts of each nucleotide measured. An important discovery—made some years before the structure of DNA was understood—is that the number of molecules of A always equals the number of molecules of T, and that the number of molecules of G always equals the number of molecules of C. How can this finding be interpreted based on the current knowledge of DNA structure?

10. Human adult hemoglobin actually contains two different proteins, called α-hemoglobin and β-hemoglobin. (α and β are the Greek letters alpha and beta.) The initial 15 nucleotides in the sense strand of the gene coding for β-hemoglobin are known to be TACCACGTGGACTGA. What corresponding nucleotide sequence is found in the antisense strand?

11. BUdR is incorporated into DNA in place of T. If a DNA strand that is undergoing replication has the nucleotide sequence CATACTTAGGAG, what sequence would be found in the partner strand synthesized in the presence of BUdR? (Use the symbol B to represent BUdR.)

12. The accompanying diagram represents a fluorescent-light view of a mitotic chromosome after two rounds of chromosome replication in the presence of BUdR. How can the pattern of bright and dim fluorescence be explained by sister-chromatid exchange?

13. In terms of the incidence of genetically caused disorders, what is the significance of the fact that chromosomes are found in homologous pairs?

14. The mutation leading to the most common form of color blindness is located on the X chromosome in humans. (The X chromosome is one of the sex-determining chromosomes; the other is the Y chromosome.) This form of color blindness is much more frequent in men than in women. Why should this be so? (Hint: Females have two X chromosomes, whereas males have only one X chromosome along with one Y chromosome; moreover, the X and Y chromosomes carry different genes.)

FURTHER READING AND REFERENCES

Avers, C. J. 1976. Cell Biology. Van Nostrand, New York. A popular textbook of cell biology.

Chaganti, R. S. K., S. Schonberg, and J. German III. 1974. A many fold increase in sister chromatid exchanges in Bloom's syndrome lymphocytes. Proc. Natl. Acad. Sci. U.S.A. 71:4508–4512. One of the first indications of elevated sister-chromatid exchange in certain syndromes.

DeWitt, W. 1977. The Biology of the Cell. Saunders, Philadelphia. Source of Figure 1.5.

Diener, T. O. 1980. Viroids. Scientific American 244:66–73. The smallest known agents of disease —essentially "naked" strands of RNA.

Dustin, P. 1980. Microtubules. Scientific American 243:66–76. On the manifold functions of these proteinaceous filaments.

Dyson, R. D. 1978. Essentials of Cell Biology, 2d ed. Allyn & Bacon, Boston. A good overview of the structure and function of cells and their organelles.

Hayflick, L. 1980. The cell biology of human aging. Scientific American 242:58–65. What causes cells to grow old?

Lake, J. A. 1981. The ribosome. Scientific American 245:84–97. A careful examination of this marvelous little factory.

Latt, S. A. 1981. Sister chromatid exchange formation. Ann. Rev. Genet. 15:11–55. A summary of current knowledge concerning sister-chromatid exchanges.

Lodish, H. F., and J. E. Rothman. 1979. The assembly of cell membranes. Scientific American 240:48–63. Discusses the structure and growth of the film that holds our cells together.

Ludmerer, K. M. 1971. Genetics and American Society: A Historical Appraisal. Johns Hopkins University Press, Baltimore. An honest history of the frequently misguided and often racist eugenics movement in the United States.

Milunsky, A. 1979. Know Your Genes. Avon Books, New York. A nontechnical introduction to genetic health.

Paterson, M. C., and P. J. Smith. 1979. Ataxia telangiectasia. Ann. Rev. Genet. 13:291–318. Review of an inherited condition that increases sensitivity to DNA-damaging chemicals.

Porter, K. R., and J. B. Tucker. 1980. The ground substance of the living cell. Scientific American 244:56–57. A careful examination reveals a gossamer lattice of filaments for support and movement of organelles.

Rothman, J. E. 1981. The Golgi apparatus: Two organelles in tandem. Science 213:1212–1219. Thorough review of how this important cellular structure is thought to work.

Therman, E. 1980. Human Chromosomes: Structure, Behavior, Effects. Springer-Verlag, New York. Contains an excellent account of mitosis.

Tzagoloff, A. 1982. Mitochondria. Plenum, New York. A thorough review of the structure and function of this complex organelle.

Wallace, D.C. 1982. Structure and function of organelle genomes. Microbiol. Rev. 46:208-240. Have mitochondria and chloroplasts evolved from prokaryotic symbionts? A review of the evidence.

Warren, S. T., R. A. Schultz, C.-C. Chang, M. H. Wade, and J. E. Trotsko. 1981. Elevated spontaneous mutation rate in Bloom syndrome fibroblasts. Proc. Natl. Acad. Sci. U.S.A. 78:3133–3137. Bloom syndrome may be due to a mutation that increases the rate of mutation of other genes.

Woese, C. R. 1980. Archaebacteria. Scientific American 244:98–122. Discusses a form of life that is neither prokaryote nor eukaryote but constitutes a kingdom of its own.

chapter 2
Chromosomes and Gamete Formation

Approximately 1 out of every 155 live-born children has some major abnormality in its chromosomes. As will be discussed in detail in Chapter 6, some of these children have physical abnormalities that can be recognized at birth; others are apparently normal at birth but suffer later impairment of physical or mental development; still others are physically and mentally normal, but their chromosome abnormality leads to a high risk of physical or mental abnormalities among their children.

This 0.6 percent (1/155) of live-born children with abnormal chromosomes is a special group because they are sufficiently normal in their development to be born alive. Many more embryos with abnormal chromosomes undergo spontaneous abortion. Indeed, about half of all spontaneous abortions are associated with major chromosomal abnormalities of the fetus.

One subject of this chapter is the number and appearance of the chromosomes in cells of normal people. There is some variation in the appearance of the chromosomes in normal individuals; the nature and extent of this variation will be noted later. Bear in mind that the immediate discussion will necessarily be about the "typical" set of chromosomes. The descriptions of chromosomes will apply to people of all ages. For the most part, they will apply to the chromosomes in the cells of both sexes, with the exception of the X and Y chromosomes, which determine sex. The descriptions will also apply to the chromosomes of people of all colors, sizes, races, nationalities, and so on; that is to say, the kinds of chromosomes in the cells of all normal human beings are essentially the same, with the exception of certain normal variants to be noted later.

Viewing Human Chromosomes

First it should be noted how chromosomes are prepared for examination and from what cells they are derived. Chromosomes are best examined in metaphase of mitosis or in the corresponding stage in meiosis—the special kind of cell division that gives rise to eggs and sperm. Cells undergoing mitosis are usually the ones examined. Contrary to what one might expect, however, the cells that are examined are not the ones that are rapidly dividing in the body. Cells of the bone marrow, lymph nodes, spleen, and other rapidly dividing tissues are generally difficult to obtain; getting samples of them requires procedures that can be painful for the patient. In some cases, even surgery is required. So the usual procedure is to obtain cells that are normally not dividing rapidly in the body and to stimulate them chemically to divide in a test tube or culture flask in the laboratory.

A typical procedure used in the study of human chromosomes is outlined in Figure 2.1. A small amount of blood is taken from the patient. (In young babies, blood is frequently obtained from a heel prick.) The blood is added to a solution containing a chemical that prevents it from coagulating, and a small amount of this is then added to a rich nutrient broth containing an ingredient called **phytohemagglutinin**, which is a chemical extracted from the juice of the red kidney bean. Phytohemagglutinin stimulates certain cells to undergo mitosis. In the sample of blood are red blood cells and several different kinds of white

blood cells. One type of white blood cell, the small lymphocyte, is the type stimulated by phytohemagglutinin to undergo mitosis. (The small lymphocyte, by the way, is the white blood cell mentioned in passing in Chapter 1 as being one of the smallest cells in the body.)

The cultures that contain the small lymphocytes in nutrient broth are incubated for about three days, during which time they divide repeatedly. Then a relatively nontoxic chemical derivative of colchicine is added to stop the cells in metaphase, and about $1\frac{1}{2}$ h later the cells are suspended in a very weak salt solution (one with a lower salt concentration than that inside the cells) and allowed to imbibe water for a few minutes to swell them, to help disperse the chromosomes, and to aid in disentangling the arms of the sister chromatids. (Figure 2.2 demonstrates the importance of colchicine and the low-salt solution in obtaining satisfactory preparations of human chromosomes.) The cells are then fixed by adding an acid-alcohol mixture to preserve them and retain their internal structures. Following this, the cells are spread out on a small glass plate or slide and allowed to dry. This drying separates the chromosomes in a cell from one another so they can be observed individually. Then the chromosomes are stained with a dye, whereupon the slides are washed, dried, and examined under a light microscope. Inevitably, the chromosomes of most cells are not suitable for study, but many cells will contain chromosomes that are well spread and suitable for examination.

Sometimes it is important to examine cells from several tissues in addition to white blood cells. In such cases, the needle of a syringe can be inserted into the marrow of certain bones, such as the breastbone, and a small number of rapidly dividing cells can be withdrawn. Cells from the lymph nodes can be obtained surgically. Or a small amount of tissue from one testicle can be removed. An-

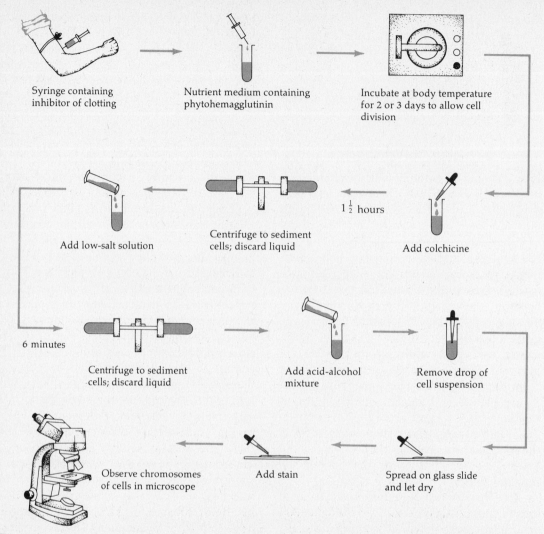

Figure 2.1 Outline of experimental procedure for viewing human chromosomes in metaphase of mitosis. Note especially the addition of colchicine (to prevent formation of the spindle) and the brief soaking in a low-salt solution (to swell the cells and disperse the chromosomes).

other approach is to surgically obtain a small piece of skin that can be cultured in the laboratory; the chromosomes of these cultured skin cells can be studied with the same techniques used to study chromosomes of white blood cells.

Studying cells from several tissues might be necessary in some cases to diagnose or check whether a patient is a chromosomal mosaic. A **mosaic** is a person who has two or more genetically different types of cells or tissues; more specifically, a **chromosomal mo-**

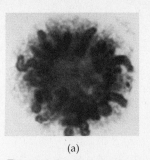
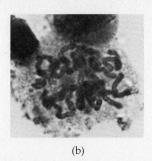
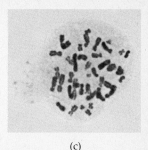

(a)   (b)   (c)

Figure 2.2 Human cells in metaphase prepared as in Figure 2.1 but *(a)* without colchicine or soaking in low-salt solution, *(b)* with colchicine but without soaking in low-salt solution, and *(c)* with both colchicine and soaking in low-salt solution. In *(a)*, the metaphase chromosomes are in close proximity to one another, held by the fibers of the spindle. In *(b)*, the spindle has been disrupted by colchicine, but the chromosomes are still jumbled and difficult to analyze. The chromosomes in the swollen cell *(c)* are much more spread out and easier to examine.

saic is an individual who has one chromosome constitution in some cells and a different chromosome constitution in other cells. The vast majority of people have exactly the same chromosome constitution in virtually all their cells, so the constitution of small lymphocytes is the same as that of brain cells, nerve cells, muscle cells, and others. In chromosomal mosaics, this is not the case. Some of the patient's cells may be normal, whereas others may have, for example, a missing chromosome or an extra one. The usual cause of chromosomal mosaicism is an error in mitosis in some cell during the development of the embryo. Only one of the daughter cells may be abnormal, or both may be. In either case, how much of the patient's body is one or the other cell type will depend on when in development the abnormal mitosis occurred and what types of tissue developed from the cells involved. The more cell types that can be examined, the more likely it is that the abnormal cells in a mosaic individual will be detected. Unfortunately, some cell types cannot be examined at all. It is virtually impossible to find out whether or not nerve cells, for example, are chromosomally normal; there is at present no convenient way to obtain and study them.

Identification of Human Chromosomes

In summary, the chromosomes of small lymphocytes in metaphase of mitosis are the ones most easily obtained and usually examined. Figure 2.3 shows the chromosome constitution of a normal female (Figure 2.3a and b) and a normal male (Figure 2.3c and d). An unarranged photograph of the chromosomes in a nucleus, as in Figures 2.3a and c, is called a **metaphase spread**; when the chromosomes in a metaphase spread are cut out and rearranged as in Figures 2.3b and d, the arrangement is called a **karyotype**.

Figure 2.3 shows that normal cells have 46 chromosomes comprising 23 homologous pairs. The homologous chromosomes match in size and shape, except for the sex chromosomes. There are two **sex chromosomes** in both males and females. In females, the two sex chromosomes match in shape and size and are truly homologous in the sense that they carry corresponding genes; this sex chromosome is designated the **X chromosome** (see Figure 2.3b). In males, the two sex chromosomes do not match in shape and size, nor do they carry corresponding genes; one of the sex chromo-

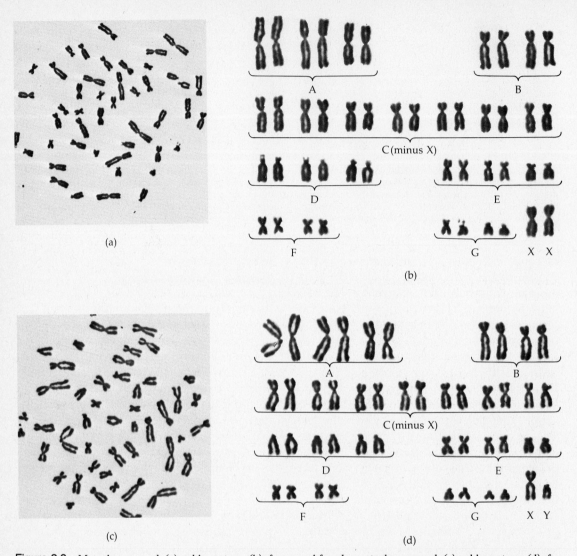

Figure 2.3 Metaphase spread *(a)* and karyotype *(b)* of a normal female; metaphase spread *(c)* and karyotype *(d)* of a normal male. Note that the only difference in the karyotypes is in the sex chromosomes: Females have two X chromosomes, and males have one X chromosome and one Y chromosome. (Identification of the X chromosome among the C-group chromosomes is somewhat arbitrary without special staining procedures.)

somes is an X (the same X chromosome as in females), and the other sex chromosome is designated the **Y chromosome** (see Figure 2.3d). In addition to the sex chromosomes—XX in females and XY in males—the cells of both sexes have 44 other chromosomes that are called **autosomes** to distinguish them from the sex chromosomes; these 44 autosomes represent 22 homologous pairs. The 44 autosomes together with the 2 sex chromosomes provide the grand total of 46 chromosomes in the cells of normal individuals of both sexes.

According to agreements reached at international conferences on nomenclature, the 22 pairs of autosomes are numbered by size from longest to shortest. A problem with this numbering system is that the lengths of the chromosomes are sufficiently variable from cell to cell that one cannot always be sure which chromosome is which. One can decide with some confidence whether a chromosome is long, medium, or short, however; and one can tell from the lengths of the arms whether a chromosome has its centromere near the middle (making it a **metacentric** chromosome), near the tip (an **acrocentric** chromosome), or somewhere between (a **submetacentric** chromosome). Based on overall length and centromere position, the chromosomes (including the sex chromosomes) can be lumped readily into seven groups, which are assigned the letters A through G (see Figure 2.3b and d).

The criteria by which chromosomes are assigned to groups A through G are listed in Table 2.1. Group A comprises the long metacentrics; there are three homologous pairs of these chromosomes, which, because of their length, are designated chromosomes 1 (the longest in group A) through 3 (the shortest in group A). Similarly, group B comprises the long submetacentrics, which are designated by the chromosome numbers 4 and 5. Group C is the largest group, consisting of chromosomes 6 through 12 plus the X chromosome. The smallest chromosomes in the complement are the short acrocentrics designated chromosomes 21, 22, and the Y (group G). The chromosome associated with **Down syndrome** (formerly called *mongolism*), a condition that will be discussed in Chapter 6, has historically been designated chromosome 21. This designation of the Down syndrome chromosome has been retained even though chromosome 21 is actually somewhat shorter than chromosome 22. The numbering of chromosomes 21 and 22 is therefore the one exception to the rule of numbering by size—number 21 being shorter than number 22.

Chromosome complements are customarily arranged and presented as karyotypes of the sort depicted in Figure 2.3b and d. When a karyotype is produced, the chromosomes are individually cut out of a photograph of a metaphase spread. The chromosomes are then separated into groups A through G, the homologous pairs of chromosomes in each group are identified, and the chromosomes are arranged in pairs and the pairs numbered from longest to shortest (except for chromosomes 21 and 22). Karyotyping hundreds or thousands of individuals can be extremely tedious, of course, but various computer-assisted methods are cur-

TABLE 2.1 DEFINITION OF HUMAN CHROMOSOME GROUPS A THROUGH G

Group	Chromosome Numbers	Distinguishing Characteristics
A	1 through 3	Large metacentrics
B	4 and 5	Large submetacentrics
C	6 through 12 plus X	Medium submetacentrics
D	13 through 15	Medium acrocentrics
E	16 through 18	Medium submetacentrics, but smaller than C group
F	19 and 20	Small metacentrics
G	21 and 22 plus Y	Small acrocentrics

rently being devised and perfected. Although the assignment of chromosomes to groups is relatively easy, distinguishing among the chromosomes within a group (and especially trying to identify the homologous pairs) is often pure guesswork. The exercise has been judged "more amiable than realistic." Actually, only five chromosomes are sufficiently different in appearance from the others to be readily identified: Chromosomes 1, 2, and 3 can be picked out, as can chromosome 16 and the Y. Fortunately, several procedures have been devised to positively identify every one of the others.

One aid to chromosome identification is called **autoradiography**. In autoradiography, a chromosome is induced to take a picture of itself. Nucleotides incorporated by the cell into the DNA of chromosomes during chromosome replication can be added to the culture medium in radioactive form. These radioactive nucleotides are unstable; they will eject electrons. An electron that hits a photographic film will cause

a spot to appear, leaving a record of its presence. Any part of a chromosome that contains the radioactive nucleotide will produce spots over the radioactive segment when overlaid with unexposed film and left in the dark to expose for several days; a chromosome that contains none of this material will not cause darkening of the film. (Figure 2.4 is an example of an autoradiograph karyotype.) The procedure of autoradiography can be used to identify individual chromosomes because each chromosome replicates at a characteristic time during interphase. If the radioactive nucleotide is added to cultures $3\frac{1}{2}$ to 4 h before the cells are harvested, then any chromosome or part of one that is found to have incorporated the radioactivity must have replicated within $3\frac{1}{2}$ to 4 h prior to metaphase, the stage in which the chromosomes are examined. Trial and error has shown that, in blood cultures, the chromosomes that replicate $3\frac{1}{2}$ to 4 h before metaphase are the last chromosomes to replicate. Thus,

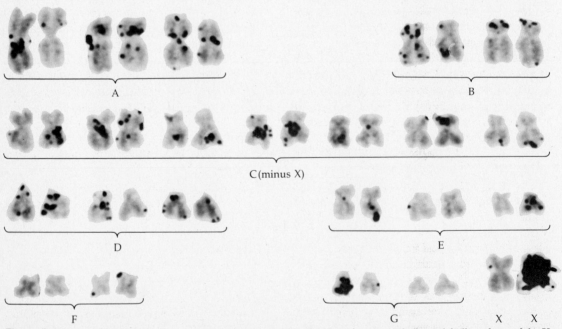

Figure 2.4 Autoradiograph of chromosomes from a normal female. Note the extremely heavy labeling of one of the X chromosomes, which is the inactive X to be discussed in Chapter 5.

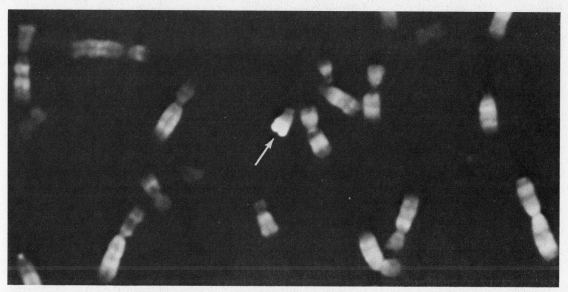

Figure 2.5 Quinacrine-induced fluorescence of human chromosomes. Note the particularly bright fluorescence of the long arm of the Y chromosome (arrow).

the late-replicating chromosomes can be identified because only they will have incorporated radioactive nucleotides that will produce spots on the film. The short arm of chromosome 5 is in this category; chromosomes 13, 18, and 21 are the last to take up radioactive label in their respective groups; and one of the X chromosomes in females is invariably late in replicating (see Figure 2.4). By extending the time of exposure to radioactive label, autoradiography has been used to reveal that chromosomes 19 and 20 are among the earliest chromosomes to replicate.

The most elegant procedures for identifying chromosomes are the ones that use special stains that bind preferentially to certain segments of the DNA in the chromosomes because these regions are particularly rich or poor in some chemical constituents. Such stains produce a visible pattern of **bands** or crosswise striations on the chromosomes. Some stains, for example **quinacrine** or **Hoechst 33258**, fluoresce brightly when examined under certain kinds of light. Chromosome

21 fluoresces more brightly than chromosome 22 when stained with quinacrine, for example; and the Y chromosome fluoresces brilliantly (Figure 2.5). The fluorescent stains mentioned above are thought to interact primarily with regions of DNA that are relatively rich in A-T nucleotide pairs (and correspondingly poor in G-C nucleotide pairs); bands in chromosomes produced by the stains are called **Q bands**. (These fluorescent stains, incidentally, are the ones mentioned in Chapter 1 that are used in conjunction with BUdR to detect sister-chromatid exchange.)

One spectacular chromosome-banding method uses the **Giemsa stain**, a mixture of two dyes. When properly used, Giemsa produces the beautiful pattern of bands, known as **G bands**, shown in Figure 2.6. Note in the figure that each chromosome has a unique pattern of bands and that homologous chromosomes can readily be identified by their identical banding patterns. Altogether, Giemsa produces about 300 G bands in the haploid set of human chromosomes. (Recall that the haploid set

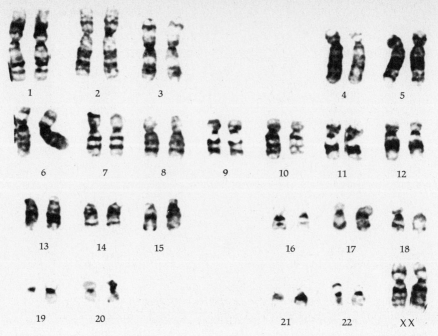

Figure 2.6 Karyotype of a human female showing the spectacular banding pattern observed with the Giemsa staining technique.

consists of 23 chromosomes, one chromosome from each of the 23 homologous pairs.) Since humans are widely believed to have between 30,000 and 90,000 genes, each G band, on the average, represents 100 to 300 genes.

In addition to Q bands and G bands, several other banding procedures have been developed. One procedure highlights the centromeric region of the chromosomes (producing **C bands**), for example; another reverses the Giemsa effect (producing **R bands**)—it stains the regions Giemsa leaves unstained and vice versa. Figure 2.7 is a composite of the various banding patterns observed in human metaphase chromosomes.

Still greater resolution of human chromosomes can be achieved by examining G bands at prophase, when the chromosomes are more extended than they are at metaphase. Two examples of prophase G banding are shown in Figure 2.8. For both chromosome 1 and the X chromosome, the metaphase G bands are

shown at the right and the prophase G bands at the left. The greater resolution of prophase banding is evident. Indeed, the number of prophase G bands in the haploid set is about 1200—four times more than the number of metaphase G bands. Each prophase G band therefore represents, on the average, 25 to 75 genes.

It should be noted here that techniques for the detailed study of human chromosomes were developed relatively recently, in the 1960s and 1970s. Indeed, human chromosomes could hardly be studied at all until the mid-1950s. Two major obstacles in the study of human chromosomes were the small number of cells at metaphase that could be examined and the tendency of the chromosomes to clump together, making it virtually impossible to be sure how many chromosomes there were, let alone to describe what they looked like. Cell division in human cells was first described in about 1880, but early counts of chromosomes

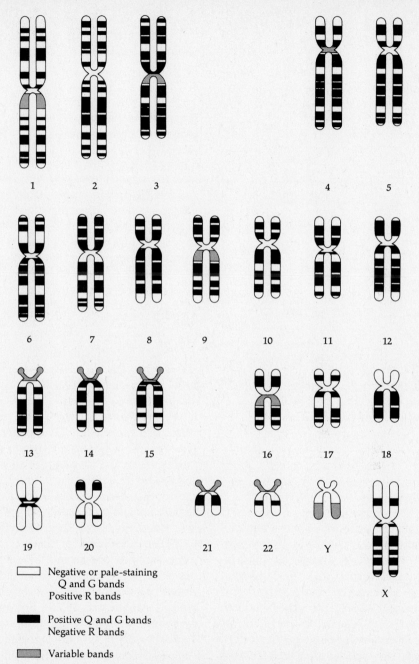

Figure 2.7   Banding pattern of human chromosomes obtained as a composite of various special staining procedures. G bands are Giemsa bands (see Figure 2.6); R bands are reverse Giemsa bands (R bands are heavy where G bands are light and light where G bands are heavy); and Q bands are quinacrine fluorescent bands (see Figure 2.5).

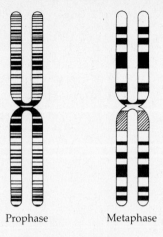

Prophase   Metaphase

Chromosome 1

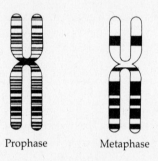

Prophase   Metaphase

X chromosome

Figure 2.8 Increase in the number of G bands observed in prophase as compared with metaphase. Only chromosome 1 and the X are illustrated, but all chromosomes show a comparable increase in the number of bands.

gave the number as 18, 24, or more than 40. Higher numbers were obtained when techniques improved, and by about 1910 the consensus had been reached that the number of chromosomes in human cells was 48. This error was perpetuated until 1956, when modern methods were first applied to the study of human chromosomes. The normal number of chromosomes in human cells was then found to be 46. Two principal techniques that finally transformed human chromosomes into objects suitable for study were the use of colchicine to increase the number of cells in metaphase that

could be analyzed and the insightful little idea to soak the cells in weak salt solution for a few minutes to swell them and untangle the chromosomes.

Normal Variation in Human Chromosomes

Some chromosomes in the human set do not look the same in every person; that is, minor variants of certain chromosomes do occur in the human population. Examples of such variants are shown in Figure 2.9. Chromosome 1 at the left of Figure 2.9a has a long arm that is noticeably longer than that of its homologous counterpart. Figure 2.9b shows chromosome 21 (right) with giant satellites. (A **satellite** is a little blob of chromosomal material attached to the short arm of a chromosome by a slender filament that stains only slightly, so that the connection between the satellite and the arm of the chromosome is nearly invisible. The filament that connects the satellite to the short arm is involved in forming the nucleoli during interphase. Note in Figure 2.7 that the chromosomes in groups D and G normally have satellites.) Figure 2.9c and d illustrate variation in the length of the long arm of the Y chromosome; the Y chromosome in Figure 2.9c is shorter than chromosome 22 (normally the Y is slightly longer), whereas the Y in Figure 2.9d is substantially longer than chromosome 22.

The prevalence of the sort of chromosome variation shown in Figure 2.9 has been examined in a number of studies, but the results of a Boston study of newborn males are typical. Among the 13,751 Boston newborns who were studied, 84 (i.e., 0.6 percent, or approximately 1 in 155) had major chromosome abnormalities of the type to be discussed in Chapters 6 and 7; these types of abnormalities are associated with severe mental or physical defects. In addition to the major chromosome abnormalities, a total of 462 (i.e., 3.36

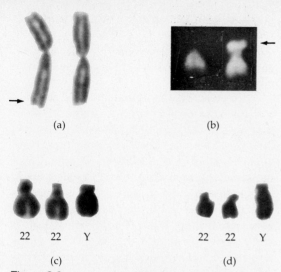

(a)   (b)

22   22   Y   22   22   Y

(c)   (d)

Figure 2.9 Some variants found in the normal chromosome complement. *(a)* An "uncoiler" chromosome 1 on the left and its normal homologue on the right; note that the long arm of the uncoiler chromosome (arrow) is about 15 percent longer than that of its homologue. *(b)* A normal chromosome 21 (stained with quinacrine) and a chromosome 21 with giant satellites (arrow). *(c)* A short Y chromosome beside a pair of chromosome 22 from the same cell for comparison; the short Y is the same length or shorter than chromosome 22—in contrast to a normal Y, which is slightly longer than chromosome 22. *(d)* A long Y chromosome beside a pair of chromosome 22 from the same cell for comparison; the long Y is much longer than chromosome 22. Each pair or set of chromosomes has been enlarged to a different scale.

percent, or approximately 1 in 30) individuals had minor chromosome variants of the type depicted in Figure 2.9. In contrast to the results of major chromosome abnormalities, infants who have these minor variants seem to be completely normal and healthy.

A summary of the normal variants found in the Boston study is given in Table 2.2. To read the table, a note on terminology is necessary. The short arm of a human chromosome is designated by the letter p (for *petite*); the long arm is designated by the letter q. A plus (+) sign following a chromosome designation indicates an addition of chromosomal material; a minus (−) sign following a designation indi-

cates a loss of chromosomal material. Thus, in Table 2.2, the symbols Dp+ and Gp+ represent addition of material (+) to the short arm (p) of a D- or G-group chromosome, respectively; that is to say, Dp+ and Gp+ are chromosomes with large satellites. Similarly, Yq+ represents a Y chromosome of the type shown in Figure 2.9d—one with an unusually long (+) long arm (q). Likewise, Yq− represents a Y chromosome of the type shown in Figure 2.9c—one with an unusually short (−) long arm (q).

As is apparent from Table 2.2, about 0.74 percent (1 individual in 135) of the Boston newborns had abnormally large satellites on either a D or a G chromosome. The Y chromosome is particularly variable—1.75 percent (1 in 57) of the males had Yq+, and 0.38 percent (1 in 263) had Yq−. All told, about 1 newborn out of 30 had some minor chromosomal variant.

Not shown in Table 2.2 is yet another normal variant—a metacentric (instead of submetacentric) C-group chromosome, which banding methods reveal to be chromosome 9. The incidence of the metacentric C is correlated with race: The variant occurs in approximately 1 out of 1350 Caucasians but in

TABLE 2.2 NORMAL VARIANTS IN HUMAN CHROMOSOMES FOUND IN A LARGE BOSTON STUDY

Type of variant	Number of cases (out of 13,751)	Percent
Dp+	51	0.37
Gp+	51	0.37
Yq+	240	1.75
Yq−	52	0.38
Other	68	0.50
Total	462	3.36

Source: Data from S. Walzer and P.S. Gerald, 1977, A chromosome survey of 13,751 consecutive male newborns, in E.B. Hook and I.H. Porter (eds.), Population Cytogenetics: Studies in Humans, Academic Press, New York, pp. 45–61.

approximately 1 out of 75 non-Caucasians. There is thus a nearly twentyfold greater incidence of the metacentric C in non-Caucasians, but it is to be emphasized that this chromosome is associated with no known physical or mental effects whatsoever.

Considering that chromosomes carry the genes, it may seem strange that certain chromosomes or parts of chromosomes can vary in size without having dramatic or even detectable effects on the individual. Several explanations for this can be offered. Perhaps the differences in length do not reflect differences in the amount of genetic information in the chromosomes as much as they reflect differences in the amount of coiling of the chromosomes; the more tightly coiled a chromosome is, the shorter it will appear. Or it may be that the variation in length occurs in parts of chromosomes that are relatively inert genetically—that is, in parts that do not carry much genetic information to begin with. It is known, for example, that the factors on the Y chromosome that determine maleness are located on the short arm of the Y, whereas the normal variation observed in the overall length of the Y is due exclusively to differences in the length of the long arm. Perhaps the long arm of the Y (and certain other chromosomal regions as well) carries so little or such a special kind of genetic information that the variation in size makes no difference.

Certain chromosomal regions, including the long arm of the Y chromosome, are often described by the term **heterochromatin**, which means that the chromatin remains relatively compacted or coiled during interphase so that the heterochromatic material can be stained with various dyes and observed during interphase. Most chromatin, recall, becomes uncoiled during interphase and is not visible in the light microscope; this type of chromatin that is not visible in interphase is called **euchromatin**. Various lines of evidence suggest that heterochromatin contains very few genes to begin with, or genes that are not expressed in the cell, so variation in the amount of heterochromatin as exemplified in the Yq− or Yq+ chromosomes would be expected to have negligible effects.

In any case, the normal types of chromosomal variation summarized in Table 2.2 show that normal individuals can have chromosomal differences; that is, there is chromosomal (i.e., genetic) variation among normal people. Chromosomal variation is just the tip of the iceberg of genetic variation, however. As will be discussed in Chapter 14, there is so much subtle variation in the DNA among individuals—variation not visible in the microscope—that no two individuals (with the exception of identical twins) are ever genetically alike.

Sexual and Asexual Inheritance

Now we must consider the special kind of cell division that is involved in the formation of gametes. This gamete-forming type of division is called **meiosis**, and it is the basis of **sexual inheritance**—that is, inheritance mediated through the union of gametes and involving two sexes. Mitosis, discussed in the preceding chapter, is the cellular basis of **asexual inheritance**—that is, inheritance from cell to cell without the involvement of gametes. Yet a third type of biological inheritance should be mentioned briefly, even though it is not of great importance in human genetics. This type is called **cytoplasmic inheritance**. As discussed in Chapter 1, some organelles in the cytoplasm have their own DNA and therefore their own genes; the mitochondria are the best examples in animals. Even though the amount of DNA in the mitochondria is minuscule compared with the amount of DNA in the nucleus, certain traits of the mitochondria are determined by the genes in mitochondrial DNA; the inheritance of these traits is said to be cytoplasmic because the genes responsible are located within the organelles in the cytoplasm. (The actual situation regarding mitochondrial inheritance is more complicated because genes in the nucleus also provide information for some of the constituents of mitochondria, so the inheritance of mitochondrial traits is due to a complex interaction of nuclear and mitochondrial genes.) Although cases of cytoplasmic inheritance are well established in several experimental organisms, few, if any, human traits are associated with cytoplasmic inheritance, although variation in human mitochondrial DNA is known to occur.

The process of sexual inheritance is more subtle than that of asexual inheritance for several reasons. With sexual inheritance, each child has two parents, and both contribute equally to the child's genetic constitution. Despite the parents' equal contributions, however, the children do not tend to resemble one parent or the other in particular, nor do the children look exactly intermediate or like a blend of the parents. Although the resemblance between parent and child is sometimes particularly striking, most children exhibit a mixture of traits or features. Some of these traits can be recognized (more or less) in the parents or other relatives; others are unique to the child. Finally, with sexual inheritance, the offspring produced are genetically different from one another; with asexual inheritance, by contrast, the daughter cells are genetically identical. The genetic diversity of offspring produced through sexual inheritance makes it seem as if sexual inheritance is completely unpredictable, almost whimsical.

The unpredictability of sexual inheritance is literally true in particular cases: What any particular unborn child will be like, whether it will exhibit one trait or another, whether it will be male or female, and so on, is impossible to predict except in rare and unusual cases or unless cells from the fetus are studied. But there is regularity in sexual inheritance nevertheless; the rules of sexual inheritance are known. These rules are statistical, involving probabilities. We can say with complete confidence that very nearly half of all children born in the world today will be female, but we cannot say what the sex of any particular unborn child will be. One might think at first that statistical rules are of no value because they cannot be used to predict particular events, but this would be wrong. For potential parents to know in advance that they have a high risk of producing an abnormal child—say, 25 or 50 percent—is to forewarn them, even though any particular child they have may be normal. Often the high risk alone may be sufficient to deter them from having children; this decision prevents a terrible anxiety during each pregnancy and it relieves society of the social and medical costs of severely malformed children.

The Strategy of Meiosis

The statistical rules of inheritance have their origin in the process of meiosis. This meiotic type of cell division has evolved to satisfy simultaneously two conflicting requirements. On the one hand, the genetic constitution of the offspring, not only humans but all species, should be similar to that of the parents. Some similarity is desirable because the very survival of the parents demonstrates that their genetic constitutions are successful, and it is generally unwise to break up winning combinations of genes. On the other hand, the genetic constitution of the offspring should not be *too* similar to the parents. Some differences are desirable because children inevitably grow up in an environment that is different in many ways from the one their parents grew up in. The survival of the parents shows that the genes carried by them are successful, but successful in the *old* environment. Sexual reproduction must equip at least some of the offspring with the ability to cope with a new and unpredictable environment and thereby allow the species to evolve, while at the same time preserving to some extent the harmoniously working genes in the parents. Meiosis satisfies both these requirements by allowing only half the genes from each parent to be passed on to a particular child, but the nature of the process is also to reshuffle the genes in such a way that each child will have a particular combination of genes unique to it alone.

In humans, the meiotic process occurs only in certain cells of the ovaries and testes. These cells are the **oocytes** of the ovaries and the **spermatocytes** of the testes. The meiotic process has often been compared to two mitotic divisions occurring in sequence. True, there are some similarities, but the differences between meiosis and mitosis are more important than the similarities. Recall that the most important event in early mitosis is the replication of each chromosome; the two sister chromatids are then separated by being pulled in opposite directions into what will become the nuclei of two daughter cells. The two daughter cells produced by mitosis each have a full diploid complement of 46 chromosomes. Meiosis is different; when complete, it results in four daughter cells, not two, and each daughter cell has a haploid set of 23 chromosomes—one representative from each of the homologous pairs. The daughter cells produced by meiosis develop into the egg or the sperm, and each carries 23 chromosomes; when the egg and sperm come together in fertilization, their fusion restores the normal diploid complement of human chromosomes: 23 pairs of homologues, 46 chromosomes altogether.

The Mechanics of Meiosis: Prophase I

Meiosis, like mitosis, begins with an interphase. This is called **interphase I**. (The stages of meiosis are named like the stages of mitosis, but because meiosis is vaguely similar to two mitotic divisions occurring in sequence, the stages are numbered I and II to distinguish them. It must be emphasized that there are critically important differences between, say, prophase of mitosis and the corresponding stages of meiosis, even though these stages are referred to by the same name.) From the standpoint of genetics, the most important event that occurs during interphase I is the replication of the DNA in the chromosomes.

After DNA replication is accomplished and the cell has mobilized for division in various other ways, the cell enters **prophase I**, marked by the first appearance of the long, threadlike chromosomes convoluted throughout the nucleus. Figure 2.10 illustrates the appearance of the chromosomes at various times during prophase I, and the accompanying photographs are from testes cells of the grasshopper *Chorthippus parallelus*. The sketches in Figure 2.10 depict prophase I in an organism that has a diploid number of four chromosomes (two homologous pairs, one shown as metacentric and the other as submetacentric); the shaded chromosomes represent those that were contributed to the individual through the sperm, and the unshaded ones came from the egg. As seen in Figure 2.10, prophase I is a very long stage comprising several noteworthy events that decisively distinguish it from anything seen in mitosis. When first seen in prophase I (see Figure 2.10a), the chromosomes are single threads, unlike their appearance in prophase of mitosis when they are clearly already doubled with their two sister chromatids intimately intertwined. Even though the DNA has been replicated in interphase I, the chomosomes themselves in prophase I of meiosis have not yet become visibly double. Throughout all of prophase I, the nucleus increases slowly in volume as the chromosomes undergo coiling so that the chromosomes appear to become progressively shorter and thicker.

An event unique to meiosis and of enormous importance occurs when the chromosomes are still quite extended. The homologous chromosomes pair up. The homologues find each other somehow and come to lie side by side, gene for gene, like zippers coming together tooth for tooth. The forces that bring about this alignment are unknown. The homologous chromosomes pair first at a few regions near the tips (see Figure 2.10b); the pairing then progresses zipperlike from these few points. This pairing of homologous chromosomes is called **synapsis**. Through the electron microscope, a special ladderlike structure composed mainly of protein can be seen lying with its rungs crosswise between the synapsed or paired regions of chromosomes; this structure, called the **synaptonemal complex**, is evidently necessary for snyapsis to occur normally. When the synapsis of each of the homologous pairs of chromosomes is completed, the nucleus will contain as many structures as there are chromosomes in the haploid set (two in Figure 2.10c). In human cells, the nucleus in the stage corresponding to Figure 2.10c has 23 structures visible in it, each consisting of two homologues intimately paired along their length. One exception in human cells should be noted, however. In males, the X and Y chromosomes do not synapse along their entire length but only tip to tip, with the tip of the short arm of the X pairing with the tip of the short arm of the Y. Throughout the period when synapsis occurs, the chromosomes continue to become shorter and thicker (compare Figure 2.10c with 2.10d).

The next noteworthy event occurs after synapsis is completed. The chromosomes finally become visibly double (see Figure 2.10e). Each chromosome is now seen to consist of a pair of sister chromatids attached to a single centromere. Thus, the synapsed homologues consist of two centromeres side by side, with each centromere having attached two sister chromatids. The whole unit is called a **bivalent**, and the chromatids attached to different centromeres are called **nonsister chromatids**. As illustrated in Figure 2.11, each bivalent consists of four chromatids (i.e., two pairs of sister chromatids) attached pairwise to two centromeres and synapsed lengthwise. The important distinction between sister and nonsister chromatids is that sister chromatids, being replicas of the same chromosome, must be genetically identical; nonsister chromatids, being replicas of different chromosomes, may be genetically different. In human cells in the stage corresponding to Figure 2.10e, the number of chromatids is 92, the number of centromeres is 46, and the number of bivalents is 23.

Within the bivalents are certain crosslike structures called **chiasmata** (the singular is **chiasma**), as illustrated in Figures 2.10e and 2.11. Chiasmata are the visible manifestations of a process called **crossing-over**, during which nonsister chromatids undergo an exchange of parts. The genetic importance of crossing-over is best made clear by analogy. The analogy is not perfect because the bivalent is unique—unlike anything else in the world. Nevertheless, for purposes of discussion, imagine two lengths of ticker tape that both carry letters and words designed to convey the same meaning. The tapes are assumed to be identical except where misprints occur (Figure 2.12a). The ticker tapes correspond to homologous chromosomes, and each one carries a single molecule of DNA as described in Chapter 1. Arrange these tapes so that they are aligned letter to letter. This arrangement represents synapsis. Now let each tape be duplicated, but be sure that each copy is still aligned letter for letter with its original. Select some point along the length of the tape and staple the copy and the original together. Do this as well to the other copy and its original, being careful to insert this second staple in the same relative position as the first. These staples represent the centromeres. This duplication and stapling produce two pairs of tapes, each pair attached to one centromere, and the tapes are aligned lengthwise. The whole structure represents a bivalent.

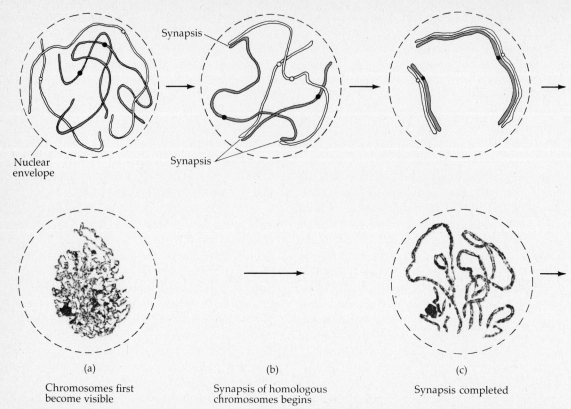

(a)   (b)   (c)

Chromosomes first   Synapsis of homologous   Synapsis completed
become visible   chromosomes begins

Figure 2.10   Principal events in prophase I of meiosis in a hypothetical cell that has two pairs of homologous chromosomes (top). Corresponding stages in testes cells of the grasshopper *Chorthippus parallelus* as viewed in the light microscope (bottom). Note particularly the pairing (synapsis) of homologous chromosomes in *(b)* and the appearance of chiasmata (crosslike structures) in *(e)*. (Photos 600–900X.)

Now the important part occurs. Proceed along the tapes until you come to some position—any position. Cut two of the tapes at exactly the same position, being careful that the two strands that are cut are attached to different centromeres (i.e., are nonsister chromatids), as shown in Figure 2.12b. Now interchange the severed ends and repair the broken strands with transparent tape (see Figure 2.12c). Be careful not to move the ends physically; leave them where they are, but take two small pieces of tape and bridge the gaps. What you have now accomplished is a splice. Two of the strands in the bivalent, the uncut ones, are identical to their parent strands. The other two

strands are spliced; each is identical along part of its length with one of the parent strands, and along the rest of its length it is identical with the other parent strand (see Figure 2.12d).

The correspondence between this paper-and-tape model and an actual chiasma is shown in Figure 2.13. Crossing-over, represented by the chiasma, corresponds to the splice between nonsister chromatids. The splicing does not involve scissors and tape, of course, and the actual mechanism is not fully worked out. But apparently there are particular proteins in the nucleus that can nick or break chromosomes. Other proteins are able to repair these breaks. When breaks occur in roughly corresponding

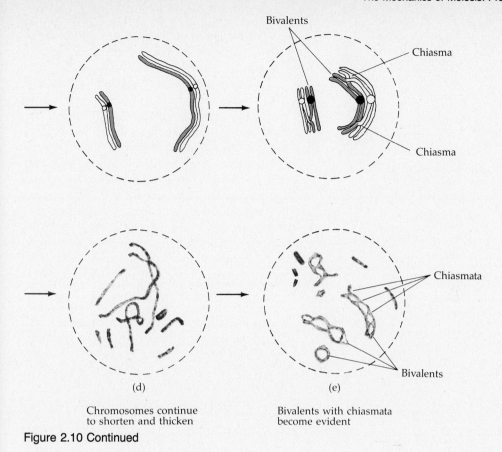

(d)

Chromosomes continue
to shorten and thicken

(e)

Bivalents with chiasmata
become evident

Figure 2.10 Continued

places in two nonsister chromatids in a bivalent, the breaks can sometimes be repaired
crosswise in the way that the tape was spliced
in the analogy; two "spliced" chromosomes
result. The breaks need not occur at exactly the
same place in the nonsister chromatids because
still another set of proteins in the nucleus is
capable of chemically degrading overlapping
segments or chemically resynthesizing the genetic information missing in gaps by replicating
the information from the corresponding region
of one of the unbroken strands. The overall
process is therefore exceedingly precise in
terms of preserving genetic information even
though the initial breaks are. not exactly

aligned. The upshot of these events is that the
spliced strands neither gain nor lose genetic
information; they merely interchange it. The
chromosome strands that are actually involved

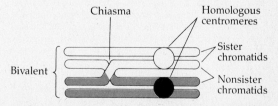

Figure 2.11   Expanded view of a bivalent late in prophase I, corresponding to Figure 2.10e. Note that the
chiasma is indicative of crossing-over between nonsister
chromatids.

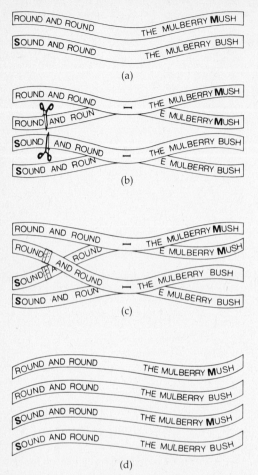

(a)

(b)

(c)

(d)

Figure 2.12 A surprisingly good analogy can be drawn between genetic information in chromosomes and letters on a ticker tape. *(a)* Two ticker tapes are paired as in synapsis; note that each tape bears a misprint (boldface letter). *(b)* The tapes have become visibly double, and breaks (cuts by scissors) occur at corresponding positions in nonsister chromatids. *(c)* The breaks are rejoined using transparent tape, forming a chiasmalike cross. *(d)* The resulting products; note that two products (the ones not involved in the splice) are like the original tapes; of the two products involved in the splice, one bears both misprints and one is entirely free of misprints.

quences of crossing-over will be discussed further in Chapter 4 and its molecular basis in Chapter 8.

Here we should emphasize that crossing-over must not be confused with sister-chromatid exchange (discussed in Chapter 1). Sister-chromatid exchanges normally have no genetic consequences because they are exchanges between genetically identical chromatids. Moreover, sister-chromatid exchanges are not associated with chiasmata. Crossing-over, by contrast, involves nonsister chromatids. Because nonsister chromatids can be genetically different, crossing-over can produce new kinds of recombinant chromosomes that are genetically different from either of the parent chromosomes (see Figure 2.12d).

The chemical events involved in crossing-over cannot be seen. Although chiasmata are known to represent crossovers, it is not known for certain when the actual break and rejoining of the chromosomes occur. Some geneticists believe that the crossover occurs in prophase I, just before the chiasmata first become visible, and a tiny amount of DNA synthesis does occur at about this time, as would be expected from a repair process. Other geneticists believe that the crossover itself actually occurs during interphase I, and a chiasma then reflects an event that occurred much earlier than prophase I. The evidence bearing on the time of crossing-over is somewhat ambiguous, but for most practical purposes, it really does not matter exactly when it occurs.

Chiasmata may have a physical function in the bivalents. Shortly after the chromosomes first become visibly double in prophase I, the nonsister chromatids in the bivalent appear to repel one another. Each pair of sister chromatids remains tightly associated, but they move apart from the other pair of sister chromatids. (See the photograph in Figure 2.13 for an example.) This repulsion is often most marked in the regions around the centromeres, because the physical moving apart of the cen-

in the splice are called **crossover strands** or **crossover chromosomes**, and they are said to have **recombined** or to be **recombinant chromosomes** because of the intermixture of genetic information between them. The genetic conse-

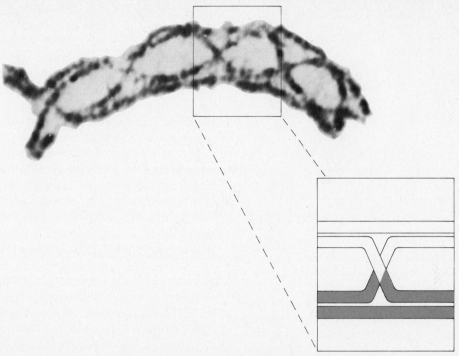

Figure 2.13   Micrograph of a bivalent with five chiasmata and diagram showing the genetic consequences of one of them (box). The bivalent here is in a stage corresponding to Figure 2.10e. (approx. 5000X.)

tromeres is easiest to see. In any case, the homologous chromosomes would probably dissociate completely if it were not for the chiasmata holding them together. Most bivalents have at least one chiasma. The shortest human bivalents (involving chromosomes of the G group) have one or two chiasmata; the longest ones (involving chromosomes of group A) have four, five, or six. The number varies from cell to cell and from bivalent to bivalent, but in males, the total number of chiasmata in all the bivalents taken together averages about 54, or 2.3 per bivalent (Figure 2.14). All bivalents usually have at least one chiasma, with the exception of the XY chromosome pair in males. No crossing-over between the X and Y occurs, and no chiasmata are observed. The two X chromosomes in females, on the other hand, behave in this respect like the auto-

somes, forming typical bivalents and exhibiting chiasmata.

The crucial sequence of events in prophase I is marked by the first appearance of the chromosomes, the pairing of homologues, the visible doubling of the chromosomes to make the bivalents four-stranded structures with chiasmata, and the apparent repulsion of homologous chromosomes in the bivalents. As the repulsion of the homologous chromosomes continues, the chiasmata sometimes seem to "slip" toward the tips of the chromosomes in much the way that a loose knot between two ropes can be slipped toward and even off the ends by pulling the ropes apart. The bivalents continue to shorten and thicken through more intense coiling toward the end of prophase I. At one point, the nucleolus begins to disappear. The spindle begins to form, and the

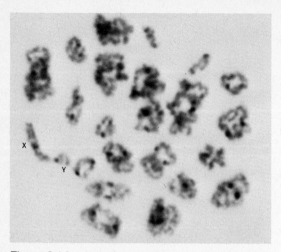

Figure 2.14 Bivalents in prophase I in a human male showing numerous chiasmata. Note the tip-to-tip association of the X and Y chromosomes.

bivalents tend to migrate to the periphery of the nucleus near the nuclear envelope. Shortly thereafter, the nuclear envelope breaks down, and this breakdown marks the end of prophase I.

The Mechanics of Meiosis: Later Events

Figure 2.15 summarizes the events that follow prophase I. The photo inserts are from testes cells of the grasshopper *Chorthippus parallelus*. In **metaphase I**, the spindle is completed and the bivalents move so that they come to lie on the equatorial plate (see Figure 2.15a). The orientation of the bivalents on the plate is usually quite precise. Each bivalent lines up so that its two centromeres lie on opposite sides of the plate, pointing toward the poles of the spindle.

Anaphase I commences when the two centromeres of each bivalent actually begin to separate as they are pulled in opposite directions (see Figure 2.15b). The pair of chromatids associated with each centromere follows

along, and the chiasmata that held the bivalents together are finally disengaged as they slip off the ends. Note the results of crossing-over in Figure 2.15b; the recombinant chromatids are those that carry partly paternal (shaded) and partly maternal (unshaded) material. As the centromeres proceed toward the poles in anaphase I, the chromatids seem to lose their attraction to each other and the arms are free to flop around somewhat, but the chromatids remain attached to the same centromere. This is the most important feature of anaphase I: The two centromeres of each bivalent are separated from each other, but the centromeres themselves do not split. Anaphase I is thus very different from anaphase of mitosis, in which the centromeres do split. The separation of homologous centromeres in anaphase I results in a halving of the chromosome number. Recall that the number of chromosomes in a cell is determined by counting the number of centromeres. Since homologous centromeres pull apart during anaphase I, each daughter cell receives the haploid number of centromeres and hence the haploid number of chromosomes.

The end of anaphase I coincides with the arrival of the chromosomes at the poles of the spindle. Only one homologue of each bivalent has gone to each pole, and in human cells, there will be 23 nonhomologous chromosomes at each pole. In **telophase I**, the chromosomes uncoil slightly and in some species a nuclear envelope begins to reform around each of the two groups of chromosomes. In many species (including humans), the cytoplasm divides during telophase I (see Figure 2.15c). In human males, the cytoplasm is divided roughly equally so that two daughter cells of equal size are formed. Each of these cells undergoes the steps that follow. In human females, the division of the cytoplasm is grossly unequal. The cytoplasm divides so that one of the daughter cells receives almost all of it. This cell is destined to

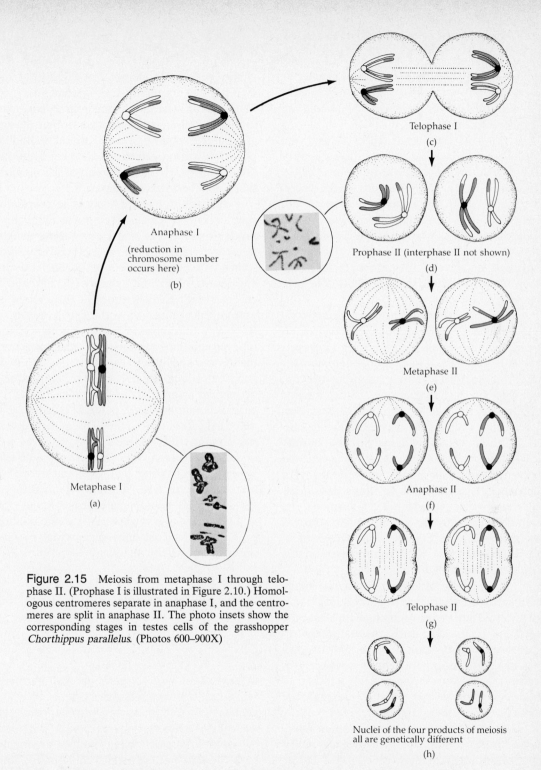

Anaphase I

(reduction in
chromosome number
occurs here)

(b)

Telophase I

(c)

Prophase II (interphase II not shown)

(d)

Metaphase I

(a)

Metaphase II

(e)

Anaphase II

(f)

Telophase II

(g)

Figure 2.15 Meiosis from metaphase I through telophase II. (Prophase I is illustrated in Figure 2.10.) Homologous centromeres separate in anaphase I, and the centromeres are split in anaphase II. The photo insets show the corresponding stages in testes cells of the grasshopper *Chorthippus parallelus.* (Photos 600–900X)

Nuclei of the four products of meiosis
all are genetically different

(h)

give rise to the egg. The other cell formed by the division is very small and lies just outside the membrane of the larger one. This small cell is called a **polar body**.

After the division of the cytoplasm, the nuclei in the daughter cells undergo a very abbreviated and incomplete interphase called **interphase II**. The chromosomes do not uncoil completely, and the nuclear envelope is not fully reformed. Most important, there is no synthesis of DNA or replication of the chromosomes during interphase II. For all practical purposes, human cells undergoing meiosis proceed almost directly from telophase I to prophase II, the next stage.

Prophase II begins what amounts to a conventional division of the chromosomes, almost like mitosis (see Figure 2.15d). However, nuclei in prophase II have the haploid number of chromosomes (23 in humans), not the diploid number. As noted earlier, the first meiotic division has reduced the chromosome number. Recall that each of the chromosomes in prophase II is associated with two distinct chromatids. These chromatids coil tightly again and become shorter and thicker. Toward the end of prophase II, the nuclear envelope disappears (in those species in which it has formed), the spindle begins to be set up again, and at **metaphase II** (see Figure 2.15e) the chromosomes line up with their centromeres on the cell's equatorial plate.

In **anaphase II**, the centromeres finally split into two parts (see Figure 2.15f). Anaphase II is thus a conventional anaphase of the sort that occurs in mitosis. After the centromeres split, each chromosome has its own independent centromere. The two new centromeres that arise from the splitting proceed to opposite poles of the spindle, dragging the arms of the chromosomes along.

In **telophase II**, the chromosomes uncoil and become diffuse, the nuclear envelope reforms, and the cytoplasm divides (see Figure 2.15g). As in telophase I, the division of the cytoplasm in males produces two daughter cells of equal size, whereas in females the division of the large cell is again unequal so that an egg and another polar body are formed. (The first polar body often undergoes the second meiotic division also, so that at the end of the meiosis, there may be three polar bodies.)

Meiosis is now complete. The spermatocyte or oocyte has divided to produce four products of meiosis, and each of these daughter cells carries the haploid number of chromosomes. The egg carries one of each of the autosomes, numbered 1 through 22, and one X chromosome. Each sperm also carries one of each of the autosomes, and it carries either an X or a Y chromosome but not both. Of the four products of meiosis in males, two carry the X and the other two carry the Y. As is apparent in Figure 2.15h, each of the products of meiosis is genetically different; they carry various combinations of paternal (shaded) and maternal (unshaded) genes. This genetic diversity among the products of meiosis is the underlying reason for the genetic diversity among the offspring that is characteristic of sexual reproduction. Earlier we mentioned that a typical human male expels about 350 million sperm in a single ejaculate; the meiotic process makes it virtually certain that each of these sperm will be genetically different!

Summary of Meiosis

At this point it may be useful to summarize the principal characteristics of chromosome behavior during meiosis.

INTERPHASE I: DNA synthesis and chromosome replication occur.

PROPHASE I: Chromosomes become visible as single strands; synapsis occurs; chromosomes become visibly double, producing bivalents within which chiasmata (the physical manifestations of crossing-over) are evident.

METAPHASE I: Bivalents align on the equatorial plate.

ANAPHASE I: Homologous centromeres pull apart, causing a reduction in chromosome number.

TELOPHASE I: Cytoplasm divides.

INTERPHASE II: Chromosomes uncoil slightly; there is no DNA synthesis or chromosome relication.

PROPHASE II: Chromosomes again coil.

METAPHASE II: Chromosomes align on the equatorial plate.

ANAPHASE II: Centromeres split and chromosomes pull apart.

TELOPHASE II: Cytoplasm divides, completing meiosis.

The stages interphase I through telophase I constitute the **first meiotic division**; the stages interphase II through telophase II constitute the **second meiotic division**.

Development of Gametes

After telophase II, the nuclei undergo what further differentiation is required to form a mature egg or sperm. In males, all the telophase II cells develop into sperm. The cells elongate and the tail of the sperm is formed (Figure 2.16). The nucleus becomes supercondensed in the sperm head, and the bulk of the cytoplasm is stripped away and degraded. What is left in the sperm head is essentially the nucleus. The head of the sperm is a sort of space capsule, and the tail is a propulsion system designed to launch and propel the nucleus toward fertilization.

The stages of meiosis in females are synchronized with events in the life cycle and the menstrual cycle. The earliest stage in meiosis—prophase I—takes place in the ovary of a female when she is still an embryo developing in her mother's uterus. When the embryo is only 4 months old, the deepest layers of the developing ovary contain many cells in the earliest part of prophase I. These cells gradually progress through prophase I. When the embryo is 7 months old, a few cells can be found in late prophase I. Late in prophase I, a curious arrest of meiosis occurs. Instead of proceeding to metaphase I, the bivalents unwind or "decondense" and the nucleus enters an interphaselike state. This state is not really interphase because none of the DNA synthesis that characterizes interphase I occurs. The cells in this interphaselike state are quiescent. In embryos 9 months old and about to be born, the ovary is heavily populated with cells in this quiescent state.

Each of the interphaselike cells is surrounded by a large number of helper cells called **follicle cells**. Each group of helper cells together with the included interphaselike cell is known as a **follicle**. The follicles are quiescent through childhood. At puberty, with the onset of menstruation and its hormonal cycles, one follicle per month is stimulated and **ovulation** (the shedding of the egg) occurs. (The synchronized rise and fall in the levels of female hormones trigger ovulation, and this is why the artificial interference with hormone levels caused by birth control pills prevents ovulation.) The stimulated follicle increases in size almost 10 times—to 4000 μm or about $\frac{3}{16}$ in—although the oocyte itself, the interphaselike cell, does not increase in size. The follicle bursts open and the oocyte is extruded from the ovary and is caught in the upper part of the fallopian tube, where it begins its journey toward the uterus. As the follicle is ripening, the chromosomes in the oocyte resume meiosis; they coil up again and undergo metaphase I, anaphase I, and telophase I. The tiny polar body is pinched off and sticks to the surface of the oocyte (see Figure 2.16). After the follicle bursts and the egg is shed, the cells go

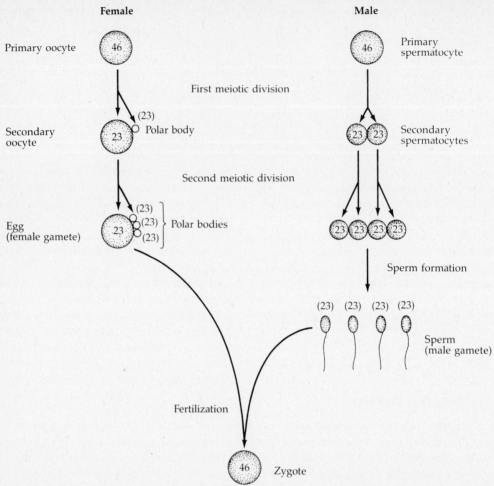

Figure 2.16 Chromosome numbers in various cell types in human reproduction, showing the grossly unequal cytoplasmic division in females leading to one large egg cell and three tiny polar bodies. The illustration is somewhat oversimplified because in human females the second meiotic division actually takes place in the fallopian tube *after* fertilization.

through prophase II and the chromosomes align for metaphase II. Metaphase II, anaphase II, and telophase II occur only after the sperm has penetrated the egg. During telophase II, the egg extrudes another polar body. The fertilized egg at this stage contains two nuclei—its own telophase II nucleus and the nucleus contributed by the sperm. The DNA in both nuclei is replicated, and the two nuclei move toward each other and fuse, producing a full diploid complement of already replicated chromosomes. The fertilized egg immediately enters prophase of mitosis, which occurs while the egg is still progressing slowly down the fallopian tube. About 3 or 4 days after fertilization, the zygote reaches the uterus, and 3 or 4 days later it becomes implanted in the uterine wall.

Earlier we mentioned that the statistical rules of sexual inheritance are the consequences of the process of meiosis. Meiosis determines how the genes will be distributed among gametes, and fertilization determines how these gametes will combine to produce individuals of the next generation. However, meiosis ordinarily is not observed directly, nor is fertilization. The genetic consequences of meiosis and fertilization are nevertheless observable in the pattern of inheritance of various traits. These patterns of inheritance—the genetic result of meiosis—are the subject of the next several chapters.

SUMMARY

1. The normal diploid complement of chromosomes in humans consists of 46 chromosomes—44 **autosomes** and 2 **sex chromosomes**. In females, the sex chromosomes are both X chromosomes; in males, one sex chromosome is an X and the other is a Y.

2. The autosomes in humans occur as 22 homologous pairs, which are assigned the numbers 1 through 22 from longest to shortest. (Chromosome 21 is slightly shorter than chromosome 22, however.) Individual chromosomes cannot be identified without special procedures, but each can be assigned to one of seven groups designated by the letters A through G. Each group contains chromosomes that are similar in size and centromere position. The longest chromosomes, which are metacentrics, constitute group A; the shortest ones, which are acrocentrics, constitute group G. The X chromosome is a C-group chromosome; the Y is a G-group chromosome.

3. Individual chromosomes or parts of chromosomes can be identified with special procedures. **Autoradiography** relies on the fact that radioactive nucleotides incorporated in DNA will emit radiation and darken photographic film. Certain fluorescent dyes interact preferentially with regions of DNA rich in A-T nucleotide pairs, so sections of chromosomes that carry such DNA are preferentially stained, producing Q bands.

4. Giemsa stain, which produces G bands, gives the highest resolution of human chromosomes. There are about 300 G bands in the haploid set of chromosomes at metaphase, and each pair of homologous chromosomes has a unique pattern of G bands. Humans are thought to have between 30,000 and 90,000 genes, so each metaphase G band represents, on the average, 100 to 300 genes. Still finer resolution of chromosomes can be gained by means of Giemsa staining during prophase, when approximately 1200 G bands are visible.

5. Normal variation among human chromosomes does occur. Approximately 1 individual out of 30 has some minor chromosome variant, such as a D-group or a G-group chromosome with abnormally large **satellites** or a Y chromosome with an atypically sized long arm. These variants usually involve parts of chromosomes that contain **heterochromatin**, a type of chromatin thought to carry few (or inactive) genes.

6. Meiosis is the basis of sexual inheritance, and it occurs only in the **oocytes** of the ovaries and the **spermatocytes** of the testes. In males, meiosis results in four haploid sperm; in females, it results in a haploid egg and three polar bodies. Fertilization restores the diploid complement of chromosomes to the zygote.

7. The major features of the first meiotic division are as follows: **interphase I** (DNA and chromosome replication); **prophase I** (synapsis and formation of bivalents containing chiasmata, which are the physical manifestations of crossing-over); **metaphase I** (alignment of bivalents on equatorial plate); **anaphase I** (separation of homologous centromeres, reducing the chromosome number); **telophase I** (cytoplasmic division). The major events in the second meiotic division are: **interphase II** (abbreviated phase, no DNA or chromosome replication); **prophase II** (chromosomes regain prophase appearance); **metaphase II** (alignment of chromosomes on equatorial plate); **anaphase II** (centromeres split and separate); and **telophase II** (cytoplasmic division). Telophase II is followed by the formation of sperm or eggs.

8. Crossing-over may be thought of as a break and reunion between nonsister chromatids. It results in **recombinant** chromosomes that carry paternal genes along part of their length and maternal genes along the other part. Crossing-over thus allows the formation of chromosomes that carry new combinations of genes.

9. The principal genetic consequence of meiosis is that all gametes will be genetically unique in the sense of carrying different combinations of genes. This diversity of gametic types is the cause of the diversity of offspring formed in **sexual reproduction**.

WORDS TO KNOW

Chromosome	Autoradiography	**Inheritance**	Chiasma
Mosaic	Q bands	Sexual	Interphase I
Metaphase spread	G bands	Asexual	Prophase I
Karyotype	Satellite	Cytoplasmic	Metaphase I
Sex chromosome	Bivalent		Anaphase I
X chromosome	Sister chromatids		Telophase I
Y chromosome	Nonsister chromatids	**Meiosis**	Interphase II
Autosome	Recombinant	Oocyte	Prophase II
Metacentric		Spermatocyte	Metaphase II
Submetacentric	**Chromatin**	Synapsis	Anaphase II
Acrocentric		Bivalent	Telophase II
A, B, C, D, E, F, G	Heterochromatin	Crossing-over	Polar body
	Euchromatin		

PROBLEMS

1. For discussion: The world's first "test-tube baby" (that is what newspaper reporters called her) was born in England in 1978. The mother could not become pregnant in the normal manner because her fallopian tubes were blocked. After treating the woman with hormones to stimulate maturation of the eggs in the ovary, physicians inserted a device called a *laparoscope* through the abdominal wall and used a needle to draw out an egg. The egg was placed in a small dish containing blood serum and other nutrients, and sperm from the husband was added. After fertilization, the zygote was transferred to another dish for several days where, by means of successive mitotic divisions, it gave rise to a small ball of cells. In the meantime, the woman was treated with other hormones to prepare the uterus. The small ball of cells was then introduced into the uterus, where it underwent further embryonic development and eventually gave rise to a healthy baby girl. Do you think the term *test-tube baby* is appropriate for this child? Do you think the procedure was justified in this case? Would you approve the use of the procedure for an unmarried woman? Would you approve if the ball of cells had been introduced into the uterus of a woman other than the egg donor? If the ball of cells had been introduced into a different woman, which woman, in your opinion, would be the child's "mother"?

2. A child born with cleft palate, small skull, malformed ears, and other major abnormalities is found to have 47 chromosomes. The extra chromosome is an acrocentric chromosome of medium size. To what group does the extra chromosome belong? What procedure could be used to identify the chromosome more precisely?

3. A physically normal girl is found to have 47 chromosomes. The extra chromosome is a C chromosome that has one G band in the short arm and five G bands in the long arm. Which chromosome is it?

4. Why do homologous autosomes have identical patterns of G bands?

5. In the somatic cells of females of all mammals, including humans, one of the X chromosomes becomes heterochromatic. What does this suggest about the activity of genes on the heterochromatic X?

6. From which parent does a male's Y chromosome derive? Which parent contributes the X?

7. The homologues of a G-group chromosome of a child and its parents are shown here. Note that one of the G chromosomes in the mother and one in the child have a large satellite. From which parent did the child receive the chromosome indicated by the single-headed arrow? Which parent contributed the chromosome indicated by the double-headed arrow?

Father

Mother

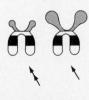

Child

8. The first meiotic division is often called the *reductional division*; the second meiotic division is often called the *equational division*. How can this terminology be explained in terms of the number of chromosomes at various stages in meiosis?

9. In which meiotic division, the first or the second, does the X chromosome separate from the Y chromosome?

10. What kind of sperm would be produced from an anaphase II nucleus in which both chromatids of the Y chromosome went to the same telophase II nucleus instead of separating? What chromosomal constitutions in the zygote would result from fertilization of a normal egg by each type of sperm?

11. In the accompanying diagram of the bivalent, *A* and *a* represent different forms of the same gene (i.e., different nucleotide sequences); likewise *B* and *b* represent different forms of another gene on the same chromosome. Which forms of the genes *A*, *a* and *B*, *b* would be found in the four telophase II nuclei from the meiosis in which this bivalent occurred? (Example and hint: The nucleus receiving chromatid 1 would carry *A* and *B*.)

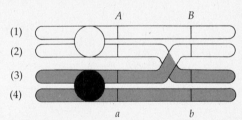

12. What is the answer to Problem 11 for the bivalent shown here?

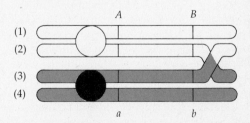

13. The homologues of a D-group chromosome in a child and its parents are shown here. The mother's chromosomes are normal. In the father, one homologue has giant satellites and the other has an addition of chromosomal material to the long arm (arrows). In the child, one homologue is normal, but the other has both giant satellites and the abnormal long arm. From which parent does the abnormal D chromosome in the child derive? What normal process in meiosis could account for the origin of the doubly abnormal chromosome? Judged on the basis of banding pattern, which D chromosome is involved?

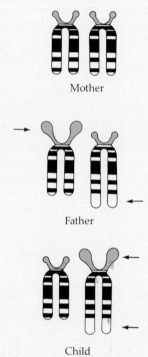

Mother

Father

Child

14. Radiation and certain chemicals can cause breaks in chromosomes, although breaks can also occur spontaneously at low frequency. The ends of broken chromosomes are often said to be "sticky" because they can rejoin with one another. Indeed, any broken end can rejoin with any other broken end, so chromosome breakage and rejoining can create a variety of types of abnormal chromosomes. (The genetic consequences of several types of chromosome abnormality will be discussed in Chapter 7.) Shown here are two types of chromosome 2. The one on the left is normal; the other is abnormal because it has (among other things) a misplaced

centromere. What event involving two chromosome breaks and subsequent rejoining could account for the creation of the abnormal chromosome from its normal counterpart? If the chromosomes shown were the homologues of chromosome 2 in an individual, would you expect the individual to have a genetic disorder? Why or why not?

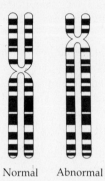

Normal   Abnormal

15. As noted in Problem 14, broken chromosome ends can rejoin in a variety of ways to produce abnormal chromosomes. Suppose an individual is found who has 45 chromosomes. One of the D chromosomes is missing, as is one of the G chromosomes. In their place, however, the individual carries the abnormal submetacentric chromosome shown here. The patient is physically and mentally normal, so it seems reasonable to conclude that the new chromosome somehow accounts for the missing ones. What event involving two chromosome breaks and subsequent rejoining could combine a D and a G chromosome into the chromosome shown? Does the new chromosome carry *all* the material from the old D and G chromosomes? If not, what material is missing? Based on the banding pattern of the new chromosome, which D and G chromosomes were involved in the rearrangement?

16. Recall that, in males, meiosis produces two X-bearing sperm and two Y-bearing sperm. Assuming that both types of sperm are equally capable of fertilization, what proportion of zygotes would be expected to be chromosomally XX and what proportion XY? Assuming an equal likelihood of survival of XX and XY zygotes, what proportion of newborns would be expected to be female? What proportion male?

17. If we designate the maternally derived chromosome in an individual as M and the paternally derived chromosome as P, then, in the absence of crossing-over, an organism with one pair of chromosomes can produce just two types of gametes because each must carry either M or P. An organism with two pairs of chromosomes can produce four types of gametes (i.e., MM, MP, PM, and PP, where, in these symbols, the first letter represents chromosome 1 and the second letter represents chromosome 2). The fruit fly, *Drosophila melanogaster*, has four pairs of chromosomes, and there is no crossing-over in the male. How many different types of sperm can a male fruit fly produce? Again ignoring the enormous diversity of gametic types generated by crossing-over, how many gametic types could be produced by an organism such as humans with 23 pairs of chromosomes?

FURTHER READING AND REFERENCES

Bloom, K. S., and J. Carbon. 1982. Yeast centromere DNA is in a unique and highly ordered structure in chromosomes and small circular minichromosomes. Cell 29: 305–317. Molecular structure of centromeres.

Comings, D.E. 1978. Mechanisms of chromosome banding and implications for chromosome structure. Ann. Rev. Genet. 12:25–46. Why do various cytogenetic procedures produce chromosome bands?

Grell, R.F., E.F. Oakberg, and E.E. Generoso. 1980. Synaptonemal complexes at premeiotic interphase in the mouse spermatocyte. Proc. Natl. Acad. Sci. U.S.A. 77:6720–6723. An argument that synapsis may precede meiotic prophase and suggesting interphase as the actual time of crossing-over.

Grobstein, C. 1979. External human fertilization. Scientific American 240:57–67. A clear discussion of the scientific issues and relevant social concerns.

John, B., and K.R. Lewis. 1965. The Meiotic System. Vol. 6, Pt. Fl. of Protoplasmatologia. Springer-Verlag, New York. A splendid review of meiosis notable for its extraordinary pictures.

Nora, J.J., and F.C. Fraser. 1981. Medical Genetics, 2d ed. Lea & Febiger, Philadelphia. A clinically oriented introduction to medical aspects of genetics.

Rhoades, M.M. 1950. Meiosis in maize. J. Heredity 41:57–67. This paper is more than 30 years old and is still one of the best descriptive accounts of meiosis to be found anywhere.

Therman, E. 1980. Human Chromosomes: Structure, Behavior, Effects. Springer-Verlag, New York. Contains an excellent account of meiosis.

Walzer, S., and P.S. Gerald. 1977. A chromosome survey of 13,751 consecutive male newborns. In Hook, E.B., and I.H. Porter (eds.). Population Cytongenetics: Studies in Humans. Academic Press, New York. pp. 45–61. Source of data in Table 2.2.

Yunis, J.J. 1976. High resolution of human chromosomes. Science 191:1268–1270. Contains diagrams like Figure 2.8 for the entire chromosome complement.

chapter 3
Mendel's Laws and Dominant Inheritance

- Some Important Terminology
- Segregation
- Dominance
- Segregation of Genes in Humans
- Homozygotes and Heterozygotes
- Simple Mendelian Inheritance in Humans
- Human Heredity: Some Cautionary Remarks
- Simple Mendelian Dominance

The mechanics of the meiotic divisions determine the statistical rules of inheritance. However, the rules of inheritance were worked out long before chromosomes had been discovered, before the details of cell division had been described, before chromosomes were known to carry the genes, and even before the overriding importance of the nucleus in inheritance had become clear. These statistical rules of inheritance were discovered by an Augustinian priest, Gregor Mendel, who carried out his most important scientific experiments in the early 1860s in a monastery in what is now Brno, Czechoslovakia, where a monument stands in his honor. (See Figure 3.1 for some memorabilia of Mendel.) Mendel studied the

pattern of inheritance of various traits in the garden pea—round versus wrinkled seeds, green versus yellow pods, and others. He was able to discern certain regularities in their pattern of inheritance. He reported these results in the journal of a local scientific society, which was rather widely disseminated in Europe; he also wrote to one of the leading plant breeders of the time, whose name was Nageli, and told him of the results. Nevertheless, the significance of Mendel's findings was completely missed by those who read his paper (if anyone did read it attentively) and by Mendel's correspondents. His elegant work lay unnoticed and unappreciated for 34 years.

Nowadays, there is a natural tendency to

Figure 3.1 *(a)* A portrait of Gregor Mendel, *(b)* part of the monastery garden in which he carried out his pea-breeding experiments, *(c)* handwritten title page of his celebrated paper, and *(d)* a page from one of his experimental notebooks.

think that Mendel was ignored because he was stuck in some backwater monastery and cut off from the world. This is far from the truth. In those days, joining the priesthood was an accepted way for a poor peasant boy like Mendel to receive a higher education. Moreover, the monastery of St. Thomas was at that time the scientific and cultural center of Brno, and Brno was the scientific and cultural center for a large part of southeastern Europe. Many of Mendel's brother monks were scientists of various sorts, so Mendel's interest in experimentation was by no means odd or unusual. Nor was Mendel cut off from the world. For 14 years he taught in the public school system of Brno, and his students recalled him as a kind and gentle man with a lively sense of humor whose teaching talent was in making complex ideas seem simple.

Mendel was therefore part of a respected scientific tradition, and isolation was not the primary cause of his neglect. More than likely, Mendel's work was ignored at the time because no one really understood it. Science often has its own tempo, and certain discoveries must come before other ones. In Mendel's case, an understanding of statistics was necessary for a proper appreciation of his work. Although a number of prominent scientists in Mendel's time were interested in determining the rules of inheritance, they did not expect these rules to be statistical. Hardly anyone even knew how to think about statistical regularities of any sort, because the intellectual framework of statistics was not laid down until the last half of the nineteenth century and the first half of the twentieth century. A mature appreciation of the unique power and beauty of statistical laws is as difficult to achieve as an appreciation of, say, cubism in art. Both share a perspective on the world that is different from the commonplace, and at first this perspective seems alien, even a little frightening. Appreciation takes time. Moreover, although the nature of inheri-

tance had caught the attention of some scientists, it was a secondary problem in the biology of the time. The principal interest was in **evolution**—the processes whereby species of organisms become progressively more adapted to their environment. Indeed, on the cold, clear winter evening of February 8, 1865, when Mendel read his paper to the 40 members of the Brno Natural History Society, there was not a single question or discussion pertaining to the paper; instead, the group discussed various issues related to evolution!

Between about 1870 and 1900, the details of cell division and the importance of chromosomes in inheritance gradually became clear. In the late 1870s, the decisive role of the nucleus in cell division and fertilization was demonstrated. Between 1880 and 1900, a whole series of investigators contributed to the understanding of mitosis and meiosis, although the most important discovery is usually considered to be the observation that meiosis reduces the chromosome number by half and that the full chromosome number of a species is restored again only at fertilization. However, the details of meiosis—that chromosomes come in homologous pairs and that bivalents represent homologues paired side by side—were not fully worked out until around 1900.

In the last years of the nineteenth century, interest in the nature of inheritance was rekindled, partly because inheritance was finally perceived to be an important element in evolution and adaptation. A number of people began performing experiments that were, unbeknown to them, very similar to Mendel's. Three men—Hugo de Vries in Holland, Carl Correns in Germany, and Erich von Tschermak in Austria—all working independently, discerned in their own plants the patterns of inheritance that Mendel had seen in his. Only when the three were ready to publish their results did they become aware of Mendel's paper; de Vries, Correns, and Tschermak all published in 1900, and all cited Mendel. With the rediscovery of Mendel's work, it became clear that all living organisms have genes—elements transmitted in inheritance—although in 1900, these were still highly abstract entities. In the early 1900s, it was realized that the genes were probably located on the chromosomes because the genes behaved in inheritance as if they were located on the chromosomes. At the same time, it also became evident that Mendel's laws of inheritance have their physical foundation in the process of meiosis. The marriage of genetics and **cytology** (the study of cell structure) began at this time.

Some Important Terminology

What were these rules of Mendel, these statistical laws? Actually there were three discoveries, covering the phenomena of dominance, segregation, and independent assortment. The discoveries pertaining to segregation and independent assortment are often referred to respectively as *Mendel's first and second laws*. The phenomenon of dominance is really not a generalization about inheritance as much as it is about the way genes act during development. A discussion of dominance will have more meaning if deferred until later; and the second law, that of independent assortment, will be taken up in Chapter 4. The nub of Mendel's theory is in the first law, the *law of segregation*. This law actually rests on several other concepts, which deserve a brief discussion. The first concept is that **hereditary traits are determined by genes** (although Mendel used the word *factor* since the word *gene* did not exist). The second is that **genes are transmitted from one generation to the next through the gametes** (the egg and sperm in animals). The third concept is that **genes come in pairs**. The idea that genes come in pairs is the most subtle of Mendel's discoveries. It introduces more nuances that the simple word *gene* can carry. We

need two new words. There is a need for a word to specify the place on a chromosome where a gene is located. The word used is **locus** (plural **loci**). A locus on a human chromosome may be specified physically by saying, for example: "This locus is so many micrometers from the tip of the short arm of chromosome 11," or "This locus is one of those present in such and such a G band on chromosome 11." A locus can also be specified genetically by saying, for example: "This locus is the location of the gene that causes sickle cell anemia when it is mutated in a certain way." Either description actually specifies two loci—the two that are located at exactly corresponding positions on the two homologous chromosomes in any one cell in an individual. The idea of a locus should convey no impression of the normality or abnormality of the gene carried there. A locus is merely a physical position on a chromosome.

We also need a word to specify the actual gene or genetic information that is present at a locus. The word for this is **allele**. Allele is one of the trickiest words in genetics. It means genes that are alike in one way but different in another. To make this clear, we must first consider what the gene at a locus is like. Recall that a gene is actually a sequence of nucleotides in the DNA at a specific locus on the chromosome, and this sequence of nucleotides is analogous to the sequence of letters on a ticker tape. The two genes that occupy homologous loci are really two instructions, either of which should enable the cell to carry out a particular function. The information is encoded in the sequence of nucleotides in the DNA. The cell has two copies of each gene or instruction; perhaps this has evolved partly to protect the cell against mistakes that may exist in one of them. This results in a benefit similar to the strategy often followed by spies and others for whom a simple misprint can have dire consequences: The same message is sent twice.

Often the genes at homologous loci carry the same sequence. They may both say "round and round the mulberry bush," to use the ticker-tape analogy in Figure 2.12. Sometimes the two copies are different. One may have the normal sequence, "round and round the mulberry bush," and the other may be aberrant in any number of ways. It may have a simple misprint: "round and round the fulberry bush" or "round and pound the mulberry bush." It may have letters missing from the end: "round and round the mul," or from the middle: "round and rd the mulberry bush." It may even have a small addition: "round and round-ound the mulberry bush." (Counterparts of each of these kinds of mistakes do occur in genes.) Sometimes both homologous loci carry an abnormal gene.

In any case, the alternative sequences of a gene, including the normal or typical ones and all the variants, are called **alleles**. Alleles are genes that are identical in the sense that they are alternative occupants of the same locus. Of course, only one allele can occupy the locus in a particular chromosome, because the nucleotides in the DNA corresponding to a particular locus must have some particular sequence; this sequence of nucleotides specifies the allele that is present in the chromosome. On the other hand, alleles are genes that are nonidentical in the sense that they carry variant forms of the genetic information of the same gene; these variant forms correspond to different nucleotide sequences or, by the ticker-tape analogy, different sequences of letters. The distinction between locus and allele is illustrated in Figure 3.2. The locus in the figure corresponds to the position on the chromosome occupied by 12 nucleotides of double-stranded DNA. (Actual loci consist of thousands of nucleotides.) Three alleles are shown: Allele 1 is the normal allele, allele 2 is an allele differing from the normal one only in the first nucleotide position (asterisk), and

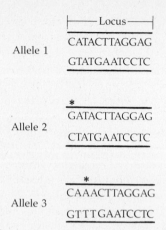

Figure 3.2 A *locus* corresponds to a segment of DNA, illustrated here as parallel lines representing the two strands with A-T and G-C nucleotide pairings between them. An *allele* refers to a particular sequence of nucleotides at a locus. Shown here are three alleles; alleles 2 and 3 differ from allele 1 at the nucleotide positions indicated by asterisks. The word *gene* can mean either "locus" or "allele" depending on the context.

allele 3 differs from the normal one only in the third nucleotide position. Although a particular locus can carry only one allele, the allele that is present can be any one of those illustrated.

To summarize, the word *gene* is a general catchall word, used fairly loosely to mean the elements that are inherited. The word *locus* means a position on a chromosome. And the word *allele* means a particular form of a gene. Both *locus* and *allele* may be used as synonyms for *gene*, but each word has its own particular shade of meaning. The Mendelian statement "Genes occur in pairs" can now be made more precise. It means "Loci occur in pairs"; that is, for each locus in a cell there is a homologous locus. This, of course, follows from the fact that each chromosome in the cell has a homologous chromosome. As noted in Chapter 2, in humans and many other animals, there is an exception to the rule that loci occur in pairs. The exception pertains to the sex chromosomes in males. The X and the Y chromosomes carry different genes. They are homologous chromosomes only in the general sense that they are both sex chromosomes. But during meiosis they pair end to end, not side by side, and no crossing-over occurs between them. The genes carried by these two chromosomes are different. Therefore the loci on the X chromosome are present only once in each cell of a male; the same goes for genes on the Y. The 2 X chromosomes in females and the 22 pairs of autosomes in both sexes do imply that each locus on these chromosomes will be present twice in each cell.

Convenience and brevity often require that the entire sequence of nucleotides or letters in a gene be designated by a single symbol. We may, for example, let the arbitrary symbol *A* stand for the entire normal sequence "round and round the mulberry bush." An aberrant or abnormal sequence of this gene might then be designated by the symbol *a*; for example, *a* might represent the sequence "round and round the fulberry bush." Using upper and lower cases of the same letter is helpful because it is a reminder that *A* and *a* are alleles, either of which may be found at a specific locus. If the locus in question is on one of the autosomes, so that there are two representatives of the locus in every cell, then both loci may carry *A*, both may carry *a*, or one locus may carry *A* and the other *a*. When this locus is described, the genetic constitution of these cells would conveniently be designated as *AA*, *aa*, or *Aa*, respectively.

This straightforward symbolism does not imply a denial of any other differences between cells of individuals. If one cell of an individual is said to carry *AA* and a cell of a different individual is said to carry *aa*, for example, this only draws attention to the one locus under consideration. The cells may have other important differences as well, but these are intentionally overlooked. This exemplifies the same kind of verbal shorthand used every day when

a person says: "Sally has red hair; Mary is a blond." Sally and Mary have innumerable other differences, but their hair color is the only point of interest to the speaker. One other use of the genetic symbols may possibly cause confusion. The genetic constitution of a *cell* may be *AA*, *aa*, or *Aa*. But if the cell is a fertilized egg and it is, for example, *Aa*, then the process of mitosis makes it almost certain that each of the cells in the person who develops from this egg will also be *Aa*. One could properly say, for instance, "The fertilized egg that gave rise to Theodore was genetically *Aa*" or "Each of the approximately 10 trillion cells in Theodore's body carries *Aa*," but both expressions are cumbersome. It is easier to say simply "Theodore carries *Aa*." It is equally correct to say "Theodore *is Aa*." The difference in terminology here is as tiny and inconsequential as that in "Theodore's head carries blond hair," "Theodore has blond hair," and "Theodore is blond."

Segregation

Mendel's **law of segregation** states that the homologous loci in an individual separate from one another, or *segregate*, during the formation of eggs and sperm, so that each gamete carries exactly one of the loci, never neither and never both. The law of segregation also states—and this is the statistical part—that in the formation of gametes, half the gametes will carry one of the homologous loci and the other half will carry the other. For Mendel to have inferred this rule from his simple pea-breeding experiments is to his everlasting honor. Once the details of meiosis were worked out, the rule became fairly obvious. An individual cell undergoing meiosis and carrying *Aa* at some pair of homologous loci must produce four telophase II nuclei; two of these nuclei must carry the chromosome with *A* on it and the other two must carry the homologous chromosome with

a on it. Thus, segregation is a direct consequence of the pairing and separation of homologous chromosomes during meiosis. In human
males, each of the telophase II cells goes on to
form a mature, functional sperm, so each spermatocyte gives rise to four sperm, two with the
A allele and two with the *a* allele. In females,
three of the telophase II nuclei are associated
with polar bodies and only one remains in the
functional egg; but which one of the nuclei
becomes the egg nucleus is a matter of chance,
so half the time it will carry *A* and half the time
a (Figure 3.3). We say that the **probability** that
a gamete carries *A* is $\frac{1}{2}$ and the probability that
it carries *a* is $\frac{1}{2}$. Note that the two probabilities
add to 1, which is just the mathematical way of
saying that each gamete must carry either *A* or
a and that these are the only two possibilities.
Of course, segregation of *A* from *a* occurs only
in *Aa* individuals. Individuals who are geneti-

cally *AA* must produce only *A*-bearing gametes; individuals who are *aa* must produce
only *a*-bearing gametes. Thus, the probability
of an *A*-bearing gamete coming from an *AA*
individual is 1, and the probability of an *a*-
bearing gamete from an *AA* individual is 0.
Likewise, the probability of an *A*-bearing gamete from an *aa* individual is 0, and the probability of an *a*-bearing gamete from an *aa* individual is 1.

Ordinarily, the genes themselves are
never seen; they are too small to observe
except in an electron microscope. Even the
chromosomes are far too small to see with the
naked eye, and most people never actually see
a typical human cell. But this is not important.
What is important is that the effects of the
genes can be observed in the appearance of the
people who carry them. Therefore, the segregation of genes can often be observed merely

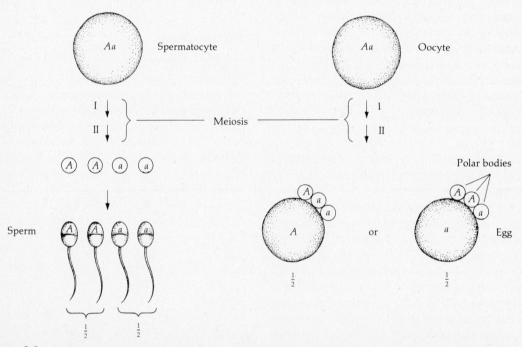

Figure 3.3 Segregation in an *Aa* heterozygote. In males, two *A*-bearing and two *a*-bearing sperm are produced. In
females, either an *A*-bearing egg or an *a*-bearing egg is produced, and each of these possibilities is equally likely.

by examining the kinds of offspring produced by a mating. Using pea plants, Mendel himself studied, among other traits, the inheritance of round seeds versus wrinkled seeds. In this comparison can be seen one of Mendel's marks of brilliance. He realized that you must always study the inheritance of *differences* between individuals. When every individual is precisely identical in some physical or mental feature, the mode of inheritance of the trait can hardly be determined because it has no pattern of inheritance—other than that all parents and all offspring have the trait. Traits that appear in every individual need not have a genetic basis at all; they might be determined by environmental factors common to all individuals.

Mendel was also careful to focus on only one or a few differences between plants and to ignore the countless other differences he must have seen. He realized as well that the differences to be contrasted must be comparable. For example, in the case of round seeds versus wrinkled seeds, he perceived that trying to determine the inheritance of round seeds versus tall height in the plants would be hopeless because the traits are not comparable. To compare them would be as meaningless as comparing pecans and pineapples; and to study the genetic basis of such a difference would lead to nonsense. The insight to know exactly what traits are to be compared and contrasted is gained by experience and, sometimes, just good luck. In many cases, the choice of characteristics to be compared is quite straightforward because the comparison is between individuals who are abnormal in some very specific way and those who are normal with respect to this same trait. In other cases, the comparison is not so obvious, and progress in genetics is often made when someone perceives a fruitful way to classify the differences among individuals; this is particularly true in human genetics.

There is an amusing and surprisingly accurate analogy to the process of segregation that may eliminate some of the abstractness and mystery associated with the process. The analogy is from the famous geneticist J. B. S. Haldane, and it appears in his book *Heredity and Politics*.[1]

Nonbiologists often find a certain difficulty in following Mendel's laws, so I am going to use analogy rather freely. In Spain every person has two surnames, one of which is derived from the father and one from the mother. For example, if you are called Ortega y Lopez you derive the name Ortega from your father, it was his father's name; and Lopez from your mother, as it was her father's name. Now I want you to imagine a strange savage nation in which everybody has two surnames; and when a child is born there is a curious ceremony by which he receives one of the surnames from one parent and one from the other, these being drawn at random by a priest, whose business it is. For example, if Mr. Smith-Jones marries Miss Brown-Robinson the children may get the name Smith or Jones from the father and Brown or Robinson from the mother; and will be called Smith-Brown, Smith-Robinson, Jones-Brown, or Jones-Robinson. If the priest draws at random those four types occur with equal frequency. A complication arises from the fact that some people may get the same name from both parents, and be called Smith-Smith, transmitting the name Smith to all their children. If you try to remember that simple scheme you will find no great difficulty in understanding the laws which govern heredity. The things which are analogous to the surnames are called genes.

For determining the kinds and frequencies of the various types of offspring expected to result from a mating, geneticists often use the sort of squares (called **Punnett squares**) shown in Figure 3.4. Use of the squares is illustrated by means of the Haldane analogy. In Figure 3.4a, Smith-Jones (the father) mates with Brown-Robinson (the mother). The father's contribution to the offspring surname is either Smith or Jones, each with probability

[1]Haldane, J.B.S. 1938. Heredity and Politics. Norton, New York, pp. 46–47.

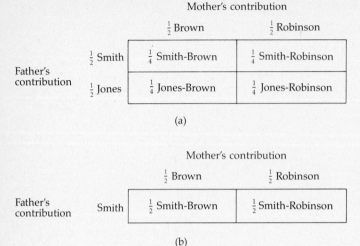

(a)

(b)

Figure 3.4   Haldane analogy of Mendelian inheritance in which each distinct allele at the locus under consideration corresponds to a distinct surname. The mating in *(a)* involves four alleles; that in *(b)* involves three alleles.

$\frac{1}{2}$; the mother's contribution is either Brown or Robinson, each with probability $\frac{1}{2}$. The possible offspring surnames are obtained by means of a sort of cross-multiplication of the father's contribution and the mother's contribution; these are shown inside the boxes of the Punnett square. Each of the four offspring surnames has a probability of occurrence of $\frac{1}{4}$ (because $\frac{1}{2} \times \frac{1}{2} = \frac{1}{4}$), which is what Haldane refers to when he says "If the priest draws at random those four types will occur with equal frequency." Segregation of surnames in a mating of Mr. Smith-Smith with Ms. Brown-Robinson is illustrated in Figure 3.4b. Because in this case all the offspring must receive the name Smith from the father, only two classes of offspring are possible, and each class has a probability of occurrence of $\frac{1}{2}$ (because $\frac{1}{2} \times 1 = \frac{1}{2}$).

Dominance

Mendel found that seeds carrying two copies of a certain allele were round and that those carrying two copies of another allele were wrinkled. The genetic constitution of the round seeds may be designated as AA and that of the wrinkled seeds as aa. When plants with these genetic constitutions are crossed, or mated, the offspring (called a **hybrid**) must receive A from one parent and a from the other. All the hybrid seeds are therefore genetically Aa. (A surname analogy for this mating is one in which Mr. Smith-Smith mates with Ms. Brown-Brown, and all the offspring are Smith-Brown.) What do these Aa offspring look like? As illustrated in Figure 3.5a, Mendel found that they were round—indistinguishable from AA seeds. The development of a round seed evidently requires only a single copy of the A allele; two copies of the A allele do not make a seed perceptibly more round than only one copy. The observation is summarized by saying that the A allele is **dominant** to the a allele; conversely, the a allele is said to be **recessive** to the A allele. The dominance of the allele that causes round seeds is only one example of **dominance**, and more about dominance will be noted later.

Dominance is an example of a rather common occurrence in genetics—that individuals who appear to be identical with respect to

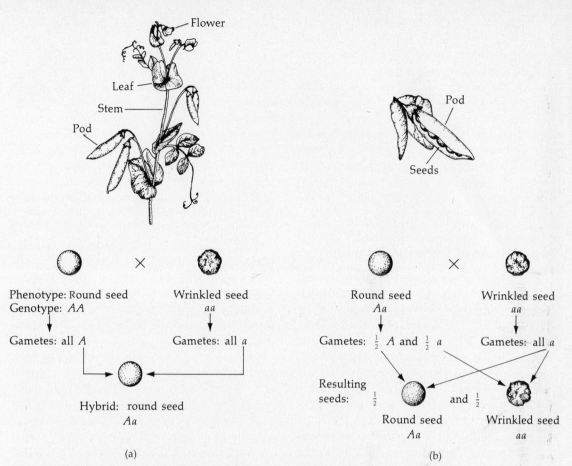

Figure 3.5 Summary of some of Mendel's results in breeding peas with round or wrinkled seeds. *(a)* The mating of a homozygous *AA* plant with a homozygous *aa* plant gives rise to heterozygous *Aa* seeds. The *Aa* seeds are round, like those in the *AA* parent; thus, *A* is dominant to *a*. *(b)* A mating of a heterozygous *Aa* plant with a homozygous *aa* plant. Because of segregation during meiosis in the heterozygous parent, half the resulting seeds will be *Aa* (and therefore round) and half will be *aa* (and therefore wrinkled).

some trait may not really be identical in the genes that underlie the trait. Because of this, it is of crucial importance to distinguish the genetic constitution of an individual from the physical appearance of the individual. The distinction is made easier to remember because there is a special terminology for it. The **geno-type** of an individual is the individual's genetic constitution; and the **phenotype** of an individu-al is the physical appearance. *AA*, *Aa*, and

aa are examples of genotypes; round and wrinkled seeds are examples of phenotypes. Mendel's observation of dominance can be summarized by saying that the phenotypes of *AA* and *Aa* seeds were round, whereas the phenotype of *aa* seeds was wrinkled.

The hybrid that results from a mating of an *AA* plant with an *aa* plant has the genotype *Aa*. If one crosses such an *Aa* hybrid with an *aa* plant, nearly half the resulting seeds will be round and the other half wrinkled (see Figure 3.5b). The proportions of the two types reflect the segregation of the A and a alleles during meiosis. Half the gametes from the Aa plant carry A, and half carry a. All the gametes from the aa plant carry a. Therefore, the seeds that arise from the mating of Aa with aa will be Aa (and therefore round) half the time and aa (and therefore wrinkled) the other half, just as a mating of Mr. Smith-Brown with Ms. Brown-Brown in Haldane's strange savage nation would lead to $\frac{1}{2}$ Smith-Brown and $\frac{1}{2}$ Brown-Brown offspring.

Segregation of Genes in Humans

Segregation of genes occurs in humans as well as in peas. However, segregation cannot easily be observed unless the trait under study is determined by the alleles at a single locus without complications due to the effects of either alleles at other loci or the environment. Such simple traits that are determined by the alleles at a single locus are known as **simple Mendelian traits**, and they are said to show **simple Mendelian inheritance**.

One example of a simple Mendelian trait that illustrates segregation in humans is the harmless trait *woolly hair*. (The woolly hair trait is rare; examples of more common simple Mendelian traits will be discussed a little later.) Woolly hair refers to a curly, fuzzy texture of the hair, superficially similar to the texture of the hair in many Negro families, although woolly hair is not exactly the same (see the family in Figure 3.6). Woolly hair is brittle and tends to break off at the tip, so it never gets very long. This trait is due to the presence of a dominant gene, which may be called W (for woolly) located on one of the autosomes (which autosome is not known). People with straight, nonwoolly hair are genotypically ww. People with woolly hair are Ww, so only one copy of the W allele is sufficient to produce the trait. (It should be emphasized that the genetic

basis of the differences in hair texture between Caucasians and Negroes and other racial groups is rather more complicated than the difference between straight and woolly hair.) The designation of W as a dominant gene uses the word *dominant* in a somewhat different sense than it was used previously. The gene A was previously said to be dominant to a because AA and Aa could not be distinguished phenotypically; both had round seeds. To retain precisely the same meaning in humans, WW and Ww would both have to produce woolly hair for W to be dominant to w. But this is one of the difficulties in human genetics—not only with respect to woolly hair but with respect to many other traits as well. Individuals who have genotype Ww are extremely rare, so they virtually always mate with people who are normal ww. In the instance of woolly hair, no cases are known in which both father and mother had woolly hair, and such a mating ($Ww \times Ww$) is the only likely origin of a genetically WW child.

Nevertheless the W gene does have one important feature in common with Mendel's A gene—only one copy of the allele is necessary to produce a detectable change in phenotype. A single copy of the A allele is sufficient to change the shape of the pea seed from wrinkled to round; likewise a single copy of the W allele is sufficient to change the texture of the hair from straight to woolly. This similarity underlies the use of the word *dominant* as it applies to the W allele, and this connotation of the word is the one always implied in human genetics. The actual phenotype of WW individuals is, of course, not known. A similar lack of knowledge applies to the analogous genotype with respect to many other dominant genes in humans. For the record, however, it should be noted that the rare individuals known who do have two copies of a dominant gene that causes a serious physical or mental defect are usually far more severely affected than those who have only one copy of the allele, and in many such

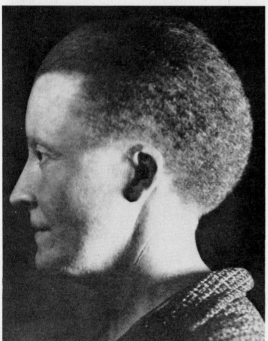

Figure 3.6 A portrait of a Norwegian family showing woolly hair. The mother (seated) has the trait, whereas the father (also seated) does not. The three older children standing in the rear all have woolly hair; the three younger standing near the front have nonwoolly hair. The photograph of a woman with uncut woolly hair demonstrates that such hair breaks off naturally before getting very long.

cases, the individual carrying two copies of the dominant allele will abort as an embryo or die very young.

Figure 3.7 is a pedigree depicting the occurrence of woolly hair in a Norwegian kindred. A **kindred** is a group of people related by marriage or ancestry, and a **pedigree** is a diagram showing the ancestral history of a group of relatives. The pedigree in Figure 3.7 uses certain conventions:

1. Matings are represented by horizontal lines connecting the individuals involved.

2. The children of a mating are represented in order of birth from left to right beneath a horizontal line connected to the parental mating.

3. Males are represented by squares, females by circles.

4. Affected individuals (in this case, affected with woolly hair) are represented by solid symbols, nonaffected by open symbols.

5. Arrows call attention to certain individuals in the pedigree, usually to the first individual who came to the investigator's attention and through whom the pedigree was discovered; in the case of Figure 3.7, however, the arrows point to the people in the family portrait in Figure 3.6.

6. In many cases, as in Figure 3.7, it is convenient to depict the successive generations (indicated by Roman numerals) as a series of concentric semicircles rather than as a series of horizontal lines in order to conserve space.

Figure 3.7 illustrates certain features that are characteristic of rare, simple Mendelian traits due to an autosomal allele. These characteristics are as follows:

1. Males and females are equally likely to be affected with the trait.

2. Affected individuals will almost always have just one affected parent. (Because the trait is

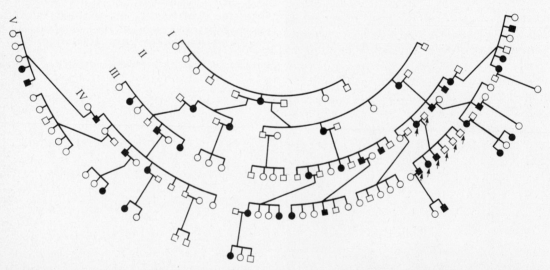

Figure 3.7 Pedigree of extensive Norwegian kindred involving the woolly hair trait. Solid symbols indicate woolly hair; open symbols indicate nonwoolly hair. Circles indicate females; squares indicate males. The Roman numerals refer to generations. Individuals designated by arrows are shown in the family portrait in Figure 3.6.

assumed to be rare, matings between two affected individuals almost never occur.)

3. An affected parent may be either male or female.

4. Half the children from an affected parent will be affected.

This last characteristic of simple dominant inheritance—that half the children of affected parents will be affected—is a result of segregation in the affected parent. Because the trait is due to a rare dominant allele, matings involving one affected parent will be between the genotypes *Ww* (the woolly haired parent) and *ww* (the normal parent). The offspring expected from such a mating are illustrated in the Punnett square in Figure 3.8. (In terms of the surname analogy, the mating is like one between Mr. Smith-Smith and Ms. Brown-Smith.) As seen in Figure 3.8, half the offspring are expected to be *Ww*, and the other half are expected to be *ww*.

Does the pedigree in Figure 3.7 exhibit half woolly haired and half nonwoolly haired offspring from matings in which one of the parents is affected? There are 20 such matings in the kindred in Figure 3.7, and these matings produced 81 offspring of whom 38 were woolly haired (*Ww*) and 43 were nonwoolly haired (*ww*). This is quite close to the 1:1 distribution of children predicted from the law of segregation (see Figure 3.8). Although the expected number is 40.5 of each kind of child, no one really expects the actual number to hit the expectation exactly. One reason is, of course, that you cannot have half a child, so the very

best fit to a 1:1 distribution of woolly and nonwoolly hair would be 40:41 or 41:40. But a more important reason why you cannot realistically expect the closest possible fit is that there is **chance variation** in the observed numbers. You cannot expect the observed numbers to hit the expectations exactly, any more than you can expect to observe exactly 50 heads in 100 flips of a coin. Indeed, even though the "expected" number of heads is 50, if exactly 50 heads were actually flipped, many people might suspect, not unreasonably, that something was fishy with the coin. (The odds of flipping exactly 50 heads are only 1 in 12.) Nevertheless, most people do expect the number of heads to be *about* 50, and indeed, 95 percent of the time the number of heads in 100 flips will be between 40 and 60. It is much the same with segregation in the case of woolly hair. Although you cannot realistically expect exactly 40.5 children of each type in the Norwegian kindred, you do expect the observed number to be fairly close to 40.5. What "fairly close" means is a matter of judgment. Most geneticists take it to mean that the difference between the observed and the expected number should be within the range that would be anticipated 95 percent of the time. In the woolly hair example, the chance of observing 81 people with a distribution as bad or worse than 38:43, when the expected number is 40.5 of each, is about 60 percent, so the observed distribution is quite comfortably in the right ball park.

Figure 3.9 illustrates the probability of obtaining various numbers of woolly haired offspring among a total of 81 offspring when it is assumed that the woolly haired trait is due to a simple Mendelian dominant. The shaded region includes those numbers that would be expected to occur by chance 95 percent of the time. As can be seen, 38 woolly haired offspring is well within this 95 percent range. The graph is based on numbers calculated from a formula that proves to be extremely useful in

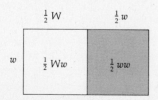

Figure 3.8 Punnett square illustrating segregation of the dominant allele *W* for woolly hair.

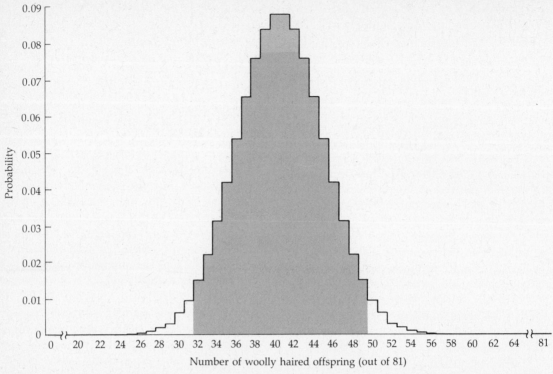

Figure 3.9 Number of woolly haired offspring possible in the pedigree in Figure 3.7 and their respective probabilities, calculated according to the formula given in the text. Shaded area denotes those possible outcomes for which the goodness of fit to a 1:1 distribution would be considered satisfactory. The actual number of woolly haired offspring in Figure 3.7 is 38.

dealing with probability problems of the sort that often arise in genetics. If we have two classes of offspring (in this case, woolly haired and nonwoolly haired) and each offspring has an equal chance of being in either class (which is a consequence of segregation in the case of woolly hair), then, among n total offspring (in Figure 3.9, $n = 81$), the probability of obtaining exactly i of one class and $n - i$ of the other is given by

$$\frac{n!}{i!(n - i)!} \left(\frac{1}{2}\right)^{i}\left(\frac{1}{2}\right)^{n - i}$$

where the symbol $n!$ (read n **factorial**) represents the product of all whole numbers from 1 to n. That is to say, $n! = n \times (n - 1) \times (n - 2)$

$\times \ldots \times 3 \times 2 \times 1$. Similarly, $i!$ is the product of all whole numbers from 1 to i, and $(n - i)!$ is the product of all whole numbers from 1 to $(n - i)$. (The quantity 0! is defined to equal 1.)

Application of this formula to cases such as Figure 3.9 when n is large entails a great deal of calculation; the probabilities of Figure 3.9 were actually calculated by a high-speed computer. For small values of n, however, the calculations can be done by hand. To illustrate the use of the formula, the family in Figure 3.6 will be considered. This family has six children; three are affected (those standing at the rear) and three are nonaffected (the boys standing beside their parents). How unusual would it be for a family of six children to contain exactly three affected and three nonaffected (a perfect 1:1 ratio)? To answer this question, we use the formula with $n = 6$ (the total number of children), $i = 3$ (the number affected), and $n - i = 6 - 3 = 3$ (the number not affected). The probability of three affected and three non-affected is thus:

$$\frac{6!}{3!3!}\left(\frac{1}{2}\right)^3\left(\frac{1}{2}\right)^3$$
$$= \frac{6 \times 5 \times 4 \times 3 \times 2 \times 1}{(3 \times 2 \times 1)\,(3 \times 2 \times 1)}\left(\frac{1}{8}\right)\left(\frac{1}{8}\right)$$
$$= 0.3125$$

That is to say, among a large number of families with six children in which one parent is woolly haired, a little more than 31 percent of the time one expects exactly three woolly haired and three nonwoolly haired children. The family in Figure 3.6, therefore, is not particularly unusual.

One further example using the formula: What is the probability of no woolly haired offspring in a family of six children with one parent affected? Here we have $n = 6$ and $i = 0$, so $n - i = 6 - 0 = 6$. The required probability is

$$\frac{6!}{0!6!}\left(\frac{1}{2}\right)^0\left(\frac{1}{2}\right)^6$$
$$= \frac{6 \times 5 \times 4 \times 3 \times 2 \times 1}{(1)\,(6 \times 5 \times 4 \times 3 \times 2 \times 1)}(1)\left(\frac{1}{64}\right)$$
$$= 0.0156$$

(Recall that $0! = 1$ and that any number raised to the power of 0 also equals 1.) Thus, only about 1.5 percent of all such families would be expected to have no woolly haired offspring.

Homozygotes and Heterozygotes

The emphasis in the examples of woolly versus straight hair in humans and round versus wrinkled seeds in peas has been on what happens when segregation occurs in only one parent. When there are two alleles at a locus, call them A and a, three genotypes will be present in the population: AA, Aa, and aa. Altogether six different kinds of matings can occur (the cross symbolizes the mating):

$$AA \times AA$$
$$AA \times Aa$$
$$AA \times aa$$
$$Aa \times Aa$$
$$Aa \times aa$$
$$aa \times aa$$

Segregation can be observed in all three matings in which one or both partners is Aa. Genotypes that carry two different alleles at a pair of homologous loci—the Aa genotype in this example—are thus the ones that undergo segregation, and geneticists have a special word for them. Genotypes such as Aa that carry two different alleles at a pair of homologous loci are called **heterozygous genotypes**, or **heterozygotes**. Conversely, the AA and aa genotypes are called **homozygous** A and **homozygous** a, respectively; either genotype is sometimes referred to as a **homozygote**. Woolly hair has been chosen as an example of segregation of genes in humans because the trait is rare, and every woolly haired person will therefore almost certainly be heterozygous. The W and w alleles will therefore undergo segregation in every mating that involves a woolly haired parent. The vast majority of woolly haired people come from matings between heterozygotes and normal homozygotes. This is characteristic of traits caused by rare dominant alleles in humans; many other examples could be cited.

Simple Mendelian Inheritance in Humans

Woolly hair is a harmless curiosity that is useful for illustration because it is a simple Mendelian trait due to an autosomal dominant allele. Woolly hair is difficult to relate to one's own hereditary endowment, however, because it is so rare. In this section, we consider some simple Mendelian traits that are sufficiently common that you may be able to identify one or more of them in yourself, among your relatives, or among your friends. Surprisingly, and perhaps paradoxically, *most phenotypic variation in humans is not due to simple Mendelian traits*. The phenotypic variation that is obvious to everyone—variation in height, weight, hair color, eye color, skin color, facial appearance, and so on—has a more complex genetic basis than do simple Mendelian traits. Traits such as height are determined by the collective or aggregate effects of alleles at tens or hundreds of loci and are also influenced by environmental factors (such as diet, in the case of weight, or exposure to sunlight, in the case of skin color). Traits that are determined by the aggregate effects of alleles at many loci are called **polygenic** or **multifactorial** traits. For multifactorial traits, the segregation of individual loci is virtually impossible to detect because the effect of any individual locus is so small when compared with the background effects of all the other loci and the environment. The background effects are like noise on a telephone line, noise so extreme that individual words become garbled or lost altogether. It is therefore important to keep in mind that *most readily observed traits are multifactorial*.

Nevertheless, a number of simple Mendelian traits involving relatively common phenotypic variation in humans have been identified (see the list in Table 3.1 and the corresponding illustrations in Figure 3.10). In spite of the relative simplicity of their mode of inheritance, these traits illustrate a number of important and complicating features that are often encountered in the study of human inheritance. Brief comments about each trait are therefore in order.

Common baldness (see Figure 3.10a). This trait is the common form of hair loss in Caucasians, made famous because President John Adams and several of his illustrious descendants had the trait. Beginning at about age 20 in males and age 30 in females, hair begins to fall out at the sides near the front

TABLE 3.1   CHARACTERISTICS OF SOME RELATIVELY COMMON SIMPLE MENDELIAN TRAITS IN HUMANS*

Trait	Illustration	Phenotype
Common baldness	Figure 3.10a	M-shaped hairline receding with age
Chin fissure	Figure 3.10b	Vertical cleft or dimple in chin
Ear pits	Figure 3.10c	Tiny pit in external ear
Darwin tubercle	Figure 3.10d	Extra cartilage on rim of external ear
Ear cerumen		Wet and sticky ear wax versus dry and crumbly ear wax
Congenital ptosis	Figure 3.10e	Droopy eyelid
Epicanthus	Figure 3.10f	Fold of skin near bridge of nose leading to almond-shaped eyes
Camptodactyly	Figure 3.10g	Crooked little finger due to too short tendon
Mid-digital hair	Figure 3.10h	Hair growth on middle segment of fingers
Phenylthiocarbamide tasting		Ability to taste phenylthiocarbamide; tasters report it as bitter
S-methyl thioester detection		Detection by smell of odiferous substances excreted in urine after eating asparagus
ABO blood group		Type A versus type B versus type O versus type AB
Rh blood group		Type Rh$^+$ (positive) versus type Rh$^-$ (negative)

*For photographs, see Figure 3.10.

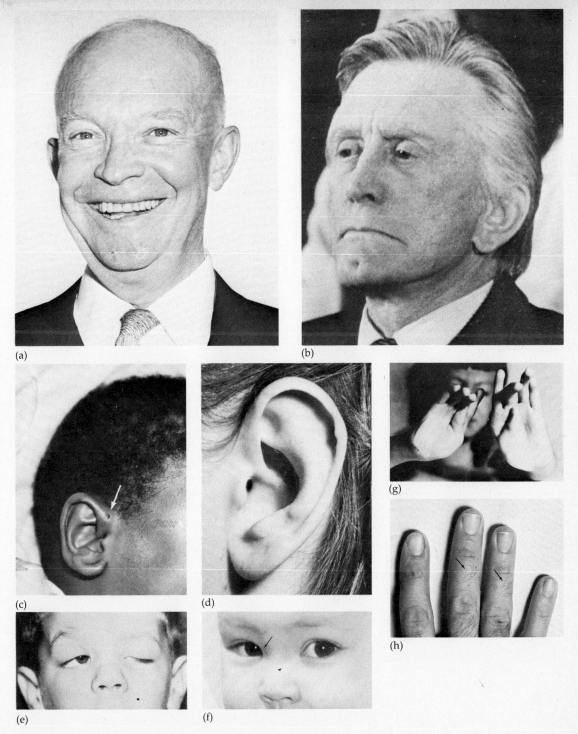

Figure 3.10 Examples of some simple Mendelian traits in humans that are relatively common: *(a)* common baldness, *(b)* chin fissure, *(c)* ear pits, *(d)* Darwin tubercle, *(e)* congenital ptosis, *(f)* epicanthus, *(g)* camptodactyly, and *(h)* mid-digital hair. See Table 3.1 for descriptions and modes of inheritance.

of the scalp, producing an M-shaped hairline. Recession of the hairline in front is usually followed by the development of a bald spot at the rear of the crown. Enlargement of the bald spot and continued recession of the front hairline lead to a final stage where only a peripheral fringe of scalp hair remains. The trait is a simple Mendelian trait due to an autosomal allele; however, the allele is *dominant* in males but *recessive* in females. The trait is most common in Caucasians: Estimates are that at least 50 percent of Caucasian men and 25 percent of Caucasian women will be affected, although the extreme progression to a remaining fringe of hair with a bald pate occurs in only about one-third of affected men (affected women typically experience less hair loss than affected men). The trait is less common among Negroes than among Caucasians, and it is rare among Orientals.

The trait of common baldness illustrates several complications. First, it is a **sex-influenced** trait, which refers to the fact that the trait is expressed differently in males and females (recall that the baldness allele is dominant in males but recessive in females). Extreme cases of sex-influenced traits, in which expression is limited to one sex, are known as **sex-limited** traits; examples of sex-limited traits in humans are growth of facial hair, which ordinarily occurs only in males, and development of breasts, which ordinarily occurs only in females. Common baldness also illustrates **variable age of onset**; individuals who have the baldness genotype have the genotype from the time of birth, but the trait does not occur until much later, and its time of occurrence is variable, beginning in the early 20s for some men but later for other men. The trait also exhibits **variable expressivity**, which means that it is expressed to different degrees in different people. The severity of expression of common baldness may vary from a slightly receding frontal hairline to an almost completely bald pate. The same genotype may be expressed differently in different individuals because of differences in **genetic background** (i.e., the genotypes at other loci that also influence the trait) or because of differences in nongenetic or environmental factors. The extreme end of variable expressivity is known as **incomplete penetrance**, in which individuals who have a genotype associated with a particular trait do not express the trait at all. Finally, common baldness illustrates **variation among populations** in its incidence. As noted, the trait is common among Caucasians, less common among Negroes, and rare among Orientals. Variation among populations is a characteristic feature of many inherited traits, and additional examples will be found in later chapters of this book.

Chin Fissure (see Figure 3.10b). The most common type of chin fissure is a perpendicular furrow in the middle of the chin, and it is due to an autosomal dominant allele. The trait is variable in expression (ranging from a chin dimple to a Y-shaped furrow), but it has almost complete penetrance in males (i.e., all males who carry the allele will have some sort of chin fissure). Females are affected only about half as frequently as males, so the trait is sex influenced. Chin fissures have a variable age of onset; although sometimes present at birth, they more commonly appear in childhood or early adulthood. Among people of German extraction, approximately 20 percent of males and 10 percent of females have the trait.

Ear Pits (see Figure 3.10c). These harmless and usually shallow pits may occur in one ear or both. Ear pits are due to an autosomal dominant allele, which is variable in expression (one ear or both affected, shallow or deep pits), has incomplete penetrance (only about half of those individuals who are heterozygous or homozygous for the dominant allele will have ear pits), and is sex influenced (twice as prevalent in females as in males). The incidence of the trait varies among populations—1 percent in American Caucasians, 5 percent in American Negroes, and said to be "very common" in Chinese.

Darwin Tubercle (see Figure 3.10d). The Darwin tubercle is a variable-sized thickening of cartilage near the upper rim of the ear, and it is probably due to an autosomal dominant allele (i.e., homozygote and heterozygote are both affected). The allele has a high penetrance and the trait is not sex influenced because it has approximately equal frequency in males and females. Prevalence varies among populations, being about 20 percent in Germany and about 50 percent in England and Finland.

Ear Cerumen In the Japanese language, the gray, dry, and brittle type of ear wax is called "rice-bran ear wax," whereas the brown, sticky, and wet type is called "honey ear wax" or "cat ear wax." The dry type is due to an autosomal recessive allele. Type of ear cerumen has remarkable variation among populations; the dry type (which occurs in homozygous recessives) has the following frequencies in various populations: Koreans (93 percent), Japanese (85 percent), American Indians (66 percent), American whites (25 percent), American blacks (5 percent). The trait illustrates two important phenomena. First, most Koreans and Japanese who have wet cerumen will actually be heterozygous for the dominant "wet" allele; they are thus heterozygous for the recessive "dry" allele. Such individuals who are heterozygous for a recessive allele are said to be **carriers** of the allele. Second, ear cerumen illustrates what is called **pleiotropy**, which refers to the fact that genes can have effects on several or many traits and not just one trait. In the case of ear cerumen, the alleles influence secretions of certain glands in the ear canal; these very same alleles

also influence the chemical makeup of secretions from certain other glands, notably those involved in perspiration. Thus, the alleles that affect ear cerumen are **pleiotropic**: They affect several traits (type of ear wax and chemical makeup of perspiration).

Congenital Ptosis (see Figure 3.10e). The word **congenital** means "present at birth," and congenital ptosis refers to a drooping eyelid usually affecting only one eye that is in most cases due to weakness in upward movement of the eyelid. The condition is usually present at birth and may in some cases require surgical correction. The trait is caused by an autosomal dominant allele with incomplete penetrance (about 60 percent of homozygous dominants and heterozygotes actually express the trait). Although the exact prevalence of congenital ptosis is unknown, it is thought to be quite common.

Epicanthus (see Figure 3.10f). *Epicanthus* refers to a condition in which the inner corner of the eye is bridged by an arching fold of skin extending from the bridge of the nose and giving the eye an almond-shaped or "Oriental" appearance. The trait is due to an autosomal dominant allele that is extremely variable in its expression. The trait often tends to disappear with age, being present in about 20 percent of 1-year-old Caucasians but in only about 3 percent of those 12 years and older. Approximately 70 percent of adult Orientals have epicanthus, but it is not clear whether the Oriental form of the trait has the same genetic basis as in other racial groups. (As a point of interest, epicanthus is a normal characteristic of the *fetus* in all racial groups.)

Camptodactyly (see Figure 3.10g). Camptodactyly is a minor hand abnormality usually expressed as an inability to completely straighten the little finger. It is due to an autosomal dominant allele with incomplete penetrance and variable expressivity, but its prevalence is not known.

Mid-Digital Hair (see Figure 3.10h). Presence of hair on the middle segment of the fingers is thought to be due to an autosomal dominant allele. Homozygotes for the recessive allele lack hair on these finger segments, even if they otherwise have abundant body hair. The prevalence of the trait is unknown.

Phenylthiocarbamide Tasting The ability to taste this chemical is due to an autosomal dominant allele usually symbolized as T; the recessive allele is symbolized as t. Thus, genotypes TT and Tt are tasters and report that the chemical is bitter, whereas tt genotypes are nontasters and cannot detect the chemical. Approximately 30 percent of Western Europeans are nontasters. Tasting ability is of some interest because it illustrates that genes can influence almost any trait, including, in this case, our sensory perceptions.

S-methyl Thioester Detection Certain odiferous substances are excreted into the urine by at least some people after eating asparagus. At one time, excretion was thought to be due to a single dominant allele, nonexcretors being homozygous recessives, but now it appears that the primary genetic variation involves the ability to smell the substances. Approximately 10 percent of Caucasians can smell the diluted asparagus-related substances.

ABO Blood Groups The ABO blood groups refer to certain constituents found on the surface of red blood cells. These constituents can be identified by means of the appropriate chemical reagents (see Chapter 13). An individual may have one of four blood types. Those individuals whose red cells have only the A substance are called type A individuals;

those with only the B substance are called type B; those with neither A nor B are said to be of type O; and those with both the A and B substances are said to be of type AB.

The ABO blood types are due to three alleles at a single locus near the tip of the long arm of chromosome 9. These alleles are designated I^A, I^B, and I^O. Individuals of genotype $I^A I^A$ or $I^A I^O$ have blood type A; those of genotype $I^B I^B$ or $I^B I^O$ have blood type B; those of genotype $I^O I^O$ have blood type O; and those of genotype $I^A I^B$ have blood type AB. Thus, I^A is dominant to I^O, and I^B is also dominant to I^O. However, because $I^A I^B$ individuals express phenotypic characteristics associated with *both* the I^A and I^B alleles, I^A and I^B are said to be **codominant**. The ABO blood group locus illustrates the phenomenon of **multiple alleles**, which refers to the occurrence of more than two alleles at a locus. The situation is genetically somewhat more complex than indicated above, because at least two types of I^A allele (designated I^{A_1} and I^{A_2}) are known.

Considerable variation in the frequency of the ABO blood types occurs among populations. Among Iowans, for example, the frequencies of A, B, O, and AB blood types are 42 percent, 9 percent, 46 percent, and 3 percent, respectively; among Chinese, the comparable frequencies are 31 percent, 28 percent, 34 percent, and 7 percent, respectively. The chimpanzee (our closest living relative) also has the ABO blood group substances, with about 87 percent of chimps having blood type A and the rest having blood type O; the I^B allele is evidently rare in chimps or does not occur at all.

Rh Blood Groups Individuals who are Rh$^+$ (Rh positive) have a certain substance present on the surface of their red blood cells, whereas Rh$^-$ (Rh negative) individuals lack this substance. (The substance is unrelated to the ones involved in the ABO blood groups.)

Presence of the Rh$^+$ substance is due to a dominant allele at a locus on the short arm of chromosome 1. We will denote the dominant allele as D and its recessive counterpart as d. Thus, DD and Dd genotypes are phenotypically Rh$^+$, whereas dd genotypes are Rh$^-$. Details of the genetics are unclear, however. The system may involve two or more alleles at each of three loci that are extremely close together on the chromosome, certain combinations of which include the D allele at one locus. Alternatively, the system may involve multiple alleles (at least eight) at a single locus, some alleles (collectively designated D) associated with the presence of the Rh$^+$ substance and other alleles (collectively designated d) with its absence. As with many other traits, there is variation in the incidence of Rh$^+$ and Rh$^-$ among populations. Among American Caucasians, the frequencies of Rh$^+$ and Rh$^-$ are about 86 percent and 14 percent, respectively. Among the Basques, a group of people who live in the Pyrenees Mountains between France and Spain, the frequencies of Rh$^+$ and Rh$^-$ are about 58 percent and 42 percent, respectively; the Basques have the highest known frequency of Rh$^-$. (For further discussion of the Rh blood groups, see Chapter 13.)

Human Heredity: Some Cautionary Remarks

It should be clear from the preceding section that determining the genetic basis of human traits can be fraught with difficulties such as incomplete penetrance, variable expressivity, variable age of onset, and sex influences. Because of these complications, some traits that may seem to be inherited may not be genetic; and some traits that may seem to be caused by the environment may nevertheless be determined by genes. Determination of the actual genetic basis of human traits therefore requires careful study and an awareness of certain pitfalls and complications. Here follows a brief compilation of some of these pitfalls.

1. *Traits that are* **familial** *(i.e., tending to "run in families") are not necessarily genetic.* Children learn to speak from their parents, for example, and they may pick up certain peculiarities in pronunciation or accent. Such peculiarities are certainly familial, but this is due to learning, not genetics. Familial traits that are transmitted from generation to generation by means of the learning process are said to be due to **cultural inheritance**.

2. *Traits caused by environmental factors may nevertheless have an underlying genetic basis.* This means that genes can determine the degree of **liability** (risk) toward the expression of environmentally triggered traits. For example, genes can cause an extreme sensitivity and violent side reactions to certain drugs, such as antibiotics. Yet, unless the individual is exposed to the drug, sensitivity will not be expressed. In this case, the drug is the environmental trigger for a genetically determined trait.

3. *Genetic traits may have no obvious pattern of simple Mendelian inheritance.* This is true because genetic traits need not be determined by the alleles at a single locus. Many traits are multifactorial (influenced by alleles at many loci). Many traits are more or less strongly influenced by the environment. Pedigrees for such traits will therefore be incompatible with simple Mendelian inheritance, but the traits are nonetheless genetic.

4. *A genetic trait may consist of several seemingly unrelated symptoms.* The cause of this is pleiotropy—the fact that genes can influence many traits at the same time. For example, the blood disorder sickle cell anemia has many seemingly unrelated symptoms, including general weakness, susceptibility to infections, fever, and recurrent pain in the joints and other body parts. However, all these symptoms are pleiotropic effects of the allele that causes the abnormal hemoglobin protein in red blood cells. (See Chapter 4 for further discussion.)

5. *Genetic traits can have different modes of inheritance in different kindreds.* Cleft palate (failure in the fusion of the halves forming the roof of the mouth) is a good example. In most kindreds in which cleft palate occurs, the trait is multifactorial. In some rare kindreds, however, the trait is due to an autosomal dominant allele.

6. *Genetic traits with well-established modes of inheritance can be mimicked by environmentally caused disorders.* For example, the kinds of malformations in the embryo that can result from certain drugs or from infection with German measles virus (rubella) may strongly resemble those caused by certain genetic disorders. Phenotypes that resemble known genetic conditions but are caused by environmental factors are known as **phenocopies**.

7. *Genetic traits may not be expressed in all individuals who have the relevant genotype.* This is true of genes with incomplete penetrance, and such examples as congenital ptosis were discussed in the previous section. Nonexpression of certain genotypes is also characteristic of sex-limited traits because, by their very nature, these traits can be expressed in only one sex.

8. *Individuals who are affected with the same genetic trait may not be affected to the same degree.* This is an important principle, and its causes include variable expressivity, variable age of onset, and sex influence on expression, all of which were discussed earlier in connection with common baldness.

9. *Genetic traits with well-established modes of inheritance may nevertheless be expressed differently in different kindreds.* This can occur because kindreds may differ in **modifier genes** (alleles at other loci that affect expression of the trait in question). Also, different kindreds may have different alleles at the **major locus** (the locus that has the major effect on expression of the trait). For example, the A substance in the ABO blood groups will be slightly different in kindreds segregating for I^{A_1} than in those segregating for I^{A_2}.

10. *Seemingly identical traits with the same mode of inheritance may nevertheless be caused by alleles at different loci.* For example, production of the pigment **melanin** is a multistep process, with each step controlled by the product of a different locus. Absence of melanin leads to **albinism**, in which affected individuals have stark white hair and pinkish skin (see Chapter 4 for a more detailed discussion). Two almost identical forms of albinism are known, both due to autosomal recessive alleles. The recessive alleles are at different loci, however, and represent defects at different steps in pigment production. Hence, some kindreds have one form of autosomal-recessive albinism, and other kindreds have the alternative form of autosomal-recessive albinism.

Simple Mendelian Dominance

At latest count, some 736 human traits were known to be simple Mendelian traits due to autosomal dominant alleles, and another 753 traits were strongly suspected to have this sort of simple Mendelian dominance as their mode of inheritance. Most of these traits are rare, usually much rarer than those listed in Table 3.1. Nevertheless, the traits are important because many of them involve inherited abnormalities of various kinds, and understanding their genetic basis permits proper **genetic counseling** (genetic advising) of parents or other relatives of afflicted individuals about what chance they may have of transmitting the genes that are responsible for the trait (or what chance they have of developing the trait themselves). Moreover, understanding the genetic basis of a trait often reveals clues about the trait that are important in developing new clinical treatments of affected individuals. A few examples of simple Mendelian dominance are discussed below; many others will be found in later chapters, and further discussion of genetic counseling will be taken up in Chapter 16.

One simple Mendelian dominant trait is of major importance in human genetics because it affects so many people. The trait is **familial hypercholesterolemia**, and it is the *single most frequent simple Mendelian disorder*. In most populations, about 1 person in 500 is heterozygous for the dominant allele. Heterozygotes for this allele have elevated levels of cholesterol in their blood serum that are untreatable by dietary restriction, early onset of atherosclerosis ("hardening of the arteries") due to deposits of fatty substances in the arteries, and a high risk of heart attack (myocardial infarction)—at least 25 times the risk in normal individuals. The allele has a pleiotropic effect leading to the development of yellowish nodules in the tendons, especially near the knuckles of the hands (see Figure 3.11 for an extreme example). These characteristic nodules have an age of onset between 30 and 40 in males (somewhat later in females), and they appear at approximately the same time as symptoms of coronary artery disease. Homozygotes for the high-cholesterol allele are much rarer than heterozygotes (the frequency of homozygotes is about one per million in most populations), and they are much more severely affected than the heterozygotes. The yellowish nodules appear earlier in life and may even be present at birth, serum cholesterol levels are greatly elevated, and death from heart disease usually occurs by age 20. To underscore the importance of familial hypercholesterolemia, there are approximately 500,000 heterozygotes in the United States alone!

Another example of simple Mendelian dominance in humans is a rare dominant gene

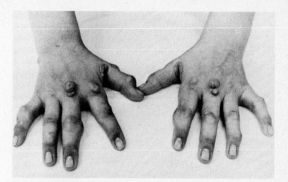

Figure 3.11   Hands of an individual with familial hypercholesterolemia showing nodules in tendons near the knuckles.

that causes a kind of dwarfism known as **achondroplasia** (Figure 3.12). Over a hundred different causes of short stature have been catalogued; this is one of them. (A person whose adult height is less than 5 ft is considered "short"; one whose adult height is less than 4 ft 4 in is considered a "dwarf.") Achondroplasia is a rare condition, affecting only about 1 in 10,000 people. One kindred that was segregating for the dominant gene causing achondroplasia—call the gene A and its normal allele a—was a Mormon family in Utah dating back to 1833 when the region was pioneered. This form of dwarfism is not incompatible with a relatively normal life. Indeed, in Utah during the late 1800s, achondroplastic dwarfs of this type were able to function well in an agrarian society, and some became community leaders and men and women of substance. Also, although the pelvis of achondroplastic women is flattened somewhat (front to back), childbearing is no problem, though it is in many other kinds of dwarfism. As in the case of woolly hair, all marriages of interest are heterozygous Aa × homozygous aa, and of 76 children born to such marriages in the Utah kindred, 34 were achondroplastic and 42 were normal, a ratio in satisfactory agreement with the expected 38:38.

Still another example of simple Men-

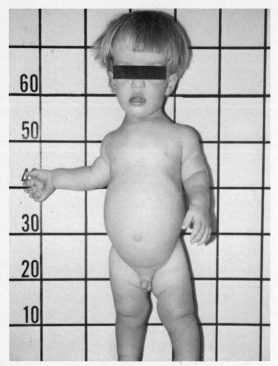

Figure 3.12   An achondroplastic dwarf.

delian dominance is **brachydactyly** or short-fingeredness (Figure 3.13a); the fingers of

affected individuals are about two-thirds the length of normal fingers owing primarily to the shortness of the outermost long bones. (There are other skeletal abnormalities in the hands as well.) This condition is of historical interest; studied in 1905, it was the first demonstration of simple Mendelian dominance in humans.

One final example of simple dominance should be mentioned because the trait is relatively common; it is **polydactyly**, or extra fingers (see Figure 3.13b). In some kindreds, the trait that occurs is a fully formed extra finger; it does no harm, so it is ordinarily not even removed. This form of polydactyly is a simple Mendelian dominant. In other kindreds, a second form of polydactyly is found, which is multifactorial. The extra finger is a tiny one, just a nub, not fully formed. The tiny extra digit, which may be next to either the thumb or the little finger, is eliminated by cutting off its blood supply with a loop of silk thread. The nub eventually degenerates, much like what happens to the remnants of the umbilical cord of a newborn baby. The two forms of polydactyly taken together are fairly common. About 1 percent of American Negroes have the trait, making it about seven times more common among blacks than among whites.

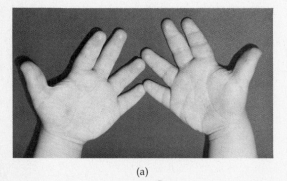

(a)

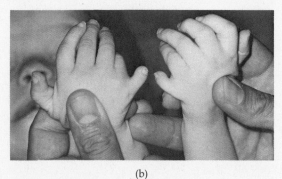

(b)

Figure 3.13   (a) Hands of a child with brachydactyly (short fingers); note that the fingers are only slightly longer than the thumb. (b) Hands of a child with polydactyly (extra fingers); the extra finger on each hand of this child is a small but well-formed digit next to the little finger.

SUMMARY

1. Alleles are alternative forms of a gene that can occupy a particular locus. The word **gene** is a general term used to refer to those elements that are transmitted in inheritance. A **locus** is the physical position of a gene on a chromosome. An **allele** is a particular form of a gene; that is, an allele is a particular sequence of nucleotides in the DNA corresponding to the gene.

2. Alternative alleles at a particular locus are often designated by upper and lower case letters, such as *A* and *a*. For an autosomal locus that has two alleles, three genetic constitutions (called **genotypes**) are possible: *AA*, *Aa*, and *aa*. The *Aa* genotype is said to be **heterozygous**; the *AA* and *aa* genotypes are said to be **homozygous** *A* and **homozygous** *a*, respectively.

3. The physical appearance of an individual is known as his or her **phenotype**. When Mendel studied the phenotypes round seeds versus wrinkled seeds in garden peas, he discovered that seeds of genotype *AA* and *Aa* were phenotypically round, whereas seeds of genotype *aa* were phenotypically wrinkled. This situation is described by saying that the *A* allele is **dominant** to the *a* allele; conversely, we could say that the *a* allele is **recessive** to the *A* allele. In human genetics, however, an allele *A* is said to be dominant to allele *a* if *Aa* heterozygotes are phenotypically different from *aa* homozygotes; in many instances in which a harmful trait is due to a dominant allele, the *AA* homozygotes will be much more severely affected than the *Aa* heterozygotes.

4. Mendel's **law of segregation** states that, during the formation of gametes, the alleles in an individual separate (segregate) from each other in such a way that each gamete will carry one or the other; moreover, half the gametes will carry one of the alleles and half will carry the other. Thus, for example, a heterozygous *Aa* individual will produce 50 percent *A*-bearing gametes and 50 percent *a*-bearing gametes. Said another way, the **probability** that a gamete from an *Aa* individual carries *A* is $\frac{1}{2}$, and the probability that it carries *a* is $\frac{1}{2}$. Segregation of alleles results from the separation of homologous chromosomes during meiosis.

5. Because of segregation in *Aa* individuals, the mating *Aa* × *aa* is expected to produce 50 percent *Aa* and 50 percent *aa* offspring. If the *A* allele is dominant, then such matings are expected to produce 50 percent of offspring having the phenotype associated with the dominant allele and 50 percent having the phenotype associated with the homozygous recessive. In any particular family, however, there need not be equal numbers of the two offspring phenotypes because of **chance variation**. The exact probabilities of various numbers of the two types of offspring can be calculated from the formula

$$\frac{n!}{i!(n-i)!}\left(\frac{1}{2}\right)^i\left(\frac{1}{2}\right)^{n-i}$$

where *n* is the total number of offspring, *i* the number of *Aa* offspring, and (*n* − *i*) the number of *aa* offspring.

6. Simple Mendelian traits are traits with inheritance that is due to the alleles at a single locus. In spite of considerable phenotypic variation for many traits in humans, most of these traits are not simple Mendelian. Many of the most easily observed traits (such as height, weight, hair color, eye color, and skin color) are **multifactorial**, which means that the traits are due to the combined effects of alleles at many loci. Many traits are also influenced by environmental factors (such as diet, in the case of weight).

7. Studies of simple Mendelian traits in humans (such as common baldness, chin fissure, ear cerumen, and others listed in Table 3.1) reveal complications in the traits or in their genetic basis. For example, genes may have **pleiotropic effects**, which refers to the fact that a gene can simultaneously influence several, even seemingly unrelated, traits. Complications in trait expression include **incomplete penetrance** (nonexpression of the trait in some individuals who have the relevant genotype), **variable expressivity** (variation in severity of expression of the trait in different individuals), and **variable age of onset** (variation in the age at which the trait appears during life). Traits may also be **sex influenced** (different expression in males and females) or even **sex limited** (expressed in only one sex). Genetic complications include **multiple alleles** (more than two alleles at the major locus affecting the trait) and **variation among populations** (incidence of the trait differing in different populations).

8. A number of additional cautions should be kept in mind regarding human heredity. The most important are (a) traits that are familial (especially behavioral traits) need not be genetic, (b) traits caused by environmental factors may nevertheless have an underlying genetic basis, and (c) genetic traits with well-established modes of inheritance can be mimicked (**phenocopied**) by environmentally caused disorders.

9. Familial hypercholesterolemia, a trait due to a simple Mendelian dominant allele, is the most frequent simple Mendelian disorder in humans. In most populations, about 1 individual in 500 is heterozygous for the allele. These individuals, in addition to other symptoms, have a greatly elevated risk of heart attack.

WORDS TO KNOW

Cultural Inheritance	Kindred	Punnett square
Probability	Phenocopy	Major locus
	Liability	Pleiotropy
Trait	Genetic counseling	Modifier gene
Phenotype		Multiple alleles
Congenital		
Familial	**Gene**	**Human Traits**
Simple Mendelian trait	Locus	Common baldness
Multifactorial (polygenic)	Allele	Familial hypercholesterolemia
Incomplete penetrance	Genotype	ABO blood groups
Variable expressivity	Homozygous	Rh blood groups
Variable age of onset	Heterozygous	Ear cerumen
Sex influenced	Dominant	Albinism
Sex limited	Recessive	Achondroplasia
Variation among populations	Codominant	Brachydactyly
Pedigree	Segregation	Polydactyly

PROBLEMS

1. For discussion: Huntington chorea is a rare, simple Mendelian disorder due to an autosomal dominant allele with complete penetrance. Heterozygotes for the allele undergo an incurable and relentless degeneration of the nervous system marked by progressive loss of walking ability, power of speech, and mental faculties. Age of onset is variable but is usually between 35 and 40; affected individuals undergo progressive deterioration but survive an average of about 15 years after onset. Among Caucasians, the estimated frequency of heterozygotes at birth is about 1 per 10,000. Because of the late age of onset, individuals with Huntington chorea usually have had their children prior to the onset of the disease. Since Huntington chorea is due to a rare, autosomal dominant allele, each child will have a 50 percent chance of being heterozygous and therefore a 50 percent chance of developing the disease later in life. A chemical procedure could be developed that permits heterozygotes to be identified long before symptoms of the disease appear. Do you think the procedure should be applied routinely to the children of people who have Huntington chorea? What possible benefits might result? What possible harm might be done? If one of your parents had Huntington chorea, would you want to know your genotype? Suppose the procedure were 90 percent accurate, so that there was a 10 percent chance of incorrectly identifying a normal person as a heterozygote. Do you think this 90 percent procedure should be used?

2. Which of the following terms refer to genotype and which to phenotype: Aa, $I^A I^B$, round seeds, blood type O, homozygous, over 6-ft tall, bald, TT, heterozygous, nontaster, wet ear wax, phenocopy?

3. One frequently hears statements such as "The inability to taste phenylthiocarbamide is an autosomal recessive trait." Although the meaning is clear, such statements are not, strictly speaking, correct. What is wrong with the statement?

4. A woman has identical twins (i.e., twins arising from a single zygote and therefore genetically identical). One twin has congenital ptosis, and the other does not. How is this possible in light of the fact that the trait is "inherited"?

5. How is it possible for two Rh$^+$ parents to have an Rh$^-$ child? Would it be possible for two Rh$^-$ parents to have an Rh$^+$ child?

6. A man homozygous for the T ("taster") allele marries a woman homozygous for the t ("nontaster") allele. What genotypes and phenotypes would be expected among their offspring and in what proportions? Draw a Punnett square for the mating.

7. An Rh$^+$ woman whose father was Rh$^-$ marries an Rh$^-$ man. What genotypes and phenotypes would be expected among their offspring and in what proportions? Draw a Punnett square for the mating.

8. A nontaster woman married to a taster man has a nontaster child. What are the genotypes of mother, father, and child?

9. A nontaster woman has a taster child. What are the possible genotypes of the father?

10. A woman of blood type O, Rh$^+$ has a child of blood type A, Rh$^-$. What are the possible phenotypes of the father for the ABO and Rh blood groups?

11. How is it possible for a male to have common baldness when neither of his parents have it? What would be the genotypes of father, mother, and son in such a case? (Use the symbols B and b for the "bald" and normal alleles, respectively.)

12. In a family of four children, one has blood type A, one B, one AB, and one O. What are the genotypes and phenotypes of the parents?

13. A woman's first husband has blood type A and their child has blood type O; her second husband has blood type B and their child has blood type AB. What are the woman's genotype and phenotype?

14. A woman with achondroplasia marries a normal man and they have four children. What is the probability that the oldest child is affected (disregarding whether the other three are affected or not)? What is the probability that exactly two children are affected?

15. A man who is heterozygous for familial hypercholesterolemia marries a normal woman and they have three children. What is the probability that exactly two children will have the trait? What is the probability that all three will have the trait? What is the probability that two *or* three of the children will have the trait? (Hint: The probability of two *or* three affected children is the sum of the probability of two affected and the probability of three affected.)

FURTHER READING AND REFERENCES

Barnes, P., and T.R. Mertens. 1976. A survey and evaluation of human genetic traits used in classroom laboratory studies. J. Heredity 67:347–352. Certain "classic" simple Mendelian traits used to illustrate human heredity are not simple Mendelian traits at all.

Bergsma, D. 1979. Birth Defects Compendium, 2d ed. Alan R. Liss, New York. This lengthy catalogue of birth defects, although primarily designed for physicians and counselors, contains a wealth of useful information.

Crow, J.F. 1979. Genes that violate Mendel's rules. Scientific American 240:134–146. A readable account of certain genes that do not segregate as expected.

Goldstein, J.L., and M.S. Brown. 1979. The LDL receptor locus and the genetics of familial hypercholesterolemia. Ann. Rev. Genet. 13:259–289. More details on this important disorder.

Hsia, V.E., Hirschhorn, K., R. L. Silverburg, and L. Godmilow (eds.). 1979. Counseling in Genetics. Alan R. Liss, New York. An up-to-date review of various aspects of genetic counseling demonstrating that much more than mere instruction is involved.

Iltis, H. 1932. Life of Mendel. Trans. by E. and C. Paul. Norton, New York. Old, but still the best available biography of Mendel.

Lison, M., S.H. Blondheim, and R.N. Melmed. 1980. A polymorphism of the ability to smell urinary metabolites of asparagus. Brit. Med. J. 281:1676–1678. An old trait reexamined.

McKusick, V.A. 1978. Mendelian Inheritance in Man. 5th ed. Johns Hopkins University Press, Baltimore. A computerized catalogue of human hereditary traits useful for many purposes.

Mohr, O.L. 1932. Woolly hair: A dominant character in man. J. Heredity 23:345–352. An extensive study of the inheritance of woolly hair, and the source of Figures 3.6 and 3.7.

Myers, R. H., J. J. Madden, J. L. Teague, and A. Falek. 1982. Factors related to onset age in Huntington disease. Am. J. Hum. Genet. 34: 481-488. Age of onset of Huntington chorea varies from 4 to 65!

chapter 4
Mendel's Laws and Recessive Inheritance

The examples of dominant inheritance of rare human traits discussed in the last chapter illustrate segregation directly because most individuals who carry the dominant allele result from matings between heterozygotes and homozygotes, and the segregation of alleles in the heterozygotes during the formation of gametes leads to a 1:1 distribution of phenotypes among the offspring. There is a class of traits in which attention should be focused on matings between two heterozygotes. These traits are caused by recessive alleles. They are conditions that are not expressed unless a person is homozygous for some allele—that is, unless two copies of some particular allele are present. Dominance and recessiveness are two sides of

the same coin. Dominant genes are dominant with respect to some other allele at the homologous locus, and this other allele is therefore recessive. Conversely, recessive genes are recessive with respect to some other allele at the homologous locus, and this other allele is therefore dominant. Ordinarily the words *dominant* and *recessive* are applied to alleles without explicit reference to which other allele the gene is dominant or recessive. This is partly by convention, partly for brevity, and partly carelessness. The allele designated as dominant or recessive is usually meant to be the abnormal allele, and the dominance or recessiveness is usually intended to mean with respect to the normal allele at the locus. The use of *abnormal* in this context should not be taken to imply "harmful" or "undesirable," especially in regard to the sort of traits listed in Table 3.1; the word is meant to imply only "atypical," "unusual," or "not common." Mid-digital hair and the AB blood type are examples of traits that are abnormal in the sense of unusual but certainly not in the sense of harmful.

Nevertheless, most of the examples of recessive inheritance to be discussed in this chapter are very serious and harmful indeed. In some cases, affected homozygous individuals can never lead a normal life and can never produce children. Consequently, the origin of seriously affected children is almost exclusively matings between two heterozygous people, people who are themselves normal (because they carry one dominant allele) and usually completely unaware that they carry the deleterious recessive allele. Unlike the traits inherited as dominants, mutation in these cases cannot be suggested as the cause of affected individuals. This is because two simultaneous mutations at corresponding loci on homologous chromosomes would be required to produce the genotype of an affected individual, and the odds against the simultaneous occurrence of events that are individually so very rare are astronomical. However, similar to the conditions inherited as dominants, traits inherited as recessives can be mimicked (phenocopied) almost perfectly by environmentally induced abnormalities or by other hereditary conditions. This pitfall emphasizes again the need for caution in the study of inherited traits.

Mechanics of Recessive Inheritance

The most important matings in the case of rare, harmful recessives are between heterozygotes, symbolically written as $Aa \times Aa$. Segregation therefore occurs in both parents. Half the eggs carry the A allele and half the a allele;

likewise for the sperm. The eggs and sperm join at random to give rise to the genotypes of the children. Children of three genotypes can be produced: AA, Aa, and aa; these are expected to occur in the proportions $\frac{1}{4}$ AA, $\frac{1}{2}$ Aa, and $\frac{1}{4}$ aa. The reason for these numbers can be seen by considering the surname analogy of the last chapter. The analogous mating is Mr. Smith-Brown with Ms. Smith-Brown. Four surnames of offspring can result: Smith-Smith, Smith-Brown, Brown-Smith, and Brown-Brown. Since the selection of offspring surnames is random, the four possibilities must be equally likely. This equal likelihood can be expressed numerically in three ways. One is to say that the ratio of the four possibilities (or the "odds," as the bookmakers would put it) is 1:1:1:1. (This is read as "one to one to one to one.") Another way is to give the proportions, which is 25 percent for each of the four possible offspring surnames. Note that the percentages associated with all the possibilities add up to 100, which is the numerical way of saying that these four offspring surnames are the *only* possibilities. Still another way to express the likelihood of the four outcomes is to express the percentages as probabilities, in this case $\frac{1}{4}$ for each of the four outcomes. Note here (as in the last chapter) that all the probabilities add up to 1.

If you designate the name Smith as A and Brown as a, then the four possible offspring surnames can be seen to correspond to the four ways that eggs and sperm can unite in the mating $Aa \times Aa$. The four possibilities are AA, Aa, aA, and aa. These possibilities are equally likely, so their ratio is 1:1:1:1 (see the Punnett square in Figure 4.1). However, the Aa and aA genotypes are identical in that each has one dominant and one recessive allele; they only seem different because the alleles are written in opposite order. The Aa and aA genotypes must therefore be lumped together. This reduces the number of outcomes to three:

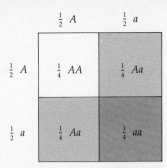

Figure 4.1 Punnett square showing segregation in a mating of two Aa heterozygotes. In both parents, half the gametes carry A and half carry a. The overall distribution of offspring is $\frac{1}{4} AA$, $\frac{1}{2} Aa$, and $\frac{1}{4} aa$.

AA, Aa, and aa. The ratio of the three is 1:2:1. Expressed as percentages, 25 percent of the children will be genotypically AA; 50 percent will be Aa; and the remaining 25 percent will be aa. Or, what is equivalent, the chance that a particular unborn child will be AA is $\frac{1}{4}$; the chance that it will be Aa is $\frac{1}{2}$; and the chance that it will be aa is $\frac{1}{4}$.

If the a allele is recessive, then only the aa individuals will be affected; the AA and Aa individuals will all be normal and indistinguishable. The ratio of normal to affected in the children of matings of $Aa \times Aa$ therefore boils down to 3:1 or, as fractions, $\frac{3}{4}$ normal and $\frac{1}{4}$ affected. This is the 3:1 ratio that many people associate with Mendelism. Note that it is a ratio of *phenotypes*. The underlying ratio of *genotypes* is still 1:2:1. One of Mendel's own experiments provides a good illustration. Mendel carried out matings of $Aa \times Aa$, where A again designates the dominant gene for round seeds in the garden pea. He observed a total of 7324 seeds from the mating; 5474 of these were round and 1850 were wrinkled—a ratio in very close agreement with the expected 5493:1831 (that is, 3:1). By subsequent breeding tests Mendel was able to show that, *among those seeds that were phenotypically round*, the ratio of AA to Aa genotypes was 1:2. Stated another way, among those seeds that received at least one A allele and were therefore phenotypically round, $\frac{1}{3}$ were genotypically AA and $\frac{2}{3}$ were genotypically Aa. These experiments verified the underlying 1:2:1 ratio of $AA:Aa:aa$.

Characteristics of Autosomal-Recessive Inheritance

In this section we consider three general characteristics of autosomal-recessive inheritance.

Pedigrees Figure 4.2 is a pedigree of one form of autosomal-recessive albinism (lack of pigmentation) that illustrates certain characteristic features of this mode of inheritance. The pedigree symbols are as described in Chapter 3, but here each person can be uniquely designated by specifying his or her generation number (I to IV in Figure 4.2) and the position (number from the left) in the generation. For example, the male in generation I is referred to as individual I-1, and his mate is designated I-2; similarly, the affected male in generation III is individual III-3, and his affected sister is individual III-6. Some characteristic features in pedigrees of traits due to autosomal recessives are as follows:

1. There will often be a **negative family history** for the trait; that is, the trait may not occur among the ancestors of affected individuals. In Figure 4.2, for example, there is a negative family history for albinism in generations I and II.

2. Males and females are equally likely to be affected with the trait.

3. Affected individuals *may* have unaffected parents. In Figure 4.2, for example, the parents of III-3 and III-6 are both unaffected, as are the parents of IV-11. If the trait in question is rare, then affected individuals will *usually* have unaffected parents (i.e., the parents will

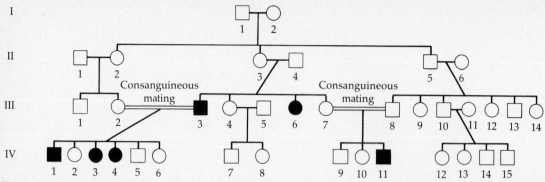

Figure 4.2   Pedigree of one form of autosomal-recessive albinism showing consanguineous matings between III-2 and III-3 and between III-7 and III-8.

be heterozygous for the recessive allele and therefore phenotypically normal).

4. Affected individuals will often arise from matings between relatives. (Matings between relatives are called **consanguineous matings**.) In Figure 4.2, for example, individuals III-2 and III-3 are first cousins, as are III-7 and III-8. The matings indicated by the labels are therefore consanguineous matings. As shown in the pedigree, **consanguinity** (i.e., genetic relationship) between mating individuals is represented by a double horizontal line connecting those involved.

Consanguinity among the parents of affected individuals is a typical feature in pedigrees involving autosomal-recessive traits, particularly rare traits. An individual who is heterozygous for a rare recessive allele can unknowingly transmit the allele to many of his or her descendants. If one of these heterozygous descendants mates with a relative, there is a significant chance that the relative will also be heterozygous; thus, the consanguineous mating will be of the type $Aa \times Aa$, say, and 25 percent of the offspring will be expected to be homozygous aa and therefore affected. On the other hand, if the heterozygous descendant mates with an unrelated individual, the unrelated individual will usually be genotypically

AA (because the a allele is rare); the nonconsanguineous mating will usually be of type $Aa \times AA$, therefore, and, although half the offspring will be heterozygous, none will be affected.

Frequency of Heterozygotes   Heterozygous individuals who carry an autosomal-recessive allele are often called **carriers**; there is an important relationship between the frequency of carriers and the frequency of affected individuals. The rule is this: *If an autosomal-recessive trait affects less than 44 percent of a population, most genotypes that carry the recessive allele will be heterozygous.* The mathematics underlying this rule will be discussed in Chapter 14, but the consequences of it are shown in Figure 4.3. The horizontal axis in the figure gives the frequency of recessive homozygotes (1 per 10 individuals, 1 per 100, 1 per 1000, and so on), and the vertical axis gives the frequency of heterozygotes (number per 500 individuals). If an autosomal-recessive trait has a prevalence of 10 percent (1 per 10), for example, the graph shows that the frequency of heterozygotes will be $\frac{216}{500}$, or 43 percent; the ratio of heterozygotes to recessive homozygotes is thus $\frac{43}{10} = 4.3$. To take another example, if the incidence of the trait is 0.1 percent (1 per 1000), the frequency of carriers

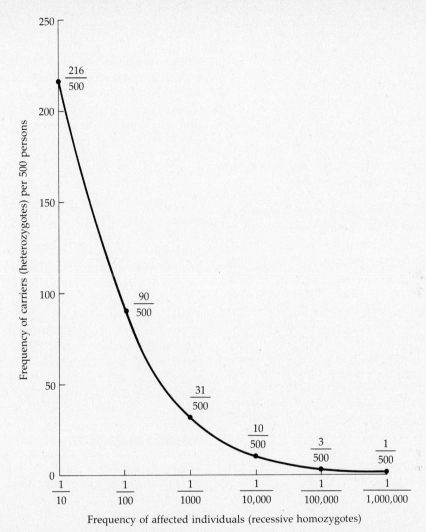

Figure 4.3 Frequency of heterozygotes for a recessive gene (vertical axis) plotted against frequency of homozygous recessives (horizontal axis). If a recessive allele is rare, there will be many more heterozygotes than recessive homozygotes.

is $\frac{31}{500}$, or 6.2 percent; the ratio of heterozygotes to recessive homozygotes in this case is $\frac{6.2}{0.1}$ = 62. In other words, heterozygotes are 62 times as frequent as affected individuals! Note that the ratio is greater in this example than in the previous one. Indeed, *the smaller the incidence of recessive homozygotes is, the greater will be the ratio of heterozygotes to recessive homozy-*

gotes. Thus, the frequency of carriers relative to the frequency of affected individuals *increases* as the incidence of the trait *decreases*.

The reason for the rule giving rise to Figure 4.3 can be seen by considering a recessive allele (call it *a*) in a gamete that is about to combine with another gamete to create a zygote. If the recessive allele is rare, then most of the time the other gamete will carry the dominant counterpart (A), and the resulting zygote will be Aa; only rarely will the other gamete also carry a and thus lead to an aa zygote. Hence, most rare recessive alleles will actually be found in heterozygous (and phenotypically normal) individuals. The cutoff of 44 percent for the incidence of the recessive trait mentioned above is the incidence at which the frequency of heterozygotes exactly equals the frequency of recessive homozygotes.

Number of Affected and Normal Offspring in Aa × Aa Matings

As noted in connection with Figure 4.1, the probability of an affected offspring from an $Aa \times Aa$ mating is $\frac{1}{4}$, and the probability of a phenotypically normal offspring is $\frac{3}{4}$. In Chapter 3 we presented a formula for calculating the overall probability of any number of affected and nonaffected offspring when the probability of each type of offspring was $\frac{1}{2}$. In this case, the probabilities of affected and nonaffected are $\frac{1}{4}$ and $\frac{3}{4}$, not $\frac{1}{2}$ and $\frac{1}{2}$. However, the formula can easily be modified: If the probability of an affected offspring from a mating is $\frac{1}{4}$, and the probability of a nonaffected offspring is $\frac{3}{4}$, then the probability that a group of n children contains exactly i affected ones and $(n - i)$ normal ones is given by

$$\frac{n!}{i!(n - i)!}\left(\frac{1}{4}\right)^{i}\left(\frac{3}{4}\right)^{n - i}$$

(Here again, $n!$ means $n \times (n - 1) \times \cdots \times 3 \times 2 \times 1$, and $0!$ is defined to equal 1.)

As an example of the use of the formula, we will calculate the probability that an $Aa \times Aa$ mating gives rise to exactly one affected and three nonaffected children. We have $n = 4$ (the total number of children), $i = 1$ (the number of affected children), and $n - i = 3$ (the number of nonaffected children). The required probability is thus

$$\frac{4 \times 3 \times 2 \times 1}{(1)(3 \times 2 \times 1)}\left(\frac{1}{4}\right)^{1}\left(\frac{3}{4}\right)^{3} = \frac{27}{64} = 0.42$$

About 42 percent of such families would be expected to have one affected and three normal children. Similarly, the probability that such a family would have no affected children is about 32 percent because, with $n = 4$, $i = 0$, and $n - i = 4$, the formula is

$$\frac{4 \times 3 \times 2 \times 1}{(1)(4 \times 3 \times 2 \times 1)}\left(\frac{1}{4}\right)^{0}\left(\frac{3}{4}\right)^{4} = \frac{81}{256} = 0.32$$

Many human traits are known to be caused by recessive genes. At the present time, 521 conditions are known to be inherited as autosomal recessives, and another 596 traits are strongly suspected to be. In the more serious of these conditions, affected individuals almost always arise from matings—frequently consanguineous matings—between heterozygotes. The heterozygous individuals in the population are phenotypically normal, but they serve as the reservoir of the abnormal allele because, as noted earlier, heterozygotes are much more frequent than recessive homozygotes.

Familial Emphysema

Emphysema refers to an overinflation and distension of the air sacs in the lungs marked by continuous shortness of breath (even when at rest), a wheezy cough, and increased blood pressure in the arteries of the lungs leading to an enlargement of the right side of the heart. Without proper treatment, the disease progresses to obstructive lung disease and eventual death due to respiratory failure or congestive heart failure. Onset of the disease usually occurs in middle age, and the lifespan of affected individuals is shortened by 10 to 30 years depending on the success of treatment.

Emphysema can be caused by environmental factors, such as heavy smoking, but one form of the disease, called **familial emphysema**, is inherited as an autosomal recessive. In most populations, the incidence of familial emphysema is about 1 in 1700 individuals; the frequency of heterozygotes, however, is much greater—about 1 in 20 individuals! The high frequency of heterozygotes is particularly important in this case because there is evidence that heterozygotes may also have an increased risk of lung disease. In one study of 103 patients with obstructive lung disease, for example, there were 5 homozygotes and 25 heterozygotes.

Familial emphysema is also known as α_1-**antitrypsin deficiency**. α_1-antitrypsin is a protein found in blood serum. Its function is to inhibit an enzyme (**trypsin**) that breaks down other proteins. The locus for α_1-antitrypsin, called the *Pi* locus, is apparently on chromosome 2, and at least 23 different *Pi* alleles have been identified. One of these alleles, designated *Pi^Z*, leads to an inactive form of α_1-antitrypsin. This allele is the one responsible for familial emphysema because *Pi^Z Pi^Z* homozygotes have an extremely high risk of developing the disease. As noted, however, *Pi^Z* heterozygotes may also have an increased risk.

Cystic Fibrosis

Among Caucasians, one of the most frequent simple Mendelian recessive diseases of childhood is **cystic fibrosis**, which is due to a recessive allele on the long arm of chromosome 5. The incidence of the condition itself (homozygous recessives) is about 1 in 2500 individuals, but about 1 in 25 individuals is heterozygous for the allele. Although relatively frequent in Caucasians, cystic fibrosis is extremely rare in other racial groups.

It is difficult to determine the mode of inheritance of cystic fibrosis and other conditions caused by rare, simple Mendelian recessives. The problem is that not all matings between heterozygotes ($Aa \times Aa$) actually produce affected children. To show that a disease like cystic fibrosis is due to a simple Mendelian recessive, one has to show that $\frac{1}{4}$ of the children from $Aa \times Aa$ matings are affected. The difficulty in showing this is that $Aa \times Aa$ matings cannot be distinguished from others in which both parents are phenotypically normal; $Aa \times Aa$ matings can be detected only if they have affected children. To see what a problem this causes, focus on families who have just two children; the proportion of such families having zero, one, or two affected children will be $\frac{9}{16}$, $\frac{6}{16}$, and $\frac{1}{16}$, respectively (using the formula given earlier). Only the latter two families would be detected, however, so the proportion of detected families who have one or two affected children would be $\frac{6}{7}$ and $\frac{1}{7}$, respectively. Among the detected families, therefore, the overall proportion of affected children is $(\frac{6}{7})(\frac{1}{2}) + (\frac{1}{7})(1) = \frac{8}{14} = 57$ percent, which is nowhere near the expected 25 percent! This problem arises because not all $Aa \times Aa$ matings are **ascertained** (discovered), and the observed proportion of affected children must be corrected for this **ascertainment bias**. The correction that should be applied is beyond the scope of this book. However, in one study of cystic fibrosis, the corrected proportion of affected children turned out to be 24.9 percent, confirming the autosomal recessive mode of inheritance.

Cystic fibrosis is characterized by the malfunction of the pancreas and other glands, resulting in the production of abnormal secretions. Its chief symptoms are recurrent respiratory infections, malnutrition resulting from incomplete digestion and absorption of fats and proteins, and cirrhosis of the liver. Patients with cystic fibrosis have an accumulation of thick, sticky, honeylike mucus in their respiratory tract, which often leads to respiratory complications. The victim becomes a prime target for secondary infections such as pneumonia or bronchitis. Cystic fibrosis is a disease of childhood, and the symptoms may appear early. The disease usually leads to death in childhood or adolescence, but the lives of affected children can be prolonged somewhat by intensive respiratory and dietary treatment. Left untreated, 95 percent of affected children will die before age 5. With treatment, the average life expectancy of affected girls is more than 12 years, and that of affected boys is more than 16 years. If treatment is begun before any appreciable lung damage has occurred, the child's chance of living to age 21 or older is now greater than 50 percent. The treatment includes a special diet, daily administration of antibiotics, administration of extracts of the pancreas of animals, and special daily lung care (sometimes including the flushing out of the lungs with an aerosol mist). The treatment is continuous and expensive, however.

Sickle Cell Anemia

Among blacks, the most common disorder inherited as a simple Mendelian recessive is **sickle cell anemia**, due to an allele on chromosome 11. The hereditary defect shows up in the red blood cells—the cells that carry oxygen from the lungs to the tissues. More specifically, the defect is in the oxygen-binding protein, **hemoglobin**, a major component of the red blood cells. (Hemoglobin is the protein that physically binds with oxygen molecules and transports them.) People homozygous for the sickle cell allele have a form of hemoglobin that tends to crystallize or stack together when exposed to lower than normal levels of oxygen. The crystallization of hemoglobin causes the entire red cell to collapse from its normal ellipsoidal shape into the shape of a half-moon or "sickle." (See Figure 4.4.) In this form the red cell cannot carry the normal amount of

oxygen. More important, the sickled cells tend to clog the tiny capillary vessels, interrupting the nutrient blood supply to vital tissues and organs. Children with the disease tire easily, may be retarded in their physical development, and tend to be susceptible to infections of all kinds—owing to their general weakened condition caused by the severe chronic anemia brought on by the reduced amount of normal hemoglobin. At intervals the affected people experience sickle cell crises marked by fever and severe, incapacitating pains in the joints, particularly in their extremities, and in their chest, back, and abdomen caused by the clogging of the blood supply to these vital areas. These painful episodes may last from hours to days to weeks. They may be provoked by anything that lowers the oxygen supply; overexertion, high altitude, respiratory ailments, and so on. Pregnancy in a woman with sickle cell anemia is particularly dangerous, not only for the mother because of the increased stress but also for the baby. The sickling of cells in the placenta can decrease or block the oxygen supply to the baby and cause a spontaneous abortion, miscarriage, or stillbirth. (Technically, the expulsion of a fetus from the time of fertilization to three months of pregnancy is an **abortion**; expulsion between three and seven months of pregnancy is a **miscarriage**; expulsion of a nonliving fetus thereafter is a **stillbirth**.)

The lifespan of people affected with sickle cell anemia is distressingly short. They frequently die in their teens or twenties, almost always before age 45. At present there is no cure for the disease, although the symptoms can be treated, transfusions can be given, and the pain can be decreased or relieved.

A confusing terminology about sickle cell anemia has come into widespread use. The carriers of the gene, the heterozygotes, are often said to have the sickle cell "trait." These people are actually quite healthy. Rare cases

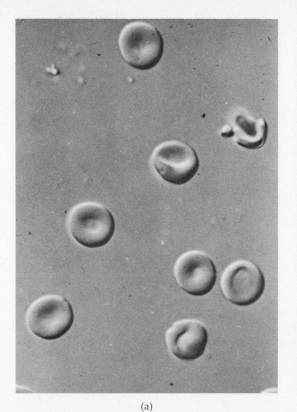

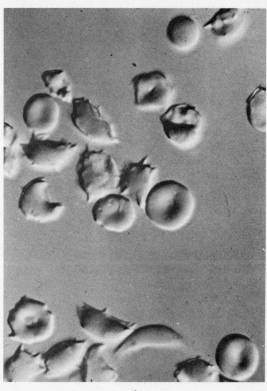

(a)   (b)

Figure 4.4 *(a)* Micrograph of red blood cells from a patient with sickle cell anemia when the cells are saturated with oxygen; under these conditions, the appearance of the cells is the same as that of normal red blood cells. *(b)* Micrograph of red blood cells from a patient with sickle cell anemia when the cells are subjected to conditions of low oxygen; note the extensive collapsing or "sickling" of the cells. With low amounts of oxygen, normal red blood cells do not collapse.

are known, however, of heterozygotes suffering symptoms of the disease brought on by extreme and prolonged reduction of oxygen in the blood. The affected people who are homozygous for the sickle cell gene are said to have the sickle cell "disease." Sickle cell anemia is the condition that afflicts *homozygous* people. These distinctions should be kept in mind to avoid confusing the condition of the heterozygotes, who are normally healthy, with that of the homozygotes, who are not.

Sickle cell anemia is extraordinarily common as compared with other serious conditions that have an equally simple inheritance. In most of West Africa, about 1 per 100 children has the disease, and about 1 person in 6 is a carrier. Among American blacks, the frequencies are lower because only part of their ancestry traces back to West Africa; about 1 per 400 children has the disease and approximately 1 person in 10 is heterozygous. This single disease causes the young to die in enormous numbers; the yearly toll worldwide has been estimated at 100,000. The reason the disease is so common is found in an infectious disease, **malaria**, and in the law of segregation.

As shown in Figure 4.5, the geographic distribution of the high frequency of sickle cell

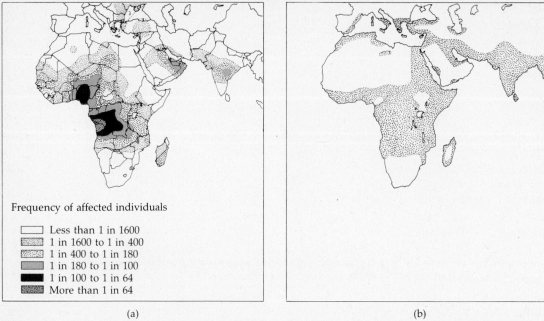

Frequency of affected individuals

☐ Less than 1 in 1600
▨ 1 in 1600 to 1 in 400
▨ 1 in 400 to 1 in 180
▨ 1 in 180 to 1 in 100
■ 1 in 100 to 1 in 64
▨ More than 1 in 64

(a)   (b)

Figure 4.5 *(a)* Map of distribution and frequency of individuals affected with sickle cell hemoglobin disease in Africa, India, the Middle East, and southern Europe. *(b)* Map of distribution of malaria caused by *Plasmodium falciparum* in about the 1920s (before extensive control programs were instituted). Note the extensive overlap of the shaded areas on the two maps.

anemia in Africa almost coincides with the geographic distribution of falciparum malaria, so named because of the protozoan parasite that causes it (*Plasmodium falciparum*). This overlap occurs because the carriers of the sickle cell allele have an enhanced resistance to the disease. The parasites causing malaria are mosquito-borne; they are transferred from bloodstream to bloodstream by mosquitos and get transported throughout the body by inserting themselves into red blood cells. The red blood cells with abnormal hemoglobin tend to sickle when infested with parasites; they clump together and are thereby effectively removed from circulation and destroyed. It may also be that the parasite cannot infect these red cells as readily as the normal ones. Still other factors may be at work, but, in any case, the heterozygous carriers of the sickle cell gene tend to be

more resistant to malaria than normal homozygotes. In one study, the risk of severe malarial infections in children who were carriers was found to be only half as great as the risk to homozygous normal children. Partly because of this, carriers of the sickle cell allele have an enhanced ability to survive and reproduce in an environment in which malaria is widespread. Inevitably, since the carriers have an advantage over even the normal homozygotes, many matings occur between two carriers, and the law of segregation foreordains that $\frac{1}{4}$ of the children will have the disease, $\frac{1}{4}$ will be homozygous normal (and therefore more susceptible to malaria than their parents), but $\frac{1}{2}$ of the children will themselves be carriers and will perpetuate the sickle cell allele because of their advantage.

Almost as if to emphasize the connection

between malaria and abnormal hemoglobin, another inherited blood disease is found in other parts of the world where malaria is common, particularly in the Mediterranean basin. In this case the abnormality is in the amount of hemoglobin. The Mediterranean condition is also inherited as a simple recessive, and homozygous recessive individuals suffer a severe and debilitating anemia that is frequently fatal in early childhood. The disease itself is called **thalassemia major**. The coextensive distribution of malaria and thalassemia was first noticed in about 1950, and this immediately led to the suggestion that the carriers might have a decreased susceptibility to malarial infections. The geographic correlation was subsequently shown to be true of sickle cell anemia and malaria in Africa as well (see Figure 4.5). The essential correctness of the interpretation of the geographic correlation is now established beyond reasonable doubt. Thalassemia major is very common in certain Italian populations, sometimes reaching an incidence of 1 per 100; up to 1 in 6 normal people are heterozygous carriers. The carriers of the gene have at worst a mild anemia known as thalassemia minor. Thalassemia is also found in appreciable frequencies among other peoples in the Mediterranean region—for example, among the Greeks, Sardinians, Armenians, and Syrians.

Tay-Sachs Disease

Like cystic fibrosis among Caucasians and sickle cell anemia among Negroes, Tay-Sachs disease is a simple Mendelian recessive that has an increased incidence in a particular population—in this case Jews. (Incidentally, many diseases are named after the person or persons who first recognized the symptoms as recurring clinical entities. In this case, Tay was a British ophthalmologist and Sachs an American neurologist; both described the disease in the

1880s, although they were not aware of its genetic basis.) **Tay-Sachs disease**, formerly called "familial infantile amaurotic idiocy," is due to a recessive allele on chromosome 15. The normal form of the gene is responsible for producing an enzyme, called **hexosaminidase A**, which functions in the breakdown of a sugary-fatty substance (**ganglioside GM₂**) in the central nervous system. The role of the enzyme in breaking down ganglioside GM_2 is outlined in Figure 4.6. Homozygotes for the Tay-Sachs allele lack a functional form of hexosaminidase A. The absence of this enzyme causes an abnormal accumulation of ganglioside GM_2 in the central nervous system, which leads to blindness, seizures, and a complete degeneration of mental and motor function. The disease symptoms are usually evident by six months of age; the disease becomes progressively more severe and is invariably fatal within the first five years of life. Among Jews from Central

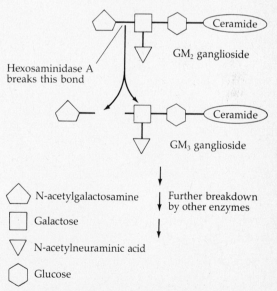

Figure 4.6   Hexosaminidase A is involved in the breakdown of ganglioside GM_2. A defective enzyme leads to accumulation of GM_2 in the central nervous system and causes Tay-Sachs disease.

Europe (Ashkenazi), the incidence of the condition is about 1 per 4000 births. This incidence is about 100 times higher than that among non-Jews or among Jews from the Mediterranean basin (Sephardic); the incidence of Tay-Sachs disease among non-Jews is about 1 per 400,000. Approximately 1 in 30 Ashkenazi Jews is a carrier of the recessive allele, which emphasizes the importance of the disease in this population.

The Complementation Test

One final example of variation among populations in the incidence of inherited disorders is albinism among the Hopi Indians of Arizona and the Jemes and Zuni Indians of New Mexico. **Albinism** results from an inherited defect in the ability of certain specialized cells to produce normal amounts of the brown-black pigment melanin. Albinos therefore lack pigmentation (Figure 4.7). Their hair, skin, and the iris of the eyes are very light or white, although their eyes look pink due to reflected light passing through the blood vessels. The condition is inherited as a simple Mendelian recessive. It is not as serious a disorder as many others, but albinos tend to have vision problems, sometimes including blindness, and are extremely susceptible to sunburn and prone to

Figure 4.7   Photograph of three Hopi girls, taken in about 1900. The girl in the middle is an albino.

skin cancer because they have no melanin to protect them from the sun's ultraviolet rays. The incidence of the condition among the Indians mentioned is about 1 in 200, and about 1 person in 8 is a carrier of the albino allele. Among people of European ancestry, by contrast, the incidence is about 1 in 40,000.

The social position of albinos in Hopi society is interesting because traditional Hopis do not recognize the condition as inherited; they believe that albinism and other abnormalities at birth are the result of some previous action or incident in the life of a parent or relative. (This is an explanation of birth defects encountered in many cultures.) The traditional explanation of albinism in Hopi children varies from case to case. One man was said to have loved a white donkey so dearly that two of his granddaughters were born white. The Indians have come to accept the high incidence of albino children as a fact of life and even to admire them. It is thought to be good luck to have one or more albinos in a village; for this reason, some Hopi women desire to have an albino baby. Albinos are thought to be clean, smart, and very pretty, although many of them never marry.

As discussed in Chapter 3, there is a second form of albinism due to a recessive allele at a different autosomal locus. This form also has an incidence of about 1 in 40,000, but it seems to be no more frequent in Southwest American Indians than in other populations. Albino Hopis almost always have the first form of albinism; they are all homozygous for a recessive allele at the same locus. That the same locus is involved is indicated by the fact that, when two albinos mate, they produce only albino children. The pedigree in Figure 4.8 is a Hopi pedigree illustrating that the children of albino parents are albinos.

The mating between III-2 and III-3 in Figure 4.8 illustrates a human example of a type of mating often carried out in experi-

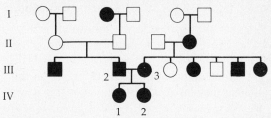

Figure 4.8 Partial pedigree of albinism in a Hopi kindred. Note that the mating between homozygotes (III-2 and III-3) produces exclusively homozygous offspring.

mental organisms. The mating is known as a **complementation test**, and its purpose is to determine whether two phenotypically similar autosomal-recessive traits are caused by alleles at the same locus. In Figure 4.8, the issue is whether the ancestors of individuals III-2 and III-3 have the same form of albinism. If III-2 and III-3 are homozygous for a recessive allele at the same locus, then (as actually happened) all their children will be expected to be albinos. To see why this is the case, let *a* represent the recessive allele for the first form of albinism and let *b* represent the recessive allele for the second form; their respective normal counterparts are designated *A* and *B*. If individuals III-2 and III-3 have different forms of albinism, then one of them will have genotype *aaBB* and the other will have genotype *AAbb*. In such a case, all their children will have genotype *AaBb* and will be nonalbino (because *a* and *b* are both recessive); the recessive genes in III-2 and III-3 are then said to **complement** or to show **complementation**, and it would be concluded that *a* and *b* are **nonallelic** (i.e., involve different loci). The children in Figure 4.8 *are* affected, however; in this case, the recessive genes in III-2 and III-3 are said to be **allelic** (i.e., involving the same locus) or to show **noncomplementation**. Therefore, because the alleles in individuals III-2 and III-3 are noncomplementing, III-2 and III-3 are homozygous for a recessive gene at the same locus;

their genotypes are both *aaBB* (i.e., homozygous for the Hopi type of albinism).

The complementation test is the experimental test of allelism. Two recessive genes that are noncomplementing are, by definition, alleles; they involve the same locus. Conversely, two recessive genes that do complement are not alleles; they involve *different* loci. An example of complementation and noncomplementation occurring in the same kindred is shown in the pedigree in Figure 4.9. The trait here is deaf-mutism (lack of ability to hear and speak). The parents of individual II-9 must both be affected with the same autosomal recessive form of deaf-mutism because the genes they carry are noncomplementing (i.e., all their children are affected). However, individual II-7 must have an autosomal recessive form of deaf-mutism that is due to a *different* locus from the form in II-9 because the genes in II-7 and II-9 show complementation (i.e., all their children are normal). Thus, the mating between II-7 and II-9 shows that there are at least two forms of autosomal deaf-mutism. (In fact, there are many forms. Many genes are involved in the ability to hear and speak, and a homozygous defect in any one of them can cause deaf-mutism.)

At this point, it might be useful to review the types of mating that can occur when there are two alleles at an autosomal locus and the genotypes of offspring that result:

$$AA \times AA \rightarrow \text{all } AA$$
$$AA \times Aa \rightarrow \tfrac{1}{2} AA \text{ and } \tfrac{1}{2} Aa$$
$$AA \times aa \rightarrow \text{all } Aa$$
$$Aa \times Aa \rightarrow \tfrac{1}{4} AA, \tfrac{1}{2} Aa, \text{ and } \tfrac{1}{4} aa$$
$$Aa \times aa \rightarrow \tfrac{1}{2} Aa \text{ and } \tfrac{1}{2} aa$$
$$aa \times aa \rightarrow \text{all } aa$$

Independent Assortment, Recombination, and Linkage

Now we must consider what happens when the segregation of *two* loci is followed simultaneously in the same individual. This is the problem addressed by Mendel's second law, the **law of independent assortment**, which asserts that the segregation of the first locus during meiosis has no influence on the segregation of the second one. Like the discovery of dominance, this generalization of Mendel's is of secondary importance compared with the law of segregation. Indeed, *independent assortment* occurs only between certain loci and not others. A principal factor in determining whether independent assortment occurs between two loci is the relative position of the loci on the chromosomes. Loci on nonhomologous chromosomes must assort independently because of the me-

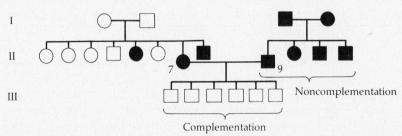

Figure 4.9 Pedigree of autosomal-recessive forms of deaf-mutism from Northern Ireland illustrating noncomplementation and complementation. The parents of II-9 were both homozygous recessives and produced only affected children; thus, the alleles in the parents of II-9 are noncomplementing (i.e., they are alleles at the same locus). Individuals II-7 and II-9 are both homozygous recessives, but all their offspring are nonaffected. In this case, the alleles in II-7 and II-9 are complementing (i.e., they are alleles at different loci).

chanics of meiosis. Individuals of genotype *AaBb* are segregating for two loci at once. If the *A* locus and the *B* locus are on different chromosomes, then the four chromatids of one bivalent will carry *AAaa* and those of the other bivalent will carry *BBbb* (Figure 4.10). The movement of the bivalents during metaphase I to the imaginary plane cutting across the cell and the orientation of the bivalents on this plane occur with total indifference to the alignment of nonhomologous centromeres. The result is that, in anaphase II, an *A*-bearing chromosome is as likely to go to the same pole with a *B*-bearing chromosome as with a *b*-bearing one; likewise, the *a*-bearing chromosome may end up at a pole with either a *B*- or a *b*-bearing chromosome (see Figure 4.10). Therefore, the *AaBb* genotype will produce four kinds of gametes—*AB*, *Ab*, *aB*, *ab*—and these will be equally likely; each of the four types will have a chance of $\frac{1}{4}$ of being present in a particular gamete. It is this 1:1:1:1 distribution of the four gametic types that characterizes independent assortment.

Something different must be taken into account when the *A* and *B* loci are on the same chromosome. It is important in the first place to know exactly which alleles in the *AaBb* genotype are linked on the same chromosome of the pair of homologues. There are two possibilities: The *A* and *B* alleles may have been inherited from one parent, *a* and *b* from the other, in which case the genotype should be written as *AB/ab*; the slash is intended to separate and identify which alleles are on which chromosome. Alternatively, the *A* and *b* alleles may have come from one parent, *a* and *B* from the other, and in this case the genotype should be denoted as *Ab/aB*. Both genotypes will produce four kinds of gametes; the *AB/ab* parent will produce *AB*, *ab*, *Ab*, and *aB*; the *Ab/aB* parent will produce *Ab*, *aB*, *AB*, and *ab*. These gametic types are genetically the same. But in each case the first two gametes

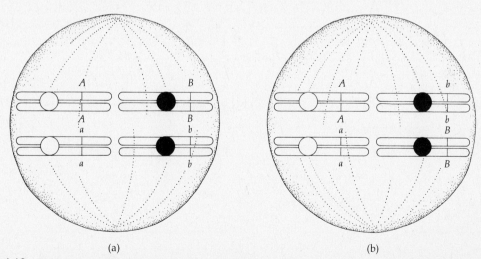

(a) (b)

Figure 4.10 Independent alignment of nonhomologous chromosomes at metaphase I of meiosis produces independent assortment of loci on nonhomologous chromosomes. Shown here are two possible alignments of nonhomologous chromosomes at metaphase I. Alignment as in part *(a)* produces two *AB* and two *ab* gametes; alignment as in part *(b)* produces two *Ab* and two *aB* gametes. Since both alignments are equally likely, the overall proportion of gametes from an *AB/ab* individual will be $\frac{1}{4}AB$, $\frac{1}{4}Ab$, $\frac{1}{4}aB$, and $\frac{1}{4}ab$. In these diagrams, the possible occurrence of crossing-over between the *A* and *B* loci and their respective centromeres has been ignored.

listed are genetically like the parental chromosomes; these gametes are known as **parental** or **nonrecombinant** gametes. The last two gametes mentioned in each case are known as **recombinant** gametes.

The particular ratio of parental to recombinant gametic types depends on the distance between the loci on the chromosome. A bivalent in which one crossover occurs between the A and B loci will, after separation at the two anaphases, be resolved into four gametes—one of each of the parental and recombinant types

[Figure 4.11(a)]. A bivalent in which no crossover occurs in the region will be resolved into four gametes—two of each of the parental types [Figure 4.11(b)]. If the A and B loci are very close together on the chromosome, few bivalents will actually experience a crossover in the region between them, so relatively few recombinant gametes will be formed. Thus, the proportion of recombinant gametes produced between two loci will increase as the physical distance between the loci on the chromosome increases.

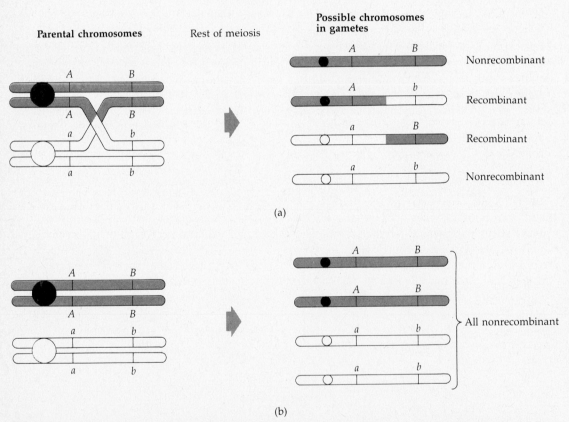

(a)

(b)

Figure 4.11 A measure of the distance between loci on the same chromosome can be obtained from the frequency of recombination. *(a)* When a crossover occurs between the loci, two gametes will be nonrecombinant and two will be recombinant. *(b)* When a crossover does not occur between the loci, all the gametes will be nonrecombinant. Thus, the frequency of recombinant gametes is proportional to the probability of a crossover between the loci, and this, in turn, is proportional to the distance between the loci on the chromosome.

The distance between loci on a chromosome can be measured and expressed in terms of the amount of recombination between the loci. The proportion of recombinant gametes produced by an individual is known as the **recombination fraction** between the loci. (The recombination fraction between two loci is calculated as the number of recombinant gametes divided by the total number of gametes.) The genetic distance between loci is expressed in terms of **map units** (so called because gene arrangements on chromosomes are typically illustrated by diagrams called **genetic maps**). *For loci that are not too far apart on the chromosome, 1 map unit corresponds to 1 percent recombination between the loci.* For example, if two loci are separated by a distance of 3 map units, there will be a 3 percent recombination between them.

Although the frequency of recombination increases as the physical distance between the loci increases, the upper limit is 50 percent, the same as observed for loci on nonhomologous chromosomes. The upper limit is reached when the loci are so far apart on the chromosome that at least one crossover almost always occurs between them. This is quite easy to see if exactly one crossover always occurs, because the crossover involves only two of the four strands of a bivalent, and therefore two of the strands will be recombinant but the other two will be parental [see Figure 4.11(a)]. This corresponds to a recombination frequency of two out of four, or 50 percent. The same is true when two, three, or more crossovers occur between the loci, but the reason is not so easy to see.

Consider, for example, what can happen when exactly two crossovers occur (Figure 4.12.). Sometimes the second crossover will involve the two chromatids that were not involved in the first crossover, and all four chromatids from such a bivalent will be recombinant [see Figure 4.12(a)]. At other times the second crossover will involve the same two chromatids as were involved in the previous crossover; in this case, the second crossover undoes the effects of the first one, and none of the chromatids from the bivalent will be recombinant [see Figure 4.12(b)]. At still other times, the second crossover will involve one chromatid that participated in the first crossover and one chromatid that did not, and in this case, two chromatids from the bivalent will be recombinant and two will be nonrecombinant [see Figure 4.12(c) and (d)]. Now, if the chromatids involved in any crossover are selected at random, as seems to be the case, then the proportions of the three types of double crossovers mentioned above will be $\frac{1}{4}$ to $\frac{1}{4}$ to $\frac{1}{2}$, respectively. Therefore the average number of recombinant chromosomes per bivalent will be $\frac{1}{4} \times 4 + \frac{1}{4} \times 0 + \frac{1}{2} \times 2 = 2$, which is the same number as obtained when only one crossover occurs between the loci. Moreover, if the chromatids involved in any crossover are selected at random, then the average number of recombinant chromatids per bivalent is two no matter how many crossovers occur between the loci. What this means is that when two loci recombine with a frequency of 50 percent (that is, when they assort independently), then you cannot be sure whether they are on different chromosomes altogether or whether they are so far apart on the same chromosome that free recombination occurs.

Loci that are on the same chromosome are said to be **syntenic**, and syntenic loci with less than 50 percent recombination are said to be **linked**. (However, the terminology is sometimes misused because syntenic loci that show 50 percent recombination are sometimes also referred to as *linked*.) One further point about linkage: The **map distance** between two loci (i.e., the number of map units separating them) can be greater than 50, which seems to imply that more than 50 percent recombination can occur. However, the map distance between

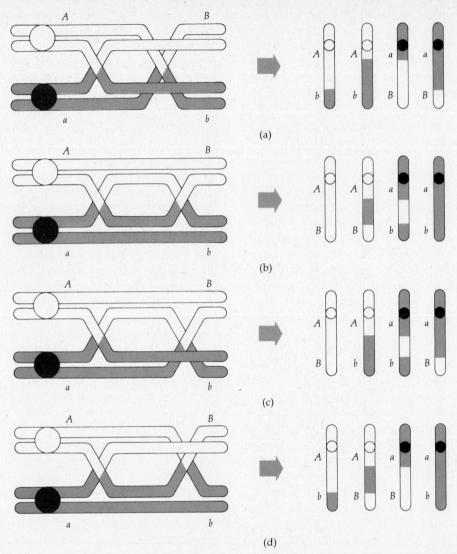

Figure 4.12 Four types of double crossovers can occur between two loci. *(a)* A four-strand double crossover (so called because all four chromatid strands are involved) produces four recombinant *(Ab* or *aB)* gametes. *(b)* A two-strand double produces four nonrecombinant *(AB* or *ab)* gametes. *(c)* and *(d)* A three-strand double (there are two types of three-strand double crossovers) produces two recombinant and two nonrecombinant gametes. If all four configurations are equally likely, then an *AB/ab* individual will produce an overall distribution of gametes of $\frac{1}{4}AB$, $\frac{1}{4}Ab$, $\frac{1}{4}aB$, and $\frac{1}{4}ab$. The general rule is that when two loci are so far apart on the chromosome that one or more crossovers almost always occur between them, then the frequency of recombination between the loci (that is, the proportion of recombinant gametes) will be 50 percent; one cannot tell from this whether the loci are far apart on the same chromosome or on two nonhomologous chromosomes.

two unlinked syntenic loci is obtained by adding together the map distances between intervening loci that are linked. Summation of many such numbers can produce map distances greater than 50, even though the actual recombination fraction between the loci would be 50 percent. This is why map distances correspond to recombination fractions only for loci that are sufficiently close together.

Linkage in Humans

The mapping of human chromosomes by means of recombination is generally a difficult matter. Nevertheless, in recent years rapid progress has been made in filling in the chromosome map of human genes (particularly by means of such methods as somatic cell genetics, which will be discussed soon). The study of linkage between loci using pedigrees is difficult because families suitable for study tend to be rare, which makes the total number of offspring available small. What is required for the study of human linkage using pedigrees is a group of parents who are double heterozygotes (like *AaBb*). If one of the alleles is rare, then finding such people will be difficult; if two of the alleles are rare, then it will be even more so. The individual's family history must also be known well enough to be able to infer whether the genotype is actually *AB/ab* or *Ab/aB*; otherwise, the recombinant chromosomes might be confused with the nonrecombinant ones. In addition, the person's mate must have a genotype that will allow the investigator to determine in each of the children which alleles were inherited on the chromosome from the doubly heterozygous parent. Finally, a large number of such families must be pooled together and the recombination frequency determined.

In spite of the problems, the study of human linkage in pedigrees has had considerable success. For example, the locus on chromosome 9 that determines the ABO blood group is known to be linked with the locus of a gene that, when mutated, leads to characteristic abnormalities of the fingernails and kneecaps and often severe kidney disease known as the **nail-patella syndrome**; the distance between the ABO and nail-patella loci is about 13 map units (i.e., there is 13 percent recombination between the loci). Similarly, the locus of the Rh blood group (chromosome 1) is linked to a locus associated with **elliptocytosis**, a rare blood disease with malformation of the red blood cells; the map distance between these loci is about 2 map units.

Figure 4.13(*a*) presents map distances (in terms of observed recombination fractions) between two pairs of linked loci. *Se* refers to the **secretor** locus, which determines whether the ABO blood group substances will be present in the saliva and other secretions; *Dm* refers to the locus of **myotonic dystrophy**, a muscular disorder; and *Lu* refers to the **Lutheran blood group**, a blood group distinct from ABO and Rh. As indicated in Figure 4.13(*a*), the map distance between *Se* and *Dm* is 4 units and that between *Se* and *Lu* is 13. However, as shown in Figure 4.13(*b*), two possible genetic maps are consistent with these data. In one map, the order of the loci is *Se-Dm-Lu* (i.e., *Dm* is in the middle); in the other map, the order is *Lu-Se-Dm* (i.e., *Se* is in the middle). Both possible genetic maps have the correct distances for *Se-Dm* and *Se-Lu* (see the boldface type). In the upper map, the inferred *Dm-Lu* distance would be 9 map units (because 13 − 4 = 9); in the lower map, the inferred *Dm- Lu* distance would be 17 map units (because 13 + 4 = 17). A decision about which genetic map is actually correct would require study of the actual amount of recombination between *Dm* and *Lu*. In any case, the data in Figure 4.13(*a*) eliminate one conceivable ordering of the loci. The order could not possibly be *Se-Lu-Dm* (i.e., *Lu* in the middle) because, with this ordering, the *Se-Dm* dis-

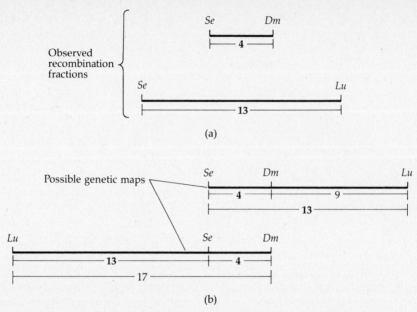

Figure 4.13   Construction of genetic maps. *(a)* Observed recombination fractions between *Se* and *Dm* and between *Se* and *Lu*. For such short distances, the percentage of recombination between two loci is equal to their distance in map units. *(b)* Two possible genetic maps, both consistent with the data in *(a)*.

tance would have to be greater than the *Se-Lu* distance, whereas, in reality, it is smaller.

Although family studies of segregation and recombination can reveal linkage, they cannot by themselves identify which of the 22 pairs of autosomes a locus is on. (The sex chromosomes are a special case.) The identification of the particular chromosome can often be accomplished by other methods. Sometimes a particular allele will be found to be inherited simultaneously with a particular chromosomal abnormality such as a giant satellite; that is, the allele will tend to be present whenever the abnormal chromosome is. (A perfect association between the allele and the chromosomal abnormality cannot be expected because of crossing-over, of course.) When such a parallelism in inheritance is found, then the locus of the allele can be inferred to be physically located on the abnormal chromosome. In these studies the relatively common and innocuous chromosome "abnormalities" such as giant satellites and unusually long short arms are extremely useful. For example, on chromosome 1 there is an "uncoiler" region found in some families. Chromosomes with this region look unusually long and certain segments stain less intensely than is normal. Many family studies have been carried out to detect particular alleles that are inherited simultaneously with this chromosome. Chromosome 1 is the best mapped human autosome. From the combined results of several methods of studying linkage, more than 30 loci have now been assigned to chromosome 1. Figure 4.14 is a genetic map of 11 loci on chromsome 1. The numbers between loci refer to the map distances (recombination fraction) between them. Note that the *map distance* between *Amy* and *FUCA* (or any other pair of loci) would be calculated as the sum of

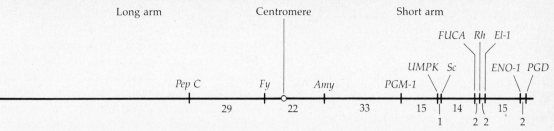

Figure 4.14   Partial genetic map of chromosome 1. The numbers between the indicated positions of the loci refer to the distance between the loci in map units. For map distances of about 15 or less, the percentage of recombination equals the map distance. For longer distances, the percentage of recombination is smaller than the map distance. Most of the map distances here have been inferred from family studies. (Abbreviations: *PepC*, peptidase C enzyme; *Fy*, Duffy blood group; *Amy*, amylase enzyme; *PGM-1*, phosphoglucomutase-1 enzyme; *UMPK*, uridine monophosphate kinase enzyme; *Sc*, Scianna blood group; *FUCA*, α-L-fucosidase enzyme; *Rh*, rhesus blood group; *El-1*, elliptocytosis-1 disease; *ENO-1*, enolase-1 enzyme; and *PGD*, 6-phosphogluconate dehydrogenase enzyme.)

the intervening distances—in this case, 33 + 15 + 1 + 14 = 63, which is larger than 50, even though the *recombination fraction* between *Amy* and *FUCA* would be 50 percent.

A powerful method of detecting linkage in humans is by means of tissue culture and **cell fusion**—the physical fusing of two somatic cells. The fused cells are called **hybrid cells**, and, because the cells divide by mitosis, their genetic study is called **somatic cell genetics**. The methods of somatic cell genetics can be used only to study genes whose presence and action can be detected in cells grown in laboratory cultures. This includes many genes that direct the production of specific enzymes (protein molecules that accelerate particular chemical reactions), but many traits cannot be detected in tissue cultures. For example, the gene that causes common baldness cannot be detected in tissue culture (at least not at the moment). The procedure for identifying which chromosome a gene is on involves the fusion of human cells with cells of another species, often mouse cells. Formation of human-mouse hybrid cells can be stimulated with special techniques, and the hybrid cells can be selected from mixtures of cells by growing the cells in environments that discriminate in favor of the hybrids. The princi-

ples involved in selecting out hybrid cells are behind many kinds of genetic studies in which rare cell types (such as particular types of mutant cells) must be isolated from heterogeneous mixtures. Indeed, these principles are so widely applied that they deserve discussion in their own right.

Principles of Mutant Hunting: Screening, Enrichment, and Selection

Consider the following problem (one that frequently arises in practice): A geneticist has a culture containing 1,000,001 cells; one cell is of particular interest because it carries some unusual mutation. How can the geneticist separate the one interesting cell from the background of 1 million uninteresting ones? This sort of problem is often encountered in genetics, and several approaches are possible. Of course, unless the mutant cell type is phenotypically distinct from the nonmutant type, the mutant type cannot be identified, let alone separated. So we must assume beforehand that the geneticist has available an **assay** (a procedure of measurement or identification) by which the mutant and nonmutant cells can be distinguished. Given an assay procedure, there are three approaches to cell separation:

1. Screening: The geneticist can **screen** (i.e., survey) the entire population by applying the assay to each of the 1 million cells individually. Although screening will eventually reveal the mutant cell, the procedure is inefficient. If the assay were rapid enough to allow the screening of one cell per second, for example, screening the entire population of 1 million cells would require 278 h—about 5 wk of hard work.

2. Enrichment: In a population that is **enriched** for a mutant cell type, the frequency of the mutant is higher than it was in the original population. In the present example, the original population has a frequency of mutants of 1 per 1 million. If the geneticist could enrich the frequency to 1 per 1000 or even 1 per 10,000, substantially less screening would be required to separate the mutant of interest. Methods of enrichment vary greatly depending on the phenotype of the mutant in question. A simple hypothetical example may illustrate the point. Suppose the mutant cell type is able to survive a treatment (with a chemical, say) that only 1 per 1000 (i.e., 10^{-3}) nonmutant cells can survive. If the geneticist treats the original 1,000,001 cells with this chemical, there will be 1000 (i.e., $10^6 \times 10^{-3} = 10^3$) nonmutant survivors in addition to the mutant, which also survives. The resulting treated population is enriched for the mutant. The frequency of the mutant cell type in the enriched population is 1 per 1000 rather than the original 1 per 1 million. To separate out the mutant from the enriched culture, only 1000 cells need be screened, which (at one assay per second) would require about $\frac{1}{2}$ h.

3. Selection: The most extreme form of enrichment is selection, in which the enrichment is so extreme that the final (**selected**) population

consists entirely or almost entirely of mutant cells. Again, selective methods vary greatly depending on the particular phenotype of the mutant. Suppose, for example, that the mutant is resistant to a chemical that kills all the nonmutant cells. Treating the original culture with this chemical will kill all but the one mutant cell. Indeed, there is now no need to assay the surviving cell at all, except to verify its phenotype. Selection is therefore an extremely powerful and efficient method for isolating rare cell types from heterogeneous populations.

The principles of screening, enrichment, and selection apply to all sorts of searches for rare phenotypes. Suppose a geneticist wishes to study a group of individuals who are heterozygous for the Tay-Sachs allele. The geneticist could assay Caucasians at random to locate the heterozygotes; this procedure would represent *screening*, and the proportion of heterozygotes found would be about 1 in 300. Alternatively, the geneticist could screen Ashkenazi Jews for heterozygosity; the Ashkenazi population represents an *enriched* population for the Tay-Sachs allele, and the proportion of heterozygotes found would now be about 1 in 30. However, the geneticist could also study parents who have had children with Tay-Sachs disease; the parents represent a *selected* population for the Tay-Sachs allele because every parent *must* be heterozygous. Thus, among the parents, the proportion of heterozygotes will be 100 percent.

Somatic Cell Genetics

Now we are in a position to discuss how human-mouse hybrid cells are obtained for the study of linkage. The most common procedure makes use of special characteristics of mutations in two genes involved in DNA synthesis. One of these genes (called *HGPRT*) codes for the enzyme **hypoxanthine guanine phosphori-bosyl transferase**. The other gene (called *TK*) codes for the enzyme **thymidine kinase**. As indicated in Figure 4.15(*a*), mammalian cells have two alternative **pathways** (routes) of DNA synthesis, which use different **precursors** (chemical predecessors); most of the DNA is produced by a so-called *major pathway*, but some is produced by a *minor pathway* involving HGPRT and TK. However, the chemical **aminopterin** prevents the major pathway from functioning, so, in an aminopterin-containing

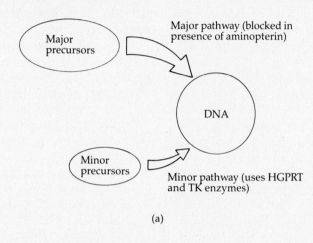

(a)

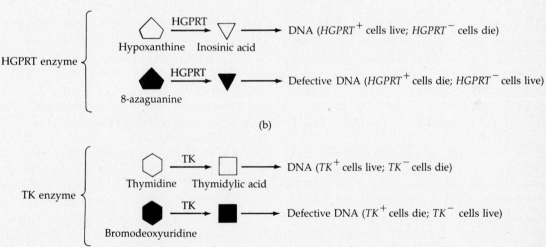

(b)

(c)

Figure 4.15 Principles underlying HAT selection. *(a)* Major and minor pathways of DNA synthesis; the major pathway is blocked by aminopterin. *(b)* Role of HGPRT in the minor pathway; in the presence of hypoxanthine and aminopterin, *HGPRT*+ cells live but *HGPRT*− cells die; in the presence of 8-azaguanine, *HGPRT*+ cells produce defective DNA and die, but *HGPRT*− cells survive. *(c)* Role of TK in the minor pathway; in the presence of thymidine and aminopterin, *TK*+ cells live and *TK*− cells die; in the presence of bromodeoxyuridine, *TK*+ cells produce defective DNA and die, but *TK*− cells survive. In HAT medium (i.e., medium containing hypoxanthine-aminopterin-thymidine), the only cells that survive are *HGPRT*+; *TK*+ cells.

medium, all the DNA is synthesized by means of the minor pathway.

The role of HGPRT in DNA synthesis is illustrated in Figure 4.15(*b*). The symbol *HGPRT*⁺ indicates cells that produce a normal HGPRT enzyme, and *HGPRT*⁻ indicates cells that produce a nonfunctional HGPRT enzyme. As indicated in Figure 4.15(*b*), appropriate manipulation of the growth medium allows one to either (1) select rare *HGPRT*⁺ cells from a background of *HGPRT*⁻ cells *or* (2) select rare *HGPRT*⁻ cells from a background of *HGPRT*⁺ cells. (Of course, any population of *HGPRT*⁺ cells will contain a tiny proportion of *HGPRT*⁻ cells because of **mutation**—a spontaneous change in a gene. Similarly, any population of *HGPRT*⁻ cells will contain a tiny proportion of *HGPRT*⁺ cells because of **reverse mutation**—mutation back to the original form of the gene.) The selective medium for *HGPRT*⁺ contains aminopterin and hypoxanthine [see Figure 4.15(*b*)]. In this medium, aminopterin knocks out the major pathway of DNA synthesis, but *HGPRT*⁺ cells can use hypoxanthine to synthesize DNA; on the other hand, *HGPRT*⁻ cells cannot use hypoxanthine so they die owing to lack of DNA synthesis. The selective medium for *HGPRT*⁻ contains **8-azaguanine** [see Figure 4.15(*b*)]. In this medium, *HGPRT*⁺ produces an abnormal precursor [shaded shape in Figure 4.15(*b*)], which is incorporated into DNA and kills the cells; *HGPRT*⁻ cells survive in this medium because they cannot incorporate 8-azaguanine; all their DNA comes from the major pathway. Hence, with an appropriate choice of growth medium, the investigator can practice **positive selection** (selection for *HGPRT*⁺) or **negative selection** (selection for *HGPRT*⁻).

Positive and negative selection can also be practiced in the case of the *TK* gene [see Figure 4.15(*c*)]. In this case, the selective medium for *TK*⁺ contains aminopterin and thymine. Aminopterin again knocks out the major path-

way, but *TK*⁺ can use the thymine to synthesize DNA via the minor pathway; *TK*⁻ cells die because they are unable to use the thymine. Selective medium for *TK*⁻ contains **bromodeoxyuridine** [see Figure 4.15(*c*)]. In this medium, *TK*⁺ cells produce defective DNA and so kill themselves, whereas *TK*⁻ cells synthesize all their DNA via the major pathway and survive.

One important implication of Figure 4.15(*b*) and (*c*) is that a medium containing hypoxanthine, aminopterin, and thymine (called **HAT medium** for the first letters of its constituents) selects for *both HGPRT*⁺ and *TK*⁺; that is, only cells that are *HGPRT*⁺ *and TK*⁺ will be able to survive in HAT medium. This consideration (along with the ones above) provides the procedure by which human-mouse hybrid cells (or hybrid cells involving other species) can be obtained. The steps in the procedure are as follows:

1. Use 8-azaguanine to select for *HGPRT*⁻ in one of the cell populations to be fused; the resulting cells will be *HGPRT*⁻; *TK*⁺.

2. Use bromodeoxyuridine to select for *TK*⁻ in the other population to be fused; the resulting cells will be *HGPRT*⁺; *TK*⁻.

3. Mix the *HGPRT*⁻ cells with the *TK*⁻ cells and place them in the HAT medium. Cells that result from human-mouse cell fusion will be *HGPRT*⁺; *TK*⁺ and will survive. Such cells will obtain their *HGPRT*⁺ gene from one of the parental cells and their *TK*⁺ gene from the other parental cell. (This sort of compensation for each other's weaknesses may remind you of the accommodation between Jack Sprat, who could eat no fat, and his wife, who could eat no lean.)

Step 3 is illustrated in Figure 4.16. Without special treatment, the frequency of cell fusion is low. However, the frequency can be enormously increased by treating the cells with

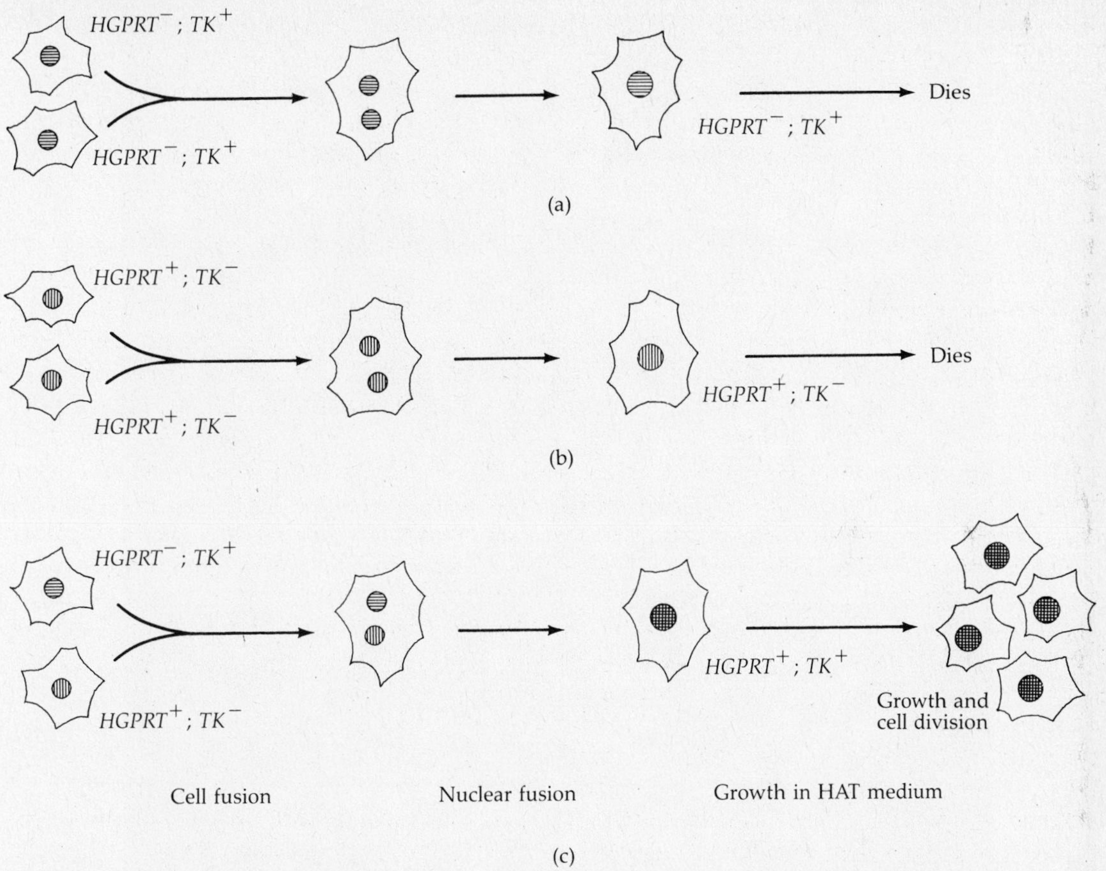

(a)

(b)

Cell fusion   Nuclear fusion   Growth in HAT medium

(c)

Figure 4.16   HAT selection of hybrid cells. One cell line is $HGPRT^-$; TK^+ and the other is $HGPRT^+$; TK$^-$. In HAT medium, the only cells that can survive are the ones that have undergone fusion. For the principles underlying the selection, see Figure 4.15.

inactivated **Sendai virus** or with **polyethylene glycol**. With such treatment the majority of cells will undergo fusion. Of course, in the mixture of fused cells there will be human-human hybrids and mouse-mouse hybrids in addition to the desired human-mouse hybrids. Nevertheless, as shown in Figure 4.16, the use of HAT medium ensures that only the human-mouse hybrids will survive.

The cells obtained by the HAT procedure are true human-mouse hybrids. They have 86 chromosomes—the full somatic complement of the mouse (40) plus the full somatic comple-

ment of humans (46). Figure 4.17 shows a quinacrine-stained metaphase spread of one such cell, and some of its human chromosomes are indicated. Although the hybrid cells perpetuate their chromosomes as they undergo successive mitotic divisions, sometimes chromosomes are lost—particularly human chromosomes. After a while, a culture of hybrid cells will become genetically heterogeneous because different cells will have lost varying numbers of human chromosomes. When individual cells from such a culture are isolated and placed in nutrient medium, each cell will give

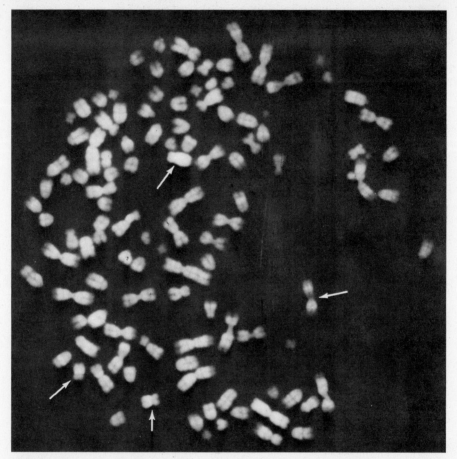

Figure 4.17 Arrows indicate several human chromosomes in a human-mouse hybrid cell stained with quinacrine.

rise to a **clone**—a group of genetically identical cells. By screening the karyotypes of the clones using such procedures as Giemsa staining (see Chapter 2), an investigator can identify which human chromosomes have been retained in the various clones. Then, upon assaying each clone for the product of a particular human gene, it can easily be determined which human chromosome carries the locus corresponding to the gene. For example, if the human gene product is present in every clone that has retained chromosome 3 but absent in every clone that has lost chromosome 3, then the locus of the

gene can confidently be assigned to chromosome 3. Examples of the successful use of this technique include the assignment of the locus of hexosaminidase A (the locus involved in Tay-Sachs disease) to chromosome 15. Indeed, more than 200 genes have now been assigned to the human autosomes, and at least two genes are known for every autosome.

An actual example of the use of hybrid cell clones to determine the chromosome that carries the gene for UMPK (**uridine monophosphate kinase**) is shown in Table 4.1. The original hybridization involved an $HGPRT^+$; TK^-

TABLE 4.1 PRESENCE OR ABSENCE OF HUMAN CHROMOSOMES AMONG EIGHT HUMAN-MOUSE HYBRID CELL LINES, AND PRESENCE OR ABSENCE OF HUMAN UMPK ENZYME*

Clone[†]	Chromosome																							Enzyme UMPK	Expected pattern
	1	2	3	4	5	6	7	8	9	10	11	12	13	14	15	16	17	18	19	20	21	22	X		
a	A	P	P	P	A	A	A	A	A	A	P	P	A	A	A	A	A	A	A	A	A	A	P	−	A
b	P	P	A	A	P	A	A	P	A	A	A	A	P	A	A	P	P	P	A	A	A	A	P	+	P
c	P	P	P	P	A	A	A	P	A	P	A	P	P	P	P	P	A	P	P	P	P	P	P	+	P
d	P	A	P	A	A	P	P	P	A	P	P	P	P	P	P	P	P	P	A	A	P	A	P	+	P
e	A	P	A	P	A	A	A	A	A	A	P	P	P	P	A	A	P	P	A	P	A	A	P	−	A
f	A	A	A	A	P	P	P	A	P	A	P	P	P	P	P	P	P	P	P	P	A	A	P	−	A
g	A	P	A	A	A	A	P	A	A	A	A	A	P	A	A	A	A	A	A	A	P	A	P	−	A
h	P	P	P	P	P	A	P	P	A	P	A	P	A	P	P	P	P	A	P	A	P	A	P	+	P

Source: Data from A. Satlin, R. Kucherlapati, and F. H. Ruddle, 1975, Cytogenet. Cell Genet. 15:146–152.

*A = absence of chromosome; P = presence of chromosome; − = absence of human enzyme; + = presence of human enzyme.

[†]All these clones have a unique combination of human chromosomes.

human cell and an $HGPRT^-$; TK^+ mouse cell. Individual clones were cultured in HAT medium, so the one human chromosome that must be retained is the one that carries the locus of HGPRT. (If this chromosome were lost, the resulting cell would be $HGPRT^-$; TK^+ and would die in the HAT medium.) Note in Table 4.1 that the one human chromosome retained in all clones is the X chromosome; the locus for *HGPRT* is thus on the X chromosome. For localizing *UMPK*, the strategy is to examine the chromosomal constitutions of the clones to identify which single chromosome has the same pattern of presence and absence (P's and A's) among the clones as the observed pattern of *UMPK* presence and absence (+'s and −'s). That is to say, one wishes to find a chromosome that has the same pattern of P's and A's as in the last column of Table 4.1. The appropriate pattern is exhibited by chromosome 1 and only by this chromosome; the *UMPK* locus must therefore be on chromosome 1.

The methods of somatic cell genetics so far discussed tell us only which loci are syntenic, not their order or their distance apart. The mapping procedure can be carried further,

however. If cells are exposed to radiation, chromosome breaks will be induced and fragments of chromosomes can be lost from the cells. Moreover, the probability that a chromosome break will occur between two loci is proportional to their distance apart. Chromosome breaks between loci that are far apart will be frequent, but breaks between loci that are close together will be rare. In consequence, close loci will tend to be found on the same chromosome fragment and will be either lost together or retained together. Conversely, loci that are far apart will often be separated by chromosome breaks, and one locus will be retained while the other is lost. These considerations provide the basis of **fragmentation mapping** of human chromosomes. From the map of chromosome 1 in Figure 4.14, for example, it is apparent that the loci *ENO-1* and *PGD* (map distance 1 unit) should tend to be inherited together—both lost or both retained. On the other hand, *PGM-1* and *PGD* (map distance 51 units) should be separated by breaks more frequently, and *Amy* and *PGD* (map distance 84 units) should be separated still more frequently.

SUMMARY

1. Matings between two heterozygotes —$Aa \times Aa$—produce a ratio of genotypes among the progeny of $\frac{1}{4}$ AA, $\frac{1}{2}$ Aa, and $\frac{1}{4}$ aa. If the a allele is recessive, then AA and Aa will be phenotypically indistinguishable, and the ratio of phenotypes among the progeny will be $\frac{3}{4}{:}\frac{1}{4}$. This 3:1 ratio of phenotypes is characteristic of autosomal recessive inheritance.

2. Pedigrees of rare traits with autosomal recessive inheritance have characteristic features including (a) there is a negative family history for the trait, (b) both sexes are equally likely to be affected, (c) affected individuals

usually have nonaffected parents, and (d) there is a relatively frequent occurrence of **consanguineous mating** (mating between relatives) among the parents of individuals that are affected.

3. Unless a trait with autosomal recessive inheritance is extremely common, there will be more heterozygotes in the population than recessive homozygotes. Indeed, for rare traits, heterozygotes will be hundreds or thousands of times more frequent than recessive homozygotes (see Figure 4.3).

4. The probability of various numbers of affected and nonaffected offspring from the mating $Aa \times Aa$ can be calculated from the formula

$$\frac{n!}{i!(n - i)!}\left(\frac{1}{4}\right)^i\left(\frac{3}{4}\right)^{n - i}$$

where n is the total number of offspring, i is the number of recessive homozygotes (affected) offspring, and $(n - i)$ is the number of non-affected offspring.

5. Among the most common autosomal recessive disorders is **familial emphysema** (α_1-**antitrypsin deficiency**), for which about 1 individual in 20 is heterozygous. The high frequency of heterozygotes is particularly important in this case because heterozygotes seem to have a higher than average risk of developing pulmonary emphysema. Like other traits, the incidence of autosomal recessive disorders can vary among populations. Among Caucasians one of the most common is **cystic fibrosis**. Among blacks one of the most frequent is the hemoglobin disorder **sickle cell anemia**, for which heterozygotes are more resistant to malaria than are both types of homozygotes. Ashkenazi Jews have a relatively high incidence of **Tay-Sachs disease** (hexosaminidase A deficiency), and Southwest American Indians have a high frequency of one form of **albinism**.

6. Allelism between two recessive genes is determined by the **complementation test**, which is carried out by mating homozygous recessives for the genes in question. If the offspring are phenotypically normal, the genes are said to be **complementing** and are judged to be **nonallelic** (involving different loci); if the offspring are affected with the trait, the genes are said to be **noncomplementing** and are judged to be **allelic** (involving the same locus).

7. Independent assortment refers to the independent segregation of alleles at two loci. If an A/a; B/b double heterozygote produces equal frequencies of the four possible gametic types (AB, Ab, aB, and ab), the loci are said to undergo independent assortment.

8. Linkage refers to a situation in which two loci fail to undergo independent assortment; linked loci are necessarily on the same chromosome. An individual of genotype AB/ab (i.e., A and B alleles on one chromosome and a and b alleles on the homologue) will produce four gametic types—two **nonrecombinant** types (AB and ab) and two **recombinant** types (Ab and aB). An individual of genotype Ab/aB will produce the same four gametic types, but the Ab and aB gametes are the nonrecombinants whereas the AB and ab gametes are the recombinants. The amount of linkage between two loci is measured by the **recombination fraction**, which equals the proportion of recombinant gametes produced by a double heterozygote. If the recombination fraction is 6 percent, for example, a double heterozygote will produce $\frac{6}{2} = 3$ percent of each type of recombinant gamete and $(100 - 6)/2 = 47$ percent of each type of nonrecombinant gamete. The maximum possible recombination fraction is $\frac{1}{2}$ (i.e., 50 percent), which corresponds to independent assortment.

9. Loci on the same chromosome are called **syntenic**. Syntenic loci that are sufficiently far apart can undergo independent assortment. A **chromosome map** is a diagram showing the arrangement of syntenic loci, and the **map distance** between loci is measured in terms of **map units**. For loci that are sufficiently close together, the number of map units between them equals the recombination fraction in percent (e.g., 6 percent recombination corresponds to 6 map units). The map distance between more distant loci is calculated by summing the map distances between intervening loci.

10. Cell cultures are particularly useful in the study of human genetics. Such methods often involve the problem of isolating rare cell types from heterogeneous cultures. Three approaches are possible: (a) **screening**, assay of individual cells from the culture; (b) **enrichment**, enhancement in the frequency of the desired cell type, followed by screening of a smaller number of cells; and (c) **selection**, enrichment to a level at which all cells or almost all cells are of the desired type, followed by assay of very few cells for verification.

11. Somatic cell genetics is the study of the inherited characteristics of somatic cells in laboratory cultures. A principal method uses **cell fusion** between human cells and cells of another species (often mouse) to produce **hybrid cells**. In the most widely used method, $HGPRT^+$; TK^- cells are mixed with $HGPRT^-$; TK^+ cells in HAT (hypoxanthine-aminopterin-thymine) medium to select for the $HGPRT^+$;

TK^+ hybrids. The $HGPRT$ and TK genes code for the enzymes hypoxanthine guanine phosphoribosyl transferase and thymidine kinase, respectively, which are involved in the minor pathway of DNA synthesis.

12. Hybrid cells sometimes lose various human chromosomes. Studies of **clones** (cultures of genetically identical cells) that have lost particular chromosomes can be used to determine which chromosome a gene is on. The product of a gene on a particular chromosome will be expressed in clones that retain the chromosome but will be absent in clones that have lost the chromosome. Irradiation of somatic cells to produce chromosome breaks is the basis of **fragmentation mapping** because loci that are close together will more frequently be found on the same chromosome fragment than will loci that are farther apart.

WORDS TO KNOW

Recessive Genes	**Independent Assortment**	Enrichment	HAT medium
1:2:1 ratio	Synteny	Selection	Fragmentation mapping
3:1 ratio	Linkage	Positive selection	
Complementation	Recombination	Negative selection	**Trait**
Noncomplementation	Recombination fraction		Familial emphysema
Nonallelic	Genetic map	**Somatic Cell Genetics**	Cystic fibrosis
Allelic	Map distance		Sickle cell anemia
Carrier	Map unit	Cell fusion	Tay-Sachs disease
Consanguineous mating		Hybrid cells	Albinism
Ascertainment bias	**Mutations**	Clone	
		$HGPRT$	
	Screening	TK	

PROBLEMS

1. For discussion: For many traits due to recessive alleles (including cystic fibrosis, sickle cell anemia, and Tay-Sachs disease) heterozygotes can be identified by appropriate chemical tests or by studies of cultured cells. Such procedures could be used to screen all prospective marriage partners to deter-

mine those in which both individuals are heterozygous. Do you think such a screening program would be warranted? What possible benefits might result? What possible harm? Which traits do you think would be the most important to screen for, and in which populations? Should such tests be mandatory or voluntary? Screening will reveal many couples in which one partner is heterozygous but the other is homozygous normal. Should such couples be informed of their genotypes? Why or why not?

2. Certain eugenic proposals include the forced sterilization of individuals who are affected with rare traits due to autosomal-recessive inheritance to eliminate the recessive allele. Moral and ethical issues aside, what does the information in Figure 4.3 imply about the effectiveness of such a program?

3. Why is consanguinity relatively frequent among the parents of individuals affected with rare autosomal recessive traits?

4. Two normal individuals have a child with cystic fibrosis. What is the genotype of the parents?

5. A woman has a brother who died in infancy from Tay-Sachs disease. What is the probability that the woman is heterozygous?

6. In a mating of two heterozygotes for a harmful recessive allele, what is the probability that a family of three children will have one affected and two nonaffected?

7. A mouse geneticist discovers two autosomal recessive genes, each of which, when homozygous, causes blindness. How could the geneticist determine whether the genes are alleles of the same locus?

8. An albino Hopi marries an albino Caucasian. What is the probability that their children will be albino? (Hint: The two forms of autosomal-recessive albinism are equally frequent in Caucasians.)

9. Why is aminopterin required in the selection of $HGPRT^+$ and TK^+ cells?

10. All the clones in Table 4.1 were selected in HAT medium as hybrid cells from a mixture of TK^+; $HGPRT^-$ mouse cells and TK^-; $HGPRT^+$ human cells. Which human chromosome carries the locus of $HGPRT$? All clones in Table 4.1 also retain the human gene for α-galactosidase. Which chromosome carries this gene?

11. The human enzyme β-galactosidase is on chromosome 3. Which clones in Table 4.1 would have human β-galactosidase?

12. If assayed for the human form of the enzyme β-glucuronidase, clones a to h in Table 4.1 would be $-, -, +, +, +, +, +,$ and $+$, respectively. Which human chromosome carries the gene for β-glucuronidase?

13. The α_1-antitrypsin locus (Pi) is on chromosome 2; the Rh locus is on chromosome 1. In the mating $Pi^A/Pi^Z;D/d \times Pi^A/Pi^A;d/d$, what genotypes of offspring are possible and in what frequencies would they be expected? Which one of Mendel's laws do these frequencies illustrate?

14. Suppose loci A and B are syntenic. What possible gametes could be formed by an individual of genotype Ab/aB? Which gametes are recombinant gametes? Which are nonrecombinant?

15. As noted in Figure 4.13, the recombination fraction between Lu and Se is 13 percent. What possible gametes would be produced by an individual of genotype $Lu\ Se/lu\ se$, and what frequencies are expected? Answer the same question for an individual of genotype $Lu\ se/lu\ Se$.

FURTHER READING AND REFERENCES

Anderson, W. F., and E. G. Diacumakos. 1981. Genetic engineering in mammalian cells. Scientific American 245:106–121. Prospects of somatic cell genetics and recombinant DNA in the treatment of disease.

Botstein, D., R. L. White, M. Skolnick, and R. W. Davis. 1980. Construction of a genetic linkage map in man using restriction fragment length polymorphisms. Am. J. Hum. Genet. 32:314–331. The theoretical requirements for this new type of mapping are explored.

Carrell, R. W., J.-O. Jeppsson, C.-B. Laurell, S. O. Brennan, M. C. Owen, L. Vaughan, and D. R. Boswell. 1982. Structure and variation of human α_1-antitrypsin. Nature 298: 329–334. Review of function of α_1-antitrypsin and the relationship between certain alleles and smoking-induced emphysema.

Cavalli-Sforza, L. L. 1974. The genetics of human populations. Scientific American 231:80–89. Excellent summary and the source of Figure 4.5.

Goodman, R. 1979. Genetic Disorders Among Jewish People. Johns Hopkins University Press, Baltimore. In a review of this book (Am. J. Human Genet. 32:471), Frederick Hecht remarks, "Everything learned about genetic disorders in Jews bears directly on our overall insight into human genetic variation."

Gordon, R., and A. G. Jacobson. 1978. The shaping of tissues in embryos. Scientific American 238:106–113. Computer-assisted study of the forces that sculpture the developing embryo.

Harper, M. E., A. Ullrich, and F. G. Saunders. 1981. Localization of the human insulin gene to the distal end of the short arm of chromosome 11. Proc. Natl. Acad. Sci. U.S.A. 78:4458–4460. Application of sophisticated mapping techniques to an important gene.

Mourant, A. E., A. C. Kopec, and K. Domaniewska-Sobczak. 1978. The Genetics of the Jews. Clarendon Press, New York. A good collection of data relevant to genetic polymorphisms in Jewish populations.

Ruddle, F. H. 1981. A new era in mammalian gene mapping: Somatic cell genetics and recombinant DNA methodologies. Nature 294:115–120. An excellent discussion of the new vistas opened by the combination of two powerful techniques.

Satlin, A., R. Kucherlapati, and F. H. Ruddle. 1975. Assignment of the gene for human UMPK to chromosome 1 using somatic cell hybrid clone panels. Cytogenet. Cell Genet. 15:146–152. Source of data in Table 4.1.

Sinnott, E. W., L. C. Dunn, and T. Dobzhansky. 1950. Principles of Genetics. McGraw-Hill, New York. Source of pedigree in Figure 4.2.

Spyropoulos, B., P. B. Moens, J. Davidson, and J. A. Lowden. 1981. Heterozygote advantage in Tay-Sachs carriers? Am. J. Hum. Genet. 33:375–380. To their question, the authors conclude no. Founder effects, genetic drift, and differential immigration are probably responsible for the high allele frequency in certain Jewish populations.

Stanbridge, E. J., C. J. Der, C.-J. Doersen, R. Y. Nishimi, D. M. Peehl, B. E. Weissman, and J. E. Wilkinson. 1982. Human cell hybrids: Analysis of transformation and tumorigenicity. Science 215:252–259. Use of human-human cell hybrids in current cancer research.

Stern, C. 1973. Principles of Human Genetics. Freeman, San Francisco. Source of Figure 4.9.

Woolf, C. M., and F. C. Dukepo. 1969. Hopi Indians, inbreeding and albinism. Science 164:30–37. A fascinating account of the social position of albinos in traditional Hopi society. Source of Figures 4.7 and 4.8.

chapter 5
The Genetic Basis of Sex

Although Mendel single-handedly worked out the statistical rules of inheritance of genes in peas, and although these rules are as valid for genes on the autosomes in humans as they are for genes in peas, the genes on the sex chromosomes in humans—the X and the Y—follow different rules. Mendel can hardly be faulted for not discovering these rules because peas and most other plants do not have sex chromosomes. Flowering plants such as peas are sexual, but both the male and female structures are present in all the flowers on the plants. Nevertheless, it is ironic that Mendel did not interpret sex as an inherited trait; if he had, he might have realized that sex itself provides one of the most convincing demonstrations of segregation.

The Sex Ratio

Human females have two X chromosomes and 22 pairs of autosomes; during meiosis these become parceled so that the egg contains one X and one of each of the autosome pairs. Males have an X and a Y chromosome plus 22 pairs of autosomes; during meiosis the X and the Y segregate and the autosomes are distributed so that half the sperm carry an X and one of each of the autosome pairs whereas the other half carry a Y along with one of each of the autosome pairs. As illustrated in Figure 5.1, if an X-bearing sperm fertilizes an egg, the resulting zygote will have two X's and 22 pairs of autosomes and will develop into a female. If a Y-bearing sperm fertilizes an egg, the zygote will have an X, a Y, and 22 pairs of autosomes and will develop into a male. The segregation of X and Y during meiosis means that a mating between a normal male and a normal female (XY × XX) is somewhat analogous to a mating between a heterozygote and a homozygote (e.g., *Aa* × *aa*).

In discussing the **sex ratio** (the ratio of males to females), one must distinguish the **primary sex ratio** (the sex ratio at fertilization) from the **secondary sex ratio** (the sex ratio at birth) because spontaneous abortion is relatively frequent and may well affect one sex more than the other and thus alter the sex ratio among survivors. Little is known about the human primary sex ratio because the earliest stages of fertilization and development are inaccessible to large-scale study. Very early spontaneous abortuses are amenable to study, but such studies present problems. First, the very earliest spontaneous abortions (including failure of the zygote to implant in the uterine wall) usually are unrecognized as such. Second, the number of such abortuses available for study is very limited. Third, spontaneous

abortuses, by their very nature, are not a random sample of the corresponding developmental stage in normal embryos. And fourth, very early embryos are exceedingly difficult to sex accurately. Theoretically, of course, the primary sex ratio should be 1:1 because of segregation of the X and Y chromosomes in the father. On the other hand, segregation in the father is not the only determinant of the primary sex ratio. It could be, for example, that the Y-bearing sperm (or the X-bearing sperm) has a slightly greater capacity to reach the site of fertilization or to participate in fertilization. It could even be the case that the X-bearing and Y-bearing sperm have different efficiencies depending on the precise physiological conditions in the mother or the father or both. In short, the primary sex ratio in humans is unknown. Nevertheless, what little evidence is available suggests that it is close to (but perhaps not exactly) 1:1.

In contrast to the limited data available about the primary sex ratio, the secondary sex ratio in humans is well documented in birth records. (Birth records with the sex of the newborn recorded were first kept in France in the mid-1700s.) Where the secondary sex ratio is known, the probability of various numbers of males and females in sibships of a specified size can easily be calculated. (A **sibship** is a group of brothers and sisters.) In making the calculation, it must be assumed that each birth is independent of the others—that is, that the sex of previously born children has no influence on the sex of an unborn child. If the secondary sex ratio were $\frac{1}{2}$ males and $\frac{1}{2}$ females, for example, the probability that a sibship of n children would consist of exactly i males and $(n - i)$ females would be given by the formula from Chapter 2 as

$$\frac{n!}{i!(n - i)!}\left(\frac{1}{2}\right)^i\left(\frac{1}{2}\right)^{n - i}$$

We will see in a moment that the secon-

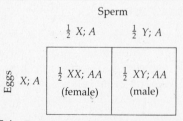

Sperm

$\frac{1}{2}$ X; A $\frac{1}{2}$ Y; A

Eggs   X; A | $\frac{1}{2}$ XX; AA (female) | $\frac{1}{2}$ XY; AA (male) |

Figure 5.1 Punnett square showing segregation of X and Y chromosomes in the male. The symbol A represents a haploid autosomal complement (i.e., one each of chromosomes 1 through 22).

dary sex ratio is not 1:1, however, and in such cases the formula must be modified. Suppose that the overall proportion of male births is m and that of female births is $(1 - m)$. Then the probability that a sibship of n children contains exactly i males and $(n - i)$ females is given by

$$\frac{n!}{i!(n - i)!} m^i (1 - m)^{n - i}$$

(The formula given for a 1:1 sex ratio is just a special case of this more general formula when $m = \frac{1}{2}$ and $1 - m = \frac{1}{2}$.)

Table 5.1 shows the calculation of the probabilities of various sex distributions in sibships of size 1 to 4. In the second column, M indicates male and F indicates female, and the third column gives the general formula for the probability applicable to any value of m. The fourth column presents the expected sex distributions (as percentages) for the case $m = 0.5$, and the final three columns pertain to actual cases. Among U.S. whites, for example, the observed secondary sex ratio has a proportion of males equal to $m = 0.5135$. (This proportion corresponds to 1055 male births for every 1000 female births.) The expected percentages of various sex distributions in sibships of size 1 to 4 for $m = 0.5135$ are indicated in the column corresponding to U.S. whites. One consequence of the slight excess of males in the secondary sex ratio is that all-male sibships will be somewhat more frequent than all-female sibships; among sibships of size 3 in U.S. whites, for example 13 percent are expected to have all males whereas 12 percent are expected to have all females.

As indicated in Table 5.1, the secondary sex ratio varies among populations, usually with a slight excess of males. The differences in the secondary sex ratio are quite small, ranging

TABLE 5.1 EXPECTED SEX DISTRIBUTIONS IN SIBSHIPS OF SIZE 1 TO 4 FOR VARIOUS VALUES OF m*

Sibship size	Sex distribution	General formula	Sex distribution, percentage†			
			$m = 0.5000$	$m = 0.5063$ (U.S. nonwhites)	$m = 0.5135$ (U.S. whites)	$m = 0.5233$ (Philippines)
1	1M:0F	m	50.0	50.6	51.4	52.3
	0M:1F	$1 - m$	50.0	49.4	48.6	47.7
2	2M:0F	m^2	25.0	25.6	26.4	27.4
	1M:1F	$2m(1 - m)$	50.0	50.0	50.0	49.9
	0M:2F	$(1 - m)^2$	25.0	24.4	23.7	22.7
3	3M:0F	m^3	12.5	13.0	13.5	14.3
	2M:1F	$3m^2(1 - m)$	37.5	38.0	38.5	39.2
	1M:2F	$3m(1 - m)^2$	37.5	37.0	36.5	35.7
	0M:3F	$(1 - m)^3$	12.5	12.0	11.5	10.8
4	4M:0F	m^4	6.2	6.6	7.0	7.5
	3M:1F	$4m^3(1 - m)$	25.0	25.6	26.4	27.3
	2M:2F	$6m^2(1 - m)^2$	37.5	37.5	37.4	37.3
	1M:3F	$4m(1 - m)^3$	25.0	24.4	23.6	22.7
	0M:4F	$(1 -)^4$	6.2	5.9	5.6	5.2

*m is the proportion of males in the secondary sex ratios.
†Some of the columns do not add to exactly 1 because of round-off error.

from 1026 males per 1000 females in U.S. nonwhites to 1097 males per 1000 females in the Philippines, but the differences are not due to chance variation because they are based on millions of births. (The designation "nonwhite," incidentally, is a U.S. Census Bureau term referring to a heterogeneous population consisting largely of blacks but also including many other ethnic groups.)

The calculations in Table 5.1 are based on the assumption that individual births are random in regard to sex. That is to say, among U.S. whites, the probability that a newborn will be male is 0.5135 irrespective of whether the newborn is the first or the tenth in a sibship and irrespective of the sexes of prior children. Whether this assumption is valid or not depends on the observed distribution of males and females in sibships of various sizes. If the observed distributions are in accord with the theoretical ones in Table 5.1, then the assumption of randomness is supported. In fact, the agreement between observed sex distributions and the theoretical ones in Table 5.1 is remarkably good, not only for sibships of four or fewer children but for larger sibships as well. Thus, *there do not seem to be tendencies for certain kinships to have boys or for certain other ones to have girls; the distribution of boys and girls in sibships seems to be random*. Many people are mildly surprised at this conclusion, for they know of one or more large kinships that consist of mostly boys or mostly girls. But a few kinships of predominantly one sex would be expected simply by the laws of chance. When one occurs, its curious sex distribution commands attention, so these kinships receive publicity in disproportion to their numbers. In short, sibships with extremely odd sex distributions are not more frequent than would be expected by chance.

The secondary sex ratio is not a constant like the speed of light, however. Not only does the secondary sex ratio vary among populations, but it varies in other ways as well. Although the effects are real, they are small in magnitude and their causes are unknown. Four correlates of variation in the secondary sex ratio have been well documented.

1. *Birth order*. The secondary sex ratio decreases with order of birth in sibships, from a high of 1066 males per 1000 females among firstborn offspring to a low of 1045 males per 1000 females among seventh and later offspring. (These and the data below pertain to U.S. whites.)

2. *Father's age*. Young fathers tend to have slightly more male offspring than older fathers; the secondary sex ratio for fathers aged 15 to 19 is about 1070 males per 1000 females, but for fathers aged 45 to 49 it is about 1049 males per 1000 females.

3. *Seasonal effects*. The secondary sex ratio changes cyclically throughout the year, from a low of 1048 males per 1000 females in February to a high of 1062 males per 1000 females in July.

4. *Temporal variation*. The secondary sex ratio changes slightly from year to year. It rose from 1056 males per 1000 females in 1935 to a high of 1063 in 1945, then dropped to 1052 in 1962 and rose again to 1060 in 1968. Peaks in the secondary sex ratio in 1945 and 1968 coincide with the Second World War and the Vietnam War, respectively, which has led some investigators to suggest that there may be a causal connection. On the other hand, male births declined steadily during the Korean War in the early 1950s.

This variation in the secondary sex ratio seems to conflict with the conclusion reached earlier that the sex of each birth is independent of previous ones and that the distribution of sexes within sibships is random. However, variation in the secondary sex ratio due to the sources discussed above is so small in magnitude that it does not significantly upset the earlier calculations.

Y-Linked Genes

Since males transmit a replica of their Y chromosome to all of their sons but to none of their daughters, pedigrees of traits due to genes on the Y chromosome (**Y-linked genes**) should be extremely simple.

1. Only males are affected with the trait.
2. Females never transmit the trait, irrespective of how many affected male relatives they may have.
3. All sons of affected males are also affected.

In spite of the relative ease with which these pedigree characteristics could be detected, few human traits have these characteristics. The principal reason for the rarity of such traits is that *the human Y chromosome carries very few genes*. (This relative paucity of Y-linked genes is also found in mice, fruit flies, and many other organisms.) Of course, one trait that is inherited along with the Y chromosome is maleness. Maleness, like the Y chromosome, is passed from father to son to grandson and so on. This mode of transmission is of great genetic importance, but it is so commonplace that its importance is easily overlooked. Moreover, as we shall see in the next chapter, individuals with abnormal sex-chromosome constitutions are phenotypically male or malelike if they have one or more Y chromosomes, but female or femalelike if they lack a Y. Thus, *the Y chromosome in humans carries the genes that trigger the embryonic development of maleness*. One Y-linked gene, apparently near the centromere, is involved in the production of a substance called the **H-Y antigen**, which appears very early in embryonic development. It is currently thought that the H-Y antigen is important in sex determination.

Are there any traits, other than maleness itself, that are determined by Y-linked genes? The inheritance of such traits could easily be detected because of their striking pedigree characteristics. Nevertheless, of the several thousand hereditary traits known, only one seems to follow the pattern of Y-linked inheritance. This is a gene for the trait **hairy ears**, which refers to the growth of stiff hair an inch or longer on the outer rim of the ears of males beyond the age of about 20 (Figure 5.2). The trait is most commonly found in men of India (incidence about 20 percent) and Israel, but it also occurs in Caucasians, Japanese, and other ethnic groups. Y-linked inheritance of hairy ears is not entirely certain because the trait is difficult to study. Not only is it extremely variable in age of onset (in some men it is not expressed until 60 or 70), but the trait is also extremely variable in its expression. Nevertheless, pedigrees of hairy ears (see Figure 5.2) do suggest Y linkage because the trait occurs only in males and is transmitted only through males. Whether the hairy ear gene is truly Y linked or not, it is clear that the Y carries few genes.

X-Linked Genes

The X chromosome, in contrast to the Y, carries as many genes as would be found on an autosome of comparable size. The genes on the X (called **X-linked** genes or, less frequently, **sex-linked** genes) seem to be a random collection of genes that are no different in their functions from those found on the autosomes. Over 100 different loci are known to be X linked and nearly 100 others are strongly suspected to be. Among these loci are the *HGPRT* locus used routinely in somatic cell genetics, loci involved in color perception and

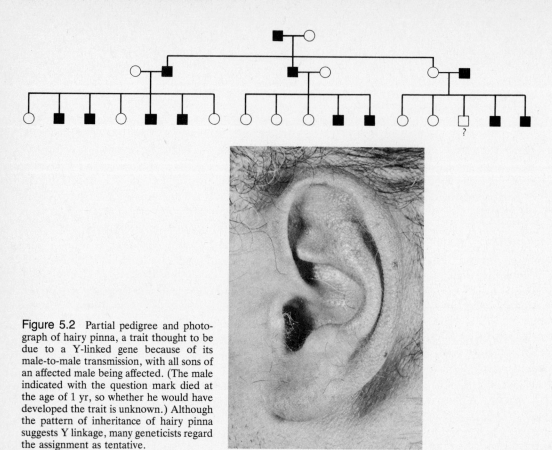

Figure 5.2 Partial pedigree and photograph of hairy pinna, a trait thought to be due to a Y-linked gene because of its male-to-male transmission, with all sons of an affected male being affected. (The male indicated with the question mark died at the age of 1 yr, so whether he would have developed the trait is unknown.) Although the pattern of inheritance of hairy pinna suggests Y linkage, many geneticists regard the assignment as tentative.

blood clotting, and a locus that, when mutated, causes a degenerative disease of muscle known as Duchenne-type muscular dystrophy.

Pattern of Inheritance The pattern of inheritance of X-linked genes is unique because males receive their X chromosome only from their mother and transmit it only to their daughters. This pattern is often called **crisscross** inheritance because an X chromosome can crisscross between the sexes in successive generations. A surname analogy for X-linked inheritance can be devised if we imagine a "strange savage nation" in which males have one surname but females have two. In this nation, a male's surname is one chosen at random from his mother, but a female always receives her father's surname in addition to one randomly chosen from her mother. Thus, for example, a mating between Mr. Smith and Ms. Brown-Robinson would produce sons named Brown or Robinson with equal likelihood and daughters named Smith-Brown or Smith-Robinson with equal likelihood.

In this analogy, of course, the names correspond to X-linked alleles. With just two alleles at an X-linked locus (call them *A* and *a*), there will be three possible genotypes in females (AA, Aa, and aa) but only two possible genotypes in males (A and a). Since males have only one X chromosome, the terms *homozygous* and *heterozygous* do not apply to their X-linked loci. Instead, males are said to be **hemizygous** for X-linked loci, and males who carry A or a are referred to as hemizygous A or hemizygous a, respectively.

If the a allele is a recessive allele associated with some disorder, then homozygous aa females and hemizygous a males will both be affected. The hemizygous a males will be affected because, having only one X chromosome, they lack a dominant allele to compensate for the recessive.

Punnett squares for the matings $AY \times Aa$ and $aY \times AA$ are shown in Figure 5.3. Here A and a represent A-bearing and a-bearing X chromosomes, respectively, and Y represents the Y chromosome. If a trait due to an X-linked recessive is rare, the matings in Figure 5.3 will be the most frequent ones in which the recessive allele is involved. Figure 5.3(a) shows that heterozygous females have 50 percent affected sons and 50 percent carrier daughters; Figure 5.3(b) shows that affected males have all normal sons and all carrier daughters.

Some Pedigree Characteristics Corresponding to the unique nature of X-linked inheritance are certain unique features in pedigrees of traits due to X-linked recessive alleles. These features are illustrated in Figure 5.4, which is a pedigree of hemophilia, a trait characterized by excessive bleeding following an injury due to the absence of an essential blood-clotting factor. To be noted in the pedigree are the following points:

1. Predominantly males are affected with the trait. Indeed, in Figure 5.4, only males are affected, which happens frequently when the trait in question is rare.

2. Affected males have phenotypically normal offspring. As noted in connection with Figure 5.3(b), however, all their daughters are carriers.

3. Affected males usually have phenotypically normal parents. This characteristic of rare, X-linked recessives occurs because most affected males arise from heterozygous (and therefore phenotypically normal) mothers.

Incidence in Males and Females In connection with Figure 4.3 we emphasized

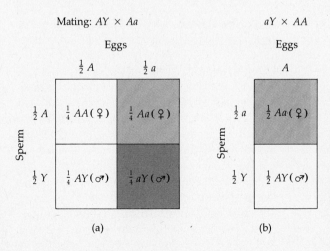

Mating: $AY \times Aa$

(a)

$aY \times AA$

(b)

Figure 5.3 Punnett squares showing inheritance of an X-linked pair of alleles, A and a. *(a)* Mating between a hemizygous A male and a heterozygous female produces $\frac{1}{4}$ homozygous A daughters, $\frac{1}{4}$ heterozygous daughters, $\frac{1}{4}$ hemizygous A sons, and $\frac{1}{4}$ hemizygous a sons. *(b)* Mating between a hemizygous a male and a homozygous A female produces $\frac{1}{2}$ heterozygous daughters and $\frac{1}{2}$ hemizygous A sons. The pattern of inheritance results from the fact that a male transmits his X chromosome only to his daughters and his Y chromosome only to his sons.

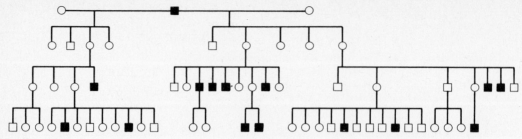

Figure 5.4 Pedigree of hemophilia in a kinship from Scotland showing several typical features of the inheritance of rare X-linked mutations. Note that only males are affected. Because males transmit their X chromosome only to their daughters, and because the rare mutation is recessive, affected fathers have normal sons and carrier daughters [see Figure 5.3(b)]. Moreover, half the *sisters* of affected males are expected to be carriers [see Figure 5.3(a)]. Carrier females are expected to have 50 percent affected sons [see Figure 5.3(a)].

that, for rare autosomal recessive alleles, heterozygotes are much more frequent than recessive homozygotes. This same relationship is true in females with respect to rare X-linked recessive alleles. That is to say, Figure 4.3 is also valid for X-linked recessives if the numbers along the horizontal axis are interpreted as referring to the incidence of recessive homozygotes among females and those along the vertical axis are interpreted as referring to the frequency of heterozygotes among females. Reference back to Figure 4.3 will emphasize again the greater frequency of heterozygotes than recessive homozygotes.

For traits due to X-linked recessive alleles, affected males occur much more frequently than affected females. This aspect of X-linked inheritance is illustrated in Figure 5.5. For a trait that affects 1 percent of females (i.e., $\frac{1}{100}$), for example, the incidence of the trait in males is 10 percent (i.e., $\frac{50}{500}$), which means that affected males are 10 times as frequent as affected females. To take another example, a rare trait that has an incidence of 1 per 1 million females will have an incidence in males of 1 per 1000 (i.e., $\frac{0.5}{500}$), so in this case affected males are 1000 times as frequent as affected females. Although the mathematical

basis for the differences in incidence between the sexes will be discussed in Chapter 14, the reason for the differences can be seen intuitively: Whereas two copies of a recessive X-linked allele are required to produce an affected female, only one copy is required to produce an affected male; and an individual is less likely to inherit two copies of a rare allele than only one copy.

At this point it is convenient to discuss several examples of X-linked inheritance.

Glucose 6-Phosphate Dehydrogenase (G6PD) Deficiency

One well-known gene on the X chromosome controls the production of an enzyme, **glucose 6-phosphate dehydrogenase** (G6PD), which is involved in carbohydrate metabolism and is important in maintaining the stability of red blood cells. Individuals who have abnormally low amounts of this enzymatic activity are prone to a severe anemia that occurs when many of their red blood cells cannot function normally and therefore break down and are destroyed. The anemia can be provoked by a number of environmental triggers such as in-

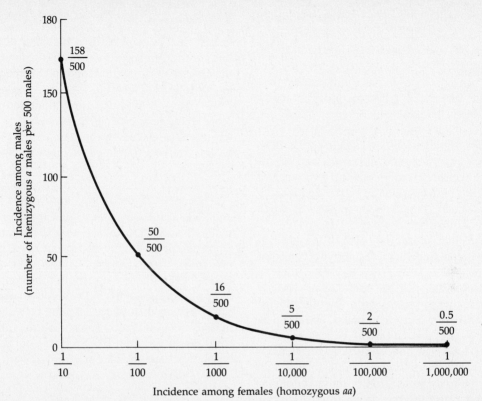

Figure 5.5 Incidence of hemizygous males for an X-linked recessive (vertical axis) plotted against incidence of homozygous recessive females (horizontal axis). When an X-linked recessive is rare, there will be many times more affected males than affected females.

haling pollen of the broad bean *Vicia faba* or eating the bean raw, in which case the illness is known as **favism**. Such individuals are also sensitive to certain drugs such as naphthalene (used in mothballs), certain sulfa antibiotics (such as sulfanilamide), or the antimalarial drug primaquine. In the absence of the offending substances these individuals are completely normal, and they recover from the anemia when the agents are eliminated. G6PD deficiency is found in high frequency in people of Mediterranean extraction (10 to 20 percent or more of males are affected) and among Asians (about 5 percent of Chinese males are affect-

ed), and it occurs in about 10 percent of black American males. As noted, the locus of G6PD is X linked, and well over 50 alleles coding variant forms of the enzyme are known. However, only a few of these alleles lead to a sufficiently defective form of G6PD to cause the drug sensitivity and anemia associated with G6PD deficiency.

Color Blindness

Almost everyone is familiar with the common form of color blindness, called **red-green color blindness**, which is inherited as an X-linked recessive condition. Several other kinds of defects in color vision are known; they differ according to which of the three pigments in the retina of the eye—red, green, or blue—is defective or present in an abnormally low amount. The most common types of color blindness involve the red or green pigments; these are collectively known as red-green color blindness, but they are not a single entity. Both conditions are X linked, however, and in different Caucasian populations the frequency of red-green color-blind males is between 5 and 9 percent. Color blindness in females is much rarer, of course, but it does occur.

Two loci on the long arm of the X chromosome seem to be involved in red-green color blindness. Defects in green vision are due to mutations at one of the loci; defects in red vision are due to mutations at the other locus. Among Western European males, about 5 percent have defects in green perception and another 1 percent have defects in red perception. Two loci are known to be involved in red-green color perception because the mutations that cause the red defects and the green defects exhibit *complementation* of the sort discussed in Chapter 4. That is to say, women who carry a red-defect allele on one X chromosome and a green-defect allele on the other have normal color vision. Such complementation is expected when two loci are involved because these women are actually heterozygous for a normal allele and a recessive mutation at each of two loci. Because of segregation, nearly half the sons of these women will have the green defect and the other half will have the red defect. Because of recombination, however, a proportion of their sons will have normal color vision and an equal proportion will have defects in both their red and green color vision. The two loci are very close to each other on the X chromosome, so the actual proportion of sons who carry recombinant X chromosomes is very small.

Hemophilia

Recall that **hemophilia** is a bleeding disorder due to an X-linked recessive that results from the excessively long time required for the blood to clot following an injury. Affecting about 1 in 7000 males, it is much rarer than red-green color blindness, yet it is probably as well known. This is partly because it affects the blood; and even though we now know that blood has no magical or hereditary properties, we still carry a vestige of the old beliefs in such expressions as "pure blood" and "blood lines" and "blood relation." A second reason hemophilia is so well known is that it occurred in many members of the European royalty who descended from Queen Victoria of England (1819–1901). She was a carrier of the gene, and by the marriages of her carrier granddaughters, the gene was introduced into the royal houses of Russia and Spain. Ironically, the present royal family of Great Britain is free of the gene because it descends from King Edward VII, one of Victoria's four sons, who was not himself affected (Figure 5.6).

Actually, there are several forms of hemophilia, and the incidence of 1 per 7000 in males applies only to the special type exemplified in some of Queen Victoria's descendents. Blood clotting is a complex process involving more than a dozen ingredients known as **clotting factors**, and different genes are responsible for producing these various clotting factors. An abnormal form of any one of these genes can lead to an inherited form of excessive bleeding. Most hereditary hemophilias are caused by genes on the autosomes, but two forms are inherited as X-linked recessives. One of these X-linked forms, called **hemophilia B** or **Christmas disease**, involves clotting factor IX, and it is extremely rare. The other X-linked form, called **hemophilia A** or **Royal hemophilia**, involves clotting factor VIII. Hemophilia A is the more common X-linked form, and it is the

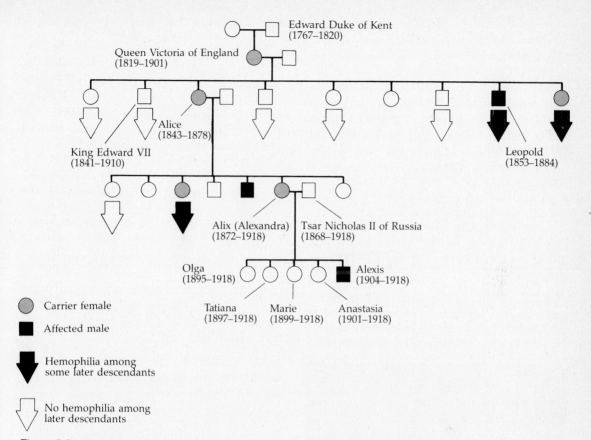

Figure 5.6   Partial pedigree of hemophilia A among the descendants of Queen Victoria, including Alexandra, Empress of Russia, and her five children.

one present in some of the European royal families.

Where Queen Victoria's hemophilia-A mutation came from is not known. She herself was a carrier, but her father was completely normal and nothing in her mother's family suggests that the gene was present. The best guess is that either the egg or the sperm that gave rise to Queen Victoria carried a new mutation. Perhaps it was in the sperm of her father, Edward Duke of Kent, because it is thought that mutations are somewhat more likely to occur in the sperm of older men, and he was 52 when Victoria was born. However it

happened, Victoria was a carrier. She had nine children (see Figure 5.6): five daughters (two were certainly carriers, two were almost certainly not, and the remaining daughter may or may not have been as she left no children) and four sons (one a hemophiliac—Leopold Duke of Albany—and three normal). In the five generations since Queen Victoria, 10 of her male descendants have had the disease. Virtually all died very young. Those who survived childhood often died in their 20s or early 30s, usually from excessive bleeding following injuries.

The most famous of Victoria's affected

male descendants is undoubtedly her great-grandson Alexis, the son of her granddaughter Alix (Empress Alexandra of Russia) and Tsar Nicholas II (see the pedigree in Figure 5.6 and the photograph in Figure 5.7). Alexis was Alexandra's firstborn son and heir to the Russian throne. Unbeknown to Alexandra, she was a carrier of the hemophilia-A allele, and her son had inherited the disease. Anna Viroubova, one of Alexandra's favorite ladies-in-waiting, recounts:

The heir was born amid the wildest rejoicings all over the Empire. After many prayers, there was an heir to the throne of the Romanoffs. The Emperor was quite mad with joy. His happiness and the mother's, however, was of short duration, for almost at once they learned that the child was afflicted with a dread disease. The whole short life of the Tsarevich, the loveliest and most amiable child imaginable, was a succession of agonizing illnesses due to this congenital affliction. The sufferings of the child were more than equaled by those of his parents, especially of his mother.[1]

The parents became preoccupied with the boy's health. At one point the Tsar observed in his diary how difficult it was to live through the worry resulting from Alexis's excessive and life-threatening bleeding from minor injuries and bruises that all children unavoidably experience. The boy survived his childhood, but some historians have argued that the Tsar's preoccupation with Alexis's health contributed to the neglect of the empire that ultimately brought on the Bolshevik Revolution in 1917. During the revolution, the Tsar, Alexandra, Alexis, and his four sisters disappeared. The fate of the family is unknown, but they are thought to have been machine-gunned to death in Ekaterinburg fortress on July 17, 1918, just 13 days before what would have been Alexis's seventeenth birthday.

Lesch-Nyhan Syndrome (HGPRT Deficiency)

As a final example of X-linked recessive inheritance we consider the rare condition known as **Lesch-Nyhan syndrome**, which is due to a deficiency of the enzyme HGPRT (recall from Chapter 4 that HGPRT refers to *hypoxanthine guanine phosphoribosyl transferase*). Persons with this condition, virtually all male, have two groups of symptoms. One results from excessive accumulation of **uric acid** in the blood. Recall from Chapter 4 that one normal function of HGPRT is to route certain cellular metabolites into DNA via the minor pathway of DNA synthesis. (A **metabolite** is any compound produced during the course of chemical processes in the cell; the totality of all these processes is called **metabolism**.) In the absence

Figure 5.7 Tsar Nicholas II of Russia, the Empress Alexandra (granddaughter of Queen Victoria), and their son, Tsarevich Alexis. Alexandra was a carrier of the hemophilia-A mutation, and her son was afflicted with the disease. For a pedigree of the family, see Figure 5.6.

[1]Anna Viroubova (Vyrubova). Memories of the Russian Court. (New York: Macmillan, 1923), p. 10.

of HGPRT, these metabolites accumulate and are eventually broken down by other enzymes into uric acid. Accumulation of excess uric acid leads to hyperuricemia (elevated levels of uric acid in the blood), bloody urine, crystals in the urine, urinary tract stones, severe inflammation of the joints (**arthritis**), and the extremely painful swelling and inflammation of joints in the hands and feet (particularly the big toe) known as **gout**. (Although gout is always associated with elevated levels of uric acid, the inheritance of most forms of gout is multifactorial; only a minority of patients with excessive uric acid and gout have HGPRT deficiency.)

Symptoms of excessive uric acid can be treated with appropriate drugs, but Lesch-Nyhan syndrome has another group of symptoms that are of unknown origin and are not alleviated by treatment for excessive uric acid. These symptoms involve the nervous system and are associated with jerky, unwilled, and uncoordinated muscular movement. The most bizarre symptom of Lesch-Nyhan syndrome is extreme aggressive behavior, most often expressed as self-mutilation of the hands and arms, usually by biting. At present, such self-mutilation can be prevented only by appropriate binding of the hands and arms. Lesch-Nyhan syndrome is an extremely serious condition. Most patients die before the age of five, and almost all die before adulthood. Though rare, the Lesch-Nyhan syndrome is important because it involves HGPRT, which allows us to understand the basis of some of its symptoms. The condition also illustrates the profound influence that genes can exert on behavior, even though the biochemical basis of this influence is not yet understood.

X Linkage and Recombination

There appears to be no recombination between the X chromosome and the Y chromosome in humans. Consequently, a male's Y chromosome is passed on intact to his sons, and his X chromosome is passed on intact to his daughters. In females, on the other hand, recombination between X-linked loci appears to occur as frequently as it does between autosomal loci. Recombination in females therefore permits the establishment of an X-chromosomal genetic map, in the same way that recombination in both sexes permits the establishment of autosomal genetic maps. Figure 5.8(a) is a small part of the genetic map of the long arm of the X chromosome showing the positions of four loci—red color blindness, green color blindness, *G6PD*, and hemophilia A. (The order of the color blindness loci is uncertain, but they are very closely linked.)

Genetic maps are not merely of academic interest but can aid in genetic counseling. In one actual case, for example, a pregnant woman who was heterozygous for hemophilia A (genotype *hemA/hemA*$^+$—the + designates the normal allele) came to a genetic counseling clinic wishing to know whether her unborn child would be affected with hemophilia. By a procedure known as *amniocentesis* (to be discussed in the next chapter), cells from the fetus were obtained, and the karyotype indicated that the woman's unborn child was a male. On the face of it, therefore, the child had a 50 percent chance of being a hemophiliac because, at present, hemophilia cannot be diagnosed in fetal cells. However, the woman was also found to be heterozygous for *G6PD* (genotype *G6PDA/G6PDB*), and *G6PD* is not only closely linked to *hemA* but also can be diagnosed in fetal cells. Family studies quickly established that the woman had inherited the *G6PDA* allele and *hemA* from one parent and the *G6PDB* allele and *hemA*$^+$ from the other parent, so her genotype was *G6PDAhemA/G6PDBhemA*$^+$. Moreover, her unborn son was found to carry *G6PDB*. This information, along with that in the genetic map in Figure

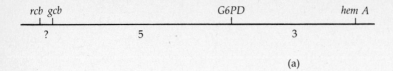

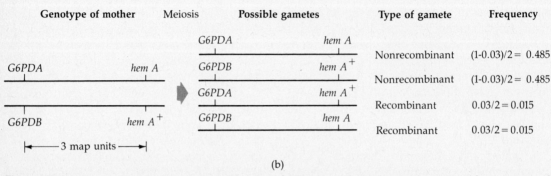

Figure 5.8 *(a)* Genetic map of part of the X chromosome. *(b)* Gametes and their frequencies expected from a woman of genotype *G6PDA hemA/G6PDB hemA⁺*. (Symbols: *rcb*, red form of color blindness; *gcb*, green form of color blindness; *G6PD*, glucose 6-phosphate dehydrogenase enzyme; and *hemA*, clotting factor VIII.)

5.8(*a*), can be used to calculate more precisely the risk of hemophilia.

The reasoning involved in the more precise calculation of risk is illustrated in Figure 5.8(*b*). It is based on the realization that, for map distances as small as the ones in this case, 1 map unit corresponds to 1 percent recombination (so 3 map units correspond to 3 percent recombination). The four possible types of gametes that the woman could produce are shown along with their expected frequencies at the right side of Figure 5.8(*b*). Among those gametes that carry *G6PDB* (the allele present in the fetus), the proportion that also carry *hemA* is $0.015/(0.015 + 0.485) = 0.03$. Thus, the risk of hemophilia in the unborn son is no longer 50 percent; the additional information about *G6PDB* indicates that the true risk is 3 percent. This risk can be calculated in another, completely equivalent, way. Since the fetus has *G6PDB*, the only way it could also have *hemA* is if the *G6PDB*-bearing chromosome had undergone recombination in the 3-map-unit region of interest. However, the probability that a *G6PDB*-bearing chromosome is a *nonrecombinant* is 97 percent, and the probability that a *G6PDB*-bearing chromosome is a *recombinant* is 3 percent. Hence, the probability that the fetus is genotypically *G6PDB hemA* (i.e., recombinant) is 3 percent. In any event, the woman was advised that her son had a relatively small risk of having hemophilia. She decided to take the risk and, for the record, gave birth to a normal son.

Dosage Compensation

In many instances, the amount of a gene product in cells is directly related to the number of copies of the corresponding gene: A cell with one gene copy will produce half as much gene product as a cell with two copies, and a cell with two copies will produce two-thirds as much gene product as a cell with three copies of the gene. Based on this simple reasoning, one would expect the amount of gene product of X-linked genes in males to be half as large as the amount in females, because males have only one X chromosome whereas females have two. This expectation is not realized in fact, however. For the X-linked locus *G6PD*, for example, a cell from a female contains the same amount of G6PD enzyme as does a cell from a male. Similarly, to take another example, a female cell contains the same amount of clotting factor VIII as does a male cell.

In all known organisms in which the sexes differ in the **dose** (number of copies) of sex-chromosomal genes, the activity of genes on the sex chromosome in one of the sexes is adjusted by a special type of regulation known as **dosage compensation**; the net effect of dosage compensation is to offset ("compensate") the sex differences in gene dosage. In mammals, dosage compensation occurs in the female, and, *in humans, dosage compensation involves the turning off of one X chromosome in each somatic cell*. That is to say, in any somatic cell of an adult female, only one X chromosome is genetically active, and only the genes on that X chromosome are working; the other X chromosome is genetically inactive ("turned off"), and the genes on that X chromosome are not expressed. However, the X chromosome that is the active one in some cells may be the inactive one in other cells. Every female is therefore a kind of **mosaic** with respect to X-linked genes. (Recall from Chapter 2 that a mosaic is an individual composed of two or more genetically different types of cells.) A normal female is a mosaic because, in approximately half her cells, only the genes on one particular X chromosome will be active; in the rest of her cells that same X chromosome will be inactive, but the genes on the other X chromosome will be expressed. (It should be pointed out that the mechanism of dosage compensation in insects such as *Drosophila* is quite different from what it is in humans and other mammals.)

The physical inactivation of one X chromosome in each somatic cell of a normal female is known as the **single active X principle** or the **Lyon hypothesis**. The actual time of the inactivation is evidently soon after fertilization, but both X chromosomes are known to be active at least through the 16-cell stage of embryonic development. At the time of inactivation, one X chromosome in each cell is chosen at random in some unknown manner, and this X chromosome is rendered inactive. The inactivation is permanent in the sense that it persists through cell division. That is to say, as cells in the embryo continue to proliferate by means of mitosis, all the descendants of a cell that have a particular X chromosome inactivated will have that same X chromosome inactivated. The time of X-chromosome inactivation need not be the same in all embryonic cells, and, indeed, in oocytes, it appears that both X chromosomes remain active throughout life.

The genetic consequences of X-chromosome inactivation are easily observed in cells from females who are heterozygous for an X-linked mutation. One experimental procedure used to study the situation is outlined in Figure 5.9, where the woman whose cells are to be studied is assumed to be heterozygous for an allele that leads to a nonfunctional form of HGPRT. The genotype of such a female is *HGPRT+*/*HGPRT−*—the + and − designating the normal and defective alleles of *HGPRT*, respectively. Half the cells in a heterozygous female are expected to have their *HGPRT+*-bearing X chromosome active, and these cells will produce the normal HGPRT enzyme. In the remaining cells, the *HGPRT−*-bearing X chromosome will be active, and these cells will produce a nonfunctional HGPRT enzyme. (In the latter case, the *HGPRT+* allele is still physically present in the cells, of course, but this allele and apparently most other genes on the same X chromo-

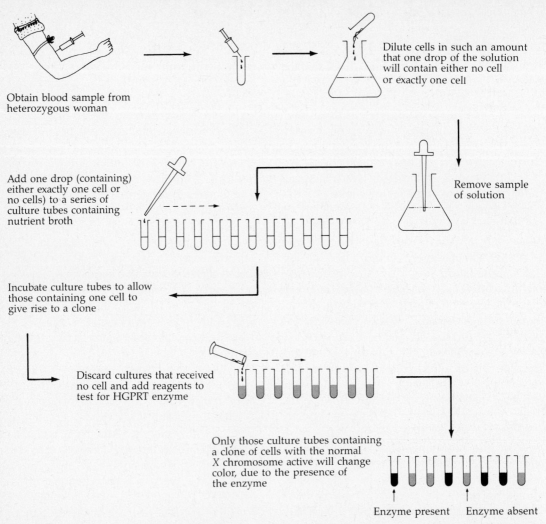

Obtain blood sample from heterozygous woman

Dilute cells in such an amount that one drop of the solution will contain either no cell or exactly one cell

Add one drop (containing) either exactly one cell or no cells) to a series of culture tubes containing nutrient broth

Remove sample of solution

Incubate culture tubes to allow those containing one cell to give rise to a clone

Discard cultures that received no cell and add reagents to test for HGPRT enzyme

Only those culture tubes containing a clone of cells with the normal X chromosome active will change color, due to the presence of the enzyme

Enzyme present   Enzyme absent

Figure 5.9   One procedure for showing that only one X chromosome is active in each somatic cell of a normal female. The woman here is heterozygous for a recessive mutation that causes a nonfunctional HGPRT enzyme to be produced. Cells from the woman are cultured one by one, each giving rise to a clone. Clones in which the normal X chromosome is active will produce the functional enzyme; clones in which the normal X is inactive will produce only the nonfunctional enzyme. Approximately half of each type of clone is found, indicating that half the cells have the normal X chromosome active and half have only the mutation-bearing X chromosome active.

some are inactive.) When cells from an $HGPRT^+/HGPRT^-$ female are separated from each other and laboratory cultures are started from single cells placed in nutrient medium, each cell gives rise to a *clone* (a group of genetically identical cells). Because of the single active X principle, half the clones are expected to produce a functional HGPRT, and the remaining clones are expected to produce a nonfunctional HGPRT. As indicated in Figure

5.9, this expectation is borne out in practice. Indeed, if the experiment in Figure 5.9 was carried out with G6PD in a female of genotype *G6PDA/G6PDB*, half the clones would express the *G6PDA* form of the enzyme and the other half would express the *G6PDB* form of the enzyme.

A female who is heterozygous for HGPRT deficiency or for G6PD deficiency will still have the functional enzyme present in her body, of course; it will be produced by the cells in which the X chromosome carrying the normal allele is active. However, because half her cells will produce only the nonfunctional form of the enzyme, she will have less functional enzyme overall than would a woman homozygous for the normal allele. For a similar reason, women heterozygous for the hemophilia-A allele have less clotting factor VIII than women who are homozygous for the normal allele, although virtually all heterozygous women have sufficient clotting factor in their blood serum to prevent hemophilia.

In certain cells of the body, the inactive X chromosome can be seen directly in the microscope. Nuclei of nondividing cells scraped off the surface layer of skin on the inside of the cheek often have a densely staining blob near the nuclear envelope [Figure 5.10(*b*)]. This blob is the coiled and inactive X chromosome, and it is called a **sex-chromatin body** or a **Barr body**. Actually, the staining procedure does not work perfectly, and for this reason the Barr body cannot be seen in all cells scraped from the inside of the cheek. Even though all the cells of normal females have one inactive X chromosome, not all the cells have a visible Barr body. The important finding is that cells from normal females frequently have a Barr body [see Figure 5.10(*b*)], whereas cells from normal males never do [see Figure 5.10(*a*)].

In the next chapter we will discuss certain relatively rare individuals who have an abnormal sex-chromosome constitution. Here we should note one important finding relative to the somatic cells of such individuals: *Irrespective of the number of X chromosomes in an individual, all X chromosomes except for one are inactivated in each somatic cell.* Thus, for example, XO females (who have only one X chromosome instead of two) have no X-chromosome inactivation, XX females have one X inactivated, XXX females have two X's inactivated, XXXX females have three X's inactivated, and so on. Consequently, cells of XO females have no Barr bodies, those of XX

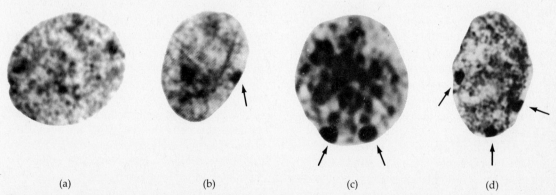

(a)   (b)   (c)   (d)

Figure 5.10   Micrographs of nuclei showing zero, one, two, or three Barr bodies (arrows). Inactive X chromosomes are tightly condensed in the nucleus and strongly absorb certain dyes, giving rise to Barr bodies. Cells that have zero, one, two, or three Barr bodies have respectively one, two, three, or four X chromosomes.

females have up to one [see Figure 5.10(*b*)], those of XXX have up to two [see Figure 5.10(*c*)], and those of XXXX have up to three [see Figure 5.10(*d*)]. As will be discussed in the next chapter, individuals with abnormal numbers of sex chromosomes have varying degrees of phenotypic abnormality, which is interesting in light of the principle that a single X chromosome is active irrespective of how many may be present. The question that arises is why such individuals should have any phenotypic abnormalities, inasmuch as all have a single active X. The answers are not known for certain. Perhaps the earliest stages of embryonic development, when all the X's in a cell are active, are critical, and an abnormal gene dosage during this period may have permanent, irreversible effects. Also, perhaps the inactivation of the X chromosomes occurs much later in some tissues than in others, and perhaps gene dosage in these tissues is critical. In addition, certain genes, such as the one on the short arm of the X coding for **steroid sulfatase**, are known *not* to be inactivated, and such uncompensated loci may be important in development.

Genes on the Y chromosome are not subject to dosage compensation, which is perhaps because there are so few genes on the Y chromosome to begin with. The Y chromosome never forms a Barr body and, presumably, all the Y chromosomes are active in males who have multiple Y's. Several active Y chromosomes seem to have little phenotypic effect, probably because of the small number of genes on the Y (see the discussion in the next chapter). A male's X chromosome is genetically active in all somatic cells, and no Barr bodies are formed. (Oddly enough, a male's X chromosome seems to be genetically inactive in spermatocytes.) The single active X principle does hold in somatic cells of males who have multiple X chromosomes, however. That is to say, males with more than one X chromosome have all except one X chromosome inactivated, and these inactivated X's are frequently visible as Barr bodies. Consequently, XXY or XXYY or XXYYY males have cells with up to one Barr body, XXXY and XXXYY males have cells with up to two, XXXXY males have cells with up to three Barr bodies, and so on. (Note that the number of Y chromosomes makes no difference in the number of Barr bodies.)

The Developmental Basis of Sex

At this point it is appropriate to discuss the embryological basis of sexual differentiation, but it will be necessary to preface the discussion with a brief overview of more general aspects of development. Along the way we will have occasion to mention one particular aspect of development—the extreme sensitivity of the process to drugs and certain other agents.

Development, of course, begins with **fertilization**, which normally occurs in one of the two **fallopian tubes**. About 4 in long and less than $\frac{1}{2}$ in in diameter, the fallopian tubes stretch from the upper corners of the uterus to the ovaries on either side; there the tubes flare out with tiny fingerlike projections to catch the egg released at the time of ovulation. The fertilized egg (**zygote**) travels down the fallopian tube toward the uterus, dividing by mitosis as it moves along. The one cell becomes two, two become four, four become eight, eight become sixteen, and so on, with all the cells sticking together in a clump (Figure 5.11). This clump of cells, called the **morula**, reaches the uterus about three or four days after fertilization.

These early cell divisions are peculiar because the cells do not grow between divisions. They merely replicate their chromosomes and divide again. The result is that the cytoplasm of the egg becomes partitioned into smaller and smaller packages, with the new cells all bunched together inside the gelatinous envelope that covered the egg.

Cell division continues in the morula

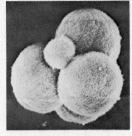

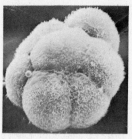

| Two cells | Four cells | Eight cells | Morula |

Figure 5.11   Scanning electron micrographs of the earliest stages in mouse development. The tiny cell visible particularly in the photograph of the four-cell stage is a polar body.

until the cells number a few thousand. The cells then begin to move in relation to one another; they push against and expand the envelope that encloses them and rearrange themselves to form a hollow sphere called the **blastocyst** (Figure 5.12). In one region of the inner wall of the blastocyst there is a thickening composed of a mass of cells known as the **inner cell mass**. The body of the embryo itself is destined to develop from this group of cells. The wall of the blastocyst will form several **embryonic membranes**, including the **amnion**, which enshroud the developing **embryo**. Once the blastocyst has been formed, it implants itself in the uterine wall; this occurs on approximately the eighth day after fertilization. From the time of implantation onward, the cells in the blastocyst must approximately double their size with each cell division. This is the first great trial of the embryo. No longer can it survive from the cytoplasm in the egg produced by the mother; the cells of the embryo must now manufacture their own cytoplasm. The relationship between the blastocyst and the mother is crucial to passing this test. The **placenta**, which develops in the region of implantation, must help transfer nutrients and oxygen from the mother's blood to the embryo and waste from the embryo back into the mother, all of this without actually intermixing the circulatory fluids of mother and child.

The development of the embryo proceeds rapidly after the blastocyst implants in the uterine wall. (Table 5.2 is a timetable of human embryonic development highlighting certain landmark events.) In the third week after fertilization, the early stages of the skeleton and nervous system are formed. By the fourth week the head of the embryo is visible and the embryo has a heart—a beating heart [Figure 5.13(a)]. The arms and legs of the embryo begin to develop; these start as tiny paddle-shaped bumps of cells that grow outward and elongate. Also in the fourth week after fertilization, the embryo acquires a tail, which later disappears. Early stages of gill development can be recognized; these are much the same as in fishes, but in humans the gills do not develop completely and they eventually recede. (The tail and primitive gills are reminders of our evolutionary ancestry.) By the fifth week of development, the chest and abdomen of the embryo are formed, fingers and toes are beginning to develop, and the eyes appear. The embryo is less than $\frac{1}{2}$ in long. By six weeks after fertilization, the embryo is still less than 1 in long. The gills, which never developed completely, are disappearing. The face and ears of the embryo are developing. The embryo begins to look like a human embryo. By the seventh week after fertilization, the embryo is unmistakably a human embryo [see Figure 5.13(b)]. Many of the internal organs have begun to take shape.

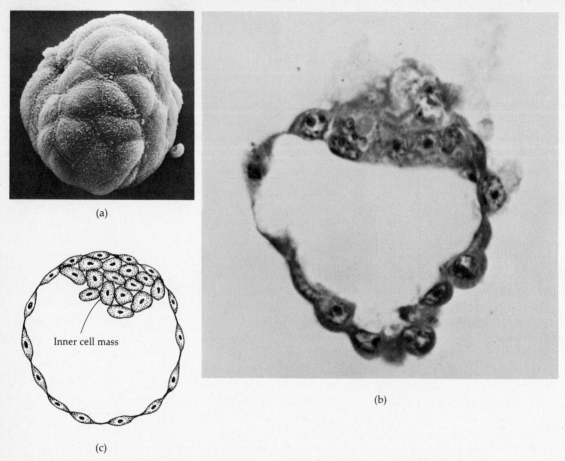

(a)

Inner cell mass

(c)

(b)

Figure 5.12   Three views of the blastocyst. *(a)*Scanning electron micrograph of mouse blastocyst. *(b)*Cross section of human blastocyst showing inner cell mass that gives rise to the body of the embryo. *(c)* Schematic of the cross section in *(b)*. In humans, the blastocyst forms about one week after fertilization.

From this point on the embryo is properly called a **fetus**. By the end of 12 weeks, the major outlines of embryonic development are almost complete [see Figure 5.13(*c*)]. The fetus is only 3 in long and weighs little more than 1 oz, yet the last months of development will be spent, in a sense, in putting on the finishing touches and, most especially, growth.

The first three months of embryonic development are the most critical because so much happens so fast. The cells in the embryo divide extremely rapidly. Different types of cells with special functions and abilities arise. (It has been estimated that humans may have up to 200 distinct cellular types, such as muscle cells, red blood cells, and so on.) Several important events occur at nearly the same time in different parts of the embryo. Everything seems to be in constant motion. Outfoldings of groups of cells occur here, infoldings there, and bulges somewhere else. Fusions of tissues happen in one place, splittings in another. Some cells migrate from up to down, some from down to up. The embryo seems intent on

TABLE 5.2   TIMETABLE OF HUMAN EMBRYONIC AND FETAL DEVELOPMENT

Week after Fertilization	Major Features of Development*
0	Fertilization
1	Egg travels down fallopian tube into uterus; begins to divide
2	Implantation of blastocyst in wall of uterus; embryonic membranes begin to form
3†	Early stages of skeleton, nervous system, and blood vessels
4†	Head, heart, and tail visible; heart begins to beat; gill-pouch rudiments; paddle-shaped rudiments of arms and legs; length 4 mm (0.2 in); see Figure 5-13 (*a*)
5†	Chest and abdomen formed; fingers and toes appear; head increases in size; eyes developing; length 8 mm (0.3 in)
6†	Face, features, and external ears developing; upper lip formed; gill rudiments disappearing; length 13 mm (0.5 in), weight 1 g (0.04 oz)
7†	Face completely developed—now resembles a human child; palate forms; length 18 mm (0.7 in), weight 2 g (0.07 oz); from now on embryo usually called *fetus;* see Figure 5-13 (*b*)
8†	Fingers distinct; testes and ovaries distinguishable; beginning of differentiation of external sex organs; beginnings of all essential internal and external structures present; length 30 mm (1.2 in), weight 4 g (0.2 oz)
12†	Limbs—including fingers, toes, nails—fully formed; external sex organs developed; sex can be determined by a trained person without microscopic examination; 77 mm (3.0 in), 30 g (1.1 oz); see Figure 5-13(*c*)
16	Movements ("quickening") begin; heart can be heard with stethoscope; hair all over body; eyebrows and eyelashes present; 190 mm (7.5 in), 180 g (6.4 oz)
21	Head hair appears; 300 mm (11.8 in), 450 g (1 lb)
25	Eyes open; 350 mm (14 in), 875 g (2 lb)
30	Fetus pink and scrawny—little fat has been deposited; if born now can survive, given special care; 400 mm (16 in), 1425 g (3 lb, 2 oz)
32	Considerable fat deposited—fetus begins to look "plump"; excellent chance of survival if born; 450 mm (18 in), 2375 g (5 lb. 4 oz)
34	Full term; skin covered with cheeselike material; head hair typically 25 mm (1 in) long; may be other hair on shoulders, but soon disappears; head still very large relative to body; 500 mm (20 in), 3250 g (7 lb. 3 oz)

*All measurements are averages; there are many wide departures from them, especially in the later months.
†Critical stages for teratogenic agents.

its business, as if in a hurry to be born. The last six months of development seem almost leisurely by contrast.

Teratogens

Because so much cell division and differentiation occur in the first 12 weeks, these are the weeks in which the embryo is most sensitive to external influences. The cells are extremely sensitive to radiation, for example, and to many drugs. Any agent that can cause malformations during development is called a **teratogen**. One dramatic example of a teratogen is the tranquilizer **thalidomide**, which was in widespread use in certain parts of Europe in the late 1950s. The drug seemed harmless enough at first, but then the effect of thalidomide on the embryos of pregnant women was recognized. When present during the first criti-

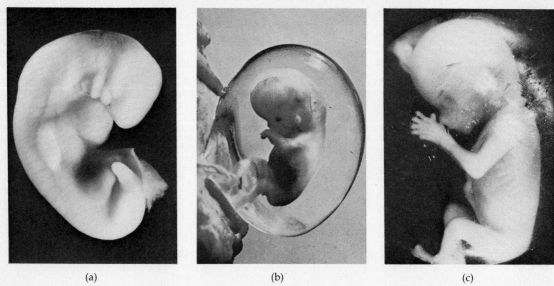

(a)   (b)   (c)

Figure 5.13   The human embryo at *(a)* 4 weeks, *(b)* 7 weeks, and *(c)* 12 weeks after fertilization. Facial features are formed by the seventh week of development; the limbs, including fingers, toes, and nails, are fully developed by the twelfth week. The ages of the embryos shown here are approximate.

cal months of embryonic development, thalidomide interferes with the normal growth of the long bones in the arms and legs (Figure 5.14). The pathetic children born to women who took the drug early in pregnancy were normal in every respect, except that they had abnormally short or absent arms or legs. Before the tragic mistake was discovered, approximately 5000 children had been maimed. Not everyone suffered, however. Those pregnant women who started to take the drug *after* the first critical months of embryonic development gave birth to normal babies.

Developing embryos are also extremely sensitive to certain kinds of infections. Of these, **rubella** or **German measles** is the best-known example. Rubella is caused by a virus, a very tiny internal parasite of cells, which multiplies inside infected cells to produce hundreds of virus particles. These particles may, in turn, infect other cells. Consequently, infection by a single virus particle can eventually lead to the

presence of millions of them in the body. Adults are not greatly threatened by German measles. Even small children are not gravely endangered. A few days of fever, red spots, and discomfort, and the infection usually runs its course.

If a woman contracts German measles in the early stages of pregnancy—in the first three months—the virus can penetrate the placental barrier and infect cells in the embryo. Such infection destroys embryonic cells and can lead to spontaneous abortion of the embryo or fetus. Among fetuses that survive the infection, a great danger is blindness—or deafness, or heart defects, or severe mental retardation, or all of these. After an adult or child has been infected with German measles virus, the body develops an immunity that prevents subsequent infection by destroying the invading virus before its cycle of reproduction can get started. In times past, young girls were encouraged to visit friends who had the disease in the

Figure 5.14 Malformation induced by the teratogen thalidomide.

hope they would contract it and thereby become naturally immune to later infection during their childbearing years. Nowadays, an antirubella vaccine is used to immunize women against the virus.

Here it is necessary to point out that infectious diseases such as German measles may have different degrees of severity in different societies. A particular infectious disease may be rather mild in a society in which the disease has been present for thousands of years, the main reason being that natural selection operates both against the most harmful forms of the disease-causing agent and against the human genotypes that are most severely affected; hence, a sort of mutual accommodation evolves. However, these same disease-causing agents, upon infecting people of a different society not previously exposed to them, may cause severe, perhaps even mortal, diseases. (Native populations in Mexico, for example, were decimated by European diseases introduced during the expeditions of Cortez and others.) This principle applies to infectious diseases of plants and animals as well.

Differentiation of the Sexes

In the latter half of the critical first 12 weeks of development, the body of the embryo takes on a visible sexual identity as its internal and external sex organs are formed. Development in the first seven weeks following fertilization is virtually identical in both male and female embryos. The sex of embryos in these early weeks cannot be outwardly distinguished. Their sex chromosomes reveal which is which, of course, but the bodies of the embryos themselves give no hint of whether they will ultimately develop as male or female. In the seventh week after fertilization, the embryo physically commits itself to one direction or the other. Sexual development then occurs very rapidly and by 12 weeks it is essentially completed. In the twelfth week a careful observer can visually determine the sex of the fetus, then only 3 in long.

Important events in sexual development do occur before the seventh week, but these events are the same in male and female embryos. The origin of the sexual structures is intimately connected with the urinary system, as might well be expected from the anatomical relationship of the two systems in adults. The gonads of the embryo, one on either side—which in females will develop into ovaries and in males into testicles—are formed as ridgelike thickenings of a region of a pair of ducts or tubes that form a sort of primitive kidney, a structure that is fully formed in the embryo by the sixth week. (The permanent kidney originates separately in development.) The embryonic gonads rapidly enlarge as the cells divide repeatedly and increase in size.

The embryonic gonad contains the **primordial germ cells**, the cells whose descendants will eventually give rise to eggs or sperm. Although the germ cells are present in the gonads in the sixth week of development and probably earlier, their actual place of origin is far removed from the gonad. Indeed, the primordial germ cells do not arise in the body of the embryo itself! Rather they arise at an early stage in the margin of the yolk sac outside the embryonic body (Figure 5.15). From there they migrate into the body of the embryo and proceed to the embryonic gonad. The germ cells from the yolk sac are genetically identical to all the other cells in the embryo and its membranes, of course, because all these cells are descendants of the fertilized egg. Yet it is remarkable how early in development the germ cells are set apart. From a genetic point of view this is not so surprising, however. Inheritance in animals proceeds fundamentally from germ cell to germ cell to germ cell down

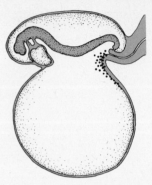

Figure 5.15 Drawing of human embryo at about $3\frac{1}{2}$ weeks after fertilization showing that the germ cells (black dots) arise outside of the body of the embryo in the margin of the yolk sac. From there, they migrate into the body of the embryo and reside in the embryonic gonad.

through the generations. A person's body is a sort of carrier or receptacle of the germ cells, although it is important to keep in mind that the genes expressed in the body are the same genes carried by the germ cells. But in a not wholly facetious sense, a hen is just an egg's way to make another egg.

By the time the embryonic gonads are formed, a second pair of ducts has developed in the embryo [Figure 5.16(*a*)]. The embryo thus has two pairs of ducts that are associated with sex. Internally, in the sixth and seventh week of development, the embryo is neither male nor female. In a certain sense it is both, because the pair of ducts that develop last—the **Müllerian ducts**—are associated with the internal sex organs of females, while the primitive kidney—the **Wolffian ducts**—is associated with the internal sex organs of males. The external parts of the body are also identical in both male and female embryos of this age. In roughly the position where the external sexual structures will be formed is a raised bump or elevation of cells that in males will develop into the penis and in females into the clitoris. Flanking the elevation of cells in the genital region of the embryo are a pair of vaguely

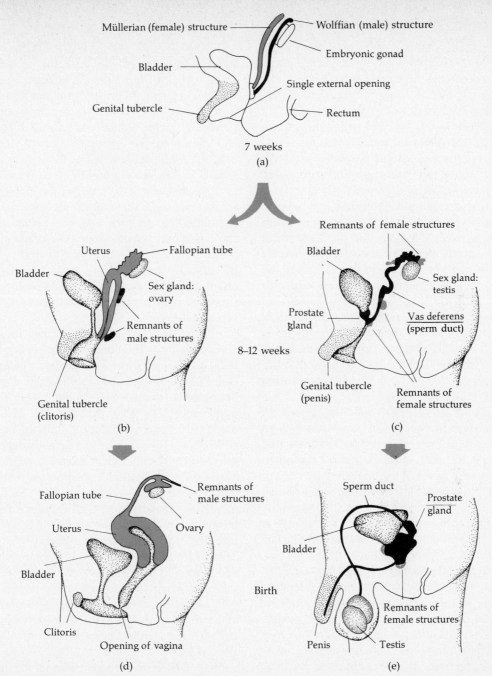

Figure 5.16 Outline of differentiation of the sex organs in humans. Until about two months after fertilization, the development of the sex organs in male and female embryos is identical; the embryo has two sets of internal sexual structures: the Müllerian ducts (female) and the Wolffian ducts (male). Presence of the male hormone testosterone in later embryonic development causes development of the male structures and degeneration of the female structures; absence of testosterone causes development of the female structures and degeneration of the male ones. Note that the penis and the clitoris develop from the same embryonic tissue, a small bump of cells called the *genital tubercle.*

outlined ridges that in males will grow out and fold toward each other and fuse, forming the scrotum; in females these ridges of tissue become the labia majora of the vagina.

Development of the two pairs of sex ducts in the embryo and formation of the subtle bumps and swellings in the embryo's genital region set the stage for subsequent sexual development. Internally the embryo is both male and female; externally it is neither. The embryo is capable of developing in either direction, male or female, and only now will it commit itself. Hereafter the sexual structures of male embryos develop in one direction and those of female embryos in another. What tips the balance of sexual development is whether the male hormone **testosterone** is secreted by the embryonic gonad. This hormone is produced in the gonads of normal males, not in normal females, and the sex-chromosome constitution of the embryo usually determines whether it will be produced. The gonads of XY embryos secrete testosterone; the gonads of XX embryos do not. (The genes involved in the synthesis of testosterone seem to be located on the autosomes; they somehow respond to the presence of the Y chromosome by becoming active, and perhaps this activation is related to the H-Y antigen.) The germ cells are not the cells that produce the hormone; this function is reserved for other cells in the embryonic gonad.

In a sense, every human embryo is actually "preprogrammed" to develop as a female. Certain individuals with a rare inherited disorder lack embryonic gonads, or their gonads degenerate at an early stage of development. None of the hormones produced by the gonads in normal people are present in these patients, and whether their sex chromosomes are XX or XY, sexually they are female. Both their external and their internal sex organs are female, although the sex organs remain infantile. They develop as females because testosterone is not present (in some cases because their tissues do

not respond to H-Y antigen or to testosterone); their sex organs remain infantile because of the absence of female hormones normally produced in the glands. A similar condition can be created experimentally in animals by surgically removing the embryonic gonads. In rabbits, for example, removal of the gonads from female embryos has no marked consequences in development. The embryo develops as a female, although the internal organs are slightly smaller than in normal females. Removal of the embryonic gonads from male rabbits has a profound effect, however, particularly if the gonads are removed very early. If they are taken out early enough, sexual development shows no signs of maleness at all, but follows a typically female pattern.

The presence of testosterone in male embryos causes the course of sexual development to deviate from femaleness to maleness [see Figures 5.16(c) and (e)]. The effect of the hormone is twofold: First, it stimulates development of the male parts of the embryo; second, it prevents differentiation of the female parts. The female parts actually degenerate, and only remnants of them are present in newborn boys. The details of how testosterone works are not yet worked out. What seems to happen is that certain cells have specific **testosterone receptors** on their surfaces that recognize the hormone and transport it into the cells. Once inside, testosterone triggers a series of actions and reactions that ultimately become reflected in certain sets of genes (those that control male development) being turned on and other sets of genes (those that control female development) being turned off.

Whatever the details of this process, the end result of testosterone activity is the development of a male embryo. The inner part of the embryonic gonad expands whereas the outer part recedes; these changes accompany the development of the embryonic gonads into testes. The parts of the Wolffian ducts that are in intimate contact with the developing testes differentiate into tubules that will serve as a conduit and a place of storage for sperm. Simultaneous with these events, the Müllerian ducts, under the influence of testosterone, gradually degenerate [see Figure 5.16(c)]. While all this is happening inside the embryo, the external genital structures are developing into the typical external sex organs of males.

In embryos that do not possess the male hormone testosterone, sexual development proceeds the way it had been preprogrammed [see Figures 5.16(b) and (d)]. In the development of a human female, the outer part of the embryonic gonad expands; the inner part recedes. The gonads differentiate into the ovaries. The upper parts of the Müllerian ducts develop into the fallopian tubes, whereas the lower parts of the Müllerian ducts fuse together. The upper portion of this fused segment becomes the uterus; the lower portion becomes part of the vagina. As this is going on, the Wolffian ducts slowly degenerate [see Figure 5.16(b)]. External female sexual structures begin to develop at about the same time.

The other structures most frequently associated with sexuality are the mammary glands. These, too, form identically in males and females until approximately the sixth or seventh week of development. The glands are first formed in the region of the nipples as tiny buds of cells that grow and penetrate into the underlying tissue. A constricted neck of tissue retains the connection between the inner part of this structure and the part near the surface, so that the whole thing looks something like an hourglass. In male embryos the connection is broken and development of the mammary glands proceeds no further. In females the connection is retained, and as development proceeds, a circular fold of tissue comes to surround the beginnings of the mammary gland, thus elevating the nipple.

Sexual development is essentially completed by the twelfth week after fertilization. Indeed, most of the major events in human development are completed by the twelfth week. The fetus is still utterly dependent on the placenta for nutrition and removal of wastes, however. If aborted, naturally or intentionally, the fetus has no hope of survival. By the sixteenth week the "quickening" occurs; the fetus begins to move. But it is still totally dependent on its mother. Not until about 30 weeks after fertilization does the fetus have a reasonable chance of surviving on its own, although in rare cases a somewhat younger fetus has survived.

Ambiguous Sex

Although most human beings are unambiguously male or female, about 1 percent have abnormalities in sexual development. These include several of the abnormal sex-chromosome constitutions to be described in Chapter 6, such as XO and XXY, but they also include a wide variety of developmental abnormalities in children who have normal sex chromosomes. Some of these conditions are known to be caused by mutations in particular genes; other conditions have extremely complex or unknown causes. Among the latter group is a particularly common class of abnormalities in sexual development known as hermaphroditism. **Hermaphrodites** are neither entirely male nor entirely female but have a mixture of male and female sex organs; that is, their sex is ambiguous. (The name *hermaphrodite* is a compound of the names of the god Hermes and the goddess Aphrodite.) A distinction should be made between the true hermaphrodites and false or **pseudohermaphrodites**. True hermaphrodites are, by definition, capable of producing *both* functional sperm and functional eggs so that, theoretically, they could fertilize themselves. Some animals, such as earthworms, and most flowering plants are true hermaphrodites. The disorder in humans is invariably pseudohermaphroditism—that is, false hermaphroditism. Indeed, pseudohermaphrodites in humans produce *neither* functional sperm nor functional eggs.

The overall incidence of pseudohermaphroditism in humans is about 0.2 percent, or 1 per 500 newborns. This makes it relatively common among abnormalities of sexual development. Pseudohermaphroditism is not a single condition, however, and its causes are diverse. Homozygosity for any one of several different autosomal recessive mutations can lead to pseudohermaphroditism, but beyond this rudimentary knowledge the genetics of the disorder are not well worked out. The trait is extremely variable from patient to patient. In some cases the internal reproductive organs are of one sex, the external organs of the other. In other cases the reproductive organs are female on one side of the body and male on the other. Often **ovotestes** are found; these are mixtures of ovarian and testicular tissue that result from the failure of either the inner or outer part of the embryonic gonad to degenerate.

Although the underlying causes of pseudohermaphroditism are diverse and generally not understood in detail, there are exceptions. Among these are the several abnormalities of the sex chromosomes to be discussed in the next chapter. Two types of inherited pseudohermaphroditism not due to sex-chromosome abnormalities are also relatively well understood, and the nature of the disorder in these cases illustrates the importance of the male hormone testosterone in sexual development. One of these types is a male pseudohermaphroditism known as the testicular-feminization syndrome; the other is a female condition known as the adrenogenital syndrome.

Testicular-Feminization Syndrome

The **testicular-feminization syndrome** is an inherited condition of chromosomally XY individuals. The mutation responsible for it is on the X chromosome, and its expression is limited to males. Affected individuals are chromosomally XY, yet they have few traces of masculinity (Figure 5.17). The external genitalia are female, a vagina is present but incompletely developed, and rarely there is a rudimentary uterus and fallopian tubes. Tissue recognizable as testicular can be found, usually in an abnormal position in the body cavity, and the Wolffian ducts, associated with male sexuality, are somewhat developed. Affected individuals are therefore chromosomally male but for the most part phenotypically female, and they are, of course, unable to bear children. Nevertheless, many do marry as women and have normal sexual relations with their husbands.

The syndrome is evidently caused by the inability of embryonic tissues to respond to testosterone. The hormone itself is produced in quantities comparable to those in normal men. Much of the detailed information on testicular feminization comes not from studies of humans but from studies of a virtually identical X-linked inherited condition in the mouse. The particular defect in affected mice is very likely a defective receptor protein whose function is to recognize testosterone at the surface of the cell membrane and transport it into the cells. Although affected mice do produce testosterone, the hormone is not incorporated into the appropriate cells. Testosterone cannot exert its triggering effect in these cells, so they do not differentiate in the normal male direction.

Adrenogenital Syndrome

The **adrenogenital syndrome** is a form of pseudohermaphroditism in humans caused in some instances by a hereditary defect in the adrenal glands. These glands, one perched atop each kidney, produce a variety of important hormones, among them **cortisol**. This hormone is produced by converting **cholesterol**, chemical

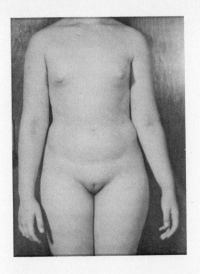

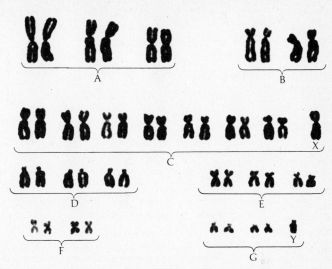

Figure 5.17 This individual is phenotypically female but chromosomally male (XY), as indicated in the karyotype. The condition is known as the testicular-feminization syndrome, due to an X-linked mutation that makes embryonic tissues unable to respond to the presence of testosterone.

step by chemical step, into the cortisol molecule (Figure 5.18). Along this kind of chemical assembly line, each step adds some atoms or takes some off or chemically rearranges the preceding molecule. Each step in this sort of assembly pathway requires the presence of a specific enzyme molecule to accomplish the chemical reaction, and the genetic information enabling the cell to manufacture each of the enzymes is present in a different gene. In the adrenogenital syndrome, one enzyme required in an intermediate step along the pathway is not functional. Thus, the chemical pathway is blocked and cortisol cannot be produced. The defective enzyme is supposed to convert one of the intermediate molecules (**progesterone**) in the pathway into the next molecule in the pathway, but the enzyme is nonfunctional due to homozygosity for a recessive mutation and the conversion cannot be accomplished. As a consequence, the intermediate molecule accumulates in cells of the adrenal gland, much as half-finished automobiles would accumulate on

an assembly line if one of the workers quit abruptly.

Progesterone and related molecules cannot accumulate forever, of course. They are gradually broken down step by step along another chemical pathway by another series of enzymes into a class of molecules that is chemically similar to testosterone; one representative molecule in this class is called **aldosterone** (see Figure 5.18). These breakdown products have an action in the body similar to that of testosterone, but not so strong. Nevertheless, the testosterone-like effects of the breakdown products cause a degree of **virilization** (masculinization) in affected females.

Females afflicted with the adrenogenital syndrome have a normal uterus and fallopian tubes; the vagina is usually small, and the clitoris may be enlarged, giving it the appearance of a penis. Externally, these people are malelike, although the reproductive organs are small and not as well formed as in normal males. The most serious effects of the syn-

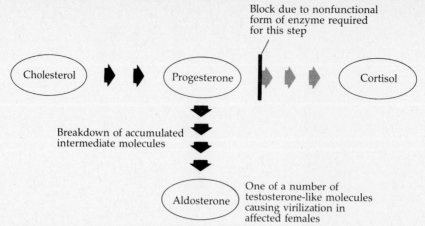

Figure 5.18 In the adrenal glands, cortisol is produced from cholesterol by means of a complex biochemical pathway; the key steps are illustrated here. Individuals with the adrenogenital syndrome have an inherited defect in one of the enzymes in the pathway, leading to a buildup in the amount of one of the intermediates (progesterone). This excess progesterone is broken down by another pathway into a number of testosterone-like molecules, one of which is aldosterone. These breakdown products mimic testosterone in their effects on the body, and they cause virilization in affected females.

drome are not the physical abnormalities, however; of greatest concern is the lack of cortisol, which can lead to death. Treatment of the condition involves continual administration of cortisol to the patient. The presence of cortisol prevents the adrenals from starting more cholesterol molecules along the chemical pathway to cortisol. This, in turn, prevents the accumulation of the intermediate molecules and their breakdown into testosterone-like forms. Finally, the absence of these testosterone-like molecules prevents further masculinization.

SUMMARY

1. The primary sex ratio is the ratio of males to females at fertilization. Although the primary sex ratio in humans is not well known, it is thought to be close to 1:1, which corresponds to the ratio theoretically expected due to segregation of the X and Y chromosomes in males.

2. The **secondary sex ratio** is the ratio of males to females at birth. A great deal is known about the secondary sex ratio because sex is one of the items typically noted in birth records. Although there is variation among populations in the secondary sex ratio, most populations have a slight excess of males.

Among U.S. whites, for example, the secondary sex ratio is about 1055 males for every 1000 females.

3. If it is assumed that successive births are independent of one another and that each newborn has a probability m of being a male, then the probabilities of various sex distributions in sibships of any given size can easily be calculated. Indeed, the probability that a sibship of size n consists of exactly i males and ($n - i$) females is given by

$$\frac{n!}{i!(n - i)!} m^i(1 - m)^{n - i}$$

The observed proportions of various sex distributions fit rather well the theoretical proportions calculated from this formula, which suggests that successive births are indeed independent (i.e., there do not seem to be inherited tendencies for certain kinships to have an excess of males and for others to have an excess of females).

4. Small but real changes in the secondary sex ratio do occur, however. The secondary sex ratio decreases slightly with birth order, decreases slightly with father's age, varies cyclically during the year, and changes somewhat from year to year. The reasons for these small changes in the secondary sex ratio are not known.

5. The Y chromosome in humans and many other organisms carries few genes beyond those responsible for the determination of maleness. One Y-linked human gene codes for the **H-Y antigen**, which is thought to be of great importance in sex determination. The pattern of inheritance of traits due to Y-linked genes is exceedingly simple: The trait occurs only in males and is transmitted only through males—from father to son to grandson, and so on.

6. The X chromosome seems to carry as many genes as an autosome of comparable size. Recessive X-linked genes will be expressed in homozygous females and in the corresponding **hemizygous** males. The pattern of inheritance of X-linked genes follows the inheritance of the X chromosome itself; namely, males receive their X chromosome from their mother and transmit it only to their daughters (**crisscross inheritance**). Pedigrees due to rare, X-linked recessive alleles therefore have certain characteristic features, including (a) primarily or only males are affected, (b) affected males have nonaffected offspring, and (c) affected males have phenotypically normal parents. Traits due to rare X-linked recessives are much more frequent in males than in females.

7. Examples of traits due to X-linked recessives include **G6PD deficiency** (characterized by drug-induced anemia), **color blindness** (two X-linked forms associated with defective red or defective green perception), **hemophilia** (two X-linked forms associated with clotting factor VIII or IX), and **Lesch-Nyhan syndrome** (deficiency of HGPRT; symptoms include gout and bizarre self-aggressive behavior).

8. Recombination between X-linked loci occurs in females just as recombination between autosomal loci occurs in both sexes. Females who are heterozygous for two loci separated by a recombination fraction r are expected to produce a proportion $(1 - r)/2$ of each of the nonrecombinant gametes and a proportion $r/2$ of each of the recombinant gametes. In the example discussed in connection with Figure 5.8(b), $r = 0.03$.

9. Dosage compensation refers to regulation of gene activity of the sex chromosomes to adjust and equalize gene dosage in the sexes. In humans, dosage compensation is accomplished according to the **Lyon hypothesis**, in which one randomly chosen X chromosome in each somatic cell in normal females is rendered genetically inactive. X-chromosome inactivation occurs early in development, and the particular X that is rendered inactive may differ from cell to cell. However, all descendants of a cell that has a particular X chromosome inactivated have that same X inactivated; so an adult female is a sort of *mosaic* for X-linked genes, with some cells having the genes on one X chromosome expressed and the rest of the cells having the genes on the other X chromosome expressed. In individuals with multiple X chromosomes, all X chromosomes but one are inactivated (**single active X**). The inactive X's can be recognized cytologically as **sex-chomatin (Barr) bodies**. The Y chromosome, in contrast to the X, does not seem to be subject to dosage compensation.

10. Human development passes through stages known as **zygote**, **morula**, **blastocyst**, **embryo**, and **fetus**. Although birth usually occurs about 34 weeks after fertilization, the beginnings of all essential internal and external structures are present after 8 weeks. The first 12 weeks of development are marked by great sensitivity to **teratogens** (agents that cause developmental abnormalities). Examples of teratogenic agents are **thalidomide** (a tranquilizer) and **rubella** (German measles virus).

11. Sexual differentiation occurs in the seventh to twelfth week of development. Prior to the seventh week, all embryos possess two sets of sex-related ducts: the female-related **Müllerian ducts** and the male-related **Wolffian ducts**. Under the influence of the male hormone **testosterone**, the Müllerian ducts degenerate and the Wolffian ducts differentiate into the internal male sexual structures. In the absence of testosterone, development is preprogrammed in such a way that the Wolffian ducts degenerate and the Müllerian ducts differentiate into the internal female sexual structures.

12. Abnormalities in sexual development include **pseudohermaphroditism** (presence of both male and female sexual structures in the same individual but neither functional eggs nor functional sperm produced), **testicular-feminization syndrome** (an X-linked gene expressed only in chromosomally XY individuals, causing nonresponsiveness to testosterone and consequent phenotypic femaleness), and the **adrenogenital syndrome** (block in conversion of cholesterol to cortisol, leading to accumulation of intermediate metabolites that are broken down into testosterone-like compounds causing masculinization of affected females).

WORDS TO KNOW

Gene	**Teratogen**	Testosterone
X-linked	Thalidomide	
Y-linked	Rubella	**Trait**
Hemizygous		G6PD deficiency (favism)
	Development	Color blindness
Sex Ratio	Zygote	Hemophilia
Primary sex ratio	Morula	Lesch-Nyhan syndrome
Secondary sex ratio	Blastocyst	(HGPRT deficiency)
	Embryo	Gout
Dosage Compensation	Fetus	Hermaphroditism
Lyon hypothesis	Amnion	Pseudohermaphroditism
Single active X principle	Placenta	Testicular-feminization
Mosaic	Müllerian ducts	syndrome
Sex-chromatin body	Wolffian ducts	Adrenogenital syndrome
Barr body	Primordial germ cells	

PROBLEMS

1. For discussion: It is theoretically feasible (but, despite occasional reports, not yet possible in practice) to separate X-bearing sperm from Y-bearing sperm by physical means, and then to use these separated sperm in artificial insemination to produce offspring of a predetermined sex. Control of the sex ratio in farm animals is a primary objective of such research. In dairy cattle, for example, females are more valuable than males; in beef cattle, by contrast, males are more valuable than females. Would you approve of research aimed at the separation of X-bearing sperm from Y-bearing sperm in humans? If such a procedure were available, can you think of any situations in which its use would be beneficial? Can you think of any ways in which the procedure could be misused? If the procedure were available, would you use it to choose the sex of your own children? Why or why not?

2. Assume for purposes of this problem that the secondary sex ratio is $\frac{1}{2}$. What is the probability that a sibship of five children will consist of all boys? Of four boys and one girl?

3. What sort of trait could be expressed *only* in males and yet not be due to a Y-linked gene?

4. A male receives his Y chromosome from his grandfather. Which grandfather?

5. A male with a single Y chromosome is found to have a Barr body. What is his sex-chromosomal constitution?

6. A female is found to have two sex-chromatin bodies. What is her sex-chromosomal constitution?

7. How many Barr bodies would be found in a phenotypic female who has the testicular-feminization syndrome?

8. A woman's father had hemophilia A. What is the probability that she is heterozygous?

9. A woman's paternal grandfather had hemophilia A. What is the probability that she is a carrier? What would the probability be if her maternal grandfather had been affected?

10. A boy of genotype *G6PDB* has a sister with genotype *G6PDA/G6PDA*. What are the genotypes of the mother and father?

11. A woman whose father had hemophilia A marries a normal man and has a son. What is the probability that the boy is affected?

12. A woman has a brother affected with Lesch-Nyhan syndrome. What is the probability that she is a carrier?

13. A man with the red form of color blindness marries a woman with the green form of color blindness. What phenotype with respect to color vision would be expected among their children?

14. What phenotypic sex would be expected in a chromosomal mosaic who is primarily XX when the embryonic gonads are XY? In a chromosomal mosaic who is primarily XY when the embryonic gonads are XX?

15. A woman of genotype *G6PDA/G6PDB* whose father had the green form of color blindness and *G6PDA* has a son who carries *G6PDA*. What is the probability that the boy is color blind? (Hint: Use the data in Figure 5.8, and remember that for genes as closely linked as *gcb* and *G6PD*, the distance in map units is the same as the recombination fraction.)

FURTHER READING AND REFERENCES

Bardin, C. W., and J. F. Catterall. 1981. Testosterone: A major determinant of extragenital sexual dimorphism. Science 211:1285–1294. The influence of testosterone on nonsexual structures.

Behnke, J. A., C. E. Finch, and G. B. Moment (eds.). 1978. The Biology of Aging. Plenum, New York. A multiauthored but readable introduction to the science of aging.

Browder, L. W. 1980. Developmental Biology. Saunders, Philadelphia. A comprehensive text dealing with a complex subject.

Edgell, C.-J. S., H. N. Kirkman, E. Clemons, P. D. Buchanan, and C. H. Miller. 1978. Prenatal diagnosis by linkage: Hemophilia A and polymorphic glucose 6-phosphate dehydrogenase. Am. J. Hum. Genet. 30: 80–84. Source of counseling case in Figure 5.8.

Edwards, A. W. F., and M. Fraccaro. 1958. The sex distribution in the offspring of 5,477 Swedish ministers of religion, 1585–1920. Hereditas 44:447–450. Finds no evidence that sex ratios "run in families."

Fuchs, F. 1980. Genetic amniocentesis. 1980. Scientific American 242:47–53. Details the procedure of aminocentesis and its uses.

Gordon, Jon W., and F. H. Ruddle. 1981. Mammalian gonadal determination and gametogenesis. Science 211:1265–1271. On the role of the X and

Y chromosomes in mammalian sexual development.

Haseltine, F. P., and S. Ohno. 1981. Mechanisms of gonadal differentiation. Science 211: 1272–1278. Excellent discussion of testosterone, testicular feminization, and the H-Y antigen.

Kurnit, D. M., and H. Hoehn. 1979. Prenatal diagnosis of human genome variation. Ann. Rev. Genet. 13:235–258. An assessment of the use of studies at the DNA level in prenatal diagnosis.

MacLusky, N. J., and F. Naftolin. 1981. Sexual differentiation of the central nervous system. Science 211:1294–1303. Gonadal hormones also influence sex-specific developmental processes in the brain.

Martin, G. R. 1982. X-chromosome inactivation in mammals. Cell 29: 721–724. What is the molecular basis of inactivation?

Ohno, S. 1979. Major Sex-Determining Genes. Springer-Verlag, New York. A short but advanced discussion of this important subject.

Oppenheimer, S. B. 1980. Introduction to Embryonic Development. Allyn & Bacon, Boston. An introduction to general aspects of the mysteries of differentiation.

Silvers, W. K., D. L. Gasser, and E. M. Eicher. 1982. H-Y antigen, serologically detectable male antigen and sex determination Cell 28:439–440. Evidence that H-Y is not the primary cause of gonad differentiation.

Wachtel, S. A. 1980. Where is the H-Y structural gene? Cell 22:3–4. The H-Y antigen is surely regulated by the Y, but is it coded on the Y?

Wilson, J. D., F. W. George and J. E. Griffin. 1981. The hormonal control of sexual development. Science 211:1278–1284. Which hormones are necessary for normal sexual development, and what do they do?

chapter 6
Abnormalities in Chromosome Number

In this chapter and the next we discuss genetic conditions due to an excess or deficiency of whole chromosomes or parts of chromosomes instead of mutations in particular genes. The importance of gene dosage in cells was emphasized in Chapter 5. Human cells have become adapted to having a proper and harmonious balance in the action of genes when each gene on the autosomes is present twice, which is related to the fact that autosomes occur in homologous pairs. The X and Y chromosomes are special cases, of course, but the dosage of X-linked genes is equalized by dosage compensation according to the Lyon hypothesis, and apparently the Y chromosome carries so few genes that dosage compensation is unnecessary.

Because of the importance of proper gene dosage, too many or too few copies of genes can upset the normal processes of development, even though the added or absent genes themselves are not necessarily mutated. These abnormalities result from gross imbalances in the number and action of genes, and a general rule is that the greater the imbalance, the more severe is the abnormality. Some imbalances are sufficiently small to have almost no effects on development. Other, larger ones are **lethal**; they lead to death of the embryo or child at an early age.

Polyploidy and Polysomy

The balance of genes can be disrupted in many different ways, and in this chapter we will consider two distinct types of abnormality in chromosome number. Recall from Chapter 1 that human gametes are **haploid**; they carry one complete set of chromosomes consisting of 22 autosomes and 1 sex chromosome. Somatic cells, by contrast, are **diploid** because they carry two haploid sets of chromosomes—46 altogether. Cells that have three haploid sets of chromosomes (totaling 69) are known as **triploids**, and those with four haploid sets (totaling 92) are called **tetraploids**. More generally, any cell that has more than two haploid sets of chromosomes is said to be **polyploid**. Polyploids involving six, eight, or even more sets of chromosomes are important in a few animals and many plants; indeed, in many flowering plants, normal individuals are polyploid. Among humans, however, the only polyploids of significance are triploids and tetraploids.

The second type of abnormality in chromosome number involves individual chromosomes instead of entire sets of chromosomes. A normal somatic cell is diploid and has 46 chromosomes. A cell with a missing chromosome would have 45, and such cells are said to be **monosomic** for the chromosome represented once instead of twice. For example, a female with only one X chromosome instead of two (i.e., an XO female) is monosomic for the X. Similarly, cells with three copies of a chromosome instead of two are said to be **trisomic** for the chromosome in question. For example, an individual with three copies of chromosome 21 is trisomic for chromosome 21; alternatively, we could say that the individual has **trisomy 21**. More generally, a **polysomic** individual has more than two copies of a particular chromosome. Among the human autosomes, the only significant level of polysomy is trisomy. For the sex chromosomes, higher levels of polysomy, such as **tetrasomy** (four copies of one of the sex chromosomes), are sometimes found.

How important are chromosomal abnormalities? Judging from their frequency alone, they are extremely important. Most people are astonished to learn that about 8 percent (1 out of 12) of diagnosed human pregnancies involve an embryo that has some major chromosomal abnormality. Most of these abnormalities lead to grossly deformed embryos that mercifully undergo **spontaneous abortion**, usually quite early in development. One might say of these embryos as the Apocrypha says of the souls of the righteous—that in the sight of the unwise they seem to die and their departure is taken for misery, but they are at peace and no torment shall touch them. Yet many such embryos do survive, the ones with less severe abnormalities. Among live-born children, about 1 in 155 (0.6 percent) have some major chromosomal abnormality. This number includes individuals with sex-chromosomal abnormalities, which are among the least severe chromosome abnormalities (these will be discussed in a few pages), but it does not include minor chromosomal variants of the sort discussed in Chapter 2 or chromosomal abnormalities that are too subtle to be detected through the light microscope.

Primary Nondisjunction

Any number of accidents can happen during meiosis that can cause an egg or a sperm to have two copies of a chromosome instead of only one; the same sort of errors can also lead to gametes that completely lack one particular chromosome. If a gamete carrying two copies of a chromosome participates in fertilization, the resulting zygote will be trisomic for that chromosome. Successful gametes that lack a particular chromosome will give rise to monosomic zygotes. Such abnormal gametes, which lead to monosomic or trisomic zygotes, result from failure of one of the pairs of chromosomes to be distributed normally to opposite poles during meiosis. The chromosomes that fail to separate (disjoin) normally are said to have undergone **primary nondisjunction** (or just **nondisjunction**).

Nondisjunction that occurs during meiosis is called **meiotic nondisjunction**. Two principal types of meiotic nondisjunction are illustrated in Figure 6.1. Figure 6.1(*a*) refers to **first-division nondisjunction**, in which chromosomes fail to separate properly at anaphase I. Sometimes, as illustrated in the figure, homologous chromosomes may synapse and undergo crossing-over, but then they both proceed to the same pole at anaphase I. (This situation is sometimes referred to as **anaphase lag** because as the other chromosomes separate, the one pair "lags" behind.) At other times in first-division nondisjunction, the chromosomes may not synapse properly or they may fall apart prematurely, and both chromosomes may then proceed to the same pole at anaphase I. Whatever the cause, if nondisjunction at the first meiotic division is followed by a normal second meiotic division, the resulting four gametes consist of two that have an extra chromosome and two that have a missing chromosome [Figure 6.1(*a*)].

Another type of meiotic nondisjunction, called **second-division nondisjunction**, is illustrated in Figure 6.1(*b*). In this case, the first meiotic division is normal, but the sister chromatids of one chromosome fail to disjoin at anaphase II. The overall result of second-division nondisjunction is one gamete with an extra chromosome, one with a missing chromosome, and two gametes that are normal.

Abnormal gametes that result from meiotic nondisjunction give rise to monosomic or trisomic zygotes. Nondisjunction can occur in mitosis, too, and in such a case it is known as **mitotic nondisjunction**. Mitotic nondisjunction results from a failure in chromosome separation at anaphase, often due to anaphase lag.

As illustrated in Figure 6.2, for the simplest case of mitotic nondisjunction, one daughter cell will be trisomic and the other will be monosomic. If mitotic nondisjunction occurs in a cell of a developing embryo, the descendants of the monosomic or trisomic daughter cells will give rise to monosomic or trisomic clones in an otherwise normal individual. Thus, mitotic nondisjunction will lead to a **chromosomal mosiac**—an individual with two or more chromosomally distinct types of cells.

Sex-Chromosomal Abnormalities

In human genetics, it is a general rule that *chromosome abnormalities involving the autosomes have more severe phenotypic effects than those involving the sex chromosomes*. This generalization has a twofold reason. First, abnormal gene dosage of the Y chromosome has a milder effect than that of an autosome because, as emphasized in Chapter 5, the Y chromosome seems to carry few genes beyond those responsible for the development of maleness. Secondly, an abnormal number of X chromosomes is rendered less harmful than an abnormal number of autosomes because of dosage compensation involving the single active X principle (see Chapter 5). Indeed, trisomy of the sex chromosomes is in certain instances so mild phenotypically that affected individuals can lead normal or nearly normal lives. On the other hand, X-chromosomal monosomy is a severe disorder, and Y chromosomal monosomy is invariably **lethal** (incompatible with life) very early in embryonic development. This contrast between the milder effects of trisomy as compared with monosomy leads to another generalization that will be discussed further in the next chapter: Generally speaking, *an increased dose of one or more genes had less severe phenotypic effects than a decreased dose of the same genes*.

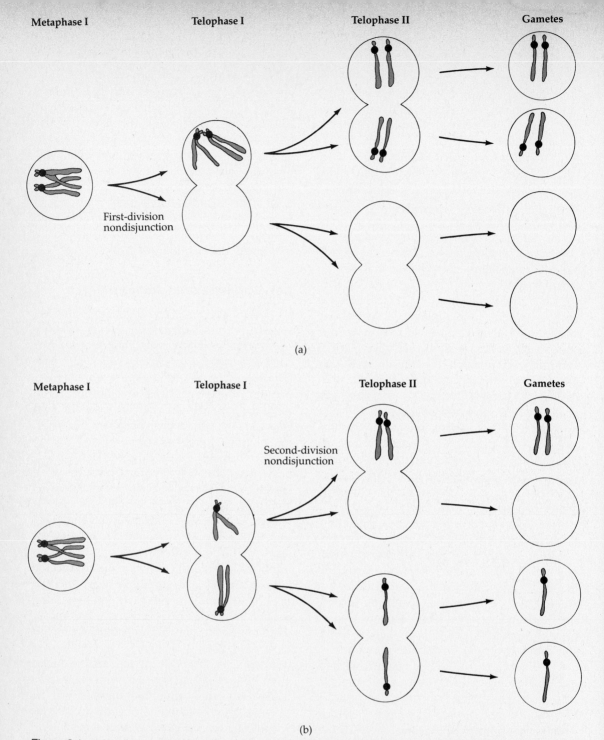

Figure 6.1 Two types of meiotic nondisjunction. *(a)* First-division nondisjunction, in which homologous chromosomes fail to disjoin at anaphase I, leads ultimately to two gametes that have an extra copy of the chromosome in question and two gametes that lack the chromosome. *(b)* Second-division nondisjunction, in which sister chromatids fail to disjoin at anaphase II, leads to one gamete with an extra copy of the chromosome in question, one gamete lacking the chromosome, and two gametes that are normal. Although anaphase is not illustrated here, nondisjunction often results from a chromosome lagging behind at anaphase.

162

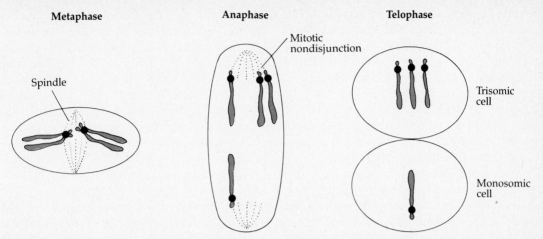

Metaphase   **Anaphase**   **Telophase**

Spindle

Mitotic
nondisjunction

Trisomic
cell

Monosomic
cell

Figure 6.2   Mitotic nondisjunction occurs when sister chromatids of one member of a pair of homologous chromosomes fail to disjoin during anaphase, often due do anaphase lag. Of the resulting daughter cells, one will be trisomic for the chromosome in question and the other will be monosomic. Subsequent mitoses of the daughter cells result in a trisomic/monosomic chromosomal mosaic.

Trisomy X

The most frequent sex-chromosomal abnormalities found among live-born children are listed in Table 6.1. Among the most common is **trisomy X**. Women with this syndrome have 47 chromosomes, including three X chromosomes (XXX). (Technically speaking, the conventions of human karyotype nomenclature require that the total number of chromosomes be listed first, the sex-chromosome constitution next (if relevant), and other chromosome abnormalities following. Thus, trisomy X females would be denoted 47,XXX, but the abbreviation XXX is a useful shorthand for our purposes.) XXX females are physically quite normal and usually able to bear children, but some have menstrual irregularities and an early onset of

TABLE 6.1   MOST FREQUENT SEX-CHROMOSOME ABNORMALITIES IN HUMANS

Sex-chromosome abnormality	Name of syndrome	Approximate frequency	Major symptoms
XXX (47,XXX)	Trisomy X "syndrome"	1 in 950 newborn females	Female; physically quite normal, but tendency toward mental retardation; fertile
XYY (47,XYY)	Double-Y "syndrome"	1 in 950 newborn males	Male; tend to be tall; behavioral effects highly controversial; fertile
XXY (47,XXY)	Klinefelter syndrome	1 in 1000 newborn males	Male; tend to be tall; sexual maturation absent, although some breast enlargement may occur; frequently associated with mental retardation; sterile
XO (45,X)	Turner syndrome	1 in 5000 newborn females	Female; short of stature, sometimes with "webbing" of skin between neck and shoulders; sexual maturation absent; mental abilities quite normal; sterile

menopause. The most striking feature of the syndrome is mental retardation, which is sometimes severe. Although the incidence of the XXX condition is about 1 per 950 newborn females, it is three times greater among the institutionalized mentally retarded. Mental ability of all kinds varies from individual to individual, of course, and some XXX women are both mentally and physically normal. These women lead completely normal lives. They are born, grow to adulthood, bear children, grow old, and die, giving no hint of their extra X chromosome. Nevertheless, on the average, XXX individuals have a high risk of mental retardation, indicating that the condition is not entirely harmless. Mental retardation is typical of individuals with extra X chromosomes: Three or more X's lead to mental retardation, and the more X's, the more severe is the disability. (Extra Y chromosomes also tend to be associated with mental defects, but the effect of the Y is rather small compared with that of the X.)

XYY and Criminality

Another sex-chromosomal abnormality associated with few physical manifestations is the XYY condition—technically 47,XYY. Affected males have 47 chromosomes; they have the full chromosome complement of normal males plus a second Y chromosome. XYY males tend to be tall, averaging over 6 ft. Beyond this, distinctive physical symptoms are practically nonexistent, so it is questionable whether the term *syndrome* really applies to this case because there is no group of characteristic symptoms. Among newborn males, the incidence of the XYY condition is thought to be approximately 1 in 950.

The behavior of XYY males has been a matter of concern and controversy since 1965, when a Scottish study uncovered an incredibly high frequency of the XYY condition among imprisoned men convicted of violent crimes. More than 50 similar studies have now been carried out in prisons in Europe and the United States, and it has held true that the frequency of XYY men confined for crimes of violence is about 10 times greater than it is among a sample of newborn males. Many of these studies have concentrated on tall men; because XYY males tend to be tall, this will automatically create a bias, and the findings must therefore be interpreted with caution. Nevertheless, the higher than expected frequency of XYY males among tall criminals has suggested to some that XYY males may undergo impaired psychosocial development, perhaps caused by their height itself, which may produce social stresses tending to evoke violent, aggressive behavior, or perhaps caused by some subtle and invisible behavioral effect of the extra Y.

One Danish study of criminality among XYY males is particularly important and revealing because it is free of many of the biases in other studies. The Danish research involved an examination of all available records on almost all men 28 to 32 years old whose height was 184 cm (72.4 in) or more and whose birthplace was Copenhagen. A total of 4139 sex-chromosomal determinations were made; among these there were 12 XYY (incidence 1 per 345), 16 XXY (incidence 1 per 259—the XXY sex-chromosomal constitution will be discussed in a few pages), and 13 individuals with other types of chromosomal abnormality. The information examined included criminal records and types of crime, socioeconomic status of the men's parents, score on a Danish army intelligence test, and an index of educational level. Parental socioeconomic status was lower among men with criminal convictions than among those without convictions, but no differences were found among XY, XYY, and XXY males. Thus, lower socioeconomic status is associated with criminality but not with sex-chromosomal constitution.

TABLE 6.2 HEIGHT, CRIMINALITY, AND INTELLECTUAL FUNCTION AMONG XY, XYY, AND XXY DANISH MALES

	Number of males	Height, cm	Army intelligence test score	Educational index
Average of all XY males	4096	187.1	43.7	1.6
XY males with no criminal record	3715	187.1	44.5	1.6
XY males with criminal record	381 (9.3%)	186.7	35.5	0.7
Average of all XYY males	12	190.8	29.7	0.6
XYY males with no criminal record	7	191.3	31.6	0.7
XYY males with criminal record	5 (41.7%)	190.2	27.0	0.5
Average of all XXY males	16	189.8	28.4	0.8

Source: Data from H. A. Witkin et al., 1976, Science 193:547–555.

Other Danish data pertaining to chromosomes and criminality are summarized in Table 6.2. Among XYY males, the rate of criminal convictions was 41.7 percent (5 out of 12), which is significantly greater than the 9.3 percent among XY males. (The term **significant** as applied to statistical tests means that an observed difference is greater than could plausibly be attributed to chance.) Among XXY males, the rate of criminal convictions (3 of 16) was somewhat greater than among XY males, but the difference is sufficiently small that it could be due to chance. The findings relative to height are in the third column of Table 6.2. On the average, XYY males are taller than tall XY males by about 3.7 cm (1.5 in), and XXY males are taller than tall XY males by about 2.7 cm (1.1 in). Tallness alone is not related to criminality, however, because men with criminal records are somewhat shorter than their noncriminal counterparts.

One of the most important findings of the Danish study is summarized in the last two columns of Table 6.2. XYY and XXY males have a significantly lower score on the Army intelligence test and a significantly lower index of educational level than do XY males. For each sex-chromosomal class, moreover, men with criminal records are substantially poorer in both measures of intellectual function than

men without criminal records. Of course, this does not necessarily imply that men with lower levels of intellectual function actually commit more crimes; they may merely be more likely to get caught. In any case, the Danish data seem to imply that the high rate of criminality among XYY males is due to their moderately impaired mental function.

Some comments are in order about the nature of the crimes committed by the XYY males. Two of the XYY males were habitual criminals. One had more than 50 convictions for petty larceny, auto theft, burglary, embezzlement, and so on, but no record of violent or aggressive behavior. The other habitual criminal, who was also mentally retarded and had spent most of his life in institutions for the mentally retarded, had convictions for arson, theft, burglary, and embezzlement, but, once again, no convictions for aggressive crimes against people. The other three XYY males were petty criminals. Two had each been convicted twice for petty theft. The other had turned in a false report about a traffic accident and had started a small fire. Overall, *the crimes of the Danish XYY males were not crimes of violence and aggression; they were mainly crimes involving property.*

The Danish study illustrates the danger of preliminary scientific reports being sensationalized by the popular press. At about the time of the original Scottish discovery of XYY males among exceptionally violent criminals, a pathological killer sneaked into a dormitory in Chicago and brutally murdered several student nurses. A newspaper claimed that the man had been "born to kill' because he was XYY. This turned out to be untrue—the Chicago killer was, in fact, XY—but the false report was widely disseminated and believed. Several proposals were made for the mass screening of newborns for the detection of XYY males to provide them with special education and intensive psychological guidance to counteract their supposed "killer instinct." It is good that the screening proposals were never carried out because, in the atmosphere of the time, they could have been harmful by creating a self-fulfilling prophecy. That is to say, if physicians and family are convinced that XYY males (or any other category of people) have severe, aggressive behavioral problems, then they may interact with the individuals in such a manner as to elicit the very behavior they expect, thereby fulfilling their own prophecy. In light of the Danish study, the rate of criminality is related more to level of intellectual function than to sex-chromosomal constitution.

Secondary Nondisjunction

In theory, individuals who are trisomic and fertile would be expected to produce 50 percent trisomic offspring because of the mechanics of the meiotic divisions. The meiotic situation in a hypothetical individual trisomic for an acrocentric chromosome is illustrated in Figure 6.3. Synapsis and crossing-over between two of the chromosomes may produce a normal bivalent at metaphase I with the extra chromosome remaining unpaired. At anaphase I the chromosomes of the normal bivalent will disjoin, and the extra chromosome will be pulled to one or the other spindle pole. If disjunction of the sister chromatids occurs normally at anaphase II, two of the resulting gametes will be normal, but the other two will carry an extra chromosome and lead to trisomic offspring. The production of gametes with an extra chromosome by trisomic individuals is known as **secondary nondisjunction**. (Secondary nondisjunction should not be confused with second-division primary nondisjunction. Secondary nondisjunction is a frequent event involving chromosomes that are trisomic to begin with; second-division primary nondisjunction, illustrated in Figure 6.1(*b*), is a rare event that occurs in chromosomally normal individuals.)

XXX and XYY individuals are fertile, and each has three sex chromosomes. Because of secondary nondisjunction, a substantial fraction of their offspring would be expected to have sex-chromosomal abnormalities. Remarkably, their offspring are chromosomally normal. Chromosome studies of cells undergoing meiosis in XYY males have provided a partial understanding of how the sex chromosomes seem to escape the theoretical consequences of secondary nondisjunction. Cells undergoing meiosis in XYY males are chromosomally XY; these cells seem to have lost or eliminated the extra Y chromosome prior to meiosis, and the sperm that result from such cells are chromosomally normal. A similar sort of chromosome elimination may occur in meiotic cells of XXX females, or perhaps XX nuclei formed as a result of secondary nondisjunction always end up in polar bodies. In any event, XYY and XXX individuals produce mostly normal gametes, and the risk of sex-chromosomal abnormalities among their offspring is correspondingly small. Consequently, most children with sex-chromosome abnormalities are born to chromosomally normal parents as a result of primary nondisjunction. Indeed, the bright fluorescence of the Y chromosome in sperm treated with quinacrine has revealed that up to 1 percent of the sperm

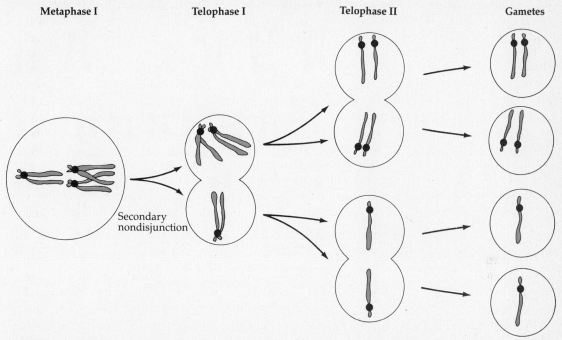

Figure 6.3   Secondary nondisjunction refers to the production of abnormal gametes by trisomic individuals. As shown here, meiosis in a trisomic individual produces two gametes that carry an extra copy of the chromosome in question and two gametes that are normal.

from normal XY males may carry two Y chromosomes. The frequency of XYY among newborn males (1 per 950) is therefore considerably less than would be expected, although the reasons for the discrepancy are not yet understood.

XXY: Klinefelter Syndrome

Although the XXX and XYY conditions have few and relatively mild physical manifestations, some sex-chromosomal abnormalities do have distinctive physical effects, chiefly affecting sexual development. Most frequent among these is the XXY condition (technically 47, XXY), which occurs in about 1 per 1000 newborn males. Approximately two-thirds of these arise from primary nondisjunction in the mother, leading to an egg carrying two X chromosomes, and there is a slightly enhanced risk of the condition among sons of older mothers. Males with the XXY chromosome constitution have a distinctive group of symptoms known as **Klinefelter syndrome** (Figure 6.4). Characteristic of the syndrome are very small testes but normal penis and scrotum. Although about half the patients have some degree of mental retardation, growth and physical development are, for the most part, quite normal, although affected males tend to be tall for their age (see Table 6.2). Puberty does not occur normally: The testes fail to enlarge, the voice remains rather high-pitched, and pubic and facial hair remains sparse. In about half the cases there is some enlargement of the breasts. Males with Klinefelter syndrome do not produce sperm and hence are sterile; up to 10 percent of men who seek aid at infertility clinics turn out to be

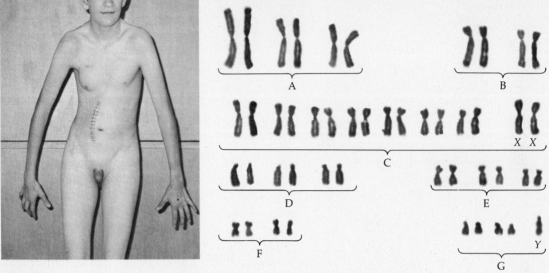

Figure 6.4   An individual with Klinefelter syndrome and the associated XXY karyotype.

XXY. As noted in connection with Table 6.2, XXY males in the Danish study seem to have moderately impaired intellectual function. However, some men with Klinefelter syndrome, though sterile, lead completely normal and even distinguished lives.

XO: Turner Syndrome

Another abnormality of the sex chromosomes, more rare than the XXY condition, is the XO condition. (The "O" used in the shorthand designation XO refers to the absence of the homologue of the X.) Individuals with this sex-chromosome constitution have 45 chromosomes altogether: 22 pairs of autosomes and a monosomic X. Thus, the XO chromosome constitution is technically designated 45,X. The frequency of XO among live-born females is about 1 per 5000, but this far under-represents the occurrence of the condition. As will be discussed later in this chapter, about 15 to 20 percent of spontaneously aborted fetuses that have a detectable chromosome abnormality are XO, making this one of the most common chromosomal abnormalities in spontaneous abortions. More than 1 percent of all recognized pregnancies in humans involves an embryo that is chromosomally XO, and approximately three-fourths of these are caused by abnormal sperm that lack a sex chromosome.

The small fraction of XO fetuses that survive develop into individuals who have **Turner syndrome** (Figure 6.5). Affected females are of short stature and often have a distinctive webbing of the skin between the neck and shoulders. There is no sexual maturation. The external sex organs remain childlike, the breasts fail to develop, the pubic hair does not grow. Internally, the ovaries are infantile or absent and menstruation does not occur. Although the IQ of affected females is very nearly normal, they have specific defects in spatial abilities and arithmetical skills.

Autosomal Abnormalities

Most autosomal trisomies—and all autosomal monosomies—are so severe that they are in-

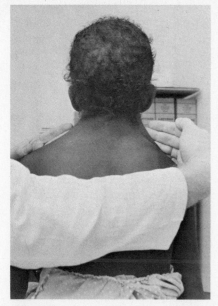

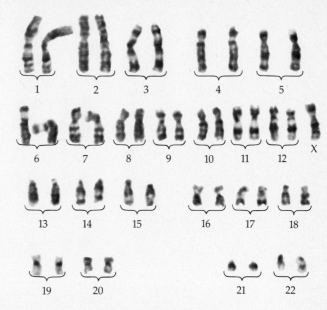

Figure 6.5   An individual with Turner syndrome and the associated XO (also symbolized 45,X) karyotype.

compatible with life. The result of the chromosomal abnormality may be failure of blastocyst implantation or spontaneous abortion of an often grossly malformed embryo or fetus. Nevertheless, three autosomal trisomies are sometimes compatible with embryonic development and birth, although the trisomic children may be severely abnormal. The incidence of the three conditions among live-born children is summarized in Table 6.3. The most common autosomal trisomy among live-born

children is **trisomy 21**, also known as **Down syndrome** after the physician who first described the condition over a century ago.

Trisomy 21: Down Syndrome

The cause of Down syndrome was for many years a complete mystery, but in 1959 it was discovered that affected children have 47 chromosomes (Figure 6.6), including three copies of chromosome 21, the smallest of the autosomes. (In the early years after the discovery that patients with Down syndrome were trisomics, it was mistakenly thought that the chromosome involved was chromosome 22; one may still find reference to "trisomy 22" in connection with Down syndrome in older books and journals. However, it is now agreed that the trisomic chromosome in all children with Down syndrome is indeed chromosome 21). As noted in Table 6.3, the correct technical designation of the Down karyotype is 47,

TABLE 6.3   MOST FREQUENT AUTOSOMAL TRISOMIES IN LIVE-BORN HUMANS

Trisomy	Name of syndrome	Approximate frequency
13 (47,+13)	Patau syndrome	1 per 5000 live births
18 (47,+18)	Edwards syndrome	1 per 6500 live births
21 (47,+21)	Down syndrome	1 per 750 live births

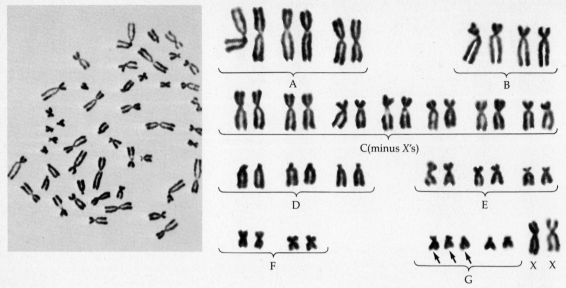

Figure 6.6   Metaphase spread and karyotype of a girl (XX) with Down syndrome showing the trisomy of chromosome 21 (arrows).

+21—the "+" preceding the "21" denoting an entire extra chromosome 21.

Children with Down syndrome are invariably mentally retarded. They are of short stature due to delayed maturation of the skeletal system, and their muscle tone is poor, leading to a characteristic facial appearance in older children and adults (Figure 6.7). Approximately 40 percent of children with Down syndrome have major heart defects. Many children with the syndrome also have epicanthus—a small fold of skin across the inner part of the eye where the tear glands are; this gives the eyes a slightly Oriental look, and the common name for the disease—**mongolism**—derives from it. (Use of the term *mongolism* for Down syndrome is now discouraged because of the unnecessary and misleading racial connotations.) Altogether there are 10 or so major symptoms of the syndrome, but the condition is somewhat variable.

Down syndrome is very frequent in all racial groups (about 1 in 750 live births). The condition has probably been present in human populations since the ancient past, long before recorded history, perhaps even before our species, *Homo sapiens*, had evolved. There is, for example, a condition in chimpanzees that is very similar in symptoms to Down syndrome. This is also caused by trisomy, and the banding pattern of the chromosome responsible is so similar to that of human chromosome 21 that the conclusion is inescapable that both chromosomes must have descended from some common ancestral chromosome in a primate population in the dim past. The incidence of Down syndrome in present-day human population is, as noted, about 1 in 750 live-born children. Down syndrome does not usually run in families and therefore most families that have had a child with Down syndrome have only a small risk of having a subsequent child with the same condition. About 97 percent of families with a trisomic-21 child are in this

Figure 6.7   All the children in these photographs have Down syndrome. As is particularly evident in *(a)* and *(c)*, the facial features characteristic of Down syndrome are not always apparent in younger children.

category. (However, as will be noted later, the risk of having a child with Down syndrome increases with the age of the mother.) The remaining 3 percent of families with an affected child have a much higher recurrence risk. In these cases one of the parents carries a chro-mosomal rearrangement that leads to a high risk of trisomy 21 in the children even though the parent is phenotypically normal. Such rearrangements will be discussed in Chapter 7. For now, concentrate on the 97 percent; in these families the trisomic-21 children are the result of primary nondisjunction leading to a gamete with two copies of chromosome 21.

Actually, the abnormal gamete in the case of Down syndrome is almost always the egg. Moreover, the risk of a pregnancy leading to a child with Down syndrome increases dramatically with the age of the mother. This effect of maternal age is shown graphically in Figure 6.8. Overall, the risk of having a child with Down syndrome is about 1 in 750, but averaging obscures the maternal-age dependence. The risk is only about 1 per 1700 in mothers 15 to 19 years old. The risk increases with age, gradually at first, so that the risk of occurrence in mothers aged 20 to 30 is about 1 per 1400; in mothers aged 30 to 35, however, it is 1 per 750. Then it shoots up (see Figure 6.8). In mothers aged 45 and over, the risk is as high as 1 in 16—about 6 percent! As a result, women over 40 have approximately 40 percent

of the children with Down syndrome, although they have only 4 percent of all children. The **recurrence risk** of Down syndrome (i.e., the risk of a subsequent child having Down syndrome when one child is already affected) also increases with the mother's age; generally speaking, the risk of a woman having a second child with Down syndrome is about three times the risk of a woman of the same age who has not previously had an affected child.

The reason for the greatly increased risk of trisomy 21 in the children of older mothers is not known. Suspicion centers directly on the meiotic process in females. The cells undergoing meiosis in the ovaries of a female embryo are arrested in late prophase I, and they stay arrested until just before the egg is ready to be ovulated. Thus a cell undergoing meiosis in a woman 45 years old has actually resumed the process after a hiatus of 45 years. This by itself may greatly increase the chance of nondisjunction or other errors. Perhaps, too, hormonal changes in older women may increase the risk of meiotic accidents. In any case, the increased risk with age is confined to the mother. The risk does not increase with the age of the father. It may seem strange that this can be stated so definitely. Older women are almost always married to older men, and you might think that the age effects would be so intertwined that you would be unable to decide what to attribute to whom. However, reliable statistical procedures have been devised to separate the age effects of mothers and fathers, and the result of the elaborate analysis is that the increased risk in older couples is due almost exclusively to the age of the mother.

Children with Down syndrome, as you might guess, have a shorter life expectancy than normal. Many have heart defects among their other multiple physical abnormalities, and they are particularly susceptible to chest infections and bronchial pneumonia. The critical year in children with trisomy 21, as in normal children, is the first year of life. The

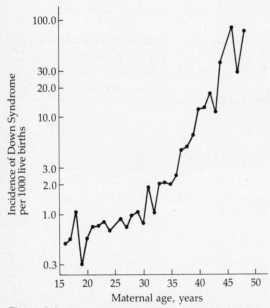

Figure 6.8 Rate of Down syndrome per 1000 live births related to maternal age. The graph is based on 330,859 live births (438 Down-syndrome births) occurring in Sweden in 1968 to 1970. The overall incidence of Down syndrome in this study was 1 per 755 live births.

average lifespan of children born with Down syndrome in 1932 was nine years. But the life expectancy of affected individuals has been steadily increasing as antibiotics and mass immunizations have decreased the risk and severity of infections. By 1947 the life expectancy was 12 years; by 1963 it was 16. Today, affected children between ages 5 and 7 have a life expectancy of up to 27 years; some survive to age 40 or 50.

Because of the increased life expectancy of individuals with Down syndrome, many of them enter the reproductive period. Males with Down syndrome are sexually incompetent and do not reproduce. Affected females are capable of reproduction, however, and approximately 50 women with Down syndrome have given birth; usually the pregnancy has resulted from rape or incest. Because of secondary nondisjunction (see Figure 6.3), one expects approximately half the children to have Down syndrome. In fact, only about 20 percent of the children are affected; the rest are completely normal. The frequency of affected children is lower than expected because fetuses with trisomy 21 frequently undergo spontaneous abortion, which will be discussed later in this chapter.

The increased life expectancy of individuals with Down syndrome has had another consequence. Today more people with trisomy 21 are living than at any time in the past, and this creates a social cost of tremendous proportions because affected individuals are often utterly dependent on others for their welfare. If an affected child is institutionalized, the social cost may be paid through taxes. But virtually everyone acknowledges that many public institutions are understaffed by undertrained and underpaid personnel. Therefore, many physicians urge the parents to raise an affected child at home and, indeed, there is a national organization of parents of Down syndrome children who do raise their children at home. The substantial medical costs must then be borne by the family. Unfortunately, there is also a psychological cost. Tragically, many people unwisely look upon the birth of an abnormal child as a stigma, a mark of Cain, and they inevitably feel remorse and guilt. The parents themselves must resist these emotions. We should long since have stopped believing with the Puritans that the birth of an abnormal child is retribution for past and hidden sins.

Concern for the welfare of society and the parents must not replace compassion for the child. An affected child can benefit from special and intensive care and training, but usually this is not delivered. Sometimes the child does not even receive the attention and stimulation required by every human being to remain alert and interested, regardless of ability. Public institutions are often too overburdened to provide the care, and the parents are not themselves trained to be able to provide it. Moreover, the retarded child is usually not capable of frolicking with other, normal children and learning from them. Indeed, small children can be especially cruel and heartless when they ridicule the physical abnormalities or mental deficiencies of others. All this spells out a very sad tale, not only for children with Down syndrome but also for countless others with various kinds of physical and mental abnormalities.

In grappling with the social issues raised by children with Down syndrome, it is all too easy to forget that the children are human beings as much in need of love, security, and affection as others. They have interests and personalities of their own, which have been poignantly described by Down-syndrome specialists Smith and Wilson:[1]

Children with Down syndrome usually take great pleasure in their surroundings, their families, their toys, their playmates. Happiness comes easily, and

[1]From D. W. Smith and A. A. Wilson, 1973, The Child with Down's Syndrome (Mongolism), Saunders, Philadelphia.

throughout life they usually maintain a childlike good humor. They are not burdened with the grown-up cares that come to most people with adolescence and adulthood. . . . Life is simpler and less complex. The emotions that others feel seem to be less intense for them. They are sometimes sad, happy, angry, or irritable, like everyone else, but their moods are generally not so profound and they blow away more quickly. . . . A child with Down syndrome, though slow, is still very responsive to his environment, to those around him, and to the affection and encouragement he receives from others.

Trisomy 18: Edwards Syndrome

Down syndrome is not the only autosomal trisomy compatible with the live birth of affected children, but it is by far the most frequent. Two other autosomal trisomies also occur in live-born children: trisomy 18 (i.e., 47, +18) and trisomy 13 (i.e., 47, +13). (see Table 6.3.) **Trisomy 18**, also known as **Edwards syndrome**, is about eight times less frequent than Down syndrome; it affects about 1 in 6500 live-born children. As with trisomy 21, the incidence of trisomy 18 increases with the mother's age, but the increase is relatively small, far less than in Down syndrome. Children with Edwards syndrome have multiple congenital abnormalities, including severe mental and physical retardation (Figure 6.9). They have an elongated skull with low-set, malformed, sometimes pointed ears; their jaw and oral cavity are small. They carry their fingers in an abnormal position, with the second finger overlapping the third. Virtually all these children are born with heart defects. Approximately 65 percent of affected newborns are female, presumably because males with the syndrome are more likely to undergo spontaneous abortion. The life expectancy of affected children also indicates a greater severity in males: The life expectancy of males is about three months; in affected females it is about nine months.

Trisomy 13: Patau Syndrome

Trisomy 13, also known as **Patau syndrome**, is equally as severe as trisomy 18. It occurs in about 1 in 5000 live births. As in other autosomal trisomies, affected children are severely retarded, both mentally and physically (Figure 6.10). Children with trisomy 13 have a small skull and eyes; the ears are often malformed and deafness is common. Most have harelip and cleft palate. Many have malformed thumbs and extra digits. Nearly 70 percent have heart defects. Patau syndrome also has a slightly increased incidence with maternal age, but again the increase is far less than the age effect in Down syndrome. The sex ratio of affected children is about 1:1. Rarely do affected children survive more than three or four months after birth.

Most cases of trisomy 18 and trisomy 13 are sporadic. They are unpredictable and do not run in families, but there are a few high-risk families. Thus the situation insofar as occurrence and recurrence are concerned appears to be much like that in Down syndrome. The abnormalities caused by the extra chromosome in all three trisomies are major and multiple, physical and mental. Each trisomy is unique, though somewhat variable. Because certain abnormalities occur regularly in trisomy 13 but only rarely in trisomy 18, for example, these criteria can be used to distinguish between the two syndromes, even though there is a considerable overlap of physical and mental defects. And unlike the abnormalities caused by single-gene mutations, the abnormalities in the trisomies cannot be traced to some single biochemical defect in the cells. Rather, the trisomy syndromes result from the cumulative action of abnormal gene dosage. Consequently there is at present no hope of effecting a "cure" for the trisomies.

The autosomal trisomies invariably involve major physical malformations and men-

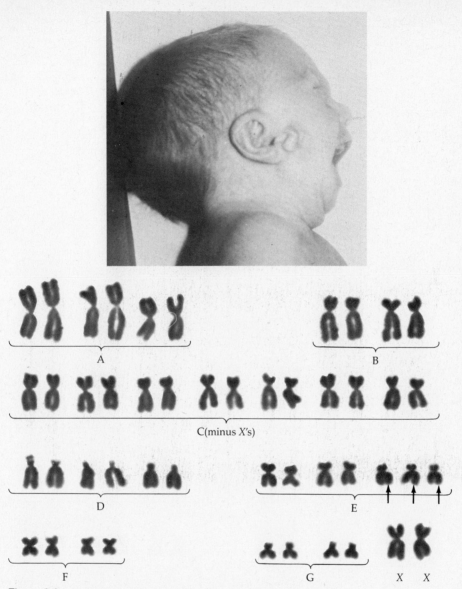

Figure 6.9   A child with Edwards syndrome (trisomy 18) and karyotype.

tal retardation and are often associated with heart defects. This is true of other major developmental disturbances as well, not only of those that are genetic in origin but also of those that are purely environmental, such as those caused by infection of the fetus with German measles virus (rubella) during the first months of pregnancy. Evidently the heart and

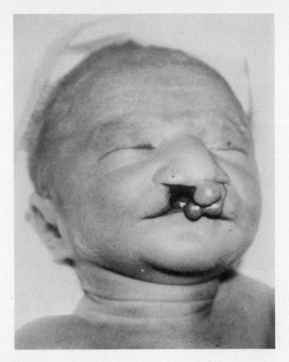

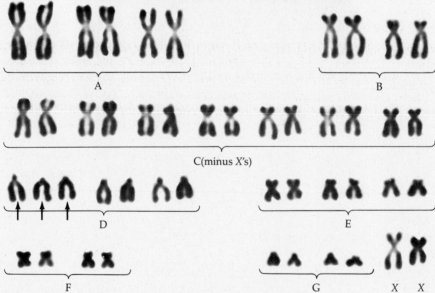

Figure 6.10 A child with Patau syndrome (trisomy 13) and karyotype.

nervous system are often affected because their embryonic development is delicate and incredibly intricate; major developmental aberrations are unlikely to leave these systems untouched. The heart is a fist-sized, four-chambered, 10-oz wonder. It beats between 60 and 200 times a minute, depending on circumstances, and its 30 million rhythmic contractions a year force nearly a million gallons of blood through almost 60,000 miles of arteries, veins, and capillaries. Little wonder that the ancients thought the center of life, personality, and emotion to be the heart! But the brain is even more impressive. In the cerebral cortex alone, the complex outer layer—the seat of the senses, voluntary muscle control, consciousness and rationality—there are nearly 10 billion nerve cells known as neurons, with each neuron supported and nourished by about 10 neuroglial cells. Each neuron has between 4,000 and 10,000 extensions connecting it to other neurons. Most of these cells and connections develop before birth, but the process actually continues until about age two. It is no surprise that major developmental abnormalities will almost always disrupt this prodigious network. To expect otherwise would be like expecting an earthquake not to interrupt telephone service! (Incidentally, brain cells in an adult cannot divide. They gradually die off and are not replaced—an inevitable part of aging. People beyond puberty lose some 10,000 neurons every day.)

Amniocentesis

Until the recent past nothing could be done about the dilemma of children born with incurable mental and physical defects. Now, in certain cases, it is possible to identify birth defects in the fetus while the fetus is still developing in the uterus. The principal method of procuring cells of the fetus suitable for laboratory study is known as **amniocentesis**.

The cells obtained in this procedure are not taken from the fetus itself, but they are genetically identical to those in the fetus. Every fetus is surrounded early in development by membranous sacs—sacs composed of cells that descend from the fertilized egg. The cluster of cells that reaches the uterus from the fallopian tube a few days after fertilization will develop into a fetus, but only certain cells in the cluster actually give rise to the body of the fetus. The rest divide and stick together as sheets of cells and move and fold so as to enshroud the fetus in the membranous sacs. The innermost membrane is known as the **amnion**. It is filled with a clear, watery fluid that cushions the fetus against mechanical shocks and prevents adhesion of parts of the fetus that might accidentally come into contact. Present in the amniotic fluid are cells that slough off from the fetus or from the amniotic membrane. These cells are, of course, genetically identical to those in the fetus. At about the fourteenth to sixteenth week of pregnancy, the amnion becomes large enough to enable insertion of a needle and removal of a sample of the amniotic fluid by piercing through the abdomen of the mother (Figure 6.11). A small amount of amniotic fluid is withdrawn, and the cells in this fluid are separated from the liquid and grown in cultures in the laboratory. The chromosomes and various biochemical or enzymatic capabilities of the cells can be examined, and in this way certain hereditary defects in the fetus can be diagnosed. If the cells are trisomic for chromosome 21, for example, this is indicative of a fetus with Down syndrome. All the known chromosomal defects can be prenatally diagnosed in this way. In addition, more than 100 other inherited disorders due to single mutant genes can be detected even though no visible chromosome abnormality is involved. For example, Tay-Sachs disease can be detected because the cultured amniotic cells lack the enzyme hexosaminidase A (see Chapter 4).

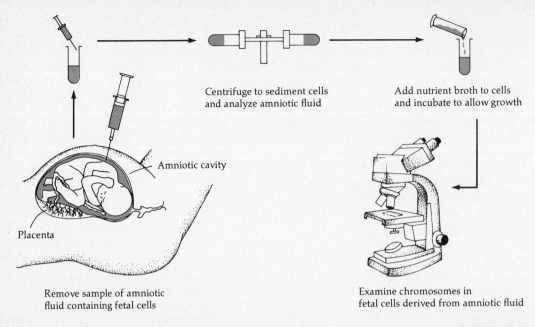

Centrifuge to sediment cells
and analyze amniotic fluid

Add nutrient broth to cells
and incubate to allow growth

Amniotic cavity

Placenta

Remove sample of amniotic
fluid containing fetal cells

Examine chromosomes in
fetal cells derived from amniotic fluid

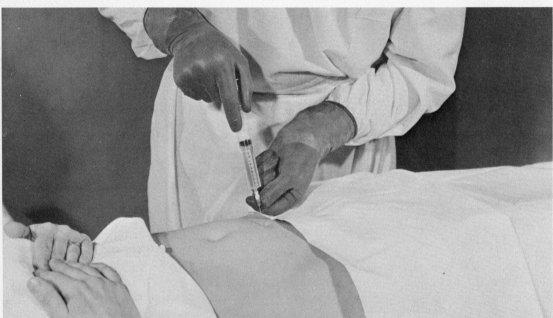

Figure 6.11 Outline of the procedure of amniocentesis in which cells from an unborn fetus are obtained and examined in the laboratory for signs of genetic disease or chromosome abnormality. The photo shows amniotic fluid being obtained from a pregnant woman.

Amniocentesis is not without its dangers, although complications arise only infrequently. (The overall risk in major health centers is about 1 percent.) Of course, great care must be exercised to avoid puncturing the fetus, but other dangers exist as well. These include the risk of infection, the danger of maternal hemorrhage, or the induction of an abortion by the procedure itself.

Thus amniocentesis will probably never become routine except in high-risk pregnancies. These are pregnancies in which the fetus is known to have a particularly high risk of having some detectable abnormality. In the case of Down syndrome, for example, the pregnancies at greatest risk are those in normal women over 35 or in the small proportion of families in which a chromosome rearrangement in one of the parents predisposes the fetus to high risk. In cases of recessively inherited conditions, the high-risk families are those in which both parents are carriers of recessive genes (the carriers of about 100 such genes can now be identified by appropriate studies of blood or other body fluids or cultured cells). When amniocentesis is carried out and the embryo is found to be normal, then there is no problem and the parents can be put at ease with regard to the trait examined. When the fetus is affected, then an abortion may be considered. The decision is left to the parents, but in the present social climate many families choose abortion over the birth of a severely malformed and mentally retarded child.

The decision whether or not to abort is usually much more agonizing when the fetus is affected with a less major abnormality or when it is not itself affected but is, like its parents, a carrier. In the latter instance the question may become one of aborting a healthy fetus, and serious moral and ethical issues must be faced. At the present time there are relatively few legal obstacles to abortion in the United States. Many thoughtful people have argued that this is the way it ought to be—that a mother should decide for herself, based on whatever grounds she considers adequate, whether to nurture or to abort the fetus; society, they argue, has no right to force a woman to carry a fetus she does not want. Many equally thoughtful people feel differently. They argue that the legality of an activity does not make it morally right; they point out that slavery was legal for centuries, although we now perceive it to have been morally wrong. If abortion is wrong, the argument goes, then it is wrong for everyone, and the legality of it is a symptom of society's disrespect for life and not an absolution of the act. It would be inappropriate to discuss in further detail here the many sincerely held points of view concerning abortion. It is no simple matter, and great wisdom is called for.

An equally difficult matter—maybe an even more explosive one—is the care of children born alive with grim, mortal, and incurable abnormalities. To what extent should extraordinary medical measures, often called "heroic" measures, be used to sustain and prolong the lives of such children? They cannot intentionally be killed, of course; infanticide is proscribed by the law, customs, and common morality of our society (although infanticide is accepted and practiced in some other societies). Plain decency requires these children to be fed and fondled and kept warm. But should their tortured lives be unnaturally prolonged? How long should a hopelessly deformed and doomed child be kept alive by medical gadgetry? And to what extent should heroic measures be used for children who are not so severely deformed? Should a child with a seriously defective heart due to Down syndrome receive heart surgery, for example? Some surgeons say no. They point out that a child otherwise normal would be completely well following successful surgery. But in cases such as Down syndrome, the patient will not be cured by surgery; the patient's life may be prolonged, but the major mental abnormalities will remain. And they argue that the well-being of the entire child and not just the heart should be of prime concern. The other side of this question is perhaps best summarized by a verse from the New Testament (Romans), that we who are strong ought to bear the infirmities of those who are weak and not live to please ourselves. The terrible dilemma applies not only to children, of course, but with equal force to the aged and to those hopelessly maimed by stroke, cancer, heart attack, or accident. The issues will be painful to resolve. They should not be ignored. Almost everyone will have to decide, if not for a child or an aged parent or an injured spouse, then at least for himself or herself should the worst come to pass. So it is perhaps worth bearing in mind the ancient Chinese proverb that if you do not think about what is distant, you will discover sorrow near at hand.

Death Before Birth

The fact is that most fetuses with major abnormalities of the autosomes do not survive to be born at all. Their development is so grossly abnormal that the pregnancy terminates early in a spontaneous abortion. The actual frequency of spontaneous abortions from all causes is a number virtually impossible to obtain. Since pregnancies are not usually diagnosed until they are two months along, abortions prior to this time are usually not detected. Very early abortions that result from failure of implantation or expulsion of the embryo within the first few weeks are often not recognized as abortions because the mother may be unaware of her pregnancy when the abortion occurs, and the abortion itself may cause her only minor discomfort. She may experience a particularly heavy menstrual discharge, for example, or perhaps she will miss a menstrual period— events that cause no particular wonder or alarm in most women because they are not

uncommon occurrences. But some unknown fraction of these are actually very early abortions.

Of recognized pregnancies—those pregnancies that last long enough to be diagnosed —about 15 percent end in spontaneous abortion (expulsion of the fetus prior to the end of the third month of pregnancy) or miscarriage (expulsion of the fetus between the third and seventh months of pregnancy). The great majority of these abortions occur before the fifth or sixth months. There is general agreement that at least as many abortions occur prior to recognition of the pregnancy as occur afterward, perhaps many more. Some studies have suggested that up to *half* of all fertilizations end in spontaneous abortions. Whatever the real answer is, spontaneous abortion in humans is an extremely common, everyday occurrence. Even the 15 percent of recognized pregnancies—nearly one out of six—represents a staggering proportion.

Table 6.4 is a composite of the results of many chromosome studies showing the chromosome constitutions that would be expected to occur among 100,000 recognized pregnancies. Results vary somewhat from study to study, so the numbers in the table should be regarded as approximate. Nevertheless, they are in the range actually found. As noted earlier, out of 100,000 recognized pregnancies, about 15,000 will undergo spontaneous abortion. These abortions represent a merciful cleansing of the population of the most drastically abnormal, grotesquely deformed fetuses. Many of these fetuses must carry genetic mutations that are incompatible with life; many others must have sustained environmentally induced damage or accidents during development. However, as the composite data in Table 6.4 indicate, approximately 50 percent of spontaneously aborted fetuses have major *chromosomal* abnormalities!

The most frequent class of abnormalities

TABLE 6.4 NUMBER AND TYPE OF CHROMOSOMAL ABNORMALITIES AMONG SPONTANEOUS ABORTUSES AND LIVE BIRTHS IN 100,000 HYPOTHETICAL RECOGNIZED PREGNANCIES*

		100,000 Recognized Pregnancies	
		15,000 abort spontaneously 7,500 chromosomally abnormal	85,000 live births 550 chromosomally abnormal
Trisomy			
A:	1	0	0
	2	159	0
	3	53	0
B:	4	95	0
	5	0	0
C:	6–12	561	0
D:	13	128	17
	14	275	0
	15	318	0
E:	16	1229	0
	17	10	0
	18	223	13
F:	19–20	52	0
G:	21	350	113
	22	424	0
Sex chromosomes			
	XYY	4	46
	XXY	4	44
	XO	1350	8
	XXX	21	44
Translocations			
	Balanced	14	164
	Unbalanced	225	52
Polyploid			
	Triploid	1275	0
	Tetraploid	450	0
Other (mosaics, etc.)		280	49
	Total	7500	550

Source: Based on data in D. H. Carr and M. Gedeon, 1977, Population cytogenetics of human abortuses, in E. B. Hook and I. H. Porter (eds.), Population Cytogenetics: Studies in Humans, Academic Press, New York, pp. 1–9; and E. B. Hook and J. L. Hamerton, 1977. The frequency of chromosome abnormalities detected in consecutive newborn studies—differences between studies—results by sex and by severity of phenotypic involvement, in E. B. Hook and I. H. Porter (eds.). Population Cytogenetics: Studies in Humans, Academic Press, New York, pp. 63–79.

*The numbers are approximate and vary somewhat from study to study. They are within the range found in most studies, though, and are close to the average values.

among chromosomally abnormal fetuses is autosomal trisomy. Approximately half of all chromosomally abnormal fetuses have autosomal trisomy. Trisomics for most autosomes are found, but all are not equally frequent. Trisomy 16 is extremely common; roughly a third of trisomic abortuses have trisomy 16. Trisomics of D- and G-group chromosomes are also quite common; roughly another third of autosomal trisomic abortuses have trisomy of a D- or G-group chromosome. Included in this category are many fetuses with trisomy 21, the ones that do not survive until birth. About three-fourths of all fetuses with trisomy 21 are eliminated by spontaneous abortion and only one-fourth are born alive. To apprehend the significance of this finding, note that if all fetuses with trisomy 21 were born alive, about 1 in 200 live-born children would have Down syndrome! The remaining third of trisomic abortuses involve autosomes other than chromosome 16 or the D- or G-group chromosomes. Why some trisomies are much more frequent than others is not really known. Part of the reason is that the rates of primary nondisjunction of the different autosomes are probably not the same; another part is that certain autosomal trisomics may be aborted very early, perhaps even before the pregnancy is recognized.

Two other types of chromosomal abnormalities account for a large fraction of chromosomally abnormal fetuses. One of the most common is the XO (i.e., 45,X) sex-chromosome constitution. Fetuses that are XO account for about 18 percent of chromosomally abnormal abortuses, but a small fraction of XO's—about 0.5 percent—do survive and have Turner syndrome. The high frequency of XO among abortuses is astonishing. Indeed, if all the XO fetuses were born alive, the incidence of the condition among live-born females would be about 3 percent!

The other frequent kind of chromosome abnormality in aborted fetuses is **polyploidy** (multiple complete sets of chromosomes, discussed at the beginning of this chapter). Triploids account for roughly 17 percent of chromosomally abnormal fetuses. A **triploid** has three complete sets of chromosomes—69 chromosomes altogether [Figure 6.12(a)]. Triploid fetuses result from an abnormal meiosis in which one of the divisions fails to progress normally, so that two full sets of chromosomes become included in the same gamete; they may also result from errors in fertilization. A polar body with its complement of chromosomes may not be extruded from the egg, for

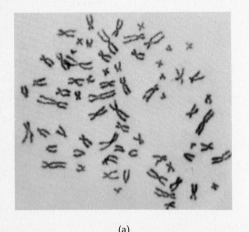

(a)

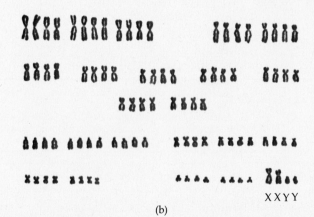

(b)

XXYY

Figure 6.12 (a) Metaphase spread of a triploid human cell (69 chromosomes). (b) Karyotype of a tetraploid human cell (92 chromosomes).

example, or it may be formed but then reenter the egg; alternatively, two sperm may both fertilize a normal egg, or one sperm carrying two complete sets of chromosomes may result in triploidy. **Tetraploid** fetuses—fetuses that have four complete sets of chromosomes totaling 92 [see Figure 6.12(*b*)]—are also rather common.

To summarize the information in Table 6.4, about 15 percent of all recognized pregnancies terminate in spontaneous abortion, and about half of these abortions involve fetuses with major chromosomal abnormalities. Among the fetuses with chromosomal abnormalities, approximately 50 percent have autosomal trisomy, 18 percent are XO, and 23 percent are triploids or tetraploids. The remaining aborted fetuses have multiple or more complex types of chromosomal abnormality. Included in the last category are fetuses that have **translocations**—chromosomes that have undergone an interchange of parts, a type of chromosome abnormality that will be discussed in the next chapter. The last category of chromosomally abnormal fetuses also includes **chromosomal mosaics**—fetuses with chromosomal abnormalities in some tissues but not in others. Chromosomal mosaics are almost always the result of mitotic nondisjunction (see Figure 6.2) in the early mitotic divisions of the fertilized egg.

There is reason to believe that Table 6.4 significantly underestimates the true incidence of major chromosomal abnormalities among human zygotes. Chromosome studies of spontaneously aborted fetuses miss those instances in which a blastocyst fails to implant in the uterine wall or in which the embryo aborts so early that the pregnancy is unrecognized. For example, one category of chromosomal abnormality that is conspicuously absent among spontaneously aborted fetuses is **monosomy**—the absence of a single chromosome. XO fetuses are monosomic for the X, of course, and they are an exception because they are so common. But only very rarely is an autosomal monosomic found. The finding of fewer monosomics than trisomics is quite unexpected because monosomics arise from the loss of a chromosome during gamete formation, whereas trisomics result from the gain of a chromosome, and in virtually all organisms chromosome loss is more frequent than chromosome gain. The likely explanation is that many human zygotes are autosomal monosomics, but these are probably aborted so early that they are hardly ever found among abortions of recognized pregnancies.

SUMMARY

1. Abnormalities in chromosome number can involve entire sets of chromosomes. The number of sets of chromosomes in a cell is indicated by the suffix **-ploid**. Thus, a cell such as an egg or a sperm that has a single set of chromosomes is **haploid**, a cell such as a somatic cell that carries two sets of chromosomes is **diploid**, a cell with three sets of chromosomes is **triploid**, one with four sets of chromosomes is **tetrapolid**, and so on. Cells with three or more sets of chromosomes are said to be **polyploid**.

2. Abnormalities in chromosome number can also involve individual chromosomes. Such abnormalities are indicated by the suffix **-somic**. For example, an otherwise diploid cell that is missing one chromosome is **monosomic** for the chromosome in question, and an otherwise diploid cell that has one extra chromosome is **trisomic** for the chromosome in question. Otherwise diploid cells that have three or more copies of a particular chromosome are said to be **polysomic**.

3. Monosomic or trisomic zygotes are

usually the result of **primary nondisjunction**—an abnormal behavior of chromosomes during meiosis that leads to gametes having two copies of a chromosome (or no copies of a chromosome) instead of exactly one. When the abnormal chromosome behavior occurs in the first meiotic division, the type of nondisjunction is called **first-division nondisjunction**; when it occurs in the second meiotic division, the type of nondisjunction is called **second-division nondisjunction**. Nondisjunction can occur during mitosis as well as meiosis, in which case it is referred to as **mitotic nondisjunction**. The result of mitotic nondisjunction is a **chromosomal mosaic**—an individual who has two or more chromosomally distinct types of cells.

4. Generally speaking, monosomy or trisomy involving the sex chromosomes is phenotypically less severe than monosomy or trisomy involving the autosomes. In all cases, however, monosomy has more extreme phenotypic effects than trisomy. For the sex chromosomes, YO (i.e., 45,Y) zygotes invariably undergo spontaneous abortion so early that the pregnancy is unrecognized; XO (i.e., 45,X) zygotes usually undergo spontaneous abortion, and the small surviving fraction has **Turner syndrome**. The XXY chromosome constitution is associated with **Klinefelter syndrome**, a condition in which affected males are taller than normal, have moderate mental impairment, fail to undergo puberty, and sometimes have breast development.

5. The sex-chromosome constitution XXX (i.e., trisomy X) has no distinctive physical symptoms. Individuals who have trisomy X have a higher than normal risk of mental retardation and may have menstrual irregularities, but many XXX women are within the normal range of mental abilities and are fertile.

6. The sex-chromosome constitution XYY also has no distinctive physical symptoms, although XYY males tend to be taller than normal males. Preliminary findings regarding XYY suggested a high rate of criminality, particularly involving crimes of violence against people. The concern over the criminality of XYY's was lessened somewhat by a later Danish study which found that, while XYY males do have higher rates of criminal conviction than their XY counterparts, the higher rate of criminality involves generally petty crimes against property rather than violent crimes against people. Moreover, the higher rates of criminal conviction may simply be due to the moderately impaired intellectual function of the XYY males.

7. XXX and XYY individuals are fertile, and theory predicts a high frequency of sex-chromosomal trisomy among their offspring due to **secondary nondisjunction**—the formation of gametes carrying an extra chromosome by individuals who are trisomic for the chromosome. However, the offspring of XXX and XYY individuals are chromosomally normal. In XYY males, the extra Y chromosome seems to be lost or eliminated from cells undergoing meiosis, but the reason for the absence of secondary nondisjunction in XXX females is not known.

8. Among the autosomes, all monosomies are incompatible with life and lead to failure of blastocyst implantation or to abortion prior to the diagnosis of pregnancy. Autosomal trisomies usually lead to abortion somewhat later during development, but three are compatible with live birth. **Trisomy 21** (also called **Down syndrome** or, less acceptably, **mongolism**) is the most frequent autosomal trisomy in live-born infants, affecting about 1 in 750 newborns. The incidence of Down syndrome increases dramatically with the mother's age to a maximum of about 6 percent in mothers aged 45 or older. In addition to other characteristic physical symptoms, children with Down syndrome are mentally retarded and many have heart defects. Nevertheless, many individuals with Down syndrome survive to adultood.

9. Trisomy 13 (Patau syndrome) and

Trisomy 18 (**Edwards syndrome**) are much more rare and more phenotypically severe than Down syndrome. Newborns with either of these syndromes usually die within the first few months of life, and virtually none survives for more than a year.

    10. In the procedure of **amniocentesis**, cells of a developing fetus are obtained by piercing the amnion with a hollow needle inserted through the abdominal wall of a pregnant woman and drawing off a small amount of fetal-cell-containing amniotic fluid. The cells can then be cultured and studied in the laboratory. All major chromosomal abnormalities can be diagnosed before birth by means of amniocentesis, and over 100 genetic disorders due to single-gene defects can be detected in fetal cells. Ordinarily, however, amniocentesis is used only in high-risk pregnancies.

    11. Chromosome studies of spontaneously aborted fetuses have revealed an astonishingly high frequency of major chromosomal abnormalities. On the average, among 100,000 hypothetical recognized pregnancies, about 15,000 will terminate in spontaneous abortion. Among these 15,000, about half will have major chromosomal abnormalities, including autosomal trisomy (50 percent of all chromosomally abnormal fetuses), X-chromosomal monosomy (18 percent), and polyploidy (23 percent triploids or tetraploids). Trisomy 16 is the single most frequent autosomal trisomy among spontaneously aborted fetuses, but D- and G-group trisomies are also relatively frequent. Indeed, about three-fourths of all trisomic-21 fetuses undergo spontaneous abortion. Because autosomal monosomics probably undergo spontaneous abortion prior to the recognition of pregnancy, chromosome studies of abortuses probably underestimate the true frequency of major chromosome abnormalities in human reproduction.

WORDS TO KNOW

Amniocentesis	**-Somic**	**Sex-Chromosomal Abnormalities**	**Syndrome**
Lethal	Monosomic		Turner
Chromosomal Mosaic	Trisomic	XXX (i.e., 47,XXX)	Klinefelter
Spontaneous Abortion	Tetrasomic	XO (i.e., 45,X)	Down
	Polysomic	XXY (i.e., 47,XXY)	Edwards
-Ploid		XYY (i.e., 47,XYY)	Patau
Haploid	**Nondisjunction**		
Diploid	Primary	**Nonlethal Autosomal Trisomies**	
Triploid	First division (primary)	Trisomy 21	
Tetraploid	Second division (primary)	Trisomy 18	
Polyploid	Secondary	Trisomy 13	
	Mitotic		

PROBLEMS:

    1. For discussion: In an actual court case, a 13-year-old boy with Down syndrome who had been institutionalized since birth had a progressively worsening heart condition requiring surgery. The boy's parents refused permission for the surgery on the ground that the surgery, if successful, would merely prolong the boy's life without alleviating his mental retardation. The physicians involved appealed to the court to take custody of the child from his original parents and, indeed, produced an unrelated couple who were willing to adopt the boy and have him live in their home. In your judgment, what are the principal medical, moral, and legal issues involved in this case? If you were the judge, how would you decide the case, and why? Would you feel differently if the child had been a newborn?

2. What kind of chromosomal constitution would result from a zygote in which the spindle failed to form during the first mitotic division but in which chromosome replication and centromere splitting did occur?

3. Why do chromosomal abnormalities involving the sex chromosomes usually have less severe phenotypic effects than those involving the autosomes?

4. Which sex-chromosomal monosomy is invariably lethal?

5. An individual has a monosomic X and trisomy 21 but no other chromosomal abnormalities. How many chromosomes does the individual have?

6. What is the distinction between meiotic nondisjunction and mitotic nondisjunction? Between primary and secondary nondisjunction? Between first-division and second-division nondisjunction?

7. If an XX zygote undergoes mitotic nondisjunction of one of the X chromosomes, what sex-chromosome constitutions would be found among cells of the resulting chromosomal mosaic?

8. If an XYY male is born as a result of primary nondisjunction, would the nondisjunction be first division or second division?

9. What sex-chromosome constitutions would be found among the four sperm produced by a spermatocyte that underwent second-division nondisjunction of the X chromosome?

10. Why is the word *syndrome* inappropriate for the XXX and XYY sex-chromosome constitutions?

11. How could a phenotypically normal male and a phenotypically normal female have a son with Klinefelter syndrome who also has the red form of color blindness?

12. Theoretically, what sex-chromosome constitutions would be expected among the offspring of an XXX female? Which ones are actually found?

13. A woman with trisomy X of genotype *G6PDA/G6PDA/G6PDB* has a chromosomally normal son. What is the probability that he carries *G6PDA*?

14. A male with the green form of color blindness and a normal female have a child with Turner syndrome who has normal color vision. In which parent did the nondisjunction occur?

15. Why are the data in Table 6.4 thought to be *conservative* estimates of the frequency of major chromosomal abnormalities in human reproduction?

FURTHER READING AND REFERENCES

Alfi, O. S., R. Chang, and S. P. Azen. 1980. Evidence for genetic control of nondisjunction in man. Am. J. Hum. Genet. 32:477–483. Certain genes, when homozygous, seem to increase the rate of nondisjunction.

Carr, D. H., and M. Gedeon. 1977. Population cytogenetics of human abortuses. In E. B. Hook and I. H. Porter (eds.). Population Cytogenetics: Studies in Humans. Academic Press, New York, pp. 1–9. One of the sources of Table 6.4.

de Grouchy, J., C. Turleau, and C. Finaz. 1978. Chromosomal phylogeny of the primates. Ann. Rev. Genet. 12:289–328. How did human chromosomes come to be the way they are?

Emery, A. E. H. 1979. Elements of Medical Genet-ics, 5th ed. Churchill Livingstone, Edinburgh. A short but fine introduction to the medical aspects of human genetics.

Hamerton, J. L. 1971. Clinical Cytogenetics. Academic Press, New York. Advanced reference with detailed descriptions of trisomy syndromes.

Hook, E. B. 1980. Rates of 47,+13 and 46 translocation D/13 Patau syndrome in live births and comparison with rates in fetal deaths and at amniocentesis. Am. J. Hum. Genet. 32:849–858. A summary of studies on the rate of occurrence of Patau syndrome.

Hook, E. B., and J. L. Hamerton. 1977. The frequency of chromosome abnormalities detected in consecutive newborn studies—differences between studies—results by sex and by severity of phenotypic involvement. In E. B. Hook and I. H. Porter (eds.). Population Cytogenetics: Studies in Humans. Academic Press, New York, pp. 63–79. One of the sources of Table 6.4.

Hook, E. B., and A. Lindsjö. 1978. Down syndrome in live births by single year maternal age interval in a Swedish study. Am. J. Hum. Genet. 30:19–27. Source of Figure 6.8.

Hsu, T. C. 1979. Human and Mammalian Cytogenetics: An Historical Perspective. Springer-Verlag, New York. An entertaining book on cytogenetics glittering with personal anecdotes of one who was there.

Martin, R. H., C. C. Lin, W. Balkan, and K. Burns. 1982. Direct chromosomal analysis of human spermatozoa: Preliminary results from 18 normal men. Am. J. Hum. Genet. 34: 459–468. A method involving *in vitro* fertilization of hamster eggs gives data consistent with those from spontaneous abortions.

Norwood, C. 1980. At Highest Risk: Environmental Hazards to Young and Unborn Children. McGraw-Hill, New York. Focuses on the dangers of such drugs as amphetamines, alcohol, and many others.

Smith, D. W., and A. A. Wilson. 1973. The Child with Down's Syndrome (Mongolism). Saunders, Philadelphia. A sympathetic account for parents, physicians, and other concerned persons.

Therman, E. 1980. Human Chromosomes: Structure, Behavior, Effects. Springer-Verlag, New York. An outstanding introduction to human cytogenetics.

Uchida, I. A., and E. M. Joyce. 1982. Activity of the fragile X in heterozygous carriers. Am. J. Hum. Genet. 34: 286–293. A form of male mental retardation associated with an X chromosome having a tendency to break.

Witkin, H. A., S. A. Mednick, F. Schulsinger, E. Bakkestrøm, K. O. Christiansen, D. R. Goodenough, K. Hirschhorn, C. Lundsteen, D. R. Owen, J. Philip, D. B. Rubin, and M. Stocking. 1976. Criminality in XYY and XXY men. Science 193:547–555. The important Danish study summarized in Table 6.2.

chapter 7

Abnormalities in Chromosome Structure

In addition to abnormalities in chromosome number such as the examples discussed in the preceding chapter, there are a great variety of structural abnormalities in chromosomes. **Structural abnormalities** refer to chromosomes having an abnormal structure; examples include chromosomes that have a portion missing or a portion represented twice. They range in size from those that are so small as to be almost undetectable in the light microscope (and probably a great many are overlooked because they lie below the limit of this form of detection) to those that are so large as to be striking and obvious. They range in seriousness from those that have a few or no phenotypic effects to those that are almost as severe as

monosomy or trisomy. Between these extremes are structural abnormalities that have no detectable effects on carriers but expose their children to great risk of severe abnormality. Among abnormal types of chromosomes that have no known deleterious effects are the variants in the normal chromosome complement mentioned in Chapter 2—variants that are found in people with normal family histories, such as chromosomes with unusually prominent satellites or variants in the length of Yq. The word *abnormal* in reference to these variants should be taken to imply "atypical" rather than "harmful"; the important thing to remember about them is that the carriers and their offspring are phenotypically normal.

Chromosome Breakage

This chapter is about chromosome abnormalities that are much more rare than those discussed in Chapter 2. The variants discussed here are harmful in humans, and they may have severe phenotypic effects in the carriers themselves or in some proportion of their offspring. These abnormal chromosomes originate from **chromosome breakage**. In their way, chromosomes are fragile objects and they sometimes break spontaneously. The incidence of chromosome breakage is greatly increased by exposure of the chromosomes to a wide variety of agents, including x-rays and certain chemicals (see Chapter 11). The effect of such agents is especially acute in cells that are actively undergoing division—embryonic cells, for example. Broken chromosomes also tend to "heal"—to undergo **restitution**. The broken ends behave as if they were "sticky" (they come together and fuse), and the restitution probably involves particular enzymes that aid in the repair process. Most of the time broken chromosomes restitute correctly, and the broken ends rejoin at the point of fracture. But sometimes they are not restituted correctly and chromosomal abnormalities result.

Fragments of chromosomes that have no centromere, which are called **acentric** chromosomes, are lost from daughter cells because they cannot be maneuvered to the poles during cell division. Generally, too, chromosomes that end up having two centromeres, which are called **dicentric** chromosomes, are lost rather quickly because there is a continual risk during cell division that the two centromeres will be pulled to opposite poles. In such a case the chromosome itself will form a bridge between the centromeres; the upshot is that either the chromosome forming the bridge will rupture somewhere in the middle and both poles will receive an abnormal, broken chromosome, or the chromosome forming the bridge will not rupture, in which case neither centromere may

travel all the way to a pole where it can be included in a telophase nucleus and the entire dicentric chromosome will be lost. In short, *chromosomal abnormalities will usually persist only if they have exactly one centromere.*

Four kinds of structural abnormalities of chromosomes are of greatest importance in human genetics. Some cells have **deficiencies** (also called **deletions**), formed when a particular segment of a chromosome is broken off and lost. Some cells have **duplications**, in which a segment of a chromosome is present in more than the normal number of copies. The two other important abnormalities involve no loss or gain of chromosomal material; they involve instead the rearrangement of parts of chromosomes. In **inversions**, for example, a chromosome is broken in two places and the middle segment is flipped end for end before restitution. In **translocations**, two nonhomologous chromosomes become broken and their terminal segments interchanged. Inversions and translocations do not usually cause phenotypic abnormalities in the carriers, but they may, especially in the case of translocations, expose the offspring of carriers to substantial risk. In this chapter, each of these four types of structural abnormality (and two others to be described later) will get a brief discussion. In these discussions it will be convenient and appropriate to use certain conventional symbols in designating human karyotypes; these are listed in Table 7.1.

Euploidy and Aneuploidy

In the last chapter we discussed individuals who have multiple complete sets of chromosomes such as triploids and tetraploids. Such polyploid individuals are said to be **euploid**, which means that the *relative* dosage of genes is not disrupted by the chromosomal abnormality. In a diploid individual, any autosomal locus is present the same number of times

TABLE 7.1 CONVENTIONAL KARYOTYPE SYMBOLS USED IN HUMAN GENETICS

A–G	Chromosome groups
1–22	Autosome designations
X, Y	Sex-chromosome designations
p	Short arm of chromosome
q	Long arm of chromosome
ter	Terminal portion: pter refers to terminal portion of short arm, qter refers to terminal portion of long arm
+	Preceding a chromosome designation, indicates that the chromosome or arm is extra; following a designation, indicates that the chromosome or arm is larger than normal
−	Preceding a chromosome designation, indicates that the chromosome or arm is missing; following a designation, indicates that the chromosome or arm is smaller than normal
mos	Mosaic
/	Separates karyotypes of clones in mosaics— e.g., 47,XXX/45,X
dup	Duplication
dir dup	Direct duplication
inv dup	Inverted duplication
del	Deletion
inv	Inversion
t	Translocation
rep	Reciprocal translocation
rob	Robertsonian translocation
r	Ring chromosome
i	Isochromosome

(twice) as any other autosomal locus. In a triploid individual, any autosomal locus is also present the same number of times (three times) as any other autosomal locus, so triploidy represents a euploid chromosomal abnormality since the dose of one gene relative to any other is the same as in a diploid—namely, 1:1. In the last chapter we also discussed monosomy and trisomy, which are called **aneuploid** chromosomal abnormalities because they *do* disrupt relative gene dosage. An individual who is monosomic for some chromosome has only one dose of the genes on the monosomic chromosome but two doses of the genes on the other chromosomes, so the relative gene dosage is upset by monosomy. Similarly, a trisomic individual has genes on the trisomic chromosome represented three times but other genes represented twice, so trisomy is also an aneuploid chromosomal abnormality.

Concepts involving gene dosage are also useful in discussing structural abnormalities of chromosomes. A structural abnormality such as an inversion, which changes only the arrangement but not the number of genes along a chromosome, could be called a *euploid* chromosomal abnormality; a structural abnormality such as a deletion, in which certain genes are missing from the chromosome, could be called an *aneuploid* chromosomal abnormality. (However, certain authors prefer to use the terms **balanced** for euploid structural abnormalities and **unbalanced** for aneuploid structural abnormalities because, historically, *euploidy* and *aneuploidy* were used to refer to abnormalities involving entire chromosomes or sets of chromosomes.) An aneuploid individual with too few copies of certain genes is said to be **hypoploid** for the genes in question, and an aneuploid individual with too many copies of certain genes is said to be **hyperploid** for the genes in question.

Duplications and Deficiencies

Figure 7.1 shows the origin of a deletion-bearing chromosome and a corresponding duplication-bearing chromosome. Figure 7.1(*a*) and (*b*) illustrate homologous chromosomes that have been broken in three places followed by repositioning of the *BC* segment from one chromosome into the gap in the other. Restitution of the broken ends produces the chromosomes shown in Figure 7.1(*c*); one chromosome carries a deletion of the *BC* region, the other a duplication of the *BC* region. An individual carrying both of the chromosomes in Figure 7.1(*c*) has a balanced (euploid) chromosomal rearrangement. However, dur-

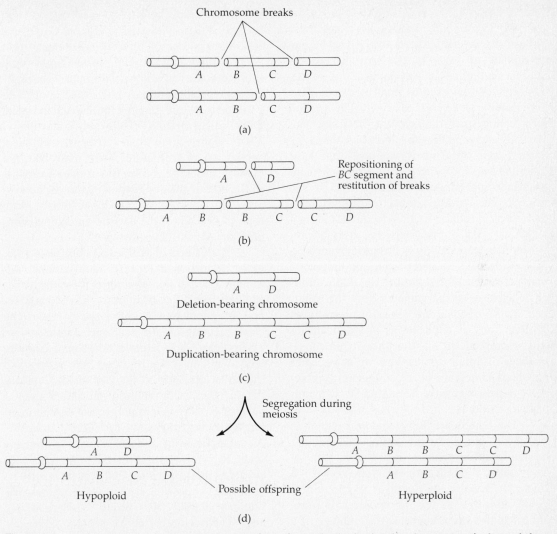

Figure 7.1 Origin of deletion-bearing chromosome (*c,* top) and duplication-bearing chromosome (*c,* bottom) from chromosome breaks indicated in *(a)* following restitution as in *(b).* Possible offspring of the individual in *(c)* are shown in *(d).*

ing meiosis, the homologous chromosomes will segregate, and half the resulting gametes will carry the deletion-bearing chromosome and half will carry the duplication-bearing chromosome. The offspring from the individual in Figure 7.1(*c*) will thus be aneuploid [See Figure 7.1(*d*)]. Those who receive the deletion-bearing chromosome will be hypoploid for the

BC region; those who receive the duplication-bearing chromosome will be hyperploid for the BC region.

The phenotypic consequences of aneuploidy are variable and depend on the length of the chromosomal region involved and on the particular genes in the region. Nevertheless, certain general characteristics of aneuploidy have emerged from an extensive study in the fruit fly, *Drosophila melanogaster*, and these general characteristics seem to apply to human aneuploids as well. In the fruit fly, experimental techniques allow duplications or deficiencies of almost any small segment of a chromosome to be created at will. The method for creating small duplications or deficiencies starts by exposing flies to x-rays and searching among their offspring for animals that have inherited translocations (chromosomes with interchanged parts). As will be discussed later in this chapter, translocations predispose their carriers to produce a high frequency of gametes carrying duplications and deficiencies. When flies carrying two different translocations are mated, gametes from one individual may partially compensate for the duplications or deficiencies in the gametes from the other, and in this manner an enormous variety of combinations of duplications and deficiencies can be created. In all, the phenotypic consequences of small duplications or deficiencies for about 85 percent of the autosomal complement in fruit flies have been studied by means of this procedure.

The results are somewhat surprising. In the first place, very few individual loci are **dosage sensitive**; that is, in only a handful of cases does a duplication or deficiency of a single locus cause the death or visible abnormality of a fly. A few particular loci cause abnormalities when hyperploid (duplicated); a few others cause abnormalities when hypoploid (deficient). In only one case does *both* hyperploidy and hypoploidy cause abnormality —in this instance, the death of the fly. The locus involved is evidently very special because it must be present in *exactly* two doses to allow survival. Unfortunately, the function of this critical locus is as yet unknown.

Aside from the handful of dosage-sensitive loci, duplications and deficiencies, if they are small, produce very minor or no visible effects at all! Any effects depend on the size of the duplicated or deficient region, and the size is usually expressed as a fraction of the total haploid complement of autosomes—that is, as a fraction of the sum of the lengths of all the autosomes in a normal gamete. Generally speaking, duplications smaller than about 1 to 2 percent of the haploid autosomal complement cause neither death nor abnormality. Those slightly larger than 1 to 2 percent cause a variety of abnormalities, but many do not cause death. The chance of survival decreases as the size of the duplication increases, however, so that flies that carry a duplication of more than 10 percent of the haploid autosomal complement almost never survive.

The effects of deficiencies are, as a rule, more severe than those of duplications. Although only a small minority of deficiencies smaller than 0.5 percent of the haploid autosomal complement will cause the death of the carrier, about half the deficiencies smaller than 1 percent of the haploid autosomal complement will cause death. Again, the severity of the effect depends on the size of the deficient region; flies with deficiencies larger than 3 percent of the haploid autosomal complement rarely, if ever, survive.

Duplications and deficiencies in humans seem to follow roughly the same pattern of effects as in *Drosophila*. Complete certainty is, of course, impossible because far less information is available. Until recently, only relatively large duplications or deficiencies could be detected in the light microscope. Now, with the advent of techniques that reveal specific banding patterns on human chromosomes, smaller aberrations can be detected, but the smallest ones that can be identified are still far larger than the smallest ones that can be seen in fruit flies. The finer level of resolution possible in fruit flies is due to a peculiar type of chromosome structure found in salivary glands and certain other tissues in fruit fly larvae. In fruit flies, homologous chromosomes undergo gene-for-gene pairing in *somatic* cells. In the salivary glands and a few other tissues, these paired and extended chromosomes undergo about 10 consecutive replications without intervening cell division; these replications result in giant chromosomes called **polytene** chromosomes, which are characterized by a spectacular pattern of bands (Figure 7.2). Substantial evidence suggests very strongly that each band (or perhaps each interband) in the polytene chromosomes corresponds to a single gene, and there are about 5000 such bands. In any event, the occurrence of polytene chromosomes in *Drosophila* permits a fine-scale resolution of small chromosomal aberrations. Unfortunately, human chromosomes are many times smaller than the giant salivary-gland chromosomes of *Drosophila*, and relatively few chromosome bands are produced with current techniques such as G banding, so that each band in a

human chromosome contains perhaps tens or hundreds of loci. Therefore a small deficiency or duplication of only a few loci in fruit flies could be detected microscopically, whereas in humans the detection would be impossible. Consequently, the effects of extremely small duplications and deficiencies in humans are virtually unknown.

The effects of larger ones in humans are known, however, and the total amount of duplication or deficiency that a human being can tolerate and still survive appears to be about the same relative percentage of the haploid autosomal complement as in *Drosophila*. Consider first deficiencies. Many children have been discovered who have partial deletions of any one of a number of chromosomes. Among these are patients who have a deletion of part of the short arm of chromosome 4, 5, or

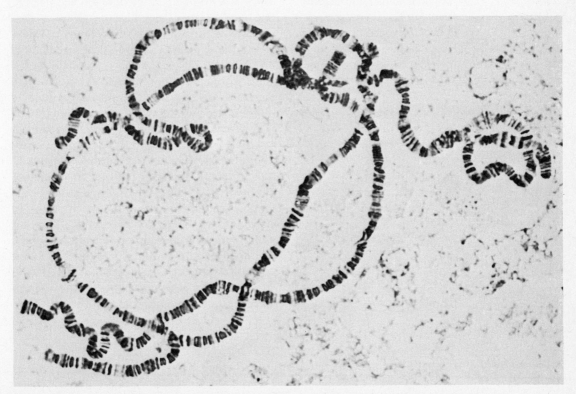

Figure 7.2   Giant polytene chromosomes found in salivary glands of the fruit fly *Drosophila melanogaster*.

18 or of part of the long arm of chromosome 13, 18, or 21. Such partial deletions are designated by a minus sign following the symbol for the chromosome arm; the deletions mentioned above would thus be designated 4p−, 5p−, 18p−, 13q−, 18q−, and 21q−, respectively (see Table 7.1). Individuals with any of these deletions suffer from various physical abnormalities, and all have varying degrees of mental retardation.

The best-known human deletion is probably that of part of the short arm of chromosome 5 (Figure 7.3). Affected people have 46 chromosomes, but part of the short arm of one of the number 5 chromosomes is missing. The abnormalities caused by the deletion are comparatively minor, except for mental retardation. A distinctive feature of the syndrome is the peculiar meowing and catlike sound of the infant's cry, which results from a malformation of the larynx. The name of the syndrome—the **cat-cry** (or **cri-du-chat**) syndrome—derives from this, although the condition is also known as **Lejeune syndrome**. The majority of affected children are female, presumably because of

increased fetal mortality of affected males. All affected children are physically and mentally retarded, and many have heart defects, a small skull with abnormally shaped ears, squinting or crossed eyes, and various other abnormalities. The condition is very rare.

The syndromes caused by deletion of part of the short arm or part of the long arm of chromosome 18 are also associated with mental and physical retardation. The physical abnormalities tend to be variable, however. Children with partial deletions of the short arm of chromosome 18 sometimes have ear and jaw malformations [Figure 7.4 (*a*)]; those with partial deletions of the long arm of 18 frequently have rather severe eye and ear defects [Figure 7.4(*b*)]. Taking into account the size of the deleted segments in these and other deletion syndromes, the maximum length of chromosome that can be deleted and still be compatible with survival seems to be about 2 or 3 percent of the autosomal complement. This is about the same relative size as can be tolerated in fruit flies.

Turning to duplications, the picture that emerges for humans is again comparable to that in fruit flies. Small duplications tend to be less severe than deletions of similar sizes. For example, children who carry a duplication of the cat-cry segment of chromosome 5 are mentally retarded, but their physical abnormalities are very mild, some being completely normal in physical appearance. Also, the trisomies of chromosome 13, 18, and 21 are hyperploid for these entire chromosomes, and whereas trisomic children can survive, at least in some cases, the corresponding monosomics (hypoploids) are invariably aborted spontaneously. Moreover, certain exceedingly rare individuals have

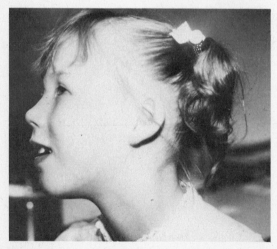

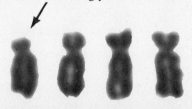

Figure 7.3 A child with the cat-cry syndrome (also called *cri-du-chat* or Lejeune syndrome) and a micrograph of the four B chromosomes. One chromosome has a deletion of part of the short arm (arrow), and banding studies reveal that the deletion is in chromosome 5.

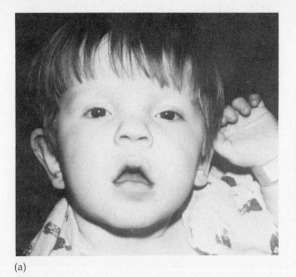
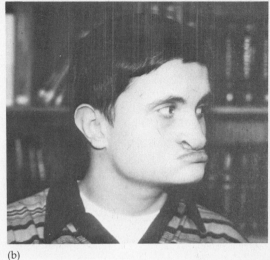

(a)   (b)

Figure 7.4 *(a)* Child with 18p− syndrome (deletion of part of the short arm of 18). *(b)* Child with 18q− syndrome (deletion of part of the long arm of 18).

double trisomy—simultaneous trisomy of chromosomes 18 and 21, for example. Since chromosomes 13, 18, and 21 constitute roughly 4, 3, and 2 percent of the length of a haploid set of autosomes, a duplication of 5 or 6 percent of the haploid autosomal complement can evidently be compatible with life, at least in a few instances.

These lengths of deletions or duplications are averages. The actual chance of survival of a child with a duplication or deficiency depends not only on the overall length of the abnormal segment but also on which part of which chromosome is involved. A particular duplication or deficiency will always cause death if it carries a dosage-sensitive locus, and the number and position of dosage-sensitive loci in humans are not known.

Unequal Crossing-Over

Chromosomes that carry duplications of a particular block of genes have the potential for creating new and different chromosomes that carry still more copies of the block. These new chromosomes result from crossing-over in an individual who is homozygous for the duplication-bearing chromosome. In such an individual, synapsis between the duplication-bearing homologues will ordinarily occur as shown in Figure 7.5(*a*). (The duplicated block of genes is indicated by the shading.) With this configuration in synapsis, crossing-over will not generate any new types of chromosomes. On occasion, however, synapsis between the duplication-bearing homologues will occur as shown in Figure 7.5(*b*), where the left-hand block of one chromosome pairs with the right-hand block of the other, with the other blocks of genes forming loops that remain unpaired. When synapsis occurs as in Figure 7.5(*b*), crossing-over within the paired block of genes [see Figure 7.5(*c*)] will generate new chromosome types, which are indicated in the anaphase I configuration at the right of Figure 7.5(*c*). Note in the anaphase configuration that two chromatids (the two not involved in the crossover) carry the original duplication; of the two chromatids that were involved in the crossover, one carries a single copy of the block of genes, and the other

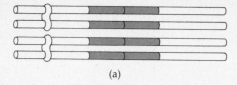

(a)

Figure 7.5 *(a)* Normal synapsis of chromosomes bearing a duplication. *(b)* Mispairing of duplicated regions. *(c)* Crossing-over within a mispaired duplication leads to the anaphase I configuration shown at the right; note that one of the chromatids involved in the unequal crossover now carries a triplication, whereas the other chromatid carries a single copy of the region.

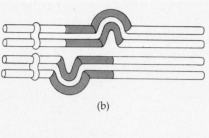

(b)

Crossing-over

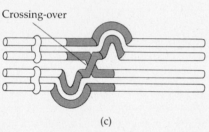

(c)

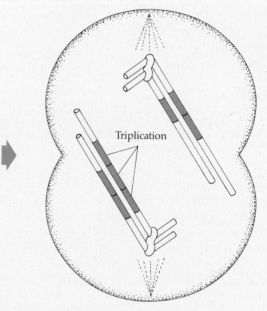

Triplication

carries a **triplication** (three copies) of the block of genes. The mispairing and crossing-over between duplicated blocks of genes that generate new chromosome types are known as **unequal crossing-over**. Of course, unequal crossing-over can occur in individuals who carry a duplication on one homologue and a triplication on the other, or in individuals who are homozygous for a triplication; the result is new chromosomes that carry still more copies of the block of genes.

Although Figure 7.5 illustrates a duplication that is large relative to the size of the chromosome, unequal crossing-over can also involve very small duplications of blocks of nucleotides along the DNA. A spectacular example of the long-term consequences of many separate events of unequal crossing-over is found in the human gene that codes for a protein called **apolipoprotein A1**. This gene carries 13 copies of a block of DNA consisting of 33 nucleotides, presumably generated by the repeated occurrence of unequal crossing-over. (Unequal crossing-over at the molecular level is discussed further in Chapter 11.)

Inversions

Duplications and deficiencies are characterized by the gain or loss of chromosomal material from the cell. Genes are actually missing or present in excess because segments of chromosomes are missing or repeated. One might therefore expect that large duplications or deficiencies would have harmful phenotypic effects. Inversions and translocations have more subtle effects, however. In these chromosomal abnormalities, only the arrangement of genes is altered; certain genes are moved and get new neighbors. Sometimes this repositioning alone causes the relocated genes to function abnormally, particularly when genes in euchromatin are moved to a position in or near heterochromatin. (See Chapter 2 for a discussion of euchromatin and heterochromatin.) Autosomal genes near the breakpoint of translocations involving the X chromosome sometimes function abnormally in females owing to a sort of "spreading effect" of X-chromosome inactivation. When a gene's function is altered by changing its position, the phenomenon is called a **position effect**. However, genes ordinarily function normally irrespective of their neighbors, so position effects are rather rare. Consequently, the carriers of inversions and translocations are phenotypically normal even though the usual arrangement of genes has been altered. This is not to say that inversions and translocations cannot cause trouble; they can and often do, particularly during meiosis.

Paracentric Inversions

Figure 7.6 depicts the events involved in the formation of an inversion. A chromosome that is broken in two places (*a*) undergoes a reversal of the broken segment (*b*), followed by restitution of the broken ends (*c*), resulting in an inverted chromosome (*d*). The normal order of genes on this chromosome is *ABCD*, but the inverted chromosome has the order *ACBD*. The inversion in Figure 7.6(*d*) is a special type of inversion called a **paracentric inversion**, so called because the inverted segment does *not* include the centromere. Note that no genes are gained or lost because of the inversion; some genes merely occur in reverse of the normal order. Aside from the possibility of position effects, the inverted chromosome might have a gene that functions abnormally if one of the breaks happened to occur in the middle of the

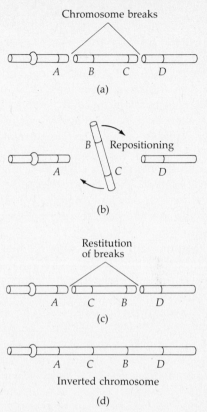

Figure 7.6 Origin of a chromosome with a paracentric inversion (*d*) following the events of breakage and reunion in (*a*) through (*c*).

gene and thereby disrupted the genetic information in the gene. Such unlucky breaks seem to be rare; even when they occur, they have little or no phenotypic effects in heterozygotes because the normal allele on the noninverted homologous chromosome will compensate. Because inversions in human chromosomes seem to be relatively rare, the great majority of individuals who carry an inversion will be heterozygous for it; that is, the homologue of the inverted chromosome will have the normal gene sequence.

Heterozygous inversions cause no problems during mitosis. Both homologous chromosomes—normal and inverted alike—replicate during interphase; both condense and coil up during prophase; both proceed independently to the equatorial plate during metaphase; the centromeres of both split during anaphase; and the sister chromatids of both are pulled in opposite directions. The nuclei formed during telophase as the daughter cells pinch apart both have 46 chromosomes. One chromosome in each daughter nucleus will have the inverted sequence; its homologue will have the normal sequence. Because mitosis is normal, both daughter cells receive exactly the correct number and kind of genes. A cell that carries an inversion can therefore give rise to completely healthy daughter cells.

Matters are not so simple for heterozygous inversions during meiosis. When synapsis occurs during prophase I, all regions of the inverted chromosome try to pair gene for gene with the corresponding regions of the normal chromosome. To accomplish this, the inverted region in one of the chromosomes must form a **loop**, as illustrated in Figure 7.7(*a*); this loop will permit gene-for-gene pairing with the homologue. Thus, except for a relatively small region around the breakpoints themselves, the normal and the inverted chromosomes can synapse all along their lengths. Because of chromosome pairing in somatic cells in *Drosophila* and other dipteran flies, such inversion loops can easily be observed in the polytene chromosomes of the salivary glands. Figure 7.8 shows an inversion loop (arrow) in salivary-gland chromosomes of the black fly, *Eusimilium aureum*.

The consequence of a heterozygous inversion depends on whether or not a crossover occurs within the inversion loop. In the event that no crossover occurs in the inversion loop, the subsequent stages of meiosis occur without mishap; two of the resulting gametes receive the inverted chromosome, and the other two gametes receive its noninverted homologue.

When a crossover does occur within the inversion loop, then, as indicated in Figure 7.7, trouble arises. Part (*a*) shows the inversion loop with a crossover at the top of the loop, and part (*b*) shows the result of this configura-

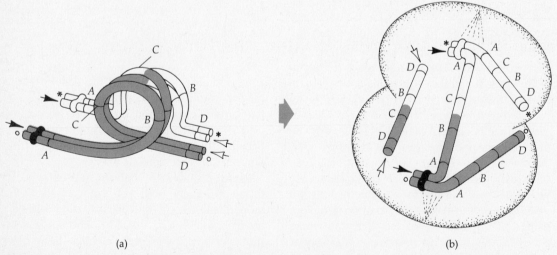

(a)   (b)

Figure 7.7   *(a)* Synapsis in an individual who is heterozygous for a paracentric inversion showing a crossover within the inversion loop. *(b)* Anaphase I configuration resulting from the crossover in *(a)*. One of the chromatids involved in the crossover is a dicentric (solid arrows); the other is an acentric (open arrows).

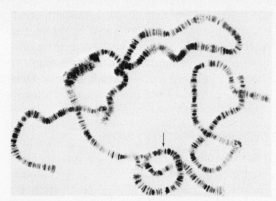

Figure 7.8 Inversion loop (arrow) in giant salivary-gland chromosomes of the dipteran black fly, *Eusimilium aureum*.

tion at anaphase I. One chromatid (its tips indicated by small circles) has the normal gene sequence, and another chromatid (tips indicated by asterisks) has the inverted sequence; these are the two chromatids that were not involved in the crossover. Of the chromatids that were involved in the crossover, one is a dicentric (solid arrows) and one is an acentric (open arrows), as can be verified by examining the corresponding chromatids in the configuration in Figure 7.7(*a*). Moreover, the dicentric chromatid is duplicated for *A* and deficient for *D*, whereas the acentric chromatid is deficient for *A* and duplicated for *D*. Of the four possible gametes formed from the anaphase I cell in Figure 7.7(*b*), one will carry the normal chromosome, one will carry the inverted chromosome, and two will have major chromosomal abnormalities as a result of the crossover. Indeed, the dicentric and the acentric often fail to be included in any telophase II nucleus, so the corresponding gametes completely lack the chromosome in question. In any event, *the chromatids involved in a crossover within the inversion loop of a heterozygous paracentric inversion become dicentric or acentric chromosomes*. Consequently, if an individual is heterozygous for a large inversion so that crossing-

over within the inversion loop will frequently occur, then there is substantial risk of major chromosomal abnormality among the offspring. In addition, among the phenotypically normal offspring (whether the inversion is large or small, and whether a crossover occurs or not), half will inherit the inverted chromosome and the other half will inherit its non-inverted homologue.

Pericentric Inversions

Figure 7.9 illustrates the formation of the second principal type of inversion, which is called a **pericentric inversion** because it *does* include the centromere. When the breakpoints of a pericentric inversion are at appropriate distances from the centromere, the inversion can alter the physical appearance of the chromosome. The chromosome in Figure 7.9(*a*) is a metacentric chromosome, for example, but its inverted derivative in part (*b*) is a submetacentric. Some pericentric inversions can therefore be detected microscopically because they change the relative position of the centromere.

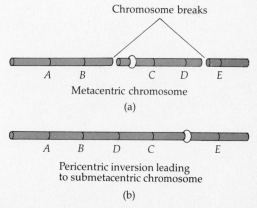

Figure 7.9 Origin of a chromosome with a pericentric inversion *(b)* following chromosome breakage illustrated in *(a)*. Note that the original chromosome is a metacentric but the inversion-bearing chromosome is a submetacentric.

As with paracentric inversions, heterozygous pericentric inversions cause no problems in mitosis because each chromosome behaves independently of its homologue. In meiosis, synapsis again produces an inversion loop [Figure 7.10(a)], and crossing-over within this loop results in abnormal chromatids [Figure 7.10(b)]. In the anaphase I configuration in Figure 7.10(b), it can be seen that one chromatid (its tips marked by small circles) is structurally normal and another (tips marked by asterisks) has the pericentric inversion; these chromatids are the ones not involved in the crossover. Of the two chromatids that were involved in the crossover, one (solid arrows) has a duplication of A and a deficiency of D, and the other (open arrows) has a deficiency of A and a duplication of D; both chromatids are **monocentric** (have a single centromere), in contrast to the situation with paracentric inversions outlined in Figure 7.7. In short, *the chromatids involved in a crossover within the inversion loop of a heterozygous pericentric inversion will carry duplications and deficiencies* and will lead to offspring who have major chromosomal abnormalities. Among the phenotypically normal offspring of an individual who is heterozygous for a pericentric inversion, half will receive the inverted chromosome and the other half will receive its noninverted homologue.

Incidence of Inversions

The prevalence of inversions in human chromosomes is largely a matter of speculation because inversions are more difficult to detect than almost any other chromosomal abnormality. In the absence of some way to recognize the linear sequence of regions along a chromosome, inversions cannot be recognized in the microscope unless they happen to be pericentric inversions that change the relative position of the centromere as illustrated in Figure 7.9. The advent of chromosome-banding techniques such as Giemsa staining makes it possible to identify inversions that are large enough to produce a recognizable alteration in the sequence of bands, but small inversions are still extremely difficult to detect. Despite the handi-

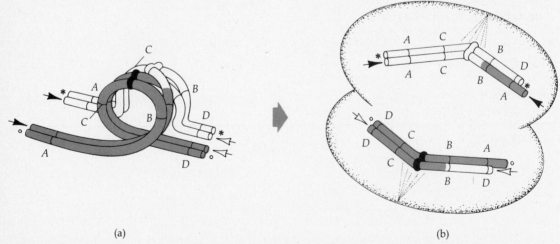

(a)

(b)

Figure 7.10 *(a)* Synapsis in an individual who is heterozygous for a pericentric inversion showing a crossover within the inversion loop. *(b)* Anaphase I configuration resulting from the crossover in *(a)*. One of the chromatids involved in the crossover has a duplication of A and a deficiency of D (solid arrows); the other chromatid has a deficiency of A and a duplication of D (open arrows).

caps in identification, inversions in most human chromosomes have been reported. Various studies suggest that perhaps 1 individual per 1000 carries a detectable inversion, but this is a very rough estimate intended only to convey the likely order of magnitude. Whatever the actual incidence of inversions may be, individuals who are heterozygous for any large inversion except an inversion of the Y chromosome have a significant risk of producing chromosomally abnormal gametes because of crossing-over within the inversion loop. (The Y chromosome is an exception because crossing-over between the X and Y does not occur.)

Reciprocal Translocations

Reciprocal translocations involve an interchange of parts between nonhomologous chromosomes. The origin of a reciprocal translocation is illustrated in Figure 7.11; the process involves a break in each of two nonhomologous chromosomes (*a*), repositioning of the terminal segments and restitution of the breaks (*b*), resulting in the reciprocally translocated chromosomes shown in (*c*). Notice that no chromosomal material is gained or lost because of the translocation; all the genes on both chromosomes will function normally provided there are no position effects and provided the breaks did not slice through the middle of a gene. Notice also that each chromosome in the translocation must be monocentric. If the repositioning and restitution had joined *Y* with *B* and *Z* with *C*, the result would be a dicentric chromosome (*ABY*) and an acentric chromosome (*ZCDE*). Neither of these chromosomes could proceed normally through mitotic cell divisions and both would very quickly be lost, perhaps resulting in the death of the cell. An

Chromosome breaks

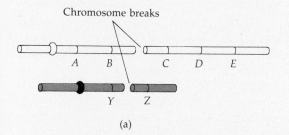

(a)

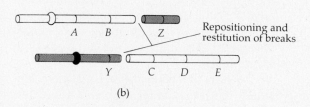

(b)

Figure 7.11 Origin of a reciprocal translocation *(c)* from breaks in two nonhomologous chromosomes *(a)* followed by repositioning of broken ends and restitution of the breaks *(b)*.

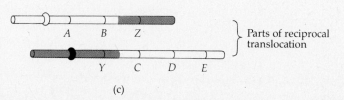

(c)

individual who carries the parts of the reciprocal translocation in Figure 7.11(*c*) would also carry the normal homologue of both chromosomes (i.e., *ABCDE* and *YZ*); such a person is said to carry a **balanced translocation** because the individual is euploid. The composite data in Table 6.4 of Chapter 6 indicate that about 1 in 500 live-born children carries a balanced translocation.

Carriers of balanced reciprocal translocations, like carriers of inversions, are not themselves phenotypically abnormal, but they often produce eggs or sperm carrying duplications and deficiencies. Again like inversions, translocations can be passed intact from generation to generation, from carrier to carrier, without being detected. On the other hand, whereas abnormal gametes arise from heterozygous inversions only if a crossover occurs in the inverted region, abnormal gametes may arise from heterozygous translocations in the absence of crossing-over.

Mitosis in cells that carry a heterozygous reciprocal translocation proceeds normally and no problems arise because each chromosome replicates and the chromatids disjoin independently of all the others. Consequently a fertilized egg that carries a balanced translocation gives rise to a phenotypically normal child. In meiosis, however, problems arise. The synapsis of the interchanged chromosomes with their normal homologues must involve all four chromosomes, forming a **quadrivalent** (Figure 7.12) rather than a bivalent. With the quadrivalent configuration of chromosomes illustrated in Figure 7.12, every gene along the chromosomes can synapse with its homologous gene across the way, except possibly for genes located near the translocation breakpoints.

The problem caused by a quadrivalent is that it has four centromeres instead of the two centromeres in a normal bivalent. During anaphase I, the centromeres of the quadrivalent

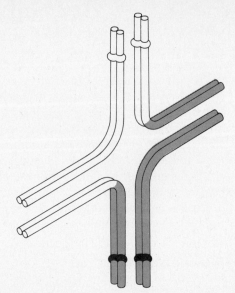

Figure 7.12   Quadrivalent formed during synapsis in an individual carrying a heterozygous reciprocal translocation. The chromosomes involved in the translocation are partly shaded and partly unshaded. Note that the quadrivalent has four centromeres; a normal bivalent has two centromeres.

will be distributed so that each telophase I nucleus will receive exactly two of them—but which two? This depends on how the quadrivalent orients itself on the equatorial plate during metaphase I, which in turn depends partly on the number and distribution of chiasmata in the arms of the quadrivalent. Since chiasmata greatly complicate the situation with reciprocal translocations without changing the essential outcomes, chiasmata have been ignored in Figure 7.12.

Which two chromosomes a gamete ultimately receives from the quadrivalent is determined by the manner of segregation of the quadrivalent in anaphase I. The quadrivalent can align on the equatorial plate in metaphase I in any of three ways, and these alignments determine the mode of segregation. The three possible orientations of the quadrivalent at

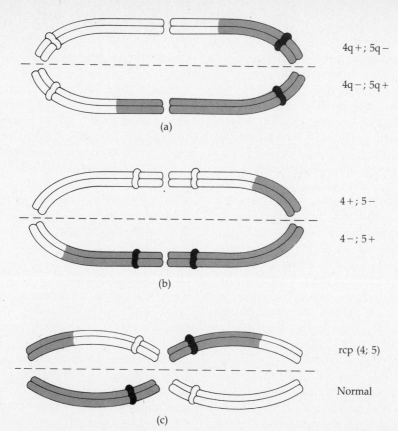

4q +; 5q −

4q −; 5q +

(a)

4 +; 5 −

4 −; 5 +

(b)

rcp (4; 5)

Normal

(c)

Figure 7.13 Possible alignments of a quadrivalent involving a heterozygous reciprocal translocation at metaphase I of meiosis. The broken line indicates the equatorial plate. For purposes of illustration, the translocation is assumed to involve the long arms of chromosomes 4 (unshaded) and 5 (shaded). Alignment *(a)* leads to adjacent-1 segregation and produces the unbalanced gametes shown at the right (duplicated for part of 4q and deficient for part of 5q, or deficient for part of 4q and duplicated for part of 5q). Alignment *(b)* leads to adjacent-2 segregation and produces gametes that are either duplicated for most of 4 and deficient for most of 5 or deficient for most of 4 and duplicated for most of 5. Alignment *(c)* leads to alternate segregation. Alternate segregation produces balanced gametes; half carry both parts of the reciprocal translocation and the other half are chromosomally normal.

metaphase I are shown in Figure 7.13, where the broken lines represent the equatorial plate. To visualize how the structure in Figure 7.12 can give rise to the orientations in Figure 7.13, recall from Chapter 2 that, late in prophase I, homologous chromosomes seem to repel each other until only their tips remain in association. Such repulsion in the quadrivalent opens out

the cross-shaped region in the center and forms a ring of four chromosomes in tip-to-tip association. The ring so formed can orient on the equatorial plate in configuration (*a*) or in configuration (*b*) of Figure 7.13. Alternatively, the ring can undergo a twist in the center and then align in configuration (*c*).

For purposes of illustration, the reciprocal translocation in Figure 7.13 is assumed to involve the long arms of chromosomes 4 (unshaded) and 5 (shaded) [symbolized as rcp(4;5) or more precisely as rcp(4q;5q)—see Table 7.1], and the symbols in Figure 7.13 refer to the types of gametes that would ultimately result from each mode of segregation. (Of course, each configuration in Figure 7.13 actually gives rise to four gametes, but in each case the two that arise from the chromosomes above the horizontal broken line will be chromosomally identical, and the two that arise from the chromosomes below the broken line will be chromosomally identical.)

The mode of segregation in Figure 7.13(*a*) is known as **adjacent-1** segregation because adjacent chromosomes having non-homologous centromeres proceed to the same pole at anaphase 1. Of the four gametes produced by adjacent-1 segregation, two will have a duplication of part of the long arm of chromosome 4 (unshaded chromosome) and a deficiency of part of the long arm of chromosome 5 (shaded chromosome), which is indicated by the symbol 4q+;5q−. The other two gametes arising from adjacent-1 segregation will have a partial deficiency of 4q and a partial duplication of 5q, symbolized 4q−;5q+. In short, *all the gametes produced by adjacent-1 segregation will have duplications and deficiencies.*

The mode of segregation in Figure 7.13(*b*) is called **adjacent-2** segregation because adjacent chromosomes having homologous centromeres proceed to the same pole at anaphase I. Gametes formed from adjacent-2 segregation will have either a duplication of most of 4 and a deficiency of most of 5 (symbolized 4+;5−), or a deficiency of most of 4 and a duplication of most of 5 (4−;5+). Here again, *all the gametes resulting from adjacent-2 segregation will have duplications and deficiencies.*

The gametes that arise from adjacent segregation (either adjacent-1 or adjacent-2) are known as **unbalanced** gametes because they

are aneuploid, and a zygote formed from such a gamete is said to carry an **unbalanced translocation**. When the interchanged parts of the reciprocal translocation are large, then zygotes receiving the unbalanced gametes will carry large duplications and deficiencies and would be expected to undergo spontaneous abortion. Indeed, the frequency of unbalanced translocations among chromosomally abnormal fetuses in spontaneous abortions is about 3 percent (see Table 6.4). On the other hand, if the interchanged chromosomal parts are small, then certain of the unbalanced combinations may be compatible with live birth; it may be noted in Table 6.4 that about 1 per 1600 live births carries an unbalanced translocation.

The mode of segregation in Figure 7.13(*c*) is known as **alternate** segregation because each gamete receives alternate (i.e., nonadjacent) chromosomes from the original quadrivalent. In alternate segregation, two gametes receive both parts of the reciprocal translocation [denoted rcp (4;5)], and the other two receive the structurally normal homologous chromosomes. The first type of gamete will lead to a phenotypically normal offspring that carries the translocation, and the second type will lead to a completely normal offspring that has no chromosomal abnormality. Thus, *with alternate segregation, all the offspring will be phenotypically normal, but half will be carriers of the reciprocal translocation.*

Theoretically, the three modes of segregation in Figure 7.13 would be expected to be equally likely, leading to $\frac{2}{3}$ duplication/deficient gametes, $\frac{1}{6}$ rcp gametes, and $\frac{1}{6}$ normal gametes. However, as noted earlier, the actual frequencies of the modes of segregation depend on the breakpoints of the translocation and on the number and position of chiasmata in the arms of the quadrivalent. Nevertheless, translocation carriers will always be expected to produce a substantial fraction of unbalanced gametes and to have 50 percent translocation carriers among their phenotypically normal offspring. If the duplications and deficiencies in the aneuploid gametes are large, the zygotes originating from them will abort very early in development, perhaps so early that the pregnancy is unrecognized. As a result, many carriers of translocations are unaware of the fact and may never know that half of their phenotypically normal offspring are also carriers.

Robertsonian Translocations

A second type of translocation is sufficiently important in human genetics that it is designated by a special term. A **Robertsonian translocation**, illustrated in Figure 7.14, involves the interchange of the long arm of one acrocentric chromosome with the short arm of a nonhomologous acrocentric chromosome [see Figure 7.14(a) and (b)], leading to a tiny metacentric chromosome consisting of both short arms along with a large metacentric or submetacentric chromosome consisting of both long arms [see Figure 7.14(c)]. The tiny metacentric chromosome frequently undergoes mitotic or meiotic nondisjunction and is lost, but its loss has no detectable phenotypic effect owing to the small amount of chromosomal material in this tiny metacentric. Such small metacentrics are sometimes discovered in screening programs, however. Indeed, in the Boston study of 13,751 consecutive newborns discussed in Chapter 2, some 5 phenotypically normal individuals were found to have such a small metacentric chromosome, for an overall incidence of 1 per 2750 newborns. Nevertheless, in most cases, the small metacentric is lost, leaving the large metacentric or submetacentric, which constitutes the Robertsonian translocation. In practical terms, a Robertsonian translocation results in a sort of "fusion" of the long arms of two acrocentrics, and for this reason a Robertsonian translocation is sometimes referred to as a **chromosomal fusion**.

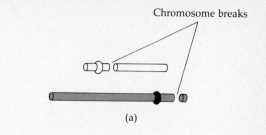

Chromosome breaks

(a)

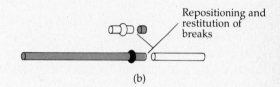

Repositioning and restitution of breaks

(b)

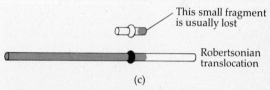

This small fragment is usually lost

Robertsonian translocation

(c)

Figure 7.14 Origin of a Robertsonian translocation *(c)* from nonhomologous acrocentrics broken as shown in *(a)* with restitution as in *(b)*. In most cases, the tiny metacentric chromosome consisting of both short arms undergoes nondisjunction and is lost.

Chapter 6 referred to certain kinships in which Down syndrome (trisomy 21) had a high risk of occurrence and recurrence. Such kinships account for about 3 percent of all cases of Down syndrome, and the reason for the high risk is the presence of a Robertsonian translocation involving chromosome 21. Figure 7.15 shows the karyotype of a female carrying such a Robertsonian translocation, this one involving chromosomes 15 and 21; note that the individual carries one normal chromosome 15, one normal chromosome 21, and the Robertsonian translocation (indicated by the arrow). In this case, as in most similar cases, the small metacentric corresponding to the short arms of chromosomes 15 and 21 has been lost.

The woman with the karyotype in Figure 7.15, although she has 45 chromosomes, is phenotypically normal. She has, however, a high risk of producing offspring with Down syndrome because of the various modes of segregation of the Robertsonian translocation. With a Robertsonian translocation, there are again three modes of alignment on the equatorial plate at metaphase I. These are illustrated in Figure 7.16, where the unshaded chromosome represents 15, the shaded chromosome represents 21, and the symbol rob(15;21) represents the Robertsonian translocation. In accordance with convention in human genetics (see Table 7.1), a + or − sign preceding a chromosome number designates a duplication or deficiency of the entire chromosome.

Segregation according to the alignment shown in Figure 7.16(a) produces gametes with either a duplication of chromosome 21 [i.e., rob(15;21); + 21] or a deficiency of chromosome 21 (i.e., −21). The first type of gamete will lead to an offspring who has Down syndrome, and the second type will lead to an embryo that undergoes spontaneous abortion. The second mode of segregation [see Figure 7.16(b)] yields gametes that have either a du-

plication of chromosome 15 [rob(15;21); +15] or a deficiency of chromosome 15 (−15); all zygotes formed from these gametes will undergo spontaneous abortion. Finally, the third mode of segregation [see Figure 7.16(c)] produces gametes that carry either the Robertsonian translocation [rob(15;21)] or the nontranslocated homologous chromosomes; offspring from the rob (15;21)-bearing gametes will be phenotypically normal carriers of the translocation, and offspring from the normal gametes will be phenotypically and chromosomally normal.

In summary, *a carrier of a Robertsonian translocation involving chromosome 21 will produce three categories of live-born offspring: (1) offspring with Down syndrome, (2) phenotypically normal carriers of the Robertsonian translocation, and (3) phenotypically and chromosomally normal offspring.* Moreover, the frequency of offspring in categories (2) and (3) will be equal. Strangely enough and for reasons that are not yet understood, the frequency of gametes leading to offspring with Down syndrome depends on which parent is the carrier. The proportion of Down-syndrome children born to carrier mothers is high—15 to 20

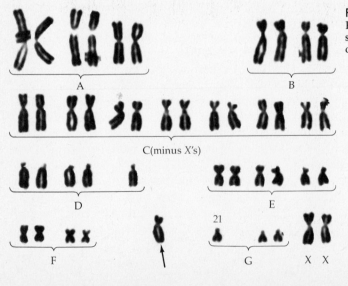

Figure 7.15 Karyotype of a woman with a Robertsonian translocation involving chromosomes 15 and 21 (arrow). Note that there are only 45 chromosomes.

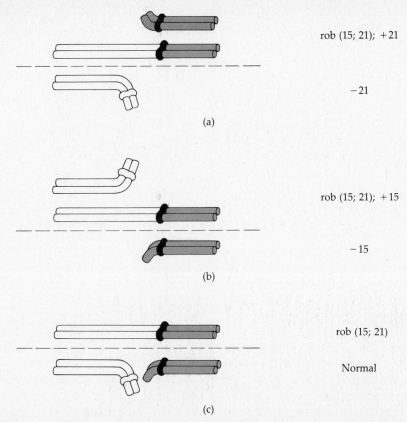

rob (15; 21); +21

−21

(a)

rob (15; 21); +15

−15

(b)

rob (15; 21)

Normal

(c)

Figure 7.16 Possible metaphase I alignments in an individual carrying a heterozygous Robertsonian translocation involving chromosomes 15 and 21. The broken line represents the equatorial plate, and the resulting gametes are indicated at the right. Gametes from alignment *(a)* have either a duplication of chromosome 21 (leading to a trisomy-21 zygote) or a deficiency of chromosome 21 (leading to a monosomy-21 zygote). Gametes from alignment *(b)* have either a duplication of chromosome 15 (leading to a trisomy-15 zygote) or a deficiency of chromosome 15 (leading to a monosomy-15 zygote). Both monosomics and the trisomy 15 will undergo spontaneous abortion, but the trisomy-21 zygote may lead to a live-born offspring with Down syndrome. The alignment shown in *(c)* produces balanced gametes, half carrying the Robertsonian translocation and the other half being chromosomally normal.

percent or more; the proportion born to carrier fathers is lower—2 to 5 percent or less. This discrepancy is true of translocations in general. Women who carry a balanced translocation seem to be 5 to 10 times more likely to produce unbalanced gametes than are men.

Further Aspects of Translocations

In addition to their importance in genetic counseling, translocations are important in other respects, two of which deserve a brief mention.

Somatic Cell Genetics In Chapter 4 we discussed the fusion of somatic cells as a central technique in the assignment of genes to human chromosomes. Once loci are known to be syntenic, fragmentation mapping is one approach to determining their sequence on the chromosome. However, hybrid cells involving translocations are also useful in determining the physical location of genes on a chromosome. The method can be illustrated by means of a hypothetical example involving a gene known to be on chromosome 15 in a hybrid-cell line having a translocation between chromosome 15 and, say, the X chromosome. As outlined in Chapter 4, loss of human chromosomes from such cell lines permits assignment of genes to chromosomes. If the gene in question behaves like an X-linked gene among clones derived from the translocation-bearing cell, then the gene must be on the segment of chromosome 15 that had been translocated to the X. Alternatively, if the gene still behaves like an autosomal gene, then it must be on the part of chromosome 15 that was not translocated to the X. Studies of a series of translocations having different breakpoints permit genes to be assigned to chromosomal regions with some precision.

Evolution Chromosomal changes occur with sufficient frequency during the course of evolution that related species rarely have identical chromosomes. (For our purposes, **evolution** may be defined as cumulative change in the genetic makeup of a population, and a **species** may be defined as a group of organisms that can interbreed and produce fertile offspring.) Indeed, some evolutionary biologists perceive chromosomal changes as playing a key role in the origin of new species. Human beings are not exempt from the forces that govern the evolutionary process, and chromosomal changes have occurred in our own evolution. Comparison of human chromosomes with those of the chimpanzee reveals at least five recognizable pericentric inversions, for example. (Humans and chimpanzees are thought to have begun their evolution as separate species approximately 10 million years ago.) Remarkably, chimpanzees have 48 chromosomes, not 46, but chromosome banding studies have allowed the chromosomal correspondences between humans and chimpanzees to be de-

duced. In the chimpanzee (*Pan troglodytes*), no chromosome has a banding pattern like that of human chromosome 2. However, the chimp has two acrocentric chromosomes not found in humans. One of these acrocentrics has a banding pattern like that of human 2p; the banding pattern of the other is like that of human 2q. The presence of these two acrocentrics is characteristic of nonhuman primates, so the inference is that human chromosome 2 arose as a fusion of the two acrocentrics during evolution. The fusion was not a typical sort of Robertsonian fusion, however. It involved the fusion of the short arms of the acrocentrics to form a dicentric chromosome, but the fusion was also accompanied by the inactivation of one of the centromeres by some unknown process. Modern humans arose from a population in which this fused chromosome had become homozygous.

Ring Chromosomes

Two other types of abnormal chromosome structure deserve a brief discussion even though they are less important than duplications, deletions, inversions, and translocations. One of these is a **ring chromosome** (i.e., a chromosome that forms a closed circle), which originates by the process illustrated in Figure 7.17. The easiest way to visualize ring-chromosome formation is to imagine that the tips of a chromosome come together and fuse. This process does not seem to occur, however. The tips of chromosomes seem to be very special structures that do not undergo fusion; in this respect, the normal tip of a chromosome is completely different from a broken end of a chromosome, because broken ends do tend to fuse. That the tips of chromosomes should be protected from undergoing fusion is perfectly understandable, of course; wholesale fusion of chromosome tips would produce chromosomal chaos—a profusion of rings and dicentrics and other abnormal structures. The special proper-

Chromosome breaks

(a)

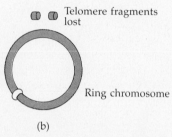

Telomere fragments lost

Ring chromosome

(b)

Figure 7.17 Origin of a ring chromosome *(b)* from breakage as shown in *(a)*, with restitution and loss of acentric telomere fragments.

ties of normal chromosome tips are designated by a special term; normal tips are called **telomeres**.

The usual manner of formation of a ring chromosome involves two chromosome breaks that occur near the telomeres of a chromosome [see Figure 7.17(*a*)]. Fusion of the broken ends then forms the ring chromosome (*b*) along with acentric fragments carrying the telomeres, which are lost from the cell during cell division.

Individuals with a ring chromosome are usually chromosomal mosaics because the ring tends to be lost. Recall from Chapter 1 that sister-chromatid exchange is a normal event in cells undergoing mitosis. In a structurally normal chromosome, sister-chromatid exchange has no extraordinary genetic consequences. Sister-chromatid exchange in a ring chromosome does have dramatic consequences, however. Figure 7.18 illustrates the sister chro-

matids of a ring chromosome during prophase of mitosis, and the chromatids are shown to have undergone sister-chromatid exchange. When the centromere splits during anaphase, the configuration in Figure 7.18 leads to a double-sized dicentric ring, which can be seen by following through the individual chromatids. As a consequence of the dicentric ring, the entire chromosome is usually lost from the cell, thus producing a chromosomal mosaic.

Isochromosomes

The final type of chromosomal abnormality to be discussed is an **isochromosome**—a chromosome that has two genetically identical arms. One mode of isochromosome formation is outlined in Figure 7.19. The normal manner of centromere splitting is illustrated in part (*a*). Rarely, however, a centromere can split incorrectly as shown in (*b*), resulting in two isochromosomes, one consisting of one arm of the original chromosome and the other consisting of the other arm. A cell with an isochromosome will have a duplication of one entire chromosome arm and a deletion of the other chromosome arm. Although isochromosomes are thought to be formed primarily by improper splitting of the centromere, they can

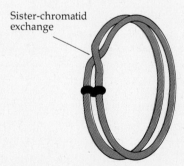

Sister-chromatid exchange

Figure 7.18 Sister-chromatid exchange during mitosis in a ring chromosome. Splitting of the centromere at anaphase leads to a double-sized ring chromosome that has two centromeres. This dicentric ring is usually lost from the cell.

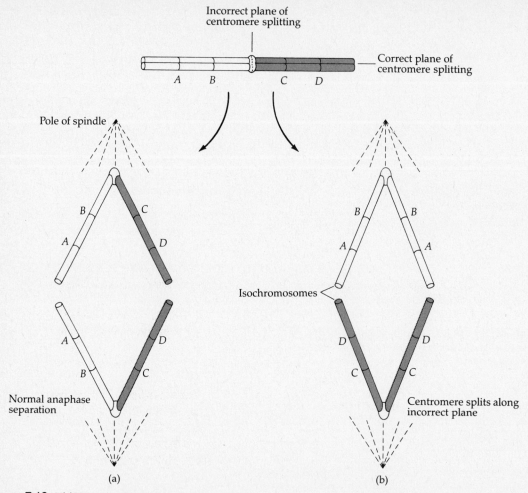

Figure 7.19 *(a)* Normal splitting of centromere at anaphase of mitosis leads to daughter cells having genetically identical chromosomes. *(b)* Splitting of centromere along an abnormal plane leads to two isochromosomes, one consisting of two copies of the left arm of the original chromosome and the other consisting of two copies of the right arm of the original chromosome. Note that each isochromosome is deficient for the opposite chromosome arm.

also be created in other ways, such as by a Robertsonian translocation involving homologous acrocentric chromosomes. In *Drosophila*, an isochromosome is usually referred to as a **compound chromosome** or an **attached chromosome**.

SUMMARY

1. Chromosome breaks sometimes occur spontaneously and can be induced by certain agents such as x-rays. Broken ends of chromosomes tend to undergo **restitution** (rejoining), but they sometimes rejoin incorrectly, creating chromosomes with abnormal structures. Chromosomes with no centromere (**acentrics**) or with two centromeres (**dicentrics**) tend to be lost during cell division, so the only types of abnormal chromosomes that are perpetuated are those that have exactly one centromere.

2. A cell is **euploid** if the relative dose of genes is the same as in a diploid; polyploid cells are euploid. A cell is **aneuploid** if the relative dose of genes is not the same as in a diploid; polysomic cells are aneuploid. Chromosomal abnormalities that change the arrangement but not the number of genes are called euploid (or **balanced**) rearrangements; chromosomal abnormalities that do change the dose of genes are called aneuploid (or **unbalanced**) rearrangements. An aneuploid cell with too few copies of certain genes is said to be **hypoploid** for the genes in question; a cell with too many copies of certain genes is said to be **hyperploid** for the genes in question.

3. A chromosome with a **duplication** carries two copies of certain genes. The phenotypic effects of a duplication depend on the size of the duplicated region and on the genes involved. Generally speaking, duplications of up to 5 or 6 percent of the haploid autosomal complement produce a variety of severe mental and physical abnormalities, but they are compatible with life. Small duplications may have virtually no phenotypic effects and, indeed, can generate chromosomes with **triplications** or even more copies of the genes by the process of **unequal crossing-over**.

4. A chromosome with a **deletion (deficiency)** has certain genes missing. As with duplications, the phenotypic effects of a deficiency depend on the size of the deficiency and the genes involved. On the whole, however, deficiencies tend to have more harmful phenotypic effects than duplications of the same genes. In humans, the largest deficiencies that are compatible with life involve 2 to 3 percent of the haploid autosomal complement; larger deficiencies are almost always lethal. One of the best-known human deficiencies is a deficiency of part of the short arm of chromosome 5 (i.e., 5p−), which is associated with the **cri-du-chat** (or **Lejeune**) syndrome.

5. Deficiencies, duplications, and other types of chromosome abnormalities can be studied in great detail in dipteran flies including the common fruit fly, *Drosophila melanogaster*. The great resolution is made possible because of a special type of giant chromosome found in salivary glands and certain other tissues of immature fly larvae. These giant chromosomes are called **polytene** chromosomes, and they result from the pairing of homologous chromosomes followed by about 10 consecutive replications without intervening cell division. The polytene chromosomes of *Drosophila* have about 5000 bands, and various lines of evidence suggest that each band (or perhaps each interband) corresponds to one gene.

6. A chromosome with an **inversion** has part of its genes in the reverse of the normal order. Inversions that do not include the centromere are known as **paracentric** inversions; those that do include the centromere are known as **pericentric** inversions. Inversions are balanced rearrangements that have few or no phenotypic effects in the carriers. However, individuals who are heterozygous for inversions can produce unbalanced gametes as a result of crossing-over within the inversion loop formed during synapsis of the inverted chromosome and its noninverted homologue. Crossing-over within the inversion loop of a heterozygous paracentric inversion produces a dicentric chromatid and an acentric chromatid; crossing-over within the inversion loop of a heterozygous pericentric inversion produces chromatids that have duplications and deficiencies.

7. A **reciprocal translocation** results from an interchange of parts between nonhomologous chromosomes. Reciprocal translocations are balanced rearrangements that usually have no detectable phenotypic effects in carriers. Heterozygotes for reciprocal translocations produce a substantial frequency of unbalanced gametes because of **adjacent** segregation from the **quadrivalent** that forms during synapsis. **Alternate** segregation, on the other hand, produces an equal frequency of balanced gametes that carry both parts of the reciprocal translocation or gametes that are chromosomally normal.

8. A **Robertsonian translocation** is a chromosome consisting of the long arms of two nonhomologous acrocentric chromosomes. The short arms of the chromosomes are usually lost without causing phenotypic harm. Robertsonian translocations involving chromosome 21 are particularly important in human genetics because one class of unbalanced gametes carries a duplication of 21q, which leads to offspring who have Down syndrome. About 3 percent of all Down-syndrome births involve a parent with such a translocation, and in these cases the recurrence risk is high.

9. A **ring chromosome** is a circular chromosome formed by the loss of both **telomere** (tip) fragments from a chromosome and the rejoining of the broken ends. Ring chromosomes tend to be lost during mitosis because sister-chromatid exchange produces doublesized dicentric rings.

10. An **isochromosome** is a chromosome that has two genetically identical arms, usually formed by misdivision of the centromere. A cell with an isochromosome has a duplication of one chromosome arm and a deficiency of the other chromosome arm.

WORDS TO KNOW

Chromosome	Gene Dosage	Duplication	Translocation
Breakage	Euploid	Deficiency	Reciprocal
Restitution	Aneuploid	Deletion	Quadrivalent
Acentric	Balanced	Unequal crossing-over	Balanced
Dicentric	Unbalanced	Triplication	Unbalanced
Polytene	Hypoploid		Adjacent-1 segregation
Fusion	Hyperploid	**Inversion**	Adjacent-2 segregation
Telomere	Dosage-sensitive locus	Paracentric	Alternate segregation
Ring		Pericentric	Robertsonian
Isochromosome		Loop	Down-syndrome associated

PROBLEMS

1. For discussion: A woman has a child with Down syndrome and is found to be a carrier of a Robertsonian translocation involving chromosome 21. She becomes pregnant again, and amniocentesis is performed. The fetus is found not to have trisomy 21. However, like its mother, the fetus is a carrier of the translocation. The mother wants to abort the fetus because, as she puts it, "I don't want my child to have to go through this terrible anxiety, and I don't want a grandchild with Down syndrome." The father is against the abortion because the fetus is phenotypically normal. He says to the mother, "After all, nobody aborted you!" In your judgment, what are the principal moral and ethical issues in this situation? If the decision about the abortion were up to you, how would you decide, and on what basis?

2. Suppose that the chromosome illustrated here breaks at the positions indicated by the arrows.

Assuming that each broken end rejoins with some other broken end, how many types of restitution are possible? Make an illustration of each possibility.

3. Suppose that the nonhomologous chromosomes shown here are broken at the positions indicated by the arrows. Assuming that each broken end restitutes with some other broken end, how many types of restitution are possible? Make an illustration of each possibility.

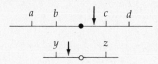

4. What is the evidence that human chromosome 2 arose by the fusion of two nonhomologous chromosomes in some ancestral species?

5. Sketch the chromosome that would result from mitotic sister-chromatid exchange in the ring chromosome shown here.

6. An isochromosome is always a metacentric. Why?

7. A submetacentric chromosome has the gene sequence $a\,b\,c \cdot d\,e$, with the raised dot representing the centromere. What would be the gene sequences of the isochromosomes formed from the submetacentric?

8. The sister of a boy with Down syndrome is phenotypically normal but has 45 chromosomes. How can this be explained?

9. Table 6.4 in Chapter 6 indicates that about 1 in 1600 live-born children carries an unbalanced translocation. What does *unbalanced translocation* mean?

10. A **tandem duplication** is a duplication in which the duplicated regions are adjacent to each other on the chromosome. For example, the duplication illustrated in Figure 7.5 is a tandem duplication. A **nontandem duplication** is one in which the duplicated regions are nonadjacent. Shown here are the homologous chromosomes in an individual who is homozygous for a nontandem duplication of gene *a*. What types of chromosomes would result from unequal crossing-over involving the duplication?

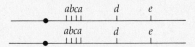

11. An individual is heterozygous for a chromosome that carries three adjacent (tandem) copies of a certain region (i.e., a tandem triplication) and another chromosome that carries four tandem copies of the same region. What possible gametes could be formed by unequal crossing-over?

12. Shown here is a pair of homologous chromosomes in an individual who is heterozygous for a large paracentric inversion; the primes represent the alleles on the inverted chromosome. Suppose that one crossover occurs in the *b–c* region and another occurs in the *c–d* region. What chromosomes would result when the double crossover is a four-strand double? When the double crossover is a two-strand double?

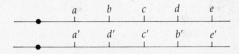

13. A chromosome with the gene sequence *a e b c d f* can arise from one with the sequence *a b c d e f* by two consecutive inversions. In which intervals would the breakpoints of the first inversion have to occur? In which intervals would the breakpoints of the second inversion have to occur?

14. Shown here are the chromosomes in an individual who is heterozygous for a reciprocal translocation involving the *cri-du-chat* segment of chromosome 5 (i.e., 5pter) and 21p. What types of gametes would result from adjacent-1 segregation? From adjacent-2 segregation? Would any of these unbalanced gametes lead to zygotes that could survive development and be born alive? Which ones, and what chromosome abnormalities would be expected?

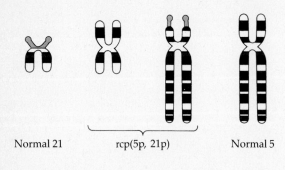

Normal 21   rcp(5p, 21p)   Normal 5

15. In the fruit fly, *Drosophila melanogaster,* the X

chromosome is an acrocentric chromosome, and isochromosomes of the X are widely used in genetic research. (However, in *Drosophila* genetics, an X isochromosome is called an *attached X* or a *compound X.*) Moreover, in *Drosophila*, XXY individuals are fertile females. Letting the symbol X̂X represent the compound X chromosome, what zygotes would be expected from the cross of an X̂XY female with a normal XY male? In *Drosophila*, individuals with no X chromosomes never survive and individuals with three X chromosomes rarely do. Taking this into account, what surviving offspring would be expected from the cross of X̂XY with XY, and in what proportions? Which sex of offspring receives the father's X chromosome? Which sex of offspring receives the father's Y?

FURTHER READING AND REFERENCES

Albertini, A. M., M. Hofer, M. P. Calos, and J. H. Miller. 1982. On the formation of spontaneous deletions: The importance of short-sequence homologies in the generation of large deletions. Cell 29: 319-328. How deletions sometimes form in *E. coli.*

Borgaonkar, D. S. 1977. Chromosomal Variation in Man, 2nd ed. Alan R. Liss, New York. A catalogue of chromosomal variants and abnormalities.

de Grouchy, J., and C. Turleau. 1977. Clinical Atlas of Human Chromosomes. Wiley, New York. A compendium of chromosomal abnormalities.

Hamerton, J. L. 1971. General Cytogenetics. Academic Press, New York. An advanced book on chromosomes and chromosome behavior in humans.

Jacobs, P. A. 1981. Mutation rates of structural chromosome rearrangements in man. Am. J. Hum. Genet. 33:44–51. The overall rate of occurrence of new detectable structural abnormalities is about 1 per 1000 gametes.

Niebuhr, E. 1978. The cri du chat syndrome. Hum. Genet. 44:227–275. A review and discussion of the 5p− syndrome.

Rowley, J. D. 1980. Chromosome abnormalities in human leukemia. Ann. Rev. Genet. 14:17–39. Discusses chromosomal involvement in leukemia.

Sandler, L., and F. Hecht. 1973. Genetic effects of aneuploidy. Am. J. Hum. Genet. 25:332–339. Compares the effects of chromosomal imbalance in fruit flies and humans.

Therman, E. 1980. Human Chromosomes: Structure, Behavior, Effects. Springer-Verlag, New York. Excellent discussion of human chromosome abnormalities.

Vogel, F., and A. G. Motulsky. 1979. Human Genetics: Problems and Approaches. Springer-Verlag, New York. An excellent advanced survey of human genetics.

Williamson, B. 1982. Gene therapy. Nature 298: 416- 418. Prospects of gene therapy and attendant ethical concerns.

Yunis, J. J. (ed.). 1977. New Chromosomal Syndromes. Academic Press, New York. Collection of articles dealing with cytogenetic techniques and chromosomal aberrations.

chapter 8
The Molecular Basis of Heredity

This is the first of several chapters that deal with the chemical nature of the gene and of the manner in which genetic information is expressed in cells. The main thread of the story is in the threadlike molecule **DNA (deoxyribonucleic acid)**. Chemically speaking, a gene corresponds to a segment of a molecule of DNA. Several important general features of DNA were introduced in Chapter 1 in connection with the discussion of sister-chromatid exchange during mitosis. A brief review of these general features of DNA is illustrated in Figure 8.1.

1. A DNA molecule consists of two strands. In Figure 8.1 (left), the molecule is depicted as a sort of ladderlike structure with long sidepieces and rungs across the middle. Using the ladder analogy, each DNA strand in the molecule is a "half-ladder" consisting of one of the sidepieces with half-rungs jutting off toward the middle. The complete ladder is formed by the side-by-side alignment of these half-ladders.

2. Each DNA strand is composed of a linear string of constituents called **nucleotides**. These are illustrated in Figure 8.1 as little ladder parts, each consisting of a section of sidepiece with a half-rung jutting off. Four nucleotides are found in DNA, which were symbolized in Chapter 1 as A (crosshatched circle in Figure 8.1), T (open circle), G (crosshatched square), and C (open square).

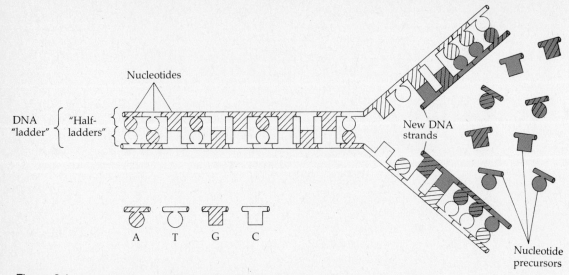

Figure 8.1 Ladder analogy of DNA structure. The fundamental units are nucleotides (A, T, G, and C); each is represented as a short section of sidepiece and an attached half-rung (square or circle). Nucleotides are attached by their sidepieces to form a half-ladder, and the double-stranded DNA molecule is composed of two such half-ladders (left). Note that A pairs only with T, and G pairs only with C. Replication is illustrated at the right, where the original DNA strands have become separated and each serves as a template for the synthesis of a complementary strand (shaded) by the one-by-one addition of nucleotides.

3. The nucleotides in the individual DNA strands that face each other across the way are **complementary**; where one strand carries an A, the other across the way carries a T; and where one strand carries a G, the other across the way carries a C. This pairing of complementary nucleotides accounts for the configuration of facing half-rungs illustrated in Figure 8.1: A hatched circle (A) is always paired with an open circle (T), and a hatched square (G) is always paired with an open square (C).

4. In the process of DNA replication, illustrated at the right in Figure 8.1, the individual strands of a DNA molecule open out and new complementary strands (represented as shaded half-ladders) are synthesized using each original strand as a **template** (a sort of "blueprint"). During the synthesis of the new DNA strands, nucleotide precursors that occur free in the cell's nucleus are added one by one to the growing strands according to the nucleotide pairing rules of A with T and G with C.

5. In any particular gene, only one DNA strand carries the actual genetic information in its nucleotide sequence; this strand is called the **sense** strand of the gene in question; we may, for purposes of illustration, assume that the upper strand in Figure 8.1 is the sense strand. The other DNA strand in the gene, which carries the complementary nucleotides of the sense strand, is known as the **antisense** strand. In the case of Figure 8.1, the antisense strand is the lower one. During DNA replication when the sense and antisense strands separate, the old sense strand serves as a template for the synthesis of a new antisense strand, and the old antisense strand serves as a template for the synthesis of a new sense strand. Consequently, the mode of DNA replication illustrated in Figure 8.1 results in two double-stranded DNA molecules, each identical in nucleotide sequence to the original unreplicated molecule.

DNA Structure: A Closer Look

Of course, DNA is not really composed of little ladder parts hooked together as illustrated in Figure 8.1. The depiction of DNA as a ladder is only an analogy. It is useful in that it conveys the essentials of DNA structure and replication in easily understood terms. On the other hand, the ladder analogy is less than ideal in that the nucleotide-pairing rules of A with T and G with C are merely arbitrary conventions relating to the little ladder parts. As will be seen in a few pages, the nucleotide-pairing rules are by no means arbitrary but are essential consequences of the molecular makeup of the DNA molecule.

Figure 8.2 provides a more detailed picture of each ladder part that makes up the molecule in Figure 8.1. Each ladder part corresponds to a particular nucleotide, but it should be noted that the nucleotides all have certain features in common. In particular, each nucleotide consists of three parts.

1. A **phosphate group** consisting of an atom of phosphorus (P) attached to three atoms of oxygen (O). One phosphate group is shown in detail in Figure 8.2(*a*), but in parts (*b*) through (*d*) the phosphate group is shown as an encircled P.

2. A molecule of the sugar **deoxyribose**, which basically has the shape of a flattened ring composed of four carbon atoms and one oxygen atom; a fifth carbon atom pokes up from the ring and is attached to the phosphate group [see Figure 8.2(*a*)]. In the sort of representations in Figure 8.2, each vertex in the rings represents an atom of carbon, but the symbol

C is not written to avoid cluttering the illustration. The deoxyribose sugar has hydrogen (H) atoms attached to the carbons at various places [see Figure 8.2(*a*)], but these hydrogens are not shown in parts (*b*) through (*d*) for the sake of simplicity. In addition, the carbon atom attached to the phosphate group is depicted in parts (*b*) through (*d*) as a kink in the line jutting up from the deoxyribose ring.

3. A molecule of a constituent called a **base**, which is attached to the opposite side of the sugar ring from which the phosphate group is attached. Four bases are found in the nucleotides that make up DNA. Each base is made up of a carbon-nitrogen (N) ring to which other atoms are attached. In the cases of **thymine** [see Figure 8.2(*b*)] and **cytosine** [see Figure 8.2(*d*)], the carbon-nitrogen ring is a single ring; these single-ring bases are known collectively as **pyrimidines**. In the cases of **adenine** [see Figure 8.2(*a*)] and **guanine** [see Figure 8.2(*c*)], the carbon-nitrogen ring is actually a fused double ring; these double-ring bases are referred to as **purines**.

For the sake of convenience, the carbon atoms of the deoxyribose sugar are numbered as shown in Figure 8.2(*b*). The carbon to which the base is attached is called the 1′ (one-prime) carbon, and the others are numbered consecutively going clockwise. Thus, the 3′ (three-prime) carbon carries a hydroxyl (−OH) group that points downward from the flat sugar ring, and the 5′ (five-prime) carbon, indicated by the kink in the line jutting up from the sugar ring, is attached to the phosphate group. Thus, a **nucleotide**—the ladder parts in Figure 8.1— actually consists of a deoxyribose sugar ring with a phosphate group attached to its 5′ carbon and one of four bases (adenine, guanine, thymine, or cytosine) attached to its 1′ carbon. The base that is attached determines which nucleotide we are dealing with:

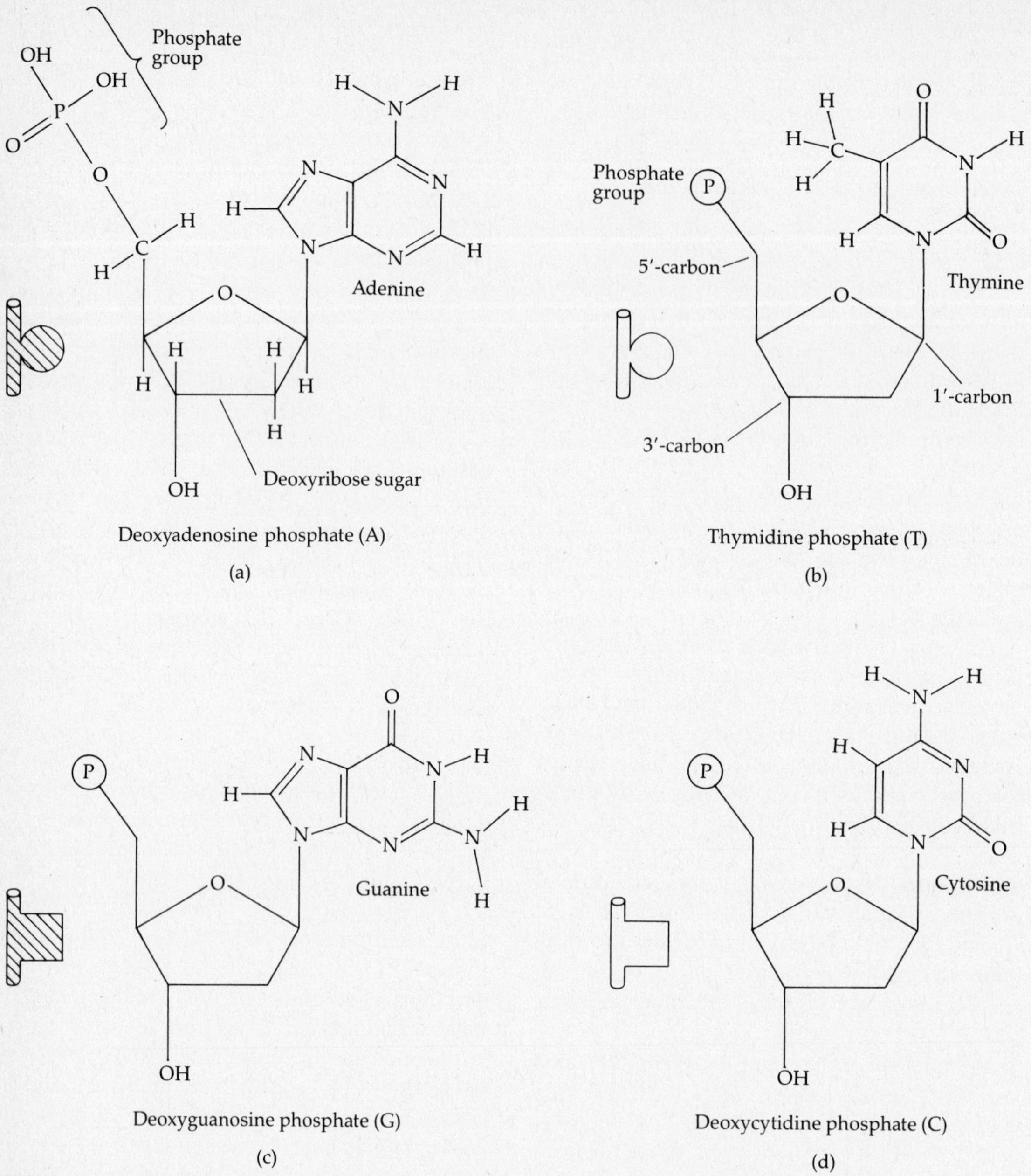
Phosphate group
Adenine
Deoxyribose sugar
Deoxyadenosine phosphate (A)
(a)
Phosphate group
5′-carbon
Thymine
1′-carbon
3′-carbon
Thymidine phosphate (T)
(b)
Guanine
Deoxyguanosine phosphate (G)
(c)
Cytosine
Deoxycytidine phosphate (C)
(d)

Figure 8.2 Chemical structure of nucleotides in DNA and their ladder-piece analogies from Figure 8-1.

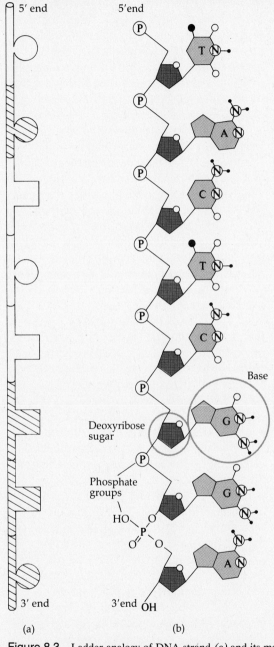

Deoxyadenosine phosphate (A) carries the base **adenine** [see Figure 8.2(*a*)].

Thymidine phosphate (T) carries the base **thymine** [see Figure 8.2(*b*)].

Deoxyguanosine phosphate (G) carries the base **guanine** [see Figure 8.2(*c*)].

Deoxycytidine phosphate (C) carries the base **cytosine** [see Figure 8.2(*d*)].

How the nucleotide ladder parts are joined together to make a single strand (half-ladder) of DNA is illustrated in Figure 8.3(*b*) along with the corresponding ladder analogy in Figure 8.3(*a*). Each nucleotide is linked through its 5′-phosphate group to the 3′ carbon of the nucleotide above. The phosphate linkage is illustrated in detail for the nucleotides at the bottom of Figure 8.3(*b*), but for the others the phosphate linkage is abbreviated as an encircled P. Also, for purposes of clarity, the chemical structures of the bases have been simplified from their more detailed versions in Figure 8.2. In Figure 8.3, each base is labeled by the first letter of its name: A, T, G, and C for adenine, thymine, guanine, and cytosine, respectively.

One feature of the structure in Figure 8.3(*b*) is particularly to be noted because it is an important characteristic of DNA structure that is not apparent in the ladder analogy in Figure 8.3(*a*). This feature is that DNA strands have a particular orientation or direction—a "top" and a "bottom"—determined by which nucleotide carries the free (unlinked) phosphate group on its 5′ carbon and which other nucleotide carries the free hydroxyl (OH)

Figure 8.3   Ladder analogy of DNA strand *(a)* and its more detailed chemical structure *(b)*. Note that each nucleotide consists of a base and a phosphate group, both attached to a deoxyribose sugar. In a DNA strand, the phosphate group of each nucleotide is chemically attached to the sugar of the nucleotide next in line. The 3′ end of the molecule is the end terminating in a sugar; the 5′ end is the end terminating in a phosphate.

group on its 3' carbon. In the case of Figure 8.3(*b*), the free 5' phosphate is on the nucleotide at the top of the DNA strand; this end of the molecule is known as the **5' end**. Similarly, the free 3' hydroxyl is on the nucleotide at the bottom of the DNA strand; this end of the molecule is known as the **3' end**. In many instances, it is important to specify the orientation of a DNA strand that has a particular sequence of nucleotides. This specification is accomplished by appending the symbol "5'" to the 5' nucleotide of the sequence and the symbol "3'" to the 3' end. Thus, for example, the DNA strand in Figure 8.3(*b*) could be written correctly as either

5'-TACTCGGA-3'

or

3'-AGGCTCAT-5'

depending on whether one wishes to copy the sequence from top to bottom or from bottom to top.

Base Pairing in DNA

One of the most important discoveries that preceded a detailed understanding of DNA structure was that DNA contains an equal number of A and T nucleotides and an equal number of G and C nucleotides. In human DNA, for example, 31 percent of the nucleotides are deoxyadenosine phosphate (A), 31 percent are thymidine phosphate (T), 19 percent are deoxyguanidine phosphate (G), and 19 percent are deoxycytidine phosphate (C). Thus, relative to their amounts, A = T and G = C. In the fruit fly, *Drosophila melanogaster*, to take another example, the nucleotides are 27.5 percent A, 27.5 percent T, 22.5 percent G, and 22.5 percent C, so, here again, A = T and G = C.

The equality of A with T and G with C in DNA was just an inexplicable curiosity until the early 1950s, when Maurice Wilkins and Rosalind Franklin at Kings College in London discovered through their x-ray diffraction studies that a complete molecule of DNA consists of two strands. (X-ray diffraction is a technique in which x-rays are transmitted through a crystal of some substance, and an atomic view of the substance is inferred from the pattern of reinforcement or interference of the transmitted x-rays.) As this information about two strands in DNA became available to James Watson and Francis Crick at Cambridge University, they began to try to construct a tinkertoy-like model of DNA. In one of their first models, the bases attached to the nucleotides stuck off toward the outside of the molecule instead of pointing inward. But they could not get such a porcupine model to work; every version was incompatible with details of the x-ray diffraction data.

If the bases point toward the interior of the molecule, how could they fit together without requiring too much space or too little space or otherwise violating the laws of physics and chemistry? According to Watson, who later wrote *The Double Helix,* a fascinating account of the discovery of the structure of DNA, he began to play with little metal cutouts of the shapes of the nucleotides to see how they could fit together. The x-ray data implied that the distance between the DNA strands was the same everywhere, which ruled out the pairing of purines (i.e., A-A, A-G, or G-G) because such pairs would be too large; it also ruled out the pairing of pyrimidines (i.e., T-T, T-C,. or C-C) because such pairs would be too small. Evidently, each pair must involve one purine and one pyrimidine—but which ones?

One happy day Watson discovered that his nucleotide cutouts could be fit together as illustrated at the right side of Figure 8.4. With the appropriate orientation of the nucleotides, adenine will pair with thymine and guanine will pair with cytosine. Watson realized that this

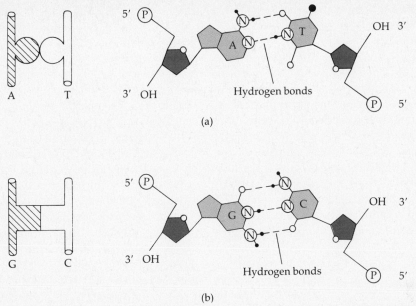

Figure 8.4   Ladder analogy and more detailed chemical structures of *(a)* A-T base pair and *(b)* G-C base pair. The base pairing is facilitated by hydrogen bonds that can form between certain atoms on the bases (broken lines). A-T base pairs form two hydrogen bonds; G-C base pairs form three hydrogen bonds.

was a key discovery for several reasons. First, it explained why, relative to their amounts, A = T and G = C. Since A and T are paired, each A must have a partner T and each T a partner A, so their amounts must be equal. Likewise for G and C. Second, it explained how the proper distance between the DNA strands is preserved: A large purine base (A or G) is always paired with a small pyrimidine base (T or C, respectively), thus giving the same spacing between the DNA strands. Third, the pairing of A with T and G with C explained how the DNA strands in a molecule are held together; the answer lies in the formation of weak chemical bonds, called **hydrogen bonds**, between the bases.

Hydrogen bonds, illustrated by the broken lines in Figure 8.4, are formed by the sharing of a hydrogen atom (symbolized by the small solid dots in the figure) by two atoms, usually nitrogen or oxygen, that have a slight negative charge. The hydrogen always has a **covalent bond** with one of the atoms (solid lines of attachment), which means that the electron belonging to the hydrogen is shared with the atom to which the hydrogen is covalently bonded. However, the other negative atom involved in the hydrogen bond exerts a slight pull on the hydrogen, too, and this second force forms the hydrogen bond. The formation of hydrogen bonds requires close physical proximity of the atoms involved, and hydrogen bonds are weak —only one-seventh to one-thirtieth as strong as covalent bonds. As illustrated in Figure 8.4, A and T form two hydrogen bonds with each other, whereas G and C form three of them. Although each hydrogen bond is weak, a DNA molecule contains so many nucleotide pairs that the collective effect is substantial. Nevertheless, in a test-tube solution containing double-stranded DNA, the strands can usually be separated by heating so as to disrupt the hydrogen bonds, although the covalent bonds in each individual strand remain intact.

One further aspect of Figure 8.4 should be highlighted. To form the A-T and G-C base pairs, the nucleotides must be in opposite orientation. In Figure 8.4(a), for example, the 5′ end of the deoxyadenosine phosphate points upward, but the 5′ end of thymidine phosphate points downward. A similar situation is apparent in Figure 8.4(b). One weakness of the ladder analogy (see Figure 8.4, left) is that this reverse orientation of the nucleotides is not immediately apparent, although the open end of each tube is intended to represent the 5′ end of the nucleotide and the blunt end to represent the 3′ end.

Duplex DNA

Once the possibilities of A-T and G-C base pairing were realized, Watson and Crick set out to build a model of DNA structure. A two-dimensional rendition of double-stranded DNA (often called **duplex DNA**) is illustrated in Figure 8.5(b) along with its ladder analogy in Figure 8.5(a). It should be emphasized that Figure 8.5(b) takes great liberties with the nature of chemical bonds. Covalent bonds are actually rather rigid because their lengths and angles can vary only within relatively narrow limits. These limits impose constraints on the actual three-dimensional structure of the molecule. In representing the structure of duplex DNA on a flat sheet of paper, the bond angles and distances have to be distorted beyond their realistic limits. Nevertheless, a representation of duplex DNA like that in Figure 8.5(b) is worthwhile because several important characteristics of DNA structure are readily seen.

The first aspect of duplex DNA to be noted in Figure 8.5(b) is that the DNA strands run in opposite directions; the 5′ end of one strand is juxtaposed with the 3′ end of the other strand. To convey this essential feature of

DNA structure, the DNA strands in a duplex are often said to have **opposite polarity** or to be **antiparallel**. A second feature of note is that a purine base (A or G) is always paired with a pyrimidine base (T or C, respectively) in the other strand; this purine-pyrimidine pairing preserves the proper spacing between the sugar-phosphate backbones of the individual DNA strands. The third and perhaps most important aspect of DNA structure is that an A in one strand is always paired with a T in the other, and a G in one strand is always paired with a C in the other; these base-pairing rules occur because of hydrogen bonding between the bases—two hydrogen bonds in the case of A-T base pairs and three in the case of G-C base pairs.

Because of the base-pairing rules of A with T and G with C, four types of nucleotide pairs can occur in duplex DNA: A-T, T-A, G-C, and C-G. There is, however, no restriction on the *sequence* of these nucleotide pairs. Figure 8.5(b) illustrates the sequence

$$5'\text{-TACTCGGA-}3'$$
$$3'\text{-ATGAGCCT-}5'$$

but any other sequence would have served as well. The unrestricted possibilities of nucleotide sequences in DNA are the feature that allows DNA to carry so much genetic information, because the genetic information in DNA resides in its nucleotide sequence. Consider, to take a simple example, a gene-sized piece of DNA consisting of 1000 nucleotide pairs. How many nucleotide sequences are possible? Because there are four possibilites for each nucleotide pair (i.e., A-T, T-A, G-C, and C-G), there are 4^{1000} possible sequences altogether. The size of the number 4^{1000} is unimaginable; it is approximately 10^{600}—a number far larger than the number of seconds that have passed since the universe began, and far larger than the number of elementary atomic particles in the universe. Little wonder that DNA can

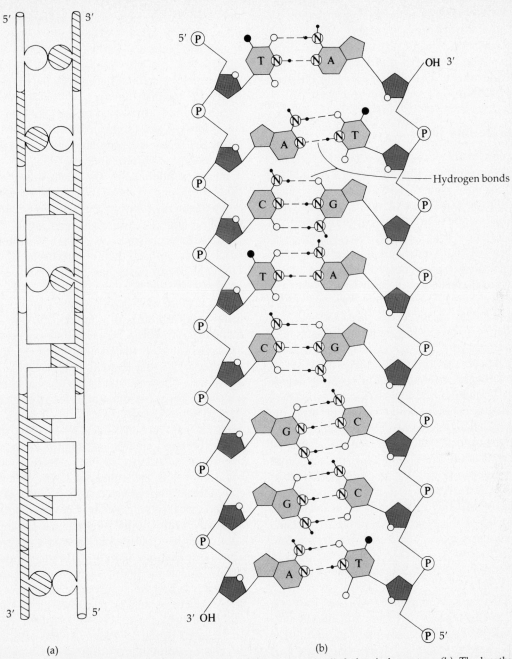

(a)   (b)

Figure 8.5 Ladder analogy of double-stranded DNA *(a)* and more detailed chemical structure *(b)*. The lengths and angles of some of the chemical bonds must be distorted to draw the configuration on a flat surface. Note that the two DNA strands have opposite polarity (i.e., 5′ to 3′ from top to bottom in the left-hand strand, 3′ to 5′ in the right-hand strand).

carry the diversity of genetic information for the millions of species of plants, animals, and microbes that exist on earth. Indeed, the diversity of life on earth expresses only a tiny fraction of the information-carrying possibilities of DNA.

The Double Helix

A more realistic representation of the three-dimensional structure of DNA is shown at the bottom of Figure 8.6. Here the sugar-phosphate backbones of the individual DNA strands can be seen to wind around one another on the outside of the DNA molecule, reminiscent of the winding of the red and white stripes on a barber pole. (The paths of the backbones can be traced easily by following the black spheres representing the phosphate atoms.) The paired bases are flat and are stacked on top of one another like a roll of coins, and the backbones wind around the outside of the roll. The three-dimensional DNA duplex in Figure 8.6 is called a **double helix**—"double" because of the two DNA strands it contains, and "helix" because the strands wind around one another in the form of a right-handed helix. The configuration of DNA illustrated in Figure 8.6 is the one originally proposed by Watson and Crick. It is known as the **B form** of DNA, and the dimensions of the molecule are well known: The diameter of the molecule is about 20 Å; each backbone makes a complete turn around the helix every 34 Å; and there are 10 base pairs in every turn, thus the thickness of each base pair is 3.4 Å.

The space-filling model of the DNA duplex at the bottom of Figure 8.6 blends into several alternative representations that highlight certain features of its structure. Immediately above the space-filling model is a representation in which the sugar-phosphate backbones are denoted as ribbons and the base pairs as thick horizontal lines. In the middle region of Figure 8.6, two base pairs have been tilted sideways (with apologies to the twisted and contorted covalent bonds) to illustrate the hydrogen bonding between A and T and between G and C (indicated by the short vertical lines between the atoms involved). At the top of the figure, the backbones of the double helix have been untwisted and the possible base pairs are symbolized by the first letter of each base; the 3' end of each DNA strand is indicated by the arrow (upper right) and the 5' end by the blunt end of the arrow (upper left).

In the years since 1953, when Watson and Crick first proposed their structure for B-form DNA, their proposal has been confirmed in virtually every important detail. The structure at the bottom of Figure 8.6 appears to be the usual conformation of duplex DNA in the living cell. However, research by x-ray crystallographers has revealed that more than 20 variations of right-handed helical DNA can also occur. These variant forms involve ones in which the bases are tilted at different angles or stacked more tightly than in the B form. Indeed, it has recently been discovered that, under certain conditions, duplex DNA can even form a left-handed helix, which has become known as the **Z form** of DNA. Whether any of these alternative configurations of DNA has biological significance is at present unclear, although recent studies suggest that the interband regions of dipteran polytene chromosomes (see Figures 7.2 and 7.8) are particularly rich in Z (left-handed) DNA. Some researchers believe that the significance of alternative DNA structures may be to enable the recognition of particular stretches of DNA by particular types of proteins. Whatever the case may be, the predominant B-form of DNA and all the variant forms obey the fundamental A-T and G-C base-pairing rules.

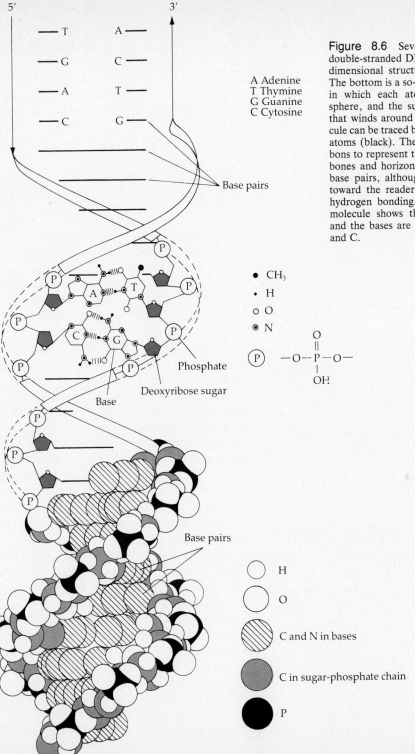

5′   3′

— T   A —
— G   C —
— A   T —
— C   G —

A Adenine
T Thymine
G Guanine
C Cytosine

Base pairs

Figure 8.6 Several representations of double-stranded DNA illustrating the three-dimensional structure of the double helix. The bottom is a so-called space-filling model in which each atom is represented as a sphere, and the sugar-phosphate backbone that winds around the outside of the molecule can be traced by means of the phosphate atoms (black). The middle section uses ribbons to represent the sugar-phosphate backbones and horizontal lines to represent the base pairs, although two base pairs (tilted toward the reader) are shown to illustrate hydrogen bonding. The upper part of the molecule shows the backbones untwisted, and the bases are represented by A, T, G, and C.

● CH₃
• H
○ O
◉ N

$$\overset{O}{\underset{OH}{P}} \quad -O-\overset{\overset{O}{\|}}{\underset{\underset{OH}{|}}{P}}-O-$$

Phosphate

Deoxyribose sugar

Base

Base pairs

○ H
○ O
▨ C and N in bases
◕ C in sugar-phosphate chain
● P

DNA Replication

Although the ladder analogy in Figure 8.1 provides an overview of the process of DNA replication that is useful and adequate for many purposes, in several respects the ladder analogy is oversimplified and potentially misleading. As a consequence, it is worthwhile in this section to examine DNA replication at a somewhat more detailed level than was possible in connection with Figure 8.1. We begin the examination with a discussion of some of the proteins involved in DNA replication.

A DNA duplex has the three-dimensional structure of a double helix (see Figure 8.6). For a duplex to be replicated, the individual DNA strands must somehow be unwound. This unwinding leads to the Y-shaped configuration shown in Figure 8.1. (In Figure 8.1, the Y is drawn on its side with its base to the left and its forks to the right.) The forks of the Y represent the individual DNA strands that are used as templates for replication, and the base of the Y represents the still-unwound part of the double helix. Such a Y-shaped configuration that occurs during DNA replication is called a **replication fork**. At least three distinct proteins are known to be involved in the unwinding process that creates the replication fork. One is called **DNA unwinding protein**; as its name implies, this protein unwinds the strands of the double helix and creates the single-stranded regions of the replication fork. This unwinding is aided by the **helix-destabilizing protein**, which binds tightly to the single-stranded regions and holds them apart. The unwinding of the double helix creates stresses and strains in regions of the helix farther down the way, but these stresses are relieved by yet another protein known as **DNA gyrase**.

The principal enzyme involved in DNA replication is **DNA polymerase**. The reaction catalyzed by DNA polymerase is illustrated in Figure 8.7. The growing end of a DNA strand is represented at the upper left (unshaded), and the nucleotide about to be added to the growing strand by DNA polymerase is at the lower left (shaded). (In this figure, the bases of the nucleotides are represented as simple shield shapes.) Note that DNA polymerase requires a free 3′ hydroxyl for the reaction to occur; this means that *DNA strands increase in length by the addition of new nucleotides to their 3′ end*. Note also that the precursor nucleotide carries *three* phosphates at its 5′ end; such a nucleotide is called a **nucleotide 5′-triphosphate** (e.g., deoxyadenosine 5′-triphosphate, thymidine 5′-triphosphate). The phosphate-phosphate bonds in the triphosphates are **high energy phosphate bonds**. In the reaction catalyzed by DNA polymerase, the innermost high-energy bond is broken and the energy so released is used to connect the innermost phosphate to the 3′ end of the growing DNA strand. Although it is not represented in Figure 8.7, DNA polymerase also requires a template strand, and each nucleotide added to the growing strand is complementary to the corresponding nucleotide in the template strand. As DNA polymerase creeps along the template strand, it adds successive nucleotides to the 3′ end of the ever-growing new strand at the rate of about 40 nucleotides per second. (Prokaryotic polymerases work much faster—up to 1000 nucleotides per second in some cases.)

Over and above its ability to elongate a DNA strand by the addition of successive nucleotides to its 3′ end, DNA polymerase has another important function, which is called its **proofreading** function. After each new nucleotide is added to the growing chain, and before proceeding to add the next one in line, the polymerase checks whether the last-added nucleotide is in base-pair agreement with the corresponding nucleotide of the template strand (i.e., it checks whether A is paired with T and G is paired with C). If the base pair is in

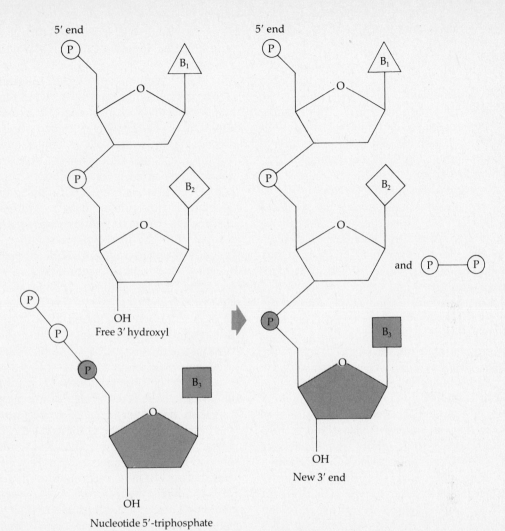

Figure 8.7   Addition of a new nucleotide to the 3′ end of a growing DNA strand (bases represented by B_1 and B_2). The nucleotide to be added (shaded) is present in the cell as a 5′ *tri*phosphate; cleavage of a high-energy phosphate bond releases energy used to attach the 5′ phosphate of the new nucleotide to the 3′ hydroxyl of the terminal sugar of the growing strand.

agreement, the polymerase goes ahead and adds the next nucleotide in line. If there is a base-pairing error (i.e., anything other than A with T and G with C), the polymerase *cleaves* the last-added (and incorrect) nucleotide from the DNA strand and replaces it with a new one according to the reaction in Figure 8.7. The ability to remove nucleotides from the end of a DNA strand makes DNA polymerase an **exonuclease** (in addition to being a polymerase). More precisely, DNA polymerase is a **3′ exonuclease** because it can cleave nucleotides off the 3′ end of a DNA strand. Thus, each base pair in a replicating DNA strand is checked twice for correctness, once just before the nucleotide is added to the growing DNA strand (when DNA polymerase functions as a polymerase) and once just after the nucleotide has been added to the strand (when DNA polymerase functions as an exonuclease). This double checking provides an extra margin of safety to greatly improve the accuracy of replication.

DNA polymerase has one significant limitation: It can add nucleotides only to the 3' end of an already established DNA strand; that is, DNA polymerase cannot *initiate* the synthesis of a new strand. How cells contend with this limitation is outlined in Figure 8.8, which illustrates the most important details of DNA replication. Part (*a*) shows the DNA "ladder" in thick gray lines, and the orientation of the antiparallel strands is indicated; this is the original DNA duplex about to be replicated. Part (*b*) shows the development of a replication fork through the activities of unwinding protein, helix-destabilizing protein, and gyrase. Part (*c*) illustrates how the cell overcomes the fact that DNA polymerase cannot initiate a new DNA strand; the cell uses a different enzyme that *can* initiate a new strand to produce a relatively short **primer** (thin black line) to which DNA polymerase can add consecutive nucleotides. However, the primer is not a DNA strand but a different type of nucleic acid called **ribonucleic acid (RNA)**, and the enzyme that produces the RNA primer is called **RNA polymerase**. (The chemical structure of RNA will be discussed in Chapter 9; for present purposes, it is sufficient to note that the structure is similar to DNA except that the sugar is different—ribose in RNA—and one of the bases is different.) Because RNA strands grow in the same direction as DNA strands (i.e., from the 5' end toward the 3' end), the RNA polymerase must initiate synthesis at the 3' end of the template strand (because the strands must be antiparallel). Thus, one primer can be

formed at the tip of a replication fork, but the other must be formed at the base [see Figure 8.8(*c*)].

After synthesis of the RNA primers, DNA polymerase elongates the RNA strands by addition of DNA nucleotides (thick black line) to the 3' end [see Figure 8.8(*d*)]. As this is proceeding, the original DNA duplex unwinds still further. Note that one of the new strands [the bottom one in Figure 8.8(*e*)] does not have a free 3' end; to continue replication from this template, a new primer must be synthesized near the base of the replication fork [see Figure 8.8(*f*)]. With this new primer on the lower strand, DNA synthesis can continue. Note that the new DNA strand at the top of Figure 8.8(*g*) is continuous but that the lower one is split into segments (in reality about 1000 nucleotides long) because of the necessity of repeated primer synthesis on this strand; these shorter DNA segments are called **Okazaki fragments** after their discoverer.

When the synthesis of an Okazaki fragment reaches the RNA primer of the previous fragment, two other enzymes come into play [see Figure 8.8(*h*)]. One of these is a **5' exonuclease**, which chops successive RNA nucleotides off the RNA primer as DNA polymerase adds successive DNA nucleotides to the strand behind; in this manner, the RNA primer is eliminated and replaced with DNA. The other enzyme involved is **DNA ligase**, which catalyzes the connection between the Okazaki fragments where they come together [see Figure 8.8(*h*)].

Continued unwinding of the original DNA duplex and synthesis as in parts (*e*) through (*h*) lead, after final elimination of the terminal primers, to the DNA duplexes in Figure 8.8(*i*). Each duplex formed by DNA replication is identical in nucleotide sequence to the original duplex. In addition, each duplex formed by DNA replication consists of one "old" DNA strand (gray line) and one newly

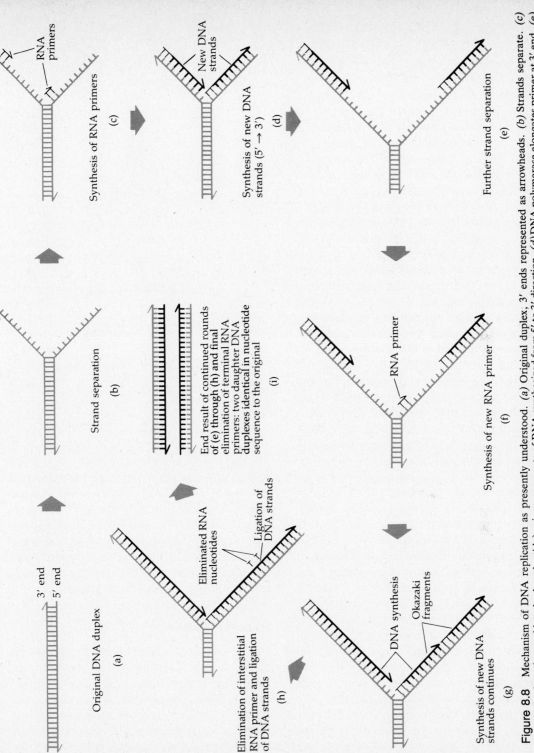

Figure 8.8 Mechanism of DNA replication as presently understood. (*a*) Original duplex, 3′ ends represented as arrowheads. (*b*) Strands separate. (*c*) Relatively short (several hundred nucleotide) primer segments of RNA synthesized from 5′ to 3′ direction. (*d*) DNA polymerase elongates primer at 3′ end. (*e*) Further separation of strands in original duplex. (*f*) New primer synthesized on one DNA strand (the bottom one as shown here). (*g*) DNA polymerase elongates new DNA strand (top) or RNA primer (bottom) at the 3′ end. (*h*) Where new DNA strand runs up against old primer, primer is eliminated, DNA synthesis fills the gap, and DNA fragments (Okazaki fragments) are ligated. (*i*) End result of continued events in (*e*) through (*h*)—two DNA molecules. Each daughter DNA molecule has one preexisting strand and one newly synthesized strand—a mode of replication called *semiconservative.*

synthesized strand (black line); this half-old/half-new mode of DNA replication is known as **semiconservative replication**.

On the whole, the ladder analogy in Figure 8.1 gives a relatively accurate description of the overall process of DNA replication, but it does gloss over many of the molecular details apparent in Figure 8.8. There is, however, an additional matter that has to be considered in regard to the replication of DNA in a eukaryotic chromosome. Human chromosome 1, to take an example, contains about 500 million nucleotide pairs of DNA. At a rate of DNA synthesis of 40 nucleotides per second (the rate for DNA polymerase), replication of a 500-million-nucleotide length of DNA, if it occurred from one end straight to the other, would require nearly 5 months! This is absurd, of course; replication of chromosome 1 occurs in a matter of hours, not months. The way this feat is accomplished is that replication is initiated by helix unwinding at many interior points of the molecule. Replication and further helix unwinding occur in both directions from these points, producing the sort of replication **bubbles** shown in Figure 8.9(*a*) and (*b*). When the rightward movement of one bubble meets the leftward movement of another, the new DNA strands are ligated (again by DNA ligase), producing the larger bubble in Figure 8.9(*c*). In short, hundreds of regions of a long DNA duplex are being replicated simultaneously, thus greatly reducing the time required for replication to be completed. Figure 8.10(*a*) is an autoradiograph of a DNA molecule from a human cell caught in the act of replication. To obtain this picture, cells undergoing DNA synthesis are bathed in a solution containing radioactive nucleotides. After a time, the DNA was extracted from the cells, spread onto a glass slide, and overlaid with photographic film. The dark spots on the film are caused by radioactivity from the radioactive nucleotides in the newly synthesized DNA strands. The black line

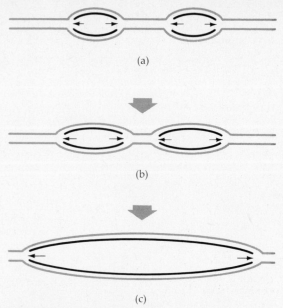

(a)

(b)

(c)

Figure 8.9 Replication bubbles in eukaryotes. A long DNA molecule begins replication at many interior points *(a)*, with replication proceeding bidirectionally *(b)*. When two replication bubbles meet, they fuse and form a larger bubble *(c)*.

underneath the DNA molecule represents 50 μm (i.e., 500,000 Å). Figure 8.10(*b*) is an interpretive drawing of the molecule in part (*a*). The black lines represent DNA strands synthesized in the presence of radioactive nucleotides; the gray lines represent nonradioactive DNA. Note the two replication forks, which proceed in opposite directions.

Recombination

One of the fundamental processes that occurs during meiosis is recombination between chromatids; it is appropriate in this section to indicate how recombination is thought to occur at the molecular level. A currently favored model of recombination is illustrated in Figure 8.11. Although there is abundant evidence that

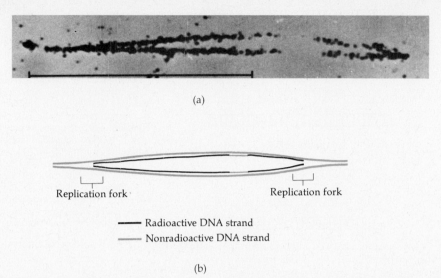

(a)

Replication fork ·   Replication fork

——— Radioactive DNA strand
——— Nonradioactive DNA strand

(b)

Figure 8.10   *(a)* Autoradiograph of a DNA molecule in a human cell caught in the act of replication. The black line represents 50μm (0.002 in). *(b)* Interpretative drawing of the molecule in *(a)*. Note similarity with Figure 8–9 *(c)*.

this model is valid for prokaryotes, its validity for eukaryotes is as yet unproven.

Figure 8.11(*a*) illustrates two DNA duplexes about to undergo recombination. The genetic information in the top duplex is assumed to correspond to alleles *A* and *B* at two loci, that in the bottom duplex to correspond to alleles *a* and *b*; use of these allele symbols may help connect Figure 8.11 with the discussion of crossing-over and recombination in Chapter 4.

The process of exchange between two DNA duplexes begins with the production of a **nick** (a break) in one of the DNA strands of each duplex [see Figure 8.11(*b*)]; these breaks are produced by particular enzymes called **endonucleases**. Beginning at the position of the nicks, the nicked DNA strands can unwind from the duplex for a short distance and then, because of complementary base pairing, can exchange pairing partners [see Figure 8.11(*c*)]. The regions of strand exchange where an unnicked strand from one duplex pairs with a nicked strand from another duplex are called **heteroduplex** regions. After formation of the

heteroduplex regions, the backbones of the DNA strands are fused together by DNA ligase [see Figure 8.11(*d*)]; after this ligation occurs, one strand from each duplex is fused with one strand from the other duplex.

The next step in the exchange process involves some topology. If the bottom duplex in Figure 8.11(*d*) is twisted end for end, then the structure in part (*e*) results. This configuration is called a **χ (chi) structure**, and the existence of such structures as intermediates in the exchange process provides strong support for the validity of the model. Figure 8.12 is an electron micrograph of an actual χ structure formed during genetic exchange in a bacterial cell.

To complete the exchange process, the χ structure must be resolved into two separate DNA duplexes. This resolution is accomplished by endonuclease nicks in two of the DNA strands followed by ligation after strand exchange, and it can be done in two distinct ways, which are indicated in Figure 8.11(*e*) by the solid and the shaded arrows. If the break-

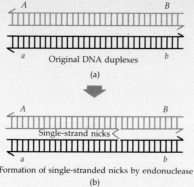

A   *B*

a   *b*
Original DNA duplexes

(a)

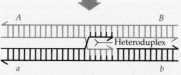

A   *B*

Single-strand nicks

a   *b*

Formation of single-stranded nicks by endonuclease

(b)

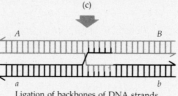

A   *B*

Heteroduplex

a   *b*

Single-strand exchange forming heteroduplex regions

(c)

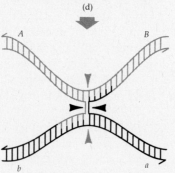

A   *B*

a   *b*

Ligation of backbones of DNA strands

(d)

Figure 8.11   Present conception of recombination at the DNA level. *(a)* Original DNA duplexes in nonsister chromatids prior to recombination. *(b)* Recombinational process initiated by single-stranded nicks. *(c)* Single strands exchange pairing partners to form heteroduplex regions. *(d)* Ligation of exchanged strands. *(e)* χ structure formed by turning lower molecule end for end. *(f)* Resolution of χ structure at positions of solid arrows leads to no recombination of outside markers (*A,a* and *B,b*). *(g)* Resolution at positions of shaded arrows leads to recombination of outside markers. Note the heteroduplex region in both *(f)* and *(g)*. While consistent with the known facts of recombination, many details of the model have not been directly confirmed in eukaryotes.

A   *B*

b   *a*

Bottom duplex in (d) turns end for end
forming χ (chi) structure

(e)

Nonrecombinant DNA duplexes   Recombinant DNA duplexes

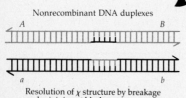

A   *B*   *A*   *b*

a   *b*   *a*   *B*

Resolution of χ structure by breakage   Resolution of χ structure by breakage
and rejoining at black arrows   and rejoining at gray arrows

(f)   (g)

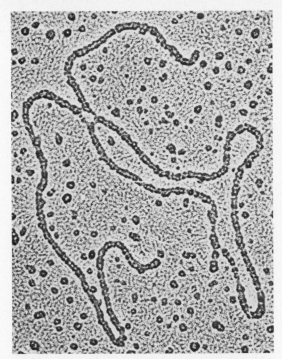

Figure 8.12 Electron micrograph of χ structure. Compare with Figure 8–11(e).

age and rejoining of the DNA strands occur at the positions of the solid arrows, the duplexes in Figure 8.11(f) result. These duplexes are both nonrecombinant; that is, one carries alleles A and B and the other carries a and b, just as did the original DNA duplexes that entered into the exchange. However, the duplexes in Figure 8.11(f) do have heteroduplex regions that result from the exchange.

The second mode of resolution of the χ structure involves breakage and rejoining at the positions indicated by the shaded arrows. To preserve the correct polarity of the DNA strands, the top strand of the upper duplex marked with A must join with the bottom strand of the lower duplex marked with b; similarly, the bottom strand of the lower duplex marked with a must join with the top strand of the upper duplex marked with B.

Exchange at the positions of the shaded arrows results in the duplexes in Figure 8.11(g). Both duplexes are recombinant (i.e., Ab or aB) and, in addition, both have small heteroduplex regions resulting from the exchange process. [Another way to appreciate how the duplexes in Figure 8.11(g) can arise from the exchange process is to imagine folding Figure 8.11(d) horizontally so that the unexchanged strands—the top and bottom ones—tilt outward and toward one another; eventually these strands will nearly touch one another, and simple breakage and reunion of the top and bottom strands will lead immediately to the duplexes in Figure 8.11(g).]

Overall, whether a χ structure will or will not lead to recombination depends on how the structure is resolved. A small heteroduplex region (actually one to several thousand nucleotides long) will always result, however. Ordinarily, such heteroduplex regions are of no genetic consequence unless the original DNA duplexes had different nucleotide sequences (i.e., carried different alleles) in the region that becomes heteroduplex. An example is illustrated in Figure 8.13(a). Here the symbol N represents any nucleotide, and paired N's in a duplex indicate properly paired nucleotides (i.e., A paired with T or G with C). Note that at one position (the center one in the figure), one duplex has an A-T nucleotide pair, which we will call "allele D" for the sake of concreteness; the other duplex carries a G-C nucleotide pair at this same position, which we will call "allele d." Figure 8.13(b) shows the heteroduplexes that result from the process in Figure 8.11, assuming, of course, that heteroduplex formation involves the region in question. Note that each heteroduplex region has one mismatched nucleotide pair—A is paired with C in the left heteroduplex, and G is paired with T in the right one. These nucleotide mismatches will automatically be resolved at the next DNA replication: Two daughter duplexes will

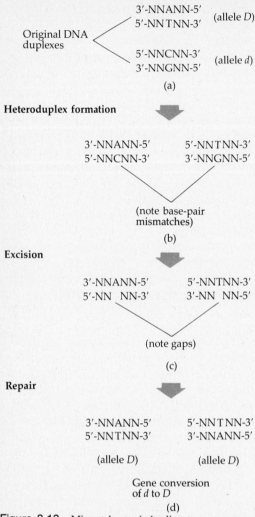

3'-NNANN-5'
5'-NNTNN-3' (allele D)

5'-NNCNN-3'
3'-NNGNN-5' (allele d)

(a)

Heteroduplex formation

3'-NNANN-5'   5'-NNTNN-3'
5'-NNCNN-3'   3'-NNGNN-5'

(note base-pair
mismatches)

(b)

Excision

3'-NNANN-5'   5'-NNTNN-3'
5'-NN NN-3'   3'-NN NN-5'

(note gaps)

(c)

Repair

3'-NNANN-5'   5'-NNTNN-3'
5'-NNTNN-3'   3'-NNANN-5'

(allele D)   (allele D)

Gene conversion
of d to D

(d)

Figure 8.13 Mismatch repair leading to gene conversion. *(a)* Original DNA molecules designated allele *D* and allele *d;* N stands for any nucleotide. *(b)* Heteroduplexes formed during recombination in the region of the sequences shown in *(a)*. *(c)* Excision of mispaired base; which base is excised is a matter of chance, and exonucleases probably widen the gap produced by excision. *(d)* Resynthesis of DNA across gap using intact strand as template corrects the mispaired base pair and can lead to gene conversion—in this case, conversion of allele *d* to allele *D*.

result from replication of the left heteroduplex, one having an A-T nucleotide pair and the other a G-C nucleotide pair; similarly, one of

the daughter duplexes resulting from replication of the right heteroduplex will have an A-T nucleotide pair and the other a G-C nucleotide pair. This resolution of the mismatches is a consequence of the fidelity of nucleotide pairing during DNA replication.

However, there is another way the nucleotide mismatches can be resolved prior to replication. By means of a process often called **mismatch repair**, either one of the mismatched nucleotides can be excised from its DNA strand and a new, correct nucleotide spliced in. The excision process is illustrated in Figure 8.13(*c*). For the sake of example, we have assumed that the bottom mismatched nucleotide in each heteroduplex is the one excised, but either nucleotide of one or both heteroduplexes could be involved. Also, the precision of excision is not really as clean as illustrated; what seems to happen is that an endonuclease cuts one DNA backbone near the site of the mismatch and that exonuclease cleaves away part of the nicked strand to produce a many-nucleotide gap that includes the originally mispaired nucleotide.

The result of resynthesis of the excised region is shown in Figure 8.13(*d*). The original A-C mismatch has been corrected to A-T (which corresponds to allele *D*), and the original G-T mismatch has also been corrected to A-T (again corresponding to allele *D*). Note especially what has occurred from a genetic point of view as a result of the mismatch repair: The original DNA duplexes carried one *D* and one *d* allele; the duplexes resulting from mismatch repair both carry the *D* allele. In effect, the *d* allele has been "converted" to the *D* allele, and this process is called **gene conversion**. The consequences of mismatch repair leading to gene conversion can be observed directly in certain fungi in which all four spores that result from meiosis (called a **tetrad**) are held together in a little sac (called an **ascus**). An individual who is heterozygous for alleles *D* and *d* would oridinarily be expected to produce two *D*-bearing and two *d*-bearing spores, and this is usually what is observed. On occasion, however, tetrads with 3*D*:1*d* or 1*D*:3*d* are observed, which are indicative of gene conversion involving heteroduplex formation and mismatch repair in two of the four chromatids.

Organization of DNA in Chromosomes

Recall from Chapter 1 that the DNA molecule in human chromosome 1 would extend for about 8 cm if it were stretched out to its full length; yet the metaphase length of chromosome 1 is about 0.001 cm—8000 times shorter than the extended length of the DNA it contains. How this prodigious feat of packaging is accomplished was at one time a complete mystery, but in recent years part of the mystery has been unraveled. The fundamental structure of the eukaryotic chromosome seems to be based on a hierarchy of coils—that is, coils of coils of coils of coils. This structure is illustrated schematically in Figure 8.14.

The extended DNA molecule found in a metaphase chromatid is illustrated in Figure 8.14(*a*); as noted earlier, it is a double helix of diameter 20 Å. The next level of chromosome structure involves the combination of DNA with certain proteins, called **histones**, to form tiny "beads" known as **nucleosomes** [see Figure 8.14(*b*)]. Each nucleosome consists of two molecules of each of four types of histones—H2A, H2B, H3, and H4; in addition to these eight histone molecules, each nucleosome also contains about 140 nucleotide pairs of DNA twisted around and through the bead. Adjacent nucleosomes are connected by a length of DNA ranging in size from 15 to 100 nucleotide pairs; strings of nucleosomes and their DNA "spacers" give extended nucleosomes a striking beads-on-a-string appearance (Figure 8.15). Within the nucleus, however, the DNA spacers are included in the nucleosomes, forming a **nucleosome fiber** with a diameter of about 100 Å [see Figure 8.14(*c*)].

At the next level of the coiling hierarchy, the nucleosome fiber coils to form a fiber 300 to 500 Å in diameter called a **solenoid** [see Figure 8.14(*d*)]. The solenoid then also coils to form a thicker fiber about 2000 Å in diameter called a **supersolenoid** [see Figure 8.14(*e*)]. Finally, the supersolenoid itself undergoes coiling to create the metaphase **chromatid** 6000 Å in diameter. Formation of these various levels of coiling is thought to be a function of yet another type of histone, called H1. Although the levels of coiling represented in Figure 8.14(*e*) and (*f*) are at present still somewhat conjectural, the model of chromosome structure does account for the remarkable packaging of an 8-cm DNA molecule in a 0.001-cm chromatid.

Repetitive DNA

In the years immediately following the discovery that the genetic material was DNA, it was a natural assumption that all the DNA in a cell carried genetic information for the manufacture of proteins. It was therefore expected that the number of genes in an organism would be related to the amount of DNA per cell in a straightforward manner. Insofar as prokaryotes are concerned, and also those viruses that infect prokaryotic cells, the expectation has been realized; there is an excellent correspondence between the number of genes and the amount of DNA. In eukaryotes, however, the correspondence breaks down. If the correspondence did hold true for eukaryotes, one would expect mammalian species to have at least as many genes—and therefore at least as much DNA per cell—as any other eukaryote. However, certain fish and amphibia have more than 10 times as much DNA per cell as any mammalian species. Moreover, especially among amphibia, closely related species may

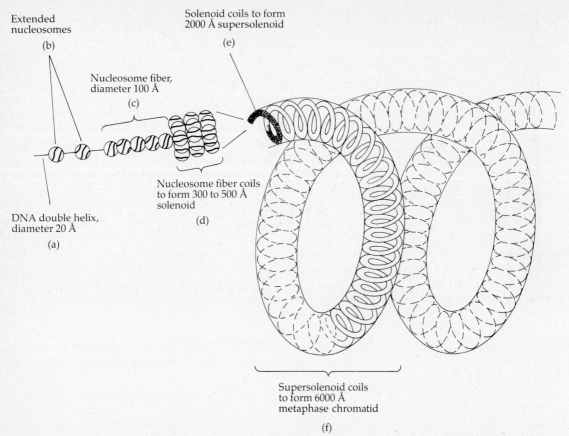

Extended
nucleosomes

(b)

Solenoid coils to form
2000 Å supersolenoid

(e)

Nucleosome fiber,
diameter 100 Å

(c)

DNA double helix,
diameter 20 Å

(a)

Nucleosome fiber coils
to form 300 to 500 Å
solenoid

(d)

Supersolenoid coils
to form 6000 Å
metaphase chromatid

(f)

Figure 8.14 One model of chromosome structure. *(a)* Extended DNA double helix. *(b)* DNA complexes with histones to form extended nucleosomes like beads on a string. *(c)* DNA spacer between nucleosomes becomes wrapped up in nucleosomes to form nucleosome fiber. *(d)* Nucleosome fiber coils to form solenoid. *(e)* Solenoid coils to form supersolenoid. *(f)* Supersolenoid coils to form metaphase chromatid. Parts *(a)* through *(d)* are well established. Details in *(e)* and *(f)* have not been directly confirmed.

differ in their amount of DNA per cell by a factor of 5 or 10.

On the whole, eukaryotic cells have far more DNA than would be expected from the number of genes they have. A haploid set of human chromosomes has about 3 billion nucleotide pairs of DNA, for example; at 1000 nucleotide pairs per gene (thought to be a reasonable estimate of the length of DNA required to code for an average protein), humans would be expected to have about 3

million genes. Yet the number of genes in humans is thought to be no greater than 100,000; that is, there is at least 30 times too much DNA. To take another example, a haploid chromosome set of *Drosophila melanogaster* contains about 160 million nucleotide pairs, which (at 1000 nucleotide pairs per gene) would correspond to 160,000 genes. Yet the number of bands in the polytene chromosomes is about 5000, and there is excellent evidence that each polytene band corresponds to one

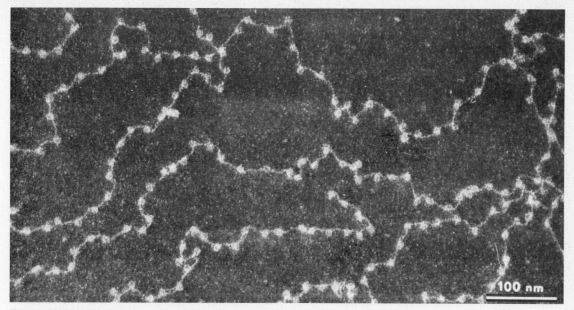

100 nm

Figure 8.15   Electron micrograph of extended nucleosomes. Compare with Figure 8.14*(b)*.

gene. Thus, in *Drosophila*, there is again about 30 times as much DNA as would be required to code for proteins.

The resolution of the discrepancy seems to be that much of the DNA in cells—perhaps most of it—does not code for proteins. Indeed, chemical and physical studies of DNA have revealed certain categories of nucleotide sequence that are not directly involved in coding for proteins. In most mammals, including humans, about 10 percent of the DNA consists of **highly repetitive** sequences; these sequences, typically 10 nucleotide pairs or shorter, are tandemly repeated up to 1 million times or more. Not only are such highly repetitive sequences found in mammals, but they also occur in most eukaryotes. The highly repetitive sequences are exceptionally simple in *Drosophila virilis*. This organism has three highly repetitive sequences. One of them is $\begin{smallmatrix}5'\text{-ACAAACT-}3'\\3'\text{-TGTTTGA-}5'\end{smallmatrix}$, which is repeated tandemly about 11 million times; another such sequence is $\begin{smallmatrix}5'\text{-ATAAACT-}3'\\3'\text{-TATTTGA-}5'\end{smallmatrix}$, repeated 3.6 million times; and the third highly

repetitive sequence is $\begin{smallmatrix}5'\text{-ACAAATT-}3'\\3'\text{-TGTTTAA-}5'\end{smallmatrix}$, repeated 3.6 million times. (Note the striking similarities among the sequences.) The sequences of highly repetitive DNA in mammals are somewhat more complex than those in *Drosophila*, but the principle is the same: Each class of highly repetitive DNA consists of a relatively short sequence repeated in tandem up to (and sometimes exceeding) 1 million times. (Repetitive DNA in humans is discussed further in Chapter 11.)

The function of highly repetitive DNA is unknown, but it is known to be located around the centromeric regions of the chromosomes. This location suggests that highly repetitive DNA plays a role in cell division. It is worth noting here that the sequence organization of highly repetitive DNA—tandem repeats of a short sequence—creates an ideal situation for the occurrence of unequal crossing-over (discussed in Chapter 7). Indeed, it is presently thought that unequal crossing-over is the principal manner in which the number of tandem repeats in a sequence of highly repetitive DNA can be decreased or increased during evolution. There must also be some mechanism that prevents the tandemly repeated sequences from becoming too different over time due to the accumulation of random mutations. (The tandem repeats are not always identical in sequence, but they are always nearly identical.) How this latter process occurs is at present unknown, but it is thought to involve something analogous to gene conversion.

In addition to the approximately 10 percent of mammalian DNA that consists of highly repetitive sequences, another 30 percent (approximately) consists of **moderately repetitive** sequences. Moderately repetitive DNA is composed of about 1500 different sequences averaging about 600 nucleotide pairs long; each sequence is repeated tens or hundreds of times in the haploid chromosome set. However, in contrast to the situation with highly repetitive DNA, the nucleotide sequences involved in moderately repetitive DNA are not repeated in tandem. Rather, moderately repetitive sequences are **interspersed** throughout the chromosomes.

The final category of DNA, which accounts for about 60 percent of the total, consists of **unique sequences**—those sequences represented just once per haploid chromosome set. The majority of such unique sequences are about 2250 nucleotide pairs long, but some are much longer. Most DNA sequences that code for proteins are represented as unique sequences. However, there are many more unique sequences than there are genes, so most unique sequences cannot be directly involved in coding for proteins.

We noted that moderately repetitive DNA sequences were interspersed along the chromosomes, which means that they are interspersed with unique sequences. Two broad patterns of interspersion can be recognized by physical studies of DNA. One pattern, called **short-period interspersion**, involves 2250-nucleotide-pair unique sequences being flanked by one or more 600-nucleotide-pair moderately repetitive sequences. The other pattern, called **long-period interspersion**, is similar, but the unique sequence involved is much longer—about 20,000 nucleotide pairs. The interspersion of unique with moderately repetitive DNA sequences suggests that moderately repetitive sequences may be involved in gene regulation, with turning on or off entire groups of genes at certain times in development or in certain tissues. Although this regulatory hypothesis is attractive and plausible, it is as yet unproven.

DNA in Prokaryotes

The fundamental double-helical structure of DNA is the same in prokaryotes as it is in eukaryotes, and the mechanisms of DNA replication and recombination are largely the same. Yet some aspects of prokaryotic DNA are strikingly simpler than in eukaryotes. Prokaryotic bacteria such as the intestinal bacterium *Escherichia coli* have just one chromosome in the form of a long molecule of double-stranded DNA with its ends joined together to form a closed circle. (The word *chromosome* is used advisedly in the context of prokaryotic DNA. Prokaryotes do not possess histones, so the nucleosome structures so characteristic of eukaryotes do not exist. In short, a prokaryotic "chromosome" is a decidedly different thing from a eukaryotic chromosome; their only common feature is the presence of DNA.) In *E. coli,* the closed circle of DNA comprises about 4 million nucleotide pairs.

The chromosome of *E. coli* and other prokaryotes has only a single **origin of replication**; that is, replication always begins at the same place and proceeds from there around the rest of the chromosome. The process of replication is illustrated in Figure 8.16, in

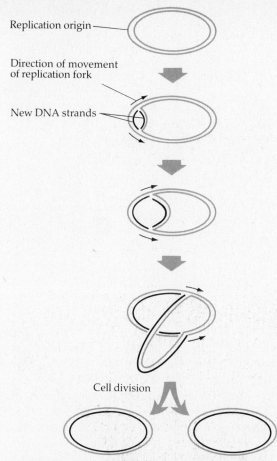

Replication origin

Direction of movement
of replication fork

New DNA strands

Cell division

Figure 8.16 Replication of circular DNA in prokaryotes. Replication begins at a single origin of replication, shown at the left. The replication bubble gradually increases in size until the entire molecule is replicated. Daughter molecules are segregated into distinct cells at the time of cell division.

of replication illustrated in Figure 8.16 also occurs in the DNA of mitochondria and chloroplasts. Their DNA is also circular, and its manner of replication suggests again an ancient prokaryotic origin of these essential organelles.)

The DNA of prokaryotes is also simpler than the DNA of eukaryotes in another respect: Prokaryotic DNA does not have highly repetitive or moderately repetitive sequences. Indeed, most of the DNA in prokaryotic cells seems to be involved directly in coding for proteins. With 1000 nucleotide pairs of DNA per protein, *E. coli*'s 4 million nucleotide pairs could code for 4000 genes. Since the organism is thought to have between 3000 and 3500 different proteins, the majority of its DNA must be directly involved in the protein-coding function.

Plasmids

In addition to its own DNA molecule, a bacterial cell may harbor one or more small, ring-shaped molecules of DNA known as **plasmids**. Plasmids are transmitted from cell generation to cell generation via the processes of DNA replication and cell division. Each plasmid has its own origin of replication, and the plasmid DNA replicates simultaneously with the chromosomal DNA. When the bacterial cell divides, the daughter cells receive the plasmid DNA along with the chromosomal DNA.

Many plasmids have a second route of transmission open to them. Part of their DNA carries genes that enable the plasmid to be transmitted from cell to cell whenever bacterial cells come into contact. (The plasmid replicates as it is transferred, so both the donor cell and the recipient cell end up with copies of the plasmid.) Because of this alternative route of transmission, many plasmids are **infectious**, not only between bacteria of the same species but also between different species and, in some

which the old DNA strands are represented as thick gray lines and the newly synthesized DNA strands as thick black lines. Note that DNA replication proceeds in a semiconservative manner in both directions from the replication origin. After replication is completed, cell division occurs in such a manner that each daughter cell receives one of the DNA molecules. (It should be mentioned that the mode cases, different genera. A plasmid that carries one or more genes that are beneficial to its bacterial host can therefore be widely disseminated among many bacterial species. Exactly this sort of dissemination has occurred in bacterial populations in response to the widespread use of antibiotics in medicine and agriculture. (**Antibiotics** are chemical agents that kill bacterial cells.) Many genes that confer resistance to particular antibiotics are located on plasmids, and these plasmids have become more frequent and widespread as a result of antibiotic use. The spread of drug resistance by means of plasmids is an even more serious problem because genes that confer resistance to different antibiotics can be on the same plasmid. Some plasmids confer simultaneous resistance to five or more antibiotics. These infectious plasmids that carry antibiotic resistance genes are known as **resistance transfer factors**.

Plasmids vary enormously in size—from small ones of a few thousand nucleotide pairs to large ones of 100 thousand nucleotide pairs or more. (Figure 8.17 is an electron micrograph of a plasmid DNA molecule.) Plasmids also vary in their **copy number**—the number of plasmids of a given type that can accumulate inside one cell. Many small plasmids have high copy numbers (20 or 30 or more copies per cell), whereas most large plasmids have low copy numbers (one or a few copies per cell). Because they can be manipulated in test tubes without being broken apart, the small, high-copy-number plasmids are of prime practical importance in the development of the arsenal of new techniques known collectively as *DNA cloning* or *recombinant DNA*.

Restriction Enzymes

Recombinant DNA refers to DNA molecules that have been artificially created by chemically splicing together DNA sequences from two

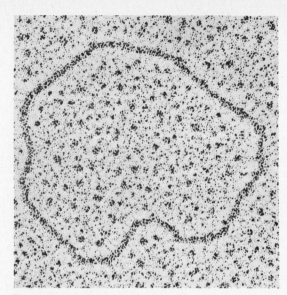

Figure 8.17   Electron micrograph of a bacterial plasmid.

or more different organisms. One technical problem that had previously hindered the study of genes directly at the DNA level was the difficulty of obtaining large amounts of DNA corresponding to a single gene. The advent of recombinant DNA has solved this problem of quantity and along the way has created powerful new methods for producing medically important substances, such as insulin.

A key ingredient in recombinant DNA technology is a class of enzymes known as **restriction endonucleases**, which are produced naturally by certain bacteria. Restriction endonucleases, often more simply called **restriction enzymes,** have the capacity to attach to a double-stranded DNA molecule only at certain places where the DNA has a specific, short nucleotide sequence. These sites of attachment are called **recognition sequences** or **restriction sites**, and each type of restriction enzyme has its own particular recognition sequence. Once a restriction enzyme attaches to its recognition sequence, it cleaves both backbones of the DNA duplex and thus breaks the molecule into two parts at the position of the restriction site. Once the DNA molecule has been cleaved, the enzyme falls off and is free to attach to its recognition sequence at some other place on the molecule.

Dozens of types of restriction enzymes are known, and each has its own particular recognition sequence. Three examples are illustrated in Figure 8.18. In the figure, the symbol N stands for any nucleotide; the only requirement is that paired nucleotides in the two strands must be complementary (i.e., A paired with T and G with C). For the restriction enzyme *Eco*RI, the recognition sequence is $\frac{3'\text{-CTTAAG-}5'}{5'\text{-GAATTC-}3'}$. (Incidentally, the names of the restriction enzymes incorporate the bacterial species of their origin; *Eco*RI and *Eco*RII come from *Escherichia coli* and *Hind*III comes from *Hemophilus influenzae*.) When *Eco*RI attaches to its restriction site, it breaks the DNA strands at the positions indicated by the arrows in Figure 8.18; the resulting fragments terminate with the sequences shown at the right side of the figure. Note that each fragment terminates with a short, single-stranded region. Moreover, the nucleotides in the single-stranded regions are complementary, which means that they can come together and be held weakly in place by hydrogen bonding. DNA fragments produced by restriction enzymes are called **restriction fragments**, and the ability of the ends of these fragments to reassociate is often described by saying that the fragments have "sticky" ends.

Restriction enzymes *Eco*RII and *Hind*III operate in a manner similar to *Eco*RI, but they have different recognition sequences. Although the recognition sequences do differ, the enzymes illustrated have one common feature; they are all **palindromes**. In the English language, a *palindrome* is a word that reads the same forward or backward, like *madam*. This same sort of symmetry characterizes palindromic nucleotide sequences. In the case of *Eco*RI, for example, the recognition sequence read from left to right in Figure 8.18 is $\frac{3'\text{-CTTAAG-}5'}{5'\text{-GAATTC-}3'}$; read from right to left it is also $\frac{3'\text{-CTTAAG-}5'}{5'\text{-GAATTC-}3'}$. As can be observed in Figure 8.18, the recognition sequences of *Eco*RI and *Hind*III are perfect palindromes, whereas that of *Eco*RII is an imperfect palindrome because of the A-T pair in its center.

DNA palindromes are important because hydrogen bonding can occur between nucleotides in the *same* DNA strand. This self-pairing feature of palindromes gives palindromic sequences the capacity to form **hairpin loops** of the sort illustrated for the *Eco*RII restriction site in Figure 8.19(*b*). Within the cell, the structure of the palindromic DNA alternates between the linear form [Figure 8.19(*a*)] and the hairpin-loop form [Figure 8.19(*b*)], but it is thought that restriction enzymes might attach to the molecule in its hairpin-loop form, although there is at present no conclusive information on the issue. Eukaryotic DNA normally contains many different palindromes; some are short (like those recognized by restriction enzymes) and others are much longer and involve tens or hundreds of nucleotide pairs. The hairpin loops formed by these palindromes are thought to provide specific sites of attachment for proteins that interact with DNA, such as proteins that are involved in the regulation of gene activity.

Recall that restriction enzymes are produced by bacteria. The enzymes are thought to be one way in which bacterial cells defend themselves against invasion by harmful foreign DNA such as plasmids or viruses that infect bacteria. When foreign DNA enters the cell, the restriction enzyme cleaves the foreign DNA at its restriction sites, and the fragments are further degraded by other enzymes. How-

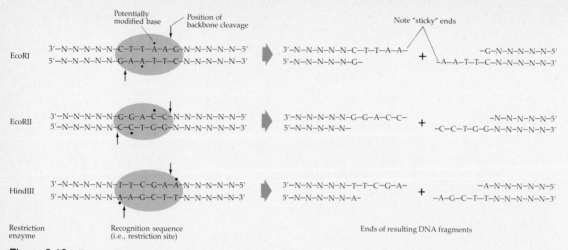

Figure 8.18 Recognition sequence (restriction site) and positions of backbone cleavage of three restriction enzymes. N represents any nucleotide. The restriction sequence is enclosed in the shaded oval, and the dots denote bases that can be chemically modified by modification enzymes to prevent cleavage. The restriction enzymes shown here all produce staggered cuts leading to restriction fragments with "sticky" (i.e., complementary) ends. Some other restriction enzymes do not make staggered cuts and so produce fragments with blunt ends.

ever, the bacteria's own DNA also contains recognition sequences of the restriction enzyme so there is the potential for self-destruction. Bacteria prevent self-destruction by means of a class of enzymes called **modification enzymes**. These enzymes chemically modify certain bases in DNA, typically by adding a methyl ($-CH_3$) group at some position in the base. When particular bases are modified in a recognition sequence, the restriction enzyme can no longer become attached to it. In Figure 8.18, the bases that can be modified to prevent recognition are indicated by dots. Thus, foreign DNA can be cleaved by restriction en-

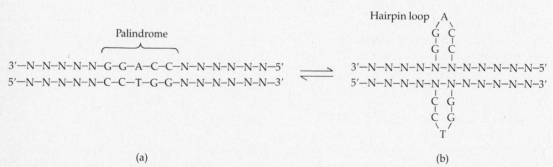

(a)   (b)

Figure 8.19 A palindrome is a DNA sequence that reads the same forward and backward—in this case 3'-GGACC-5'. Palindromes can undergo intrastrand pairing to form hairpin-loop structures. In a cell, a palindrome can thus alternate between the structures in (a) and (b).

zymes because it is not modified, but the bacteria's own recognition sequences are protected by modification.

Restriction Mapping and DNA Sequencing

When a piece of duplex DNA is cut by a restriction endonuclease, the resulting fragments (called *restriction fragments*) can be separated according to size by placing the fragments in slots at the edge of a slab of a jellylike material (usually **acrylamide** or **agarose**) and subjecting the gel to an electric current for several hours (Figure 8.20). This procedure, which is called **electrophoresis**, separates the DNA fragments by size because smaller fragments move more rapidly in response to the electric field. Usually the DNA is combined with a dye that fluoresces in response to ultraviolet light. After electrophoresis, the several size classes of DNA are visible as discrete bands when viewed with ultraviolet light (see Figure 8.20).

The fragments produced by one restriction enzyme can be removed from the gel and digested with a different restriction enzyme. The size of the fragments produced by the second enzyme reveals the positions of the recognition sequence of the second enzyme relative to those of the first. For example, if an *Eco*RII restriction fragment is cut into two fragments of equal size by *Eco*RI, then we know that somewhere on the DNA molecule there is an *Eco*RI site flanked by two equally distant *Eco*RII sites. When such studies are carried out systematically with a large number of different restriction enzymes, the relative positions of all the restriction sites in a particular piece of DNA can be determined. The resulting diagram showing the locations of the restriction sites is known as a **restriction map**. Figure 8.21 is a restriction map of a segment of mouse DNA that includes the mouse α-globin gene. (α-globin is one of the constituents of hemoglobin.)

Generating a restriction map is often a preliminary to the ultimate level of DNA study —**DNA sequencing**, which refers to the determination of the actual nucleotide sequence in a molecule. Restriction enzymes are important tools in DNA sequencing because they provide fragments of convenient length with known

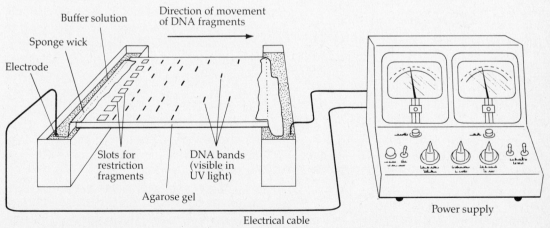

Figure 8.20 Typical setup for electrophoresis of DNA fragments.

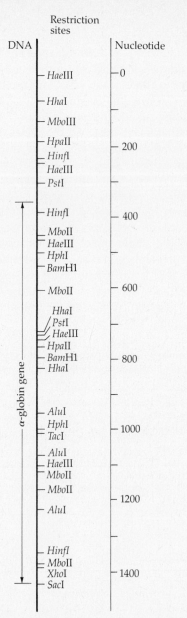

Figure 8.21 Restriction map of mouse α-globin gene showing positions of restriction sites for several restriction enzymes.

the method involves heating a restriction fragment to cause the strands to dissociate and then separating the strands by gel electrophoresis. For purposes of illustration, in Figure 8.22 we assume that the strand to be sequenced is actually 3'-ATGCATGC-5'. The next step in the procedure is to add a radioactive deoxyadenosine nucleotide to the 3' end of the strand to be sequenced. This radioactive A is represented as A* in Figure 8.22, and in practice the addition reaction is catalyzed by an enzyme.

The solution containing the now-radioactive DNA strand is next divided into four aliquots, and each aliquot is treated with a particular chemical reagent. One reagent breaks the DNA strand at any position where an A appears. The chemical reaction is such that the A is destroyed in the process of breakage. However, the reaction is carried out very gently, so most of the A's remain intact; only an occasional A is attacked and the DNA strand is cleaved at that point. A second reagent attacks G's and breaks the strand; a third attacks C's; and the fourth attacks either T's *or* C's. Again, these reagents destroy the nucleotide at the point of breakage, but they are carried out gently to allow most of the nucleotides to remain intact.

Each of the four reagents produces a unique set of fragments of the DNA strand. These fragments are separated according to size by electrophoresis. Those fragments that carry the radioactive 3' terminus are identified by autoradiography, in which a photographic emulsion is placed over the gel and radioactive emissions cause darkening of the overlying film. When the film is developed, dark bands appear over those positions occupied by fragments that have the radioactive 3'-A* (see Figure 8.22).

The nucleotide sequence of the DNA strand can be inferred directly from the pattern of radioactive bands. The most rapidly migrating band consists of single nucleotides, the next fastest is a dinucleotide, the next a trinucleo-

terminal sequences. One method of DNA sequencing in widespread use is illustrated in Figure 8.22. Single strands are sequenced, so

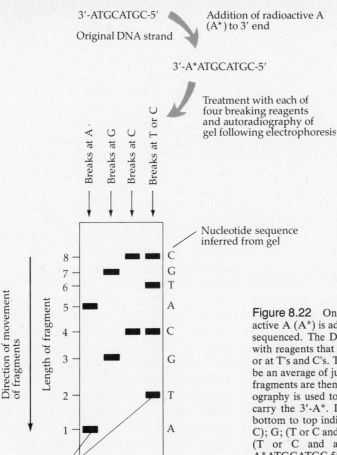

3'-ATGCATGC-5'   Addition of radioactive A
                  (A*) to 3' end
Original DNA strand

3'-A*ATGCATGC-5'

Treatment with each of
four breaking reagents
and autoradiography of
gel following electrophoresis

Breaks at A
Breaks at G
Breaks at C
Breaks at T or C

Nucleotide sequence
inferred from gel

Direction of movement
of fragments

Length of fragment

8   C
7   G
6   T
5   A
4   C
3   G
2   T
1   A

A*
3'

Bands on
autoradiograph
resulting from
radioactive A*

Figure 8.22 One method of DNA sequencing. A radioactive A (A*) is added to the 3' end of the fragment to be sequenced. The DNA is split into four parts and treated with reagents that cleave the strand at A's, at G's, at C's, or at T's and C's. Treatment is so controlled that there will be an average of just one break per molecule. The broken fragments are then run side by side on a gel, and autoradiography is used to locate the positions of fragments that carry the 3'-A*. In this case the pattern of bands from bottom to top indicates the sequence A; (T or C but not C); G; (T or C and also C); A; (T or C but not C); G; and (T or C and also C); thus, the sequence is 3'-A*ATGCATGC-5'. In practice, fragments as small as those shown here migrate so fast that they run right off the gel, but sequences farther back in the fragment can be obtained. The initial sequence of this fragment would be determined by sequencing an overlapping restriction fragment.

tide, and so on (see Figure 8.22). To determine the nucleotide next to the A*, one simply identifies which reagent has created a fragment consisting of a single nucleotide; in the case of Figure 8.22, this is the A-breaking reagent, so the sequence begins as 3'-A*A. The next nucleotide is determined from the fragment carrying two nucleotides; in Figure 8.22 the fragment appears when the strand is broken at T's or C's. However, the band does not appear with the C-breaking reagent, so the next nucle-

otide must be a T. Thus, the initial sequence is 3'-A*AT. The trinucleotide fragment appears in the G-breaking column, so the next nucleotide is G—that is 3'-A*ATG. The tetranucleotide fragment appears both in the C-breaking column and in the T- or C-breaking column, so this fragment must have been produced by the breaking of a C. The sequence so far is thus 3'-A*ATGC, and the rest of the sequence can be read directly from the gel as shown in Figure 8.22.

In actual practice, sequences of up to 300 nucleotides can be read off a single gel. To sequence a long molecule of DNA, the sequences of smaller restriction fragments are first determined. Then the sequences of overlapping fragments produced by a different restriction enzyme are determined. The sequences that overlap can be identified by direct comparison, and in this manner the nucleotide sequence of the entire molecule can be pieced together. DNA sequencing is technically straightforward, accurate, and rapid—so much so that DNA-sequence information is at present accumulating much more quickly than its significance can be comprehended.

Recombinant DNA

Restriction analysis and DNA sequencing both require relatively large amounts of a particular DNA fragment. The method for obtaining such amounts involves splicing the fragment of interest into a bacterial plasmid. As the plasmid replicates inside a living bacterial cell, the fragment of interest is replicated too. When the bacterial cell divides, all its offspring receive the artificially created plasmid. Within a large clone of bacteria descended from a single, plasmid-carrying cell, there may be 10^{15} (a thousand trillion) or more replicas of the plasmid, each carrying the DNA fragment of interest. All that is required to recover the DNA fragment is to purify the plasmid DNA from the clone.

One procedure for obtaining recombinant DNA is outlined in Figure 8.23. (Alternative procedures are similar in principle but vary in such respects as the use of DNA from bacteria-infecting viruses rather than plasmids.) Figure 8.23 illustrates the production of recombinant DNA from a eukaryote. Part (*a*) shows eukaryotic DNA fragments (labeled A and B) produced by a restriction endonuclease, and the "sticky" single-stranded ends (illustrated in detail in Figure 8.18) are to be noted.

Part (*a*) also shows the circular DNA of a plasmid that has one restriction site for the restriction enzyme in question; when cleaved, the plasmid will have the structure illustrated, and its sticky ends will be complementary in nucleotide sequence to those of the eukaryotic fragments. As a consequence of the complementary nucleotide sequences of the sticky ends, the single-stranded regions of the fragments and the plasmids will form hydrogen bonds when mixed together, as illustrated in Figure 8.23(*b*). The gaps in the DNA backbones are then permanently sealed by the addition of DNA ligase [see Figure 8.23(*c*)]. The plasmids in Figure 8.23(*c*) are artificially created plasmids that have been rendered larger than the original one by the addition of a fragment of eukaryotic DNA. In principle, the procedure of plasmid cutting and insertion is not very different from the cutting and insertion that a jeweler does to enlarge the size of a ring. However, eukaryotic DNA will have many restriction sites along its length, so there will be many more different restriction fragments than just the A and B shown in the figure. Thus, the collection of recombinant plasmids in Figure 8.23(*c*) will often consist of a highly heterogeneous collection of plasmids that have different restriction fragments.

Figure 8.23(*d*) through (*f*) illustrates how individual plasmids from this heterogeneous collection can be separated out. The first step is called **transformation**, in which the plasmids are introduced into plasmid-free bacteria. The conditions of the transformation are so arranged that each bacterial cell receives one or no plasmids, so there is the problem of how the plasmid-infected cells are to be separated from the uninfected ones. This problem is solved by the use of a plasmid that carries an antibiotic-resistance gene. When a mixture of plasmid-carrying and plasmid-free bacteria is treated with the antibiotic in question, only those cells that carry the plasmid will survive. This procedure represents one more application of the

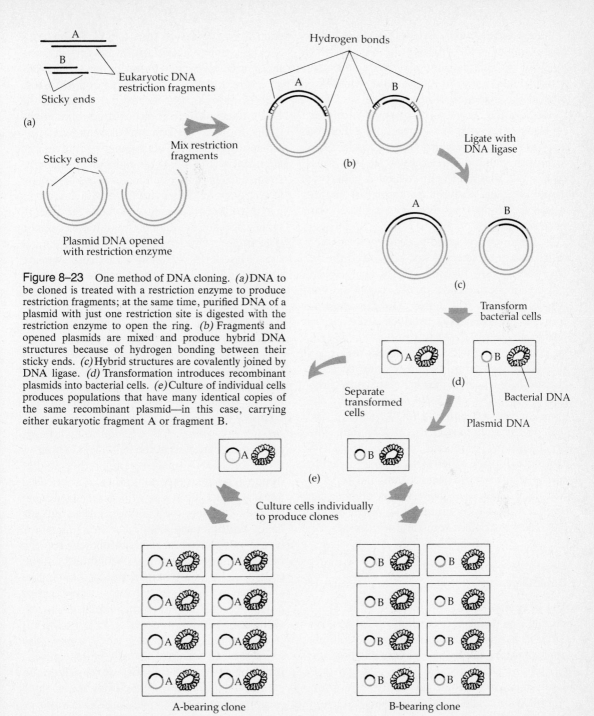

Figure 8–23   One method of DNA cloning. *(a)* DNA to be cloned is treated with a restriction enzyme to produce restriction fragments; at the same time, purified DNA of a plasmid with just one restriction site is digested with the restriction enzyme to open the ring. *(b)* Fragments and opened plasmids are mixed and produce hybrid DNA structures because of hydrogen bonding between their sticky ends. *(c)* Hybrid structures are covalently joined by DNA ligase. *(d)* Transformation introduces recombinant plasmids into bacterial cells. *(e)* Culture of individual cells produces populations that have many identical copies of the same recombinant plasmid—in this case, carrying either eukaryotic fragment A or fragment B.

principle of **selection** discussed in Chapter 4. For example, if the original plasmid carried a gene for resistance to tetracycline, then, after transformation, the only cells that would survive in the presence of tetracycline would be those that had received the plasmid. Ordinarily, the mixture of cells is spread out evenly on an agar surface containing the antibiotic. Division of the antibiotic-resistant cells soon leads to a small clump of cells (called a **colony** or **clone**) at each position on the agar surface where an original antibiotic-resistant cell had been deposited.

Once the plasmid-bearing transformed cells have been selected and separated on the agar surface [see Figure 8.23(e)], each colony can be transferred individually to a liquid culture medium in which the cells can undergo successive divisions to produce large clones [see Figure 8.23(f)]. At this point, all the cells in any one clone will have plasmids that carry the same restriction fragment, but different clones may have different restriction fragments. In Figure 8.23(f), for example, the clone on the left carries the A fragment and that on the right carries the B fragment. The procedure for producing recombinant DNA molecules is often called **DNA cloning**. Because of the small size of bacteria and the rapidity with which they can divide, DNA cloning can provide virtually unlimited quantities of any particular restriction fragment of interest.

Applications of DNA Cloning

The development of the technology of DNA cloning has caused a revolution in genetics because it has opened the way for a direct biochemical analysis of genes and of the manner in which the expression of genes is controlled. With traditional approaches, information about a gene's structure and function had to be inferred from studies of the ultimate product coded by the gene (usually a protein,

such as α-globin) or from the phenotypic effects caused by mutations in the gene. A vast amount of important information about gene structure and function has been obtained from such traditional approaches, but studies of the DNA itself are often more direct and to the point. Indeed, one new approach to genetics—an approach still in its infancy—involves the chemical alteration of cloned DNA fragments to produce precisely defined changes in nucleotide sequence. When these altered fragments are introduced back into the cell, the phenotypic effects of the engineered mutation can be determined. This new approach stands genetics on its head. In the past, geneticists tried to understand the chemical nature of a mutation by studies of its phenotypic effects; with engineered mutations, geneticists can now study the phenotypic effects of mutations whose chemical nature is already known.

It should not be assumed, however, that direct studies of DNA can completely supplant traditional approaches to genetic analysis. Genes whose expression is regulated by a complex network of interactions involving the products of other genes are not easily accessible to direct analysis at the DNA level because, at present, DNA analysis is limited to studies of one gene at a time, not groups of interacting genes. In addition, as will be discussed in Chapter 16, there is an important class of traits called **multifactorial traits** that are determined by the joint effects of the environment and the interactions of many genes. With multifactorial traits (examples are height, weight, skin color, and various aspects of behavior), the effect of any single gene may be so small as to be swamped by environmental effects or the effects of other genes. In such cases, a gene that influences the trait can scarcely be recognized, let alone studied at the DNA level. Thus, the genetic analysis of multifactorial traits or of complex phenotypes dependent on networks of interacting genes must still be carried out through more traditional approaches.

In addition to causing a revolution in genetics and molecular biology, the technology of recombinant DNA has caused a revolution in pharmacology—so much so that a number of specialized genetic-engineering companies are thriving, and many huge pharmaceutical and chemical companies have eagerly expanded their research departments to encompass the new technology. The promise of recombinant DNA is, of course, that bacteria carrying an appropriate DNA fragment might produce the gene product corresponding to the fragment. If this product was of medical or other importance, the bacteria could be used to produce it in abundance and relatively inexpensively. The promise of recombinant DNA has already been achieved in several noteworthy instances. Bacterial production of human insulin is a prime example. In the past, the insulin used in the treatment of diabetes had to be extracted from animal pancreases, and it was scarce and expensive. In addition, many diabetics developed allergic reactions to the animal form of the protein. Recombinant DNA techniques have now been used to create bacterial strains that produce human insulin in large amounts, and this insulin has been used clinically with no reported adverse effects. Insulin production is one of the cardinal practical achievements of recombinant DNA, but other successes are human **interferon** (a protein thought to be important in the body's defense against viral infection) and human **growth hormone** (a protein needed for successful treatment of children with a rare disease called pituitary dwarfism; it is otherwise obtainable in active form only from the pea-sized pituitary gland of human cadavers). These examples of the practical uses of recombinant DNA are only the early successes. Many other examples, some equally dramatic, are sure to follow.

The blessings of recombinant DNA are not entirely unmixed, however. From the very beginnings of the research, some people, including a few scientists themselves, have had a vague sense of unease. In some cases, these feelings were motivated by a general philosophical belief that one ought not to tamper with living things, and these feelings were enhanced by sensationalist press reports about scientists "creating new forms of life" with allusions to Doomsday Bugs and Frankensteins. For others, the source of anxiety was not the creatures that might be unleashed by design; it was the potential catastrophes that might be triggered by accident. These people pointed out that *E. coli* (then and still a popular bacteria for DNA cloning) is a normal resident of the intestinal tract of all humans and other mammals. If the wrong genes were cloned in *E. coli* by accident (an often-used hypothetical example involved the genes for tumor induction from human cancer-causing viruses), and if the *E. coli* clone carrying these genes escaped from the laboratory, the escaped clone might be able to invade the intestine and replace the normal bacteria. In such an event, an epidemic of awesome proportions might ensue. On the other hand, almost everyone agreed that the dangers of recombinant DNA were potential, not imminent. Nevertheless, critics of the procedures argued that the risks had not been properly assessed. They urged caution, circumspection.

Sympathetic to the need for caution, a group of 11 prominent and influential scientists banded together in the summer of 1974 and urged a worldwide moratorium on several lines of research involving recombinant DNA that were judged to be potentially hazardous. Four scientists in the group were themselves directly involved in the research they agreed to suspend. The moratorium on the potentially harmful research was followed by an international conference held February 24 to 27, 1975, at the rustic, former YMCA camp of Asilomar on the California coast. The conference included 86 American and 53 foreign scientists from 15 countries, and its purpose was to discuss the research and try to reach agreement on how to minimize the potential dangers. The conferees at Asilomar, which included attorneys and public health officials as well as scientists, decided that there were certain experiments for which the risks outweighed the benefits. Among these were experiments to clone DNA from known disease-causing organisms or from organisms that produce toxins or venoms. The Asilomar group urged that such experiments be banned. The group also decided that other experiments could be classified according to their degree of risk and that each degree of risk could be minimized with appropriate levels of physical and biological containment of the recombinant clones. **Physical containment** refers to the laboratory setting of the research—to such issues as whether it is open to unauthorized access and whether proper facilities for working with potentially dangerous organisms are available and used. The highest degree of physical containment applies to laboratories in which the organisms are maintained and manipulated in glove boxes and not directly handled by anyone. Such facilities were originally developed for the study of such extremely pathogenic organisms as the ones that cause smallpox, anthrax, or botulism. **Biological containment** refers to the organism that harbors the recombinant DNA. The lowest level of biological containment involves the normal laboratory strain of *E. coli*; the highest level involves strains carrying certain mutations that render them so weak and feeble and so nutritionally fastidious that they are virtually unable to survive outside of the laboratory. The conferees at Asilomar decided that each allowable experiment could be carried out with levels of physical and biological containment appropriate to its degree of risk. The Asilomar conference was followed by others, both in the United States and abroad, with the result that the Asilomar recommendations, somewhat modified, are now embodied in regulations that embrace all government-supported research. Other countries have followed suit with similar regulations. These regulations are periodically reexamined and revised (although some types of experiments are still forbidden), and there is now widespread agreement that the risks of DNA cloning have been appropriately weighed and minimized.

Perhaps the greatest risk of recombinant DNA research has been insufficiently discussed. It is not the risk that some harmful bacteria will escape from the laboratory and cause an epidemic. It involves the manner in which the products of such clones are to be used. Drug abuse is one of the most serious problems faced by today's society; the main offenders are tobacco, alcohol, and tranquilizers. Will the miracle products of DNA cloning be used more responsibly than other drugs? Brain biochemists have recently discovered entire classes of mind-bending small-protein hormones; some obliterate pain and others cause feelings of elation and well-being. These small proteins will be easy to produce in large quantity with DNA cloning. How are such drugs to be used? It is hoped that they will be used responsibly, but one is naggingly reminded of Aldous Huxley's *Island*, in which the inhabitants willingly maintain themselves in a state of drug-induced euphoria, oblivious to their political and spiritual oppression.

SUMMARY

1. DNA occurs in cells as a double-stranded (**duplex**) molecule, and many of the important structural characteristics of DNA are adequately illustrated by a ladder analogy. In this analogy, duplex DNA is represented as a ladder consisting of two long sidepieces connected at intervals by rungs. Each strand in the duplex molecule is represented as a sidepiece with half-rungs jutting off, and the ladder is completed by bringing corresponding half-

rungs on the two sidepieces into contact. When DNA replicates, the two half-ladders separate, and each one is used as a **template** (pattern) for the step-by-step creation of a new partner half-ladder.

2. Actually, each DNA strand consists of a linear sequence of chemical constituents called **nucleotides**. A nucleotide is composed of three parts: (a) a sugar ring (deoxyribose sugar in the case of DNA), (b) a phosphate group attached to the 5′ carbon of the sugar ring, and (c) one of four chemical groups called **bases** attached to the 1′ carbon of the sugar ring. Four bases are found in DNA: **adenine, guanine, thymine,** and **cytosine**. The first two of these are known collectively as **purines**, the second two as **pyrimidines**. These four bases determine the four nucleotides that occur in DNA: **deoxyadenosine phosphate** (symbolized **A**), which carries adenine; **deoxyguanosine phosphate** (**G**), which carries guanine; **thymidine phosphate** (**T**), which carries thymine; and **deoxycytidine phosphate** (**C**), which carries cytosine.

3. Nucleotides are linked together by means of the phosphate group of one becoming attached to the 3′ carbon of the sugar ring of the next nucleotide in line. The alternating sugar-phosphate-sugar-phosphate pattern so produced provides the backbone of the DNA strand, and each sugar carries a base that juts off the backbone. (These protruding bases correspond to the half-rungs in the ladder analogy.) One end of the backbone of a DNA strand terminates with a phosphate group; this is called the **5′ end** of the strand. The other end of the strand terminates with a free hydroxyl (−OH) on the 3′ carbon of the sugar ring; this end is called the **3′ end** of the strand.

4. Duplex DNA is formed when two DNA strands come together so that their bases meet in the middle (forming the ladder). The strands are brought together by weak chemical bonds called **hydrogen bonds** that form between corresponding bases in the two strands.

However, adenine will form hydrogen bonds only with thymine, and guanine will form hydrogen bonds only with cytosine. Because of these base-pairing restrictions, the nucleotides in the two strands of duplex DNA are **complementary** in the sense that, wherever one strand carries an A, the other across the way will carry a T, and wherever one strand carries a G, the other across the way will carry a C. In addition, hydrogen bonds can form between the bases only if the two strands are **antiparallel**; that is, the strands must run in opposite directions so that the 5′ nucleotide of one strand is paired with the 3′ nucleotide of the other strand.

5. The normal, three-dimensional configuration of duplex DNA is a **double helix,** in which the sugar-phosphate backbones wind around the outside of the molecule in a right-handed helix, and the paired bases are stacked on top of one another in the interior. This double-helical configuration is 20 Å in diameter. Each backbone makes a complete turn around the helix every 34 Å, and there are 10 base pairs per complete turn. This right-handed helix is known as the **B form** of DNA, but many alternative configurations of duplex DNA are also known. Most involve minor alterations in the double helix in which the bases are somewhat tilted or more tightly stacked. However, under certain conditions, duplex DNA can be induced to assume more bizarre configurations such as ones with the bases pointing outward or ones in which the helical coil is left-handed (the **Z form** of DNA). Whether any of these alternative forms of DNA is of biological importance is at present unknown.

6. DNA replication involves the activities of several enzymes; at least three act in such a way as to unwind the double helix and produce a Y-shaped **replication fork**. A key enzyme in DNA replication is **DNA polymerase**, which uses the **high-energy phosphate bonds** in **nucleotide 5′-triphosphates** to add successive nucleotides to the 3′ end of an established DNA strand. Of course, each added nucleotide is complementary (i.e., A with T and G with C) to the corresponding nucleotide in the template strand. DNA polymerase also has a **proofreading** function in that it can detach the last-added nucleotide in the event of nucleotide mismatch and reattach the correct nucleotide. Because DNA polymerase cannot initiate a new DNA strand, a **primer** strand composed of RNA is produced by an **RNA polymerase**, and DNA polymerase elongates the primer at its 3′ end. Because DNA polymerase can add to only the 3′ end of a strand, only one new strand (the one growing in the 5′ to 3′ direction) can be synthesized continuously. The other new strand is synthesized in relatively small fragments (**Okazaki fragments**), which are attached to one another by **DNA ligase**. Long DNA molecules in eukaryotes initiate replication at many points along their length, and the **replication bubbles** so produced have two replication forks. When DNA replication is completed, each daughter molecule consists of one newly synthesized strand and one original strand; this mode of replication is called **semiconservative**.

7. Recombination between the DNA molecules in nonsister chromatids is initiated by single-strand **nicks** in each duplex. The duplexes unwind from the position of the nicks, and the nicked strands exchange pairing partners to produce a region of **heteroduplex DNA**. After the nicks are resealed by DNA ligase, the DNA molecules can shift in position to create a **χ (chi) structure**, which can be resolved into its separate duplexes in two ways: one leaves alleles that flank the heteroduplex **nonrecombinant**, and one leaves flanking alleles **recombinant**. Noncomplementary nucleotides in the heteroduplex region can be corrected by a process of excision and resynthesis known as **mismatch repair**. Mismatch repair can lead to the apparent conversion of one allele into another. This **gene conversion** can be observed directly by studying the four products of meiosis in certain fungi that hold the four products together in the form of **tetrads**. A heterozygous A/a fungus will normally produce $2A$:$2a$ tetrads; gene conversion is indicated by $3A$:$1a$ or $1A$:$3a$ tetrads.

8. DNA is organized into chromosomes by means of a hierarchy of coils upon coils. The lowest level of structure involves an association between the DNA duplex and certain **histones**, which combine to form a sort of beads-on-a-string structure composed of **nucleosomes** (the beads) and connecting spacer DNA (the string). When the spacer DNA is not extended between the nucleosomes, the beads-on-a-string structure becomes a **nucleosome fiber** (diameter 100 Å). This fiber coils to form a **solenoid** (diameter 300 to 500 Å), which also coils to form a **supersolenoid** (diameter about 2000 Å). Finally, the supersolenoid coils to produce the 6000-Å **chromatid** that can be observed in metaphase.

9. The genetic information that enables a cell to manufacture a particular kind of protein is encoded in the sequence of nucleotides in the corresponding gene. In eukaryotes, however, only a relatively small fraction of DNA seems to code for proteins. Certain categories of non-protein-coding DNA have been detected by physical and chemical studies of **repetitive DNA**. One category, amounting to about 10 percent of the DNA in most mammals, is called **highly repetitive** DNA; it consists of short nucleotide sequences (usually less than 10 nucleotide pairs) that are repeated in tandem up to 1 million times or more. The function of highly repetitive DNA is not known, but its location near the centromeric regions of the chromosomes suggests that it plays a role in cell division. A second category of DNA consists of **moderately repetitive** sequences. These sequences account for about 30 percent of the total DNA and average about 600 nucleotide pairs in length; there are more than 1000 such sequences, each repeated tens or hundreds of times. These sequences are **interspersed** with other DNA sequences and scattered throughout the chromosomes. The final major category of DNA consists of **unique** sequences—sequences represented once in each haploid chromosome set. Most unique sequences are two to several thousand nucleotide pairs long, but some are much longer. Although some of the unique-sequence DNA carries the genetic information for proteins, much of it must have other functions. About 60 percent of the total DNA consists of unique sequences, but there are many more unique sequences than would be required for protein-coding purposes. In prokaryotes, by contrast, repetitive sequences corresponding to highly repetitive and moderately repetitive DNA do not occur, and most of the unique-sequence DNA seems to have a protein-coding function.

10. Plasmids are small, circular molecules of DNA that exist inside prokaryotic cells. These molecules undergo replication and are transmitted to the daughter cells during cell division. Some plasmids are also **infectious** because they carry genes that enable their transfer from one host cell to another. One class of infectious plasmids known as **resistance transfer factors** is particularly important in medicine because they carry genes that confer resistance to one or more **antibiotics**.

11. Restriction endonucleases are enzymes that attach to particular nucleotide sequences of duplex DNA (their **recognition sequences** or **restriction sites**) and cleave both strands of the duplex at or near these sites. Recognition sequences are often **palindromes** —sequences of duplex DNA that are identical when read in either direction—and it is thought that palindromes may facilitate recognition by their ability to form **hairpin loops**. In addition, the backbone cuts produced by restriction enzymes are often staggered, which creates fragments that have short, single-stranded termini that are complementary in nucleotide sequence and so can form hydrogen bonds with

one another. Such complementary termini are called **sticky ends**. Restriction enzymes are produced by bacterial cells, which prevent destruction of their own DNA by virtue of **modification enzymes** that chemically modify certain bases in the recognition sequence and so prevent cleavage.

12. Study of DNA fragments produced by a group of different restriction enzymes can be used to construct a **restriction map** of the DNA molecule in question. A restriction map is a diagram showing the relative positions of restriction sites along a DNA molecule. Restriction fragments of convenient length can then be used in **DNA sequencing** to determine the nucleotide sequence of the DNA. One popular method of DNA sequencing involves the chemical breakage of a single DNA strand at each of the four nucleotides in turn, and the resulting fragments are separated by **DNA electrophoresis**. The nucleotide sequence of the DNA strand can be inferred directly from the pattern of bands found on electrophoresis.

13. Recombinant DNA (or **DNA cloning**) refers to DNA molecules that have been artificially created by chemically splicing together DNA fragments from two or more different organisms. One method for producing recombinant DNA involves the use of the sticky ends of a DNA fragment produced by a restriction enzyme. These ends will form hydrogen bonds with the ends of a plasmid that has been cut open in one place by the same enzyme, and the inserted piece can be permanently joined to the plasmid with DNA ligase. The recombinant plasmid can then be reintroduced into a living cell by means of **transformation**, and the resulting transformed cell can be allowed to divide repeatedly to create a large, genetically identical population (a **clone**). Many replicas of the original DNA fragment can be recovered from the clone by isolating the plasmid DNA.

14. Uses of recombinant DNA range from the production of large quantities of a particular DNA sequence for use in restriction mapping or DNA sequencing to practical applications in the creation of bacterial clones that produce gene products of medical importance, such as insulin, growth hormone, and interferon. In the early years of recombinant DNA research, some anxiety arose about its potential hazards. It is now widely agreed that the potential risks have been minimized by appropriate standards of **physical containment** and **biological containment**. Less fully discussed is whether the miracle drugs that will come out of future recombinant DNA research will be used wisely and responsibly.

WORDS TO KNOW

Deoxyribonucleic Acid (DNA)
Double helix
Duplex
Antiparallel strands
5' end
3' end
Antisense strand
Sense strand
Highly repetitive
Moderately repetitive
Unique sequence
Hydrogen bond
Covalent bond

Nucleotide
Complementary pairing
Phosphate group
Deoxyribose sugar
Base

Base
Adenine
Thymine
Guanine
Cytosine
Pyrimidine
Purine

DNA Cloning
Restriction enzyme
Restriction site
Restriction fragment
Restriction map
Modification enzyme
Plasmid
Recombinant DNA

Replication
Unwinding
Fork
Bubble
Template
Primer
DNA polymerase
Proofreading
Okazaki fragments
DNA ligase
Semiconservative

Recombination
Nick
Endonuclease
Exonuclease
Heteroduplex
χ (chi) structure

Mismatch repair
Gene conversion

Chromosome
Histone
Nucleosome
Nucleosome fiber
Solenoid
Supersolenoid
Chromatid

PROBLEMS

1. For discussion: Shortly after the first DNA-cloning methods were developed, the news was reported in some newspapers, magazines, and television in such statements as "Scientists have succeeded in creating new forms of life that have never before existed." In light of Figure 8.23, do you think such statements accurately convey what the methods do? In what ways might such statements be misinterpreted?

2. How many different DNA sequences of exactly three nucleotide pairs are possible?

3. How can heat separate the individual strands of a DNA duplex from each other without breaking chemical bonds within the strands themselves?

4. What daughter duplexes would result from replication of the following DNA molecule?

$$5'\text{-AGCGTTCACC-}3'$$
$$3'\text{-TCGCAAGTGG-}5'$$

In each daughter molecule, indicate which strand is the old strand and which the newly synthesized one. What is this mode of replication called?

5. A normal human gamete contains about 2.89×10^9 nucleotide pairs of DNA. In the normal B configuration of duplex DNA, each nucleotide pair occupies 3.4 Å of length. What length of DNA is found in a human gamete, assuming all of it is in the B form? (Note: 1 Å = 1×10^{-8} cm or 3.94×10^{-9} in.)

6. The accompanying sequence is a portion of the DNA sequence of a human α-globin gene. Complete the sequence. (The sequence shown is from S. A. Liebhaber, M. Goossens, and Y. W. Kan, 1980, Proc. Natl. Acad. Sci. U.S.A. 77:7054–7058.)

```
5'-G A A A C T G G G T G G C C G A T T C G T C C C G C C G G G A C A C T G G-3'
3'- G A T G G C G G G G G C G C G G G A A C G G T T T A T T T T G G G T C A -5'
```

7. How could gene conversion of *D* to *d* occur in the heteroduplexes illustrated in Figure 8.13?

8. Consider an organism whose DNA consists of 25 percent A, 25 percent T, 25 percent G, and 25 percent C. What would be the average distance in nucleotides between the restriction sites of an enzyme like *Alu*I (5'-AGCT-3'), whose restriction site consists of four nucleotides? What would be the average distance between *Asu*I (5'-GGGCC-3') sites? Between *Eco*RI (5'-GAATTC-3') sites? Between *Acy*I (5'-GRCGYC-3') sites, where R = any purine and Y = any pyrimidine?

9. What kind of hairpin-loop structure could be formed by the following DNA sequence?

5'-ATGCTTTGAAAGCAT-3'
3'-TACGAAACTTTCGTA-5'

10. The accompanying DNA sequence is part of the sequence of a virus called M13 mp7 from nucleotides 6201 through 6300. This sequence is unusually rich in restriction sites. For each of the following restriction enzymes (restriction sites in parentheses), find whether they occur in the sequence and, if so, where. For convenience, represent the site of occurrence as the number of the nucleotide that corresponds to the 5' end of the restriction site.

*Alu*I	(5'-AGCT-3')
*Bam*HI	(5'-GGATCC-3')
*Eco*RI	(5'-GAATTC-3')
*Sal*I	(5'-GTCGAC-3')
*Pst*I	(5'-CTGCAG-3')
*Xho*II	(5'-RGATCY-3'),
	R = purine, Y = pyrimidine
*Hind*III	(5'-AAGCTT-3')

```
  6201      6210      6220      6230      6240      6250      6260      6270      6280      6290      6300
   |         |         |         |         |         |         |         |         |         |         |
5'-CACACAGGAAACAGCTATGACCATGATTACGAATTCCCCGGATCCGTCGACCTGCAGGTCGACGGATCCGGGGAATTCACTGGCCGTCGTTTTACAACGT-3'
```

11. If the DNA duplex shown here is cleaved with *Eco*RI and inserted into an *Eco*RI site in a plasmid, what will be the sequence of the resulting cloned fragment? (The restriction sequence of *Eco*RI is illustrated in Figure 8.18.)

5'-AGGCGAATTCGACCCTTTCAAGAATTCTCT-3'
3'-TCCGCTTAAGCTGGGAAAGTTCTTAAGAGA-5'

12. In connection with DNA cloning (see Figure 8.23), what is the process of transformation? In what way does transformation prove that DNA is the genetic material?

13. A DNA fragment is labeled and treated as in Figure 8.22, and the resulting autoradiograph is shown here. What is the nucleotide sequence of the fragment?

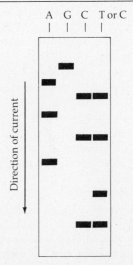

14. An alternative method of DNA sequencing uses so-called dideoxynucleotides, which are nucleotides lacking their 3′-hydroxyl group. In this method, the strand to be sequenced is used as a template for the synthesis of a complementary strand by means of DNA polymerase, and a small amount of one of the four possible dideoxynucleotides is added to the reaction mixture. Wherever a dideoxynucleodite is incorporated into the growing DNA strand, replication is terminated, and a series of DNA fragments of varying length is thus produced. Why does replication terminate when a dideoxynucleotide is incorporated? Should the dideoxynucleotides be added to the reaction mixture as 5′ monophosphates or 5′ triphosphates; why?

15. In the dideoxy procedure mentioned in Problem 14, one ingredient in the reaction mixture is a short DNA strand that can form a duplex with the strand of interest. This short DNA strand serves as a primer of replication. Why is a primer necessary?

FURTHER READING AND REFERENCES

Abelson, J. 1980. A revolution in biology. Science 209:1319–1321. A brief overview of the potentials of recombinant DNA.

Bauer, W. R., F. H. C. Crick, and J. H. White. 1980. Supercoiled DNA. Scientific American 243:118–133. In many forms of DNA the double helix itself forms a higher-order helix.

Berg, P. 1981. Dissections and reconstructions of genes and chromosomes. Science 213:296–303. Nobel prize lecture by one of the originators of recombinant DNA.

Cocking, E. C., M. R. Davey, D. Pental, and J. B. Power. 1981. Aspects of plant genetic manipulation. Nature 293:265–270. Discussion of the prospects and problems of recombinant DNA in the improvement of crop plants.

Gilbert, W. 1981. DNA sequencing and gene structure. Science 214:1305–1312. Nobel prize lecture by the inventor of the sequencing technique discussed in the text.

Gilbert, W., and L. Villa-Komaroff. 1980. Useful proteins from recombinant bacteria. Scientific American 242:74–94. Why and how recombinant DNA has produced an ongoing revolution in pharmacology.

Hentschel, C. C., and M. L. Birnstiel. 1981. The organization and expression of histone gene families. Cell 25:301–313. A thorough review of this important class of genes.

Hopwood, D. A. 1981. The genetic programming of industrial microorganisms. Scientific American 245:91–102. Recombinant DNA techniques in making certain useful microbes even more useful.

Kornberg, A. 1978. Aspects of DNA replication. Cold Spring Harbor Symp. Quant. Biol. 43(1):1–9. Excellent overview of bacteriophage DNA replication leading off a two-volume symposium devoted to DNA replication and recombination.

Kornberg, R. D., and A. Klug. 1980. The nucleosome. Scientific American 244:52–64. The discovery of the elementary subunit of chromatin structure is chronicled.

Leder, P., J. N. Hansen, D. Konkel, A. Leder, Y. Nishioka, and C. Talkington. 1980. Mouse globin system: A functional and evolutionary analysis. Science 209:1336–1342. The piecewise construction of eukaryotic genes and some suggestions concerning their mode of evolution.

Liebhaber, S. A., M. J. Goossens, and Y. W. Kan. 1980. Cloning and complete nucleotide sequence of the human 5′-α-globin gene. Proc. Natl. Acad. Sci. U.S.A. 77:7054–7058. Complete sequence of one of the two human α genes.

Liebhaber, S. A., M. Goossens, and Y. W. Kan. 1981. Homology and concerted evolution at the α1 and α2 loci of human α-globin. Nature 290:26–29. Comparison of the duplicated human α-globin genes and the α-pseudogene.

Lilley, D. M. J., and J. F. Pardon. 1979. Structure and function of chromatin. Ann. Rev. Genet. 13:197–233. Details of chromatin are reviewed.

Nishioka, Y., and P. Leder. 1979. The complete sequence of a chromosomal mouse α-globin gene reveals elements conserved throughout vertebrate evolution. Cell 18:875–882. Source of Figure 8.21.

Nordheim, A., M. L. Pardue, E. M. Lafer, A. Möller, B. D. Stollar, and A. Rich. 1981.

Antibodies to left-handed Z-DNA bind to interband regions of *Drosophila* polytene chromosomes. Nature 294:417–422. A real surprise, and a discovery of great potential importance.

Novick, R. P. 1980. Plasmids. Scientific American 243:102–127. A good overview of plasmids and their way of "life."

Potter, H., and D. Dressler. 1976. On the mechanism of genetic recombination: Electron microscopic observation of recombination intermediates. Proc. Natl. Acad. Sci. U.S.A. 73:3000–3004. Discusses a currently favored model of recombination at the molecular level.

Richards, J. (ed.). 1978. Recombinant DNA. Academic Press, New York. An informal account of the early years of the recombinant DNA debate.

Sanger, F. 1981. Determination of nucleotide sequences in DNA. Science 214:1205–1210. Nobel prize lecture by the inventor of one sequencing technique.

Seidel, G. E., Jr. 1981. Superovulation and embryo transfer in cattle. Science 211:351–358. Modern reproductive management in cattle husbandry.

Singer, M. F. 1982. SINEs and LINEs: Highly repeated short and long interspersed sequences in mammalian genomes. Cell 28:433–434. Structure and possible functions of certain classes of repetitive DNA.

Smith, G. R. 1981. DNA supercoiling: Another level for regulating gene expression. Cell 24:599–600. On the potential importance of DNA structures beyond the duplex structure.

Wang, A. H-J., G. J. Quigley, F. J. Kolpak, J. L. Crawford, J. H. van Boom, G. van der Marel, and A. Rich. 1979. Molecular structure of a left-handed double helical DNA fragment at atomic resolution. Nature 282:680–686. Contains space-filling models and other diagrams of left-handed DNA.

Wang, J. C. 1982. DNA topoisomerases. Scientific American 247:94–109. About an enzyme that may be important in replication and which can change the topological form of DNA molecules, linking them, or even tying them in knots.

Watson, J. D. 1976. Molecular Biology of the Gene, 3rd ed. W. A. Benjamin, Menlo Park, Calif. A classic textbook superbly produced and well written.

Watson, J. D. 1980. The Double Helix. Norton, New York. An entertaining account of the discovery of DNA structure with commentary, reviews, and original papers.

chapter 9

Transfer of Genetic Information: DNA to RNA to Proteins

Genetic information is carried in the nucleotide sequence of DNA, but proteins and a few kinds of RNA are the day-to-day molecular workhorses in the cell. Indeed, much of the genetic information in DNA becomes encoded in the nucleotide sequence of RNA and is precisely the information that enables the cell to manufacture particular kinds of proteins. Structurally, DNA is a relatively simple and monotonous molecule, much the same everywhere along its length. Its sequence of nucleotides is not monotonous, of course, but the structure of the double helix itself is. Although portions of the double helix may sometimes assume alternative helical configurations, or small regions may unwind and form more complex three-dimensional shapes such as hairpin loops, DNA is not nearly as versatile as protein molecules in forming complex three-dimensional molecular structures. (The relatively limited repertoire of DNA conformation may nevertheless be important in determining when certain genes will or will not become active.) Much of the biochemical machinery in the cell is involved in converting the genetic information in DNA into the structure of proteins. In eukaryotic cells, however, DNA is confined to the nucleus (except for the tiny amount that occurs in certain organelles); most proteins are located in the cytoplasm. The processes by which genetic information in the nucleus guides the manufacture of particular proteins in the cytoplasm are the principal subjects of this chapter.

The Structure and Versatility of Proteins

Protein molecules have complex, convoluted, three-dimensional structures. One example is illustrated in Figure 9.1, which depicts the three-dimensional structure of hemoglobin, the oxygen-carrying protein in red blood cells. Such complex structures are revealed by the laborious technique of x-ray diffraction, which can be applied to virtually all types of large molecules, including DNA and RNA as well as proteins. One major conclusion from diffraction studies of proteins is that each kind of protein molecule has a different three-dimensional structure from every other kind. This variety of shapes is what allows proteins to carry out so many different functions in the cell; each type of protein molecule assumes a specific shape that enables it to carry out its specific function in the cell. Genetic information in DNA is like a set of instructions for making a carpenter's tools. The instructions are verbal and structurally monotonous; the sequence of nucleotides in DNA corresponds to a sequence of letters and words on paper. The tools made from these instructions—mallets, awls, clamps, planes, chisels, and so on—are structurally very different because of their diverse tasks. The "tools" in the cell are its proteins.

Fortunately, proteins have an underlying simplicity that permits certain fundamental aspects of their structure to be comprehended, even though the actual three-dimensional structures of relatively few proteins have been studied in detail. Proteins are composed of linear strings of constituents—their "building blocks"—known as **amino acids**. There are only 20 commonly occurring types of amino acids, although a typical protein molecule may consist of 100 or more amino acids hooked together. Moreover, most amino acid molecules have an important structural similarity; they can be represented in the manner shown in Figure 9.2. Three aspects of the structure in

Figure 9.1 Three-dimensional structure of adult hemoglobin molecule which contains two each of two types of polypeptide chains, called alpha (α) and beta (β). The heme groups that bind with oxygen are shown as small rectangles.

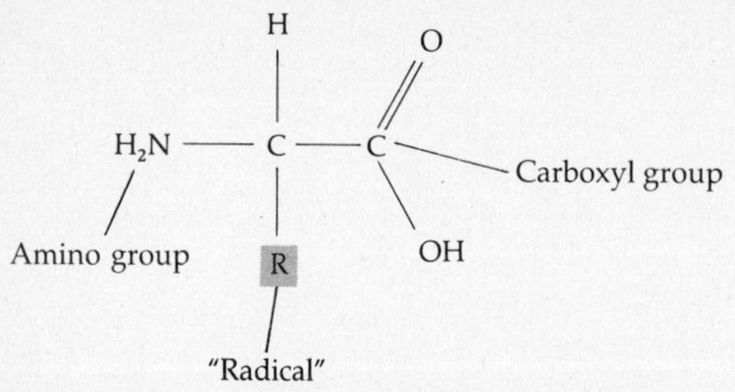

Figure 9.2 Generalized structure of an amino acid having a central carbon attached to an amino group, a carboxyl group, and a "radical" (shaded). The structure of the "radical" differs from amino acid to amino acid and gives each its unique chemical characteristics.

Figure 9.2 should be noted. One end of the amino acid carries an **amino group** ($-NH_2$), and this group defines the **amino end** of the amino acid; the other end carries a **carboxyl group** ($-COOH$), which defines the **carboxyl end** of the amino acid. The group denoted R (for "**radical**") in the shaded square in Figure 9.2 can have many different structures depending on the particular amino acid in question; use of R in this manner is a convenience for representing a generalized amino acid without having to specify its entire structure in detail.

A list of the 20 commonly occurring amino acids and their molecular structures is given in Figure 9.3. In each case, the R group for the particular amino acid is indicated by shading. Two commonly used abbreviations for each amino acid are also given. One abbreviation represents each amino acid by a single upper-case letter, like G for glycine, S for serine, or D for aspartic acid; these single-letter abbreviations are particularly useful for depicting lengthy amino acid sequences. The other abbreviation represents each amino acid by a group of three lower-case letters, like gly for glycine, ser for serine, or asp for aspartic acid; these three-letter abbreviations are convenient because each one calls to mind the name of the corresponding amino acid.

As can be observed in Figure 9.3, glycine is the simplest amino acid; the R represents a lone hydrogen atom. The R groups of certain amino acids—such as glycine, serine, threonine, and tyrosine—tend to mix well with water because they form hydrogen bonds with water molecules; the R groups of other amino acids—such as alanine, valine, leucine, and phenylalanine—tend to repel water. The R groups of lysine, arginine, and histidine tend to become positively charged under the condtions usually found in cells; the R groups of aspartic acid and glutamic acid tend to become negatively charged. These properties of amino acids are particularly important in maintaining the convoluted three-dimensional structure of protein molecules.

Inborn Errors of Metabolism

Although the role of amino acids as the building blocks of proteins is of greatest concern in this chapter, individual types of amino acids are sometimes important. One well-known and important example involves the amino acid **phenylalanine**, which when present in abnormally high concentrations in body fluids causes the disease **phenylketonuria** (or **PKU**). Phenylketonuria is inherited as an autosomal recessive and its incidence is about 1 per 10,000 newborns; the disease is characterized by severe mental retardation and light skin pigmentation and hair color (Figure 9.4). Phenylketonuria is one of many conditions caused by inherited defects in particular enzymes. These inherited enzyme defects are known as **inborn errors of metabolism**; in many instances, such as Tay-Sachs disease (discussed in Chapter 4) or Lesch-Nyhan syndrome (discussed in Chapter 5), the precise nature of the biochemical defect is known.

Figure 9.5 is an overview of the biochemical pathways involved in phenylalanine metabolism. Phenylalanine is one of the approximately 10 amino acids that mammals cannot

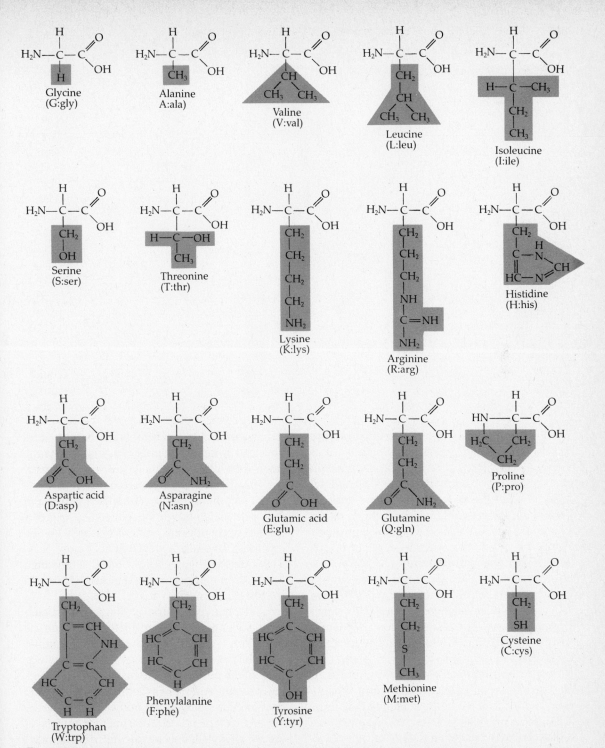

Figure 9.3 Structures of the 20 commonly occurring amino acids, laid out as in Figure 9–2. Note particularly the various types of R groups. Proline is an exceptional amino acid in that its R group is attached back to its own amino group. Each amino acid has two abbreviations, one letter and three letters.

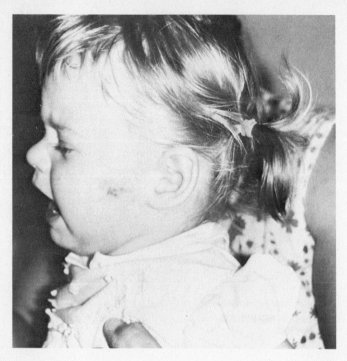

Figure 9.4   A 13-month old child with phenylketonuria.

produce from chemical precursors; these are known as **essential amino acids** because it is essential for health that they be supplied in dietary sources. Normally, excess phenylalanine in the diet is converted into another amino acid (**tyrosine**), which is then either converted into harmless breakdown products or used in other ways, such as in the synthesis of the skin pigment **melanin** (see Figure 9.5). In the case of phenylketonuria, the inborn error of metabolism is an inherited defect in the enzyme **phenylalanine hydroxylase**, which normally catalyzes the conversion of phenylalanine to tyrosine. Children with a defect in this enzyme are unable to convert phenylalanine into tyrosine, and this inability results in the accumulation of phenylalanine in the tissues and fluids of the body, including the brain. The excess phenylalanine is broken down into other products, including such harmful ones as **phenylketones**. These harmful breakdown products in some unknown way cause severe

damage to a child's developing nervous system and lead to severe mental retardation. At the same time, the abnormally low levels of tyrosine in affected children often result in a reduced amount of pigment because the pigment molecule melanin is produced from tyrosine by a series of chemical alterations (actually involving more steps than depicted in Figure 9.5). The reduced pigmentation leads to the light complexion and hair color that also characterize phenylketonuria.

If phenylketonuria is diagnosed in children soon enough after birth, they can be placed on a special diet low in phenylalanine. The child is allowed only as much phenylalanine as can be used by the body in manufacturing proteins, and phenylalanine is prevented from accumulating in the body. By this special dietary treatment, the detrimental effects of excess phenylalanine on mental development are almost completely avoided. Several years after birth, when the nervous system is fully

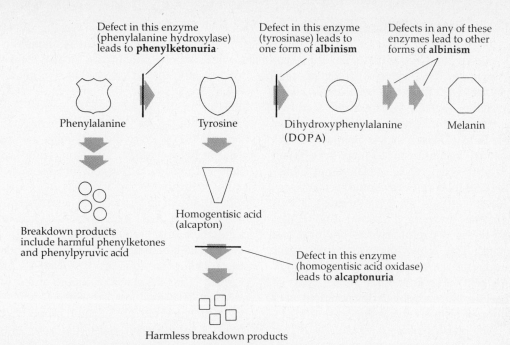

Figure 9.5   Overall pattern of metabolism of phenylalanine showing several inborn errors of metabolism including enzyme defects leading to phenylketonuria, alcaptonuria, and forms of albinism.

developed, the affected child can return to a normal or nearly normal diet with no ill effects. The treatment is so effective that, in most parts of the United States today, all newborn children are routinely examined for the chemical signs of phenylketonuria.

Since the dietary treatment of phenylketonuria began in the 1950s, a generation has passed. Children who would have been doomed to severe mental retardation have passed through childhood on the low-phenylalanine diet and have become seemingly normal and healthy adults. It is in this generation that a sad corollary of an otherwise overwhelming success story has become evident. When women who have an inborn error in phenylalanine hydroxylase attempt to have children of their own, well over half the children are born with severe and incurable mental retardation. At first it seemed reasonable to hope that the risk might be lessened by placing

the mother on a low-phenylalanine diet while pregnant. So far, the treatment seems to have had little or no effect, although experiments along these lines are continuing and there are some hopeful signs of progress. The only present recourse is proper genetic counseling of the women to inform them of the awesome risk in hope they may choose such alternatives to childbirth as adoption.

Figure 9.5 also illustrates several other inborn errors of metabolism. The first step in the conversion of tyrosine to melanin involves an enzyme, **tyrosinase**, which converts tyrosine into a substance called **dopa**. A defect in this enzyme is associated with one inherited form of **albinism** (see Chapter 4 for a discussion); defects in other steps along the way lead to other inherited forms of the trait.

A final example of an inborn error of metabolism is the relatively harmless trait **alcaptonuria**, which is of historical importance because it was the first inherited metabolic defect discovered in humans; the inherited nature of alcaptonuria was discovered by Archibald Garrod in 1902, who was far ahead of his time in realizing the biochemical implications of Mendelian genetics. Alcaptonuria is caused by an inherited defect in an enzyme (**homogentisic acid oxidase**) that is involved in the breakdown of tyrosine (see Figure 9.5). When this enzyme is defective, the substance **homogentisic acid** (also called **alcapton**) accumulates and is excreted into the urine because lack of the enzyme prevents it from being further broken down. Alcaptonuria has one dramatic symptom: The urine, upon standing, turns black (due to the natural oxidation of homogentisic acid). The only other known effect of the trait is the occasional development of a mild form of arthritis due to the deposition of a dark pigment in cartilage.

Amino Acids and Polypeptides

A principal function of phenylalanine, tyrosine, and the other amino acids in Figure 9.3 is their role as constituents of protein molecules. Various numbers of each of the amino acids can be linked together in any order to give rise to a specific protein. The manner in which the amino acids are linked is illustrated in Figure 9.6. Figure 9.6(*a*) shows a chain of four amino acids with R groups denoted R_1 through R_4. The heavy lines represent the covalent bonds that link each amino acid to the next one; these linking bonds are called **peptide bonds**. Note that one end of the amino acid chain (the left end in Figure 9.6) carries a free amino group; this group defines the *amino end* of the chain. The other end of the chain (the right end in Figure 9.6) carries a free carboxyl group, so it is called the *carboxyl end* of the chain.

At the right of Figure 9.6(*a*) is a fifth amino acid (radical R_5) that is about to be added to the growing chain. The new peptide bond is always formed between the carboxyl group of the old chain and the amino group of the new amino acid. In the formation of the bond, a molecule of water (H_2O) is released, which accounts for the apparently missing HOH (i.e., H_2O) from each pair of linked amino acids. Addition of the amino acid in Figure 9.6(*a*) leads to the five-amino-acid molecule in Figure 9.6(*b*). The amino end of the chain is the same as before, but the carboxyl end is contributed by the last-added amino acid. Although only five amino acids are shown in Figure 9.6, such a chain would ordinarily contain 100 or more amino acids.

A chain of amino acids linked together as shown in Figure 9.6 is properly called a **polypeptide**. The sequence of amino acids in a polypeptide constitutes its **primary structure**. For a polypeptide to function properly, the chain of amino acids must fold and contort itself into a specific three-dimensional shape, but this folding is an automatic consequence of the amino acid sequence (i.e., the primary structure) of the polypeptide. Under conditions found in the cell, some amino acids have positive charges in their R groups; others have negative charges. R groups of unlike charges tend to attract one another; those of like charges repel. Supplementing the effects of charged amino acids are the uncharged ones that tend to form hydrogen bonds with water or with each other. Still other amino acids repel water and tend to reach an interior position in the three-dimensional structure of the molecule where little water is present. Some regions of the backbone of a polypeptide chain may form into a helical shape (such helical regions constitute the **secondary structure** of the polypeptide), and R groups in one part of the chain may interact with nearby neighbors. The result of all these forces is that polypeptide chains coil and twist and fold automatically until they attain a relatively stable configuration (see Figure 9.1, for example).

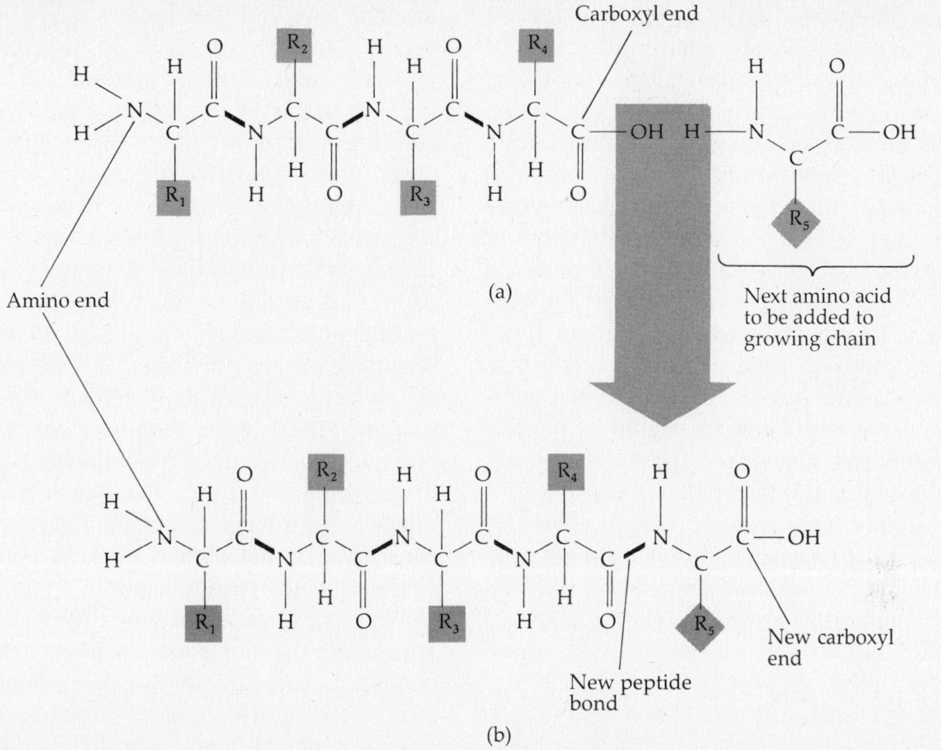

Figure 9.6 Polypeptides consist of amino acids hooked together by means of peptide bonds (heavy lines). *(a)* Polypeptide in process of synthesis with the next amino acid about to be added to its carboxyl end. *(b)*Elimination of water molecule (OH and H in shaded arrow) accompanies formation of new peptide bond.

We emphasize the word *relatively* in the phrase *relatively stable configuration*. Everything inside the cell is in constant motion; protein molecules are no exception. Parts of every protein molecule are constantly and spontaneously unfolding and then folding up again, like a spastic worm. The "structure" of a protein really means the three-dimensional form it assumes most of the time, and this relatively stable three-dimensional conformation constitutes the **tertiary structure** of the molecule.

Several particular amino acids are especially important in determining the folding of polypeptide chains into precise three-dimensional configurations. **Proline**, for example, causes bends or kinks in the backbone of the molecule because, unlike other amino acids, its nitrogen atom is covalently linked back into its R group and therefore has restricted freedom of movement. Another amino acid of particular importance in determining three-dimensional structure is **cysteine**. The $-SH$ groups dangling from two cysteines in different parts of the polypeptide chain (or even in different polypeptide chains) can be linked together by jettisoning their hydrogens and joining together covalently as $-S-S-$. These $S-S$ or **disulfide** bridges hold the molecule relatively rigid in its folded configuration. Indeed, if only the disulfide bridges are broken by chemical means, the protein will usually unfold and be unable to carry out its proper function. This unfolding that leads to loss of function is called **denaturation**, and it should be pointed out here that certain kinds of mutations are of great usefulness in genetic analysis because the defective protein (usually an enzyme) is susceptible to denaturation. The mutations are called **temperature-sensitive mutations**; they code for proteins that denature and become defective at one temperature (called the **restrictive** temperature), but they do not denature and therefore remain functional at another, usually lower, temperature (called the **permissive** temperature). Temperature-sensitive mutations are useful in genetic analysis because they permit the study of mutations that affect proteins that are essential for survival—mutations that are lethal (unable to survive) at the restrictive temperature but able to survive at the permissive temperature. (See Chapter 11 for further discussion.)

As noted earlier, a *polypeptide* is a chain of amino acids. A **protein** is a polypeptide-containing molecule that carries out some cellular function. In some cases, a protein consists of a single polypeptide chain. More commonly, two or four or more identical polypeptide chains must aggregate to produce a functional protein molecule. Indeed, sometimes two or more *different* polypeptide chains must aggregate to form the functional protein. Such aggregations of polypeptide chains constitute the **quaternary** structure of the protein; a good example of quaternary structure is found in the hemoglobin molecule depicted in Figure 9.1. The functional hemoglobin molecule consists of four polypeptide chains. Two of these (the lightly stippled chains) are identical and are both coded by the α-globin gene. The other two chains (the darkly stippled ones in Figure 9.1) are the **beta (β) chains**; the β chains are also identical and are coded by the β-globin gene. (Incidentally, the mutation that causes sickle cell anemia is in the β-globin gene.) The α and β chains—two of each—interact to create four very precise holes, each of which becomes occupied by a small iron-containing molecule called **heme** (small rectangles in Figure 9.1). Each heme is capable of binding and carrying one molecule of oxygen from the lungs to other parts of the body.

Although the three-dimensional structure of a polypeptide chain is an automatic consequence of its amino acid sequence (i.e., its primary structure), the precise details of the folding and aggregation are so complicated that they can be predicted for only short and simple artificial polypeptides. The sad truth is that knowledge of the amino acid sequence of a polypeptide does not tell much about its three-dimensional structure; that is a separate and much more difficult problem. However, since the folding of the amino acid chain is automatic (it takes place without special "folding" enzymes), the question of how the genetic information in the nucleotide sequence of DNA becomes expressed in the three-dimensional structure of a protein can be rephrased as the much simpler problem of how the information in DNA is expressed in the amino acid sequence (primary structure) of a polypeptide chain. This process involves two rather distinct steps: **transcription**, which is the synthesis of an RNA molecule using the genetic information in DNA; and **translation**, the manufacture of a polypeptide chain from information in the RNA molecule. In each of these processes the molecular structure of RNA is itself a key to understanding.

The Structure and Synthesis of RNA: Transcription

The molecular structure of **RNA (ribonucleic acid)** is very much like that of DNA, with three important exceptions. First, the sugar in RNA is **ribose** rather than deoxyribose [see Figure 9.7(a)]. Note the extra hydroxyl (−OH) on the

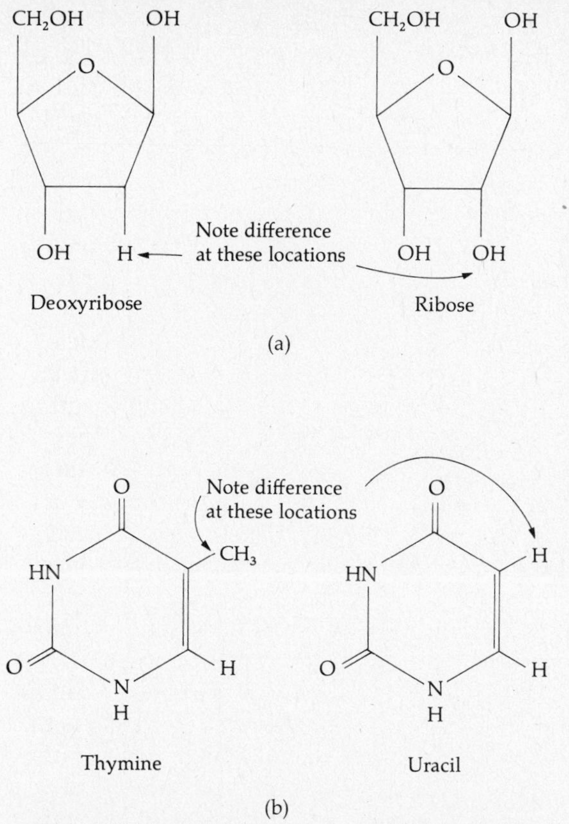

Note difference at these locations

Deoxyribose

Ribose

(a)

Note difference at these locations

Thymine

Uracil

(b)

Figure 9.7 Two major differences between DNA and RNA. *(a)* The sugar deoxyribose in DNA is replaced with ribose in RNA. *(b)* The base thymine (T) in DNA is replaced with uracil (U) in RNA.

2′ carbon of ribose, which makes it somewhat more difficult for two complementary strands of RNA to form a double helix. Second, most **ribonucleotides** (in contrast to deoxyribonucleotides) contain the bases adenine, guanine, cytosine, and **uracil (U)** instead of adenine, guanine, cytosine, and **thymine**. Uracil (U) plays the same role in RNA as thymine (T) plays in DNA; that is, since the structures of uracil and thymine are very similar [see Figure 9.7(*b*)], uracil undergoes base pairing with adenine just as thymine pairs with adenine. The third important difference between

DNA and RNA is that whereas DNA is usually found as a duplex consisting of two complementary strands, RNA is usually found in the cell as a single-stranded molecule. On the other hand, even though RNA is a single-stranded molecule, it can fold back onto itself to form hairpin-loop structures in which complementary bases can pair (i.e., A with U and G with C).

The nucleotide sequence of an RNA molecule is specified by the nucleotide sequence in the corresponding length of a DNA strand. The transfer of sequence information from a strand of DNA to a molecule of RNA is accomplished in the process of *transcription*, which is illustrated in Figure 9.8. A key enzyme involved in transcription is called **DNA-dependent RNA polymerase**; it is a different type of RNA polymerase than the one used for primer synthesis during DNA replication (see Chapter 8), but it will be convenient for present purposes to refer to the transcription enzyme as **RNA polymerase**. At the beginning of transcription, RNA polymerase binds with the **sense strand** (i.e., the information-containing strand) of DNA at a region known as the **promoter**, which is associated with a nucleotide sequence called the RNA polymerase **recognition site**. Once the polymerase becomes associated with its recognition site (this is facilitated by certain other proteins), it moves along the sense strand in the 3′ to 5′ direction until it reaches another nucleotide sequence called the RNA polymerase **binding site**, where it attaches more strongly. (The 3′ to 5′ direction along the DNA sense strand is often called the **downstream** direction, so we could say that the polymerase moves downstream from the recognition site to the binding site.) The polymerase continues its movement downstream, and a few nucleotides from the binding site, transcription actually begins.

Figure 9.8 illustrates an RNA molecule (called an **RNA transcript**) caught in the act of

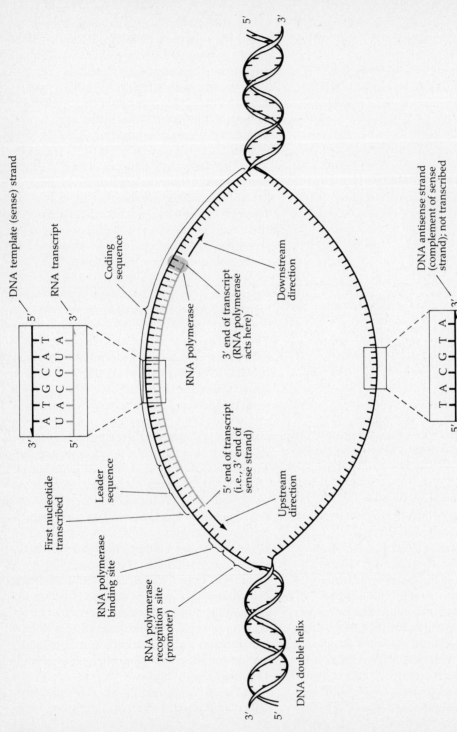

Figure 9.8 Transcription. A portion of the DNA duplex opens out; RNA polymerase interacts with its recognition site (promoter), binds at its binding site, and synthesizes an RNA molecule (transcript) complementary in base sequence to the sense strand of DNA. Note that the sense strand is transcribed in the 3' to 5' direction as the RNA molecule is synthesized in the 5' to 3' direction. The antisense DNA strand is not transcribed.

being synthesized. The first ribonucleotide in the transcript represents the 5′ end of the molecule. As the polymerase moves downstream, it adds one new ribonucleotide at a time to the 3′ end of the ever-elongating transcript. Each added ribonucleotide is complementary to the corresponding nucleotide in the DNA template; that is, an A, T, G, or C in the DNA causes the addition of a U, A, C, or G, respectively, to the RNA (see the expanded region of the template and transcript in Figure 9.8). In this manner, the nucleotide sequence information in DNA becomes transcribed into RNA. Because the bases in the RNA transcript are *complementary* to the bases in the *sense* strand of DNA, the bases in the RNA transcript will be *identical* to those in the *antisense* strand of DNA (except, of course, that T in DNA is replaced with U in RNA). This relationship between the bases in the RNA transcript and the bases in the antisense strand of DNA is illustrated in the expanded boxes in Figure 9.8; the base sequence of the transcript (upper box) is identical to the base sequence of the antisense strand (lower box), except for the occurrence of U instead of T in RNA. For this reason, DNA sequences are often presented in terms of the sequence of the antisense strand, oriented with its 5′ end toward the left. In terms of the antisense strand (or the RNA transcript itself), the upstream direction means toward the 5′ end (i.e., toward the left) and the downstream direction means toward the 3′ end (i.e., toward the right).

A **coding sequence** in an RNA transcript is a sequence of nucleotides that specifies a corresponding sequence of amino acids in a polypeptide. (How this specification occurs will be the subject of the next section.) As noted in Figure 9.8, upstream of the first coding sequence in an RNA transcript is a **leader sequence**, which provides a site of attachment for **ribosomes**—the particles on which polypeptide synthesis occurs. In Figure 9.8, transcription

will continue in a rightward direction until a termination sequence is encountered by the RNA polymerase. The nature of these termination sequences is not well understood at present, but proper termination does seem to involve other proteins that interact with the polymerase. When termination occurs, the polymerase detaches from the DNA and the RNA transcript is released. However, prior to termination, the 5′ end of the transcript may peel off the template strand and form a sort of "tail." This displacement makes room for other RNA polymerases to follow along behind the first one. In this manner, many RNA transcripts of the same DNA strand can be produced almost simultaneously. Figure 9.9 is an electron micrograph showing the many transcripts being produced from a gene in the spotted newt, *Triturus viridescens*.

Two actual DNA sequences involved in the initiation of transcription are shown in Figure 9.10. Sequence (*a*) is part of a gene called *lacZ* in the bacterium *Escherichia coli*. (The function and regulation of *lacZ* will be discussed in Chapter 10.) As indicated in the figure, the *lacZ* promoter comprises nucleotides 1 through 84, and it is directly adjacent to the next gene upstream, which is called *lacI*. The RNA polymerase interaction site extends from nucleotide 1 through nucleotide 72, and it is adjacent to the RNA polymerase binding site 5′-TATGTTG-3′. All polymerase binding sites in prokaryotes have a similar sequence; the most commonly occurring one is 5′-TATAATG-3′; and such a binding sequence is often called a **TATA box** or a **Pribnow box**. Six nucleotides downstream from the Pribnow box, transcription begins. The first 38 nucleotides of the transcript constitute the leader sequence, and within the leader is a special sequence 5′-AGGA-3′ called the **Shine-Dalgarno** sequence, which is involved in the binding of the transcript to ribosomes. The coding sequence of the *lacZ* transcript pro-

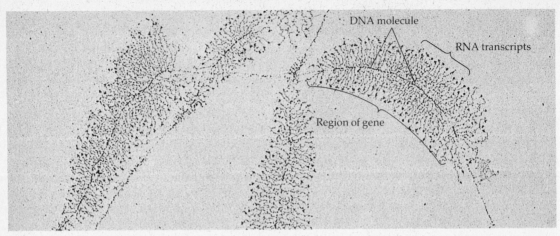

Figure 9.9 Electron micrograph of a DNA molecule caught in the act of transcription. The DNA shown codes for ribosomal RNA in the spotted newt, *Triturus viridescens*. Note that many RNA transcripts are produced simultaneously because many polymerases move in tandem along the DNA.

ceeds rightward from nucleotide 123; note that each group of three adjacent nucleotides codes for one amino acid, for reasons that will be discussed in a few pages.

Sequence (*b*) in Figure 9.10 is part of the mouse α-globin gene. The leftward extent of the RNA polymerase interaction site is not known, but the polymerase binding site 5′-TATAAG-3′ occupies positions 343 through 348; eukaryotic promoters all have similar binding-site sequences. A eukaryotic binding-site sequence is often called a **TATA box** or a **Goldberg-Hogness box**. Several nucleotides downstream from the TATA box, transcription begins. Nucleotide 373 is called a cap site, and it is important for ribosome binding. (The nature of the cap site will be discussed later in connection with eukaryotic RNA processing.) The α-globin coding sequence begins at nucleotide 405 and, again, each group of three adjacent nucleotides codes for one amino acid.

The α-globin gene contains two long and nearly perfect palindromes that may be important in polymerase interaction, RNA processing, or ribosome binding. The first comprises nucleotides 296 through 338, the second nucleotides 374 through 399. These palindromes permit the DNA to form foldback **secondary structures**, as illustrated in Figure 9.11. Indeed, since the rightmost palindrome is transcribed into RNA, the RNA itself can assume a foldback secondary structure involving this palindrome. Figure 9.11 is just one example of how secondary structures in DNA or RNA can be important in gene regulation. It is of interest to note in this connection that the *lacZ* leader sequence also contains a nearly perfect palindrome comprising nucleotides 85 through 105.

The details of polymerase interaction sites, binding sites, transcription initiation, palindromes, secondary structures, and all the rest must not overshadow the central and most important feature of transcription. The essential thing to remember is that during transcription RNA molecules are produced from the sense strand of a DNA moelcule and that the nucleotide sequence of an RNA transcript is complementary to the nucleotide sequence of the DNA sense strand. Transcription is the first step in gene expression in all organisms—prokaryotes and eukaryotes alike. The next

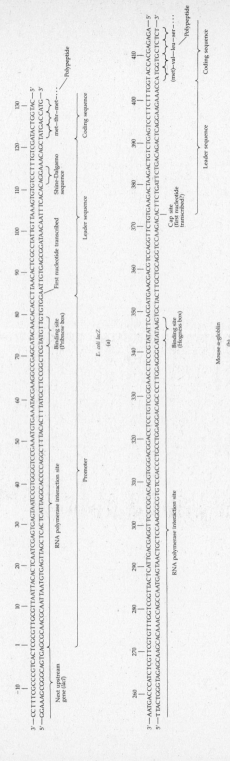

Figure 9.10 Upstream sequences important in the transcription of (a) *lacZ* gene in *Escherichia coli* and (b) α-globin gene in the mouse. Note the RNA polymerase interaction site, the RNA polymerase binding site, and the leader sequence that precedes the actual coding sequence of each polypeptide. The overall organization of the human α-globin genes is similar to that of the mouse, though there are many differences in actual sequence. (The human sequence in Problem 6 in Chapter 8 corresponds to nucleotides 338 through 410 in the mouse.)

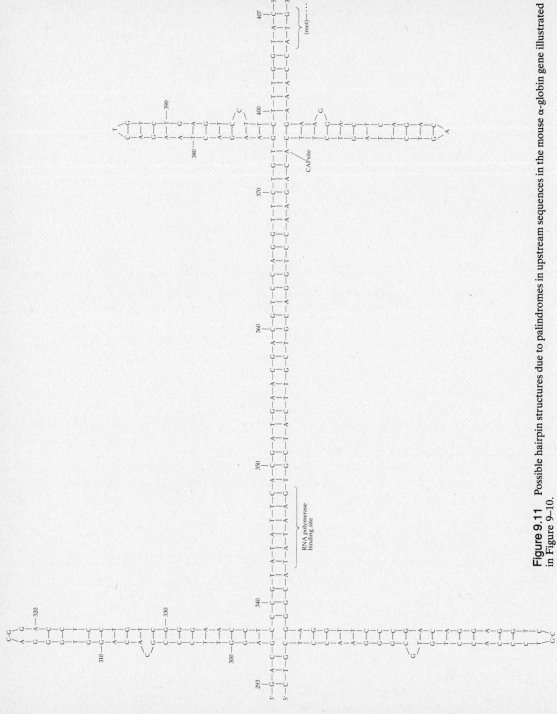

Figure 9.11 Possible hairpin structures due to palindromes in upstream sequences in the mouse α-globin gene illustrated in Figure 9–10.

step in gene expression is very different in prokaryotes and eukaryotes, however. In prokaryotes, which have no nucleus, ribosomes can attach to the RNA transcript immediately and begin polypeptide synthesis. Indeed, the usual situation in prokaryotes is that the ribosomes attach to the 5′ end of the transcript and begin polypeptide synthesis even before the 3′ end of the transcript has been produced! In eukaryotes, by contrast, the RNA transcript is not immediately accessible to ribosomes because the DNA is in the nucleus whereas the ribosomes are in the surrounding cytoplasm. The RNA cannot direct polypeptide synthesis until it is released from the nucleus. However, prior to their release, eukaryotic RNA transcripts undergo a profound chemical alteration within the nucleus, which is called **RNA processing**. Once the processed RNA is released into the cytoplasm, it directs polypeptide synthesis in a manner very similar to the situation in prokaryotes. Because of this similarity, it will be convenient to discuss polypeptide synthesis next and defer discussion of RNA processing until a later section of this chapter.

Polypeptide Synthesis: Translation

Most genes produce RNA transcripts that specify a sequence of amino acids to be hooked together in the making of a polypeptide. This process, which uses the information in the nucleotide sequence of an RNA molecule to dictate the amino acid sequence of a polypeptide, is known as *translation*. The RNA molecules whose nucleotide sequences become translated into polypeptides are called **messenger RNA's** (or mRNA's). In prokaryotes, the primary RNA transcript corresponds to the mRNA because the primary transcript is the RNA molecule that is translated. In eukaryotes, however, the mRNA is the RNA that is released from the nucleus; this differs markedly from the primary transcript because of the extensive processing that the primary transcript undergoes. Not all RNA transcripts end up as mRNA's, however; that is, the RNA transcribed from a number of genes is not translated into polypeptides. These genes—the genes for ribosomal RNA and transfer RNA—produce transcripts that undergo elaborate chemical alteration in the cytoplasm and become essential ingredients in the cellular apparatus used in the translation of mRNA produced from other genes. Thus, ribosomal RNA and transfer RNA function in translation, but they themselves are not translated.

The essence of translation is that the 5′ end of the mRNA molecule becomes associated with a ribosome in the cytoplasm, and the ribosome slowly moves down the whole length of the mRNA molecule; the genetic information in the mRNA is translated as the ribosome goes along by the addition of amino acids one by one to an ever-lengthening polypeptide chain. The ribosome "reads" the information in the mRNA molecule word by word; each "word" consists of three adjacent ribonucleotides. Each such sequence of three ribonucleotides triggers the addition of a particular amino acid to the growing polypeptide chain, and the amino acids added to the growing chain are brought to the ribosome by molecules of transfer RNA. With this brief overview of translation as background, it is now necessary to discuss in more detail some components of the translational apparatus.

The Translational Apparatus

Ribosomal RNA (rRNA), as its name implies, is associated with the ribosomes. These tiny particles, found by the thousands in the cytoplasm of most cells, are the sites of polypeptide synthesis; translation of mRNA molecules takes place on the ribosomes. As one might expect of structures able to perform such a prodigious task as polypeptide synthesis, ribosomes are very complex little particles, each containing one copy of each of three distinct types of RNA molecule and some 50 different kinds of proteins. Ribosomes exist in the cytoplasm as two distinct and separate subunits that come together only when an mRNA is actively being translated. Each of the 50 or so different proteins in a ribosome is derived from a separate mRNA molecule transcribed from a distinct gene, like any other protein. Quite remarkably, rRNA and the ribosomal proteins are able to join together spontaneously, even in a test tube, to form functional ribosomes ready for translation. Therefore no special genes are required to aid the assembly of ribosomes from their protein and rRNA components.

Ribosomal RNA is transcribed from corresponding genes in the DNA, and to ensure that rRNA will be produced in sufficient quantities, these genes are **redundant**; some 100 or more copies of the rRNA genes are present in the chromosomes. Although the redundant genes for one class of rRNA are dispersed among several chromosomes, those for the other two classes are repeated in tandem, although there are untranscribed "spacer" sequences between the repeating units. Active production of rRNA from these clustered genes results in a local accumulation of rRNA near the site of its synthesis in the nucleus, and this local accumulation is associated with what becomes visible in the microscope as the nucleolus.

Transfer RNA (tRNA) molecules are the molecules that actually "translate" the nucleotide sequence of mRNA into the amino acid sequence of a polypeptide. One might regard the ribosome as the site of translation—the "reader"—whereas the tRNA's are the "dictionary." During translation, the ribonucleotides in the coding sequence of the mRNA are read from one end to the other in groups of three; each group is called a **codon** and each codon specifies a particular amino acid to be inserted into the polypeptide chain at the corresponding position. Examples of codons are the sequences 5'-UUU-3', 5'-UAU-3', 5'-AUG-3', and 5'-CCC-3'. Specific transfer RNA's recognize the codons and bring the correct amino acids into line, and each amino acid is then linked onto the growing polypeptide chain.

Transfer RNA's, of which there are dozens of different kinds, contain 70 to 80 ribonucleotides. Each kind of tRNA molecule is transcribed from a different gene, but because tRNA is required in large quantity, these genes are often redundant, usually with many tandem copies of the corresponding DNA sequence. The tRNA's undergo extensive modification in the cytoplasm prior to participation in translation through the addition of a few ribonucleotides to one end of the molecule and the chemical modification of certain bases within the molecule. The fully tailored tRNA is able to fold back on itself in a sort of cloverleaf configuration to allow base pairing to occur between certain complementary sequences. This structure is shown schematically for phenylalanine tRNA from yeast cells in Figure 9.12(a), although the three-dimensional configuration of the molecule is slightly more complex [see Figure 9.12(b)]. All known tRNA's have this same overall structure: four base-paired regions and three loops of unpaired bases. The business end of the molecule is the bottom loop in the figure, which contains a specific sequence of three bases known as the **anticodon**. By base pairing, the anticodon is able to recognize the corresponding three-base codon in the mRNA. Each of the many kinds of tRNA has a unique anticodon that distinguishes it from all other tRNA's.

The other end of a tRNA molecule ready to participate in translation has an amino acid attached to its 3' terminus. For example, the 3' end of the tRNA in Figure 9.12 would have

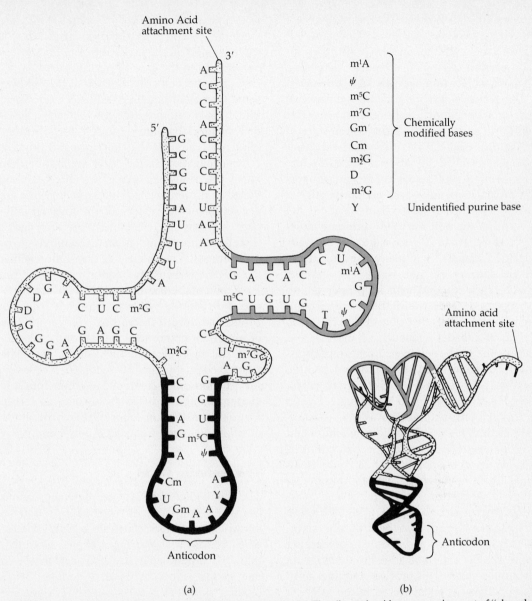

Figure 9.12 Structure of phenylalanine tRNA from yeast cells. *(a)* The ribonucleotide sequence in a sort of "cloverleaf" configuration; note a number of chemically modified bases and the black loop containing the anticodon and the gray loop. *(b)* The same tRNA molecule in the three-dimensional form it assumes, where the bars indicate hydrogen-bonded base pairs. The gray and black regions correspond to those in part *(a)*.

phenylalanine attached to it. Transfer RNA's with the same anticodon always become coupled with the same amino acid, and this coupling is carried out by 20 or more "activating enzymes" called **aminoacyl-tRNA synthetases**, each of which is specific for one amino acid. Thus, in translation, a codon on the mRNA specifies an anticodon according to the rules of base pairing, and the tRNA bearing this anticodon is arranged in advance to carry the correct amino acid at the other end. In this way, a particular three-base codon in mRNA specifies a particular amino acid.

All in all, well over 100 different molecules are required to translate any mRNA molecule correctly. The many constituents of the ribosomes are necessary, as are the tRNA's, each charged with its proper amino acid. The first step in translation is the recognition of mRNA by the ribosome. This step involves the leader sequence of the mRNA, although the leader sequence itself is not translated. The actual "start" signal for polypeptide synthesis is the codon 5'-AUG-3', which corresponds to the amino acid **methionine**. When this codon is encountered, methionine is established as the first amino acid at the amino end of the polypeptide. Then the second codon is read and its corresponding amino acid is attached to the methionine. The third codon is read, and its corresponding amino acid is attached to the second amino acid. Then the fourth codon is read, and the fifth, and sixth, and so on, with each additional codon causing another amino acid to be added to the growing polypeptide chain. This process continues until a "stop" signal is encountered in the mRNA; the "stop" signal can be any of the codons 5'-UAG-3' (called the **amber** terminator), 5'-UAA-3' (called the **ochre** terminator), or 5'-UGA-3' (sometimes called the **opal** or **umber** terminator). (Any of these stop codons could correctly be called a **terminator** codon.) When the ribosome encounters a terminator codon, the finished polypeptide is released from the

ribosome. After release of the polypeptide, the methionine with which translation started is often clipped off the finished polypeptide by an enzyme, as in the case of α-globin.

This account of translation glosses over some important details. We neglected to mention certain cytoplasmic proteins called **initiation factors** that are required for ribosome binding to the mRNA. There are also proteins called **elongation factors** that are required to move the ribosome along the mRNA. And there are proteins called **release factors** that are necessary for the release of the finished polypeptide from the ribosome. On the whole, therefore, translation is an exceedingly complex process. The essence of translation is that the mRNA is translated codon by codon from the 5' end of the coding sequence toward the 3' end. Polypeptide synthesis proceeds from the amino end toward the carboxyl end at the rate of about 15 amino acids per second. A molecule of, say, hemoglobin (the β chain) is completed in roughly 10 seconds. Moreover, several ribosomes can move simultaneously in tandem along an mRNA molecule, as shown for rabbit mRNA in Figure 9.13. Such chains of ribosomes are often called **polysomes**.

The Genetic Code

The detailed correspondence between codons in mRNA and the amino acids they specify is known as the **genetic code**; the genetic code for all 64 codons is shown in Table 9.1. Note first that all 20 amino acids are signaled by at least one codon, most by several codons. AUG (coding for methionine) is the "start" signal for translation, and AUG codons also serve to insert methionine at interior positions in the polypeptide chain. Codons UAG, UAA, and UGA correspond to "stops" in translation. The codon assigments in the code were worked out by a variety of techniques; one of the most straightforward involved the use of artificially synthesized mRNA's of known base sequence.

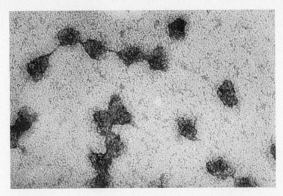

Figure 9.13 Electron micrograph of an mRNA molecule being translated simultaneously by several ribosomes moving along it in tandem. This mRNA and group of ribosomes, taken from a rabbit cell, is active in hemoglobin synthesis.

Such mRNA's can be translated in a test tube by an appropriate mixture of ribosomes, tRNA's, amino acids, and so on, and the amino acid sequence of the polypeptide produced can be compared with the known sequence of bases in the mRNA. (By jimmying the ribosomes with large amounts of magnesium ion, translation can be made to begin with any codon and not just with AUG, so AUG is not necessary to initiate translation using this technique.) An mRNA sequence 5'-UUUUUUUUUUUU · · · -3', for example, will be translated as phe-phe-phe-phe · · · —a long chain of phenylalanine molecules; thus, 5'-UUU-3' must be a codon for phenylalanine.

Most amino acids are denoted by more than one codon; that is, the genetic code has synonyms. Notice that most synonymous codons are identical in their first two bases and differ only in the third position. This is because the anticodons in tRNA's do not base-pair with complete fidelity in the third position. Many anticodons can distinguish pyrimidines (U or C) from purines (A or G) in the third codon position, but they cannot distinguish U from C or A from G in this position. The weak fidelity of base pairing in the third codon position is evidently due partly to the structure of the anticodon itself and to how the tRNA adheres to the ribosome during translation. One popular hypothesis accounting for the third-base rule is called the **wobble hypothesis**, according to which the anticodon base that pairs with the third codon base is free to move ("wobble") somewhat on the ribosome and thus form base pairs with any pyrimidine or any purine, depending on the anticodon.

The genetic code is **universal**, which means that every living thing—from bacteria to slime molds to peach trees to marmosets to humans—uses exactly the same code in translation. The most compelling evidence on this point derives from experiments in which the translational machinery of bacteria—ribosomes, tRNA's, and all the rest—are combined in a test tube with mRNA coding for hemoglobin from rabbits or any other mammal. The bacterial machinery faithfully translates the mammalian mRNA into hemoglobin. (Some differences in the translational apparatus of different organisms have been noted, however. The ribosomes of bacteria, for example, are smaller than ribosomes in more highly evolved plants and animals. Also, species differ in the relative amounts of various tRNA's. Nevertheless, these organisms all use the same genetic code.)

Evidently the genetic code evolved early in primitive organisms as long as 3 billion years ago, and it remained intact in the subsequent spectacular evolutionary radiation of microorganisms, plants, and animals. This evolutionary conservatism is not surprising. Once a genetic code is established, it is virtually impossible to alter the codon assignments significantly, because such a change in the code would simultaneously change all the proteins produced, and most of these alterations would result in polypeptides unable to function properly. However, there is one known exception to the universality of the genetic code. The translational apparatus of human **mitochondria**

BLE 9.1 THE GENETIC CODE

		Second nucleotide in codon			
	U	**C**	**A**	**G**	
U	UUU phe ⎫ F UUC phe ⎭ UUA leu ⎫ L UUG leu ⎭	UCU ser ⎫ UCC ser ⎪ S UCA ser ⎬ UCG ser ⎭	UAU tyr ⎫ Y UAC tyr ⎭ UAA (stop) UAG (stop)	UGU cys ⎫ C UGC cys ⎭ UGA (stop) UGG trp W	
C	CUU leu ⎫ CUC leu ⎪ L CUA leu ⎬ CUG leu ⎭	CCU pro ⎫ CCC pro ⎪ P CCA pro ⎬ CCG pro ⎭	CAU his ⎫ H CAC his ⎭ CAA gln ⎫ Q CAG gln ⎭	CGU arg ⎫ CGC arg ⎪ R CGA arg ⎬ CGG arg ⎭	
A	AUU ile ⎫ AUC ile ⎬ I AUA ile ⎭ AUG* met M	ACU thr ⎫ ACC thr ⎪ T ACA thr ⎬ ACG thr ⎭	AAU asn ⎫ N AAC asn ⎭ AAA lys ⎫ K AAG lys ⎭	AGU ser ⎫ S AGC ser ⎭ AGA arg ⎫ R AGG arg ⎭	
G	GUU val ⎫ GUC val ⎪ V GUA val ⎬ GUG val ⎭	GCU ala ⎫ GCC ala ⎪ A GCA ala ⎬ GCG ala ⎭	GAU asp ⎫ D GAC asp ⎭ GAA glu ⎫ E GAG glu ⎭	GGU gly ⎫ GGC gly ⎪ G GGA gly ⎬ GGG gly ⎭	

First nucleotide in codon (5′ end)

Note: The table shows the correspondence (the genetic code) between three-base codons in mRNA and the amino acid inserted into the polypeptide. Amino acids are represented by two types of abbreviations; a three-letter and a single-letter abbreviation. The asterisk (*) on the codon AUG (methionine) denotes that this codon is the "start" signal for protein synthesis. Codons UAA, UAG, and UGA are termination (stop) signals.

uses a form of the code that differs from Table 9.1 in several respects. In mitochondria, codon 5′-AUA-3′ codes for methionine instead of isoleucine, and 5′-UGA-3′ codes for tryptophan instead of termination. In addition, 5′-AGA-3′ and 5′-AGG-3′ are termination codons instead of codons for arginine. Interestingly, the mitochondrial code follows the wobble rule with no exceptions; in the third codon position, U is always equivalent to C, and A is always equivalent to G.

The Mechanics of Translation

The genetic code specifies the exact correspondence between the coding sequence of an mRNA and the polypeptide produced. As an example, consider the mRNA coding sequence 5′-AUGUUUGUGGGCCC-3′, a sequence that would have been transcribed from a DNA sense strand that has the sequence 3′-TACAAAACACCCGGG-5′. The first five codons in the mRNA are therefore AUG, UUU, UGU, GGG, and CCC. Ordinarily, of course, the AUG codon would not be at the 5′ end of the mRNA; the 5′ end would carry a leader sequence important in initiating translation, but the leader sequence itself would not be translated. Initially, the mRNA and two ribosomal subunits join together along with methionine (met)-carrying tRNA (at the AUG codon) and phenylalanine (phe)-carrying tRNA (at the UUU codon next in line). The configuration can be represented schematically as shown in Figure 9.14(*a*). Note that the ribosome has *two* binding sites for tRNA; the left-hand site is often called the **P** (for **peptidyl**) binding site, and the other is referred to as the **A** (for **aminoacyl**) binding site. In Figure 9.14(*a*), the tRNA's are drawn as simple hairpins, forgetting about the side loops. Also, the anticodons are shown as exact base-pair complements of the codons. This is not strictly correct, especially with respect to the base that pairs in the third codon position, which is often one of the chemically modified bases mentioned in connection with Figure 9.12. Thus, Figure 9.14(*a*) takes some liberties with the anticodons.

When the mRNA-ribosome-tRNA complex is established as in Figure 9.14(*a*), four things happen almost simultaneously. First, the met (methionine) is cleaved from its tRNA. Second, the methionine is connected chemically to the next amino acid, in this case phe (phenylalanine), by a peptide bond such as the ones illustrated in Figure 9.6. Third, the tRNA for methionine is released from the ribosome and becomes free in the cell; there it becomes recharged with another molecule of methionine by its associated activating enzyme. Fourth and finally, the ribosome shifts along the mRNA to the next three-base codon, in this case UGU. This shifting moves the polypeptide-bearing phe-tRNA from the A site to the P site, and it thereby frees the A site for occupation by the next tRNA, in this case cysteine (cys)-tRNA. The resulting configuration is shown in Figure 9.14(*b*).

Then the same things happen: The met-phe is clipped off the tRNA for phenylalanine and attached by a peptide bond to cys, the tRNA for phenylalanine falls free, and the ribosome moves on to the next codon—GGG in the example. At this stage the configuration is as shown in Figure 9.14 (*c*); a more detailed version of the configuration is shown in Figure 9.15. The number 1 in Figure 9.15 represents

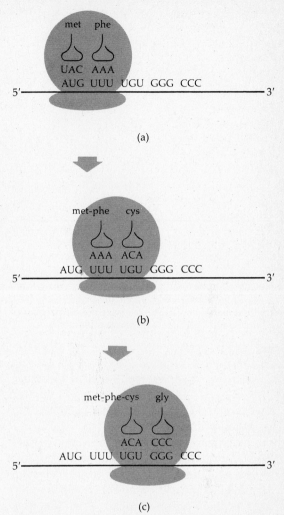

(a)

(b)

(c)

Figure 9.14 Translation involves movement of the ribosome (shaded shape) codon by codon in the 5′ to 3′ direction along the mRNA. Correct amino acids are incorporated by means of base pairing between mRNA codon and tRNA anticodon. At each stage, two tRNA's (hairpin shapes) are associated with the ribosome. One of them (the left one as illustrated here) carries the ever-growing polypeptide chain; the other tRNA carries the next amino acid to be added to the polypeptide.

the tRNA for phenylalanine that has just been released; 2 represents the polypeptide-bearing tRNA for cysteine that has just been shifted to

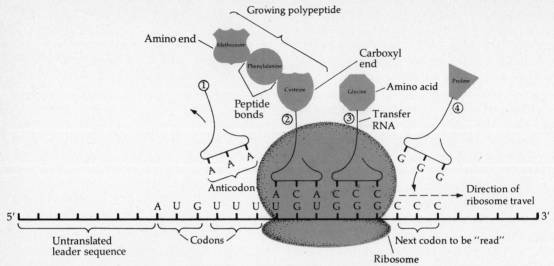

Figure 9.15   An mRNA molecule at one instant during translation. The tRNA for phenylalanine (1) has just been released from the ribosome. The cysteine of the growing polypeptide will now be detached from its tRNA (2), and a peptide bond will form with the glycine attached to its tRNA (3). The ribosome will then shift to the right one codon, tRNA 2 will be released as tRNA 3 (carrying the polypeptide) comes to occupy the left-hand position, and the proline tRNA (4) will come to its place at the right-hand position.

the P site; 3 represents the tRNA for glycine that has just occupied the A site; and 4 represents the tRNA for proline waiting in line for the ribosome to shift to the next codon (CCC) and again vacate the A site.

The same events of peptide bond formation and ribosome shifting happen over and over again during translation. With each codon translated, a specific amino acid is added to the ever-lengthening polypeptide. The first five amino acids of the polypeptide coded by the mRNA in Figures 9.14 and 9.15 would be met-phe-cys-gly-pro, corresponding to codons AUG, UUU, UGU, GGG, and CCC. The string of amino acids grows until one of the codons UAG, UAA, or UGA is encountered in the mRNA. These codons, recall, mean "stop." At this point, the polypeptide attached to the last tRNA remaining on the ribosome is clipped off, and the polypeptide, which began folding into its three-dimensional form as it

was being synthesized, becomes free to carry out its function in the cytoplasm, perhaps in conjunction with other polypeptides.

Overlapping Genes

The part of an mRNA molecule that codes for a polypeptide is often called a **cistron**. Since the base sequence of an mRNA is determined by the base sequence in the corresponding DNA, one could as well define a cistron as the DNA sequence that codes for a particular polypeptide. The concept that one coding region of DNA codes for one and only one polypeptide is referred to as the **one gene–one polypeptide hypothesis**, and this hypothesis has proven immensely fruitful in genetics because it is almost universally true.

Although the one gene–one polypeptide hypothesis is almost always valid, several instances are now known in which it is incorrect.

These exceptions, all of them in prokaryotes or viruses, involve DNA sequences that code for two polypeptides. One well-known example occurs in the virus φX174, which infects cells of *E. coli*. (A virus that infects bacterial cells is called a **bacteriophage** or, more simply, a **phage**.) The DNA of φX174 contains 5386 bases (i.e., 5.386 kilobases, a **kilobase** being a length of DNA containing 1000 bases). Its base sequence is known completely, and the part of it extending from base number 390 to base 848 is shown in Figure 9.16. The sequence illustrated is that of the antisense strand because, as noted earlier, the RNA transcript has the same base sequence as the antisense strand (except that U replaces T).

The sequence in Figure 9.16 is of interest because it codes for two polypeptides designated D and E. Moreover, the coding sequence for E is contained entirely within the coding sequence for D; that is, the *D* gene and the *E* gene are **overlapping**. As indicated in the figure, translation of the D polypeptide initiates with the AUG codon at positions 390 to 392. (The amino acids here are denoted by their conventional single-letter abbreviations.) Translation of D proceeds in the normal fashion codon by codon all the way to the ochre terminator at positions 846 to 848. The D polypeptide contains 152 amino acids, but the codons for only 20 of these are shown. However, translation can also initiate *internally* at the

AUG at positions 568 to 570; initiation at this codon produces the E polypeptide, which is terminated by the opal codon at positions 841 to 843. Altogether, the E polypeptide comprises 91 amino acids, only 10 of which are illustrated.

Note especially that, in the region of overlap (bases 568 through 840), the amino acid sequences of the D polypeptide and the E polypeptide are *different* even though they are translated from the same stretch of RNA. This difference can occur because translation of D and translation of E involve different reading frames. The **reading frame** of translation refers to which bases occupy the first codon positions as translation proceeds. For the D polypeptide, the first codon positions are the bases 390, 393, 396, 399, 402, and so on all the way to 567, 570, 573, 576, and 579, and to the end; for the E polypeptide, by contrast, the first codon positions are 568, 571, 574, 577, 580, and so on; that is, the reading frame for E is shifted one base to the right relative to the reading frame for D. The difference in the amino acid sequence of D and E therefore arises because the *first* codon positions in the E reading frame are the *second* codon positions in the D reading frame. Figure 9.16 illustrates yet another example of overlapping genes in φX174. The terminal sequence UAAUG is read in D's reading frame as UAA to terminate translation, but the AUG is also used to initiate

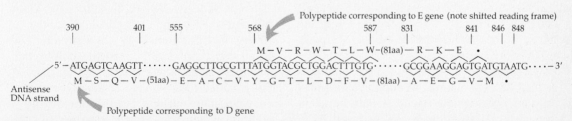

Figure 9.16 A case of overlapping genes in bacteriophage φX174. The antisense strand shown has the same base sequence as the RNA transcript from the corresponding sense strand (except that U in RNA replaces T in DNA). The mRNA is translated in two distinct reading frames as shown to generate the D and E polypeptides.

translation of another polypeptide called J; that is, the termination codon of D overlaps the initiation codon of J. Altogether, φX174 contains eight pairs of overlapping genes. The reason for this highly exceptional sort of genetic organization is thought to be coding efficiency—to pack as much coding information into as few bases as possible to keep the DNA molecule small enough to be packaged within the tiny virus particle.

Eukaryotes: Split Genes and RNA Processing

Overlapping genes represent some very fancy genetic footwork for an organism as small and relatively simple as φX174, but in one respect the situation is simple: The polypeptides and their coding regions in the DNA are **colinear**, which means that adjacent amino acids are always coded by adjacent triplets of bases in the DNA. At one time it was widely assumed that colinearity was a characteristic of all genes —that coding regions were uninterrupted. Although this generalization is true of prokaryotes, it has become clear in recent years that many, if not most, genes in eukaryotes are much more complex. The complexity arises because the coding region of a eukaryotic gene is often **split** into smaller regions, sometimes into a dozen or more smaller regions, and these are not adjacent in the DNA. The split-gene organization of the mouse α-globin gene is depicted in Figure 9.17. Across the top of the figure is the base sequence of the antisense DNA strand from base 405 to 1191. Lower down is the base sequence of the coding portion of the α-globin mRNA and the amino acid sequence of α-globin itself. As can be seen, initiation of translation begins with the AUG codon corresponding to bases 405 to 407 in the DNA; the methionine (M) with which transla-

tion begins is later cleaved off the polypeptide, so it is enclosed in parentheses. Translation proceeds in the normal manner until amino acid 30 (E: glutamic acid). Up to this point the gene and polypeptide are colinear; adjacent amino acids are coded by adjacent triplets of bases in the DNA.

After amino acid 30, coded by DNA bases 495 to 497, comes a surprise. One would have expected the next amino acid (R: arginine) to be coded by DNA bases 498 to 500. In fact, the first two positions are coded by bases 498 to 499, but the third position is coded by base 622! The entire region between bases 500 and 621—a sequence 122 bases long—does not appear in the mRNA. Such a sequence that interrupts the coding region of a gene but does not appear in the mRNA has come to be called an **intervening sequence**, or an **intron**, and the 122-base intron in α-globin is symbolized **IVS1**. Coding sequences of DNA that do appear in the mRNA are called **coding regions** or **exons**, and the 96-base exon of α-globin is symbolized **CR1**.

The second coding region of mouse α-globin (CR2) extends from base 623 through 826 and codes for amino acids 32 through 99. Then there occurs a second intervening sequence (IVS2) of 134 bases. And there follows the third coding region (CR3) between positions 961 and 1089, which codes for amino acids 100 through 141 and the ochre terminator UAA. Trailing at the 3′ end of the mRNA is a 101-base sequence transcribed from the DNA but not translated, and attached to the A (transcribed from base 1191) is a stretch of 40 to 50 consecutive A's called the **poly-A tail**. This poly-A tail is not transcribed from DNA but is added to the 3′ end of the transcript by enzymes in the nucleus. The function of poly-A addition is uncertain, but it may increase the stability of the mRNA. In any case, most eukaryotic mRNA's have poly-A tails consist-

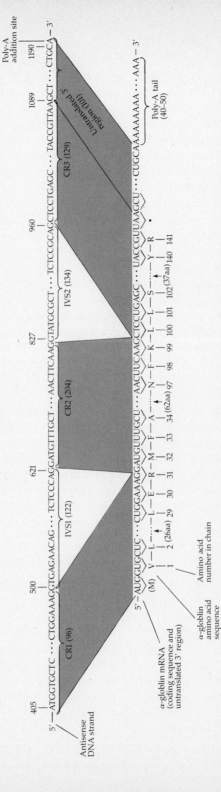

Figure 9.17 Intervening sequences (introns) in the mouse α-globin gene. Sequences corresponding to IVS1 and IVS2 do appear in the α-globin RNA transcript, but they are removed by a highly precise splicing process that brings the coding regions (CR1, CR2, and CR3) into juxtaposition in the processed mRNA. The human globin genes also have three exons (coding regions) and two introns.

ing of tens or hundreds of consecutive A's, although for some unknown reason the histone mRNA's are not polyadenylated.

Intervening sequences can be detected even without the sort of detailed sequence comparison outlined in Figure 9.17. The method involves the formation of a heteroduplex between an mRNA and the corresponding strand of DNA from which it was transcribed. When duplex DNA is mixed with complementary mRNA under the appropriate conditions, the mRNA will displace the antisense DNA strand and form a mRNA-DNA heteroduplex with the sense DNA strand due to hydrogen bonding between the bases. The displaced antisense strand forms a loop (called an **R loop**) that can be identified in the electron microscope. If the mRNA comes from a single coding region, the R loop will appear as in Figure 9.18(*a*). On the other hand, if the mRNA comes from two coding regions interrupted by an intervening sequence, a double R-loop structure will be formed [see Figure 9.18(*b*)], with the intervening sequence remaining as a DNA-DNA duplex. Such R-loop analysis provided the first evidence for the existence of intervening sequences. Figure 9.19 shows the R-loop structure formed between a region of viral DNA and one of its transcripts.

RNA Splicing

Figure 9.20 illustrates the manner of processing of eukaryotic RNA transcripts. For the moment, concentrate on the coding regions (shaded) and intervening sequences (unshaded) in the middle of the DNA [see Figure 9.20(*a*)]. During transcription, the coding regions and the intervening sequences are transcribed, so the resulting primary transcript [see Figure 9.20(*b*)] contains base sequences corresponding to the exons (coding regions) and the introns (intervening sequences). Since the aggregate size of the introns can be larger and sometimes much larger than the aggregate size of the exons, the primary transcript can be much larger than the fully processed mRNA. Indeed, the primary transcripts of eukaryotes can be as large as 60 kilobases (i.e., 60,000 bases, abbreviated 60 kb). Actually, primary transcripts from different genes can be extremely variable in size, and for this reason the large primary transcripts have come to be called **heterogeneous nuclear RNA** (or **hnRNA**) —"nuclear" because they are found only in the nucleus. During RNA processing in the nucleus, the intervening sequences are removed from the transcript by **splicing**; that is, the transcript is cut exactly at the exon-intron junctions, and the exons are ligated together, leaving the excised introns free in the nucleus [see Figure 9.20(*c*)]. The splicing is catalyzed by certain enzymes, and it is exquisitely precise, but the splicing signals are not understood. Many, but not all, intervening sequences have a GT at their 5′ end and an AG at their 3′ end (for two examples see Figure 9.17), but this GT/AG rule is surely not the only factor involved in splicing because GT · · · AG sequences are also present within coding regions (compare bases 408 to 409 and 498 to 499 in Figure 9.17). At present, the other signals that may be involved in splicing are unknown.

Functions of Intervening Sequences

The role of intervening sequences in eukaryotic genetics and evolution is unclear. Part of the problem is that such sequences have so recently been discovered that critical experiments have not yet been carried out with a sufficiently wide variety of genes in a diversity of eukaryotic organisms. Another part of the problem is that intervening sequences may have several functions, or some may have one function and others another function. Although there are

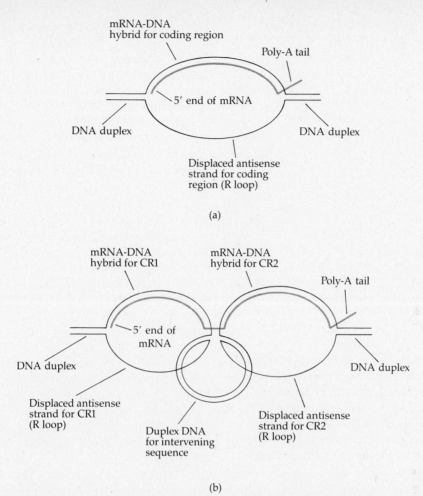

Figure 9.18 Configurations expected when mRNA is hybridized with corresponding duplex DNA. *(a)* An R loop formed by displacement of the antisense DNA strand by the mRNA; this single R-loop configuration will occur with unsplit genes because all the DNA sequence will be present in the mRNA. *(b)* Double R-loop configuration expected when a gene has one intervening sequence; each coding region displaces its corresponding antisense strand, but the region corresponding to the intervening sequence cannot be displaced because this sequence does not occur in the mRNA.

many opinions about the possible functions of intervening sequences, three broad views are under active discussion.

1. *Regulation*: This view holds that intervening sequences are important in regulating gene expression, perhaps by influencing when an

mRNA will be released from the nucleus, or perhaps by influencing the location of its release into the cytoplasm. This hypothesis is supported by the results of an experiment in which a portion of hemoglobin DNA was chemically ligated to a gene from a mammalian virus. When the intervening sequence is left in

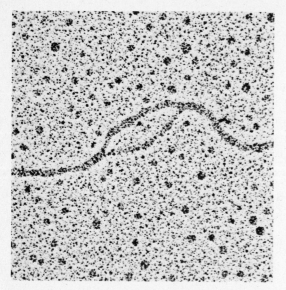

Figure 9.19   R-loop structure formed by hybridization between a viral gene and its RNA transcript.

its proper orientation, the hemoglobin fragment can be found in virus-infected cells. When the ligation is such as to reverse the orientation of the intervening sequence, on the other hand, the hemoglobin fragment is not produced because the primary transcript is unstable and breaks down in the nucleus.

2. *Modular evolution*: This view holds that intervening sequences are leftovers from past evolution. A major premise is that polypeptide chains have an internal division of function, with some regions doing one thing and other regions another. These functionally distinct regions are often called **modules** or **domains**. The modular evolution hypothesis is that each module is coded by a distinct coding region separated from others by an intervening sequence. This sort of organization would create an ideal situation for the evolutionary creation of new types of polypeptides because recombination between the intervening sequences of different genes could create virtually limitless combinations of modules, and some of these

combinations might produce new polypeptides of use to the organism. Support for this hypothesis comes from the observation that the middle coding region of the hemoglobin genes codes for the part of the polypeptide that carries the heme (see Figure 9.1). This sort of module-coding region correspondence is found in several other polypeptides as well, but it is by no means universal. The gene for chicken **ovalbumin** (an egg white protein), for example, is split into seven coding regions, but these do not correspond in any obvious way with functionally distinct parts of the polypeptide.

3. *Acceleration of sequence divergence*: This hypothesis is based on the observation that new genes often evolve by gene duplication and sequence divergence. **Sequence divergence** refers to the dissimilarity in the base sequence of related genes. Once a duplication occurs, one of the copies is free to change because the other copy will still carry out the original function. Over the course of evolution, as each generation of organisms replaces the previous one, mutations will occur. Many of these mutations will be lost from the population by chance because their carriers may leave no descendants. A lucky few mutations that are not harmful may increase in frequency by chance, and after a long succession of lucky increases in frequency may even come to replace the original form of the gene. (This process of chance changes in gene frequency is called **random genetic drift**, and it will be discussed further in Chapter 15.) A precious few mutations may actually be beneficial to their carriers and enable them to produce more offspring than those who carry the normal form of the gene; in such a case, each succeeding generation will have a disproportionate representation of the mutation, and eventually the mutation will replace the nonmutant form of the gene in the population. (This process of gene substitution is called **selection**, and it will also be discussed in Chapter 15.) In any event, random genetic

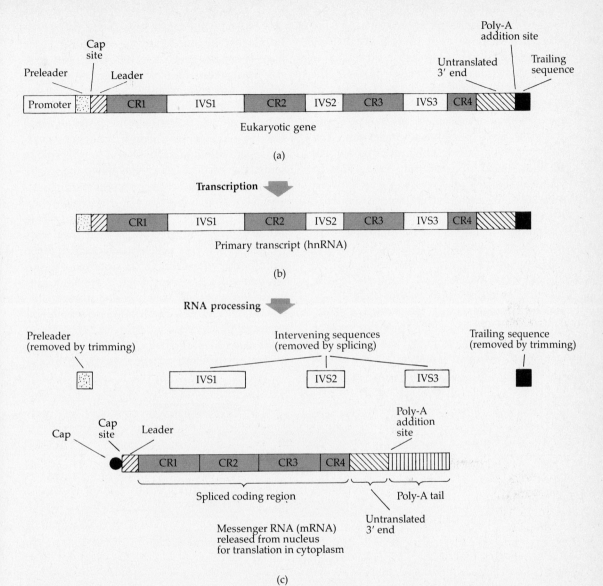

Figure 9.20 Generalized structure of eukaryotic gene. *(a)* The DNA has sequences corresponding to promoter, preleader, leader, coding regions (several), intervening sequences (several), untranslated 3' end, and trailing sequence. *(b)* Primary transcript includes all the sense-strand DNA sequences except the upstream promoter sequences. *(c)* RNA processing includes removal of preleader and trailing sequence, splicing out of intervening sequences, addition of cap at 5' end, and addition of poly-A tail to 3' end.

drift and selection are two forces that act to promote sequence divergence of genes.

On the other hand, there are also proc-

esses that act to prevent sequence divergence. For example, the many copies of the ribosomal RNA genes are virtually identical in sequence. The mechanisms by which sequence divergence of duplicate genes is prevented are not understood in detail, but two processes have been proposed. One process is thought to involve heteroduplex formation between the DNA strands of different copies followed by mismatch repair of the sort discussed in connection with gene conversion in Chapter 8; this process is also called **gene conversion**. The other process involves unequal crossing-over and is outlined in Figure 9.21. Shown in part (*a*) are mispairing and crossing-over between duplicate genes that have undergone sequence divergence (shaded and unshaded rectangles). One product of the unequal crossover will be the triplication shown in part (*b*). In a later generation, crossing-over between the triplication and the duplication will again reduce the number of gene copies to two, but these will be identical in sequence (*c*). By chance or selection, the nondiverged duplication can replace the diverged one shown in (*a*); this process has been called **concerted evolution**.

But if gene conversion or concerted evolution can prevent sequence divergence, then how can new genes evolve? Here is where intervening sequences are thought to enter the picture. Since intervening sequences do not code for amino acids, they are free to change (except, of course, for the splicing signals), and they are known to undergo sequence divergence extremely rapidly. Once the intervening sequences of duplicate genes have undergone sufficient sequence divergence, the pairing necessary for gene conversion or concerted evolution will no longer occur, and the duplicate genes will then be free to go their own way. Evidence for the acceleration-of-divergence hypothesis comes from the α-globin and β-globin genes. Their coding regions are sufficiently similar that the conclusion is inescapable that they arose as a duplication of a common ancestral globin gene an estimated 500 million years ago; in particular, the mouse β-globin coding regions are *identical* to those of the

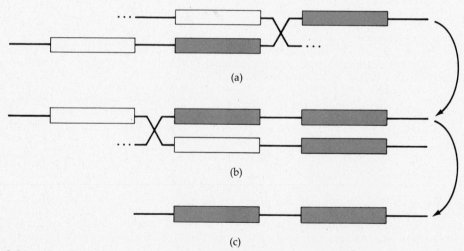

(a)

(b)

(c)

Figure 9.21   Maintenance of sequence identity of tandemly duplicated genes can occur by two rounds of unequal crossing-over. *(a)* Nonsister chromatids carrying the duplication paired unequally during meiosis; the shaded and unshaded rectangles represent a nucleotide sequence difference that has occurred in one copy of the duplication. The unequal crossover produces a triplication as one of its products. *(b)* The triplication produced in part *(a)* is shown as being heterozygous with the duplication; a second crossover generates the duplication shown in part *(c)*, in which both copies of the gene have an identical sequence. Sequence differences can also be eliminated by gene conversion-like processes.

α-globin coding regions at 232 out of 411 nucleotide sites. However, the intervening sequences are so different that their relationships can no longer be ascertained. On the other hand, humans have a duplication of the α-globin gene (these genes are called α_1 and α_2), and they are virtually identical in nucleotide sequence, including the intervening sequences.

The Cap Structure

Although the splicing of RNA to excise intervening sequences is surely the most spectacular aspect of RNA processing in eukaryotes, it is only one part of the story. The other parts are also illustrated in Figure 9.20. The left-hand part of (a) shows the promoter sequence, which permits RNA polymerase recognition and binding but is not itself transcribed. Transcription of the beginning of the gene involves the sequences called the preleader, the cap site, and the leader [see Figure 9.20(b)]. The cap site is a single nucleotide; in the mouse α-globin gene the cap site is number 373, and it is the first nucleotide transcribed; there is no "preleader" in this case. In other cases there may be a preleader upstream from the cap site, but, during RNA processing, the preleader is removed by an exoribonuclease enzyme in a process called **trimming**, and a special ribonucleotide, called a **cap**, is added to the 5′ end of the cap site. The capping ribonucleotide varies somewhat in its structure, but it always involves a modified ribonucleotide attached to the 5′ end in an unusual orientation. The cap on the mouse α-globin mRNA is illustrated in Figure 9.22. In this case the modified ribonucleotide is **7-methyl guanosine**, which is an ordinary guanosine that has been modified by the addition of a methyl group (shaded circle); the unusual linkage is the 5′-to-5′ linkage illustrated rather than the typical 5′-to-3′ linkage. In addition, the adenine and the ribose at the 5′ end of the RNA are modified by methylation. Capping of the mRNA seems to be essential for ribosome attachment in eukaryotes. This requirement makes it difficult to imagine that eukaryotes could have overlapping genes in the sense of φX174, as translational initiation cannot occur internally.

The other aspect of RNA processing illustrated in Figure 9.20 is polyadenylation of the 3′ end, which was discussed earlier in connection with α-globin mRNA in Figure 9.17. In polyadenylation, part of the 3′ end of the primary transcript may be trimmed back to the poly-A addition site, and to this site the poly-A tail is attached. After RNA processing by splicing, trimming, capping, and polyadenylation, the product is transported from the nucleus to the cytoplasm, where it finally plays its role as mRNA. The leftover parts of the primary transcript remain behind in the nucleus, where they may perform some as yet unknown function.

Pseudogenes

One aspect of eukaryotic genetics that has been discovered through direct sequence analysis of DNA is the existence of pseudogenes. **Pseudogenes** are genes that have an obvious sequence similarity to other genes but are not expressed; that is, they are "silent." An example is a mouse α-globin pseudogene. This gene has undergone sufficient sequence divergence from the normal α that, if it were transcribed and translated, it would lead to an abnormal polypeptide. Whether this α pseudogene is transcribed is not known, but it is certainly not translated. Indeed, the mouse contains a second α pseudogene that is not even transcribed because its RNA polymerase binding site has changed in sequence; this second α pseudogene is of interest because by some unknown process its intervening sequences have been neatly and precisely deleted. Humans have

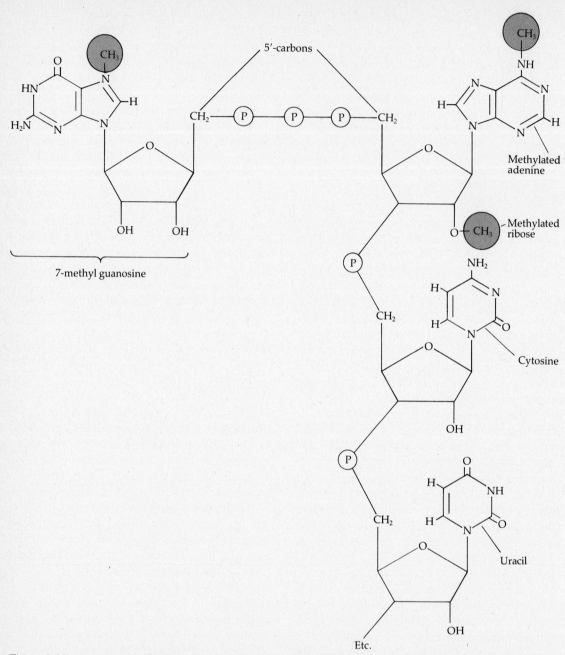

Figure 9.22   One type of cap structure found at the 5′ end of eukaryotic mRNA's. The mRNA is shown on the right, the cap on the left. Note that the cap is a modified guanosine attached by means of a 5′ triphosphate to the 5′ end of the mRNA. Various other groups on the 5′ ribose and base can also be modified. Such cap structures are involved in recognition between mRNA's and ribosomes.

pseudoglobin genes, too. In addition to the normal α-gene duplication, there is an α pseudogene about 4 kb (i.e., 4000 bases) upstream from the duplication. Moreover, while humans have only one functional β-globin gene, there are two β-pseudogenes.

How widespread pseudogenes will turn out to be is still unknown, and their functions are even more mysterious. Possible explanations range from views in which the pseudogenes are remnants of evolutionary mistakes (ancient duplications that were simply inactivated by mutation rather than deleted) to views in which pseudogenes are genes caught in the act of evolving (temporarily out of commission but perhaps one day to be reactivated to carry out a novel function) to views in which pseudogenes, while not expressed as polypeptides, play an essential role in the expression of their normal counterparts.

SUMMARY

1. Proteins are composed of chains of **amino acids** called **polypeptides**. The nucleotide sequence of most genes is used to specify the sequence of amino acids in a single type of polypeptide, and this **primary structure** (i.e., amino acid sequence) of the polypeptide determines how it will fold into its three-dimensional configuration (**tertiary structure**). However, several polypeptides from the same gene, or even polypeptides from different genes, may aggregate to form the **quaternary structure** of a finished protein.

2. Inborn errors of metabolism refer to mutations that lead to defective enzymes that disrupt vital chemical processes in the cell or organism. Classical examples of such inborn errors involve enzymes in the metabolism of the amino acids **phenylalanine** and **tyrosine**. One enzyme (**phenylalanine hydroxylase**) normally converts phenylalanine into tyrosine; when this enzyme is defective, harmful amounts of phenylalanine build up and result in the severe form of mental retardation called **phenylketonuria**. Other mutations in phenylalanine-tyrosine metabolism lead to various forms of **albinism** or to the relatively harmless black-urine condition **alcaptonuria**.

3. Genetic information in DNA is used to specify the sequence of amino acids in a polypeptide by means of a two-step process in prokaryotes and a three-step process in eukaryotes. The first step is always **transcription**, in which the sequence of nucleotides in the sense strand of DNA is used as a template for the production of a complementary sequence of **ribonucleotides** in a molecule of **RNA (ribonucleic acid)** by means of base pairing. The final step is always **translation,** in which the RNA directs the sequence of amino acids to be incorporated into a polypeptide. Eukaryotes have a third step between transcription and translation, which is called **RNA processing.**

4. The structure of RNA is similar to that of single-stranded DNA except that (a) the sugar in RNA is **ribose**, and (b) the base **uracil** in RNA plays the role of thymine in hydrogen bonding with adenine.

5. The key enzyme in transcription is **RNA polymerase**. Transcription begins when the polymerase attaches to a DNA strand at a region called the **promoter**, which consists of a polymerase recognition sequence and a binding site. The polymerase travels in the $3'$ to $5'$ direction along the sense strand of DNA, initiating a new RNA strand (called a **transcript**) a few nucleotides **downstream** (i.e., in the $5'$ direction of the sense strand) from the binding site. New ribonucleotides are added one by one to the $3'$ end of the elongating transcript, and this process ends when the polymerase encounters a termination signal farther downstream. In prokaryotes, the primary transcript consists of a **leader sequence**, important for ribosome binding to initiate translation, a **coding sequence**, which carries the base sequence that specifies the amino acids during translation, and a 3' **untranslated region** at the 3' end. The primary transcript in prokaryotes corresponds to **messenger RNA (mRNA)**, the RNA molecule actually used in translation. In eukaryotes, the mRNA is produced in the nucleus from the primary transcript by means of RNA processing.

6. Translation begins when a **ribosome** particle attaches to the leader sequence of an mRNA and moves along the mRNA in the 3' direction until it reaches the coding sequence. From there, the ribosome moves in exact steps of three bases at a time. Such groups of three bases are called **codons**, and each specifies a particular amino acid to be added to the polypeptide chain. The amino acids themselves are brought to the ribosome by molecules of **transfer RNA (tRNA)**. Each tRNA has at one end a three-base **anticodon** that can hydrogen bond with the complementary codon, and at the other end it has an attached amino acid that corresponds to the codon. The correspondence between codons and the amino acids they encode is called the **genetic code**. Codon 5'-AUG-3' stands for methionine, and this is always the first (**amino terminal**) amino acid in a new polypeptide chain (although the methionine is frequently removed from the finished polypeptide later on). When the 5'-AUG-3' codon and its tRNA are aligned, the ribosome brings in the tRNA corresponding to the second amino acid and attaches the methionine to this amino acid by means of a **peptide bond**. The tRNA for methionine becomes detached from its amino acid in the process and falls free from the ribosome. The ribosome then shifts to the next codon, brings in the appropriate tRNA, attaches the growing peptide to this amino acid, and lets the previous tRNA fall free. This process continues codon by codon

and amino acid by amino acid until one of the three **terminator codons** is encountered. These three are 5'-UAG-3', 5'-UAA-3', and 5'-UGA-3', and each will cause the polypeptide to be detached from the last remaining tRNA and be released into the cytoplasm.

7. The genetic code is nearly **universal**. Virtually any translational apparatus, from any organism whatsoever, will translate the same coding sequence in the same way. The code also has synonyms—codons that code for the same amino acid. Most of the synonymous codons differ in the third base position: Where one has a U, the other will have a C (i.e., both pyrimidines), or where one has an A, the other will have a G (both purines). The equivalence of pyrimidines (and purines) in the third codon position is thought to be due in part to the lack of rigidity at this position when the tRNA pairs with the codon; the tRNA can **wobble** somewhat, and this reduces the fidelity of base pairing, with any pyrimidine in the third codon position being suitable, or any purine. The translational apparatus of human **mitochondria** uses a slightly altered genetic code, different in several instances from the standard code. Whereas the standard code permits two departures from the wobble rule of U-C equivalence and A-G equivalence in the third codon position, the human mitochondrial code permits no exceptions to the rule.

8. While it is almost always true that a particular coding region of an mRNA codes for a unique polypeptide, several exceptions in prokaryotes are known. In these instances, ribosomes can attach to the middle of the coding region and initiate translation with a nearby 5'-AUG-3' codon. Translation of this second polypeptide then proceeds in the normal fashion. If the **reading frame** (i.e., the first base of each codon) is the same in the second polypeptide as in the first, then the second polypeptide will have the same amino acid sequence as part of the first polypeptide. However, and this is the most interesting case, when the reading frames of the two polypeptides are shifted, then the amino acid sequences of the two polypeptides may be entirely different. The region of DNA that carries such a doubly translated coding sequence is said to have **overlapping genes**. So far, overlapping genes have been found mainly in bacterial viruses (**bacteriophages**), which have to pack a large amount of coding information into a small amount of DNA.

9. In prokaryotes, the coding region of an mRNA corresponds to a continuous stretch of DNA. In eukaryotes, however, the coding region of a gene may be **split** into several smaller pieces between which noncoding regions are found. The smaller coding regions are called **exons** and the noncoding regions in between are called **introns** or **intervening sequences**. Some genes are split into a dozen or more exons, and the total length of the introns may be substantially greater than the total length of the exons. In the transcription process, both introns and exons are transcribed to produce a large precursor of mRNA called **heterogeneous nuclear RNA (hnRNA)**. As part of RNA processing, the introns are spliced out of the RNA to produce a molecule in which the separate exons lie side by side. This splicing is carried out by enzymes in the nucleus, but the nature of the splicing signals is unknown.

10. The function of intervening sequences is completely mysterious. Some believe that their presence is essential to the normal regulation of gene expression. Others believe that introns flank exons that represent functionally distinct amino acid sequences (**modules** or **domains**); recombination between introns of different genes could therefore create many combinations of exons, and some of these may produce novel proteins of benefit to the organism and so be retained in evolution. In this view, intervening sequences are leftovers from the way genes evolve. Still another view is that introns, by undergoing high rates of **sequence divergence** (i.e., sequence differentiation) in evolution, act to accelerate the rate of sequence divergence of nearby exons. Such acceleration by introns would involve reducing the amount of sequence homology required for such processes as **gene conversion** or **unequal crossing-over** that might otherwise hinder the sequence divergence of duplicate genes.

11. In addition to removal of intervening sequences by splicing, RNA processing in eukaryotes involves changes at the 5′ and the 3′ ends of the transcript. To the 5′ end of the molecule is added a **cap**—a modified ribonucleotide attached to the 5′ end in an atypical fashion; the cap seems to be necessary for ribosome attachment and the initiation of translation. The 3′ end of the transcript is usually **polyadenylated**—tens or hundreds of consecutive A's are added to the 3′ end; this poly-A tail is thought to increase the stability of the mRNA. After splicing, trimming, capping, and polyadenylation, the RNA transcript finally becomes an mRNA and is released from the nucleus to participate in translation.

12. Pseudogenes are "silent" genes that are similar in base sequence to some other gene but are not expressed. Some pseudogenes may be transcribed but the RNA is not processed or translated; other pseudogenes are not even transcribed. All known pseudogenes have undergone sequence changes that would result in a defective polypeptide if the pseudogene were expressed. In humans, in addition to a duplication of the normal α-globin gene, there is an α-globin pseudogene; although humans have only one functional β-globin gene, there are two β-globin pseudogenes. Pseudogenes thus seem to be common, but their function, if any, is unknown. Speculations range from pseudogenes being thought of as essential for the normal regulation of expression of their normal counterparts, to views in which pseudogenes are thought to be presently nonfunctional but caught in the act of evolving.

WORDS TO KNOW

Ribonucleic Acid (RNA)

Ribose
Ribonucleotide
Uracil
7-methyl guanosine

Protein

Amino acid
Amino end
Carboxyl end
Peptide bond
Polypeptide
Primary structure
Tertiary structure
Denaturation

Transcription

Promoter
Goldberg-Hogness box
 (Pribnow box, TATA box)

RNA polymerase
RNA transcript
Heterogeneous nuclear RNA
 (hnRNA)
RNA processing
Cap site
RNA splicing
Poly-A tail

Syndrome

Inborn error of metabolism
Phenylketonuria
Alcaptonuria
Albinism

Translation

Messenger RNA (mRNA)
Coding sequence
Leader sequence
Transfer RNA (tRNA)
Aminoacyl-tRNA synthetase

Ribosomal RNA (rRNA)
Ribosome
Codon
Anticodon
Genetic code
Initiation codon
Terminator codons

Gene

One gene–one polypeptide
 hypothesis
Cistron
Split
Intervening sequence (intron)
Exon
Temperature-sensitive mutation
Conversion
Overlapping gene
Pseudogene

PROBLEMS

1. For discussion: Some people seem to believe that inherited disorders are worse than environmentally caused disorders because they think that inherited disorders are untreatable. How does the case of phenylketonuria (PKU) bear on this point of view? Would you consider the dietary treatment of PKU to be a "cure"? Most newborns in the United States are now routinely tested for PKU with a rapid and simple test. Since 1962, when screening on a wide scale began, about 20 million infants have been screened, and about 2000 cases of PKU have been discovered in this manner. The screening program has neverthless been criticized, mainly on two counts. First, some of the discovered cases of PKU are actually "false positives"—normal infants who for unknown reasons have high levels of phenylalanine in the blood but not PKU. These children are placed on the low-phenylalanine diet, which is at least restrictive and has been alleged to be nutrition-

ally harmful in normal children. Second, most states that require PKU testing provide no genetic counseling to parents or to affected children at appropriate ages. Do you think such criticisms are valid? Are they sufficiently strong that screening should be terminated? If not, what might be done to increase acceptance of screening? What does PKU illustrate about the complexity of inborn errors of metabolism? What does it illustrate about the social applications of genetic knowledge?

2. What unique feature of the 5′ end of a eukaryotic mRNA would distinguish it from a prokaryotic mRNA?

3. What aspect of RNA structure promotes the formation of foldback loops of the sort found in transfer RNA's?

4. Shown here is a partial metabolic pathway for the synthesis of arginine in the fungus *Neurospora crassa*, which was important in the development of

Judson, H. F. 1979. The Eighth Day of Creation. Simon & Schuster, New York. A fascinating account of the growth of molecular biology.

Lerner, M. R., and J. A. Steitz. 1981. Snurps and scyrps. Cell 25:298–300. The possible importance of small nuclear ribonucleoproteins (snRNPs, pronounced "snurps") and small cytoplasmic ribonucleoproteins (scRNPs, pronounced "scyrps") in RNA processing is explored.

Lewin, B. 1980. Gene Expression, Vol. 2: Eucaryotic Chromosomes, 2nd ed. Wiley, New York. An excellent summary of eukaryotic gene expression.

Little, P. F. R. 1982. Globin pseudogenes. Cell 28:683–684. Speculations on the origin and evolution of pseudogenes.

Michelson, A. M., and S. H. Orkin. 1980. The 3′ untranslated regions of the duplicated human α-globin genes are unexpectedly divergent. Cell 22:371–377. Nucleotide sequences of α_1-globin and of an α-thalassemia gene.

Modiano, G., G. Battistuzzi, and A. G. Motulsky. 1981. Nonrandom patterns of codon usage and of nucleotide substitutions in human α- and β-globin genes: An evolutionary strategy reducing the rate of mutations with drastic effects? Proc. Natl. Acad. Sci. U.S.A. 78:1110–1114. Codon usage is not at all like what would be expected by chance alone.

Nishioka, Y., and P. Leder. 1979. The complete sequence of a chromosomal mouse α-globin gene reveals elements conserved throughout vertebrate evolution. Cell 18:875–882. Source of Figure 9.10(b), 9.11, and 9.17.

Rich, A., and S. H. Kim. 1978. The three-dimensional structure of transfer RNA. Scientific American 238:52–62. A detailed look at this important class of molecules.

Scriver, C. R., and C. L. Clow. 1980. Phenylketonuria and other phenylalanine hydroxylation mutants in man. Ann. Rev. Genet. 14:179–202. Detection and treatment of PKU and some related conditions.

Sharp, P. A. 1981. Speculations on RNA splicing. Cell 23:643–646. As the author notes, "In this void of knowledge [about the process of RNA splicing], it is fun to daydream about what the process might be like."

Stanbury, J. B., J. B. Wyngaarden, D. S. Frederickson, J. L. Goldstein, and M. S. Brown. 1983. The Metabolic Basis of Inherited Disease, 5th ed. McGraw-Hill, New York. Distinguished compendium of inborn errors of metabolism.

Stent, G. S., and R. Calendar. 1978. Molecular Genetics, 2nd ed. Freeman, San Francisco. An advanced textbook covering virtually all aspects of molecular genetics in a readable format.

Sussman, J. L., and S. H. Kim. 1976. Three-dimensional structure of a transfer RNA in two crystal forms. Science 192:853–858. Source of data in Figure 9.12.

chapter 10
The Regulation of Gene Expression

Every cell in the body is an organized, structured, integrated whole. Cells are not watery little bags of DNA, RNA, protein, lipid, carbohydrate, and all the other constituents. Cells are alive. They are themselves little "organisms" that have different structural parts—nucleus, mitochondria, ribosomes, and so on—and they regulate their own **metabolism**, which is the aggregate of all the physical and chemical processes that maintain a cell in the living state. A cell undergoing mitosis, for example, proceeds through a carefully orchestrated sequence of events marked early on by DNA synthesis, then chromosome coiling, spindle formation, and finally division itself.

The sequence and duration of these events are not haphazard, but precisely regulated and controlled.

By **regulation** is meant the ability of cells to express genes at only certain times. For example, *Escherichia coli*, a bacterial species that normally lives in the mammalian gut, carries genes with products that enable the cells to derive energy from **lactose** (milk sugar); these genes become active only when lactose is present in the bacteria's environment, which is to say that the genes are regulated. Genes in human cells are regulated, too, but the processes of regulation in human cells are still poorly understood. Not every gene is expressed in every human cell, and, as a cell proceeds through its life, certain genes become turned on whereas others become turned off. The ability of cells to orchestrate which of their genes will be expressed and which will not be expressed is the central reason cells are able to differentiate—to become specialized—as a human embryo develops from a fertilized egg. It is fundamentally a difference in regulation that makes a brain cell different from a liver cell, and a spleen cell different from a muscle cell. Consequently, an understanding of the regulation of gene expression is a key to understanding our own development and aging.

Gene expression is a complex process and there are many **control points**—places in the process at which gene expression could be regulated. Many of these control points are listed in Table 10.1, and, as one can see, the list begins with possible changes in the DNA itself. Other control points occur at the level of transcription, RNA processing, transport of mRNA, translation, and even after translation. In this chapter, examples of regulation at each of these levels will be discussed. An important point to keep in mind is that there is not a single, universal process of regulation. There are many possible control points for genes, and different genes can therefore be regulated in different ways. Moreover, a population does not evolve in such a way that its regulatory processes will be elegant, clever, or simple (though many such processes are). Populations evolve to survive and they are opportunistic. If a cumbersome system of regulation should occur by chance and work tolerably well, there will be further evolution for refinement and effectiveness, but not necessarily for simplicity. On the whole, regulation seems to encompass a hodgepodge of diverse, ad hoc processes, each of which is retained because it happens to do the job. Some of these processes are beautiful in their simplicity; others are bewilderingly complex and poorly understood.

TABLE 10.1 SOME POSSIBLE CONTROL POINTS IN EUKARYOTIC GENE REGULATION

I. Regulation by DNA changes
    A. Base modification
    B. Amplification or diminution
    C. Transposition
    D. Inversion
    E. Splicing

II. Regulation of transcription
    A. Positive control
    B. Negative control
    C. RNA polymerase and helper proteins

III. Regulation by RNA processing
    A. Splicing and transcript stability
    B. Capping, polyadenylation, and transport

IV. Regulation by translational control
    A. Masked or untranslated mRNA's
    B. Attenuation and mRNA conformation
    C. Stability of mRNA's

V. Regulation at the posttranslational level
    A. Polyproteins
    B. Activation
    C. Inhibition
    D. Stability of proteins

Regulation of Hemoglobin Genes in Human Development

Precise regulation of the expression of the genes for **hemoglobin**—the oxygen-carrying protein in the blood—occurs during human development. Humans do not have a single type of hemoglobin; they have several. As mentioned in Chapter 9, every hemoglobin molecule consists of four polypeptide chains. In the molecules that constitute the bulk of hemoglobin in the adult, two of these chains have an amino acid sequence called α (**alpha**), and the other two have an amino acid sequence called β (**beta**). The majority of adult hemoglobin can therefore be represented by the polypeptide formula $\alpha_2\beta_2$. Of course, the α chain is the product of the α-globin gene (which is actually a duplication), and the β chain is the product of the β-globin gene. (The mutation that causes sickle cell anemia is in the β-globin gene.) Adults have also a third hemoglobin chain, the δ (**delta**) chain, which is β-like and encoded by the δ-globin gene. The δ-globin gene is not as strongly expressed as the β-globin gene, however; for every 100 α chains in the adult, there are 97 β chains but only 3 δ chains, so adult hemoglobin is 97 percent $\alpha_2\beta_2$ and 3 percent $\alpha_2\delta_2$.

Each stage of human development has its characteristic types of hemoglobin, presumably because each type of hemoglobin has its own oxygen-carrying capability, and the genes expressed at any one developmental stage are appropriate to the availability of oxygen at that stage. The earliest embryos produce two other hemoglobin chains (Figure 10.1)—The ζ (**zeta**) chain, which is α-like and the ε (**epsilon**) chain, which is β-like. There are actually two ζ-globin genes (a duplication like the α-globin situation), and in earliest embryonic development the predominant hemoglobin has the polypep-

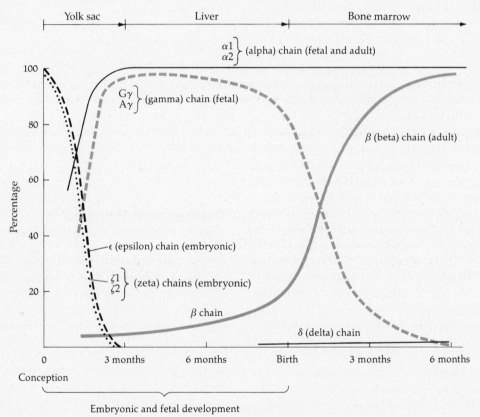

Figure 10.1 Graph showing regulation of the production of hemoglobin polypeptide chains and the site of synthesis during human development. Each type of polypeptide chain arises as the product of a different gene. The vertical axis denotes the proportion of hemoglobin molecules containing the various chains. Since each hemoglobin molecule contains four chains, the proportions at any age add up to more than 100 percent. Note the ε and ζ chains, found only in embryos, the fetal γ chain, and the small quantity of δ chain. By six months, a baby has essentially the adult complement of hemoglobin—97 percent $\alpha_2\beta_2$ and 3 percent $\alpha_2\delta_2$.

tide formula $\zeta_2\epsilon_2$. Later embryos and fetuses begin a hemoglobin transition by initiating expression of the α-globin genes and still another duplicate pair—the $^G\gamma$ (G-**gamma**) and $^A\gamma$ (A-**gamma**) pair—in liver cells. The γ chains are identical polypeptides except at amino acid position 136, where $^G\gamma$ carries glycine but $^A\gamma$ carries alanine. Since the γ chains are β-like, late embryos have three types of hemoglobin—namely, $\zeta_2\epsilon_2$, $\alpha_2\epsilon_2$, and $\zeta_2\gamma_2$. By the end of embryonic development, the embryonic globins ζ and ϵ have all but disappeared, and the fetal hemoglobin consists largely of $\alpha_2\gamma_2$ (see Figure 10.1).

At about the fifth month of development, there begins a gradual decline in the quantity of the γ chain, and expression of the β chain begins as the focus of hemoglobin synthesis switches from the liver to the bone marrow. The decrease in γ chains is almost exactly compensated by the increase in β chains. The δ chain is not produced until just before birth, and it is present in only small amounts. At birth most of the hemoglobin is $\alpha_2\gamma_2$, with the rest largely $\alpha_2\beta_2$ except for a trace of $\alpha_2\delta_2$. The exchange of β for γ then occurs more rapidly; at about two months, the proportions of $\alpha_2\beta_2$ and $\alpha_2\gamma_2$ have been reversed from what they were at birth. The production of β instead of γ continues, and slightly more δ is produced, until at six months the child has essentially its adult complement of hemoglobin. (Incidentally, a mutation is known that affects the control of the expression of γ. In an individual who carries this mutation, γ continues to be produced in adult life. The condition is very mild, and affected individuals are, for all practical purposes, completely healthy.)

The occurrence of embryonic and fetal hemoglobin distinct from adult hemoglobin clears up a minor mystery about sickle cell anemia. The sickle cell mutation causes a defective β chain, called β^S, to be produced. Individuals who are homozygous for this muta-

tion have adult hemoglobins $\alpha_2\beta^S_2$ and $\alpha_2\delta_2$; these individuals are severely anemic, and the $\alpha_2\beta^S_2$ tends to cause sickling of the red blood cells when the amount of oxygen is low. But the amount of oxygen at the placenta *is* rather low, so one wonders why homozygous embryos do not asphyxiate in utero. The answer is that the embryo and fetus do not rely on the β chain for their hemoglobin; β chains are not produced in large amounts until *after* birth. In embryos homozygous for the sickle cell gene, all the β^S chains are defective, but the ϵ and γ chains are completely normal because they are coded by different genes.

Regulation by DNA Modification

The regulation of expression of the human globin genes in Figure 10.1 is a preeminent example of regulation because it involves genes being turned on (α, β) or off (ζ, ϵ, γ) at different times in development and in different tissues (fetal liver, adult bone marrow). How this regulation occurs is not understood, but recent evidence suggests that it may involve the **modification** of certain strategically located cytosines in the DNA (control point IA in Table 10.1). In mammalian DNA, 2 to 7 percent of the cytosine bases in DNA are modified by an enzyme that creates **5-methyl cytosine** by the addition of a methyl group (-CH$_3$) to the base. (5-methyl cytosine has a methyl group at the same relative position as does thymine—see Figure 8.2.) Most of the 5^{m}C (i.e., 5-methyl cytosine) bases are found in the dinucleotide sequence 5'-CG-3', but lower levels of methylation of cytosines in other sequences can also occur. Cytosine methylation of DNA can be studied by means of certain pairs of restriction enzymes that have the same restriction site; one will cut the site only if a cytosine is unmodified and the other will cut the site whether modified or not. For example, the enzyme *Hpa*II will cut the sequence $\begin{smallmatrix} 5'\text{-CCGG-}3' \\ 3'\text{-GGCC-}5' \end{smallmatrix}$

only if the internal cytosine is unmethylated; the enzyme *Msp*I will cut at this sequence even in the presence of methylation. Thus, an unmethylated restriction site will produce the same restriction pattern with *Hpa*II as with *Msp*I, but the pattern will be different if the internal cytosine is modified.

Figure 10.2 shows the organization of globin genes in human DNA. There are two clusters of such genes: the cluster of β-like genes, which is located on chromosome 11, and the cluster of α-like genes on chromosome 16. The entire β-like cluster occupies nearly 55 kb (i.e., 55,000 base pairs), and the α-like cluster occupies about 25 kb. Note that there are two functional α genes (α1 and α2), two functional ζ genes (ζ1 and ζ2), and two functional γ genes (^{G}γ and ^{A}γ). In addition, there is an α pseudogene (ψα1) and two pseudo-β genes (ψβ1 and ψβ2). The direction of transcription is indicated, and it is of interest that the order of

genes in both clusters is the same as their order of expression during development.

The numbers beneath the β-like map in Figure 10.2 show the positions of 16 potentially methylated cytosines whose methylation can be detected with appropriate restriction enzymes. Interesting patterns emerge when the extent of base modification is compared with the expression of the genes. In cell types such as sperm or fetal brain that do not express hemoglobin, virtually all the cytosines in Figure 10.2 are modified, with the sole exception of site 7, which is methylated in sperm and brain but not in some other nonexpressing cells. Relative to fetal brain or adult liver cells, fetal liver cells have less methylation, particularly at sites 2, 4, 6, and 9. This undermethylation correlates well with expression of the γ chains in fetal liver but not in fetal brain or · adult liver. Similarly, hemoglobin-producing bone marrow cells from adults are

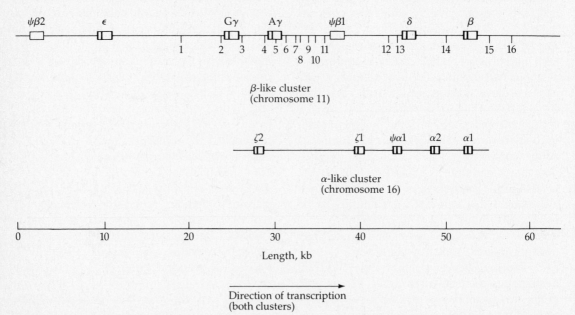

Figure 10.2 Organization of human β-globin-like gene cluster (chromosome 11) and α-globin-like gene cluster (chromosome 16). Note the two β pseudogenes (ψβ1 and ψβ2) and the α pseudogene (ψα1). The numbers represent possible sites of cytosine modification.

undermethylated relative to cells from bone marrow that do not express hemoglobin, particularly at sites 12 and 15, which are near the δ and β genes. Thus, there is some connection between relative lack of methylation and gene expression, but methylation cannot be the whole story. The complication comes from the placenta and from certain tissue culture cells that have uniformly low levels of methylation at all sites yet fail to express hemoglobin. At present, it appears that undermethylation is just one of several prerequisites for gene expression.

Gene Amplification or Diminution

Certain genes are controlled by increases or decreases in their number (process IB in Table 10.1). An example of **gene amplification** occurs in oocytes of the African clawed toad, *Xenopus laevis*; it involves amplification of the DNA that codes for ribosomal RNA (this class of DNA is symbolized as rDNA). The first suggestion of gene amplification was the finding that *Xenopus* oocytes contain hundreds of nucleoli that are not associated with the chromosomes. Recall that nucleoli are the sites of active rRNA synthesis, and they are usually closely associated with the chromosomal location of the rDNA; each such chromosomal site is said to carry a **nucleolus organizer**. The rDNA genes of *Xenopus* are clustered, and there is only one such cluster per haploid set of chromosomes. Thus, somatic cells have two nucleoli, yet oocytes can have up to 1000. Measurements of the amount of rDNA in somatic cells and oocytes revealed at least a 1500-fold increase in rDNA in the oocytes. The principal technique for such measurements is **nucleic acid hybridization**, which relies on the fact that two single-stranded nucleic acids (either DNA or RNA) will form hydrogen bonds and form a duplex molecule only if they are complementary or nearly complementary in

base sequence. In the case of rDNA, purified DNA from somatic cells or oocytes is separated into single strands by heating and then mixed with radioactive rRNA; the amount of rRNA that hybridizes (or **anneals**) with the DNA is then a measure of the amount of rDNA in the two types of cells.

Actually, there are three types of rRNA in eukaryotic cells. These are called 5S RNA, 18S RNA, and 28S RNA. (The S refers to the rate of sedimentation of the molecule in a centrifuge; larger S values are associated with larger molecules.) In *Xenopus,* as in other eukaryotes, only the 18S and 28S genes are clustered, and there are about 450 copies of each gene in such a cluster; the 5S genes are located elsewhere, but each haploid complement contains about 20,000 5S genes. In oocytes, only the 18S and 28S genes are amplified, presumably because there are enough 5S genes without amplification. The reason for rDNA amplification in oocytes seems to be the need for oocytes to produce enormous numbers of ribosomes to support protein synthesis early in development. The amplified rDNA functions only in oocytes, and the extra nucleoli do not reappear after meiosis. Lest it be thought that gene amplification is unique to *Xenopus* or to rDNA, it should be mentioned that certain genes that code for eggshell proteins in *Drosophila* are amplified more than 10-fold during oogenesis.

The opposite side of the coin of gene amplification is **gene diminution**, in which certain genes are reduced in number or entirely eliminated during development. The best-known example of chromosome elimination occurs in the horse intestinal roundworm, the nematode *Parascaris equorum*, which was studied by cytologists prior to 1900. This organism is a diploid, and its haploid chromosome number is $n = 1$; that is, zygotes have two large chromosomes [Figure 10.3(*a*)]. The terminal portions are heterochromatic and

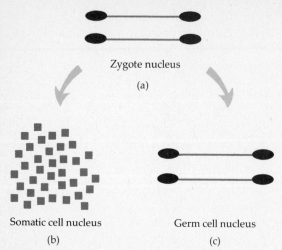

Zygote nucleus

(a)

Somatic cell nucleus

(b)

Germ cell nucleus

(c)

Figure 10.3 Chromosome diminution in the nematode *Parascaris equorum*. The heterochromatic chromosome termini in the zygote *(a)* are eliminated in somatic cells as the euchromatic part of the holocentric chromosomes becomes fragmented *(b)*, but the zygote chromosomes are retained intact in the germ line *(c)*.

shaped like a club [black portions in Figure 10.3(*a*)]. However, in cells of the embryo that are destined to become somatic cells, these terminal portions break apart from the chromosomes and are lost during cell division because they lack a centromere; in addition, the euchromatic part of the chromosomes [see Figure 10.3(*a*), shaded] undergoes fragmentation into dozens of small parts, each of which is perpetuated through cell division because it behaves as if it had a centromere of its own. [A chromosome like the shaded part of the ones in Figure 10.3(*a*) whose fragments behave as if each carried a centromere is said to have a **diffuse centromere** or to be **holocentric**.] During development, the somatic complement of *Parascaris* appears as in Figure 10.3(*b*). On the other hand, in cells destined to form the **germ line** (i.e., the lineage of cells that ultimately form gametes), the large chromosomes remain unbroken and intact [see Figure 10.3(*c*)].

A more recent example of DNA diminution during development involves the ciliated protozoan *Oxytricha*. This organism consists of a single cell, yet it has two types of nuclei. One is called the **macronucleus**, and it corresponds to the nuclei in somatic cells of higher eukaryotes; the other is called a **micronucleus**, and it is involved in meiosis and fertilization and so corresponds to the nuclei of germ cells in higher eukaryotes. Figure 10.4 gives information on the DNA in the macronucleus and micronucleus, but some background is necessary to discuss the data. The experiments in Figure 10.4 are known as **reassociation** experiments. They are carried out by heating purified DNA fragments so that the duplex molecules separate into their individual strands; the mixture is then held at a temperature allowing reassociation of the complementary strands, and at various times the proportion of double-stranded DNA is measured by determining how much ultraviolet light is absorbed. (Duplex DNA is a good absorber of ultraviolet light.) A graph is then made as in Figure 10.4, with the vertical axis showing the proportion of DNA remaining single stranded at any time, and the horizontal axis representing the logarithm of the initial amount of DNA (C_0) times the time—this latter quantity called C_0t (pronounced "**cot**"). If all the DNA sequences in the original mixture are represented the same number of times—that is, if each sequence is unique in the sense of Chapter 8—then the difference between the C_0t value corresponding to 90 percent single-stranded DNA and the value corresponding to 10 percent single-stranded DNA should be about two log units (i.e., two intervals along the horizontal axis in Figure 10.4). This expectation holds good for *Escherichia coli*, which has almost no repetitive DNA sequences, but it also holds for macronuclear DNA from *Oxytricha*. However, micronuclear DNA requires much longer to

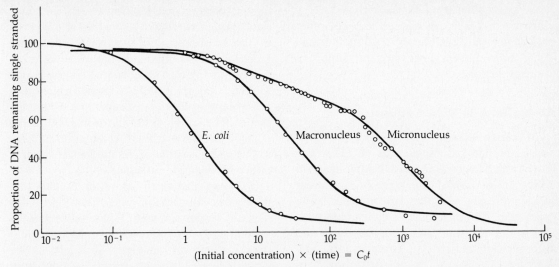

Figure 10.4 DNA diminution in the ciliated protozoan *Oxytricha*. In this experiment, DNA molecules are first separated into their individual strands and then allowed to hybridize (or renature) with one another. The shape of the curve for macronuclear DNA parallels that for *E. coli* DNA, thus indicating that there are few repetitive DNA sequences. The renaturation curve for micronuclear DNA is elongated, indicating that micronuclear DNA has and retains repetitive sequences.

reassociate from the 90 percent to the 10 percent level—roughly three units on the horizontal axis. This result means that DNA sequences in micronuclear DNA are not equally abundant, which implies that micronuclear DNA has repetitive sequences of the sort discussed in Chapter 8. Indeed, the reassociation curve for the micronuclear DNA in Figure 10.4 corresponds to about 5 percent highly repetitive DNA, 25 percent moderately repetitive DNA, and 70 percent unique-sequence DNA. However, as noted earlier, the macronucleus does not contain these repetitive sequences.

Curves like those in Figure 10.4 also convey the **complexity** of the DNA, which refers to the total amount of DNA that carries distinct base sequences, discounting repetitions of any distinct sequence. The complexity is related to the C_0t value corresponding to 50 percent reassociation. This value for *E. coli* is 1.46, for macronuclear DNA it is 18.9, and for

micronuclear DNA it is between 157 and 1096. To put the matter another way, macronuclear DNA has a complexity some 12.9 (i.e., $\frac{18.9}{1.46}$) times that of *E. coli*, and the complexity of micronuclear DNA relative to *E. coli* is between 108 ($\frac{157}{1.46}$) and 751 ($\frac{1096}{1.46}$). Since the DNA of *E. coli* contains 4.2 Mb (Mb = **megabase**, which equals 1 million base pairs), the macronuclear complexity is about 54 Mb (4.2 × 12.9), and that of the micronucleus is between 450 Mb (4.2 × 108) and 3150 Mb (4.2 × 751). Thus, not only has the macronucleus lost repetitive DNA, but it has lost a great deal of sequence information as well.

Such extreme examples of DNA diminution as *Parascaris* and *Oxytricha* are known only in nematodes, protozoa, crustaceans, and some insects. The process has not been discovered in other eukaryotes, particularly mammals, which leads to the belief that DNA diminution is not significant in other eukaryotes, except perhaps on a small scale. On the other hand, DNA can change in ways other than large-scale amplification or diminution. These more subtle changes are the subject of the next section.

Alterations in DNA Sequence

Transposition The genetic information in certain base sequences of DNA can move from one position on a DNA molecule to a different position on the same or a different DNA molecule and thereby influence gene expression. This movement is called **transposition**, and a dramatic example is the genes that determine mating type in baker's yeast, *Saccharomyces cerevisiae*. In this organism, haploid cells can undergo fusion to produce a diploid, which then undergoes meiosis to produce haploid spores. The ability of cells to fuse is determined by a locus called *MAT* (for mating type). Two mating types can occur—α or *a*—and only dissimilar cells can fuse (i.e., *a* with α). The genetic information for the *a* mating type resides in a DNA sequence at the *MAT* locus; thus, cells of mating type *a* are of genotype *MATa*. Similarly, the α genetic information resides in a different DNA sequence, and cells of mating type α are genotypically *MAT*α.

In some types of yeast, mating types are stable; they remain the same generation after generation except for rare mutations. *Saccharomyces cerevisiae* is different. Its mating type can switch back and forth between *a* and α, sometimes as rapidly as every cell division. For many years this ability to switch was a mystery, but now it has become clear that it involves transposition.

The underlying genetic situation for most laboratory strains of mating type α is illustrated at the top of Figure 10.5. The figure shows part of the third chromosome and the location of genes *his4*, *leu2*, and *thr4*, which are in-

volved in amino acid synthesis. At the center is the *MAT* locus, and in a strain of mating type α, this is occupied by the α DNA sequence (stippled). Two other loci are also indicated: *HML* and *HMR*. These, too, carry the *a* or α sequence information, but the information in these **casettes** is not expressed; it is merely stored. The strain shown has α in *HML* and *a* (shaded) in *HLR*—the usual situation in laboratory strains, but each casette locus can carry either mating-type sequence.

When the yeast spore in Figure 10.5 undergoes cell division, one of two things can happen. Either the mating-type sequences remain in place and the resulting products exhibit the α mating type (see Figure 10.5, left), or the *a*-type genetic information can transpose from where it is and displace the alternative sequence at *MAT* (see Figure 10.5, right). The resulting products will thus carry the *a*-type sequence at the *MAT* locus and express the *a* mating type. In a later generation, of course, the α genetic information can transpose to *MAT* and switch the mating type back again to α.

This transposition phenomenon has been proven directly at the DNA level using the fact that the α sequence is about 200 base pairs larger than the *a* sequence. Since sequences homologous to *HML*, *MAT*, and *HMR* have been separately cloned, these clones can be used as specific chemical **probes** (i.e., devices) to identify the type of information (*a* or α) resident at each locus. The type of experiment involved is called a **Southern blot**. In a Southern blot experiment, restriction fragments of cellular DNA are subjected to electrophoresis to separate them by size. Then the gel is covered with an equal-sized sheet of a special type of filter paper, and the DNA fragments are transferred (blotted) onto the paper, where they stick in place. If the DNA were visible on the paper, the pattern of bands would be a perfect mirror image of those on the original

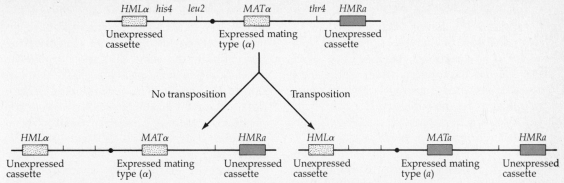

Figure 10.5 Mating type interconversion in baker's yeast, *Saccharomyces cerevisiae*. Shown above is a small portion of chromosome 3 including the expressed mating-type locus (*MAT*) and unexpressed mating-type information in two cassettes (*HML* and *HMR*); in this case, the mating type is α. In the event that the *a* information in *HMR* does not displace the α information in *MAT*, the mating type remains α (lower left). However, if the *a* information in *HMR* does displace the α information in *MAT*, the mating type converts to *a* (lower right).

gel. The DNA bands are actually invisible, but the probe can be used to bring into prominence only those bands that contain a particular DNA sequence. The trick is to treat the filter so as to denature the DNA strands and chemically link them onto the surface. Then the filter is bathed in a solution that contains radioactive single-stranded DNA representing the cloned sequence. Finally, the filter is carefully washed to remove all the radioactive DNA except the strands that cling to the filter because they have hybridized with their nonradioactive counterparts. The washed filter is autoradiographed, and the position of the radioactivity reveals the position to which the DNA fragments complementary to the probe had migrated in the original gel.

Analysis of yeast DNA by Southern blots using clones of *HML* and *HMR* will reveal which sequence (*a* or α) is present at each locus, because α is about 200 base pairs longer than *a*, and therefore restriction fragments that carry α will migrate more slowly in the gel than those that carry *a*. A strain such as the one in Figure 10.5 would be revealed to carry *HMLα* and *HMRa*, proving not only the existence of the *HML* and *HMR* loci but also the presence

of both α and *a* in unexpressed form. The critical experiment involves comparing the *MAT* locus in the unswitched and switched cell types at the bottom of Figure 10.5. In this case, Southern blots using radioactive *MAT* DNA reveal that the *MAT*-bearing fragments in the α-mating-type cell are about 200 base pairs longer than the *MAT*-bearing fragments in the *a*-mating-type cell, which means that switching of mating type is accompanied by physical transposition of genetic information to the *MAT* locus.

Although there is no doubt that the regulation of mating-type genes in yeast involves unexpressed casettes and transposition of genetic information, a number of questions remain. It is not known, for example, what happens to the genetic information that formerly occupied the *MAT* locus when the transposition occurs. Nor is the mechanism of the transposition clear. It is known that the sequence that transposes also remains at its original place. For example, the switched cell in Figure 10.5 remains *HMRa*. However, the transposition could involve replication of the sequence in question followed by incorporation at the *MAT* locus by some sort of recombination. At the other extreme, the transposition of genetic information could involve some sort of one-directional gene conversion (one-directional because the *MAT* locus is always the converted one). Perhaps the largest unanswered question is how common casettes and transposition actually are in the regulation of gene expression in eukaryotes as a whole and for other genes in yeast.

An Invertible Switch A remarkable example of genetic regulation at the DNA level occurs in the bacterium *Salmonella typhimurium*, and it involves cells being able to switch back and forth between two different protein constituents in their flagella. One protein is called H1, its alternative is called H2, and cells expressing H1 or H2 are said to be in the H1 phase or the H2 phase, respectively. (These flagellar proteins should not be confused with eukaryotic histone proteins; the only similarity is the symbols used to designate them.) Alternation between these two phases is known as **phase variation**, which is regulated by means of a 970 base-pair (bp) sequence of DNA located adjacent to the *H2* gene (Figure 10.6). As shown in Figure 10.6(*a*), the 970-bp sequence carries a gene called *hin,* the function of which will be discussed in a moment, and also a promoter sequence that promotes transcription in the direction indicated by the arrow. In addition, the 970-bp sequence is flanked by a 14-bp inverted repeat; that is, the sequence indicated by the left-pointing arrow reads

5′-AAGGTTTTTGATAA-3′
3′-TTCCAAAAACTATT-5′

whereas the sequence indicated by the right-pointing arrow reads

5′-TTATCAAAAACCTT-3′
3′-AATAGTTTTTGGAA-5′

which are the same sequences but in reverse order.

When the 970-bp sequence is situated as shown in Figure 10.6(*a*), transcription begins immediately downstream of the promoter, and transcription and translation of two adjacent genes occur. The *H2* gene leads to the expression of H2 flagella. The gene denoted *rh1* is a different sort of protein; it binds with a region of the H1 gene and causes **repression** of transcription. Thus, when *rh1* is expressed, *H1* is repressed.

Because of the 14-bp inverted repeat, however, the region in question can form the structure shown in Figure 10.6(*b*). Recombination in the 14-bp stem of the loop (the function of the *hin* gene product is to catalyze recombination at this site) leads to the configuration in Figure 10.6(*c*). Here the promoter points in the opposite direction as before and will no longer promote transcription of the *H1–rh1* region. [The *hin* and promoter sequences are written upside down in Figure 10.6(*c*) because the recombination event in Figure 10.6(*b*) will have placed these sequences on the other strand of the DNA duplex.] When the promoter is in the configuration of Figure 10.6(*c*), *H2* and *rh1* remain unexpressed, and, because of the absence of the *rh1* gene product, the *H1* gene will be expressed and will lead to type H1 flagella. At a later time, of course, recombination within the inverted repeat can again switch flagellar type back to H2. The sort of 970-bp element in Figure 10.6 represents one type of **controlling element**—in this case a **recombinational switch**. Eukaryotes have controlling elements, too, most of which have repeated sequences at their ends (see Chapter 11). On the other hand, it is not yet known whether eukaryotes regulate any genes by the sort of recombinational switch discussed for *Salmonella*.

DNA Splicing Chapter 9 emphasized the importance of **RNA splicing** for the removal of intervening sequences during RNA processing in eukaryotes. In at least one case,

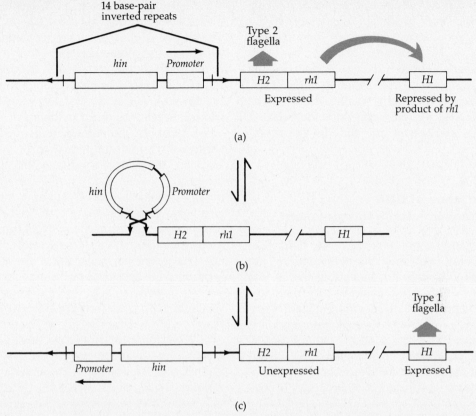

Figure 10.6 Recombinational switch controlling flagellar type in *Salmonella typhimurium*. The switch carries a gene (*hin*) involved in site-specific recombination and a promoter, and it is flanked by a 14 base-pair inverted repeat. In the orientation shown in *(a)*, the switch promotes transcription of *H2* (leading to type 2 flagella) and *rh1* (leading to repression of the type 1 flagellar gene, *H1*). However, pairing and recombination involving the inverted repeat, catalyzed by the product of *hin*, can occur as shown in *(b)*. The recombination event reverses the orientation of the switch, *H2* and *rh1* remain unexpressed, and expression of *H1* leads to the production of type 1 flagella.

eukaryotes are able to splice together distinct segments of *DNA* during development. The case in question involves certain proteins called **antibodies** that are important in the **immune system**—the processes by which the body recognizes and removes such invading foreign substances as viruses or bacteria. Details of the **DNA splicing** that occurs in the creation of antibody genes will be discussed in Chapter 13. For present purposes, it need only be remarked that the polypeptide constituents of antibodies consist of two parts: a **constant** region, which can have any one of a small number of possible amino acid sequences, and a **variable region**, which can have any of a large number of possible amino acid sequences. The enormous number of different antibody polypeptides that can exist is due in part to the manner in which antibody genes are created during development. In the differentiation of cells capable of producing antibody, any of a small number of DNA sequences that code for the constant part of the final polypeptide can become spliced onto any of a large number of different DNA sequences that code for the variable part of the final polypeptide. By this splicing process, which is called **combinatorial joining**, a large number of fused constant-region/variable-region combinations become possible, as discussed further in Chapter 13. However, combinatorial joining occurs only in somatic cells; in the germ line, the relevant DNA sequences do not undergo rearrangement.

Importance of DNA Rearrangements in Development

The overall importance of DNA rearrangements in eukaryotic development is not known. One difficulty in finding out is that some DNA changes, including base modification, amplification, transposition, and inversion, are reversible; no genetic information is lost from the cell as a result. Other types of changes in DNA are irreversible, however; genetic information is irretrievably lost. Irreversible changes include such processes as diminution and DNA splicing, and it is possible to experimentally address the issue of how important such irreversible changes may be.

The very concept of a somatic cell free from irreversible DNA changes entails that the cell's nucleus should be able to support embryonic development. That is to say, the somatic cell nucleus should be able to substitute for the zygote nucleus of a newly fertilized egg and guide the egg successfully through development. Indeed, in plants, such nuclear replacement is unnecessary; when properly cultured, adult plant cells from differentiated tissues of carrot, tobacco, and other plants are able to divide and give rise to completely normal seedlings and fertile adults. In plants, therefore, irreversible DNA changes do not seem to accompany normal development—at least in the adult tissues examined.

In animals the experimental procedures

are more difficult because nuclear replacement (called **nuclear transplantation**) is necessary. The animal egg is a marvelously complex cell with regional differences in its cytoplasm. After fertilization, when the chromosomes replicate and the cells divide, the cytoplasm of the daughter cells is not identical because each has a different region of cytoplasm from the original egg. The different cytoplasm evidently triggers the initial differentiation of the cells, and later in development, interactions between differentiated cells lead to still further differentiation. The embryo eventually comes to have lung cells, liver cells, brain cells, kidney cells, blood cells, and many more types of cells—each type differentiated, distinct from the others. The protein hormone insulin, for example, is produced in specialized cells in the pancreas but not elsewhere. It has been estimated that vertebrates have at least 200 different cellular types.

In vertebrates, the first nuclear transplantation success was achieved in the frog, *Rana pipiens*, and later in the African clawed toad, *Xenopus laevis*, mentioned earlier in connection with rDNA amplification. The experimental protocol is outlined in Figure 10.7. It begins with the destruction (by ultraviolet light) of the nucleus in an unfertilized egg; at the same time, the nucleus from a tadpole intestinal cell is removed by microsurgery. The intestinal cell nucleus is then carefully injected into the cytoplasm of the nucleus-deficient egg. In most cases the procedure does not work; the egg with the transplanted nucleus does not develop. The reason for the failures seems to be that the surgery is so difficult that there is often mechanical damage to the nucleus or the egg or both. However, in about 1 percent of cases, the intestinal nucleus does successfully take over development and gives rise to a normal, fertile animal. Differentiated tadpole intestinal nuclei therefore seem to have undergone no irreversible DNA changes.

Similar nuclear transplantation experi-

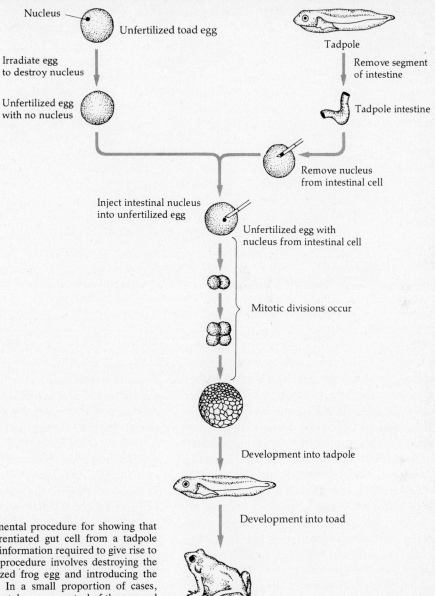

Figure 10.7 Experimental procedure for showing that the nucleus of a differentiated gut cell from a tadpole contains all the genetic information required to give rise to a complete toad. The procedure involves destroying the nucleus of an unfertilized frog egg and introducing the intestinal cell nucleus. In a small proportion of cases, the transplanted nucleus takes over control of the egg and guides the development of a mature toad.

ments are much more difficult in mammals because mammalian eggs are much smaller than those of amphibians. Success has nevertheless been achieved in the mouse by means of the procedure outlined in Figure 10.8. The

first step is the isolation of blastocysts from the donor parent (gray mouse—see Figure 5.12 for close-up views of the blastocyst). After blastocyst isolation and removal of the membranous envelope, cells from the inner cell mass—the

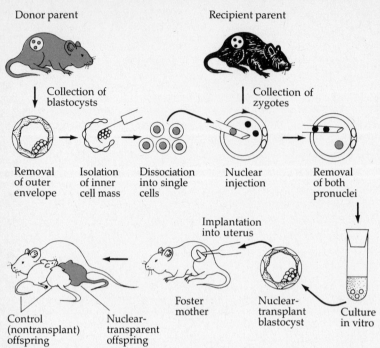

Donor parent

Recipient parent

Collection of blastocysts

Collection of zygotes

Removal of outer envelope

Isolation of inner cell mass

Dissociation into single cells

Nuclear injection

Removal of both pronuclei

Implantation into uterus

Control (nontransplant) offspring

Nuclear-transparent offspring

Foster mother

Nuclear-transplant blastocyst

Culture in vitro

Figure 10.8 Nuclear transplantation in the mouse. A nucleus from the inner cell mass of a gray strain of mice is introduced into the fertilized egg of a black strain, and the male and female pronuclei in the egg are removed. The nuclear-transplant egg is then cultured to the blastocyst stage and introduced into the uterus of a white foster mother, where, if successful, it gives rise to a gray mouse.

part of the blastocyst destined to give rise to the embryo itself—are isolated and dissociated into single cells. In the meantime, newly fertilized eggs are obtained from the recipient mouse (black) prior to the fusion of the **male pronucleus** (the haploid nucleus contributed by the sperm) and the **female pronucleus** (the haploid nucleus contributed by the egg itself). Then a nucleus from a single cell from the inner cell mass is carefully removed and transplanted into the zygote; with the same tiny needle, the male and female pronuclei originally in the zygote are both removed. The transplanted cells are cultured in vitro, where about 35 percent of them divide and produce new blastocysts. These blastocysts are then introduced into a foster mother (white mouse), where further development proceeds. In the first experiment of this kind, 3 normal baby mice were obtained from 16 implanted blastocysts; these 3 mice grew into fertile adults, and

their own coat color as well as the coat-color genes transmitted to their offspring proved that they had indeed arisen from the nucleus of the donor mouse. Thus, mammalian blastocyst cells from the inner cell mass are able to support embryonic development. It is interesting, though, that nuclei from blastocyst cells not associated with the inner cell mass cannot support development.

Early embryonic cells not only have nuclei capable of supporting development, but they can also be disassociated and nevertheless reaggregate properly. That is to say, cells from two or more genetically different preblastocyst embryos can be disassociated and mixed together and introduced into the uterus of a foster mother. This mixture of cells will reorganize and form a normal blastocyst that gives rise to a normal mouse, but the mouse is a mosaic because cells from genetically distinct embryos have contributed to its makeup. Such mixed mice are known as **chimeric** or **allophenic** mice. Figure 10.9 is a picture of a chimeric mouse derived from a mixture of embryonic cells from a black strain and a white strain. It is worth noting that chimeric mice with cells from two or three different embryos can readily be created, but three seems to be the limit; it is as if the inner cell mass derived from at most three earlier embryonic cells.

Two principal issues are raised by the success of mammalian nuclear transplants. The first is a social issue. A successful nuclear transplant represents a clone of the donor organism, and the technical procedures would not be much more difficult in humans than they are in mice. Widespread cloning in humans raises the specter of Aldous Huxley's *Brave New World*, where children with prescribed qualities befitting their future assignments to suitable castes in the social and economic hierarchy are made to order in uniform batches of thousands by genetic technicians and mass-

produced in test tubes in baby hatcheries. It provokes visions of a world in which evil men like Adolf Hitler could perpetuate and multiply their genotypes by the thousands. However, research in human cloning is not now being carried out, nor may it ever be. The research is directed toward other mammals, and the ability to clone farm animals would have tremendous benefits. The genetic segregation that occurs during meiosis is both a help and a hindrance to an animal breeder. In the early stages of breed improvement, segregation is helpful because it creates many different combinations of genes, and the best animals can then be chosen for breeding. Once a truly prize animal is encountered, however, segregation becomes a nuisance. The natural offspring of a prize animal will not, in general, be as good as the parent: first, because one parent contributes only half the genes to each offspring and, second, because the beneficial combination of genes in the parent is broken

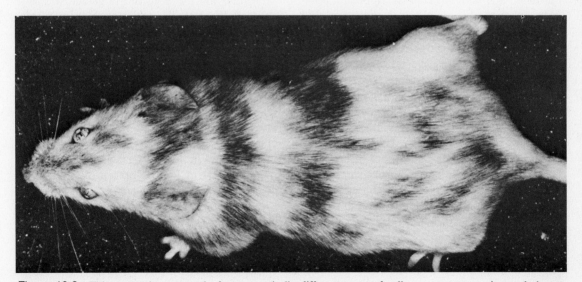

Figure 10.9 This mouse is composed of two genetically different types of cells: one type gave rise to dark-coat pigmentation and the other type to light-coat pigmentation. The mouse was produced artificially by combining cells from very young embryos of two different strains of mice (one with dark-coat pigmentation and the other with light) and implanting the mixed group of cells into the uterus of a female mouse where it then developed normally, giving rise to the individual shown here.

up and reshuffled by recombination and segregation. Both these problems could be avoided by cloning the prize animal, to the great benefit of breeders and the world meat supply. The animal technology might then be applied to humans, with the only obstacles being conventions of decency and law. Nevertheless, there will always be mavericks willing to defy convention or break the law, and some may be clever enough to succeed. There is no foolproof defense against such violations, only the hope that decent men and women involved in any large-scale human cloning project would quietly sabotage the procedures or blow the whistle.

The second issue raised by mammalian nuclear transplantation relates back to regulation. Since only embryonic cells have been studied, it could be that some adult somatic cells have undergone such irreversible DNA changes that their nuclei would be unable to support development. Experience with amphibians and plants suggests the opposite, however. In spite of their physical and functional differences, most cells in the body seem to have an identical set of genes. Differentiation thus results from differences in gene regulation in the various types of cells, not from irreversible changes in DNA. Despite their genetic similarity, a great variety of cell types can arise in the embryo by the expression of only a restricted group of genes in each type—in much the way that a great variety of different chords can be played on a piano by depressing different groups of keys. How the coordinated regulation of large groups of genes can occur is one of the great unsolved mysteries of genetics.

Regulation of Transcription

The preceding discussion of DNA changes during development brings us to item II of Table 10.1—regulation of transcription. Regulation at the level of transcription is thought to

be of major importance in eukaryotic gene expression, but here we will consider the prototypic case of transcriptional control—the lactose-utilizing genes of *E. coli*—not only because the situation is understood in detail but also because its general features are thought to occur in eukaryotes as well.

Lactose is a sugar composed of two parts: a molecule of glucose and a molecule of galactose covalently bound together. To be able to grow on lactose a bacterial cell must do two things. First it must transport lactose molecules into the cell, because cells are not naturally permeable to lactose; then it must cleave the lactose into its glucose and galactose constituents. The glucose can be used for energy immediately, because glucose is the common currency of metabolism. A number of enzymes forming a metabolic pathway are always present for chemically degrading glucose to extract from it some of the energy in its chemical bonds. The metabolism of galactose requires a little finagling. A molecule of galactose is first subjected to a precise series of chemical alterations and is then shunted into the glucose pathway for the subsequent degradation.

Each of these processes—entry of lactose into the cell and its initial cleavage into glucose and galactose—requires its own specific protein. (The enzyme that cleaves lactose is known as β-**galactosidase**, so called because the chemical bond between glucose and galactose in the lactose molecule is a β-galactoside bond; the transport of lactose across the cell membrane requires a β-galactoside **permease**.) These proteins are required only when lactose is present in the medium surrounding a bacterial cell. In the absence of lactose, these proteins have no function. Although lactose is a fairly common sugar, it is not invariably present in the environment where *E. coli* lives—namely, in the gut. Consequently, cells that waste energy in producing β-galactosidase and permease in the absence of lactose will be at a disadvantage in competition with cells that can produce these proteins only when they are needed. This imposes a natural selection in favor of cells that can regulate the genes for these enzymes. *E. coli* has evolved a way of regulating these genes; it does so by controlling their transcription. Actually, there are two levels of transcriptional control of the lactose genes—positive control and negative control—and it will be convenient to discuss these separately.

Positive Control **Positive control** at the level of transcription refers to a situation in which a protein or other molecule must bind to a promoter region of DNA for transcription to be possible. Positive control of the lactose genes in *E. coli* is illustrated in Figure 10.10. The organization of the genes on the bacterial DNA is shown in part (*a*). Four coding sequences of DNA are involved: *lacZ* codes for the β-galactosidase enzyme, *lacY* codes for the permease, and *lacA* codes for another enzyme (β-galactoside **transacetylase),** whose function in lactose metabolism is not known. These coding sequences are adjacent to one another, which illustrates one important distinction in genetic organization between prokaryotes and eukaryotes. Whereas genes involved in a single metabolic pathway in prokaryotes are usually tightly linked or even adjacent, such genes in eukaryotes are usually not clustered and often even reside on different chromosomes. The fourth coding region in Figure 10.10(*a*) is *lacI,* which codes for a protein called a **repressor.** The *lacI* gene is unregulated, so it always produces a small amount of repressor protein (about 10 molecules per cell), which is depicted in Figure 10.10 by the shaded circle; this repressor protein can bind to the DNA at a site called the **O** site (O for **operator**); the function of this binding will be discussed in a few paragraphs. The entire DNA sequence from the right end of *lacI* to the left end of *lacZ* was given in Figure 9.10(*a*); the operator region

encompasses bases 78 through 113 as they are numbered in that figure.

Upstream from the operator is a binding site for another regulatory protein called the **catabolite activator protein** (or **CAP**). The CAP-binding site includes bases 1 through 39 in Figure 9.10(*a*). The **promoter** region for the lactose genes comprises bases 1 through 87, so it completely overlaps the CAP site and partially overlaps the operator.

The catabolite activator protein is the regulatory protein that mediates positive control of the lactose enzymes because RNA polymerase cannot bind to the promoter unless the CAP is bound to the CAP site. Whether the CAP will bind to the CAP site is determined by the intracellular amount of another molecule called **cAMP** (for **cylic adenosine monophosphate**); the amount of cAMP is determined, in turn, by the amount of glucose in the bacteria's environment. The structure of the cAMP molecule will be illustrated later in this chapter. For now, the important point is the relationship between the cAMP level and glucose.

Glucose is a preferred source of carbon and energy for *E. coli*, and its availability is related in some as yet unknown manner to the intracellular level of cAMP. When glucose is available to cells, levels of cAMP decrease, and this decrease prevents the formation of a CAP-cAMP complex. When CAP is not bound to cAMP, it cannot bind with the CAP site, and transcription of the *lacZ, Y*, and *A* genes will not be possible [see Figure 10.10(*a*)]. However, when glucose is not available, the level of cAMP increases. The high cAMP enables binding to CAP, and the CAP-cAMP complex [shaded shield in Figure 10.10(*b*)] can then bind to the CAP site of a number of genes coding for enzymes that metabolize alternative sugars. Lactose is one of these alternatives, and the CAP-cAMP complex binds to the CAP site of the *lac* region and thereby permits transcription [see Figure 10.10(*b*)]. Although

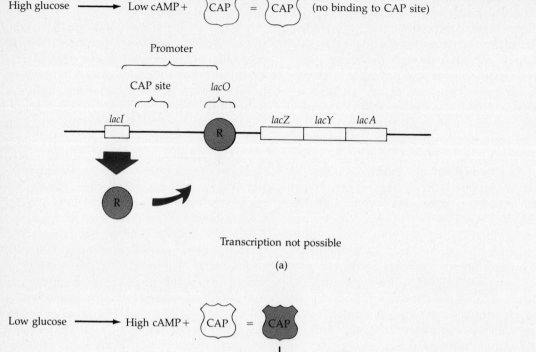

Transcription not possible

(a)

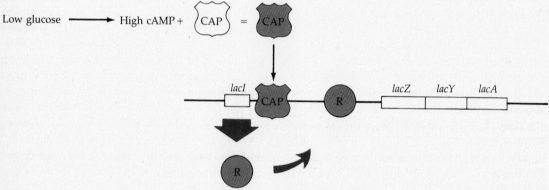

Transcription possible

(b)

Figure 10.10 Role of CAP (catabolite activator protein) in positive regulation of the lactose operon in *E. coli*. *(a)* In the presence of high glucose levels, the cell reduces the amount of cAMP, which is then unavailable for binding with CAP and no CAP complexes with the CAP site in the promoter; under these conditions, transcription of the lactose operon cannot occur. *(b)* In the presence of low glucose, enough cAMP is available to bind with CAP and complexing to the CAP site does occur; under these conditions, transcription *can* occur, but whether or not it does depends on the lactose repressor protein (R).

the binding of CAP-cAMP to the CAP site makes transcription possible, it does not make it inevitable. Whether transcription actually occurs depends on the availability of lactose, which is signaled by the state of the repressor protein, coded by *lacI*.

Negative Control   The repressor protein mediates negative transcriptional control of the lactose-related genes. **Negative control** of transcription refers to a situation in which a protein or other molecule must be *unbound* to the promoter region for transcription to occur. How the lactose repressor works is illustrated in Figure 10.11. Parts (*a*) and (*b*) both assume that glucose is unavailable, so that the CAP-cAMP complex can bind to the CAP site and make transcription possible. However, if lactose (small black squares) is not available, the repressor protein binds to the operator region and prevents transcription [see Figure 10.11(*a*)]. When lactose is available, on the other hand [see Figure 10.11(*b*)], the repressor binds with the lactose instead of with the operator; in this case, transcription will occur. Any molecule that binds with a repressor is called an **inducer**. The inducer of the lactose genes is not lactose itself, as indicated in the figure, but rather a closely related molecule called **allolactose**. Allolactose is produced from lactose by a sort of side reaction of β-galactosidase; this enzyme is present in small amounts in all cells because regulation is a little "leaky." As indicated in the figure, the coding regions of *lacZ, Y*, and *A* are transcribed in a single large messenger RNA known as a **polycistronic mRNA**. (Recall from Chapter 9 that a **cistron** is a DNA sequence that codes for a single polypeptide.) All three proteins—β-galactosidase, permease, and transacetylase—are translated from this same transcript. The group of adjacent cistrons (*lacZ, lacY*, and *lacA)* that are transcribed together constitutes a bacterial **operon**. (Such grouping of genes into operons is unique to prokaryotes; eukaryotes do not produce such polycistronic mRNA's.) Once the mRNA's produced in transcription are translated, the enzymes metabolize the available lactose. When the supply of lactose is exhausted, the repressor protein again becomes free of lactose and is able to bind once more to the operator and prevent transcription.

Positive and negative transcriptional control mediated by regulatory proteins is thought to be an important mechanism of regulation in eukaryotes, although the details are undoubtedly different from those involved in the lactose operon of *E. coli*. Such regulatory proteins would necessarily be able to bind with DNA, and two broad categories of such proteins are known. One category comprises the **histones** (discussed in Chapter 8), which are particularly rich in the amino acids lysine and arginine. The other category consists of **acidic proteins**, which are particularly rich in aspartic and glutamic acid. Both histones and acidic proteins are probably involved in gene regulation, but the regulation imposed by the histones is thought to be quite nonspecific; that is, histones probably have a general effect on DNA rather than a specific effect on particular genes. Histones are not particularly attractive candidates for specific gene regulation because of their lack of diversity. There are, for example, only five major classes of histones, and the relative amounts of these are the same in cells of different tissues and even cells of different organisms—a situation one would certainly not expect of regulatory molecules involved in the specialized activities of differentiated cells. Moreover, the amino acid sequence of a major histone protein, (H4), in cattle is identical to the sequence found in peas, except for two amino acids. So it is hard to see how proteins of such little diversity could account for the precise regulation of specific genes.

The acidic proteins are better candidates as specific regulators. These proteins are exceedingly diverse, for one thing, and different tissues produce different classes of acidic proteins. Metabolically active tissues—those actively producing proteins from many genes—also produce more of the acidic proteins than do relatively inactive tissues. In addition, the

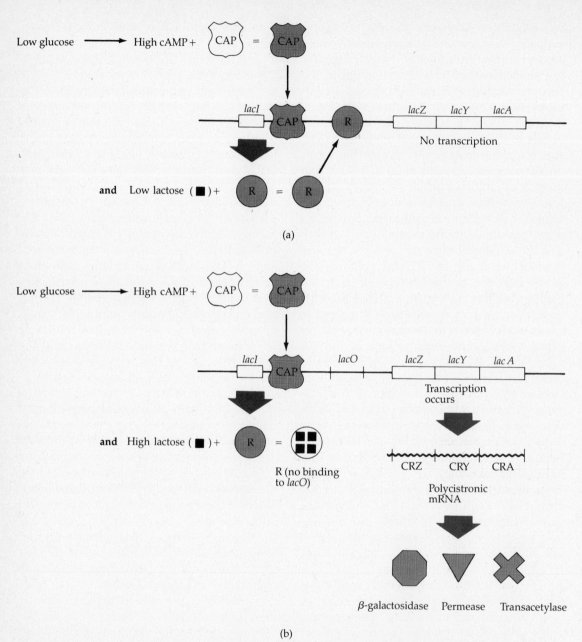

Figure 10.11   Role of lactose repressor protein (R) in negative regulation of the lactose operon in *E. coli*. Both *(a)* and *(b)* assume that CAP is bound with the CAP site, so transcription is possible. *(a)* In the presence of low lactose (black squares), not enough is available to bind with the repressor, so the repressor binds with the operator site (*lacO*) and prevents transcription. *(b)* In the presence of high lactose, repressor binds with the lactose (actually allolactose) and is thereby unable to bind with *lacO*; transcription and translation of the lactose-metabolizing genes then occur.

synthesis of particular classes of acidic proteins is associated with gene activity, and some of these proteins are known to bind to specific regions of DNA. All these things would be expected of molecules that can stimulate transcription. Although suspicion centers on acidic proteins as regulators of specific genes, much more remains to be learned. Certain hormones that stimulate transcription may exert their effects by interacting with these proteins—but what do they do? Acidic proteins (and histones, too) are known to be modified during the cell cycle by the addition of phosphate or other groups, and not all are modified at the same time or to the same extent. Perhaps hormones and other regulators act indirectly on transcription through effects on protein modification, but no one knows for sure.

An example of regulation at the level of transcription in a eukaryote involves the induction of synthesis of **ovalbumin**, a major egg white protein, by the female steroid hormone **estrogen** in the chicken. The essentials of the process are illustrated in Figure 10.12. Cells of the egg-laying duct of chickens contain a protein called **estrogen receptor protein**, indicated by the black crescent in the figure. When estrogen (gray circle) is released by certain glands during the reproductive cycle, the hormone forms a complex with the receptor pro-

tein and is brought into the cell [parts (*a*) and (*b*)]. The steroid-receptor protein complex is then transported into the nucleus [see Figure 10.12(*c*)], where, under the influence of certain acidic proteins associated with the DNA, the steroid-receptor complex binds to specific promoter sites in the DNA. This binding, in turn, stimulates transcription, and specific ovalbumin RNA transcripts appear in the nucleus; these transcripts are processed into ovalbumin mRNA and transported back to the cytoplasm [part (*d*)], where they are translated to produce the ovalbumin protein. Thus, the principal effect of estrogen is the stimulation of transcription. It is worthy of note at this point that the male steroid hormone testosterone seems to exert its effects on target cells in the same manner—via the stimulation of transcription.

RNA Polymerase and Helper Proteins

It is evident from the preceding discussions that many different proteins can interact with regions of DNA and either promote or prevent transcription. However, transcriptional control may also be mediated by **RNA polymerase** itself or by other proteins that interact with the polymerase. Indeed, eukaryotes have three principal types of RNA polymerase: **polymerase I**, **polymerase II**, and **polymerase III**. Poly

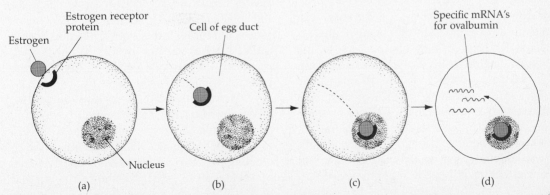

Figure 10.12 Hormonal activation of ovalbumin genes. Estrogen at the cell surface binds with estrogen receptor protein (*a*) and is brought into the cytoplasm (*b*) and then into the nucleus (*c*), where it promotes transcription of the ovalbumin gene (*d*).

merase I is associated with the nucleolus, and its principal function seems to be the transcription of ribosomal RNA; polymerase II is thought to be the enzyme that transcribes most genes, such as globin or ovalbumin; and polymerase III seems mainly involved in the transcription of small RNA's such as 5S ribosomal RNA and tRNA. In addition, polymerases require other factors for the proper initiation and termination of transcription. These "helper proteins" are not well characterized in eukaryotes, but in *E. coli* two such factors have been extensively studied. In *E. coli,* a particular protein called **sigma** (σ) binds to RNA polymerase and aids in recognizing the proper starting point for transcription. Once transcription has been initiated, σ is released from the polymerase and becomes free to bind to a different polymerase molecule to help it get started. *E. coli* has another helper protein called **rho** (ρ), which binds to RNA polymerase and is necessary in some instances for the polymerase to recognize the transcription-termination signal. (Some termination signals can be recognized by the polymerase without ρ, however.) It seems likely that there are analogous helper proteins for RNA polymerase in eukaryotes as well.

Regulation by RNA Processing

As discussed in Chapter 9, eukaryotes can regulate gene expression by means of the extensive RNA processing that occurs in the nucleus. This level of regulation is not available in prokaryotes. For many eukaryotic genes, an essential step in RNA processing is the removal of intervening sequences by splicing. Of course, removal of intervening sequences is essential to make the relevant coding regions adjacent in the final mRNA, but there is evidence that intervening sequences may have functions of their own. In one case involving a gene in the mitochondrial DNA of yeast, an intervening sequence is translated,

and the resulting polypeptide is a splicing enzyme that removes the intervening sequence from its own transcript and from other similar transcripts. Cases in which intervening sequences code for their own splicing enzymes are thought not to occur in nuclear DNA, however, because no process is known whereby an RNA transcript could be translated within the nucleus. On the other hand, intervening sequences are necessary for transcript stability, at least in some instances.

One case in which transcript stability requires an intervening sequence occurs in the monkey cell virus **simian virus 40 (SV40)**. This virus has a chromosome of circular duplex DNA, and its entire 5226-base sequence was one of the first long sequences to be established. The virus carries very few coding sequences, but some are for **early proteins** (produced early after infection) and others are for **late proteins** (produced later after infection). Among the late proteins is VP1, a major constituent of the viral **capsid** (coat). Transcription of SV40 genes occurs in the infected cell's nucleus, and the primary transcript for VP1 comprises nucleotides 260 through 2590 (approximately); during RNA processing, the entire region comprising nucleotides 451 through 1382 is removed by splicing. The fully processed VP1 mRNA is illustrated at the left side of Figure 10.13, where it can be seen that the splicing process joins the leader sequence (bases 260 through 450) with the coding region for VP1 (bases 1423 through 2511); the deleted sequence 451 through 1382 is indicated by the inverted triangle.

Whether splicing is a requirement for VP1 expression has been studied in an artificially created SV40 that precisely lacks the 451 to 1382 base sequence. The procedure used in creating the virus is illustrated at the right side of Figure 10.13. Part (*a*) shows the normal SV40 DNA, and recognition sites for the restriction enzymes *Hpa*II and *Eco*RI near nucleotides 265 and 1700, respectively, are

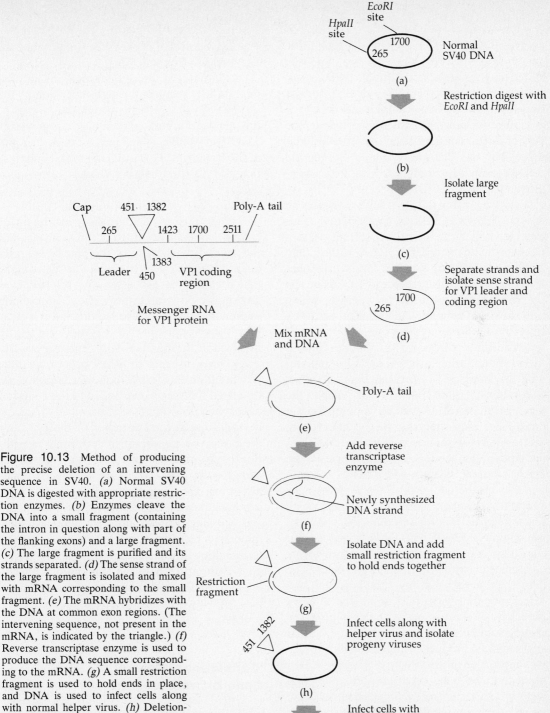

Figure 10.13 Method of producing the precise deletion of an intervening sequence in SV40. *(a)* Normal SV40 DNA is digested with appropriate restriction enzymes. *(b)* Enzymes cleave the DNA into a small fragment (containing the intron in question along with part of the flanking exons) and a large fragment. *(c)* The large fragment is purified and its strands separated. *(d)* The sense strand of the large fragment is isolated and mixed with mRNA corresponding to the small fragment. *(e)* The mRNA hybridizes with the DNA at common exon regions. (The intervening sequence, not present in the mRNA, is indicated by the triangle.) *(f)* Reverse transcriptase enzyme is used to produce the DNA sequence corresponding to the mRNA. *(g)* A small restriction fragment is used to hold ends in place, and DNA is used to infect cells along with normal helper virus. *(h)* Deletion-bearing progeny viruses are isolated and studied.

321

indicated. Digestion of SV40 DNA with these enzymes yields two fragments (*b*), and the larger one can be separated by gel electrophoresis (*c*) as described in Chapter 8. Then the individual strands of the duplex are disassociated and the sense strand for the leader and VP1 coding region is isolated (*d*). This strand is mixed with the VP1 messenger RNA, and the mRNA forms a heteroduplex with the complementary ends of the DNA (*e*).

The next step in the construction uses an enzyme known as **reverse transcriptase**. This enzyme, which is coded by certain viruses whose genetic material is RNA rather than DNA, uses RNA as a template and produces DNA; that is, reverse transcriptase acts in just the reverse way as RNA polymerase, which uses a DNA template and makes an RNA copy. In the SV40 construction, reverse transcriptase adds successive nucleotides to the DNA strand using mRNA as the template, with the direction of synthesis being counterclockwise in Figure 10.13(*f*). Of course, the newly synthesized DNA strand cannot contain nucleotides corresponding to the intervening 451 to 1382 base sequence, as the mRNA does not contain this sequence. After synthesis, the DNA strand is isolated, and its ends are held together by the addition of a complementary strand from a small restriction fragment, which forms heteroduplex regions that bridge the gap [see Figure 10.13(*g*)]. This DNA is then used to transform monkey cells that had previously been infected with a "helper" SV40 virus. The helper virus supplies any functions that the deletion-bearing virus may be unable to carry out, and, in the course of the infection, the single-stranded DNA bearing the deletion is replicated to produce a deletion-bearing duplex molecule; this molecule is then replicated many times and will be represented among the progeny viruses. Progeny viruses that carry the precise deletion [see Figure 10.13(*h*)] can then be studied individually or in combination with other types of helper viruses to determine what

effect the deletion of the intervening sequence has had on virus function.

One might think that deletion of the intervening sequence would not hinder the production of VP1 protein. On the face of it, the artificially created deletion simply saves SV40 the trouble of splicing the sequence out of the transcript. However, the actual finding is that the deletion-bearing viruses do not produce VP1. Indeed, messenger RNA corresponding to VP1 is not produced. Transcription of the region including the leader and VP1 coding sequence seems to occur normally, but the transcript that lacks the intervening sequence breaks down so rapidly in the nucleus that it cannot be processed into mRNA. It is not known why transcripts without intervening sequences break down so rapidly, but it is clear from this experiment that RNA splicing can be involved in the regulation of gene expression.

Other aspects of RNA processing may also regulate gene expression. In Chapter 9, we discussed the special **cap** added to the 5′ end of eukaryotic mRNA's (see Figure 9.22). Actually, there are several varieties of the cap, and these may be important in regulating translation of the mRNA. Similarly, the **poly-A tail** added to the 3′ end of eukaryotic mRNA's may also affect gene expression; different mRNA's can have poly-A tails of very different length, and the length of the tail may influence such important features of mRNA as its resistance to breakdown in the cytoplasm. Finally, transport of the mRNA to the cytoplasm is almost certainly an important level of gene regulation, but very little is known about the transport process.

Translational Control of Gene Expression

Transport of mRNA to the cytoplasm does not guarantee translation, as several levels of regulation are still possible. For example, unfertilized sea urchin eggs contain mRNA's that are untranslated. These untranslated mRNA's are known as **masked mRNA's**, and they can remain translationally inactive for months. Yet, minutes after fertilization, these masked mRNA's become unmasked and begin to be translated. Another example of untranslated mRNA occurs in *Drosophila melanogaster*. Recall that the DNA for eggshell proteins in this organism is amplified during oogenesis. Concomitant with DNA amplification, mRNA's corresponding to the eggshell proteins appear in the cytoplasm but remain untranslated. These untranslated mRNA's are then broken down. Only later in oogenesis does the translatable form of eggshell protein mRNA appear.

The mechanisms of translational regulation in eukaryotes are not well understood, but in prokaryotes they frequently involve the three-dimensional conformation of the mRNA. A well-known example of translational regulation in prokaryotes involves the phenomenon of **attenuation**, which occurs in operons that contain the genes for amino acid synthesis. Regulation of the operon for tryptophan (trp) synthesis in *E. coli* is illustrated in Figure 10.14; there are actually two levels of control. The genes *trpE*, *trpD*, *trpC*, *trpB*, and *trpA* all code for enzymes involved in tryptophan synthesis, and they are organized on the bacterial chromosome as illustrated in Figure 10.14(a). The gene *trpL* codes for a small polypeptide, the function of which will be discussed in a moment, but the sites labeled **transcription pause site** and **attenuator** should be noted. The first level of control of the tryptophan operon involves transcriptional control mediated by the trp repressor protein, indicated in Figure 10.14(a) by the open circle. When tryptophan levels in the cell are high, the repressor protein binds with tryptophan, producing an active form of the repressor (shaded circle), which then binds to the operator and renders transcription impossible. By

means of the repressor-tryptophan interaction, high levels of tryptophan prevent transcription of the tryptophan-producing genes.

When the intracellular levels of tryptophan are low [see Figure 10.14(b)], the repressor protein does not bind to the operator, and transcription of the operon begins. However, slightly downstream of the *trpL* sequence is the *transcription pause site*, where the RNA polymerase pauses. Whether transcription will continue much beyond the pause site depends on whether or not the *trpL* polypeptide is translated, and this dependence of continued transcription on translation constitutes the attenuation process. The *trpL* polypeptide is extremely short; it consists of just 14 amino acids, but numbers 10 and 11 in the chain are both tryptophan. The frequency of $\frac{2}{14}$ tryptophans is very high because typical *E. coli* proteins contain, on the average, only about 1 tryptophan for every 100 amino acids, and the 2 tryptophans in the *trpL* polypeptide occur near the translation-termination condon. As illustrated in Figure 10.14(b1), if the level of charged trp-tRNA in the cell is so low that the trp codons cannot be translated, then the ribosome "stalls" at those codons. (A **charged tRNA** is a tRNA that is linked to its corresponding amino acid—in this case tryptophan.) When the ribosome stalls due to lack of charged trp-tRNA, transcription continues beyond the transcription pause site to include *trpEDCBA*; thus, the enzymes required to synthesize tryptophan will be produced [see Figure 10.14(b1)]. On the other hand, when the level of charged trp-tRNA is high enough that *trpL* can be translated, then the RNA conformation charges and the RNA polymerase continues transcription only to the *attenuator site* [see Figure 10.14(b2)]; thus, high levels of charged trp-tRNA prevent the production of enzymes that function to produce more tryptophan.

Attenuation is an example of translation regulating transcription through changes in the

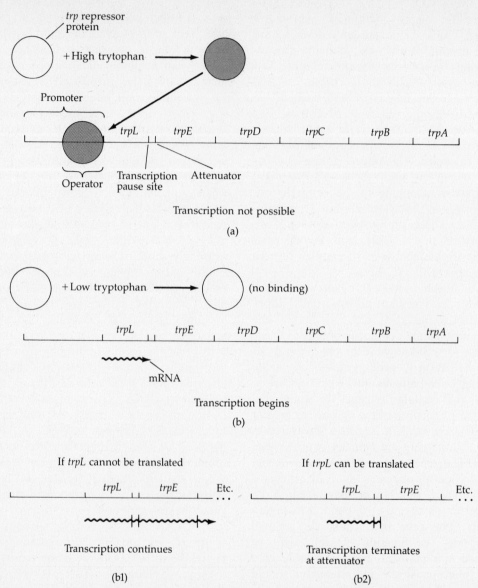

Figure 10.14   Attenuation of transcription in the tryptophan operon in *E. coli.* *(a)* The *trp* repressor protein (circle), in the presence of high tryptophan, becomes complexed with the tryptophan and binds the operator, making transcription impossible; thus, the tryptophan-synthesizing enzymes will not be produced. *(b)* With low levels of tryptophan, the *trp* repressor remains unbound with the operator and transcription of the *trpL* region occurs, but it stops at the transcription pause site. *(b1)* If the level of tryptophan is so low that *trpL* cannot be translated, transcription continues and the tryptophan-synthesizing enzymes are produced. *(b2)* If the level of tryptophan is sufficiently high that *trpL* can be translated, transcription continues only to the attenuator and then terminates.

conformation of the RNA. An example of translation of a leader polypeptide regulating further translation of the same mRNA is illustrated in Figure 10.15. Part (*a*) shows a region of an mRNA produced by a plasmid found in cells of *Staphylococcus aureus*. This plasmid confers resistance to the antibiotic **erythromycin**, but synthesis of the enzyme responsible for the resistance is regulated by erythromycin itself at the level of translation. Erythromycin is an antibiotic because it binds with ribosomes and interferes with polypeptide synthesis. The plasmid-encoded mRNA in Figure 10.15 codes for a methylation enzyme that methylates certain bases in the ribosomal RNA and thereby renders the ribosomes resistant to erythromy-

cin. SD1 in Figure 10.15 represents a Shine-Dalgarno sequence; recall from Chapter 9 that such a sequence is a ribosome-binding sequence. Downstream from SD1 is the coding sequence for a leader polypeptide of 19 amino acids, and the initiation (AUG) and termination (UAA) codons are indicated. Farther downstream is another Shine-Dalgarno sequence (SD2), and the AUG codon is the initiation codon for the 243-amino-acid methylation enzyme. The heavy black arrows labeled 1 through 4 represent inverted repeats in the mRNA: Region 1 is complementary to region 2, region 2 is complementary to region 3, and region 3 is complementary to region 4.

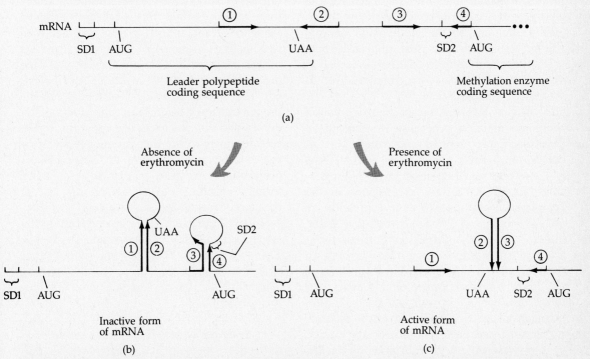

Figure 10.15   Regulation of erythromycin-methylating enzyme by mRNA conformation in *Staphylococcus aureus*. (*a*) mRNA sequence corresponding to part of a plasmid coding for the enzyme. SD1 and SD2 represent Shine-Dalgarno sequences, and regions of homology and their orientations are denoted by the numbered arrows. (*b*) Configuration in the absence of erythromycin, 1 paired with 2, and 3 with 4; SD2 is hidden and the methylation enzyme cannot be translated. (*c*) Configuration in the presence of erythromycin, 2 paired with 3; in this configuration the methylation enzyme is translated.

In the absence of erythromycin [see Figure 10.15(*b*)], translation of the leader polypeptide occurs normally, ribosomes do not accumulate in the leader coding sequence, and the mRNA assumes a form in which region 1 pairs with region 2 and region 3 pairs with region 4. Translation of the methylation enzyme does not occur in this configuration because the SD2 sequence is sequestered in the 3-4 loop. However, the presence of low levels of erythromycin [see Figure 10.15(*c*)] causes a buildup of ribosomes in the leader coding sequence owing to improper translation caused by the antibiotic; this ribosome buildup makes region 1 inaccessible to pairing with region 2, and the configuration in which region 2 pairs with region 3 occurs instead. With this configuration, the Shine-Dalgarno sequence for the methylation enzyme becomes available for ribosome binding, so the methylation enzyme will be translated. This sort of translational regulation could occur in eukaryotes as well as prokaryotes, but eukaryotic examples are not presently available.

A final level of translational regulation should be mentioned because it seems to be particularly important in eukaryotes. This involves the **stability** of cytoplasmic mRNA, which refers to the rate at which mRNA molecules are broken down by enzymes in the cytoplasm. Regulation of gene expression by mRNA stability is possible because a stable mRNA can be translated hundreds or thousands of times before being degraded, but an unstable mRNA can be translated only tens or hundreds of times before breaking down. An interesting example of translational regulation by means of mRNA stability involves the production of **casein** proteins in mammary glands. In the presence of the hormone **prolactin**, mammary glands synthesize a substantial amount of casein; when prolactin is withdrawn, casein synthesis is much reduced. This reduc-

tion is due in part to a decreased stability of the corresponding mRNA in the absence of the hormone.

Posttranslational Regulation

Polyproteins   Now we can come to section V of Table 10.1—regulation at the posttranslational level. One type of posttranslational regulation in eukaryotes involves **polyproteins**—several proteins produced by the posttranslational cleavage of a single, long polypeptide. An example of processing of a polyprotein is illustrated in Figure 10.16 for **insulin**, a protein produced in certain cells in the pancreas that is important in regulating the body's glucose metabolism. The first polypeptide produced is called **preproinsulin**; there are four regions within this molecule: a **signal sequence** (at the amino terminal end) and then regions B, C, and A. Certain key disulfide bridges, formed by covalent linkages between the sulfur atoms of two cysteines, are indicated. The signal sequence is essential for the proper transport of preproinsulin across membranes and into the secretory apparatus of the cell. Removal of the signal sequence by cleavage at the arrow in Figure 10.16(*a*) produces **proinsulin**, a partially processed but inactive form of insulin. Further processing of the molecule by cleavage at the arrows in Figure 10.16(*b*) removes the C region and leads to the active form of insulin illustrated in part (*c*). The expression of insulin and other polyproteins can therefore be regulated by controlling the rate of processing of the precursor polypeptide.

Enzyme Activation   Some enzymes in the cell are present in an inactive state, and *activation* of these enzymes requires other cellular events. An example illustrated in Figure 10.17 involves regulation of **glycogen phos-**

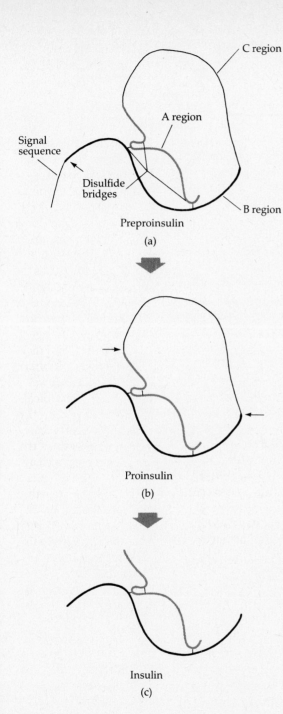

Preproinsulin

(a)

Proinsulin

(b)

Insulin

(c)

C region

A region

Signal
sequence

Disulfide
bridges

B region

phorylase kinase (GPK) by the hormone **epinephrine** in liver cells. **Glycogen** is a principal storage carbohydrate in liver cells, and GPK catalyzes the breakdown of glycogen into usable sugars. GPK is inactive in liver cells, but its activity is induced by a chain of events beginning with the binding of epinephrine to certain receptor sites on the cell surface [see Figure 10.17(*a*)]. This binding stimulates **adenylate kinase**, which produces cyclic adenosine monophosphate (cAMP) from adenosine triphosphate (ATP); the structures of ATP and cAMP are illustrated in Figure 10.18. The cAMP then activates another enzyme, a **protein kinase** [represented as a square in Figure 10.17(*b*)], which in turn activates GPK (circle). The GPK then catalyzes glycogen breakdown.

Inhibition   Some proteins are normally in an active state but can be rendered inactive (**inhibited**) by binding with other molecules in the cell. An example is the lactose repressor protein in Figure 10.11. In the absence of lactose (actually the related molecule allolactose), the repressor protein binds to the operator; in the presence of allolactose, the repressor protein becomes bound with the allolactose and in this bound state is unable to interact with the operator. The binding site for the operator in the repressor protein is different from the binding site for allolactose, however; the binding with allolactose causes a change in the conformation of the operator-binding site, and this change renders interaction with the operator impossible.

The interaction between the lactose re-

Figure 10.16 Posttranslational processing of insulin polyprotein. *(a)* Preproinsulin contains a signal sequence (important in secretion), a B region, a C region, and an A region. *(b)* The signal sequence is first cleaved off [at arrow in *(a)*], but the protein is still inactive. *(c)* The C region is then removed [at arrows in *(b)*], and the A and B regions form the active insulin molecule.

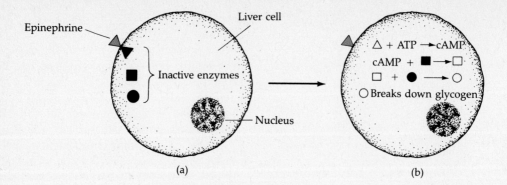

▲ , △ Inactive and active adenylate cyclase

■ , □ Inactive and active protein kinase

● , ○ Inactive and active glycogen phosphorylase kinase

Figure 10.17 Activation of liver glycogen phosphorylase kinase by epinephrine. *(a)* Epinephrine at cell membrane interacts with adenylate cyclase. *(b)* Activated adenylate cyclase (open triangle) produces cAMP, which activates a protein kinase (open square), which, in turn, activates glycogen phosphorylase kinase (open circle).

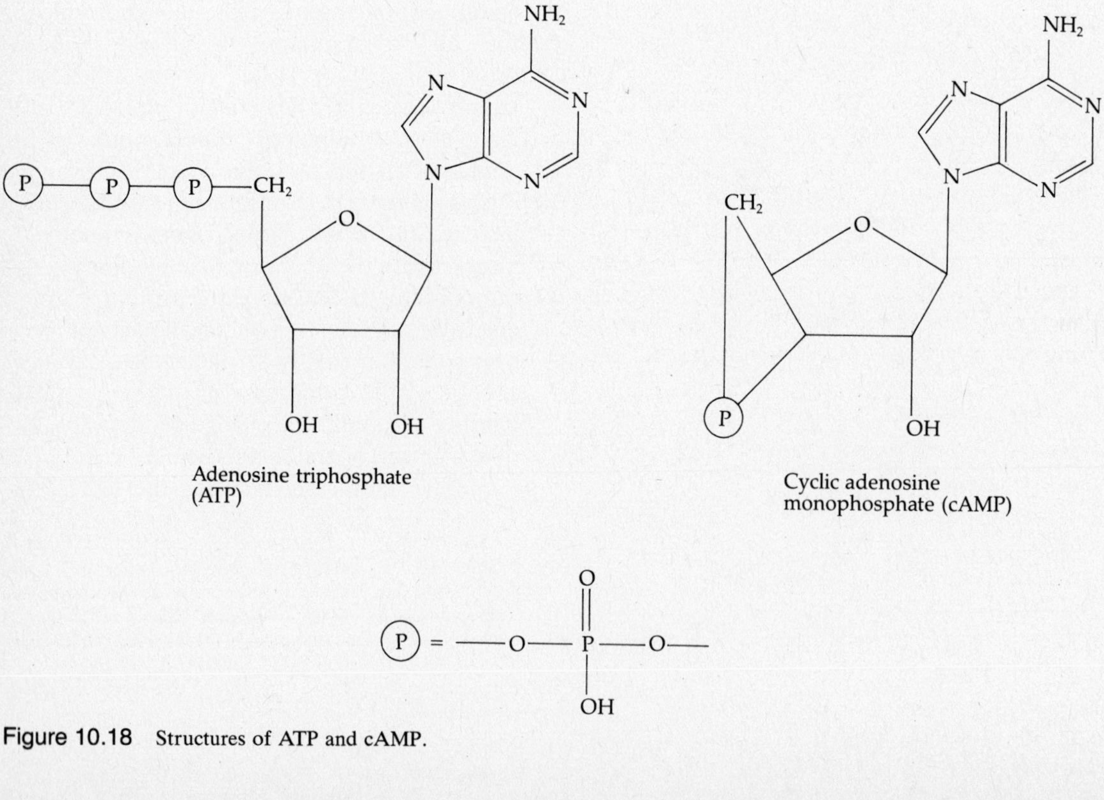

Adenosine triphosphate
(ATP)

Cyclic adenosine
monophosphate (cAMP)

$$(P) = -O-\overset{\displaystyle O}{\underset{\displaystyle OH}{\overset{\|}{P}}}-O-$$

Figure 10.18 Structures of ATP and cAMP.

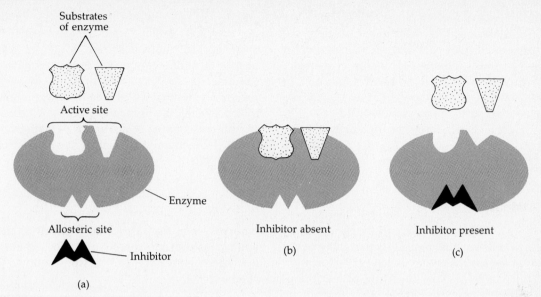

Substrates
of enzyme

Active site

Enzyme

Allosteric site

Inhibitor

(a)

Inhibitor absent

(b)

Inhibitor present

(c)

Figure 10.19 Allosteric inhibition. *(a)* Schematic of an enzyme, its substrates, and inhibitor. Note that the substrates and inhibitor bind to different sites on the enzyme. *(b)* In the absence of inhibitor, the enzyme is active. *(c)* In the presence of inhibitor, substrate-binding sites change conformation and the enzyme becomes inactive.

pressor protein and allolactose is an example of **allosteric regulation**—regulation of the activity of an enzyme or protein by a molecule with no structural similarity to the normal substrate. An enzyme subject to allosteric inhibition is illustrated in Figure 10.19(*a*). The shaded shape represents the enzyme, and its **active site** (the site for substrate binding) is indicated as being complementary in shape to the **substrates** (stippled). (A substrate of an enzyme is any molecule that is chemically transformed by the enzyme's catalytic activity.) The enzyme also has an **allosteric site**, which is complementary in shape to the **inhibitor** (black). In the absence of inhibitor [see Figure 10.19(*b*)], the active site assumes its normal conformation, permitting entry of its substrates, so the enzyme will be active. In the presence of inhibitor [see Figure 10.19(*c*)], the conformation of the active site changes to exclude the substrates, so the enzyme will be inactive.

Allosteric regulation is a widespread form of posttranslational control in both prokaryotes and eukaryotes. In many instances involving biochemical pathways for the synthesis of such molecules as amino acids, the first enzyme in the pathway is allosterically inhibited by the end product of the whole pathway—for example, an amino acid. Consequently, when the amino acid has been produced by the pathway in sufficient amount for the cell's needs, the amino acid will inhibit its own further production by inactivation of the first enzyme. Allosteric inhibition of the first enzyme in a pathway by the end product of the pathway is called **feedback inhibition**.

Protein Stability A final level of control of gene expression involves protein stability—controlling the rate at which proteins are broken down by enzymes in the cytoplasm. Some proteins are broken down quite rapidly and must be resynthesized continuously to remain present in the cell; other proteins are broken down slowly and can remain active for long periods. Although regulation at the level of protein stability is probably important in prokaryotes and eukaryotes, few details are known about the processes involved.

Despite the foregoing discussion of the many levels of regulation that are possible, we still know little of how the tens of thousands of genes in a cell are simultaneously regulated. Yet they are regulated, and the upshot is that most normal individuals are more or less healthy most of the time. But genes can **mutate** (become altered) and the products of such mutated genes may be unable to function normally, causing the individuals who carry them to suffer discomfort or disease or perhaps even death. How mutations occur is the subject of the next chapter.

SUMMARY

1. Regulation refers to the processes by which cells are able to express particular genes at the appropriate times. Since gene expression is a complex process involving transcription, RNA processing, translation, and others, gene expression has many **control points**—places in the process at which control over gene expression can be exerted.

2. The regulation of **hemoglobin** synthesis is an example of gene regulation during development. Each hemoglobin molecule consists of two α (**alpha**)-like polypeptides and two β (**beta**)-like polypeptides. Humans have two α-like polypeptides, called ζ (**zeta**) and α; both chains are coded by duplicate genes, so there are two ζ genes ($\zeta1$ and $\zeta2$) and two α genes ($\alpha1$ and $\alpha2$) in addition to an unexpressed pseudogene for α ($\psi\alpha1$). These genes are arranged on chromosome 16 in order of their expression during development, the ζ genes being upstream from the α genes; the ζ genes are expressed in early embryonic development, whereas the α genes are expressed primarily in the fetus and adult. Humans have four β-like polypeptides—ϵ (**epsilon**), γ (**gamma**, coded by two nearly identical genes called $^G\gamma$ and $^A\gamma$), δ (**delta**), and β (there are also two β pseudogenes—$\psi\beta1$ and $\psi\beta2$). The expressed β-like genes are arranged on chromosome 11 in order of their expression, ϵ being expressed in embryos, γ in the fetus, and δ and β in the adult. However, δ expression is poor, so most adult hemoglobin consists of two α and two β chains ($\alpha_2\beta_2$).

3. DNA modification, particularly by the addition of methyl ($-CH_3$) groups to certain cytosines (**methylation**), is one control point of gene expression, and it has been implicated in hemoglobin regulation. In particular, active hemoglobin genes seem to be less methylated than inactive ones, although methylation by itself is only one aspect of hemoglobin regulation.

4. Some organisms can regulate gene expression by means of DNA **amplification** (increasing the amount of DNA corresponding to certain genes) or DNA **diminution** (decreasing the amount of DNA corresponding to certain genes). In the African clawed toad, *Xenopus laevis*, the genes for producing ribosomal RNA are greatly amplified in oocytes, and amplification of certain eggshell protein genes also occurs in *Drosophila*. As for diminution, the nematode *Parascaris equorum* loses terminal parts of the chromosomes in somatic cells, but the chromosomes remain intact in germ-line cells. The unicellular protozoan *Oxytricha* undergoes DNA diminution in its **macronucleus** (corresponding to a somatic cell nucleus in other organisms) but not in its

micronucleus (corresponding to a germ-cell nucleus). DNA sequences that are absent in the macronucleus include repetitive sequences, as the rate at which separated DNA strands from the macronucleus **reassociate** is indicative of the absence of repetitive sequences. In addition, the **complexity** of macronuclear DNA measured by means of reassociation is at least 10 times less than that of the micronucleus.

5. Regulation can also be mediated by DNA **transposition** (change of location). In baker's yeast (*Saccharomyces cerevisiae*), for example, mating type (*a* or α) is determined by which DNA sequence (*a* and α) is at the mating-type locus (*MAT*). Yet both *a* and α sequences can be present in unexpressed **cassettes**, and these unexpressed sequences can transpose to the *MAT* locus and displace the sequence previously present at *MAT*. Thus, mating type can switch from *a* to α and back again to *a* by means of such transpositions.

6. Regulation can also be achieved by means of **recombinational switches**. An example of a recombinational switch is a DNA sequence in *Salmonella typhimurium* that controls flagellar type. This switch carries a promoter sequence flanked by a small inverted repeat within which recombination can occur. In one orientation, the promoter promotes transcription of the *H2* flagellar gene along with a gene that represses the H1 flagellar type. In the other orientation, the promoter promotes in the opposite direction, so *H2* and the repressor are not expressed, thus leading to the expression of the H1 flagellar type.

7. Genes can be created during development by means of **DNA splicing**. An example is the DNA splicing that brings together the **constant region** and the **variable region** of **antibody** molecules, which are key elements in the immune system. This splicing process, called **combinatorial joining**, can generate a large number of antibody types because many alternative constant regions and variable regions exist in the DNA, and any constant region can be spliced with any variable region.

8. The importance of irreversible DNA changes during development is still unclear, but it can be tested by determining whether nuclei of differentiated cells can support embryonic development. In plants such as carrot and tobacco, cells from an adult plant can be induced by appropriate conditions to divide and undergo development into a normal plant; thus, irreversible DNA changes do not seem to be significant. In amphibians, nuclei of tadpole intestinal cells can by **nuclear transplantation** be introduced into egg cytoplasm, and a fraction of such transplanted nuclei will support normal development. In mice, transplanted nuclei from the **inner cell mass** of the **blastocyst** can sustain normal development into fully fertile adults; this experiment represents the first **cloning** of a mammal, with all its attendant moral and ethical issues should it be applied to humans. Previous experiments had shown that cells of disrupted mouse embryos that are combined and implanted into a female's uterus will reorganize and give rise to a morphologically normal offspring. Such offspring are called **allophenic** mice; they are **chimeras** (genetic mosaics) because they have two or more genetically distinct types of cells. On the whole, there is no convincing evidence that irreversible DNA changes play a major role in development.

9. Gene expression can also be regulated at the level of transcription, and two broad categories of transcriptional control can be distinguished. In **positive control**, a protein or other molecule must bind to a promoter region for transcription to be possible. In **negative control**, a protein or other molecule must *not* bind to a promoter region for transcription to be possible. Both types of control are exemplified in the cluster of adjacent genes (an **oper-**

on) involved in the metabolism of **lactose** in *E. coli*. Positive control is mediated by **catabolite activator protein (CAP)**. In the presence of low levels of **cyclic adenosine monophosphate (cAMP)**, which accompany the availability of **glucose** in the environment, CAP cannot bind the *lac* promoter and transcription cannot occur; with high levels of cAMP, a CAP-cAMP complex forms, which does bind to the promoter and makes transcription possible. Negative control of the lactose genes is mediated by the **lactose repressor protein**. In the absence of lactose, the repressor binds to the DNA at a site called the **operator** (which overlaps the promoter) and renders transcription impossible. In the presence of lactose (actually **allolactose**), the repressor becomes complexed with the allolactose and is unable to bind the operator; in this case transcription can occur. In eukaryotes, positive transcriptional control is thought to be more important than negative control, and a certain class of DNA binding proteins called **acidic proteins** is implicated in the process. Transcriptional control is exemplified in the case of the gene for egg white **ovalbumin**, the transcription of which is induced by the hormone **estrogen** when it is transported into the cell nucleus by **estrogen receptor protein**. Transcription can also be regulated by the type of **RNA polymerase** involved and by certain proteins that help to initiate or terminate transcription; examples in *E. coli* are **sigma** (an initiation helper) and **rho** (a termination helper).

10. RNA processing is an important control point in eukaryotes but not in prokaryotes. In **simian virus 40 (SV40)** containing a gene whose intervening sequence has been precisely excised by recombinant DNA techniques, the transcript of the gene breaks down in the nucleus and is not properly processed into mature mRNA; in this instance (and several others), intervening sequences play an essential role in gene expression. Regulation at the level of RNA processing may also be exercised by the type of cap structure and by the presence or length of the poly-A tail.

11. Regulation also occurs at the level of **translation**. Unfertilized sea urchin eggs contain large amounts of **masked** (untranslated) mRNA's, and untranslated mRNA's for eggshell proteins in *Drosophila* accumulate in the cytoplasm as the corresponding DNA is being amplified in the cytoplasm. Bacterial operons for amino acid synthesis have translational control of transcription by means of **attenuation**. Initial transcription transcribes only a short sequence coding for a leader polypeptide and pauses at a **transcription pause site** slightly downstream. The leader polypeptide is relatively rich in the amino acid coded by the operon, and if there is enough available amino acid to correctly translate the leader, transcription continues only to the **attenuator** site a little farther downstream and then terminates. Otherwise, transcription of the entire operon occurs. Translational control of translation is exemplified by the **erythromycin**-induced **methylation** enzyme in *Staphylococcus aureus*. The mRNA for the methylation enzyme carries a short leader-polypeptide sequence upstream from the enzyme, and the conformation of the mRNA depends on the presence or absence of erythromycin. In the presence of the antibiotic, the conformation makes the methylation-enzyme sequence accessible to ribosomes, and the enzyme will be translated. In the absence of the antibiotic, the methylation-enzyme sequence becomes inaccessible to ribosomes. Translational control can also be mediated by the stability of the mRNA, as evidenced by the mRNA for **casein** in mammary glands, which is unstable in the absence of the hormone **prolactin**.

12. Posttranslational control involves regulation at the level of the translated polypeptide. In some cases, a polypeptide must be cleaved into several parts before becoming biologically active, as in the case of **insulin**, in which two polypeptide segments are removed from the original longer polypeptide (**pre-proinsulin**). In other instances, long polypeptides are cleaved into two or more parts, each of which is a different enzyme in its own right. Such polypeptides that are cleaved into two or more functional parts are called **polyproteins**.

13. Some proteins are produced in an inactive state and must be **activated** by other events in the cell. Examples are **glycogen phosphorylase kinase** (GPK), which breaks down **glycogen** (a storage carbohydrate) in liver cells. Activation of GPK is ultimately due to such hormones as **epinephrine**, which activates **adenylate kinase** to produce cAMP, which then activates a **protein kinase**, which finally activates GPK. Other proteins are produced in an active state but can be inactivated (**inhibited**) by other molecules in the cell. In many synthetic pathways, the ultimate product of the pathway inhibits the first enzyme in the pathway, thus regulating its own synthesis; this sort of regulation is known as **feedback inhibition**.

14. Many proteins have **allosteric** sites that are distinct from their primary **substrate-binding** site. Small molecules can become bound to the allosteric site and influence the protein's activity. Examples are the lactose repressor, which when bound with allolactose cannot bind the operator, and the tryptophan repressor, which when bound to tryptophan does bind with its operator. Feedback inhibition is another example of **allosteric regulation.**

15. Despite the many known examples of regulation at virtually all possible control points, little is known about the relative importance of the various control processes in eukaryotes.

WORDS TO KNOW

DNA	**RNA Transcription and Processing**	**Lactose**	**Enzyme**
Modification	Repression	Operon	Activation
5-methyl cytosine	Positive control	Operator	Inhibition
Amplification	Negative conrol	Promoter	Allosteric regulation
Diminution	Acidic proteins	Repressor	Feedback inhibition
Hybridization	RNA polymerase	Inducer	Substrate
Reassociation	Reverse transcriptase	*lacZ* gene	
Complexity	RNA splicing	*lacY* gene	**Hemoglobin**
Transposition	Masked mRNA	*lacI* gene	Alpha (α)
Casettes		CAP	Beta (β)
Southern blot	**Translation**	cAMP	Gamma (γ)
Recombinational switch	Attenuation	Polycistronic mRNA	Delta (δ)
Splicing	Polyprotein		Epsilon (ϵ)
			Zeta (ζ)

PROBLEMS

1. For discussion: Are there any circumstances in which you think the production of a human clone would be justified? Are there any circumstances in which mass cloning would be justified? (Let your imagination run free and consider such possibilities as long-distance space travel and the aftermath of a nuclear holocaust.)

2. Attenuation in *E. coli* involves a coupling

between translation and continued transcription. For what reason would this type of regulation not be expected to occur in eukaryotes?

3. Distinguish between the CAP discussed in positive regulation of the lactose operon in *E. coli* and the cap discussed in connection with eukaryotic mRNA in Chapter 9.

4. Distinguish between a polyprotein and a polycistronic messenger RNA. In what way can a polyprotein in eukaryotes serve the same regulatory function as a polycistronic mRNA in prokaryotes?

5. What is the principal difference between combinatorial joining as occurs in antibody production and the sort of splicing that occurs in the expression of, for example, hemoglobin?

6. What DNA sequence would be produced from the following RNA sequence by the action of reverse transcriptase?

5'-AUUCGUAACUUCGGGACCC-3'

Reverse transcriptase requires deoxynucleotide triphosphates to create the DNA transcript. In light of this fact, does reverse transcriptase move from right to left along the molecule shown here, or from left to right?

7. In connection with DNA electrophoresis of restriction fragments, what is a Southern blot and what is a probe? How can a Southern blot and a probe be used to determine the size of the DNA fragment that carries a particular nucleotide sequence?

8. Restriction enzyme *Hpa*II cleaves DNA at the positions of the arrows wherever this sequence occurs, but only if the C's marked with an asterisk are not methylated.

$$\downarrow^*$$
5'-CCGG-3'
3'-GGCC-5'
$$_{*}\uparrow$$

Enzyme *Msp*I cleaves the same sequence at the same places irrespective of cytosine modification. What DNA fragments would be produced by *Hpa*II digestion of the following sequence? By *Msp*I digestion? (In the sequence shown, the asterisks denote methylated C's.)

5'-ACCGGTCCGGGACCGGATCCGG-3'
3'-TGGCCAGGCCCTGGCCTAGGCC-5'

9. What would happen as a result of recombination in Figure 10.6 if the 14-bp repeat (arrows) were direct repeats (i.e., had the same orientation) instead of inverted repeats as shown?

10. In the invertible switch in Figure 10.6(*a*), what is the function of the *hin* gene? How would a mutation in the *hin* gene affect the ability of a cell to change its flagellar type? Consider what would happen if the *hin* gene had a temperature-sensitive mutation enabling the *hin* gene product to function at 30°C but not at 37°C; could flagellar switching occur at 30°C? At 37°C? Why?

11. Considering the positive and negative controls over transcription of the *lac* operon in *E. coli* (see Figure 10.10 and 10.11), would a mutant cell that is unable to produce CAP respond to lactose in the medium by being induced to produce high levels of lactose-degrading enzymes?

12. Suppose the lactose repressor in Figure 10.11 is genetically defective so that it is unable to bind to the inducer (allolactose). Would lactose-degrading enzymes be produced in the presence of lactose? Why or why not?

13. Two kinds of mutations are known that lead to the production of lactose-degrading enzymes even in the absence of lactose. (Such constantly producing strains are said to be **constitutive**.) One type of mutation occurs in the *lacI* gene and renders the repressor unable to bind with the operator. The other type of mutation occurs in the *lacO* region and so alters its structure that normal repressor cannot bind to it. Explain in each case why the cells produce high levels of lactose-degrading enzymes in the absence of lactose. If cells with such mutations were cultured in glucose, what would you expect to happen to the levels of lactose-degrading enzymes?

14. As noted in connection with Figure 10.14, the *trpL* coding region in *E. coli* contains two tryptophan codons. Suppose mutations occurred in such a way that both of these tryptophan condons were changed into leucine codons. What effect would you expect these mutations to have on the process of attenuation?

15. Many metabolic pathways are regulated by feedback inhibition, in which the first enzyme in the pathway interacts allosterically with the end product of the entire pathway and so becomes inhibited when sufficient end product has been produced. Give a reason why the first enzyme in the pathway is the one inhibited rather than the last enzyme in the pathway.

FURTHER READING AND REFERENCES

Adhya, S., and S. Garges. 1982. How cyclic AMP and its receptor protein act in *Escherichia coli*. Cell 29:287–289. Review of this potent regulatory molecule.

Astell, C. R., L. Ahlstrom-Jonasson, M. Smith, K. Tatchell, K. A. Nasmyth, and B. D. Hall. 1981. The sequence of the DNAs coding for the mating-type loci of *Saccharomyces cerevisiae*. Cell 27:15–23. The complete sequence of *a* and α.

Bank, A., J. G. Mears, and F. Ramirez. 1980. Disorders of human hemoglobin. Science 207:486–493. Discussion of hemoglobin abnormalities with emphasis on gene regulation.

Brown, D. D. 1981. Gene expression in eukaryotes. Science 211:667–674. Scholarly and readable review of gene regulation.

Darnell, J. E., Jr. 1982. Variety in the level of gene control in eukaryotic cells. Nature 297:365–371. The most frequent level of control seems to be transcriptional.

Davidson, E. H., and J. W. Posakony. 1982. Repetitive sequence transcripts in development. Nature 297:633–635. What are the possible functions of moderately repetitive DNA sequences?

Ehrlich, M., and R. Y.-H. Wang. 1981. 5-methylcytosine in eukaryotic DNA. Science 212:1350–1357. More detail on modified cytosines and their potential importance.

Groudine, M., R. Eisenman, and H. Weintraub. 1981. Chromatin structure of endogenous retroviral genes and activation by an inhibitor of DNA methylation. Nature 292:311–317. Further evidence of the role of methylation in gene expression.

Gruss, P., C.-J. Lai, R. Dhar, and G. Khoury. 1979. Splicing as a requirement for biogenesis of functional 16S mRNA of simian virus 40. Proc. Natl. Acad. Sci. U.S.A. 76:4317–4321. An element in vitro mutagenesis experiment discussed in the text in connection with Figure 10.13.

Hicks, J., J. N. Strathern, and A. J. S. Klar. 1979. Transposable mating type genes in *Saccharomyces cerevisiae*. Nature 282:478–483. Discussed in connection with Figure 10.5.

Horinouchi, S., and B. Weisblum. 1980. Post-transcriptional modification of mRNA conformation: Mechanism that regulates erythromycin-induced resistance. Proc. Natl. Acad. Sci. U.S.A. 77:7079–7083. Discussed in connection with Figure 10.15.

Huehns, E. R., N. Dance, G. H. Beaven, F. Hecht, and A. G. Motulsky. 1964. Human embryonic hemoglobins. Cold Spring Harbor Symp. Quant. Biol. 29:327–331. Source for some of the data in Figure 10.1.

Illmensee, K., and P. C. Hoppe. 1981. Nuclear transplantation in *Mus musculus*: Developmental potential of nuclei from preimplantation embryos. Cell 23:9–18. The first successful cloning of a mammal. Source of Figure 10.8.

Illmensee, K., and L. C. Stevens. 1979. Teratomas and chimeras. Scientific American 240:120–132. Discusses one way to experimentally produce genetic mosaics (chimeras).

Lauth, M. R., B. B. Spear, J. Heumann, and D. M. Prescott. 1976. DNA of ciliated protozoa: DNA sequence diminution during nuclear development in *Oxytricha*. Cell 7:67–74. Source of Figure 10.4.

Maniatis, T., E. F. Fritsch, J. Lauer, and R. M. Lawn. 1980. The molecular genetics of human hemoglobins. Ann. Rev. Genet. 14:145–178. The molecular arrangement and expression of the globin genes.

McKay, D. B., and T. A. Steitz. 1981. Structure of catabolite gene activator protein at 2.9 Å resolution suggests binding to left-handed B-DNA. Nature 290:744–749. Evidence that left-handed DNA may be biologically important.

Nevins, J. R., and M. C. Wilson. 1981. Regulation of adenovirus-2 gene expression at the level of transcriptional termination and RNA processing. Nature 290:113–118. Transcripts produced early in infection terminate prematurely relative to those produced late in infection.

Perutz, M. F. 1978. Hemoglobin structure and respiratory transport. Scientific American 239: 92–125. How the molecule clicks back and forth between two alternative structures and so carries oxygen or carbon dioxide.

Proudfoot, N. J., M. H. M. Shander, J. L. Manley, M. L. Gefter, and T. Maniatis. 1980. Structure and *in vitro* transcription of human globin genes. Science 209:1329–1336. Concerns the organization and regulation of the globin genes in humans. Data used in Figure 10.2.

Rosenberg, M., and D. Court. 1979. Regulatory sequences involved in the promotion and termination of RNA transcription. Ann. Rev. Genet. 13:319–353. The focus of this detailed review is on prokaryotes.

Simon, M., J. Zieg, M. Silverman, G. Mandel, and R. Doolittle. 1980. Phase variation: Evolution of a controlling element. Science 209:1370–1374. Material discussed in connection with Figure 10.6.

van der Ploeg, L. H. T., and R. A. Flavell. 1980. DNA methylation in the human γδβ-globin locus in erythroid and nonerythroid tissue. Cell 19:947–958. Data used in Figure 10.2.

Stroynowski, I., and C. Yanofsky. 1982. Transcript secondary structures regulate transcription termination at the attenuator of *S. marcescens* tryptophan operator. Nature 298:34–38. On the importance of RNA secondary structures.

Wigler, M. H. 1981. The inheritance of methylation patterns in vertebrates. Cell 24:285–286. Cells have ways to transmit their methylation pattern from generation to generation.

Yanofsky, C. 1981. Attenuation in the control of expression of bacterial operons. Nature 289:751–758. A thorough review of this method of regulation.

chapter 11
Mutation

Genetics involves three fundamental processes. The first is **gene transmission**, which refers to the manner in which genes are passed from generation to generation and in eukaryotes is exemplified by Mendel's laws. The second is **gene expression**, which refers to the processes by which genes exert their effects on cells and organisms. The third fundamental process is **mutation**, which refers to heritable changes in the genetic material. All living organisms undergo mutation. Indeed, mutation is thought to be essential to the long-term survival of any biological species, because a species without mutation would be unable to acquire new genes that enable it to adapt to a changing environment and would consequently become extinct. Mutation is a two-edged sword, however. It is a **random process** in the sense that mutations occur independently of the effects they may have on their carriers. Although mutation is necessary for the long-term survival of a species, mutations that have beneficial effects on their carriers in terms of survival and reproduction are very rare. Most mutations are harmful, or at best have no effect on phenotype. Yet any species must accommodate this proportion of harmful mutations to benefit in the long run from the much smaller proportion of favorable mutations that occur.

In practice, new mutations are recognized by their effects on phenotype—by the occurrence of one or more offspring that have an unexpected phenotype in a mating. Alternatively, the procedures of enrichment and selection discussed in Chapter 4 can be used to recover new mutations. Whatever the method of their discovery, individuals with unusual phenotypes must be studied individually to determine whether the phenotype is actually due to mutation. Unusual phenotypes can have many causes other than mutation. For example, if the trait in question is multifactorial, rare and unexpected phenotypes can arise from segregation and recombination producing rare combinations of alleles that are brought together in fertilization; in such cases, the unusual offspring phenotype will not be transmitted in any simple manner even though the phenotype is genetic in origin. Another reason for caution in attributing unusual phenotypes to mutation is the possibility of **phenocopy**—the occurrence of phenotypes that resemble particular inherited disorders but are actually caused by environmental factors. (See Chapter 3 for further discussion of phenocopies.) For example, a newborn who suffers from a prolonged lack of oxygen during delivery may sustain brain damage with symptoms very similar to those of a child who is homozygous for a simple Mendelian mutation that causes mental retardation, but in the case of oxygen deprivation the phenotype is due to phenocopy and not mutation. In short, the hypothesis that a new mutation has occurred must always be verified by observation or experiment.

Somatic Mutations and Germinal Mutations

Unusual phenotypes can sometimes be due to simple Mendelian mutations and yet not be transmitted to the offspring of the mutant individual. (A **mutant** individual is one who has a phenotype caused by a recognized mutation.) The apparent paradox of non-

transmission relates to an important distinction between somatic mutations and germinal mutations. A **somatic** mutation is one that occurs in a somatic cell of the body; a **germinal** mutation is one that occurs in a germ cell of the ovaries or testes. A somatic mutation may affect the phenotype of the individual in which it occurred, but the mutation will not be transmitted to future generations because it cannot be included in the gametes. Individuals in which a somatic mutation has occurred will be genetic **mosaics**; they will have two genetically different types of cells. In a mosaic, tissues arising from nonmutant cells will be normal, whereas those arising from the original mutant cell will be mutant. In humans, among the easiest somatic mutations to detect are those that affect eye color. A person may have a brown sector in an otherwise blue eye, for example, or even one blue eye and one brown eye. In some instances a mosaic individual may have mosaic ovaries or testes, particularly when the mutation occurs very early in embryonic development, and in these instances the mutation can be transmitted to future generations.

Mutations that occur in or include the germ cells, or those that occur directly in the sperm or eggs, are the most important in genetics and evolution because of their transmissibility. It should be emphasized that mosaicism is highly unlikely except in the one generation in which the new mutation occurs. Once a mutation has been transmitted through a sperm or egg, the resulting offspring will not be mosaic because all its cells will carry the mutation.

Traits acquired during the lifetime of an individual are not mutations and they are not inherited. Toil as we may to master knowledge, our children are born ignorant. Exercise and sweat as we do to develop muscles, our children are born weak. Skills in art, science, athletics, and all other acquired abilities rest with their possessors in the grave. This is not to deny that certain *propensities* may be inherited. There may well be inherited abilities toward, say, music. But the practice and perfection of these abilities are still lonely, individual tasks. There is a good side to this story, however. The tragic accidents that afflict people during their lives are not transmitted either. Disease, accidental loss of eyes or limbs, surgical mutilations—all these and other acquired disabilities are burdensome to those afflicted, but they are not transmitted to the children.

Types of Mutations

Mutations can occur in any gene. They can therefore affect any part of the body at any time in life, from embryonic development to old age. Mutations have specific effects related to the gene involved. For example, a mutation at the testicular-feminization locus causes the many separate symptoms of the testicular-feminization syndrome, but such a mutation never leads to sickle cell anemia. Similarly, a mutation of the Tay-Sachs locus causes Tay-Sachs disease but not cystic fibrosis. On the other hand, mutations in different genes that affect the same organs or the same biochemical pathways can lead to similar symptoms. Dozens of different genes, when mutated, can cause deafness, for example, and even more mutations can result in mental retardation.

Sometimes it is useful to classify mutations according to their effects. One such classification is based on **age of onset**—the age at which the phenotype associated with the mutation typically appears. At one extreme of age of onset are mutations that are expressed early in life, such as Tay-Sachs disease and cystic fibrosis, in which the phenotype is present at birth or can even appear during embryonic development. At the other extreme are mutations expressed later in life, such as the

mutation leading to premature baldness. A well-known serious disorder caused by a mutation expressed relatively late in life is **Huntington disease** (or **Huntington chorea**). This rare dominant mutation causes an incurable and relentless degeneration of the nervous system, usually beginning some time in middle age and progressing rapidly, so that in a few years the victim, once healthy and vigorous, becomes bedridden and helpless. Famed folk singer Woody Guthrie, who died in 1967, was afflicted with this tragic disorder.

Mutations can also be classified according to the severity of their phenotypic effects. The most severe mutations are, of course, **lethal**, which means that they are incompatible with life; Tay-Sachs disease is a lethal disorder. Equally severe from an evolutionary point of view (but not from a social point of view) are **male-sterile** or **female-sterile** mutations, which render the affected individual unable to reproduce; in some instances of mutations that cause sterility, the affected individual may be completely normal except for the inability to produce functional gametes. At the other end of the spectrum of severity are mutations that seem to have no marked effect on the ability of individuals to survive and reproduce. Included in this category are mutations like those listed in Table 3.1 in Chapter 3—common baldness, chin fissure, ear pits, Darwin tubercle, type of ear cerumen, and so on.

Some mutations are **conditional mutations**, which means that they are expressed only in certain environments. In *Drosophila*, for example, it is easy to find **temperature-sensitive lethals**—mutations that cause death at one temperature (called the **restrictive temperature** —typically 25°C for *Drosophila*) but are compatible with life at a different temperature (called the **permissive temperature**—typically 18°C for *Drosophila*). The usual interpretation of temperature-sensitive mutations is that they alter the amino acid sequence of a key protein or enzyme in such a way that the protein's three-dimensional configuration is sensitive to temperature. At the low permissive temperature, the protein can assume an approximately normal configuration and carry out its function in the cell, but at the high restrictive temperature, the protein unfolds (**denatures**) and is unable to carry out its function. In humans, the G6PD deficiency discussed in Chapter 5 is a sort of conditional mutation because the associated anemia occurs in response to such environmental triggers as eating raw fava beans or inhaling naphthalene fumes. Conditional mutations can occur in virtually any gene that codes for a protein; such mutations are widely used in genetic studies ranging from studies of viruses that infect bacteria to human cells in culture.

Some mutations are recognized by their effects on the expression of other mutations. Chief among these are **enhancer** mutations, which intensify the phenotypic effect of some other mutation; and **suppressor** mutations, which diminish the phenotypic effect of some other mutation. Enhancers and suppressors are extreme examples of **modifier** mutations—mutations that modify the phenotypic expression of other mutations. The extremes of modification are illustrated by an X-linked mutation in *Drosophila melanogaster* known as *forked*. By itself, the *forked* mutation leads to shortened and gnarled bristles sharply bent at the ends. Another mutation, located on chromosome 2, interacts with *forked* and leads to an extreme expression of the bristle phenotype; this mutation is known as *enhancer of forked*. Still another mutation, X-linked but at a different locus than *forked* itself, so diminishes the effect of *forked* that the bristles are virtually normal; this mutation is called *suppressor of forked*. In most instances of modifier mutations, the molecular basis of the modification is unknown. An exception is a type of suppressor called a *nonsense* suppressor, although it will be convenient to defer discussion of this until later in this chapter.

To classify mutations by their effects is only one way of looking at them. Mutations can also be subdivided by whether they are dominant or recessive. Recall that, in human genetics, a dominant mutation expresses itself when heterozygous (that is, in single dose); a recessive mutation is expressed only when homozygous (that is, in double dose). Examples of dominant mutations are those that cause achondroplastic dwarfism, certain forms of polydactyly, and Huntington disease; examples of recessive mutations are those causing cystic fibrosis, Tay-Sachs disease, albinism, phenylketonuria, and hemophilia.

The vast majority of newly arising mutations are recessive. Only one copy of a normal allele is usually sufficient for health and vitality, so in the case of harmful recessive mutations, the heterozygotes are normal and only the homozygotes are affected. In the case of dominant mutations, the heterozygotes are affected. However, in most instances involving harmful dominant mutations, the phenotype of individuals who are homozygous for the dominant is unknown because harmful dominants are usually very rare in the population and homozygotes seldom, if ever, arise. When the homozygotes can be recognized, as in familial hypercholesterolemia discussed in Chapter 3, the dominant homozygotes are usually more severely affected than the heterozygotes. This finding of greater severity of expression in dominant homozygotes is common in laboratory animals such as *Drosophila* and mice; in many cases the homozygotes are lethal, so they do not even survive. There is little reason to think that humans are exceptional in regard to the greater severity of expression in individuals who are homozygous for harmful dominants.

Although classifying mutations as dominant or recessive is useful for certain purposes, it is admittedly rather crude. Dominance and recessiveness are not so much actual properties of genes as they are reflections of the way the phenotype is examined. Consider, for example, the sickle cell mutation β^S and its normal allele β. We can inquire which of the three genotypes—β^S/β^S, β^S/β, and β/β—causes severe anemia and premature death. Only an individual with β^S/β^S is so affected, and by this criterion β^S is recessive, exhibiting its effect only when homozygous. We may also ask whether the red blood cells can be made to undergo sickling with very low oxygen tension. The cells from both β^S/β and β^S/β^S people will sickle, so by this criterion β^S is dominant, its effect showing up when heterozygous. We may also ask what is the proportion of abnormal hemoglobin in the red blood cells. β/β has essentially 0 percent, β^S/β^S has 100 percent, and β/β^S has some amount in between. By this criterion neither β nor β^S is dominant. This example shows that dominance and recessiveness have meaning only with reference to a particular way of examining the phenotype. As ordinarily used, the term recessive mutation means a mutation for which heterozygotes (or carriers) do not have the disorder characteristic of the mutation. Recessive genes are seldom recessive in all respects, however, and in most cases careful examination of phenotypes will reveal subtle differences between carriers and noncarriers. Indeed, much current medical research is devoted to finding criteria by which homozygous normal individuals and heterozygous carriers of particular mutations can be distinguished from one another.

The Genetic Basis of Mutation

The most natural classification of mutations is based on the sort of changes that occur in the chromosomes or DNA. Heritable changes in chromosomes include translocations, inversions, deficiencies, duplications, and isochromosomes, the genetic consequences of which

were examined in Chapter 7. Heritable changes in chromosomes also include alterations in chromosome number, such as trisomy, monosomy, and triploidy, which were discussed in Chapter 6. Many of these changes are associated with gross aneuploidy—upsets in relative gene dosage—and are often lethal in the embryo and not transmitted from generation to generation. All these chromosomal changes are heritable from cell to cell by means of mitosis, however, and this qualifies them as mutations. Chromosomal changes that are visible in the light microscope are necessarily rather large; otherwise, they could not be seen. Consequently, most visible chromosomal aberrations involve a deficiency or duplication of many genes simultaneously, or a change in the relative positions of large blocks of genes. But mutations occur at a much more subtle level, too, when small changes occur in the nucleotide sequences of DNA. These changes are not visible in the microscope, but they may have profound phenotypic effects in the individuals who carry them. Such subtle changes in DNA are far more common than visible chromosomal abnormalities among mutations that occur spontaneously.

Mutations at the level of DNA include the substitution of one base for another or the addition or deletion of bases in the molecule. If an addition or deletion is large enough, the alteration may be visible through the microscope; thus, the dividing line between mutations called chromosomal changes and changes at the level of DNA is sometimes fuzzy.

One relatively common kind of mutation is a simple **base substitution**, which occurs when one base pair at some position in a DNA duplex is replaced with a different base pair. If a base substitution occurs in a region of the DNA molecule that codes for a polypeptide, then the mRNA transcript produced from the gene will carry an incorrect base at the corresponding position; when this mRNA is translated into a polypeptide chain, the incorrect base may cause an incorrect amino acid to be inserted in the polypeptide chain. Hence, a base substitution in DNA can lead to an **amino acid substitution** in the corresponding polypeptide.

What effect an amino acid substitution might have on the functioning of a polypeptide depends on where in the chain the substitution occurs and on which amino acid is replaced by which other one. Sometimes a substitution may occur in a part of the molecule that can tolerate the change without undergoing a marked change in overall three-dimensional structure; the function of the molecule will then be quite normal. At other times the substitution replaces an amino acid, such as leucine, with a chemically similar one, such as isoleucine, and the net effect on the polypeptide may again be small. At still other times a base substitution in DNA leads to no amino acid substitution at all; this can happen because the genetic code has synonyms, so several different codons can denote the same amino acid in translation (see Table 9.1). Codons 5'-UUA-3', 5'-UUG-3', 5'-CUU-3', 5'-CUC-3', 5'-CUA-3', and 5'-CUG-3' all code for leucine, for example. Mutations that cause slight or no changes in the functional capacity of the polypeptide have no obvious phenotypic effects; individuals who carry such mutations are normal or very nearly normal. Certain of these changes can be detected by a special procedure known as **protein electrophoresis**, however. The procedure of protein electrophoresis is similar to DNA electrophoresis discussed in Chapter 8. A major difference is that, whereas the conditions of DNA electrophoresis are usually adjusted so that molecules are separated on the basis of size, the conditions of protein electrophoresis are usually adjusted so that molecules are separated largely on the basis of charge. Consequently, if an amino acid substitution causes a change in the overall electric charge of the protein molecule, then the normal and mutant polypeptides can be distinguished because one of them will move faster across the jellylike slab when subjected to the electric field.

Protein electrophoresis has shown that many loci in humans, perhaps up to 30 percent, have two and sometimes more alleles that are widespread in the population. The proportion of loci with two or more commonly occurring alleles in populations of insects and other invertebrates is even higher, perhaps up to 60 percent. The estimates of 30 and 60 perent are somewhat uncertain, however. They refer to the proportion of loci for which common alleles involve base substitutions that lead to a difference in the overall electric charge of the protein, since protein electrophoresis detects mainly charge differences. Thus, if one includes base substitutions that do not alter the charge of the protein molecule, the overall proportion of loci having two or more common alleles may be greater than the estimates. On the other hand, the estimates of 30 and 60 percent are based on a rather small number of loci as compared with the total number in any organism, so the estimates could be too high. In any event, such subtle genetic variation of the sort detectable by protein electrophoresis is widespread. Because all genotypes that carry such common alleles seem to be equally healthy, the notions of "normal" and "mutant" alleles lose their meaning. It is much the same with common variants of hair color, eye color, and other traits. One certainly could not consider brown hair a "normal" phenotype and all other hair colors "mutant." So, too, with common genetic differences expressed at the protein level. In these cases all the common alleles are to be considered "normal"; several *different* normal alleles of a gene often occur in populations.

The base substitutions of concern to most people are those that cause dramatic changes in the functional ability of the affected protein and therefore cause inherited disorders. One well-known example is the mutation responsible for sickle cell anemia. Recall that the major adult hemoglobin is a molecule with four polypeptide chains: two α chains and two β chains. The sickle cell mutation is in the first coding region of the gene that codes for the β chain. The β chain consists of 146 amino acids, and the sequence (from the amino end of the molecule) is val-his-leu-thr-pro-glu-glu-lys- . . . ; the three dots represent the other 138 amino acids. (The names of the amino acids in the chain are presented using the three-letter abbreviations in Figure 9.3.) The sickle cell mutation leads to an amino acid substitution at exactly one position in the chain—position number 6—at which a valine (val) replaces a glutamic acid (glu). The β^S chain therefore has the sequence val-his-leu-thr-pro-val-glu-lys- . . . . This seemingly trivial change profoundly alters the molecule. Molecules of $\alpha_2\beta_2^S$ tend to crystallize or become stacked like long rods under conditions of little oxygen; this causes the red blood cells to assume their characteristic sickle or half-moon shapes. Sickled cells tend to become rigid, clump together, and lose their ability to squeeze through the tiny capillaries (which are narrower than the diameter of normal red blood cells). This, in turn, often leads to clogging of these microscopic vessels, resulting in loss of the blood supply to the tissues nourished by them. Thus, one wrong amino acid in one polypeptide chain can cause numerous and varied abnormalities throughout the entire body (Figure 11.1).

Missense, Nonsense, and Frameshift Mutations

Various types of mutations can conveniently be discussed in relation to the β-globin gene. Recall from Chapter 10 that the human β-globin gene is split into three coding regions interrupted by two intervening sequences. Figure 11.2(a) illustrates the nucleotide sequence of the beginning of the first coding region. The sense strand is at the top, and transcription occurs from left to right (from the 3' end of the sense strand to the 5' end). The spaces that separate adjacent triplets of bases do not exist in the actual molecule, of course; they are present in Figure 11.2 only for convenience of discussion. Figure 11.2(b) shows the corresponding portion of the β-globin mRNA produced after transcription and RNA processing. Although it is not shown in the figure, the next upstream codon from GUG is the AUG initiation codon, which codes for methionine. Figure 11.2(c) shows the first eight amino acids at the amino terminal end of normal β-globin; the initial methionine with which translation begins is not represented in the finished polypeptide because it is enzymatically removed from the chain. The underlined nucleotides represent the restriction site for the restriction enzyme *Mst*II, which cleaves the DNA at this position.

As noted in the previous section, a **base substitution** mutation is one in which a base pair in a DNA duplex is replaced with a different base pair. A base substitution that leads to an amino acid substitution in the corresponding polypeptide is called a **missense** mutation, and there are two principal types of missense mutations. One is illustrated in Figure 11.3. Part (a) again shows a portion of the first coding region of the β-globin gene, and part (b) shows the result of a base substitution in which the normal ^T_A pair indicated is replaced with an ^A_T pair. In this substitution, a pyrimidine base in the sense strand (T) is replaced with a purine base (A), and a purine base in the antisense sense strand (A) is replaced with a pyrimidine base (T). Any base substitution in which a purine (A or G) is replaced with a pyrimidine (T or C), or in which a pyrimidine is replaced with a purine, is

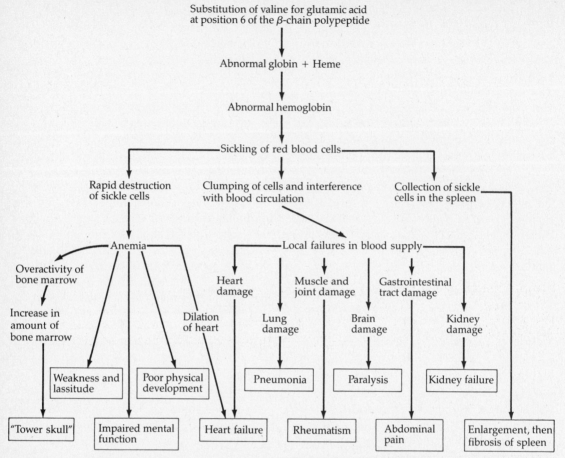

Figure 11.1   Manifold pleiotropic effects arising from the sickle cell hemoglobin mutation.

called a **transversion**. The mutation in Figure 11.3 is thus a transversion mutation. Transcription and RNA processing of the mutant DNA lead to the mRNA shown in Figure 11.3(c); note that the sixth codon, which normally reads GAG, now reads GUG. Translation of this mRNA leads to a substitution of valine for glutamic acid at the sixth position in the polypeptide (black arrow), as can be verified by examination of the genetic code in Table 9.1 in Chapter 9. The transversion mutation illustrated in Figure 11.3 is indeed the actual molecular change responsible for β^s hemoglobin.

The β^s mutation also obliterates the *Mst*II restriction site, so *Mst*II will not cleave the β^s DNA at this position. This molecular difference provides a rapid and convenient method for *in vitro* diagnosis of sickle cell anemia.

Figure 11.4 illustrates the second type of missense mutation, which is called a **transition** mutation because it involves the replacement of one purine (in this case G) with another purine (in this case A) or one pyrimidine (in this case C) with another pyrimidine (in this case T). In the transition mutation in Figure 11.4, a $_G^C$ base pair in the original molecule is

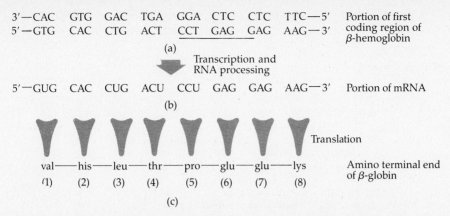

3′—CAC	GTG	GAC	TGA	GGA	CTC	CTC	TTC—5′	Portion of first
5′—GTG	CAC	CTG	ACT	CCT	GAG	GAG	AAG—3′	coding region of
								β-hemoglobin

(a)

Transcription and
RNA processing

| 5′—GUG | CAC | CUG | ACU | CCU | GAG | GAG | AAG—3′ | Portion of mRNA |

(b)

| val | his | leu | thr | pro | glu | glu | lys | Amino terminal end |
| (1) | (2) | (3) | (4) | (5) | (6) | (7) | (8) | of β-globin |

(c)

Figure 11.2 *(a)* DNA sequence coding for the first eight amino acids of human β-globin. Small gaps separate adjacent codons for ease of reading. The sense strand is at the top. *(b)* Corresponding mRNA sequence. *(c)* Amino terminal end of β-globin polypeptide. [The first amino acid is actually methionine, but it is later cleaved off the molecule and so its codon is not shown in *(a)* and *(b)*.] The sequence in *(a)* is used in Figures 11.3, 11.4, 11.5, 11.7, and 11.8 to illustrate various types of mutations.

replaced with a ^T_A base pair. The corresponding portion of the mRNA is shown in part (c), and in this case the original GAG codon in position 6 is altered to AAG. Translation of the mutant

mRNA produces a β-globin polypeptide in which the normal glutamic acid at position 6 is replaced with a lysine. This particular mutation is known as the $β^C$ mutation. Since the $β^C$

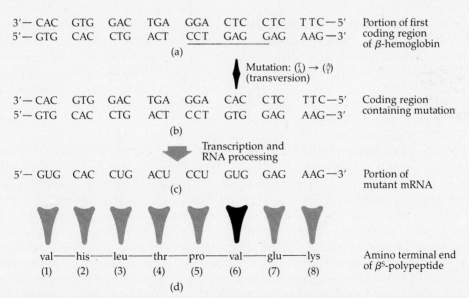

3′— CAC	GTG	GAC	TGA	GGA	CTC	CTC	TTC—5′	Portion of first
5′— GTG	CAC	CTG	ACT	CCT	GAG	GAG	AAG—3′	coding region
								of β-hemoglobin

(a)

Mutation: $\binom{T}{A} \rightarrow \binom{A}{T}$
(transversion)

| 3′— CAC | GTG | GAC | TGA | GGA | CAC | CTC | TTC—5′ | Coding region |
| 5′— GTG | CAC | CTG | ACT | CCT | GTG | GAG | AAG—3′ | containing mutation |

(b)

Transcription and
RNA processing

| 5′— GUG | CAC | CUG | ACU | CCU | GUG | GAG | AAG—3′ | Portion of |
| | | | | | | | | mutant mRNA |

(c)

| val | his | leu | thr | pro | val | glu | lys | Amino terminal end |
| (1) | (2) | (3) | (4) | (5) | (6) | (7) | (8) | of $β^S$-polypeptide |

(d)

Figure 11.3 Molecular basis of the $β^S$-hemoglobin mutation. *(a)* Normal DNA sequence. *(b)* Mutated sequence with ^A_T base pair replacing ^T_A base pair at position 17. *(c)* Corresponding mRNA. *(d)* Corresponding polypeptide with valine replacing normal glutamic acid at position 6. The mutation in question is a transversion mutation because it involves the substitution of a purine (A) for a pyrimidine (T).

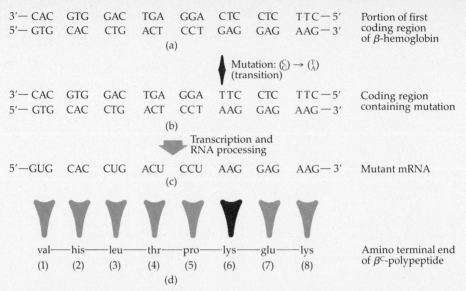

Figure 11.4 Molecular basis of the β^C-hemoglobin mutation. *(a)* Normal DNA sequence. *(b)* Mutated sequence with ^T_A base pair replacing ^C_G base pair at position 16. *(c)* mRNA from mutant sequence. *(d)* Amino terminal end of β^C polypeptide with lysine replacing normal glutamic acid at position 6. The mutation is a transition because one pyrimidine (T) replaces another pyrimidine (C).

mutation is relatively common in areas of the world where malaria is prevalent, it is thought that heterozygotes for β^C may have some protection against malaria, as is certainly the case with β^S.

Not all base substitutions are missense mutations. Figure 11.5 illustrates a type of mutation called a **nonsense** mutation. A nonsense mutation is one that creates a chain-terminating codon [i.e., UAA (ochre), UAG (amber), or UGA (opal)] in a coding region. The name *nonsense mutation* may seem odd, but it comes from the fact that chain-terminating codons are at times referred to as *nonsense codons*. In the hypothetical example in Figure 11.5, the transversion ^C_G to ^A_T leads to a UAG terminating codon at the sixth position in the mRNA. Thus, translation of this mRNA leads to a truncated polypeptide because translation terminates at the UAG codon (black arrow).

In prokaryotes such as *E. coli* and in lower eukaryotes such as yeast, certain mutations have the ability to suppress nonsense mutations. These nonsense-suppressing mutations are called **nonsense suppressors**, and they involve mutations in genes for transfer RNA (tRNA) that reduce the faithfulness of translation so that chain-terminating codons are sometimes misread. The essential situation with nonsense suppression is illustrated in Figure 11.6. Part *(a)* shows the amber nonsense mutation discussed in connection with Figure 11.5, and the truncated polypeptide is indicated. Part *(b)* shows how suppression can occur in the presence of a nonsense-suppressor mutation in a gene for serine tRNA. Such a mutant tRNA will correctly translate serine codons, but, on occasion, will misread a chain-terminating codon (in this case UAG) as a serine codon. Consequently, in some fraction of attempts at translation of the mutant

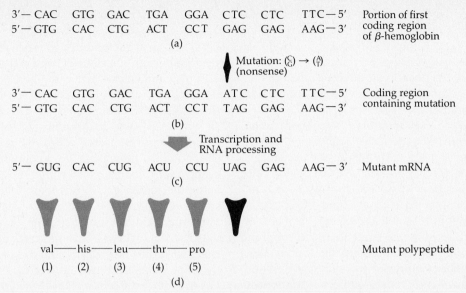

Figure 11.5 A nonsense mutation creates a chain-terminating codon, here illustrated with a hypothetical example involving the β-globin gene. *(a)* Normal sequence. *(b)* Mutant sequence with $_T^A$ replacing $_G^C$ at position 16. *(c)* Corresponding mRNA; note UAG (amber) terminator codon. *(d)* Mutant polypeptide consists of just five amino acids because translation terminates at UGA.

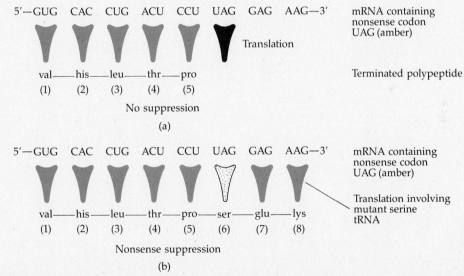

Figure 11.6 Suppression of nonsense mutations can occur by virtue of compensatory mutations in tRNA genes that sometimes misread the nonsense codon and insert an amino acid. *(a)* mRNA from a gene containing a nonsense (UAG) mutation in the sixth codon; translation terminates at position 6. *(b)* Same mRNA in a genetic background containing a mutant serine-tRNA gene that can insert serine at UAG; in this case misreading by the serine tRNA (stippled arrow) allows translation to continue. Such nonsense suppressors are so far known only in bacteria and yeast. In some fraction of attempts at translation, the mutant serine tRNA inserts serine at the UAG codon.

mRNA, the UAG codon will be misread as serine (stippled arrow), and translation will be able to proceed normally beyond this point. If the serine-containing polypeptide is able to function, the phenotypic effects of the original nonsense mutation will have been suppressed by the nonsense suppressor.

Of course, nonsense suppressors can occur in tRNA genes other than serine tRNA. For example, if a UAG suppressor occurs in a gene for glycine tRNA, a UGA codon will sometimes be misread as a glycine codon, and glycine will be inserted into the polypeptide at this position. It should also be emphasized that a particular nonsense suppressor will suppress only one of the chain-terminating codons; that is, an **amber suppressor** will suppress only the UAG (amber) terminator, an **ochre suppressor** will suppress only the UAA (ochre) terminator, and an **opal suppressor** will suppress only the UGA (opal) terminator. Of course, a nonsense suppressor will occasionally misread nor-

mal termination codons, too; for example, an amber suppressor will occasionally misread a UAG codon at the end of a normal gene and the result is an abnormally long polypeptide. However, organisms that carry nonsense suppressors are able to survive in spite of the occasional misreading of normal genes.

Not all base-substitution mutations lead to missense or nonsense codons. Because the genetic code (see Table 9.1) contains many synonymous codons, some base substitutions leave the amino acid sequence of the corresponding polypeptide unaltered. Such mutations are said to be **silent**; an example is illustrated in Figure 11.7. In this case, a $\frac{C}{G}$-to-$\frac{T}{A}$ transition at the third position in codon 6 of the β-globin gene leads to an mRNA that has GAA as its sixth codon. However, GAA is a synonymous codon for glutamic acid, so the amino acid sequence of the resulting polypeptide will be completely normal. As a brief inspection of the genetic code will reveal, most

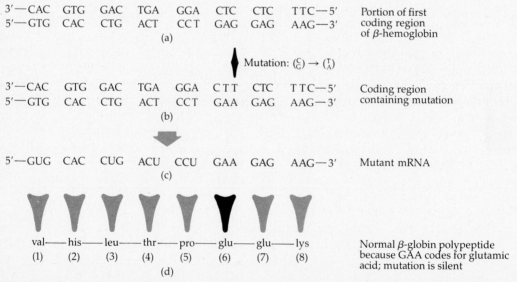

Figure 11.7   Silent mutations change one codon into a synonymous codon and thus lead to no change in amino acid sequence. *(a)* Normal DNA sequence. *(b)* Mutant sequence with $\frac{T}{A}$ instead of $\frac{C}{G}$ at position 18. *(c)* mRNA; note GAA codon replacing normal GAG. *(d)* Polypeptide has normal amino acid sequence because both GAA and GAG code for glutamic acid.

silent substitutions in coding regions would be expected to involve transitions in the third position of a codon. The situation may be very different for intervening sequences, however. Since a particular length and base sequence of an intervening sequence does not seem to be essential for proper gene function, many mutations in intervening sequences—transitions, transversions, even inversions, deletions, or insertions—may be silent.

A final example of a type of mutation that has a relatively simple molecular basis is known as a **frameshift** mutation. A frameshift mutation alters the reading frame of an mRNA during translation and thereby greatly alters the amino acid sequence of the corresponding polypeptide. Frameshift mutations are caused by deletions or additions of a small number of nucleotides in a coding region; since the genetic code consists of triplets of nucleotides, any deletion or addition of a number of nucleotides other than an exact multiple of three will cause a shift in reading frame. An example of a frameshift mutation associated with a single-nucleotide deletion is illustrated in Figure 11.8. Part (*a*) is again a portion of the first coding region of the human β-globin gene, and the left side of the figure illustrates the normal course of transcription and translation of this sequence. As in previous figures involving this gene, small gaps are introduced in part (*c*) to show the normal triplet reading frame. The right side of the figure shows the consequence of a deletion of the indicated $\frac{A}{T}$ nucleotide pair. When this deletion-bearing sequence is examined in terms of its triplet reading frame, it can be seen that all triplets beyond the second one are different from those in the normal sequence. This lack of correspondence occurs, of course, because the single-nucleotide deletion causes the triplet reading frame beyond the deletion to be shifted one nucleotide to the left. The extensive lack of correspondence due to the frameshift mutation is also evident in the

mRNA, and the resulting amino acid sequence in the polypeptide is hardly recognizable as a β-globin sequence. Such out-of-frame translation will continue along the mutant mRNA until a termination codon is encountered.

Mutations Resulting from Unequal Crossing-Over

Recall from Chapter 7 that unequal crossing-over involving duplications can lead to an increase or decrease in the number of copies of the region. When this process occurs at the molecular level, it creates new types of DNA sequences that qualify as mutations. An example of unequal crossing-over creating new mutations is found in the human β-globin gene and is illustrated in Figure 11.9. Part (*a*) shows part of the DNA sequence coding for amino acids 83 through 98, which is 18 nucleotides upstream from the second intervening sequence. The boxes indicate two nearby regions that have a perfect eight-nucleotide homology that can act as a duplication. Part (*b*) shows the coding sequence for amino acids 88 through 103; it has been shifted to the left relative to part (*a*) to show how the regions of homology can mispair (shaded boxes). As before, the gaps in the DNA sequence indicate the translational reading frame.

Figure 11.9 (*c*) and (*d*) show the DNA sequences resulting from crossing-over in the misaligned shaded boxes. Sequence (*c*) combines the left part of (*a*) with the right part of (*b*). This sequence is deleted for the nucleotides that code for amino acids 91 through 95, and it is of some interest that this deletion, called the **Gun Hill deletion**, is found in certain rare individuals. Sequence (*d*) is the complementary crossover product to sequence (*c*); this sequence carries a duplication of the nucleotides that code for amino acids 91 through 95. (Literally hundreds of mutations affecting

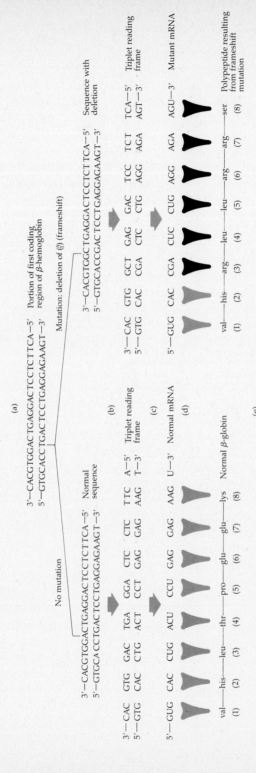

Figure 11.8 A frameshift mutation results from a small insertion or deletion that alters the downstream translational reading frame. Here a hypothetical frameshift mutation in the β-globin gene is illustrated. Left side: the normal situation; right side: the mutant situation. (*a*) Normal DNA sequence. (*b*) Deletion of $\frac{A}{T}$ base pair at position 8. (*c*) DNA sequences separated into codons by gaps for ease of reading. (*d*) Corresponding mRNA's. (*e*) Translation; note that all codons (and thus amino acids) downstream from and including position 3 are altered as a result of the frameshift mutation (black arrows).

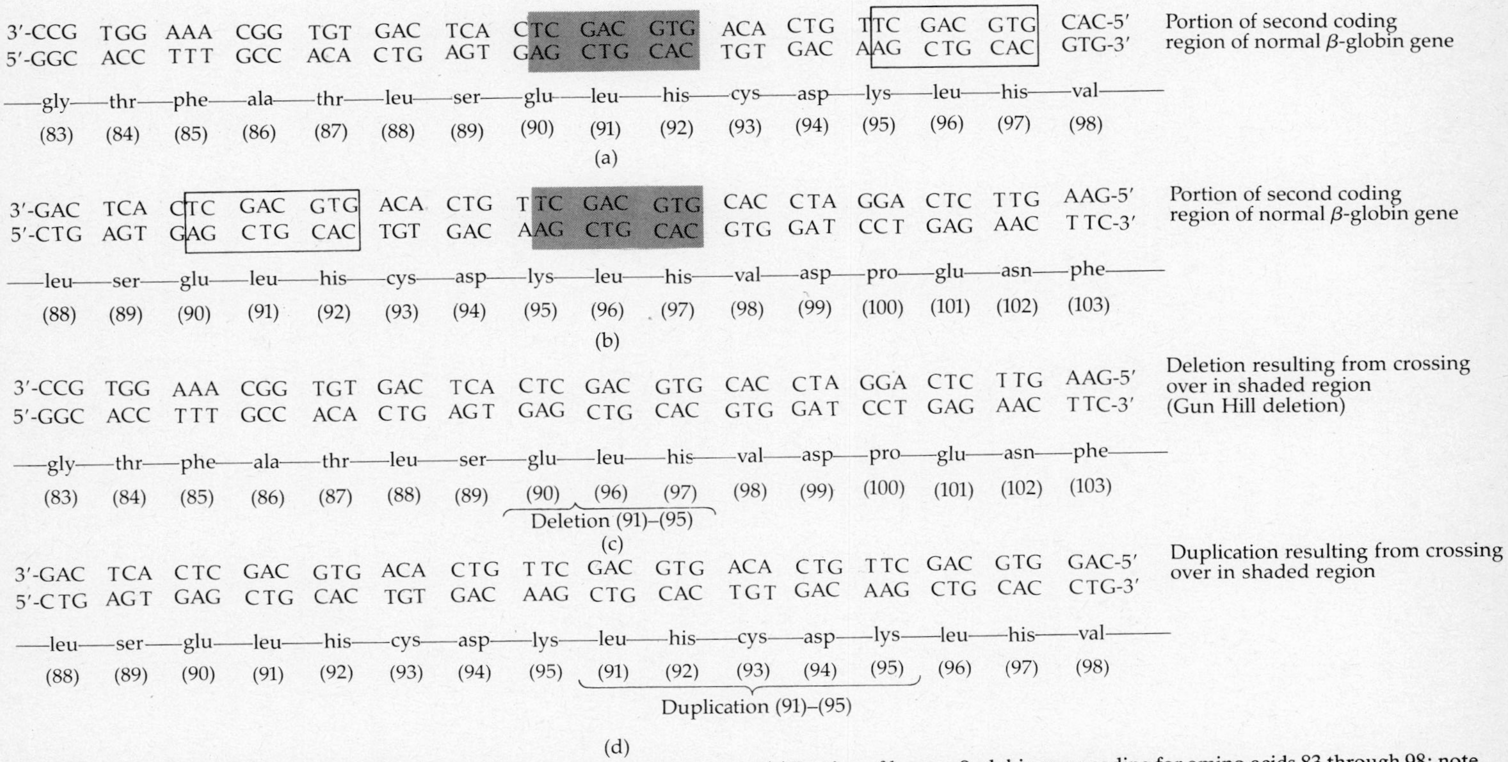

Figure 11.9 Deletions and duplication can result from unequal crossing-over. (*a*) Portion of human β-globin gene coding for amino acids 83 through 98; note the small duplicated sequence (rectangles). (*b*) Portion of human β-globin gene with duplicated regions (rectangles) misaligned with those in (*a*). Crossing-over in the shaded rectangle of the sequences misaligned as in (*a*) and (*b*) generates two new sequences. (*c*) One product carries a deletion and thus is missing amino acids 91 through 95; this product is known as the Gun Hill deletion. (*d*) The other product carries a duplication of amino acids 91 through 95.

(a) Portion of second coding region of normal β-globin gene

```
3'-CCG  TGG  AAA  CGG  TGT  GAC  TCA  CTC [GAC  GTG] ACA  CTG [TTC  GAC  GTG] CAC-5'
5'-GGC  ACC  TTT  GCC  ACA  CTG  AGT  GAG [CTG  CAC] TGT  GAC [AAG  CTG  CAC] GTG-3'
    gly  thr  phe  ala  thr  leu  ser  glu  leu  his  cys  asp  lys  leu  his  val
   (83) (84) (85) (86) (87) (88) (89) (90) (91) (92) (93) (94) (95) (96) (97) (98)
```

(b) Portion of second coding region of normal β-globin gene

```
3'-GAC  GTG  ACA  CTG  TTC [GAC  GTG] CAC  CTA  GGA  CTC  TTG  AAG-5'
5'-CTG  CAC  TGT  GAC  AAG [CTG  CAC] GTG  GAT  CCT  GAG  AAC  TTC-3'
    leu  his  cys  asp  lys  leu  his  val  asp  pro  glu  asn  phe
   (91) (92) (93) (94) (95) (96) (97) (98) (99)(100)(101)(102)(103)
```

(c) Deletion resulting from crossing over in shaded region (Gun Hill deletion)

```
3'-CCG  TGG  AAA  CGG  TGT  GAC  TCA  CTC |  GAC  GTG  CAC  CTA  GGA  CTC  TTG  AAG-5'
5'-GGC  ACC  TTT  GCC  ACA  CTG  AGT  GAG |  CTG  CAC  GTG  GAT  CCT  GAG  AAC  TTC-3'
    gly  thr  phe  ala  thr  leu  ser  glu    leu  his  val  asp  pro  glu  asn  phe
   (83) (84) (85) (86) (87) (88) (89) (90)   (96) (97) (98) (99)(100)(101)(102)(103)
                                        └─ Deletion (91)–(95) ─┘
```

(d) Duplication resulting from crossing over in shaded region

```
3'-GAC  TCA  CTC  GAC  GTG  ACA  CTG  TTC  GAC  GTG  ACA  CTG  TTC  GAC  GTG  CAC-5'
5'-CTG  AGT  GAG  CTG  CAC  TGT  GAC  AAG  CTG  CAC  TGT  GAC  AAG  CTG  CAC  GTG-3'
    leu  ser  glu  leu  his  cys  asp  lys  leu  his  cys  asp  lys  leu  his  val
   (88) (89) (90) (91) (92) (93) (94) (95) (91) (92) (93) (94) (95) (96) (97) (98)
                         └──────── Duplication (91)–(95) ────────┘
```

TABLE 11.1 DISORDERS OF HUMAN HEMOGLOBIN

Type of hemoglobin	Signs and cause
S	Homozygosity for β^S mutation (sickle cell anemia)
C	Homozygosity for β^C mutation
Lepore	δ-β chain fusion caused by unequal crossing-over in misaligned δ-β region
Anti-Lepore	Triplication consisting of normal δ gene–δ-β fusion gene–normal β gene; reciprocal crossover product of the unequal crossover leading to Lepore hemoglobin
Constant Spring	Elongated α chain due to missense mutation in terminator codon
Gun Hill	Shortened β chain due to deletion eliminating amino acids 91–95
Unstable hemoglobin	Increased blood destruction; associated with at least 70 mutations, most affecting β chain
Methemoglobulinemia	Reduced ability to transport oxygen; associated with several mutations, some affecting α, others β
Hemoglobin-induced erythrocytosis	Anemia due to increased oxygen affinity of hemoglobin causing stimulation of red blood cell production and consequent increased destruction; associated with at least 20 mutations
α-thalassemia	Reduced synthesis of α chain; principal cause is deletion of one or more α-gene copies; *hydrops fetalis* is a lethal disorder associated with absence of α chain
β-thalassemia	Reduced synthesis (β^+ form) or no synthesis (β^0 form) of β chain; some patients have β deletions; at least one β^0 form involves a mutation at the splice junction at the downstream end of the second intron
Hereditary persistence of fetal hemoglobin	Continued synthesis of γ chains; postulated deletion of DNA sequence responsible for $\gamma \rightarrow \beta$ switch

human hemoglobin are known, and several of the most important types are summarized in Table 11.1).

Rates of Mutation

Missense, nonsense, and frameshift mutations are the simplest changes that occur in DNA. Other, more complex, mutations include deletions, duplications, inversions, and, as discussed in a few pages, insertions. Certain questions about mutations inevitably arise: Where do mutations come from? How do mutations happen? How frequently do mutations occur? Relative to these issues it is important to distinguish between newly arising mutations and preexisting mutations. Most mutations in an individual are not the result of new mutations in that person; most such mutations are inherited from the previous generation, and they may have originated many generations in the past. There is a vast reservoir of genetic damage in normal, apparently healthy individuals, but it is concealed because the mutations are recessive and heterozygous and therefore hidden by their corresponding normal alleles. In every generation, each of these mutations segregates in sperm and eggs, and each runs the risk of becoming homozygous if it meets a gamete that carries a similar mutation at the same locus. In most instances the recessive mutations do not become homozygous, however. Rather they are passed from heterozygous parent to heterozygous child, from generation to generation, seldom giving warning of their presence. In addition to these recessive mutations, there are the many dominant mutations that have incomplete penetrance and are therefore not always expressed. These, too, can be transmitted from generation to generation. And on top of all this are the offspring of parents who themselves have inherited disorders. Color blindness, hemophilia, sickle cell anemia, Huntington disease, and many other inherited disorders do not always prevent reproduction by affected individuals. Many, and indeed most, mutations present in individuals in any generation are inherited from the previous generation.

The ultimate source of mutation is the spontaneous, unpredictable change of a normal gene into a mutant form. These mutations are said to be **spontaneous** because they occur in normal genes in individuals who have had no previous contact with radiation or mutation-causing chemicals. There is a small amount of radiation everywhere on earth, of course, and many exotic chemicals are present in the environment. Nevertheless, only a few spontaneous mutations can firmly be attributed to these sources. Spontaneous mutations are random and unpredictable. The famous "Royal hemophilia" among the descendants of Queen Victoria probably arose as a spontaneous mutation in the queen herself or in the germ cells of her father or mother (see Chapter 5).

Many spontaneous mutations seem to arise from errors that occur during the replication of DNA, and some of these are due to certain rare chemical configurations that the bases can assume. For example, Figure 11.10(a) shows the normal pairing configuration between adenine (A) and thymine (T). Note that the amino (NH_2) group attached to the number 6 atom in the purine ring has two hydrogen atoms associated with it. On rare occasions, one of these hydrogens can shift position and become associated with the number 1 atom of the ring; this shift gives rise to the rare **imino** form of adenine, and, as illustrated in Figure 11.10(b), this form of adenine can undergo hydrogen bonding with cytosine. The configurations of adenine shown in the figure are known as **tautomeric forms**. Although any particular adenine can shift back and forth between the normal and the imino form, the normal **amino** form is strongly favored so that, at any one instant, almost all the adenines will be in the normal configuration. The rare adenines that are in the imino form during DNA replication are nevertheless important because this form pairs with cytosine. At the next replication the adenine will almost surely have returned to its normal configuration, so the ^A_C base pair will be resolved in the daughter molecules as ^A_T and ^G_C. Thus, the tautomeric shift of adenine to its enol form results in the change of a normal ^A_T base pair to a mutant ^G_C base pair—a *transition mutation* in the terminology of Figure 11.4.

All the bases in DNA can undergo tautomeric shifts. An example involving thymine is illustrated in Figure 11-10(c) and (d). Part (c) shows the normal configuration of thymine, but the hydrogen associated with atom 1 in the pyrimidine ring can shift to become associated with the oxygen attached to atom 6; this tautomeric shift converts the normal **keto** form of thymine into the rare **enol** form, and the enol form can pair with guanine. The ^T_G base pair so formed will be resolved at the next replication into a ^T_A base pair in one daughter molecule and a ^C_G base pair in the other daughter molecule. The resulting mutation is again a transition mutation.

One feature of DNA polymerase has the effect of reducing the amount of mutation due to mispaired bases in their rare tautomeric configurations. Recall from Chapter 8 that DNA polymerase has two capabilities: It is a 5′

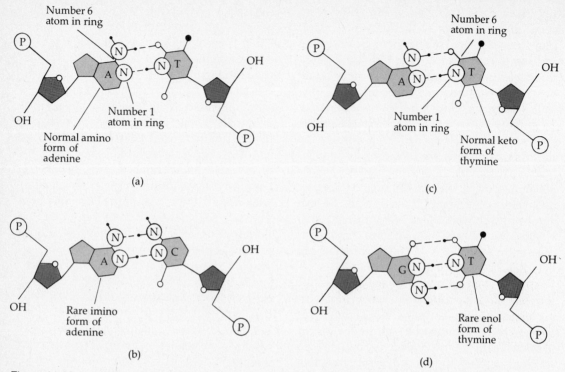

Figure 11.10   Base mispairing due to tautomeric shifts. *(a)* Normal AT base pair. *(b)* A can pair with C when A is in its rare imino form. *(c)* Normal AT base pair. *(d)* T can pair with G when T is in its rare enol form.

→ 3′ polymerase, which means that it elongates DNA strands at their 3′ end; but it is also a 3′ → 5′ exonuclease, which means that it can cleave nucleotides from the 3′ end of a DNA strand. The exonuclease function of DNA polymerase is often called its **proofreading** function. After each new nucleotide is added to a growing DNA strand, the enzyme checks again or "proofreads" the nucleotide to determine whether it forms a proper base pair. If it does not, the enzyme uses the exonuclease capability to remove the last-added nucleotide from the DNA strand. For example, if adenine in the template strand is in its imino form, the polymerase will add a cytosine nucleotide to the daughter strand; but the adenine will al-

most instantaneously shift back to the amino form, and an improper ^A_C base pair will be recognized by the polymerase during proofreading, so the incorrect cytosine nucleotide will be cleaved off again.

Another relatively frequent cause of spontaneous mutation is **deamination**, a rare chemical occurrence in which an amino group ($-NH_2$) becomes reactive with water and is replaced with oxygen ($=O$). Figure 11.11(*a*) shows that deamination of cytosine results in uracil. Since uracil undergoes hydrogen bonding with adenine, deamination of cytosine eventually would lead to a ^G_C-to-^A_T transition. However, uracil is not a normal constituent of DNA, and cells have an enzyme that can detect its presence. This enzyme is known as **DNA-uracil glycosidase**, and its function is to excise uracil-containing nucleotides from DNA strands; the gap left in the strand is then filled in with the correct nucleotide by other repair enzymes. By means of the excision and repair of DNA strands that contain uracil, the amount of mutation due to the deamination of cytosine is much reduced.

In Chapter 10 we mentioned that certain bases, particularly cytosine, are sometimes found in DNA in a chemically modified form. The most common modification of cytosine involves the addition of a methyl group ($-CH_3$) to the number 5 atom in the pyrimidine ring. Such a modification results in the base **5-methyl cytosine**, which is illustrated in Figure 11.11(*b*). Deamination of 5-methyl cytosine is not as easily corrected as deamination of cytosine, because deamination of the methylated base produces thymine [see Figure 11.11(*b*)], not uracil, and thymine is not subject to removal by DNA-uracil glycosidase. Thus, 5-methyl cytosines would seem to be particularly prone to undergoing transition mutations and this has been shown to be the case for the *lacI* gene in *Escherichia coli*. Recall from Chapter 10 that *lacI* codes for the repressor of the lactose operon. Within this gene there are certain **hotspots**—sites in the gene at which spontaneous mutations occur much more frequently than at other sites. These hotspots turn out to be the positions of 5-methyl cytosines. However, in strains that do not methylate certain cytosines, the hotspots associated with these cytosines no longer occur.

The essential situation regarding deamination of unmethylated and methylated cytosine is summarized in Figure 11.12. Part (*a*) shows deamination of an unmethylated cytosine. The result of the change depends on whether the uracil nucleotide is removed be-

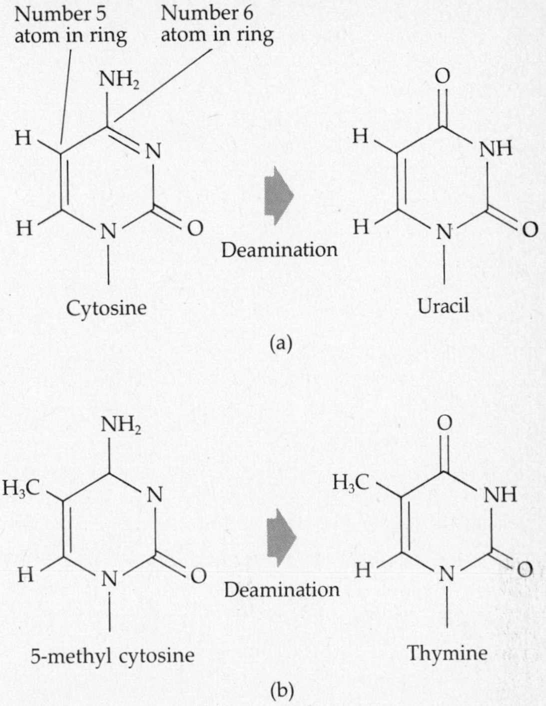

Figure 11.11   (*a*) Deamination of cytosine produces uracil. (*b*) Deamination of 5-methyl cytosine produces thymine.

fore or after the next replication. If removed before replication (top), no mutation results; if removed after replication (bottom), a ^G_C-to-^A_T transition will have occurred in the daughter duplex that originally contained the U. In the case of deamination of 5-methyl cytosine, illustrated in part (*b*), the uracil-removing enzyme plays no role in error correction, and a round of DNA replication leads to the ^G_C-to-^A_T transition. Of course, the ^G_T mismatch is susceptible to repair by other enzymes, such as the ones involved in gene conversion (see Chapter 8), but the mismatch cannot be corrected by

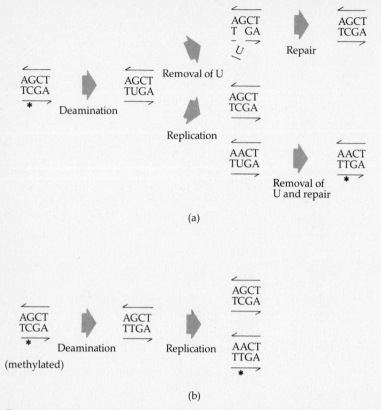

(a)

(b)

Figure 11.12   *(a)* Deamination of cytosine produces uracil in DNA, which can be removed and repaired by enzymes in the nucleus. If repaired before replication (top), no mutation results; if repaired after replication (bottom), a mutation results because the unrepaired U causes incorporation of A during replication. *(b)* Deamination of 5-methyl cytosine produces thymine in DNA, which cannot be removed. After replication, one daughter molecule will carry a mutation.

DNA-uracil glycosidase. For this reason, 5-methyl cytosines are hotspots for mutations in the *lacI* gene, but whether hotspots in other genes or in eukaryotes have the same chemical basis is not yet known.

Cellular functions such as the proofreading capability of DNA polymerase, the uracil-removing capability of DNA-uracil glycosidase, the mismatch repair capability of other enzymes, and still other DNA repair capabilities all serve to reduce the occurrence of spontaneous mutations below the level that would be expected from purely chemical considerations. The upshot is that spontaneous mutations occur only rarely. In spite of their rarity, though, spontaneous mutations are frequent enough that their rate of occurrence can be measured. Measuring **mutation rates** in such experimental organisms as bacteria, yeast, fruit flies, mice, and cultured cells is straightforward. The procedure in the case of yeast and bacteria is merely to introduce a known number of cells into a selective medium that permits only those cells that carry a particular new mutation to grow; the mutation rate is then calculated from the proportion of surviving cells. In fruit flies and mice, mutation rates can be measured by means of special breeding programs that allow certain kinds of mutations such as those with visible effects or recessive lethals to be identified. The measurement of mutation rates in humans is not so straightforward, however. Human beings are not noted for their willingness to arrange their matings for the convenience of geneticists. On the other hand, people do tend to keep detailed marriage and medical records, especially in industrialized countries, and hospital records are a vital source of genetic information.

In some instances the mutation rates of specific genes in humans can be measured directly. Dominant mutations that cause specific abnormalities not mimicked by nongenetic abnormalities or by mutations in other genes can be studied with this in mind, but the penetrance of the mutations must be nearly complete; that is, virtually all individuals who carry the mutation must express the trait for the measurement of the rate to be accurate. One disorder that fulfills nearly all these requirements is the type of dwarfism known as **achondroplasia**, which was mentioned in Chapter 3. In a study of 94,075 Danish newborns, 8 exhibited the trait. Since the parents of these children were normal, one can assume no dominant alleles were present in them, so the affected children must carry new mutations. The 94,075 children represent 188,150 sets of chromosomes (i.e., $2 \times 94,075$). In 8 of these chromosome sets a new mutation to achondroplasia occurred, so the estimated rate is $\frac{8}{188,150}$, which is about 0.00004 or 4×10^{-5} per generation. (This number is likely to overestimate the mutation rate of a single gene because at least two different genes can mutate to cause achondroplasia.)

The rates of occurrence of other dominant mutations, and of X-linked and autosomal recessives, can be estimated indirectly. By measuring the loss in reproductive ability of affected individuals, one can estimate how many mutations must occur every generation to maintain the frequency of the disorder at its observed level in the human population. This calculation grows out of the simple observation that if the frequency of the disorder is neither increasing nor decreasing, then the number of mutant genes eliminated from the population by the reduced reproduction of affected individuals must be exactly counterbalanced by the number of new mutant genes introduced into the population by new mutations. The mutation rates of several genes have been estimated in this manner, and the rate averages between 1 per 100,000 (i.e., 10^{-5}) and 1 per million (i.e., 10^{-6}) genes per generation (see Table 11.2). It must be emphasized that this range represents an average, a "ballpark" estimate. The mutation rate actually varies from gene to gene, and the average may be biased on the high side because genes that mutate at exceptionally low rates are not included among those studied. Despite these reservations, the estimate that mutations occur spontaneously in 10^{-5} to 10^{-6} genes per generation seems to be fairly reliable. It also agrees well with information derived from laboratory animals.

A rate of 10^{-5} to 10^{-6} mutations per gene per generation can be interpreted in two ways. One way is to focus on a particular locus and inquire how many normal alleles at this locus undergo mutation in any one generation. Imagine a collection of between 100,000 and 1 million sperm produced by a man homozygous for the normal allele at, say, the phenylketonuria locus. Then a mutation rate of 10^{-5} to 10^{-6}

TABLE 11.2 ESTIMATES OF SPONTANEOUS MUTATION RATES IN HUMANS

Trait	Estimated number of spontaneous mutations per 100,000 gametes per generation
Epiloia	0.8
Aniridia	0.5
Microphthalmus	0.5
Wardenberg syndrome	0.4
Facioscapular muscular dystrophy	0.5
Pelger anomaly	0.9
Myotonia dystrophica	1.6
Myotonia congenita	0.4
Huntington disease	0.2
Retinoblastoma	0.4
Neurofibromatosis	13–25
Hemophilia	2.7
X-linked muscular dystrophy	5.5

Note: Estimates of mutation rates in humans can be obtained by a variety of methods. In many cases the estimates are quite imprecise, yet most estimates are between 10^{-5} and 10^{-6} mutations per gene per generation. Except for hemophilia and X-linked muscular dystrophy, all the traits listed are due to autosomal dominant mutations.

means that, on the average, one sperm will actually carry an allele of this locus that has undergone a spontaneous mutation. The same reasoning would apply to a collection of 100,000 to 1 million eggs from homozygous normal women. This interpretation suffers from the fact that the estimated mutation rate is only an average. The calculation is in error to the extent that the phenylketonuria locus or whatever locus is under consideration may have an atypical mutation rate that deviates from the average for all loci.

A second interpretation of the meaning of the human mutation rate is obtained by focusing on all the genes in a single cell at once. If a human sperm or egg contains, say, very approximately, 50,000 genes, then what percentage of these gametes will carry a gene that

has undergone a spontaneous mutation? The answer is somewhere between $50,000 \times 10^{-5} = 0.50 = 50$ percent and $50,000 \times 10^{-6} = 0.05 = 5$ percent. In other words, roughly somewhere between 5 and 50 percent of all sperm or eggs will carry at least one new mutation. This may seem like a large number, but it is actually only a small fraction of the mutations already present in the sperm and eggs of normal individuals.

Insertion Mutations and Transposable Elements

For many years it was thought that most spontaneous mutations involved relatively simple changes at the nucleotide level—transitions, transversions, or perhaps an occasional small insertion or deletion or inversion. Recombinant DNA techniques have changed this picture completely. It is now clear that many spontaneous mutations involve massive changes at the nucleotide level, changes such as the insertion of thousands of nucleotides right in the middle of a gene. In many cases the nucleotide sequence of the large insertion can be recognized as one of a **family** (or group) of similar or identical sequences present in 20 or 30 or more copies and **dispersed** (scattered) throughout the genome. In *Drosophila melanogaster*, for example, one such family of sequences is called **copia**; the copia sequence is about 5 kb long and it is found at about 30 widely scattered chromosomal locations (Figure 11.13). What is unique about copia and other analogous **dispersed repeated gene families** (as they are sometimes called) is that the sequences are able to change location within the genome; that is, they are able to move from one place in a DNA molecule to a different place in the same molecule or to a different molecule altogether. Because of their mobility, such sequences are often called **transposable elements**. Transposable elements seem able to

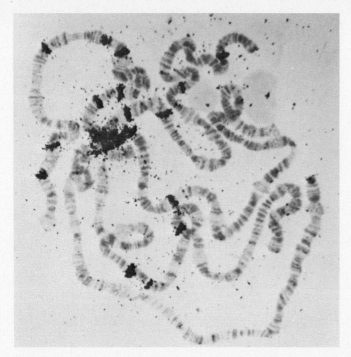

Figure 11.13 Autoradiograph showing many sites of hybridization of copia to *Drosophila* salivary-gland chromosomes.

become inserted at virtually any site in a DNA molecule; when one transposes into a site in an existing gene, the insertion often disrupts the gene's ability to function. Transposition thus becomes a source of spontaneous mutation.

The prototype transposable elements are found in prokaryotes, and eukaryotic transposable elements have many similarities with their prokaryotic counterparts. In *E. coli*, for example, a number of transposable elements called **insertion sequences** (or **IS sequences**) are a normal constituent of the bacterial DNA; the number of copies of each IS sequence varies from one or two for the sequence designated IS5 to five to eight for IS1. All known insertion sequences have one common feature: Their terminal sequences are repeats, usually inverted but sometimes direct. This aspect of structure is illustrated for the element IS1 in Figure 11.14. The entire element consists of 768 nucleotide pairs, but only the 23 pairs at each end

are shown in detail. Note that the left end is an almost perfect inverted repeat of the right end; the sequences fail to match at only 5 of the 23 sites (arrows). Such near identity of flanking sequences is a common feature of transposable elements in both eukaryotes and prokaryotes, but in some cases the sequences are direct repeats rather than inverted repeats. The copia element in *Drosophila* has a nearly perfect 276-nucleotide-pair direct repeat at its ends, for example; it is curious that each direct repeat has a less perfect 17-nucleotide-pair inverted repeat at *its* ends.

When two insertion sequences of the same kind are sufficiently close together on the bacterial DNA, they can become mobilized together; the whole structure, consisting of the two IS sequences and the DNA between them, can therefore transpose as a unit. These larger transposable elements flanked by insertion sequences are called **transposons**, and they play a

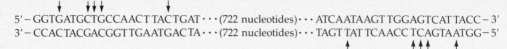

5′ – GGTGATGCTGCCAACT TACTGAT · · · (722 nucleotides) · · · ATCAATAAGT TGGAGTCAT TACC – 3′
3′ – CCAC TACGACGGT TGAATGAC TA · · · (722 nucleotides) · · · TAGT TAT TCAACC TCAGTA ATGG – 5′

Figure 11.14 Sequence of *E. coli* insertion sequence IS1 showing nearly perfect inverted repeat of the 23 base pairs at its ends. Nonmatching base pairs are indicated by arrows.

key role in microbial evolution. In Chapter 8 we mentioned bacterial resistance transfer factors—infectious plasmids that carry genes conferring resistance to one or more antibiotics. In many cases, an antibiotic-resistance gene on a plasmid is actually part of a transposon. Evidently, a newly evolved gene for antibiotic resistance can become mobilized by flanking insertion sequences and thereby become a new transposon. When the new transposon inserts into an infectious plasmid, it can become widely disseminated among many bacterial species, even some distantly related species. Indeed, as the plasmid spreads, it may even accumulate additional antibiotic-resistance genes by the insertion of other antibiotic-resistance transposons. Resistance transfer factors that carry genes conferring resistance to six or more different antibiotics are known to occur; these are thought to have arisen by the successive insertion of transposons.

Figure 11.15 illustrates the structure of a well-studied bacterial transposon, designated Tn5, which codes for resistance to the antibiotic **neomycin**. The entire 5.7-kb element consists of a 2630-bp (bp stands for **base pairs**) unique sequence flanked by 1534-bp inverted repeats (heavy lines). The wavy lines beneath the transposon depict the origin and direction of transcription; the circled P represents the triphosphate at the 5′ end of the mRNA. The shaded rectangles depict the polypeptides produced from the transcripts. The right-hand inverted repeat codes for two polypeptides—a case of overlapping genes; the polypeptides are translated in the same reading frame, but one is initiated at a different site from the other. One or both of these polypeptides are associated with the **transposase** function; that is, one or both are involved in catalyzing the transposition of Tn5. The left-hand inverted repeat of Tn5 is also transcribed, but the polypeptides produced are virtually nonfunctional because translation is prematurely terminated owing to an ochre (UAA) nonsense codon present at the position indicated. The G-to-T transversion that creates the ochre terminator is the only difference between the inverted repeats, but it is important because the very same mutation creates the promoter for the neomycin-resistance gene.

Each transposable element has its own characteristic **rate of transposition**—that is, the probability of undergoing transposition each generation—and the rates vary greatly from approximately 10^{-3} to 10^{-5}. These rates are very much greater than the rates of spontaneous nucleotide substitutions, and they provide many insertion mutations with the curious property of **instability**, which refers to the tendency of a mutated gene to undergo further mutation. As noted earlier, the rate of spontaneous mutation from a normal gene into a mutant form ranges from 10^{-5} to 10^{-6} per generation; this rate is often called the **forward** mutation rate because it involves mutation from a normal to a mutant allele. **Reverse** mutation refers to the opposite situation, in which a mutant allele undergoes another mutation that converts it back into the normal form. Rates of reverse mutation are typically very

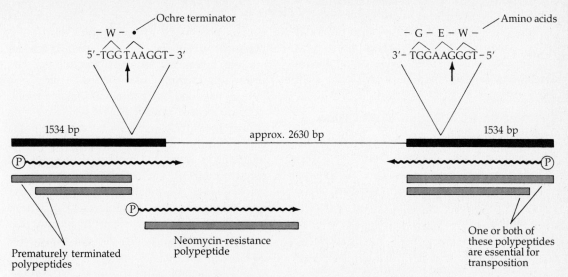

Figure 11.15 Structure of bacterial transposon Tn5. The neomycin-resistance gene is flanked by a nearly perfect 1534-bp inverted repeat. The one mismatch in the inverted repeats is indicated by the arrows, which, in the left repeat, creates an ochre terminator (UAA) and simultaneously creates the neomycin-resistance promoter. The two complete polypeptides from the right inverted repeat represent a case of overlapping genes, but these are translated in the same reading frame.

much lower than rates of forward mutation—lower by a factor of 100 or 1000 or more. Why this is the case can be appreciated by considering a simple base-substitution mutation. A normal allele can be rendered mutant by any of a large number of possible substitutions because there is a wide scope for amino acid substitutions or the creation of nonsense codons; most of these changes can render the corresponding protein nonfunctional or poorly functional. For the reverse situation, the number of possible mutations that can restore the function of an already mutant allele is strictly limited; in many instances, the allele's function can be restored only by a reverse substitution involving the identical nucleotide responsible for the original forward mutation. The many possibilities for forward mutation contrasted with the restricted possibilities for reverse mutation accounts for the large discrepancy in the rates of forward and reverse mutation.

Mutations that result from the insertion of a transposable element into a gene have an easy avenue for reversal, however. All that is required is that the transposable element undergo another transposition from its site in the gene. As mentioned above, rates of transposition can be 10^{-4} per generation or higher, so many insertion mutations will have rates of reverse mutation many orders of magnitude larger than would be expected from simple base-substitution considerations. Such relatively easily reverted mutations are known as **unstable** mutations; the discovery of unstable mutations in corn and *Drosophila* provided the first indications of transposable elements in eukaryotes. The origin and functions of eukaryotic transposable elements are unknown, but two items are worth mention. One is that many repeated gene families are transcribed and translated; indeed, in *Drosophila*, the most abundant mRNA's in tissue culture cells are transcripts of such sequences. The second item of note is that the class of moderately repetitive DNA sequences discussed in Chapter 8 may consist largely of families of transposable elements.

Radiation As a Cause of Mutation

Although many mutations occur spontaneously, mutations can also be induced by certain kinds of radiation and chemicals. Agents that can cause mutations, called **mutagens**, are of great concern because even a small increase in the mutation rate from these sources is amplified into an enormous number of unnecessary birth defects because of the large number of genes that can be mutated and the large size of the human population. It has been estimated, for example, that if the mutation rate were increased by 20 percent for one generation and then allowed to slip back to its normal level, the consequences of that temporary increase in the United States alone would be about 400,000 new, severe genetic defects spread over many generations, with perhaps 40,000 new genetic defects being expressed in the first generation. This is a staggering biological and social cost, and it should be emphasized that corresponding genetic benefits from new mutations are so rare as to be virtually nonexistent. Mutations occur randomly, and practically no mutation—of either spontaneous or induced origin—is beneficial. The kinds and frequencies of mutations that occur are not determined by what will or will not be desirable in the cell. A cell that has a defective protein has no way of "sensing" that the protein is defective, so it cannot, on the basis of recognition of a defective protein, cause a corrective mutation in the gene encoding the protein to restore the function. The chance that a random mutation should actually prove beneficial has been likened to the chance that an intoxicated monkey with a screwdriver could produce a finer ad-

justment of an already accurate watch. Of course, a very few extremely rare new mutations are beneficial and do occur; such favorable mutations are indispensable for the continuing evolution of a species.

X-rays and other forms of ionizing radiation are potent mutagens. These penetrating rays pierce through cells and along their paths cause electrons to be ejected from their usual locations in atoms and molecules; the affected electrons dart off and collide with other molecules, causing more damage, and many of the ions (charged atoms or molecules) produced by molecules that lose electrons become extremely reactive. In the nucleus these ions may combine with bases in the DNA and cause mistakes in base pairing during replication, leading ultimately to base substitutions; they may also sever the sugar-phosphate backbones of DNA and thereby cause physical breaks in the chromosomes (see Figure 11.16, for example).

A basic unit of radiation is the **curie (c)**, which corresponds to an amount of material that produces 3.7×10^{10} radioactive emissions per second. Another unit, the **roentgen (r)**, pertains to the number of ionizations produced by radioactive emissions; 1 r corresponds to an amount of ionizing radiation sufficient to produce two ionizations per cubic micron of water or living tissue. Although the roentgen is widely used as a measure of radiation, it is not strictly applicable to nonionizing radiation such as ultraviolet light, which is far less penetrating than ionizing radiation and does not produce ionizations. In biology, radiation dose is often calculated according to the amount of energy absorbed by living tissue; this absorbed-energy unit is called a **rad**; for ionizing radiation, 1 rad is slightly greater than 1 r. Radiologists often use a different measure that is convenient for comparing the damage done to human tissue by different kinds of radiation; this unit is the **rem**, with 1 rem defined as the quantity of

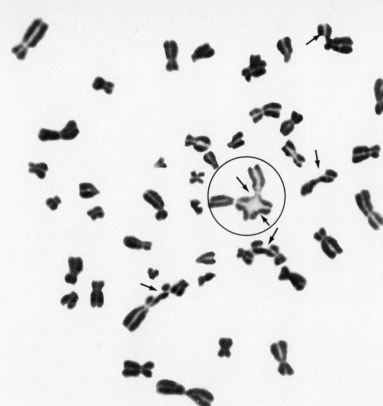

Figure 11.16 Human chromosomes showing breaks (arrows) induced by 270 r (roentgen) of x-rays. The circled structure arose from interchange and restitution of broken chromatids in two chromosomes; the positions of the breaks before restitution are denoted by arrows within the circle. A dose of only 20 r of x-rays is sufficient to produce one visible chromosome break per cell in human chromosomes in tissue culture.

radiation sufficient to produce the same damage to human tissue as 1 rad of high-voltage x-rays. Thus, for x-rays and other forms of ionizing radiation, the roentgen, the rad, and the rem are nearly equivalent amounts of radiation.

Some exposure to radiation is unavoidable because there are certain background sources such as cosmic rays, radioactive elements in the earth and on its surface, and radioactive elements in food that are incorporated into the body's tissues. The average American receives about 0.05 rem/yr from cosmic rays, about 0.06 rem/yr from terrestrial radiation, and about 0.02 rem/yr from ingested radioactive elements. The inescapable background radiation is thus about 0.13 rem/yr, or roughly 6 rem between birth and age 45.

X-rays are mutagenic, and the frequency of mutations induced by moderate doses of x-rays is strictly proportional to the dose; the more x-rays a cell receives, the more mutations and chromosome breaks will be produced. The proportionality between x-ray dose and mutation rate for X-linked recessive lethals in *Drosophila* is illustrated in Figure 11.17; the **dose-response relationship** is a straight line, and there is about a 3 percent increase in the mutation rate for every 1000 r of x-rays. The x-ray doses in Figure 11.17 are very large

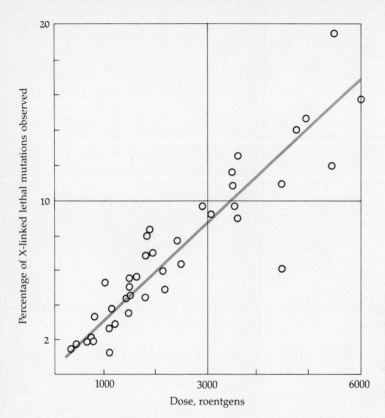

Figure 11.17 Summary of studies on the induction of lethal mutations in the X chromosome of *Drosophila* by x-rays. Note that most of the points fall close to the straight line. Thus, the rate of induction of new lethal mutations is proportional to the dose of x-rays over the range of doses shown here.

because adult *Drosophila* can survive such massive amounts of x-rays. Adult humans, by contrast, are much more sensitive to x-ray damage; doses larger than 450 r usually cause the death of the exposed individual. In human cells in tissue culture, exposure to 20 r of x-rays induces at least one visible chromosome break per cell.

The simple straight-line relationship between mutation frequency and x-ray dose illustrated in Figure 11.17 breaks down at very high doses because, at such high doses, many cells are killed outright and many survivors have two or more mutations. The dose-response relationship is also more complex with regard to translocations and inversions and other chromosomal abnormalities that require two breaks; the reason is that appreciable numbers

of cells with two or more breaks do not occur until relatively large doses of x-rays are incurred. At the other end of the spectrum, when the x-ray dose is very small, the dose-response relationship is difficult to determine because, at low doses, the overall mutation rate is not very different from the spontaneous mutation rate. Although it is generally believed that the proportionality between mutation rate and x-ray dose extends down to very small doses, so that any amount of radiation would be expected to increase the mutation rate, the matter is by no means settled. In addition, the amount of mutational damage produce by x-rays depends on whether the radiation is **acute** (delivered all at once) or **chronic** (delivered as a series of smaller doses). Acute doses are generally three or four times more effective in producing mutation; the difference apparently is that chronic irradiation gives the cells more opportunity to repair certain kinds of genetic damage in the DNA.

Radiation that hits the testes and ovaries —the **gonadal dose**—is obviously of greatest importance to future generations. Radiation also affects somatic cells, however. People cannot usually survive acute doses over 450 r, and 100 r will produce **radiation sickness**—a condition marked by nausea, vomiting, headache, cramps, diarrhea, loss of hair and teeth, decrease in the number of blood cells, and prolonged hemorrhage. Dividing cells, such as those in the bone marrow, are generally more sensitive to radiation than are nondividing cells. People accidentally exposed to high doses of radiation generally experience for some time after exposure a much lowered resistance to infectious disease. This is evidently because the rapidly dividing white blood cells, which play a central role in immunity, are very sensitive to x-rays. Certain cancers are also treated with ionizing radiation, often the gamma rays from radioactive cobalt; the reason is that cells in a tumor are rapidly dividing and therefore more sensitive to radiation than the surrounding cells, which are not dividing. The embryo, loaded with rapidly dividing cells, is also very sensitive to x-rays; this is why women who are pregnant (or who might be) must not casually allow themselves to be exposed to ionizing radiation.

A convenient measure of the mutagenic effects of radiation is the **doubling dose**—the dose that induces a frequency of mutation exactly equal to the spontaneous frequency so that the overall frequency is doubled. This is a useful measure despite the fact that the relative proportions of specific kinds of mutations caused by x-rays are somewhat different from those arising spontaneously; x-rays generally produce more deletions and more lethal mutations, for example. The doubling dose in mice

is about 40 rad, and it is thought to be about the same in humans. The question is: How much radiation do we receive over and above natural background radiation? Certain people such as x-ray technicians are exposed to radiation in their work; their exposure is limited by federal law to less than 0.3 rad/wk. Most people receive far less radiation. A significant source of radiation is the radioactive debris trapped in the upper atmosphere from previous above-ground atom bomb tests. This debris leaks slowly back to earth, and it exposes everyone to about 0.008 rem/yr. This exposure may not seem large, but it is about 6 percent of the background level, and the cumulative dose from birth to age 45 due to this source alone is about 0.4 rem.

For most people, the most important single source of ionizing radiation other than background is medical and dental x-rays. On the average, the exposure from these sources is about 0.073 rem/yr, but there is wide variation from individual to individual. People who are chronically ill and in frequent need of diagnostic x-rays receive doses many times larger than the average, whereas others may never have had an x-ray in their lives. As always, the gonadal dose is the most important for future generations, and this risk is greatest in x-rays of the abdominal region. With proper shielding of the reproductive organs, the gonadal dose is negligible for dental x-rays or for x-rays of the skull, neck, upper chest, and limbs. On the average, a male in the United States will receive a testicular dose of about 1.3 rem from diagnostic x-rays from birth to age 45; the ovarian dose to an average female would be less, about 0.3 rem, since the ovaries are inside the body cavity and therefore more protected. Although the gonadal dose from dental and medical x-rays is moderate, it should be emphasized that unnecessary x-rays are always to be avoided.

Another mutagenic form of radiation is

ultraviolet light—a short-wavelength component of ordinary sunlight that causes the skin to tan or, more painfully, burn. Ultraviolet rays are not very penetrating, so the damage they do directly is usually confined to cells in the skin. Ultraviolet light affects DNA in at least two ways. The bases of DNA tend to absorb these wavelengths and become energized, which increases the incidence of mispairing. Ultraviolet light can also cause chemical crosslinks to form between adjacent thymines in a DNA strand. These linked thymines constitute a **thymine dimer**; the formation of a thymine dimer is illustrated in Figure 11.18(*a*). Thymine dimers are ordinarily repaired by a process called **excision repair**, which is illustrated in

Figure 11.18(*b*). Essentially a cut-and-patch process, repair of a thymine dimer begins when a particular **excision enzyme** recognizes the dimer and cleaves the DNA backbone to remove the dimer. Exonuclease enzymes then widen the gap by removing additional nucleotides, and repair enzymes take over and restore the original base sequence by resynthesizing the excised piece using the intact strand as a template.

Ultraviolet light is usually of no concern genetically because it does not penetrate to the germ cells in the testes or ovaries. Ultraviolet light does damage the genetic material in the living cell layers in the skin, although this damage is often repaired by the appropriate

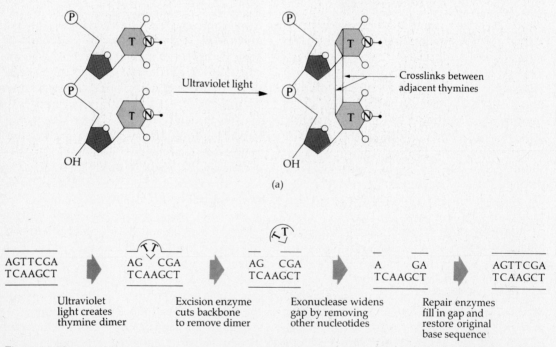

(a)

Figure 11.18 *(a)* Ultraviolet light can cause chemical crosslinks (covalent bonds) between adjacent thymines in a strand of DNA, as shown here. Recall that the rings of the bases in DNA are rather flat and are stacked on top of one another; thus, the atoms shown crosslinked are in close physical proximity. *(b)* Mechanism of repair of crosslinked thymines involving excision, widening of the gap, and resynthesis.

enzymes. However, some individuals have a rare disorder known as **xeroderma pigmentosum**, which is inherited as an autosomal recessive. Patients with xeroderma pigmentosum are extremely sensitive to ultraviolet light, and in childhood or early teens most of them develop multiple skin cancers (Figure 11.19), which usually lead to death before age 20. Although there appear to be several forms of the disorder, it is now known that at least some xeroderma pigmentosum patients have defective DNA repair systems so they are unable to repair the damage induced by ultraviolet light. Xeroderma pigmentosum provides one of the many links between mutation and cancer. Other links will be discussed in a few pages.

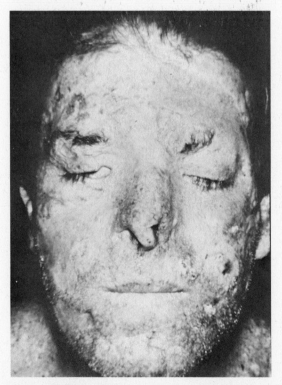

Figure 11.19   This person has xeroderma pigmentosum and thus has multiple skin tumors.

Chemicals As Mutagens

Hundreds of chemicals are known to be mutagenic; several examples are shown in Figure 11.20. The agent **nitrous acid** causes deamination; that is, it removes NH_2 groups from a base such as cytosine or adenine and substitutes an oxygen atom. The role of spontaneous deamination of cytosine in mutagenesis was discussed earlier in this chapter in connection with Figures 11.11 and 11.12. Figure 11.20(a) shows the production of uracil from a nitrous-acid-induced deamination of cytosine. As noted earlier, uracil-containing nucleotides can be removed from DNA by the enzyme DNA-uracil glycosidase. Uracils that escape removal cause mutations because U pairs with A, so the effect of the deamination is to create a $\frac{G}{C}$-to-$\frac{A}{T}$ transition. It should be emphasized again that cytosines that are methylated at their number 5 position are potent targets for deamination-induced mutation. Methylated cytosines are hotspots for such mutations because the deamination of 5-methyl cytosine produces thymine (see Figure 11.11), which is a normal constituent of DNA and so is not removed by any enzyme; since T pairs with A, deamination of 5-methyl cytosine produces a $\frac{G}{C}$-to-$\frac{A}{T}$ transition. Nitrous acid also deaminates adenine, producing a base that pairs with cytosine; deamination of adenine thereby causes $\frac{A}{T}$-to-$\frac{G}{C}$ transitions.

Nitrous acid exemplifies a class of chemicals that are mutagenic because they attack and chemically alter bases in the DNA. Another class of mutagens consists of **base analogues**; these substances are chemically similar to the bases and can be incorporated into DNA in place of one of the normal constituents. Base analogues are mutagenic because they mispair more frequently than do the normal bases and so increase the mutation rate. An example of a base analogue is **5-bromouracil**, illustrated in Figure 11.20(b). This base analogue is incorporated into DNA in place of thymine, and it has

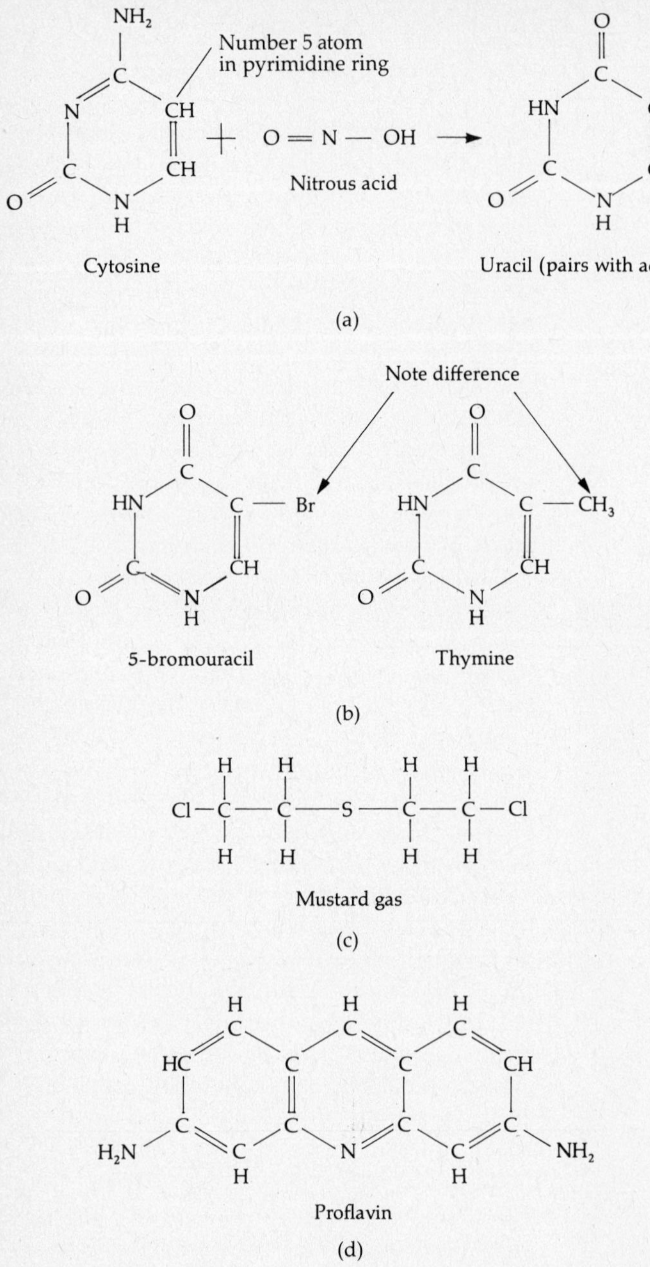

(a)

(b)

Mustard gas

(c)

Proflavin

(d)

Figure 11.20 Some chemical mutagens. *(a)* Nitrous acid causes deamination and converts cytosine to uracil; see also Figure 11.11. *(b)* 5-bromouracil is incorporated into DNA in place of thymine but often pairs with guanine; see also Figure 11.10. *(c)* Mustard gas is an alkylating agent and so a potent mutagen. *(d)* Proflavin inserts itself between the bases of a DNA molecule, causes distortions of the molecule during replication, and so leads to single-base deletions or insertions (frameshift mutations.)

the same keto and enol tautomeric configurations that thymine does [see Figure 11.10(c) and (d)]. However, 5-bromouracil is much more likely than thymine to assume the enol form; in this configuration the base analogue pairs with G instead of A, so 5-bromouracil causes ^A_T-to-^G_C transitions. Incidentally, the deoxynucleotide of 5-bromouracil is **bromodeoxyuridine (BUdR)**, which was discussed in Chapter 1 as causing differential fluorescence of sister chromatids in connection with the study of mitotic sister-chromatid exchange. If you drink coffee or cola beverages, you may be interested to know that caffeine was once thought to act as a base analogue and to be weakly mutagenic in yeast and bacteria; more recent tests have failed to confirm its mutagenicity, however.

A third class of mutagens consists of **alkylating agents**, of which **mustard gas** [see Figure 11.20(*c*)] is an example. Alkylating agents are among the most potent mutagens known. They are highly reactive chemicals that become attached to the bases in DNA and so cause errors in replication, although various repair enzymes can evidently undo some of their effects. Mustard gas was one of the poison gases used systematically against people during World War I. Both sides had canisters ready for the same purpose in World War II, but it was never used because neither side wanted to be the first to unleash it.

Yet another class of mutagens consists of the **acridines**, of which **proflavin** [see Figure 11.20(*d*)] is an example. Acridines interact with DNA by becoming inserted between the bases of the molecule; the distortions produced by the intercalated acridines cause errors in DNA replication. The errors typically involve one nucleotide being skipped or an extra nucleotide being inserted, which corresponds genetically to a single-base deletion or a single-base insertion, respectively. Acridines thus cause frameshift mutations because, when the deletion or insertion occurs in a coding region of DNA, the translational reading frame of the mRNA will be shifted one nucleotide to the right or to the left, respectively, which will lead to an incorrect amino acid sequence (see Figure 11.8).

The Ames Test

The testing of chemicals for mutagenic activity has been greatly aided by the development of a procedure called the **Ames test** after its inventor. The organisms used in the Ames test are special strains of the bacterium *Salmonella typhimurium*. These strains are highly sensitive to mutagens because they are defective in the excision-repair system for correcting DNA damage, and they also have cell membrane defects that make them highly permeable to chemicals; some tester strains also carry a resistance transfer factor that enhances mutagen sensitivity. The Ames test relies on certain mutations in the *histidine* operon, which contains the genes for the synthesis of the amino acid histidine. Some tester strains have a known base-substitution mutation, so mutagens that cause base substitutions will revert these mutations and so permit growth on media lacking histidine. Other tester strains have a known frameshift mutation, so frameshift mutagens will revert these and permit growth on media lacking histidine. In addition, the chemical to be tested is first mixed with an extract of liver, typically rat liver, because many chemicals are not themselves mutagenic but are converted into mutagenic forms by enzymes in the liver. (The liver is the primary organ for metabolizing foreign chemicals in mammals.)

A typical Ames test involves mixing the suspected mutagen with the liver extract and with about 10^9 cells of one of the tester strains. This mixture is then spread (**plated**) onto a Petri dish (the **plate**) containing histidine-less medium, and the plates are incubated for two days. Each colony that appears on the agar surface is a clone consisting of the descendants of a single cell whose histidine mutation had been reverted. The number of colonies appearing on the experimental plates with mutagenized cells is compared with the number appearing on control plates with non-mutagenized cells; a significant increase in the reversion rate is taken as evidence of the mutagenic activity of the suspected chemical. In practice, suspected mutagens are tested in a series of concentrations, and the dose-response relationships are determined. These dose-response curves are almost always linear (see Figure 11.21 for some examples).

The Ames test is simple, rapid, inexpensive, and exquisitely sensitive. Some chemicals can be detected to be mutagenic in amounts as small as 10^{-9}g, and a condensate of as little as $\frac{1}{100}$ of a cigarette can be shown to be mutagenic. Moreover, the test is quantitative. Chemicals need not be classified simplistically as "mutagenic" or "nonmutagenic." They can be classified according to their potency as mutagens because more than a million-fold range in potency can be detected in the *Salmonella* test.

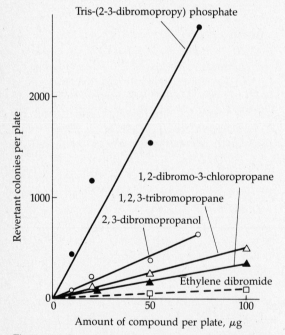

Figure 11.21 Straight-line dose-response relationship obtained with various chemical mutagens in the Ames test. The broken line indicates that the concentration of the chemical was 10 times that indicated on the horizontal axis.

Mutagens and Carcinogens in the Environment

The Ames test is only one of a group of short-term tests now favored for the rapid screening of chemicals. Other tests measure the potency of chemicals to induce mutations in *Drosophila*, to induce mutations in cultured animal cells, or to induce chromosome breaks in laboratory animals such as mice or rats. The need for screening potential mutagens and **carcinogens** (agents that cause cancer) is acute. Over 50,000 untested synthetic chemicals are already in use, and about 1000 new ones are produced commercially each year. In the past, only a small fraction of these chemicals were adequately tested prior to use, partly because thorough testing of the carcinogenic potential of a single substance in rodents is expensive in terms of both capital (up to $500,000) and time (about 3 yr). Such tests are also not feasible on an appropriate scale. For example, a widespread carcinogen that caused cancer in just 1 percent of all people in the United States would result in more than 2 million new cases of cancer. Yet to detect such a potency in test animals would require an experiment involving 10,000 rats or mice; this scale is 100 times the number of animals usually involved in such studies. Partly as a consequence of such difficulties, and partly out of ignorance, carelessness, callousness, and greed, our environment has been contaminated with many substances that are now known to be carcinogenic or mutagenic. Among a group of 168 Canadians, for example, almost all were found to have in their tissues significant quantities of chlorinated hydrocarbons such as DDT, DDE, PCB, BHC, oxychlordane, heptachlor epoxide, and dieldren. All are carcinogens. More ominously, all these chemicals are found in significant quantities in human milk.

An example of the damage of inadequate testing involves a chemical known as tris-BP. This substance was used as a flame retardant in children's polyester pajamas during 1972 to 1977. Then it was discovered that tris-BP is a potent mutagen in *Salmonella* and *Drosophila*; it interacts with human DNA and damages mammalian chromosomes; it was found to be a carcinogen in experimental tests in rats and mice; and it was found to be capable of causing sterility in laboratory animals. Moreover, the substance was shown to be absorbed through the skin, and its breakdown products could be detected in the urine of children who were wearing the treated sleepware. Unfortunately, before all this was known, more than *50 million* children were exposed to the chemical through contact with their nightclothes. The price to be paid for the tris-BP mistake is not known. Cancer has a latent period of 20 to 30 yr before it is expressed, so any increase in the cancer rate will not be detectable until the 1990s or later. Any mutagenic consequences will take generations to be revealed.

The development of rapid and inexpensive procedures such as the *Salmonella* system for detecting mutagens has eased the testing crisis somewhat. Of particular good fortune and importance is that tests for mutagenicity have proven to be extremely useful for identifying carcinogens as well. In tests involving hundreds of chemicals in *Salmonella*, for example, over 90 percent of all known carcinogens were also mutagenic. Similarly, a very high proportion of mutagens were also carcinogenic. In addition, there is a good correlation between a chemical's potency as a mutagen and its potency as a carcinogen. It therefore seems likely that some weak mutagens are in fact carcinogens but were improperly classified as noncarcinogens because of inadequate numbers of test animals. In short, most carcinogens are mutagenic, and most mutagens are carcinogenic. This discovery implies that many carcinogens produce their effects by direct interactions with DNA, and it establishes yet another link between mutation and cancer. From the standpoint of environmental health, the high

mutagen-carcinogen correspondence means that tests as simple as the *Salmonella* test can be used to identify virtually all mutagens and most carcinogens prior to their commercial use. Because the test is so rapid and inexpensive, suspects among the 50,000 synthetic chemicals already in use can readily be examined. In one case, for example, an antibacterial food additive called AF-2 used widely in fish and soybean products in Japan was found to be extraordinarily mutagenic in *Salmonella*. Subsequent tests showed it to be carcinogenic in mice, rats, and hamsters, although previous tests for carcinogenicity had had negative results. Use of AF-2 as a food additive has, of course, been prohibited, but this does not undo the 9 yr in which it was used in large amounts. As with tris-BP, it is too early to know the effects of the error.

When a mutagen, carcinogen, or **teratogen** (an agent causing birth defects) escapes notice and is released for use, it might be imagined that the agent would quickly be recognized by its effects on people. This is far from the truth. Unless the agent is exceptionally potent or causes an otherwise unusual disease, detection is extremely difficult. Two examples illustrate the point. Cigarette smoking became suspect as a carcinogen because it causes lung cancer, an otherwise relatively uncommon type of cancer. Even so, it took decades to accumulate convincing evidence because of the time delay involved in carcinogenesis. The teratogen thalidomide, which is, incidentally, not itself teratogenic but is converted into a teratogen by enzymes in the body, was identified because it causes an absence of arms and legs, an otherwise exceedingly rare birth defect (see Chapter 5). Even so, it took 5 yr to track down the agent responsible for the defect. Detection is exceptionally difficult with mutagens because the mutations they cause are relatively nonspecific; that is, a general increase in the mutation rate among humans would create a diversity of genetic defects. Although a few of these might be unique, most would be abnormalities already well known. There would simply be slightly more of everything—more achondroplasia, more hemophilia, more cystic fibrosis, more schizophrenia, more mental retardation, more diabetes, and so on. No single birth defect might be striking enough or common enough to draw attention to itself and flag the fact that something had gone wrong. For this reason, a doubling or tripling of the mutation rate might go unrecognized for decades; a smaller increase might never be detected. But the price of even a minor increase in the mutation rate is measured in hundreds of thousands of children who are needlessly crippled, deformed, or mentally deficient. This discussion emphasizes the importance of rigorous testing of chemicals before they are put into widespread use.

SUMMARY

1. A **mutation** is a heritable change in the genetic material. Mutations can involve any level of genetic organization, from gross changes in chromosome number such as polyploidy or polysomy, to aberrations in chromosome structure such as inversions, translocations, and deletions, to fine-scale changes at the molecular level such as **base substitutions** (the substitution of one nucleotide for another in the DNA). Mutations can occur in any cellular type; the most important categories are **somatic** mutations, which occur in somatic cells and are not transmitted to future generations, and **germinal** mutations, which occur in germ cells and can be transmitted.

2. Mutations can be classified in many ways depending on the purpose of the classification. One classification is based on the time of life when a mutation is expressed; those with an early or late **age of onset** are expressed relatively early or late in life. **Huntington disease** is an example of a severe genetic disorder with a relatively late age of onset. The severity of phenotypic effects provides another means of classifying mutations; **lethal** mutations are incompatible with life, and **sterile** mutations render affected individuals unable to reproduce. In experimental genetics, an important class of mutations is **conditional** mutations, which are expressed under certain environmental conditions (the **restrictive** conditions) but are not expressed under other conditions (the **permissive** conditions). In *Drosophila*, for example, **temperature-sensitive** mutations are typically expressed at 25°C (restrictive) but not at 18°C (permissive); most temperature-sensitive mutations are thought to involve amino acid substitutions that render the corresponding protein susceptible to **denaturation** (unfolding) and inactivation at the restrictive temperature. Certain mutations can be classified according to their effects on the expression of other mutations; such **modifier** mutations fall into two principal categories: **enhancers**, which amplify the expression of another mutation, and **suppressors**, which diminish the expression of the other mutation. Mutations are also sometimes classified as **dominant** or **recessive** with respect to other alleles at the locus; although useful for some purposes, this classification depends on the level of sensitivity with which the phenotype is examined; a mutation that is recessive with respect to some criterion may have incomplete or complete dominance with respect to another criterion.

3. Mutations can conveniently be classified according to their genetic basis—whether they involve gross changes in chromosome number or structure or more subtle changes at the DNA level. An important category of molecular-level mutation is **base substitutions**, which occur when one nucleotide in the DNA

is replaced with a different nucleotide. The effect of a base substitution depends on where in the gene the substitution occurs and which nucleotide is substituted for which other one. Relative to intervening sequences, the effects of base substitutions are at present unknown. Relative to coding regions, base substitutions may leave the corresponding amino acid sequence unaltered (**silent** substitutions) because one codon may be changed into a synonymous codon. Other substitutions lead to nonsynonymous codons, and the corresponding polypeptide undergoes an **amino acid substitution** (the replacement of one amino acid with another). Many amino acid substitutions can be detected by means of **protein electrophoresis** because they alter the rate of movement of the protein in response to an electric field. Among humans and most other organisms, many enzymes have two or more commonly occurring electrophoretic forms that differ by one or more amino acid substitutions due to base substitutions in the corresponding coding regions. The overall proportion of loci that exhibit such genetic variation is highly uncertain because the number of loci so far studied is small relative to the total number in the organism. Most naturally occurring alleles associated with electrophoretic variation have very small or no effects on health and vitality. The sickle cell mutation, which results in a substitution of valine for glutamic acid at position 6 in the 146-amino-acid β chain of hemoglobin, is a prominent exception because it is harmful.

4. Base substitution mutations that lead to amino acid substitutions are called **missense** mutations, and the sickle cell mutation is an example. Two broad categories of base substitution mutations can conveniently be distinguished: **transitions** involve the substitution of one pyrimidine (C or T) for another pyrimidine, or the substitution of one purine (A or G) for another purine; **transversions** involve the substitution of a pyrimidine for a purine or vice versa.

5. Not all base substitutions are silent or missense mutations. Some substitutions create new chain-terminating codons within coding regions and thereby cause premature termination of polypeptide synthesis during translation. Such substitutions are called **nonsense** mutations. In experimental organisms such as yeast and bacteria, there is a class of mutations called **nonsense suppressors**; these are mutations in tRNA genes that cause misreading of the genetic code in such a way that the amino acid corresponding to the tRNA is sometimes inserted into the polypeptide at the site of an otherwise chain-terminating codon, so complete translation of an mRNA with an internal chain terminator can sometimes occur. **Amber** suppressors misread the amber (UAG) codon, **ochre** suppressors misread the ochre (UAA) codon, and **opal** suppressors misread the opal (UGA) codon.

6. Frameshift mutations involve the addition or deletion of a number of nucleotides, but not an exact multiple of three, in a coding region of DNA. Such mutations alter the translational reading frame of the mRNA and thus change the entire amino acid sequence downstream from the site of the mutation. Additions or deletions of an exact multiple of three nucleotides do not alter the reading frame and are not frameshift mutations.

7. Regions of DNA that are sufficiently similar in nucleotide sequence to undergo base pairing can generate deletions and duplications as a result of **unequal crossing-over** in the mispaired region. Unequal crossing-over is thought to be the source of the **Gun Hill deletion** in hemoglobin β.

8. Spontaneous mutations occur in the absence of known **mutagens** (agents that cause mutations). Spontaneous mutations sometimes occur when a base undergoes a rare shift in its atomic configuration leading to an alternative

tautomeric form of the base. These rare tautomeric forms often undergo improper base pairing. For example, the common **keto** form of thymine pairs with adenine, but the rare **enol** form can pair with guanine. Rare tautomeric forms exist only briefly; they quickly convert back to the normal forms. Thus, if an improper nucleotide is inserted into a growing DNA strand as a result of a tautomeric shift, rapid conversion back to the normal form will create a base-pair mismatch, and the improper nucleotide can be removed from the strand by the **proofreading** (i.e., exonuclease) capability of DNA polymerase. Some spontaneous mutations are due to **deamination**—the removal of amino groups (NH_2) from certain bases, particularly cytosine. Deamination of cytosine produces uracil, which pairs with adenine. Many of these uracils are corrected by **DNA-uracil glycosidase**, which removes uracil-containing nucleotides from DNA. However, deamination of 5-methyl cytosine results in thymine, which cannot be corrected by the uracil-removing enzyme. For this reason, 5-methyl cytosines in DNA are associated with **hotspots** —sites in a gene that are particularly susceptible to spontaneous mutation. The error-correcting enzymes that interact with DNA reduce the spontaneous mutation rate below what it would otherwise be. On the whole, the spontaneous mutation rate in humans ranges between about 10^{-5} and 10^{-6} mutations per locus per generation; that is, between 1 in 100,000 and 1 in 1,000,000 genes at a locus will undergo spontaneous mutation each generation. These rates are averages, however, and there is wide variation from locus to locus.

9. Recent findings suggest that many spontaneous mutations actually result from insertions of long (several kilobase) sequences of DNA in or near the locus in question. The inserted sequences are often members of families of similar or identical sequences that are found scattered at many chromosomal sites throughout the genome (**dispersed repeated multigene families**). These sequences are able to move from one chromosomal site to another, and they are called **transposable elements**. The prototype of transposable elements occurs in prokaryotes, where certain sequences, called **insertion sequences**, exist in multiple copies and are able to change position in the bacterial DNA. When two insertion sequences of the same kind are sufficiently close together and flanking a bacterial gene, they can transpose as a unit and carry the bacterial gene along. Such a complex transposing unit is called a **transposon**. In bacteria, many transposons carry genes for antibiotic resistance. Indeed, transposons are thought to be key elements in the spread of antibiotic resistance in bacteria because antibiotic-resistance transposons can transpose to infectious plasmids and thereby become widely disseminated. A common feature of transposable elements in both prokaryotes and eukaryotes is that the nucleotide sequences at the ends of the elements form direct or inverted repeats. The 768-bp element IS1 in *E. coli* is flanked by a 23-bp inverted repeat, for example, and the 5-kb **copia** element in *Drosophila* is flanked by a 276-bp direct repeat.

10. Mutations caused by the insertion of transposable elements are typically **unstable**, which means that the rate of **reverse** mutation (mutation back to the normal form of the gene) is very much higher than typical rates of reverse mutation. Generally speaking, the rate of reverse mutation is lower by a factor of 100 or 1000 than the rate of **forward** mutation (mutation from a normal to an abnormal form). For insertion mutations, the rates of reverse mutation may equal or even exceed the rates of forward mutation.

11. Induced mutations occur in response

to known mutagens. One potent mutagen is **ionizing radiation**, of which x-rays are a prime example. X-rays cause breaks in DNA molecules and in chromosomes. For low doses of x-rays, the **dose-response relationship** is a straight line; that is, the number of mutations is strictly proportional to the dose. At higher doses, the dose-response relationship is more complex. In addition, the amount of mutation caused by a given dose of x-rays depends on whether the dose is **acute** (delivered all at once) or **chronic** (split into a series of smaller doses). Only a small proportion of spontaneous mutations can be attributed to background radiation. In humans, the **doubling dose** for x-rays (the amount required to double the mutation rate) is thought to be about 40 rad; the level of background radiation we receive in a 45-year reproductive lifetime averages about 6 rad.

12. Ultraviolet light is an example of a nonionizing radiation that is mutagenic. Most damage is done to the skin because the light is not very penetrating, but the energy is absorbed by DNA and causes chemical crosslinks between adjacent thymines (**thymine dimers**). These are usually repaired by means of a cut-and-patch process known as **excision repair**. However, certain patients with the recessive disorder **xeroderma pigmentosum** have a defective excision-repair system and are particularly prone to ultraviolet-light-induced skin cancers.

13. Certain chemicals are potent mutagens. Examples are **nitrous acid**, which causes deamination of bases and therefore mispairing; **base analogues** (such as **bromodeoxyuridine**, or BUdR), which are incorporated into DNA but mispair more frequently than their normal counterparts; **alkylating agents** (such as **mustard gas**), which chemically react with DNA and so cause mutations; and the **acridines** (such as **proflavin**), which become inserted between adjacent bases in DNA and so cause single-base additions or deletions, thus leading to frameshift mutations.

14. Detection of mutagens has been greatly aided by the development of the **Ames test**, which uses certain highly mutagen-sensitive strains of *Salmonella typhimurium* that contain known types of mutations in the *histidine* operon. Suspected mutagens are usually mixed with an extract of liver enzymes, because many chemicals are not themselves mutagenic but are converted into mutagenic forms by liver enzymes. The Ames test is highly sensitive (as little as $\frac{1}{100}$ of a cigarette can be shown to be mutagenic), and over a million-fold range of mutagenic potency can be assessed quantitatively.

15. Screening for mutagens, **carcinogens** (agents that cause cancer), and **teratogens** (agents that cause birth defects) is particularly important because more than 50,000 untested synthetic chemicals are already in use and more than 1000 new ones are produced commercially each year. Potential carcinogens are particularly difficult to screen because the tests are expensive and lengthy and must often be based on a limited number of animals. The Ames *Salmonella* test has proven to be useful as a preliminary test of carcinogens because, in the Ames test, most carcinogens are mutagenic and most mutagens are carcinogenic. Examples of widely used chemicals that have proven to be mutagenic and carcinogenic are tris-BP, used for 6 yr in the United States as a flame retardant in children's polyester pajamas, and AF-2, used for 9 yr as a food additive in fish and soybean products in Japan. Prescreening of potential mutagens is essential because a mutagen in widespread use might not produce sufficiently striking effects that it could even be discovered, yet hundreds of thousands or millions of new, harmful mutations would result.

WORDS TO KNOW

Mutation

Somatic
Germinal
Mosaic
Lethal
Sterile
Age of onset
Conditional
Modifier
Forward
Reverse
Frameshift
Spontaneous
Rate
Hotspot
Deamination
DNA-uracil glycosidase

Base substitution

Transition
Transversion
Amino acid substitution
Silent

Missense
Nonsense
Nonsense suppressor
Tautomeric forms

Chemical mutagenesis

Nitrous acid
Deamination
Base analogue
Alkylating agent
Acridine
Ames test
Carcinogen
Teratogen
Mutagen

Transposable element

Dispersed repeated gene family
Insertion sequence
Transposon
Unstable mutation

Radiation

X-ray
Curie
Roentgen
Rad
Rem
Dose-response relationship
Acute dose
Chronic dose
Gonadal dose
Doubling dose
Ultraviolet light
Thymine dimer
Excision repair

Syndrome

Xeroderma pigmentosum
Huntington disease

PROBLEMS

1. For discussion: In your view, what is the proper role of the federal government in testing or verifying the safety of chemical food additives? At present, tests are carried out under the auspices of the manufacturer to provide data to the appropriate federal authorities. Some critics argue that research supported by a manufacturer has a built-in conflict of interest, and unfavorable results may be suppressed. Other critics complain that the required tests are too time-consuming and expensive and should be replaced with faster, cheaper tests, even if these do not involve mammals. Still other critics argue that the government should stay out of the picture entirely because manufacturers of food additives who through negligence market a harmful additive may be legally liable for those individuals who can prove they were harmed by it. What is your opinion about these various issues?

2. In what sense can the mutation responsible for phenylketonuria be considered a "conditional mutation"? What do conditional mutations imply about the importance of both genotype and environment in producing particular phenotypes?

3. Distinguish between forward and reverse mutation rates. Considering base substitution mutations, why is the forward mutation rate expected to be much larger than the reverse mutation rate?

4. Would base substitutions due to tautomeric shifts be transitions or transversions? Why?

5. Recall from Chapter 9 that much of the redundancy in the genetic code is due to third-position "wobble," in which U and C are translated equivalently and A and G are translated equivalently. Knowing this, would you expect most "silent" nucleotide substitutions to involve transitions or transversions, and why?

6. Problems 7 through 13 pertain to the DNA sequence shown here. Assuming that transcription of the sequence occurs from left to right, what is the sequence of the corresponding RNA? What is the amino acid sequence of the corresponding polypeptide? (Note: The single-letter amino acid abbreviations reveal a message.)

```
       3   6   9  12  15  18  21  24
       |   |   |   |   |   |   |   |
5'-GGGGAGAACGAGACAATCTGCTCG-3'
3'-CCCCTCTTGCTCTGTTAGACGAGC-5'
```

7. For this and Problems 8 through 11, write the DNA sequence in Problem 6 *after* the mutation has occurred, the RNA sequence, and the amino acid sequence of the corresponding polypeptide. Then classify the mutation according to the following categories:

1. Transition
2. Transversion
3. Missense
4. Nonsense
5. Frameshift
6. Silent

(Note: More than one of these categories may be appropriate for each mutation.) Determine the consequences of a substitution of ^G_C for the ^C_G at position 14.

8. What would be the result of a substitution of ^T_A for the ^C_G at position 9?

9. Determine the result of an insertion of ^T_A between positions 2 and 3.

10. What would happen if the ^G_C at position 2 were deleted?

11. What would be the result of a substitution of ^T_A for the ^G_C at position 10?

12. Suppose the mutant DNA sequence in Problem 11 were in a cell having a mutant leucine tRNA gene that sometimes misread the amber terminator. What possible polypeptides would result from transcription and translation? Suppose the mutant leucine tRNA sometimes misread the ochre terminator instead of the amber terminator. What polypeptides would result from transcription and translation in this type of cell?

13. Most sites of cytosine methylation involve the sequence 5'-CG-3'. How many such sequences occur in the DNA molecule depicted in Problem 6? (Remember that there are two strands to consider.) If all the 5'-CG-3' sequences in Problem 6 were methylated, which base pairs would correspond to "hotspots" of mutation relative to the mutagen nitrous acid? For each of the 5'-CG-3' sequences in turn, what base substitution mutations would result from deamination and replication? What polypeptide sequences would correspond to these mutations?

14. Suppose that the promoter-carrying DNA sequence illustrated in Figure 10.6 in Chapter 10 were actually a transposable element. What would be the effect of this element being inserted immediately upstream from a gene? Would the orientation of the element matter?

15. Shown here are two homologous DNA sequences, each of which has a frameshift mutation. Sequence (*a*) has a deletion of base pair number 3; sequence (*b*) has a ^G_C insertion between positions 7 and 8. Assuming that transcription occurs from left to right, what polypeptide would correspond to sequence (*a*)? To sequence (*b*)? Suppose recombination between (*a*) and (*b*) occurred between base pairs 5 and 6. What DNA sequences would result? What would be the corresponding polypeptides? (Note: Use the single-letter amino acid abbreviations.)

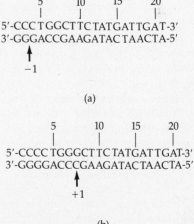

```
        5      10     15     20
        |       |      |      |
5'-CCCTGGCTTCTATGATTGAT-3'
3'-GGGACCGAAGATACTAACTA-5'
   ↑
  −1
```

(a)

```
        5      10     15     20
        |       |      |      |
5'-CCCCTGGGCTTCTATGATTGAT-3'
3'-GGGGACCCGAAGATACTAACTA-5'
              ↑
             +1
```

(b)

FURTHER READING AND REFERENCES

Ames, B. N. 1979. Identifying environmental chemicals causing mutations and cancer. Science 204:587–593. An excellent review of all aspects of the problem. Source of Figure 11.21.

Baird, M., C. Driscoll, H. Schreiner, G. V. Sciarratta, G. Sansone, G. Niazi, F. Ramirez, and A. Bank. 1981. A nucleotide change at a splice junction in the human β-globin gene is associated with β⁰-thalassemia. Proc. Natl. Acad. Sci. U.S.A. 78:4218–4221. The molecular basis of one of the β-thalassemias is discussed.

Cairns, J. 1981. The origin of human cancers. Nature 289:353–357. Expresses the view that most human cancers may be the result of genetic transpositions.

Calos, M. P., and J. H. Miller. 1980. Transposable elements. Cell 20:579–595. Excellent review of transposable elements, mainly in prokaryotes.

Campbell, A. 1980. Some general questions about movable elements and their implications. Cold Spring Harbor Symp. Quant. Biol. 45(1):1–9. A general introduction to transposable elements heading a two-volume symposium.

Cohen, S. N., and J. A. Shapiro. 1980. Transposable genetic elements. Scientific American 242:40–49. Discusses these "jumping genes" in prokaryotes and eukaryotes.

Crow, J. F., and C. Denniston. 1981. The mutation component of genetic damage. Science 212:888–893. Discusses the assessment of genetic damage expected from an increased mutation rate.

Echols, H. 1981. SOS functions, cancer, and inducible evolution. Cell 25:1–2. Evidence on the relationship between mutations and cancer.

Jackson, J. A., and G. R. Fink. 1981. Gene conversion between duplicated genetic elements in yeast. Nature 292:306–311. A clever demonstration of the effects of gene conversion.

Jones, R. W., J. M. Old, R. J. Trent, J. B. Clegg, and D. J. Weatherall. 1981. Major rearrangement in the human β-globin gene cluster. Nature 291:39–45. An unusual intrachromosomal crossover simultaneously creates a large inversion and two deletions.

Klein, G. 1981. The role of gene dosage and genetic transpositions in carcinogenesis. Nature 294:313–318. Cancer can result from DNA rearrangements that upset normal regulation.

Marotta, C. A., J. T. Wilson, B. G. Forget, and S. W. Weissman. 1977. Human β-globin messenger RNA. III. Nucleotide sequence derived from complementary DNA. J. Biol. Chem. 252:5040–5053. Source of data in Figure 11.9.

McCann, J., and B. N. Ames. 1976. Detection of carcinogens as mutagens in the Salmonella/microsome test: Assay of 300 chemicals: Discussion. Proc. Natl. Acad. Sci. U.S.A. 73:950–954. Discusses public health aspects of mutagen screening.

McCann, J., E. Choi, E. Yamasaki, and B. N. Ames. 1975. Detection of carcinogens as mutagens in the Salmonella/microsome test: Assay of 300 chemicals. Proc. Natl. Acad. Sci. U.S.A. 72:5135–5139. Detailed results of many chemicals.

McElheny, V. K., and S. Abrahamson (eds.). 1979. Banbury Report—1. Assessing Chemical Mutagens: The Risk to Humans. Cold Spring Harbor Laboratory, Cold Spring Harbor, N.Y. Proceedings of a conference designed to provide factual data relevant to this important subject.

McKnight, S. L., and R. Kingsbury. 1982. Transcriptional control signals of a eukaryotic protein-coding gene. Science 217:316–324. Beautiful use of engineered mutations in genetic analysis.

Nagao, M., T. Sugimura, and T. Matsushima. 1978. Environmental mutagens and carcinogens. Ann. Rev. Genet. 12:117–159. Technical aspects of testing are reviewed.

Neel, J. V., and W. J. Schull. 1954. Human Heredity. University of Chicago Press, Chicago. Source of Figure 11.1.

Orkin, S. H., S. C. Goff, and R. L. Hechtman. 1981. Mutation in an intervening sequence splice junction in man. Proc. Natl. Acad. Sci. U.S.A. 78:5041–5045. In this patient with α-thalassemia, a 5-bp deletion was found within the 5′ splice junction of the first intervening sequence.

Rothstein, S. J., R. A. Jorgensen, K. Postle, and W. S. Reznikoff. 1980. The inverted repeats of Tn5 are functionally different. Cell 19:795–805. Discussed in connection with Figure 11.15.

Rothstein, S. J., and W. S. Reznikoff. 1981. The functional differences in the inverted repeats of Tn5 are caused by a single base pair nonhomology. Cell 23:191–199. Discussed in connection with Figure 11.15.

Spradling, A. C., and G. M. Rubin. 1981. *Drosophila* genome organization: Conserved and dynamic aspects. Ann. Rev. Genet. 15:219–264. Reviews types of DNA including transposable elements and their possible roles in development and evolution.

Strickberger, M. W. 1976. Genetics, 2nd ed. Macmillan, New York. Source of Figure 11.1.

Strobel, E., P. Dunsmuir, and G. M. Rubin. 1979. Polymorphisms in the chromosomal locations of elements of the 412, copia and 297 dispersed repeated gene families in *Drosophila*. Cell 17:429–439. Different strains of *Drosophila* have transposable elements at different locations.

Upton, A. C., 1982. The biological effects of low-level ionizing radiation. Scientific American 246:41–49. Outlines the hazards of ubiquitons natural and artificial sources.

Weatherall, D. J., and J. B. Clegg. 1982. Thalassemia revisited. Cell 29:7–9. Molecular basis of thalassemias.

Weisburger, J. H., and G. M. Williams. 1981. Carcinogen testing: Current problems and new approaches. Science 214:401–407. How can mutagen and carcinogen testing be improved?

chapter 12
Viruses and Cancer

Human beings play host to an enormous variety of other living things. Like other organisms, our bodies are attractive homes for all manner of parasitic plants, animals, and microbes. Various fungi, tiny mites, and bacteria eke out their existence in habitats between our toes, in our hair or teeth, under our nails, or in small folds of our skin. Many other parasites find homes in our insides, particularly in the gut. Included here are roundworms and tapeworms, for example, and many single-celled organisms such as the species of amoeba that causes amoebic dysentery and the organism responsible for "traveler's diarrhea." Many types of bacteria live in the gut; most are completely harmless most of the time and some may be beneficial in digestion. (Normal diges-

tion in cattle is aided by bacteria in their digestive tracts, and termites are utterly dependent on their intestinal microbes to digest cellulose.) But some bacteria invade other parts of the body—tuberculosis bacillus and plague bacillus, for example—and they are not nearly so innocuous.

Our bodies are also prey to even smaller forms of life—viruses. These tiny things (one hesitates to call them "organisms" because they cannot reproduce outside living cells) are able to subvert the biochemical machinery of cells to accomplish their own ends. This chapter is about viruses, including certain viruses that can induce genetic changes in cells that lead to them becoming cancerous. This chapter is also about the ability of certain viruses to transport genes from one cell to another—an ability that might someday be harnessed to perform a sort of "genetic surgery."

Viruses

Viruses are submicroscopic parasites of cells, and hundreds of different kinds are known. Generally speaking, viruses are **host specific**; that is, a particular kind of virus is able to attack only the cells of one or a few related species of microorganisms, animals, or plants. Indeed, some viruses are able to infect only cells of certain tissues that carry on the cell surface specific molecules to which the virus can attach. Viruses may be relatively innocuous when they infect one type of cell but extremely harmful when they infect another. One example of a virus with such a dual personality is **rubella**, or German measles virus, which was discussed in Chapter 5. In most populations in which German measles occurs, it is a rather mild disease, seldom accompanied by serious complications. But the virus is able to cross the placenta of pregnant women and attack embryonic cells, causing severe birth defects. A second example is **herpesvirus**, actually a family of viruses. Some types of herpesvirus live in the subsurface layer of skin in humans or in the protective and insulating sheath of cells surrounding certain nerves. Herpesvirus is usually harmless, and several percent of all normal individuals are carriers of it. The carriers occasionally suffer a herpes outbreak, a characteristic type of runny cold sore around the lips or nose, but the outbreak usually subsides in a few days and the sore heals. What keeps the virus in check, or at least subdued, is the body's immune system; the general function of the immune system is to seek out, recognize, and destroy cells, tissues, virus particles, or large molecules that are "foreign" in the sense that they are not part of the body itself. In particular, herpesvirus can-

not usually invade the interior of the body but is confined to cells where it does little real harm, such as cells in the skin. On rare occasions, however, the virus is able to circumvent immune control, and some of its prime targets are cells of the central nervous system. Upon infecting the optic nerve, herpesvirus can cause blindness. Most runny cold sores are produced by a type of herpesvirus called type I. More serious is herpesvirus type II, which is associated with **genital herpes**—a venereal disease with symptoms such as swelling, itching, and throbbing in the genital area followed by local inflammation and an eruption of painful blisters. Although type I herpes usually lives above the waist, about 20 percent of genital herpes cases are due to type I virus.

Genital herpes is at epidemic proportions in the United States. Between 5 and 10 million Americans are infected, which is more than the combined total of all curable types of venereal disease. Unlike such venereal diseases as gonorrhea and syphilis, sexual contact is not necessary for the spread of genital herpes. Complications of genital herpes include an increased risk of urogenital cancer in long-term infections. Particularly serious is life-threatening **neonatal herpes**—herpes infection of the newborn through contact with the mother's genital tract. Complications of neonatal herpes include herpes of the eye, which causes blindness, and herpes of the brain membranes, which causes death. Genital herpes is a public health problem of immense proportions because there is as yet no cure.

Some viruses are relatively harmless in one host but cause serious disease in another. Herpesvirus again provides an example. This virus, which normally infects humans, can also infect rabbits and mice; in these animals, it almost always causes severe damage to the central nervous system. Another example is a virus called B virus; in monkeys, it produces an infection similar to that of herpesvirus in humans, but when B virus infects humans it is fatal.

Many human diseases are of viral origin. Among these are the common cold, measles, mumps, chicken pox, smallpox, influenza (flu), polio, yellow fever, and encephalitis, and it is now generally agreed that certain viruses can cause malignant tumors. The viruses that cause yellow fever, Rocky Mountain spotted fever, and encephalitis are in a special category of viruses that alternate between blood-sucking arthropods (usually insects such as mosquitoes or related animals such as ticks) and vertebrates. The arthropod bites and sucks blood from the vertebrate, and in so doing picks up the virus with the blood. The virus infects the insect or tick, and when the arthropod bites another vertebrate, some viruses are transferred back to the vertebrate bloodstream. Most varieties of arthropod-borne virus are innocuous, but, of course, yellow fever virus and some others are exceptions. The alternation of yellow fever virus between mosquitoes and humans is one reason so much emphasis has been placed on mosquito control in many parts of the world to reduce the incidence of yellow fever.

Characteristics of Viruses

Although viruses are small and can usually be examined only through the electron microscope, they do have an internal structure. The structure consists basically of the viral genetic material (DNA or, in many cases, RNA) surrounded by a protective protein coat called a **capsid**. The capsid of most viruses that infect animal cells is surrounded by still another coat, called the **envelope**, composed of lipid, protein, and carbohydrate, which is acquired as the viruses exit from the host cell by budding through special regions of the host cell membrane. These aspects of viral structure are illustrated for herpesvirus in Figure 12.1.

The genetic material of all viruses is nucleic acid, but the nucleic acid may be DNA or RNA, and it may be double stranded (two complementary strands held together by base pairing) or single stranded. The nucleic acid may be linear or circular, and there may be one molecule or several. The exact nature of the genetic material depends on the particular virus. Herpesvirus and the tadpole-shaped virus called **lambda** (symbolized λ), which infects the intestinal bacterium *Escherichia coli*, have as their nucleic acids one linear, double-stranded DNA molecule. Polyoma virus, one of the cancer-causing viruses, has a circular, double-stranded DNA molecule. Polio virus and influenza virus both have as their genetic material single-stranded RNA's, but whereas polio virus contains only one RNA molecule, influenza virus contains eight different RNA molecules. The nucleic acid molecules of viruses are hundreds or thousands of times shorter than those in cells, so viruses have relatively few genes. Polyoma virus, one of the smallest viruses, has six genes; influenza virus has eight genes. Even so, these viruses pack an awesome punch because they contain within the few genes sufficient information to take over a living cell. The virus λ, being relatively large and complex, has sufficient genetic material to have roughly 50 genes; λ contains about 50 kb of DNA, and its nucleotide sequence has been completely determined.

Viruses vary enormously in size and shape. Some are small and nearly spherical, shaped like a cut diamond with 20 identical triangular faces. Polio virus, for example, is about 300 Å in diameter. The "average" cell is roughly 10 μm (100,000Å) in diameter, so it has roughly 333 times the diameter of polio virus. Since the volume of a sphere increases as the cube of the diameter, however, the volume of a typical cell is some 40 million times the volume of the virus. To draw a rough analogy, the size of a virus stands in the same propor-

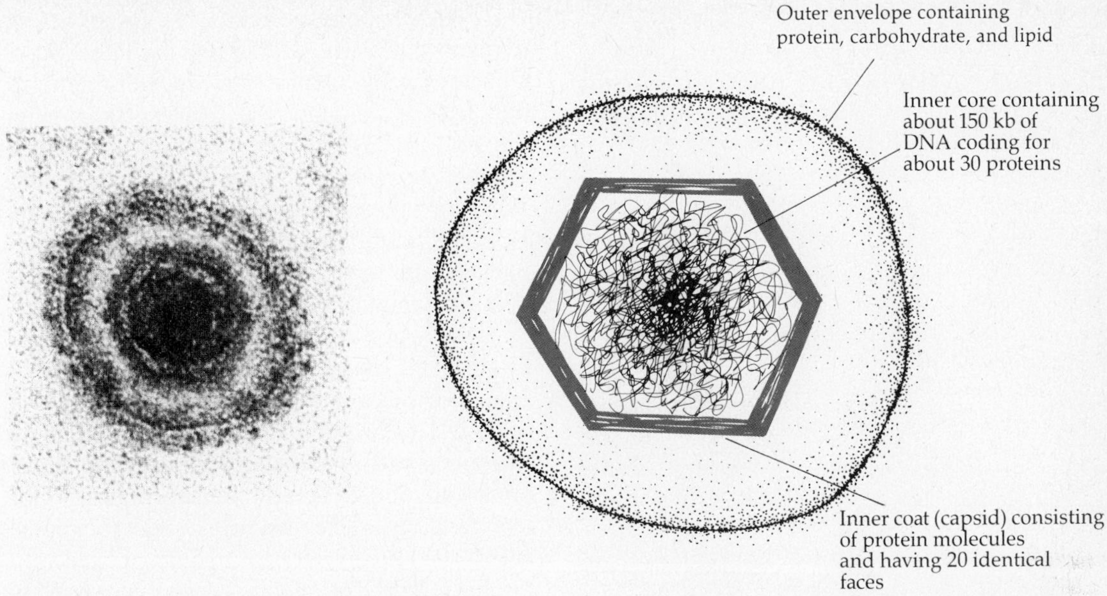

Outer envelope containing protein, carbohydrate, and lipid

Inner core containing about 150 kb of DNA coding for about 30 proteins

Inner coat (capsid) consisting of protein molecules and having 20 identical faces

Figure 12.1 Electron micrograph of herpesvirus and interpretative sketch of its structure. The diameter of the virus is 1300 to 1600 Å.

tion to the size of an average cell as the size of a thimble in relation to the size of an average bedroom.

Influenza virus is roughly spherical; it has a diameter of about 1000 Å, so the volume of an average cell is about 1 million times larger. These viruses have a curious appearance as seen through the electron microscope (Figure 12.2). Protruding out from their surface layer composed mainly of lipid are some 600 or more "spikes" made of a complex of carbohydrate and proteins. Some of these spikes are evidently the structures that recognize cell surfaces and permit infection, because viruses with their spikes stripped off are not infectious.

Among the largest viruses are the pox viruses, which are ovoid or brick-shaped and are almost as large as a small bacterial cell. Smallpox virus is ovoid, for example, and it has a diameter along its long axis of about 2500 Å. Even though these are large particles, an average animal cell is still some 65,000 times larger. Among the smallest viruses at the other end of

the size scale are polyoma viruses, which are **oncogenic** (cancer causing) in animals, including humans. Polyoma viruses are very small indeed, having a diameter of about 50 Å. Other virus shapes and sizes also occur. Tobacco mosaic virus, which infects tobacco plants, is roughly cylindrical, some 180 Å across and 3000 Å long. The *E. coli* virus λ is almost tadpole-shaped (Figure 12.3); it is about 2000 Å from "head" to "tail" and about 500 Å in diameter in the region of the head.

Viruses are consummate parasites, totally unable to reproduce outside living cells. A major reason for their total reliance on cells is that viruses lack ribosomes, transfer RNA's, and much of the other machinery required for protein synthesis; viruses also lack mitochondria, chloroplasts, and other cellular machinery required for capturing or converting energy. The most important thing a virus *does* contain is its genetic material, which in some viruses is DNA and in others is RNA. The genetic material of the virus carries informa-

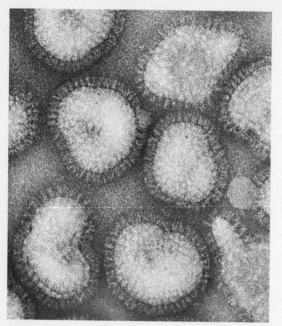

Figure 12.2 Electron micrograph of an influenza virus magnified about 200,000 times. The "halo" that surrounds each virus particle is formed by molecules projecting out from the viral surface.

tion for the manufacture of viral proteins. Some viral proteins function in replication of the viral nucleic acid, others form the protective protein coat of the virus, and still others are key elements in establishing mastery over the internal workings of a cell. When a virus enters a living cell, it can steal control of the cell from the cell's own nucleus and subvert the cellular machinery into such tasks as making viral proteins and replicating the viral nucleic acid. The products of this synthesis are then assembled into new offspring viruses.

Lambda: A Viral Parasite of *Escherichia coli*

Because of some similarities in the life cycle of the virus λ, which infects *E. coli*, and those of

many other viruses, λ has been the object of much study. The first step in the infection of a bacterium by λ is the adherence of the tip of the tail of λ onto the cell surface [Figure 12.4(*a*)]. Then, somewhat like a hypodermic needle releasing its contents, λ injects its DNA into the cell [Figure 12.4(*b*)]. (It should be noted here that most animal viruses do not infect cells by injecting their DNA in this manner; animal viruses generally enter cells directly by passing through the cell membrane, often aided by the tendency of some cells to engulf viruses and other small particles.) Once the λ DNA is inside the bacterial cell, either of two things can happen, depending on the metabolic state of the bacterium and probably other factors as well: The virus may undergo what is called the *lytic cycle*, or it may be spliced into the bacterial DNA and be replicated passively

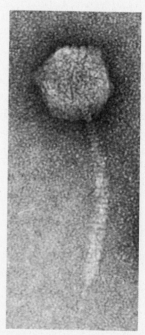

Figure 12.3 Electron micrograph of λ, a virus that infects *E. coli*. The diameter of the head is about 500 Å.

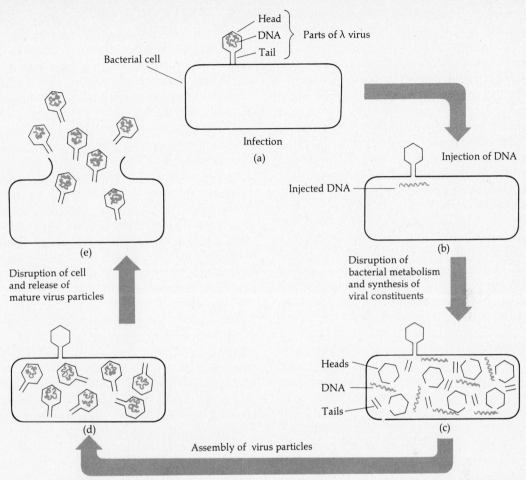

Figure 12.4 Lytic cycle of bacteriophage λ, showing how the virus absorbs to the cell *(a)*, injects the DNA *(b)*, and establishes control leading subsequently to the production of viral constituents *(c)*, which are then assembled into mature virus particles *(d)* and released as the cell is split open *(e)*. The DNA of the virus actually circularizes upon entry and remains so during replication; it is shown here as a linear molecule for clarity.

along with the bacterial DNA as the cell divides.

In the **lytic cycle** of λ, illustrated in Figure 12.4, the viral DNA takes over the biochemical machinery of the bacterial cell; it induces this machinery to make viral proteins, the DNA of the virus is replicated, and the DNA and protein components are assembled into whole viruses [see Figure 12.4(c) and (d)].

The bacterial cell then breaks open (e), which is the "lysis" of the lytic cycle, and the approximately 100 progeny viruses inside are liberated. (The exit of λ viruses from infected cells is not typical of the exit of animal viruses from their host cells; animal viruses often push outward through the cell membrane one by one and the host cell is not burst.)

Some of the genes involved in the take-over of the bacterial cell by the λ virus are illustrated in the λ genetic map in Figure 12.5. The horizontal line in the figure represents the linear duplex DNA molecule in the head of the mature phage. (A **phage** or **bacteriophage** is a virus that infects bacterial cells.) The linear molecule actually has short (12 bp) single-stranded regions at each end. These 12-bp overhanging strands are complementary in base sequence, so they can pair with one another and create a circular DNA molecule in which genes S and R are brought into juxtaposition with genes A and W. Indeed, the first thing that happens upon entry of λ DNA into the bacterial cell is that the DNA circularizes by means of its 12-bp **cohesive ends,** and these ends become covalently linked by bacterial DNA ligase. All subsequent events involve this circular molecule, although Figure 12.4 illustrates it in linear form for the sake of clarity.

Bacteriophage λ is organized into four **operons** (transcriptional units), three of which are involved in the lytic cycle. Immediately after the DNA becomes circular, transcription of two of these operons occurs and gives rise to two **early mRNA's**, designated "left" and "right" in Figure 12.5. When these mRNA's are translated, some of the proteins prevent normal bacterial functions from being carried out. Other proteins are involved in the production of new virus particles, such as the products of genes O and P, which are involved in replication of the viral DNA. At some point about midway in the lytic cycle, mRNA from a third transcriptional unit (**late mRNA**) is produced. Transcription begins near gene S and proceeds toward gene R (see Figure 12.5) and, since the DNA is circular, continues clockwise through genes A, W, B, and so on. The late mRNA codes for head proteins (genes A through F), tail proteins (Z through J), and several other proteins. After the head and tail proteins have been produced, the viral DNA becomes encapsulated in the head proteins, the

tails become attached, and a specific enzyme (**lysozyme**, coded by gene R) causes the bacterial cell to **lyse** or burst open. Lysis liberates the 100 or so virus particles that have been produced, and these are then able to infect other bacterial cells. The entire lytic cycle of λ occurs in less than 1 h.

Since bacteriophage λ has genes, mutations in these genes are expected. Many mutations of genes in λ have been found, and they are of great interest to geneticists. Such mutations can be studied by means of the same principles as are involved in genetic studies in eukaryotes. Recall from Chapter 4 that a fundamental test in genetics is the **complementation test**, which determines whether two mutations can complement each other's defects and so permit normal function of the organism. If the mutations do complement, they are inferred to be in different genes; if they fail to complement, they are inferred to be in the same gene. In bacteriophage, a complementation test is carried out by infecting a bacterium simultaneously with two different phage, each carrying a mutation. If the infection is productive and the cell lyses, the mutations are complementing and so are in different genes; if the infection is nonproductive, the mutations are noncomplementing and so are in the same gene.

A second fundamental test in genetics is the study of **recombination**, which was discussed at the chromosomal level in Chapter 4 and at the DNA level in Chapter 8. Recombination can be studied in bacteriophage and other viruses because, when two virus particles, each carrying a different mutation, infect the same cell, among the progeny viruses may be found some that contain neither mutation and some that contain both. These recombinant viruses result from breakage and rejoining of two viral DNA molecules at a position between the sites of the two mutations. The occurrence of recombination in viruses permits

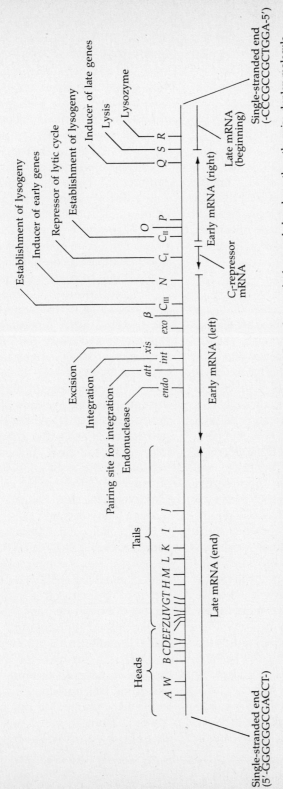

Figure 12.5 Genetic organization of λ DNA showing location of various genes. Note the homologous single-stranded ends on the otherwise duplex molecule.

mutations of the viral DNA to be located genetically or **mapped** in relation to the position of other mutations, much as chromosomes can be mapped in higher organisms. Such studies lead ultimately to the sort of map shown in Figure 12.5. In short, the genes of viruses can be studied in much the same way as the genes of other species. Indeed, some of the fundamental concepts of mutation, complementation, and recombination were originally worked out in viruses because of their small size, their rapid life cycle, their ease of culture, and their availability in enormous numbers.

Lysogeny

Bacteriophage λ is one of a class of phage called **temperate** phage, which have an alternative to the lytic cycle. (Bacteriophage that do not have this alternative life cycle are said to be **virulent**.) The alternative to the lytic cycle is called **lysogeny**, and it involves a process in which the DNA of the virus becomes incorporated into the DNA of the bacterium without disrupting the metabolism of its host. Insertion of λ DNA into the bacterial DNA requires several products of the early mRNA's, and continued maintenance of λ in this inserted state requires the product of the C_I gene (see Figure 12.5). The C_I gene product is a repressor protein that functions in two ways: (1) It represses transcription of all other λ operons and thus prevents initiation of the lytic cycle, and (2) it confers upon the bacterium an immunity to infection by other λ particles because other λ DNA's will be repressed immediately upon entering the cell.

The physical integration of λ is illustrated in Figure 12.6. After the λ DNA has become circularized, a special region denoted *att* is brought into close proximity with a corresponding region on the bacterial DNA called the λ **attachment site** [see Figure 12.6(*a*)]. As shown in the figure, this attachment site is between the genes for galactose utilization (*gal*) and biotin synthesis (*bio*). The juxtaposition of the two molecules evidently involves base pairing because both the *att* site in λ and the λ attachment site in the bacterium have the identical sequence of 15 nucleotides shown in the figure. Pairing of the two DNA molecules is followed by a site-specific recombination within these 15 nucleotides, which is catalyzed by the product of the λ *int* gene [see Figure 12.6(*b*)]. This recombination event between the phage and bacterial DNA results in the physical insertion of the λ DNA into that of the host [see Figure 12.6(*c*)]. In its inserted state, λ DNA is said to constitute a **prophage**.

Once inserted, the prophage is carried passively along by the bacterium. Every time the bacterial DNA replicates, the λ DNA is replicated, and every time the bacterial cell divides, each daughter cell receives a copy of the viral DNA as part of its genetic endowment. The physiology of the bacterium is not disrupted by the presence of the viral DNA linked to its own DNA; the bacterium divides just as rapidly and grows just as well as one not infected by the virus. Indeed, the bacterium benefits by the association because it is now immune to infection by other λ particles. This phage-bacterial association may persist through many bacterial cell divisions.

The molecular determinants of whether a bacteriophage will undergo the lytic cycle or lysogeny are quite complex, but the determination depends ultimately on whether the C_I repressor is produced in sufficient amount to repress the other operons. The C_I repressor is an example of a protein that exhibits **autogenous regulation**; that is, it regulates its own synthesis. In the case of C_I, the repressor protein, in addition to repressing other operons, stimulates transcription of its own operon and thus continues to be produced. Some "pump priming" is necessary to get the C_I operon started, and this is the function of the

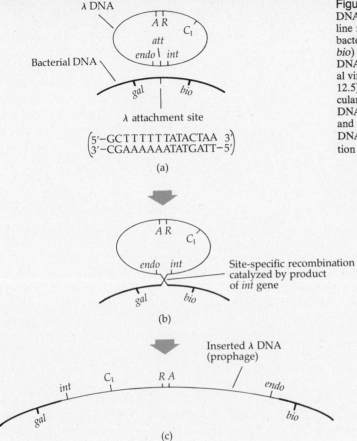

(a)

(b)

Inserted λ DNA
(prophage)

(c)

Figure 12.6 Integration of λ DNA into the DNA of *E. coli* during lysogeny. The heavy line represents a segment of the long, circular bacterial DNA; two bacterial genes (*gal* and *bio*) are indicated. The light line represents λ DNA and shows the relative positions of several viral genes (same gene symbols as in Figure 12.5). During the integration process, the circularized λ DNA aligns with the bacterial DNA at the λ attachment site (*a*). A breakage and reunion (recombination) between the two DNA molecules occurs (*b*), leading to the insertion of the λ DNA into the bacterial DNA (*c*).

C_{II} and C_{III} genes in the early mRNA's, the products of which stimulate the initial transcription of C_I. However, the effectiveness of the C_{II} and C_{III} gene products is determined by their interactions with certain molecules in the host cell, and in this manner the physiological state of the host ultimately determines whether the phage will undergo the lytic cycle or lysogeny.

There is always a small chance that integrated prophage will excise from the bacterial DNA and proceed to enter the lytic cycle. The likelihood of this happening depends on a variety of conditions, including the nutritional state of the host, and it may also be greatly enhanced by treating the bacterial cells with ultraviolet light. The excision process is roughly the reverse of the integration process (Figure 12.7). The regions of the virus and host DNA that were in proximity during integration are again brought together, and the viral DNA loops into a circle, producing a molecule that looks like a figure 8 except that the loop on one side, the bacterial DNA, is much larger than the loop on the other side [see Figure 12.7(*b*)]. The products of the phage *int* and *xis* genes now catalyze breakage and rejoining of the molecules at the position of the cross in the

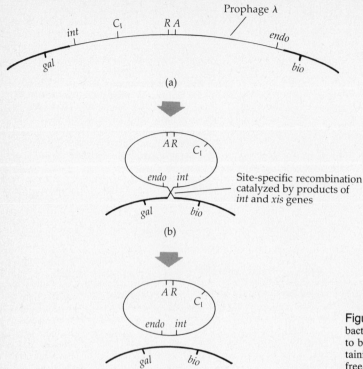

Figure 12.7 Excision of λ DNA from the bacterial host DNA. *(a)* Prophage λ attached to bacterial DNA. *(b)* Formation of loop containing λ DNA and recombination. *(c)* λ DNA freed from bacterial DNA. (Compare with Figure 12.6.)

figure 8, and the viral DNA becomes free, as if it had just been injected into the cell and become circularized [see Figure 12.7(c)]. A lytic cycle now takes place and, less than 1 h later, the bacterium lyses and 100 or more λ virus particles are liberated.

Transduction: Genetic Surgery by Viruses

Usually the process of Figure 12.7 by which prophage λ DNA is excised from the DNA of *E. coli* proceeds without a hitch, but occasionally it is slightly inexact. In these cases, the viral DNA picks up one or a few bacterial genes in its loop as it excises, leaving behind in the bacterial DNA a few viral genes. The process of defective excision is illustrated in Figure 12.8. Instead of the attachment sites of λ and the bacterium coming into proximity to form the figure 8, the pairing sites are displaced and bacterial genes become associated with the λ loop and viral genes with the bacterial loop [see Figure 12.8(b)]. Recombination in the paired region then produces a loop of λ DNA with certain viral genes replaced with genes from the host; conversely, the host chromosome has some of its genes replaced with genes from the virus [see Figure 12.8(c)]. Although Figure 12.8 illustrates how λ can pick up the *gal* genes, displacement of the loop to the other side would allow it to pick up the *bio* genes.

When such a defective phage as the one illustrated in Figure 12.8 infects a new bacterial host, it carries along the *gal* genes it now

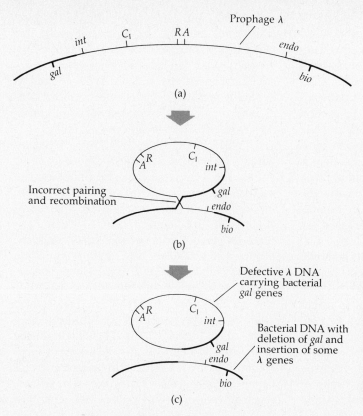

(a)

(b)

Incorrect pairing and recombination

(c)

Defective λ DNA carrying bacterial *gal* genes

Bacterial DNA with deletion of *gal* and insertion of some λ genes

Figure 12.8 Defective excision of λ leading to a virus capable of transduction. *(a)* Prophage λ attached to bacterial DNA. *(b)* Defective loop formation and recombination. *(c)* λ DNA freed from bacterial DNA. Note that the viral DNA carries certain bacterial sequences (in this case *gal*) and is deficient for certain viral sequences (in this case a region including *endo*). (Compare with Figure 12.7.)

contains. These *gal* genes can replace those on the host chromosome by means of recombination events flanking the *gal* region, so λ can mediate the transfer of *gal* (or *bio*) genes from one bacterium to another. Such bacteriophage-mediated genetic exchange is known as **transduction**, and the process has proven to be an important tool in the study of bacterial genetics. Actually, λ is an exceptional sort of transducing phage because it can carry only the few bacterial genes flanking the λ attachment site; such phages that can transduce only a few genes are called **specialized transducing phage**. More useful in some studies is a class of phage called **generalized transducing phage**, which can transduce any bacterial gene. An example of a generalized transducing phage in *E. coli* is

P1 phage. This bacteriophage does not become incorporated into the bacterial DNA during lysogeny; instead it replicates as a self-replicating plasmid. Moreover, during its lytic cycle, the host DNA is degraded into fragments that can mistakenly be packaged into mature phage particles and so become vehicles for transduction. Indeed, bacteriophage P1 can transfer up to 2 percent of the bacterial DNA from one cell to another.

Although the biology of phage λ was once a highly specialized subject allocated to professional "lambdologists," the phage has recently become important in many branches of molecular genetics because it is a popular vector for the cloning of prokaryotic and eukaryotic DNA molecules. (See Chapter 8 for a general discussion of molecular cloning.) A number of special strains of λ are available for cloning; each has conveniently located restriction sites that flank a gene that is nonessential for the lytic cycle, such as *int* or C_I. This nonessential restriction fragment is removed and replaced with a comparable restriction fragment from the DNA to be cloned. Then either the hybrid phage is packaged in vitro using purified λ head and tail proteins, or it is circularized with DNA ligase and used in DNA transformation of *E. coli*. In either case, the cloned DNA fragment is amplified by replication of the hybrid phage as it undergoes the lytic cycle in infected cells.

Before leaving the subject of λ, it should be noted that its life cycle is exceptional in several respects. It can undergo lysogeny and mediate transduction—a talent that is rather rare even among viruses that infect *E. coli*. Most types of bacterial viruses can carry out only the lytic cycle. Although the lytic cycle of λ is typical of viruses that infect bacteria, viruses that infect human cells and cells of other mammals are not quite the same. Mammalian viruses can establish control over the metabolism of infected cells, but they differ from λ in how the viruses penetrate host cells and how the progeny viruses exit.

Influenza Virus

An example of a virus that infects human cells is **influenza virus**—a virus whose envelope is essentially a lipid-protein sphere that has a large number of carbohydrate-protein projections or **spikes** (see Figure 12.2). Inside the envelope is the protective protein coat and the viral genetic material, in this case eight single-stranded RNA molecules, along with a few additional proteins that play a role in establishing infection. Influenza virus attacks cells of the respiratory tract—nose, throat, lungs—and causes such symptoms as coughing, runny

nose, fever, and a general aching. The virus is transmitted from person to person in the tiny water droplets expelled in coughs and sneezes. Actually, there are several types of influenza, called types A, B, and C. Types B and C generally produce a milder infection than does A, and major influenza epidemics are almost always due to varieties of type A virus.

With modern medical care, influenza alone is rarely fatal. However, especially in the young, the weak, and the aged, influenza can weaken the patient and make him or her more susceptible to other, more severe infections. As a result, the complications of influenza can often be fatal. A major influenza epidemic occurred in 1917 and 1918, killing 20 million people, including 548,000 in the United States. The most recent major worldwide epidemic was in 1968—the "Hong Kong flu"—which killed 51 million people, 20,000 in the United States alone. The spread of the 1968 epidemic is depicted in Figure 12.9. The strain of virus responsible for the disease arose in Southeast China in July 1968. From there it spread, reaching California in October 1968 and New York in December. The official designation of the virus responsible for Hong Kong flu is A/Hong Kong/1/68 (H3N2)—A because it is a type A virus, Hong Kong for where it was first isolated, 1 because it was the first of its kind isolated, and 68 for the year of its discovery. The symbol H3N2 in the strain designation refers to the components of the spikes that are recognized by the immune system; H stands for **hemagglutinin**, which makes up one type of spike, and N stands for **neuraminidase**, which makes up the other type of spike.

Influenza virus infects susceptible cells by adhering to their surface and literally passing through the cell membrane [Figure 12.10(*a*), (*b*), and (*c*)]. What enters the cell is the viral RNA and associated proteins; the envelope of the virus is left behind at the cell membrane. The RNA and proteins of the virus make their

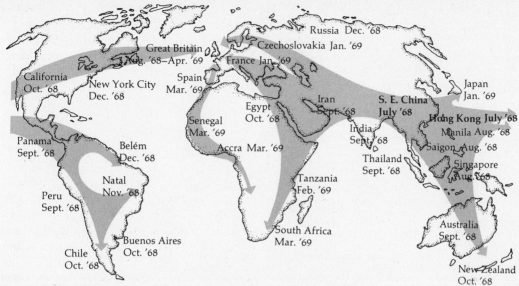

Figure 12.9   Map showing the spread of Hong Kong flu in 1968 and 1969.

way to the cell nucleus [Figure 12.10(*d*)], where they establish control over the cell and cause it to manufacture more viral RNA and more viral proteins [Figure 12.10(*e*)]. When cells in tissue culture are infected, 10 to 12 h later extensive disintegration and fusion of cells occurs, and synthesis of cellular DNA, RNA, and proteins can no longer be detected. Eventually the infected cell dies. In some strains of influenza virus, a small fraction of cells may be able to survive the initial cycle of virus infection. These cells will continue to produce virus particles, sometimes for days, without undergoing significant changes in their own behavior or growth.

Once influenza virus establishes an infection in a cell, the viral RNA is replicated in the nucleus. (The viral material entering the cell includes specific proteins that function in replicating the RNA.) Each of the eight RNA's is transcribed into a single mRNA coding for one viral protein (influenza thus has only eight genes and eight proteins), and the viral pro-

teins are produced on ribosomes in the cytoplasm. Proteins destined to become part of the viral envelop migrate to the cell membrane, while other viral proteins migrate to the nucleus [see Figure 12.10(*f*)]. There, in the nucleus, an RNA-protein complex is formed; this will eventually become the interior of the mature virus particle. The RNA-protein complex leaves the nucleus and makes its way to the cell membrane, where the carbohydrate-protein hemagglutinin and neuraminidase spikes of mature virus particles have already become attached. The RNA-protein complex is included in a small bud that pinches off from the rest of the membrane [see Figure 12.10(*g*)]. This bud forms the mature virus particle (*h*), and its lipid-protein envelope is composed partly of constituents that are cellular, not viral, in origin. It may be that other parts of the virus are also cellular in origin. In any event, the first progeny viruses from infected cells are released 3 to 4 h after infection.

From the point of view of the influenza

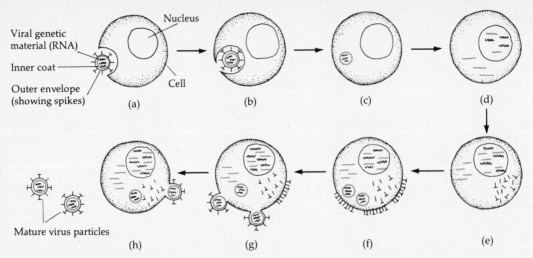

Figure 12.10   Life cycle of influenza virus showing how the virus enters the cell [*(a)* and *(b)*], shedding its envelope in the process *(c)*. Viral proteins (gray lines) and RNA (wavy lines) make their way to the nucleus *(d)*, where control over the cell is established and viral RNA is replicated. Viral coat proteins, produced in the cytoplasm, migrate to the nucleus *(e)*, where a protein-RNA complex is formed, which then migrates back to the cytoplasm *(f)*. In the meantime, other viral proteins, which will form part of the outer envelope of the virus, are synthesized in the cytoplasm and become attached to the cell membrane [*(e)* and *(f)*]. Then the protein-RNA complex buds off the infected cell at regions where the viral "spikes" have previously been attached [*(g)* and *(h)*].

virus, the process of infection is very inefficient. Many of the virus particles produced are defective because they lack one or more of the pieces of viral RNA or some of the viral proteins. Between 100 and 1000 virus particles must be applied to a culture of sensitive cells to obtain one productive infection. In a healthy human being or other organism, the virus faces an additional problem: the immune system. The hemagglutinin and neuraminidase spikes on the viral envelope provide the virus with a sort of "signature"; these spikes have certain three-dimensional configurations of atoms that the body's immune system recognizes as belonging to a foreign invader. The configurations recognized by the immune system are known as **antigenic determinants**, and the immune system responds to foreign antigenic determinants by producing specific proteins that can recognize the determinants and be-

come attached to them (see Chapter 13). These proteins are known as **antibodies**, and they aid in the destruction of invading viruses. A virus actually carries many different antigenic determinants, with each one individually recognized by the immune system. A particle, such as a virus, that carries antigenic determinants and stimulates the immune system is known as an **antigen**.

Viruses rendered inactive—unable to infect or to reproduce in cells—may still carry viral antigenic determinants that stimulate the immune response, and individuals exposed to such inactive viruses produce antibodies against them. Subsequent invasions by active viruses are less likely to lead to full-blown infections because antibodies to combat the viruses will already be present. (Influenza vaccinations are based on this principle.) Similarly, an infection caused by the active virus itself stimulates the immune response so that the virus cannot easily reinvade the same person once he or she has recovered. Such people are said to be **immune**. An influenza virus that sweeps through a population in an epidemic will leave in its wake a great number of immune individuals. Then the same strain of virus can no longer produce an epidemic in that population because the proportion of people who are immune is too high; sufficient numbers of nonimmune people do not come into contact for rapid transmission of the virus from person to person.

Acquired immunity to influenza is the principal reason influenza is usually not a significant problem in public health for a few years following a major epidemic. However, the antigens present on the envelope of the virus are encoded by viral genes, and these genes mutate like any other genes. Eventually there arises a mutated form of the virus that has sufficiently altered antigenic determinants to escape destruction by the immunity in the population. This new strain of virus is free to sweep through the population in another epidemic.

An **epidemic** is a localized outbreak of disease, and influenza epidemics are usually due to mutated viruses that have minor alterations in their antigenic determinants. These minor changes in antigenic type are said to be due to **antigenic drift**. At intervals of 10 to 30 yr there is an influenza **pandemic**—a worldwide epidemic. The antigenic difference in viruses responsible for pandemics are much greater than those that occur with antigenic drift. Such markedly different viruses are said to have undergone an **antigenic shift**. Antigenic shifts are thought to arise when a human virus and an animal virus infect the same animal because the progeny viruses can carry many possible combinations of RNA molecules of human or animal origin. Some of these

progeny viruses are able to infect humans, but their antigens may be very different because they are coded for by RNA molecules of nonhuman viruses. For example, in strain A/Hong Kong/1/68 (H3N2), mentioned earlier in connection with Hong Kong flu, seven of the eight viral RNA molecules seem to be of human-virus origin; the other molecule—which codes for hemagglutinin—seems to be from a nonhuman virus, very possibly a bird or a horse.

There is evidence that the antigenic change–epidemic–immunity cycle of influenza virus may involve periodic recurrences of past viral types. That is to say, the antigenic type of the virus obtained after some years of antigenic drift and antigenic shift may be very similar to the antigenic type that the virus possessed long ago. From the point of view of the virus, the antigens it started out with years ago are just as good as new ones if almost everyone with antibodies against the original antigens is dead —say, 40 or 50 yr after the original epidemic. Some very old people who developed antibodies against strains of virus current in the 1890s carry antibodies that are able to recognize and combat many of the "newest" viral strains cropping up today.

In the past, one had to await the appearance of a new antigenic type of influenza virus to study its features and culture it in the laboratory to obtain the antigens used in inoculations against the virus. However, the accelerated evolution of viral antigenic types has now been achieved in the laboratory, and the strains evolved in the laboratory can be used to provide antigens for inoculations that can be undertaken before the corresponding viral type has appeared spontaneously in the human population. For example, the London flu virus was obtained in the laboratory 2 yr before its actual appearance in London in the early 1970s. By these laboratory procedures, epidemiologists can try to stay several years ahead of the varieties of influenza virus instead of being chronically 1 yr behind.

Cancer

Viruses are responsible for many human diseases, and one of the most widespread and insidious of these is **cancer**, an unrestrained growth of cells. Not all cancers are known to be associated with viruses, but certain ones are. The evidence that certain viruses cause cancer is very strong, especially in laboratory animals such as mice, but conclusive proof is exceedingly difficult to obtain. For example, although it is certainly true that cells infected with certain types of virus are prone to become cancerous, this could be because the viral infection renders the cells more susceptible to another, nonviral agent that is the actual "cause" of the cancer. (It is important to emphasize in this connection that most cancers are thought to be caused by environmental agents; cigarette smoking is the most important single cause. Cancer is not one disease; although we use a single word to describe "cancer," there are at least 100 different varieties of the disease. There also seems to be genetic variation among people in their susceptibility to cancer-causing agents.)

Cells in the body of an adult are not, for the most part, actively dividing. They are metabolically active, which is to say alive, but they are not always in the process of dividing. Only a few cell types such as white blood cells and cells in the bone marrow are continuously producing daughter cells. In most tissues, only a few percent of cells are engaged in mitosis, just enough to replace cells that may die or become defective. Conditions can occur, however, that stimulate normally quiescent cells to divide. At the site of a wound, for example, the cells are somehow stimulated to undergo rapid division, and this is partly what allows the wound to heal.

One phenomenon of considerable importance for maintaining the cells of the body in a practically nondividing state is known as **contact inhibition**. This simply means that cell division tends to be inhibited by cell-to-cell contact. In tissue cultures of animal cells, for example, normal cells will continue to grow and divide until they form a layer of cells across the bottom of the culture dish. When the number of cells in this layer becomes such that the cells begin to bump into one another, the rate of cell division decreases dramatically. More than this, the cells that come into physical contact align themselves with respect to each other, giving to the layer of cells a regular pattern of lines and whorls [Figure 12.11(a)].

Cancer may arise from a single cell that has undergone a genetic change, making it insensitive to the mechanisms such as contact inhibition that ordinarily control the rate of cell division. This cell begins to divide rapidly, and since the change in the cancer cell is hereditary, the daughter cells also divide, unaffected by the mechanisms that control cell division in normal cells. The growing mass of cells pushes normal cells aside and steals nutrients from the surrounding tissue. Eventually this process results in a whole group of malignant (cancerous) cells—a tumor. (Not all, or even most, tumors are malignant. Sometimes a group of cells proliferates and forms a tumor, but then cell divisions cease. These tumors are called *benign* in contrast to *malignant* tumors.) If a malignant tumor is detected sufficiently early, it can sometimes be removed surgically with little risk of recurrence. Cancer can also be treated with radiation or with drugs that preferentially kill dividing cells, and recent methods involve an attempt to turn the full force of the immune system against the malignant cells.

Whereas cells are normally confined to

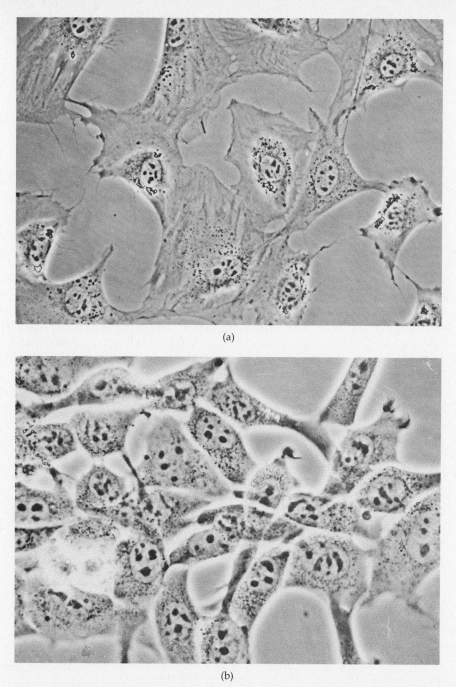

(a)

(b)

Figure 12.11 *(a)* Normal cells, sensitive to contact inhibition, show an organized pattern of growth in tissue culture. *(b)* Transformed cells, insensitive to contact inhibition, show a disorganized arrangement.

certain characteristic tissues—kidney cells are not found in the brain, for example—the cells in a malignant tumor are not so confined. Cells that slough off from the tumor can make their way via the bloodstream or lymphatic system to other tissues, where they may invade and cause a tumor in the new location. (The sloughing off of tumor cells and their invasion of other tissues is known as **metastasis**. Some types of cancer, such as lung cancer and breast cancer, metastasize quite soon after they first arise, which is part of the reason these cancers are so insidious and must be detected early to give reasonable prospects of cure. Other cancers, such as prostate cancer in older men, metastasize much later or hardly at all.) Tumors grow and divide at the expense of the surrounding tissues. Left untreated, and in some cases even if treated, cancer will finally lead to tumors in one or more vital organs, causing death through failure of the organ to function.

In tissue culture, cells from malignant tumors do not respond to contact inhibition. The layer of cells in the dish becomes ever thicker because the cells continue to divide, and the cells do not orient regularly when they come into contact but tend to climb over each other. The arrangement of cells in the culture dish therefore appears disorganized and chaotic [see Figure 12.11(b)].

The membranes of cancer cells are not normal. They develop characteristic antigens. These can stimulate the production of antibodies against themselves, and the immune system, by attacking the cancer cells, plays an important role in the body's defense against cancer. Current evidence suggests that cancerous cells arise frequently, perhaps even daily, in normal people and that these cells are usually destroyed or kept in check by the immune system. This point of view is buttressed by the fact that people afflicted with certain hereditary defects in the immune system are particularly prone to malignancies. The development of a malignancy requires two events in any case: the alteration of a normal cell into a cancerous one and a failure of the immune system to attack the tumor effectively.

The causes of various types of cancer are generally unknown. Since there are many kinds of cancers that arise in many types of tissue, there may be as many different causes. Most cancers are said to arise "spontaneously," which is a graceful way to say their cause is unknown. But there is a suspicious link between mutations and cancer.

Genetics of Cancer

Certain rare cancers are inherited. Among these are xeroderma pigmentosum, inherited as an autosomal-recessive condition associated with malignancies of the skin (see Chapter 11), and **retinoblastoma**, an autosomal-dominant condition marked by tumors in the retinas of the eyes. Inherited cancers are individually quite rare, however. Most common cancers are not known to be inherited, and the risk of cancer in close relatives of cancer victims is only slightly higher than the risk in nonrelatives. Nevertheless, the more than 50 inherited forms of cancer do establish a connection between mutations and cancer.

Another link between cancer and mutated genes comes from the relationship between carcinogens and mutagens discussed in Chapter 11 in connection with the Ames *Salmonella* test—the fact that most cancer-causing agents are mutagenic and most mutation-causing agents are carcinogenic. For some years this link could not be established clearly because many carcinogens seemed not to be mutagenic in cultured cells. Then it was discovered that these carcinogenic chemicals are not themselves carcinogens but are modified by chemical processes in the body, particularly by enzymes in the liver, and are converted into their active, carcinogenic forms. These active modified forms of the carcinogens are also the forms that are mutagens. Enzymatic activation is the reason that chemicals to be examined in the Ames test are first mixed with an extract containing liver enzymes.

A final link between altered genetic material and cancer is found in the relationship between tumors and chromosomal abnormalities. For example, patients with a particular type of leukemia (**chronic myelogenous leukemia**) frequently contain blood cells that have a particular chromosomal abnormality in which part of the long arm of chromosome 22 is deleted (i.e., 22q−) and attached to the long arm of chromosome 9 (i.e., 9q+); the 22q− part of this translocation is often called the **Philadelphia chromosome** after its city of discovery. Many other cancers are also associated with chromosomal abnormalities of various types, but the associations are usually not as pronounced as that between chronic myelogenous leukemia and the Philadelphia chromosome. However, it is not yet clear whether tumor cells become chromosomally abnormal after they become cancerous, whether the chromosomal abnormality occurs first and induces the cancerous growth, or whether some unknown factor predisposes cells both to cancer and to the development of chromosomal abnormalities. Although absolute proof is lacking, all these links taken together strongly implicate genetic components in cancer, and it is widely believed that many "spontaneous" cancers arise by somatic mutations.

Polyoma: A DNA Tumor Virus

There is also strong evidence that some viruses can cause normal cells to become cancerous. Among these viruses and providing a prominent example is **polyoma**, one of a whole group of oncogenic (cancer-causing) viruses whose genetic material is DNA. Polyoma is among the smallest of viruses (Figure 12.12). It consists of an outer protein coat that surrounds a double-stranded, circular molecule of DNA. The viral DNA is small (5292 bp—its sequence is completely known) and codes for six proteins. The normal host of polyoma is the mouse, and it is widespread in both wild and laboratory mice. The virus does not seem to cause tumors in its wild host species, but it does cause them in newborn mice of some laboratory strains and in newborn rats, rabbits, and hamsters. Although polyoma is not known to infect humans, a closely related virus (**SV40**) does.

The body's first line of defense against polyoma and other viruses is antibodies that may be present or formed that have specificities against the viral coat. A second line of defense against viruses is found in a class of proteins called **interferons**, which render cells immune to virus infection. The details of how interferon works and its effectiveness are not fully understood. A major obstacle in research has been obtaining sufficient amounts of interferon

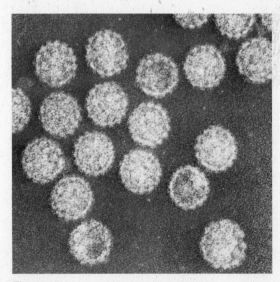

Figure 12.12 Polyoma virus magnified 270,000 times.

for study. Large amounts are now available thanks to DNA cloning, and laboratory tests and clinical trails are under way. In any case, if all or some of the virus particles can overcome barriers set up by the immune system and interferon, they can stick to the surface of susceptible cells. (The process of infection by polyoma has been studied primarily in tissue culture using susceptible cells of mice or hamsters.) The cells ingest the viruses by forming vacuoles around them: The cell membrane surrounds the virus, and the cup formed by the membrane around the virus is pinched off toward the interior of the cell. This brings the virus inside the cell (Figure 12.13).

Once inside the cell, one of three things involving the virus takes place. First, the infection may be abortive. In **abortive infection**, the process of infection stops in its early stages for

reasons not entirely understood; the polyoma virus is digested by enzymes and the cell goes on as before [see Figure 12.13(a)]. Abortive infections are the most common kind with polyoma, and this is also true of influenza and many other viruses. A second possible sequence of events constitutes polyoma's **lytic cycle**. Between 1 and 3 percent of the engulfed virus particles will enter the lytic cycle upon infection, and almost all the remainder will undergo abortive infection.

It will be convenient to discuss the events of polyoma's lytic cycle in connection with the diagram in Figure 12.14. The innermost circle represents the circular DNA of the virus numbered in arbitrary units from 0 to 100, and unique restriction sites for some enzymes (*Eco*RI, *Bam*I, and *Bgl*I) are indicated. (A **unique** restriction site in a molecule is a restric-

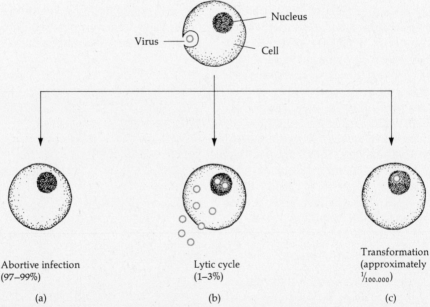

Figure 12.13 The possible outcomes of infection by polyoma. *(a)* In abortive infection, which occurs in 97 to 99 percent of infections, the virus is destroyed by the cell. *(b)* In the lytic cycle (1 to 3 percent of infections), the virus takes over the cell, leading to the production and liberation of numerous new virus particles. *(c)* In transformation, which occurs in about 1 in 100,000 infections, the infected cell does not produce new viruses, yet it retains the virus and becomes transformed in a manner similar to many types of cancer cells.

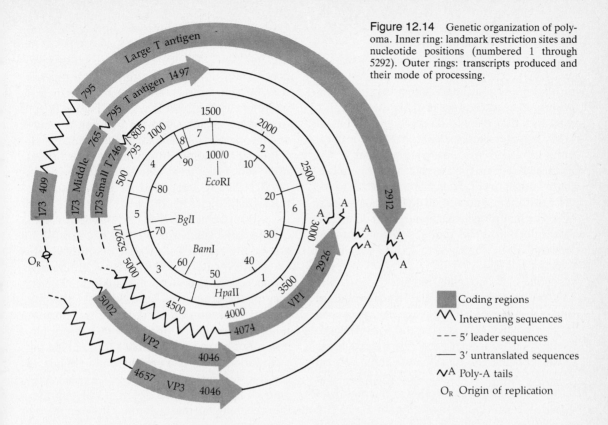

Figure 12.14 Genetic organization of polyoma. Inner ring: landmark restriction sites and nucleotide positions (numbered 1 through 5292). Outer rings: transcripts produced and their mode of processing.

Coding regions
Intervening sequences
5′ leader sequences
3′ untranslated sequences
Poly-A tails
O_R Origin of replication

tion site that occurs exactly once.) The enzyme *Hpa*II cleaves polyoma DNA into eight fragments, which are shown in the ring with segments numbered 1 through 8. The outer edge of the ring shows the nucleotide numbers of the DNA, from 1 through 5292 (numbering begins arbitrarily at one of the *Hpa*II sites).

The RNA's produced by the virus are shown in the outer three rings of Figure 12.14. Coding regions are denoted by shaded boxes, intervening sequences that are spliced out of the transcripts are indicated by jagged lines, 5′ leader sequences by dashed lines, untranslated 3′ sequences by solid lines, and poly-A tails by A's. The beginning and end nucleotides in each coding region are indicated by numbers, and the symbol O_R stands for **origin of replication**—

the place on the molecule at which DNA replication begins. As can be seen, polyoma transcripts are extensively spliced, and there are several overlapping genes. The overlapping regions of the large T, middle T, and small T antigens are all translated in the same reading frame, as are the overlapping regions of VP2 and VP3. However, the mRNA for the carboxyl terminus of VP2 and VP3 overlaps that for the amino terminus of VP1, and these are translated in offset reading frames. (Proteins VP1, VP2, and VP3 form parts of the viral coat; the functions of large T, middle T, and small T antigens will be discussed in a few paragraphs.)

In the lytic cycle, the viral DNA takes over cellular metabolism. Some 10 to 12 h after infection, the virus-coded T antigens begin to appear. DNA replication commences 12 to 15 h after infection, and shortly thereafter the late mRNA's (coding for VP1, VP2, and VP3) are produced and translated. Infected cells undergo pronounced changes in their own metabolism, including enhanced activity of the cellular DNA polymerase responsible for viral replication, increased synthesis of ribosomal RNA, increased rate of protein synthesis, increased rate of transport of six-carbon sugars across the cell membrane, and altered cell surface properties. The first progeny viruses appear 20 to 25 h after infection. Virus production continues for approximately 36 to 48 h, after which the infected cell dies.

The third possible outcome of infection of a cell by polyoma is called **transformation**. In this case no new virus particles are produced by the infected cell, but the cells do undergo changes (a "transformation") similar to those of cancer cells. The cells do not exhibit contact inhibition in tissue culture, and when injected into a live animal, there follows, in a matter of weeks, the development of a tumor at the site of inoculation. Small pieces of such tumors can be transplanted into other animals and soon give rise to full-blown tumors. However, animals that have developed an immune response to polyoma virus by being exposed to the virus are found to be resistant to the transplantation of polyoma-induced tumors because the transplanted tumor pieces are destroyed by the resistant animal's immune system. The antigens responsible for this immune response are called **transplantation antigens**; they are the three T antigens (small, middle, and large) in Figure 12.14. Transformation of cells by polyoma is rare. Roughly 1 in 100,000 virus particles will lead to transformation of the host cell [see Figure 12.13(c)], although mutant strains of polyoma are known that are incapable of inducing transformation.

In the process of transformation, the viral DNA becomes incorporated into the DNA of the host cell. The details of the integration process are not known, although the process may be analogous to the manner in which λ integrates into *E. coli* DNA (see Figure 12.6), or perhaps analogous to the manner in which eukaryotic transposable elements change position in the genome. In any case, transformed cells can have from one to several copies of viral DNA integrated at sites scattered throughout the chromosomes.

Transformed cells do not produce the late mRNA associated with VP1, VP2, and VP3, nor do progeny viruses appear. However, the early mRNA associated with the T antigens is produced and translated in transformed cells. One of these antigens—middle T—is the one primarily responsible for inducing transformation. How middle T brings about transformation is not known, but middle T protein has an enzymatic activity of a **protein kinase**; that is, it can transfer the terminal phosphate of an ATP onto an amino acid in a different polypeptide. It is becoming increasingly clear that an important mechanism of intracellular regulation occurs through activation or deactivation of particular enzymes by means of phosphorylation of certain of their amino acids. It therefore seems plausible that the middle T antigen could disrupt cellular control and induce transformation by means of its protein kinase activity. This possibility is even more plausible in light of findings from another transforming virus discussed in the next section. In any case, transformation is accompanied by the appearance of new, viral-coded T antigens on the cell surface. The immune system now gets a second chance to defend against the cancerous cells by attacking cells with the virus-specified antigens. If this attack fails, the transformed cell will continue to proliferate and cause a tumor.

Retroviruses, Endogenous Viruses, and Selfish DNA

This section is about tumor viruses with genetic material that consists of a single-stranded RNA molecule. Such viruses are usually small, consisting of their RNA surrounded by a protein capsid in an outer lipid-containing envelope composed partly of cellular constituents. They infect sensitive cells by passing through the membrane, much like polyoma, and inside the cell the protein coat is removed to release the RNA. One of the enzymes brought into the cell by the virus is a **reverse transcriptase**, which produces a DNA strand using the viral RNA strand as a template. A complementary DNA strand is then produced by other enzymes, forming a duplex DNA molecule, which then becomes covalently closed to create a circular DNA containing the viral genetic information. This circular DNA is called a **provirus**, and it proceeds to the nucleus, where it is incorporated into the host DNA. Because of their capacity to produce DNA from RNA, RNA tumor viruses are often called **retroviruses**.

Retroviruses have a very different life cycle from DNA viruses such as polyoma, not only because of their RNA genetic material and reverse transcriptase, but also because retroviruses do not have a lytic cycle. All successful retrovirus infections involve integration of the provirus into the host DNA. Moreover, provirus integration does not necessarily upset normal functions of the host cell. The provirus DNA is transcribed, the RNA is processed, and virus-specified proteins are produced in the cytoplasm. Some of these proteins form the capsid around viral RNA transcript. Other proteins associated with the viral envelope migrate to the cell membrane and are inserted in it. Then, similar to the situation with influenza, the core particles bud through the cell membrane at the regions tagged with viral envelope protein, thus becoming mature progeny viruses. As reproduction of the retrovirus is going on, the host cell can be behaving normally and even undergoing successive cell divisions. Unlike the case with polyoma, cells with integrated provirus do not necessarily die. Thus, a single infected cell can produce thousands of progeny retroviruses.

Many kinds of retroviruses are known; they infect such animals as chickens, mice, rats, hamsters, cats, pigs, deer, monkeys, and baboons. Indeed, many vertebrate species contain multiple copies of provirus DNA as a constituent of the normal genome. Viral genetic information can therefore be transmitted from parent to offspring through the germ line, but it can also be transmitted from individual to individual by means of infection with virus particles. Such normally occurring integrated proviruses are known as **endogenous viruses**. Although typically present in 5 to 50 copies per haploid genome, endogenous viruses are usually not expressed. Expression of the viral genetic information seems to be controlled by the host cells, and there can be a prolonged latent period. Sometimes infectious virus particles are produced in the absence of disease; at other times production is associated with disease (cancer). Endogenous viruses seem to persist in species for long periods of time, as closely related species tend to have closely related endogenous viruses. On occasion, cross-species transmission seems to occur. For example, one endogenous virus in domestic cats has been found to be more similar to a virus in primates than to viruses in other cats, as if a primate-to-cat transmission had occurred sometime in the distant past. Endogenous viruses may be an example of what has been called **selfish DNA**—DNA that maintains itself in a genome by virtue of its own intrinsic characteristics including the ability to integrate and transpose. Transposable elements (Chapter 11) are also often considered examples of selfish DNA.

Avian Sarcoma Virus and Cancer

From the standpoint of human health, retroviruses are important because they can cause malignant transformations. One well-studied example of a cancer-causing RNA virus is the chicken virus called **avian sarcoma virus**. (A **sarcoma** is a tumor of connective tissue.) A genetic map of the RNA of avian sarcoma virus is shown in Figure 12.15. Structurally it resembles a eukaryotic mRNA in having a cap at the 5' end and in being polyadenylated at the 3' end. Highlighted at the top of the figure is a 20-nucleotide direct repeat occurring just downstream from the cap and just upstream from the poly-A tail. Such repeated sequences are reminiscent of transposable elements (see Chapter 11), with which retroviruses share several other characteristics such as presence in the genome in multiple copies.

Avian sarcoma virus has four genes coding for four proteins. These are shown in Figure 12.15 along with the approximate number of nucleotides in each gene. (The entire RNA molecule consists of about 9700 nucleotides.) The gene designated *gag* codes for the capsid protein, *pol* codes for reverse transcriptase, and *env* codes for the virus-specified protein of the outer envelope. (*pol* and *env* are actually overlapping genes.) The gene of interest in transformation is the one designated *src*.

The *src* gene codes for a protein kinase, and it is thought to induce transformation by means of a poorly understood process outlined in Figure 12.16. Initially, a *src*-containing transcript is produced from the integrated provirus. After RNA processing in the nucleus, the *src* mRNA is released into the cytoplasm, where it is translated. The *src* protein kinase, which phosphorylates other proteins, then initiates an unknown series of cellular events (indicated by the question mark), which ultimately has manifold effects on cellular metabolism. An example involves an enzyme called **sodium-potassium ATPase** (Na^+-K^+ ATPase), which provides energy for the membrane "pump" that regulates the balance between sodium and potassium ions in the cell. The *src*-triggered events somehow decrease the activity of the ATPase, so the pump operates less efficiently. This decrease in efficiency alters the transport of glucose across the cell membrane and increases the rate of glucose breakdown by glycolysis or fermentation (also called **anaerobic respiration** because the process does not require free oxygen), which accounts for the 60-year-old observation that cancer cells frequently have an abnormally high level of anaerobic respiration. At the same time, a multitude of other cellular changes occur, including altered tubules and filaments and other

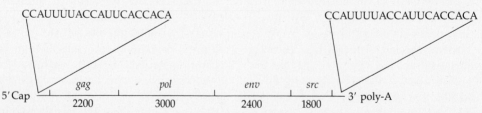

Figure 12.15 Genetic map of avian sarcoma virus. Note 5' cap, 3' poly-A tail, and 20-bp direct repeat near the ends. Genes *gag* and *pol* are translated in different reading frames. Genes *pol* and *env* actually overlap and are translated in different reading frames.

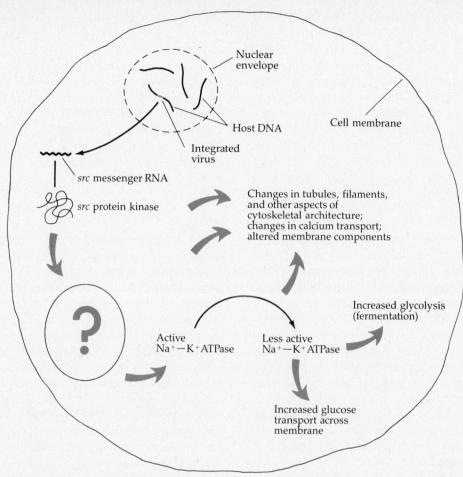

Figure 12.16 The *src* gene of avian sarcoma virus codes for a protein kinase. By an as yet unknown mechanism (question mark), this enzyme produces manifold changes in cellular metabolism, including changes in glucose transport and increased glycolysis associated with many (but by no means all) tumors.

aspects of cytoskeletal architecture, changes in calcium transport in and out of the cell, and alterations in membrane components (see Figure 12.16). These other changes may well lead to the initiation of cell division and release from contact inhibition, which is to say malignancy.

Remarkably, normal chicken cells possess a protein kinase that can act in the same way as the *src* enzyme, but the chicken enzyme

is produced in very small amounts. Introduction of avian sarcoma virus increases the amount of phosphorylation by 50 to 100 times because of the contribution from the *src* gene, and this increase is sufficient to initiate the events. Even more remarkable is the finding that the *src* gene is almost identical to the normal chicken gene that codes for the corresponding protein kinase. This finding, along with the other similarities between retroviruses and transposable elements, suggests that transforming genes like *src* were once normal cellular genes that have become mobilized (transposable). Of course, once mobilized on a retrovirus, the gene can be reintroduced into the genome by provirus integration, and there is evidence that some pseudogenes have been formed by this sort of gene hopping.

It is still unclear how important protein kinases are in the occurrence of other types of cancer. Many cancers exhibit an increased amount of anaerobic respiration, and some (but not all) have high levels of phosphate incorporation as would be expected from excess amounts of protein kinase. It should also be pointed out that no cancer-causing *human* virus has as yet been discovered, although it seems almost certain that they exist. In any event, studies with avian sarcoma and related viruses are revealing for the first time the molecular basis of at least some types of malignant transformations.

SUMMARY

1. Viruses are submicroscopic parasites of cells that are unable to reproduce outside cells. Most viruses are **host specific**, which means they are able to infect just one or a few related species of bacteria, animals, or plants. Although some viruses seem to be relatively harmless, some can cause serious disease. Examples of disease-causing viruses are **rubella** (German measles, birth defects in embryos), the **herpesvirus** family (cold sores, **genital herpes** venereal disease associated with birth defects and very possibly urogenital cancer), **influenza**, mumps, chicken pox, smallpox, encephalitis, the common cold, and certain types of cancer.

2. Viruses are extraordinarily diverse. Their genetic material may be DNA or RNA, single stranded or double stranded, linear or circular, and with one molecule or several. The genetic material is packaged inside a protective protein coat called a **capsid**, and, in animal viruses, the capsid is surrounded by an **envelope** composed of lipid, protein, and carbohydrate, which is acquired as the virus exits through the host cell membrane. Viruses vary enormously in size. Some, like the pox viruses, are almost as large as a small bacterial cell and have enough nucleic acid to code for hundreds of proteins; at the other end of the size spectrum are viruses like **polyoma** (a mouse tumor virus), which has only six genes.

3. The virus **lambda** (λ), which infects cells of *Escherichia coli*, has provided several fundamental concepts related to viral function. This bacterial virus, or **bacteriophage**, has two alternative life cycles: the **lytic cycle**, in which the virus takes over control of the host metabolism and produces about 100 progeny viruses that are released when the cell undergoes **lysis** (bursting open), and **lysogeny**, in which the λ DNA is integrated into the host DNA by a site-specific recombination event between circularized λ DNA and the host DNA. Whether infection will result in the lytic cycle or lysogeny depends in part on the physiological state of the host because two products of **early mRNA** (the mRNA produced early after infection) interact with host molecules and can stimulate transcription of the operon containing the gene C_I. The C_I gene codes for a repressor that represses other λ operons, but it exhibits **autogenous regulation** (self-regulation) in that it stimulates its own transcription. Thus, integrated λ DNA, called **prophage**, can be maintained indefinitely because C_I represses other λ operons. If the amount of C_I repressor falls below a certain level, the prophage can exit from the host DNA by another site-specific recombination event and undergo the lytic cycle. In the lytic cycle, production of the early mRNA's is followed by the production of a **late mRNA**, which codes for head proteins, tail proteins, and **lysozyme** (the lysing enzyme). The lytic cycle–lysogeny options make λ a **temperate phage**, in contrast to a **virulent phage**, which has only a lytic cycle.

4. Sometimes the λ prophage undergoes defective excision and incorporates a piece of host DNA in place of some of its own DNA. A bacterial cell infected with such a virus can incorporate a gene or genes from the previous host by ordinary recombination. Such gene transfer mediated by a virus is called **transduction**. λ is an example of a **specialized transducing phage** because it can mediate transduction of only host genes near the λ **attachment site** (the site of prophage insertion). Some other transducing phage are **generalized transducing phage** because they can mediate the transduction of any host genes.

5. λ has become a popular vector for the cloning of bacterial and eukaryotic DNA fragments. Special strains of λ have been created for cloning. These typically have conveniently located restriction sites flanking genes that are nonessential for the lytic cycle. Hybrid phage are created by replacing these nonessential genes with an appropriate restriction fragment to be cloned.

6. Influenza virus is a virus that infects the respiratory tract of birds, many mammals, and humans. The virus consists of eight molecules of single-stranded RNA (each corresponding to a different gene) inside a protein capsid surrounded by a lipid-protein envelope festooned with virus-coded **spikes** of either **hemagglutinin** or **neuraminidase**. These spikes are **antigenic**—they elicit an immune response—and animals produce **antibodies** (immunity proteins) that recognize and attack the virus. Influenza outbreaks occur when the proportion of immune individuals is sufficiently low that the virus can spread from individual to individual. Local outbreaks (**epidemics**) are usually the result of **antigenic drift**—minor changes in the hemagglutinin or neuraminidase antigens due to mutations in the corresponding genes.

Worldwide epidemics (**pandemics**), such as the pandemic of Hong Kong flu in 1968, are usually due to **antigenic shifts**—major changes in the viral antigens due to recombination between viruses of human and nonhuman origin. Antigenic changes in influenza virus may have a tendency to be cyclical, because some current viral strains are antigenically similar to strains that were prevalent around the turn of the century.

7. Cancer refers to a heterogeneous collection of diseases characterized by unrestrained cell division and by an inability of cells to respond to **contact inhibition** of division by neighboring cells. Although most cancers are thought to be induced by environmental agents, there are nevertheless links between mutations and cancer. First, more than 50 inherited forms of cancer are known, all of them rare; examples are xeroderma pigmentosum (a simple Mendelian recessive) and **retinoblastoma** (retinal tumors associated with a simple Mendelian dominant). Second, most carcinogens have been found to be mutagenic, and most mutagens have been found to be carcinogenic. Third, certain cancers are associated with chromosomal abnormalities, such as **chronic myelogenous leukemia**, which has a striking association with the **Philadelphia chromosome** (a 22q− deletion, but actually one part of a translocation). These connections between mutations and cancer suggest that at least some environmentally induced cancers may arise from somatic mutations.

8. Some cancers are caused by viruses, but no clear-cut **oncogenic** (cancer-causing) human virus has as yet been isolated. Indeed, many well-studied cancer-causing viruses do not seem to cause cancer in their natural hosts. For example, **polyoma** virus is widespread in wild mouse populations, its natural host, but it does not seem to induce tumors; yet polyoma causes cancer in newborn mice of various laboratory strains and in newborn rats, rabbits, and hamsters.

9. Polyoma is an example of a DNA tumor virus. Its genome consists of 5292 bp of circular, double-stranded DNA, and it codes for six proteins; three "early" proteins (large T, small T, and middle T antigens) and three "late" proteins (VP1, VP2, and VP3—constituents of the viral capsid). Three outcomes of infection of cells with polyoma are possible. The infection may be **abortive** (failed), which is the most common outcome. In 1 to 3 percent of infections, polyoma undergoes its **lytic cycle**, in which progeny viruses are produced and the infected cell invariably dies. In a small proportion of cases (i.e., about 10^{-5}), polyoma induces the **transformation** of normal cells into cancer cells. In transformation, from one to several copies of polyoma DNA become incorporated into the host genome. Only the early proteins are produced, and the middle T antigen is thought to be the one primarily responsible for transformation. (One function of middle T antigen is as a **protein kinase**—an enzyme that is able to attach phosphate groups to certain amino acids in other protein molecules.) Transformed cells express the polyoma T antigens as well as certain unusual host antigens, and these provide the immune system with a last opportunity to attack the transformed cells.

10. **Retroviruses** are single-stranded RNA viruses coding for a **reverse transcriptase** enzyme that produces a DNA strand using the viral RNA strand as a template shortly after infection. Following synthesis of the complementary DNA strand and circularization, the circular DNA (called the **provirus**) becomes integrated into the host genome. The integrated provirus need not disrupt host cell functions, and the integrated provirus can be transcribed and lead to the formation of progeny viruses. Thus, an infected host cell can survive and undergo division and release thousands of progeny retroviruses. Retroviruses always become integrated in successfully in-fected cells; these viruses do not possess a conventional lytic cycle.

11. Endogenous viruses are integrated proviruses that occur as a normal constituent of the host genome. Typically having 5 to 50 copies per haploid genome, endogenous viruses are widespread in vertebrate species, including primates. Being a normal constituent of the genome, endogenous viruses can be passed from generation to generation through the germ cells. Although provirus expression is controlled by the host and provirus genetic information is usually not expressed, endogenous viruses do carry the genetic information for producing mature retrovirus particles. Provirus expression may have a prolonged latent period, and cells that actively produce virus particles are sometimes cancerous and sometimes not. Cross-species transmission of endogenous viral DNA sometimes seems to occur, as illustrated by one such virus in the domestic cat, which is very primatelike in its characteristics. Endogenous viruses may be an example of **selfish DNA**—DNA that maintains itself in a genome by virtue of integration and transposition.

12. Many retroviruses can cause malignant transformations; one example is **avian sarcoma virus**, which causes **sarcomas** (connective tissue tumors) in birds. The RNA of avian sarcoma virus is about 9700 nucleotides long, carries a cap at its 5' end, is polyadenylated at its 3' end, and has a 20-nucleotide direct repeat abutting the cap and the poly-A tail. It has four genes: *gag* (a capsid protein), *pol* (reverse transcriptase), *env* (an envelope protein), and *src* (a protein kinase responsible for malignant transformation).

13. Transformation induced by avian sarcoma virus involves a series of events initiated by the *src* kinase. These events have manifold biochemical effects on cells, including alterations in the membrane, in cell structure, and in calcium transport. One such effect is a decreased efficiency of the enzyme **sodium-potassium ATPase**, which leads to decreased efficiency in the operation of the membrane pump that regulates the intracellular balance of sodium and potassium. This sodium-potassium imbalance is associated with increased transport of glucose and a higher than normal level of glucose metabolism by means of **anaerobic respiration** (breakdown of glucose without the use of free oxygen).

14. The *src* gene is almost identical to a normal chicken gene that codes for the same protein kinase, which suggests that *src* may originally have been acquired from the host. Other retroviruses carry genes that are similar if not identical to their normal host counterparts. These similarities, together with structural similarities between proviruses and transposable elements, suggest that cancer-causing retroviruses may represent transposable elements gone awry.

WORDS TO KNOW

Virus	Lambda	Influenza	Retrovirus
Host specific	Lytic cycle	Spikes	Reverse transcriptase
Herpesvirus	Lysis	Epidemic	Provirus
Capsid	Lysogeny	Pandemic	Endogenous virus
Envelope	Cohesive ends	Antigenic drift	Avian sarcoma virus
Oncogenic	Attachment site	Antigenic shift	Protein kinase gene
Bacteriophage	Prophage		Selfish DNA
Temperate	Specialized transduction	**Polyoma Virus**	
Virulent	Generalized transduction	SV40	**Syndrome**
		Abortive infection	Retinoblastoma
		Lytic cycle	Chronic myelogenous leukemia
		Transformation	

PROBLEMS

1. For discussion: Would you support a proposal to require all prospective marriage partners to submit to a test for genital herpes as they are now required to submit to a test for syphilis? What possible benefits would come from such a requirement? What possible harm?

2. What is the distinction between a viral capsid and a viral envelope?

3. In a famous experiment carried out in 1952 by A. D. Hershey and M. Chase, bacteriophage were simultaneously labeled with two types of radioisotopes—P^{32} (to label the nucleic acid) and S^{35} (to label the proteins). After infection, viral material remaining at the cell surface was detached and discarded, and the radioisotopes inside the infected cells were analyzed. The principal finding was that most of the radioactivity inside infected cells was due to P^{32}. This experiment was widely hailed as showing that the genetic material of the virus was its nucleic acid. How does this conclusion emerge from the data presented above?

4. In extracting DNA from *E. coli*, an early step involves addition of lysozyme to the cells. What is the purpose of adding lysozyme?

5. Distinguish between a temperate and a virulent bacteriophage. What can a temperate phage do that a virulent phage cannot?

6. What are endogenous viruses and what characteristics do they share with copialike transposable elements of the sort discussed in Chapter 11?

7. In connection with influenza virus, what is the difference between antigenic drift and antigenic shift? What is the difference between an epidemic and a pandemic?

8. A polyoma virus is approximately spherical and has a diameter of about 50 Å. An average cell is also approximately spherical and has a diameter of about 10 μm. What is the relationship between the volume of polyoma virus and the volume of an average cell? (The volume of a sphere is given by $V = \frac{4}{3} \pi r^3$, where r is the radius; 1 Å = 10^{-4} μm.)

9. Bacteriophage λ is often considered a sophisticated type of transposon. What features does λ share with transposons such as Tn5 (see Figure 11.15) to justify this comparison? (In thinking about this problem, it may help to be aware that, when the host bacterium has a deletion of the λ attachment site, the viral DNA can integrate at other places in the host DNA.)

10. A λ geneticist has six mutations and wishes to determine how many different genes are represented. She thus carries out complementation tests, in which cells are coinfected with phage of each of two mutant types. If a normal lytic cycle follows coinfection, the mutations complement and are thus in different genes. If a lytic cycle does not ensue, the mutations fail to complement and are thus in the same gene. Shown here are the results of coinfection, where + indicates complementation and − indicates noncomplementation. (The numbers 1 through 6 denote the individual mutations.) How many genes are represened by the six mutations?

	1	2	3	4	5	6
1	−	+	+	−	−	+
2		−	+	+	+	−
3			−	+	+	+
4				−	−	+
5					−	+
6						−

11. You are given a small vial containing a bacteriophage known to mediate transduction in *E. coli*.

How could you determine experimentally whether it is a specialized transducing phage or a generalized transducing phage?

12. Two genes that are sufficiently close together can undergo cotransduction, which means that they can be carried together in the same bacteriophage particle. The frequency of cotransduction between two genes is the proportion of cells transduced for one of the genes that are simultaneously transduced for the other. The following are the frequencies of cotransduction among three genes in *E. coli*:

metC–tolC	28%
uxaA–tolC	6%
metC–uxaA	0%

What is the order of these three genes along the *E. coli* DNA?

13. Cotransduction frequencies of three *E. coli* genes are as follows:

uxaA–ebgA	40%
tolC–ebgA	32%

Use these data and those in Problem 12 to infer the order of the genes *ebgA*, *metC*, *uxaA*, and *tolC*.

14. The accompanying pedigree involves a male with xeroderma pigmentosum. In what way is the pedigree typical of rare disorders due to an autosomal-recessive allele?

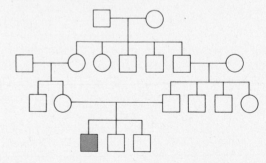

15. A normal woman has a brother affected with xeroderma pigmentosum. What is the probability that she is a carrier?

FURTHER READING AND REFERENCES

Air, G. M. 1981. Sequence relationships among the hemagglutinin genes of 12 subtypes of influenza A virus. Proc. Natl. Acad. Sci. U.S.A. 78:7639–7643. Antigenic drift and antigenic shift at the nucleotide sequence level.

Bishop, J. M. 1981. Enemies within: The genesis of retrovirus oncogenes. Cell 23:5–6. Do retroviruses steal their cancer-causing genes from cellular DNA?

Bishop, J. M. 1982. Oncogenes. Scientific American 246: 80–92. These cancer-causing genes were first found in viruses, but they are also constituents of vertebrate cells.

Cairns, J. 1980. Cancer: Science and Society. Freeman, San Francisco. An overall view of what is known about cancer.

Chedd, G. 1975. Closing in on the flu. Science Year: The World Book Science Annual. Field Enterprises Educational Corp., Chicago. Source of Figure 12.9.

Coffin, J. M. 1980. Structural analysis of retrovirus genomes. In J. R. Stephenson (ed.). Molecular Biology of RNA Tumor Viruses. Academic Press, New York. Source of data in Figure 12.15.

Croce, C. M., and H. Koprowski. 1978. The genetics of human cancer. Scientific American 238:117–125. Identification of genetic factors involved in malignancy.

Devoret, R. 1979. Bacterial tests for potential carcinogens. Scientific American 241:40–49. Details on how the tests work and what kind of genetic damage is detected.

Doolittle, W. F., and C. Sapienza. 1980. Selfish genes, the phenotype paradigm and genome evolution. Nature 284:601–603. A philosophical look at the implications of selfish DNA.

Epstein, S.S., and J.B. Swartz. 1981. Fallacies of lifestyle cancer theories. Nature 289:127–130. Challenges the often-held view that lifestyle is a major cause of cancer and that people who develop cancer have brought it on themselves.

Favera, R. D., E. P. Gelmann, R. C. Gallo, and F. Wong-Staal. 1981. A human *onc* gene homologous to the transforming gene (*v-sis*) of simian sarcoma virus. Nature 292:31–35. A transforming gene in a virus is a normal constituent of human DNA.

Fields, S., G. Winter, and G. G. Brownlee. 1981. Structure of the neuraminidase gene in human influenza virus A/PR/8/34. Nature 290:213–217. The 1413-base-pair sequence of the gene is presented.

Griffin, B. E., E. Soeda, B. G. Barrell, and R. Stader. 1980. Sequence and analysis of polyoma virus DNA. In J. Tooze, (ed.). DNA Tumor Viruses. Cold Spring Harbor Laboratory, Cold Spring Harbor, N.Y., pp. 831–896. Source of Figure 12.14.

Hayward, W.S., B. G. Neel, and S. M. Astrin. 1981. Activation of a cellular *onc* gene by promoter insertion in ALV-induced lymphoid leukosis. Nature 290:475–480. Avian leukosis virus can integrate adjacent to a cellular gene and initiate transcription from a viral promoter leading to cancer.

Johnson, A. D., A. R. Poteete, G. Lauer, R. T. Sauer, G. K. Ackers, and M. Ptashne. 1981. λ repressor and cro—components of an efficient molecular switch. Nature 294:217–223. The ingenious molecular circuitry involved in λ regulation.

Kitamura, N., B. L. Semler, P. G. Rothberg, G. R. Larsen, C. J. Adler, A. J. Dorner, E. A. Emini, R. Hanecak, J. J. Lee, S. van der Werf, C. W. Anderson, and E. Wimmer. 1981. Primary structure, gene organization and polypeptide expression of poliovirus RNA. Nature 291:547–553. The 7433-nucleotide virus encodes one long polypeptide that is cleaved into 12 viral proteins.

Nash, H. A. 1981. Integration and excision of bacteriophage λ: The mechanism of conservative site-specific recombination. Ann. Rev. Genet. 15:143–167. Details on the precision of integration and excision.

Nicolson, G. L. 1979. Cancer metastasis. 1979.

Scientific American 240:66–76. This process is the real life-threatening danger in most cancers.

Perera, F., and C. Petito. 1982. Formaldehyde: A question of cancer policy? Science 216:1285–1291. A case study of scientific and political issues in evaluating and regulating a possible cancer-causing substance.

Ptashne, M., A. Jeffrey, A. D. Johnson, R. Maurer, B. J. Meyer, C.O. Pabo, T. M. Roberts, and R. T. Sauer. 1980. How the λ repressor and cro work. Cell 19:1–11. Details on the molecular interactions that control the phage life cycle.

Reddy, V. B., B. Thimmappaya, R. Dhar, K. N. Subramanian, B. S. Zain, J. Pan, P. K. Ghosh, M. L. Celma, and S. M. Weissman. 1978. The genome of simian virus 40. Science 200:494–502. Presents all 5226 nucleotides.

Roizman, B. 1979. The organization of the herpes simplex virus genomes. Ann. Rev. Genet. 13:25–57. A detailed review of the genetics of herpes virus.

Schimke, R. N. 1978. Genetics and Cancer in Man. Churchill Livingstone, Edinburgh. A slim but authoritative introduction to genetic aspects of cancer.

Simons, K., H. Garoff, and A. Helenius. 1982. How an animal virus gets into and out of its host cell. Scientific American 246:58–66. The process is described in illuminating detail.

Temin, H. M. 1979. Viral oncogenes. Cold Spring Harbor Symp. Quant. Biol. 44(1):1–7. An overview of cancer-causing viral genes in the first of two volumes devoted to the subject.

Temin, H. M. 1980. Origin of retroviruses from cellular movable genetic elements. Cell 21:599–600. A provocative hypothesis about the origin of retroviruses.

Varmus, H. E. 1982. Form and function of retroviral proviruses. Science 216:812–820. Details of several RNA tumor viruses are reviewed.

Webster, R. G., and W. J. Bean, Jr. 1978. Genetics of influenza virus. Ann. Rev. Genet. 12:415–431. Details of influenza structure and genetics, including antigenic drift and antigenic shift.

Weinberg, R. A. 1980. Origins and roles of endogenous retroviruses. Cell 22:643–644. Where do such retroviruses come from?

Wiley, D. C., I. A. Wilson, and J. J. Skehel. 1981. Structural identification of the antibody-binding sites of Hong Kong influenza hemagglutinin and their involvement in antigenic variation. Nature 289:373–378. Detailed molecular structure of hemagglutinin.

chapter 13
Immunity and Blood Groups

We live in an ocean of hostile microorganisms such as the viruses associated with influenza and cancer, and immunity is a sort of life jacket that keeps us afloat. The immune response to invading pathogens, as found in humans and other mammals, is one of the sublime achievements of evolution. It is a system that recognizes and destroys viruses, bacteria, and transformed cancerous cells; at the same time it can identify our own normal tissues and refrain from attacking them. Unfortunately, the immune system is not perfect. Its failings are proclaimed by dysentery, influenza, and cancer; by allergies, rheumatism, and rheumatic fever; and by many degenerative disorders of aging. On the other hand, the immune system is so effective that it presents a major hindrance to the transplantation of tissues from one person to another. Even though a great deal is known about the immune system, the nature of immunity is complex and so sophisticated that it still defies complete description. The study of immunity is revealing new principles in biology while providing dramatic new procedures in the fight against disease.

The Immune Response

Immunity is rooted in a trillion or so white blood cells called **lymphocytes** and in related blood cells known as **macrophages**. Lymphocytes, which die and are replenished throughout life at the rate of about 10 million cells per minute, are produced by the division of certain **stem cells** in the mushy, reddish bone marrow where many blood components are manufactured. Once formed, these lymphocyte precursors undergo further differentiation in either of two distinct ways. About half the lymphocytes pass through or under the influence of the thymus—a small organ underlying the breast-bone in children which grows until puberty and then gradually shrinks, virtually disappearing by adulthood. Lymphocytes that differentiate under the influence of the thymus are called **T cells** or **thymus-dependent cells**; these cells constitute one arm of the immune system. (T cells are the cells usually studied in routine chromosome analysis because they can easily be stimulated to divide by such agents as phytohemagglutinin.) The other half of the lymphocytes undergo a different sort of processing, many details of which are still not understood. In birds, this processing occurs in an organ called the **bursa of Fabricius**, an organ not found in mammals. An equivalent processing occurs in mammals, but it seems to take place in bone marrow and perhaps in the spleen. The lymphocytes that undergo this alternative type of processing are known as **B cells** or **bone-marrow-derived cells**; B cells constitute a second arm of the immune system. Figure 13.1 is a photograph of a differentiated lymphocyte (a T or a B cell) from a normal mouse.

T cells circulate in the bloodstream, whereas B cells are concentrated in the lymphatic system, which consists of the spleen (the main blood filter), the lymph nodes (local lymph filters), and clumps of cells associated with the gut, respiratory tract, and urogenital tract (including cells in the tissues of the appendix, tonsils, and adenoids). (See Figure 13.2 for the anatomical relationships between some of the principal constituents of the immune system.) The immune system thus monitors all foreign invaders that come into the body through either normal body openings or wounds, and all invaders in the blood or lymphatic system. The T cells are known as agents of **cellular immunity**; they include **cytolytic** (killer) T cells that recognize and attack such alien intruders as bacteria and fungi, and they can even attack such abnormal cells as transformed cancer cells. Dividing rapidly and surrounding the foreign invaders, T cells produce a local inflammation. They can release poisons or toxins to kill the intruders, and they release chemical signals that summon legions of large scavenger white blood cells (macrophages) that devour and digest the invaders.

The other arm of the immune system, which relies on B cells, attacks invaders indirectly. Alien cells carry immunity-stimulating substances called **antigens** on their surface. These antigens trigger B cells to divide and secrete protein molecules called **antibodies**, which circulate in the blood and stick to the antigens of the foreign invaders. The antigen-antibody complex marks the invaders for destruction by initiating the sequential activation of more than a dozen blood proteins, collectively designated **complement**, which results in lethal damage to the membrane of the invading cell. Macrophages and other scavenger cells then finish the destruction. Because antibodies circulate freely in the blood, B cells are known as the agents of **humoral immunity**.

The crux of immunity is the recognition of foreign antigens by T cells and B cells and the stimulation of response. Normal recognition and response require **cell cooperation**—the interaction of cell types. Appropriate response of cytolytic T cells requires interaction with a class of T cells called **helper T cells**. Appropriate response of B cells requires interaction not only with helper T cells but also with a macrophage like cell sometimes called an **antigen-presenting macrophage** because it seems to attach to the antigen and "present" it to the lymphocyte. Details of cell cooperation aside, a multitude of substances are antigenic (that is, able to elicit an immune response). Indeed, almost anything as big or bigger than a protein molecule is antigenic. Viruses, bacteria, and foreign cells are therefore antigenic, as are most protein molecules, large carbohydrates, and some nucleic acids introduced from outside the body. On the other hand, practically no small molecules are able to elicit an immune response.

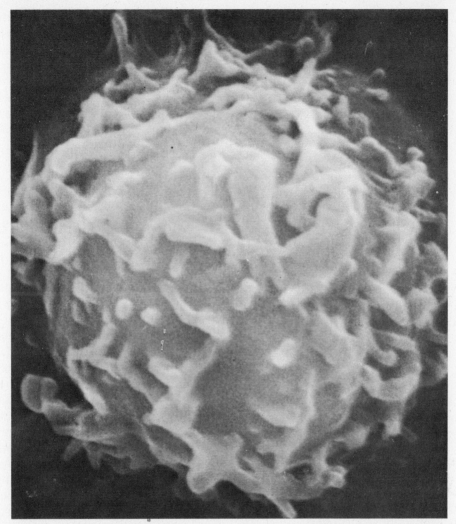

Figure 13.1   Scanning electron micrograph of a differentiated lymphocyte from a normal mouse.

Although most small molecules do not stimulate antibody production, antibodies that react with small molecules can often be stimulated if the small molecule is first linked with a larger antigen, such as a protein. In this context, the small molecule is called a **hapten** and the larger antigen a **carrier**. The role of the carrier is to interact with the antigen-presenting macrophage, which presents the hapten portion of the molecule to the B cell.

Clonal Selection

What is it about an antigen that elicits an immune response? It is the detailed relief, the texture, of its surface. All large molecules, including those on the surface of viruses, bacteria, and cells of higher organisms, fold into precise three-dimensional configurations, providing a sort of landscape with identifiable features. In much the way that every human

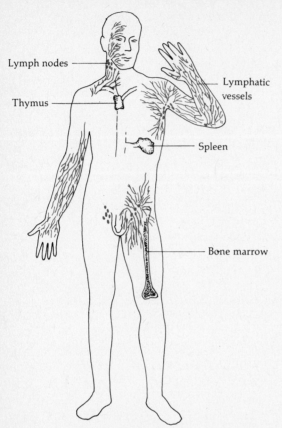

Lymph nodes

Thymus

Lymphatic vessels

Spleen

Bone marrow

Figure 13.2   The relationships of some principal components of the immune system—lymph nodes and lymphatic vessels, thymus, spleen, and bone marrow.

face can be recognized by its distinct surface features—the precise shape of the ears, the exact arrangement of skin creases around the eyes, the shape and placement of blemishes on the nose—a number of distinct surface features of an antigen can be recognized by their three-dimensional configuration of atoms. Each separate, recognizable feature of an antigen is known as an **antigenic determinant** or an **epitope**; since as few as four amino acids in a protein can provide a distinct antigenic landmark (epitope) on its surface, large molecules or cells may have tens or hundreds of different antigenic determinants. The only requirement for a material to be antigenic is that it have at

least one surface landmark sufficiently distinct from substances normally present in the body to be recognized as foreign. Because large molecules and cells have many antigenic determinants, the vast majority of such substances from sources outside an individual have at least one antigenic determinant different from those normally present in the body; these identify the substances as foreign. Moreover, a typical antigen has many different antigenic determinants and so elicits the production of many different antibodies, each type of antibody in response to one type of antigenic determinant on the antigen. Even many small molecules have a sufficiently distinctive surface relief to be recognized by antibodies, but as noted earlier, they do not typically elicit an antibody response unless coupled with a macrophage-stimulating larger antigen (the carrier).

So antigenic determinants (epitopes) are the features of antigens that are recognized and attacked by the immune system. However, each B cell or T cell is able to recognize only one antigenic determinant. A B cell has on its surface about 100,000 identical **receptor sites** whose surface relief exactly matches the specific antigenic determinant recognized by the cell; a T cell has about 10,000 receptor sites. The match between receptor site and antigenic determinant is complementary; they match like lock and key (Figure 13.3). When an antigen gains entrance into the body, therefore, not all T cells and B cells swing into action. The only lymphocytes that respond are the ones with receptors that happen to fit, or "recognize," an antigenic determinant on the antigen.

When the body is invaded by a foreign antigen, each antigenic determinant on the intruder stimulates the B cells and T cells that respond specifically to it. A B cell whose surface receptors match one of the antigenic determinants becomes combined with the antigen, and this combining, along with functions provided by macrophages and helper T cells, stimulates the B cell to undergo mitosis. The

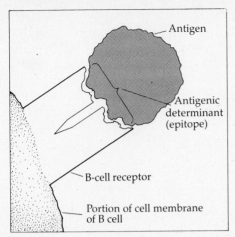

Figure 13.3 The complementary fit of an antigenic determinant and a receptor site on the surface of a B cell.

cells produced by successive cell divisions constitute a **clone** because they all derive from the same parental cell and are thus genetically identical. The stimulation process is called **clonal selection** because the B cell that gives rise to the clone is selected from among billions of B cells by the antigenic determinant it fits.

The process of clonal selection is outlined in Figure 13.4. An antigenic determinant on an antigen (shaded shape) combines with the complementary surface receptor of a B cell and stimulates the cell to undergo division. (Note that B cells with receptors that are not complementary to the antigenic determinant are not stimulated.) From the stimulated B cell, a clone of cells is produced. Within this clone are cells that differentiate and begin to secrete antibodies—proteins able to recognize the same antigenic determinant that caused stimulation. These antibody-manufacturing cells are known as **plasma cells**, and their endoplasmic reticulum is covered with ribosomes and greatly distended by accumulated antibody molecules. An active plasma cell produces and secretes about 2000 identical antibody molecules every second.

Other cells of the clone do not become plasma cells, however. They cease dividing or divide very slowly. The cells in a clone that become quiescent can be induced to resume division and give rise to mature plasma cells by a second assault of the antigen. Thus, the immune response to a repeated infection or to a booster shot is much faster and produces more antibody. The readiness to mount such a secondary immune response is commonly called **immunity**. The initial or primary response of the immune system is rather slow, taking several days. This slowness is why people may fall prey to such fast-acting diseases as influenza if not previously exposed. However, one exposure to a particular antigen, such as a strain of influenza virus, stimulates antibody production, which wards off subsequent attacks by the same or very similar antigens. Once stimulated, the memory of the immune system may persist for decades. As mentioned in Chapter 12, certain people exposed to a particular strain of influenza virus in the 1890s can still produce antibodies against that strain. Such long-term memory in the immune system is a function of T cells.

Varieties of Antibody

A plasma cell makes only one kind of antibody, but because the body has plasma cells from thousands of different clones, thousands of different antibodies can be produced. Although each type of antibody is unique, antibodies can be classified into five general groups: IgG, IgM, IgA, IgD, and IgE, where the Ig stands for **immunoglobulin**. Table 13.1 outlines some characteristics of the antibody classes. In molecular structure, antibodies are composed primarily of protein with a relatively small amount of carbohydrate attached. The fundamental unit of all antibodies consists of two long or **heavy** polypeptide chains and two short or **light** polypeptide chains. Each antibody class has its own characteristic type of heavy chain, which is designated by a lower-

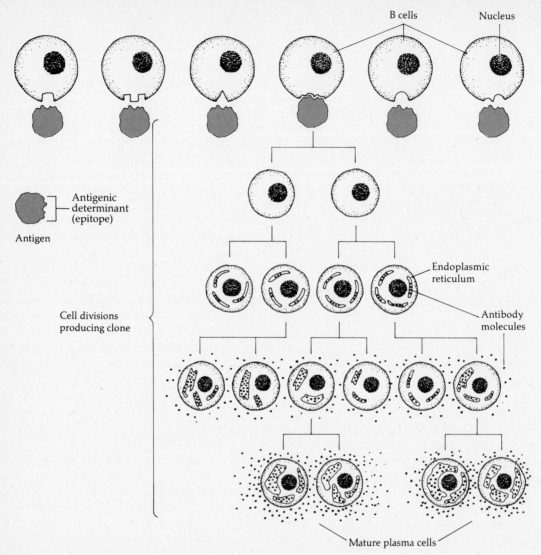

Figure 13.4   Clonal selection. An antigenic determinant on an antigen (shaded shape) combines with the complementary surface receptor of a B cell and stimulates the cell to undergo division. (Note that B cells whose receptors are not complementary to the antigenic determinant are not stimulated.) From the stimulated B cell, a clone of cells is produced. In this clone are cells whose endoplasmic reticulum is markedly distended; these cells, called *plasma cells*, produce and release large numbers of antibody molecules that attack the antigenic determinant on the antigen. Some cells in the clone do not immediately become plasma cells but cease dividing; these cells provide a more immediate response to a second assault of the same antigen.

case Greek letter (γ, μ, α, δ, ε) corresponding to an antibody **class** (G, M, A, D, E). Heavy chains range in size from about 500 amino acids (γ) to about 700 amino acids (ε). Although each antibody class has its own heavy chains, they all contain light chains of just one of two classes, called κ and λ, consisting of about 220 amino acids.

TABLE 13.1   SOME CHARACTERISTICS OF HUMAN IMMUNOGLOBULINS

Class	IgG	IgM	IgA	IgD	IgE
Heavy (H) chains	γ	μ	α	δ	ε
Light (L) chains	κ, λ	κ, λ	κ, λ	κ, λ	κ, λ
Number of subclasses	4	2	2	—	—
Serum concentration, mg/ml	8–16	0.5–1.9	1.4–4.2	<0.04	<0.007

Source: E. S. Golub, 1981, The Cellular Basis of the Immune Response, 2nd ed., Sinauer Associates, Sunderland, Mass., p. 215.

IgG is the workhorse of immunity and is the prevalent type of antibody in the blood serum. Each IgG molecule consists of two heavy chains and two light chains and thus has the structure $\gamma_2\kappa_2$ or $\gamma_2\lambda_2$ (never $\gamma_2\kappa\lambda$ because the light chains in an antibody are always the same). However, there are four nearly identical genes that can code for the γ chains in any one molecule. These four γ-chain genes lead to four **subclasses** of IgG; each subclass is determined by the particular γ chain present in the antibody. In addition, there are normal allelic variants of these and some other antibody genes, which are inherited in simple Mendelian fashion. These normal variants of antibody genes are known as **allotypes**. Although IgG is the most concentrated antibody class in blood serum, IgG molecules can also cross the placenta. Indeed, IgG is the only class of antibody able to cross the placenta.

Although IgG antibody has been the most intensively studied, a few words about the others are in order. IgM is a large and powerful antibody specialized to combat certain kinds of antigens. It seems to be involved in **rheumatoid arthritis**—a disorder of the immune system marked by swollen, stiff, and painful joints. IgM molecules contain 10 heavy chains (μ) and 10 light chains (either κ or λ), but there are two μ-chain genes that give rise to two IgM subclasses. IgA is the antibody found in the gut, respiratory tract, urogenital tract, and other mucous membranes. IgA molecules can contain a variable number of heavy chains and light chains, but all the heavy chains are of the α type and all the light chains are of either the κ or λ type. Again, distinct genes for the α chain produce distinct subclasses of IgA. IgD and IgE appear in much smaller amounts of blood serum than do the other classes of antibody. Both classes consist of two light and two heavy chains; the heavy chain is δ in IgD and ε in IgE. The IgE antibody seems to be involved in allergic reactions. Although the functions of IgD are not well known, this class of antibody is thought to be involved in the recognition process.

It is important to point out that the surface receptors of B cells and T cells are themselves antibodies. That is to say, B cells have surface antibodies, thought to be primarily IgM and IgD, which have the same antigen-recognizing specificity as the antibodies the B cell itself is capable of synthesizing. Thus, when a B cell is stimulated by an antigen, it will produce antibodies that react with the antigen. Although B-cell receptors seem to be normal antibodies consisting of heavy and light chains, T-cell receptors seem to be a different sort of "antibody" consisting of only heavy chains.

A Closer Look at IgG

As noted earlier, the fundamental unit of antibody structure consists of two heavy polypeptide chains and two light polypeptide chains. This structure is illustrated for the IgG molecule in Figure 13.5, where it can be observed

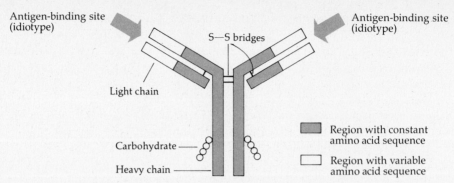

Figure 13.5 An IgG molecule showing the basic Y-shaped structure formed of two heavy and two light polypeptide chains held together by S—S bridges. Note the two antigen-binding sites; they are identical. The great diversity of types of antibodies arises from the variable amino acid sequences found in the parts of the heavy and light chains associated with the antigen-binding sites.

that the two heavy and the two light chains are arranged in the shape of a Y—actually a double Y like this: 𝖄 . The longest lines represent the heavy chains, which extend from the tines of the Y down into the stalk. The short lines in the tines of the Y represent the light chains. The carbohydrate part of the molecule, which differs for each class of antibody, is attached to the heavy chains in the stalk of the Y, and the whole structure is held together by -S-S- bridges between the polypeptide chains (see Figure 13.5). The two heavy chains in any particular molecule are identical, as are the two light chains—a result of the fact that each plasma cell has only one set of genes actively making a particular antibody. The Y-shaped antibody molecule has two sites that bind with the corresponding antigenic determinant (epitope). These sites, located at the tips of the tines of the Y, are called **antigen-binding sites** and establish the antibody's **idiotype**. (The antigen-binding site of an antibody, being a sort of reflection of an epitope on an antigen, has antigenic determinants of its own. An antigenic determinant located within the antigen-binding site of an antibody molecule is called an **idiotype**.)

Although Figure 13.5 illustrates the structure of IgG, a similar structure is found in all classes of antibody, although, of course, each class has its own heavy chain (see Table 13.1) and type of carbohydrate. IgG, IgD, and IgE consist of one such Y-shaped unit, IgA consists of a variable number, and IgM consists of five Y-shaped units joined at the bottoms of their stalks. Since the antigen-binding sites of an antibody molecule are on the arms of the Y, the five Y-shaped units in an IgM molecule provide 10 powerful sites, enabling the antibody to bind to foreign particles such as viruses or cells. The stalk portion of the Y is recognized by other components of the immune system, such as complement or killer macrophages, which allows these components to recognize and attack antibody-coated antigens.

A large number of uniform samples of different IgG molecules have now been isolated and their amino acid sequences determined. (The source of large samples of a particular type of antibody is usually a patient with a tumor that affects the immune system and destroys normal controls, thereby causing an enormous overproduction of one particular type of antibody.) The amino acid sequences of both the heavy and light chains in the upper half of the tines of the Y are different in every type of IgG antibody examined! These **variable sequences** (see Figure 13.5)—approximately

115 amino acids long—give each antibody its ability to recognize a specific antigenic determinant. Because every type of IgG molecule has a different amino acid sequence in its variable region, every type of IgG molecule will have a different surface relief in this part of the molecule. The antigen-binding sites of the antibody molecule have a surface landscape that just fits, like lock and key, the surface landscape of an antigenic determinant (Figure 13.6). The lower halves of the light chains are not variable, however; their amino acid sequences are identical (**constant**) in all IgG molecules with light chains that derive from the same gene. Likewise the **constant** (i.e., not variable) three-quarters of the heavy chains have identical amino acid sequences in all molecules with heavy chains that derive from the same gene.

Gene Splicing and the Origin of Antibody Diversity

All normal individuals are able to produce at least 1 million (10^6) antibodies that differ in their antigen-binding specificity and hence in their amino acid sequence. How many genes are involved in coding for this extensive repertoire of amino acid sequences? At one extreme, every possible light chain and every possible heavy chain would correspond to one conventional gene in the germ line. In this **germ-line theory** of antibody diversity, there would have to be at least 1000 (10^3) light-chain genes and 1000 (10^3) heavy-chain genes; random combinations of light chains and heavy chains could then produce the necessary 10^6 antibodies ($10^3 \times 10^3 = 10^6$). At the other extreme, there could be relatively few germ-line antibody genes, with antibody diversity being created anew during each individual's lifetime by somatic mutation of these genes or by some sort of recombinational process; this view, which emphasizes relatively few germ-line genes, is called the **somatic theory** of antibody diversity.

Application of recombinant DNA technology to the issue of antibody diversity has now revealed that neither the germ-line theory nor the somatic theory is completely correct, and neither is completely wrong. Relative to the constant (C) regions of the light and heavy

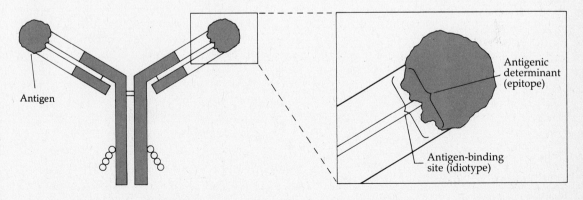

Figure 13.6   An IgG antibody molecule bound with antigen and showing the complementary fit of the antigen-binding site and the antigenic determinant. The B cell in Figure 13.3 is shown as being stimulated by this same antigen. The stimulated B cell gives rise to plasma cells producing antibodies such as the one shown here, which is able to recognize the same antigen.

chains, there is virtually no variation. Humans, for example, have 10 genes for the constant portion of the heavy chains (four coding for alternative forms of the γ chain of IgG, two for the μ chain of IgM, two for the α chain of IgA, and one each for the δ chain of IgD and the ε chain of IgE) and three genes for the constant portion of the light chains (one coding for κ and two for λ); some additional genes for the constant portions may yet be discovered, but the number is not likely to be large. The problem of the origin of antibody variability pertains to the variable (V) regions of the polypeptide chains, and a major source of variation has turned out to be unprecedented. The variable region of antibody genes is pieced together by **DNA splicing**. For the light chains,

for example, there are an unknown but relatively large number (i.e., about 250) of DNA segments that code for most of the polypeptide. Any one of these variable (V) regions can be spliced onto any one of a small number (i.e., two to five) of **joining** (J) regions to create an intact coding sequence for the variable part of a light chain. This splicing process for the light chain is called **V-J joining**, and it is outlined for a mouse κ chain in Figure 13.7. Part (*a*) illustrates the arrangement of κ-related sequences along the DNA in a mouse chromosome. The many V regions are symbolized V1, V2 and so on, and it is to be noted that each V region has its own leader sequence (stippled). Farther downstream are the J regions (light shading); the mouse has five (per-

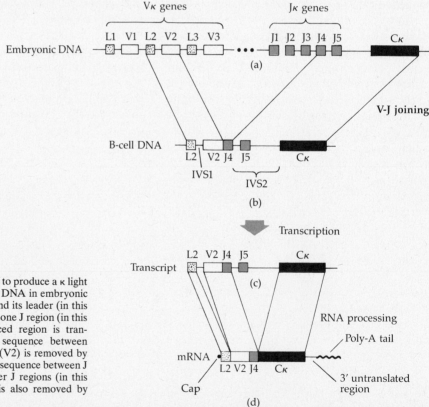

Figure 13.7   V-J joining to produce a κ light chain. (*a*) Organization of DNA in embryonic cells. (*b*) One V_L region and its leader (in this case L2-V2) is spliced with one J region (in this case J4). (*c*) Entire spliced region is transcribed. (*d*) Intervening sequence between leader (L2) and V region (V2) is removed by RNA splicing; intervening sequence between J region (J4), including other J regions (in this case J5), and C regions is also removed by RNA splicing.

haps only four) J regions associated with the κ chain. Still farther downstream is the sequence for the constant part of the κ chain (dark shading). In a B cell, illustrated in part (*b*), the DNA has been rearranged by V-J joining, which in this example has brought L2-V2 into juxtaposition with J4. Now transcription occurs [part (*c*)], and in the transcript there are two intervening sequences that will be removed during the RNA processing. One of these intervening sequences [IVS1 in Figure 13.7(*b*)] separates the leader from the V-coding region. The other (IVS2) separates the spliced V-J region from the C region. Note that, in this example, the J5 region is removed from the transcript as part of IVS2. The reason this process can work properly is that each J region is followed in the DNA by a splicing-junction sequence (see Chapter 9). Thus, all J sequences downstream from the V-J region will be removed during RNA processing. If the example in Figure 13.7 involved the joining of V2 with J1, then J2 through J5 would be removed during processing. In any case, the processed mRNA [see Figure 13.7(*d*)] reflects a piecemeal gene containing a leader (L), three adjacent coding regions (V2, J4, and Cκ in this example), and a 3′ untranslated region; this mRNA will be translated into a complete light chain.

Creation of a heavy-chain gene by DNA splicing is slightly more complex than creation of a light-chain gene; the situation in the mouse is illustrated in Figure 13.8. Part (*a*) shows the arrangement of sequences along the DNA in germ-line cells. Again there is a large number of V regions, each preceded by its own leader. Downstream of the V regions is a set of **diversity** (D) regions and farther downstream a set of joining (J) regions. (These V, D, and J regions are distinct from the V and J regions of light chains; that is, a light chain can never include a V or a J region from a heavy chain, and vice versa. Indeed, the κ and λ light chains

each have their own V and J regions.) Finally, farther downstream from the heavy-chain J regions are the coding sequences for the constant portions of the heavy chains—Cμ, Cδ, Cγ³, and so on. Note that each C gene is separated into four coding regions separated by intervening sequences; each of these coding regions corresponds to a distinct **domain** (functionally distinct part) of the heavy chain—an exon-intron relationship found in some other genes as well (see Chapter 9).

DNA splicing to create a heavy-chain variable region is illustrated in Figure 13.8(*b*). In the case of heavy chains, any V region and its leader are spliced onto any D region, which is, in turn, spliced onto any J region. This splicing process is called **V-D-J joining**; in the example in Figure 13.8, V-D-J joining connects L2-V2 with D1 and D1 with J3. Transcription [part (*c*)] is followed by RNA processing [part (*d*)], in which all intervening sequences are removed. Here again, as in the case with light genes, each J region is followed by a proper splicing sequence so that all J regions following the V-D-J sequence are removed. The mRNA in Figure 13.8(*d*) would be translated as a complete μ chain with variable part V2-D1-J3 and constant part Cμ. The corresponding B cell would thus produce IgM.

V-J joining (light chain) and V-D-J joining (heavy chain) are often referred to collectively as **combinatorial joining**. This sort of DNA splicing is a powerful process in generating antibody diversity. If there were 250 V regions for the κ chain, for example, then with 5 J regions there could be $250 \times 5 = 1250$ different κ chains from combinatorial joining alone. Similar arguments apply with more force to the heavy chains because three regions (V, D, and J) are involved in joining. With as few as 250 V_H regions, 10 D_H regions, and 4 J_H regions, for example, there would be $250 \times 10 \times 4 = 10,000$ possibilities. Thus, combinatorial joining is by itself a principal source of anti-

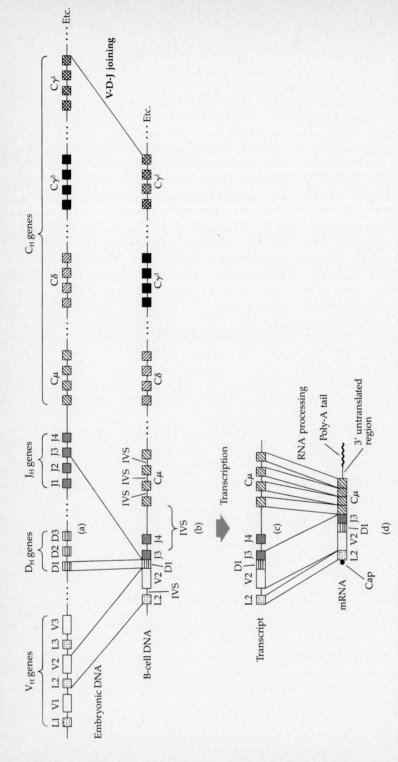

Figure 13.8 V-D-J joining in the origin of mouse IgM. (*a*) Organization of DNA in embryonic cells. (*b*) One V_H region and its leader (in this case L2-V2) is spliced onto one D region (D1 in this example), which is, in turn, spliced onto one J region (here J3). (*c*) The spliced DNA is transcribed. (*d*) During RNA processing several intervening sequences are removed including the one separating the leader sequence (L2) from the variable region (V2), the one between the J region (J3) and the constant region (Cμ), which also contains other J regions (in this case J4), and several in the constant region.

body diversity, but it is not the whole story. The DNA splicing process can also create altered codons at the splice junctions, providing an additional source of variation. In addition, heteroduplex formation and mismatch repair between related V regions can create still more variability. Thus, at the present time, it appears that antibody variability arises mainly from combinatorial joining of a moderately large number of variable-region genes with a much smaller number of D and/or J regions. This source of variation is augmented by sequence alterations at the splice junctions and perhaps by recombinational processes.

Figure 13.8 illustrates the DNA in a cell

producing IgM. All B cells pass through this IgM-producing stage of their existence; the IgM becomes associated with the cell membrane and serves as the antigen receptor. However, once stimulated by antigen, a B cell can undergo a **class switch**, in which antibody production is switched from the IgM type to any of the other classes. Class switching involves yet another DNA splice, and the process is outlined in Figure 13.9. Part (*a*) shows the DNA in the IgM-producing cell in Figure 13-8(*b*). After antigen stimulation, the DNA that codes for the variable part of the heavy chain can be spliced onto a DNA region that codes for a different constant part. In the example in Fig-

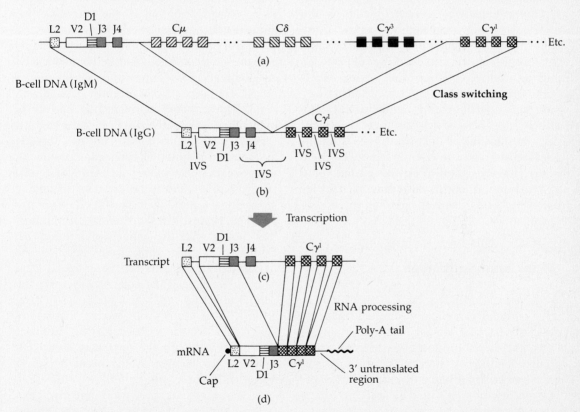

Figure 13.9 Class switching from IgM to IgG in the mouse. *(a)* B-cell DNA as in Figure 13.8*(b)*. *(b)* Splicing brings the V-D-J region into proximity with a new constant region (in this case Cγ¹). *(c)* Transcription. *(d)* Removal of intervening sequences analogous to that in Figure 13.8*(d)*.

ure 13.9(*b*), the class switch involves the new constant region $C\gamma^1$. After the splicing, transcription and RNA processing occur as before, except that the mRNA in part (*d*) will be translated as a γ chain and thus the cell will produce IgG. The antigen-binding specificity of the IgG will be the same as the previously produced IgM because this specificity resides in the variable portion of the heavy chain, which remains intact during the class switch. (It should be noted that class switching involves only the heavy chains; both the IgM-producing and the IgG-producing cells in Figure 13.9 would produce identical light chains.)

The light-chain genes are on one chromosome and the heavy-chain genes are on another. Since humans and other mammals are diploid, any individual has two copies of the light-chain genes and two copies of the heavy-chain genes. In this context it is reasonable to inquire whether combinatorial joining affects both light-chain-bearing chromosomes and both heavy-chain-bearing chromosomes. The answer appears to be yes, but there is a catch: Only one rearranged light-chain gene is actually expressed, and likewise for the heavy-chain gene. This restriction of expression has come to be known as **allelic exclusion**, but the detailed processes involved are not yet understood.

Monoclonal Antibodies

We have mentioned that large antigens such as virus particles carry many different antigenic determinants. An animal exposed to such a virus therefore produces a heterogeneous group of antibodies, and each type of antibody has a specificity for one antigenic determinant. In many instances, it would be useful to have a homogeneous antibody, such as would be produced by a single clone of B cells, as this antibody would react with one and only one antigenic determinant. For example, two closely related but distinct viruses will have many antigenic determinants in common, but each will have a few unique antigenic determinants not found on the other. If a homogeneous antibody against one of these unique antigenic determinants was available, the viruses could easily be distinguished because one would combine with the antibody and the other would not. Another use of homogeneous antibody is in mapping the distribution of certain antigenic determinants on the cell surface; since antibodies can easily be made fluorescent by the appropriate laboratory procedure, one need only determine the distribution of fluorescent areas on the cell surface; these will precisely reflect the distribution of the corresponding antigenic determinant.

The homogeneous antibody produced by a single clone of B cells is called a **monoclonal antibody**, and procedures for procuring monoclonal antibodies have recently been developed. The principal problem that had to be overcome was that normal B cells do not survive in laboratory culture, but this obstacle has been surmounted by application of some of the principles of somatic cell genetics discussed in Chapter 4. The procedure for obtaining monoclonal antibody is illustrated in Figure 13.10. It begins with the stimulation of a normal mouse with the antigen of interest (*a*); clones of B cells arise in response to the antigenic determinants on the antigen, so at some later time when spleen cells from the animal are removed, a few of them—in fact, a tiny proportion—will secrete antibodies against the antigenic determinant of interest. These few cells are the important ones, and they are represented as the black circles in Figure 13.10(*b*).

A second part of the procedure involves a type of tumor called a **myeloma** [see Figure 13.10(*c*)]. Myelomas are B-cell tumors that have the ability to divide in cell culture, but

Figure 13.10 Production of monoclonal antibodies. *(a)* Mouse stimulated with antigen. *(b)* Some B cells from spleen produce the antibody of interest (black circles). *(c)* Myeloma cells. *(d)* Growth of myeloma cells in 8-azaguanine selects for *HGPRT*⁻ cells. *(e)* Mixture of fused and unfused cells. *(f)* Growth in HAT medium selects for fused cells, some of which secrete the antibody of interest (half black circle). *(g)* Individual cells are separated and cultured. *(h)* Clones, one of which produces the antibody of interest (in this case clone 4).

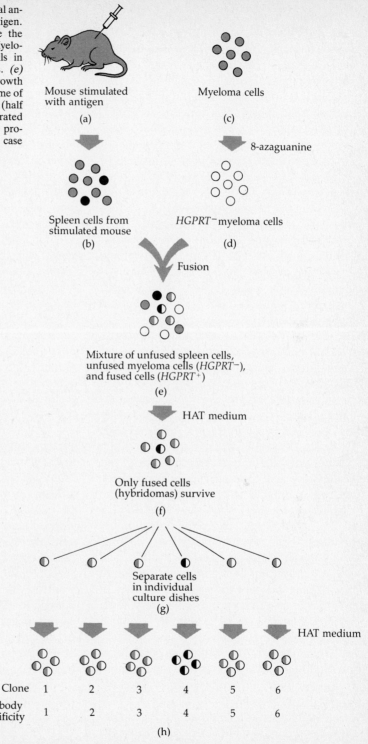

Mouse stimulated
with antigen

(a)

Myeloma cells

(c)

8-azaguanine

Spleen cells from
stimulated mouse

(b)

HGPRT⁻ myeloma cells

(d)

Fusion

Mixture of unfused spleen cells,
unfused myeloma cells (*HGPRT*⁻),
and fused cells (*HGPRT*⁺)

(e)

HAT medium

Only fused cells
(hybridomas) survive

(f)

Separate cells
in individual
culture dishes

(g)

HAT medium

| Clone | 1 | 2 | 3 | 4 | 5 | 6 |
| Antibody specificity | 1 | 2 | 3 | 4 | 5 | 6 |

(h)

many myelomas produce no antibody of their own. Myeloma cells are first grown in the presence of 8-azaguanine to select mutants that are defective in the enzyme hypoxanthine guanine phosphoribosyl transferase (HGPRT). (The principle involved is that $HGPRT^+$ cells produce defective DNA in the presence of 8-azaguanine and so commit a sort of suicide— see Figure 4.15.) Once the $HGPRT^-$ mutants are obtained in Figure 13.10(d), these are mixed with the normal spleen cells (b) under conditions that promote cell fusion. Three types of cells will be found in the mixture (e): (1) unfused spleen cells, (2) unfused myeloma cells, which remain $HGPRT^-$, and (3) fused spleen-myeloma cells (called **hybridomas**), which are rendered $HGPRT^+$ by the normal spleen cell gene.

The next step is to grow the cell mixture in HAT medium (HAT medium consists of hypoxanthine, aminopterin, and thymine—see Figure 4.15). This medium selects out the hybridomas [see Figure 13.10(f)] because unfused spleen cells cannot divide in culture in any case and because $HGPRT^-$ cells cannot synthesize DNA in HAT medium. Hybridoma cells not only are able to grow in culture, but they are also able to secrete the antibody from whatever normal B cell was involved in the fusion. Among the hybridomas, therefore, a small proportion will secrete the one antibody of interest.

The rest of the procedure is routine. The antibody-producing hybridomas are separated into individual culture dishes (g) and grown in HAT medium so that each original hybridoma cell produces a clone of genetically identical cells, all producing exactly the same antibody (h). There follows the tedious procedure of determining which clone is producing the monoclonal antibody of interest, and in the case of Figure 13.10 it turns out to be clone 4. Once identified, the monoclonal antibody can be obtained in almost unlimited quantity for as long as desired, because hybridoma cells survive in culture indefinitely.

Breakdowns of Immunity

Certain disorders are known in which affected individuals lack T-cell function or B-cell function or both. Patients with such **immunodeficiency diseases** are able to survive only with modern antibiotics and, in certain cases, only in germ-free artificial environments. The medical history of such patients highlights the division of labor in the immune system. Patients who lack T-cell function accept mismatched skin transplants (rejection is a T-cell function) and fall prey to viral infections, whereas patients who lack B-cell function accept mismatched blood transfusions and are prone to bacterial infections. Table 13.2 summarizes the characteristics of several important immunodeficiency diseases. Immunodeficiency diseases can sometimes be remedied by transplants of bone marrow (to provide normal lymphocyte precursors), but great care must be taken to genetically match the transplant donor and recipient because the transplanted tissue might otherwise begin to attack the recipient!

All human beings are genetically unique, and this uniqueness includes various aspects of the immune response. No two persons have exactly the same immune response. Some people can respond to certain antigens while others cannot, or some may respond strongly while others respond weakly. Innocuous antigens like wool, egg albumin, cat fur, or house dust do not bother most people, yet they elicit speedy and sometimes disastrous responses from the immune systems of people who are allergic to them. It is common experience that certain people are more susceptible than others to allergies, colds, poison ivy, and many other agents, and such susceptibilities may run in families. However, the mode of inheritance of these susceptibilities involves many genes and environmental factors working simultaneously, and it is therefore very complex.

The immune system is a potent weapon against foreign tissues. It would be equally effective against one's own tissues were it

TABLE 13.2   IMMUNODEFICIENCY DISEASES

Disorder*	Signs	Inheritance*
Common variable immunodeficiency	Reduced number of plasma cells; deficiency of IgG, IgA, IgM, IgD, and IgE	Heterogeneous; includes AR, AD, and multifactorial forms
Agammaglobulinemia, X-linked, Infantile (Bruton agammaglobulinemia)	Absent or severely diminished peripheral and lymphoid tissues; absence of tonsils and adenoids; few or no B cells	XL
Severe combined immunodeficiency (Swiss-type AR agammaglobulinemia)	Immunologic stem-cell disorder; thymus absent or reduced in size; absent T-cell functions, often low IgA and IgM levels	AR
X-linked severe combined immunodeficiency (Swiss-type XL agammaglobulinemia)	Immunologic stem-cell disorder; lack of B-cell and T-cell function	XL
Immunoglobulin A deficiency (isolated IgA deficiency)	Abnormally low levels of IgA in serum and secretions; incidence about 1 in 600 births	Heterogeneous; at least one AR form
Thymic agenesis (DiGeorge syndrome)	Absence of thymus; severely impaired T-cell function	AR (?)

*AR denotes autosomal recessive, AD autosomal dominant, and XL an X-linked gene.

not for **immunological tolerance**—the process whereby the immune system, especially T cells, comes to recognize and to tolerate one's own antigens. Exactly how immunological tolerance comes about is not known. At some time during the embryological development of the immune system, those parts of the system that would be programmed to attack the specific antigenic determinants present in an individual's body are somehow destroyed or made inactive. What remain functional are only those lymphocytes that recognize "nonself" or foreign antigenic determinants. If a foreign antigen is introduced into a young animal just before maturation of the immune system, the animal will often recognize this antigen as "self" and tolerance to it will develop; reinvasion of the animal by the same antigen at some later time will not elicit an immune response.

Sometimes immunological tolerance

fails, and the immune system begins to attack one's own tissues—a condition known as **autoimmune disease**. Rheumatoid arthritis and several other diseases of aging are known or strongly suspected to involve a partial breakdown of immunological tolerance and are thus examples of autoimmune diseases. A particularly serious example in children is **rheumatic fever**. Rheumatic fever is caused by a bacterium, and certain antigens on the bacterium are similar to those on the heart. The bacterial antigens are different enough from heart antigens to stimulate antibody production; at the same time, they are similar enough to heart antigens that the antibodies produced against them will also attack the tissues of the heart. This is the cause of the heart damage that frequently occurs in children afflicted with rheumatic fever.

Why did humans and other mammals

evolve an immune system that would tolerate "self" antigens and reject "nonself" antigens, that would reject even the antigenic determinants of exotic laboratory chemicals that no T cell or B cell would meet in millions of years? One plausible answer is found in the theory of **immune surveillance**, which suggests that the immune system evolved not only to fight foreign invaders but also to "police" the cells and tissues of the body and quell any "insurrections" (mutant clones) that might arise. According to this theory, newly mutated cells with new antigenic determinants are continually occurring in our bodies, being recognized as "nonself" by our immune system, and being destroyed. The theory of immune surveillance thus suggests that cancer represents a twofold failure: first, the origin of a transformed clone of cells, and second, failure of the immune system to destroy the clone or hold it in check. The theory is supported by the observation that individuals with immunodeficiency diseases are particularly prone to cancer. Indeed, some normal individuals with cancer *do* make antibodies against their cancer cells, but the cancer cells secrete molecules that neutralize the antibodies, presumably by filling up the antigen-binding sites.

The principle that the immune system responds to nonself antigens but does not respond to self antigens is true only within certain limits. Indeed, it is now known that self-recognition is *essential* in the processes of cell cooperation involving macrophages, T cells, and B cells. In particular, cell cooperation can occur only between cells that share certain surface antigens. These self-recognition antigens are products of a locus called the **major histocompatibility complex** (or **MHC**), which is involved in a multitude of immunity-related traits (Table 13.3). The importance of self-recognition may be more fundamental than its role in cell cooperation. Indeed, self-recognition may be at the heart of the

TABLE 13.3 TRAITS CONTROLLED BY THE MAJOR HISTOCOMPATIBILITY COMPLEX

Serologically detected cellular alloantigens
Transplantation antigens
Cell-mediated lympholysis target antigens
Mixed lymphocyte reaction
Immune responses
Graft versus host reaction
Serologically defined I antigen (Ia antigens)
T cell:B-cell interactions
Hybrid resistance
Tumor virus susceptibility
Serum serological protein
Testosterone levels
Complement levels
Liver cyclic adenosine monophosphate levels

Source: E. S. Golub, 1981, The Cellular Basis of the Immune Response, 2nd ed., Sinauer Associates, Sunderland, Mass., p. 70; after D. C. Shreffler and C. David, 1975, Adv. Immunol. 20:125–195.

overall regulation of the immune system. Recall that the *idiotype* of an antibody is the set of antigenic determinants of its antigen-binding site. An idiotype thus represents a sort of mirror image of the antigenic determinant. An antibody directed against the idiotype of another antibody (i.e., an **anti-idiotype antibody**) will thus recreate in its own idiotype the three-dimensional configuration of the original antigenic determinant. Anti-idiotype antibodies are therefore capable of stimulating cells of the immune system that can recognize the original antigen. It is now known that antibodies directed against an antigen also stimulate production of anti-idiotype antibodies, which stimulate other cells that recognize *their* antigen-like idiotype, and so on in an ever-widening network of idiotype and anti-idiotype responses that amplifies the original antigenic stimulation and brings the full force of the immune system to bear on the invading antigen. This complex network of regulatory interactions has become known as the **network hypothesis**.

Transplants

The same powers that make the immune system a potent weapon against the onslaught of microorganisms make it a serious barrier to the success of organ transplants and tissue grafts. Human cells and tissues are antigens, of course. Their antigenic determinants are on the cell surfaces and are genetically determined by the alleles at more than a dozen loci known as **histocompatibility loci**. Histocompatibility antigens are produced according to the general rule that each allele at a locus produces a specific antigen. A homozygous locus produces tissue antigens corresponding to whichever allele is homozygous; a heterozygous locus produces two different tissue antigens corresponding to the two different alleles on the homologous chromosomes. Since each of the dozen-plus histocompatibility loci has many alleles in the human population, each of us has a unique combination of alleles; thus the combination of antigenic determinants on our cells is also unique. When any tissue is transplanted from a donor to an unrelated recipient, the recipient's lymphocytes will recognize the nonself antigens and set out to reject the tissue. Two exceptions in which no immune barrier arises should be mentioned. Tissues from one part of an individual's body can be transplanted to another part without fear of rejection, as in the now commonplace skin grafts used in the treatment of victims of severe burns and in the hair transplants performed for cosmetic purposes. Identical twins, because they are genetically identical, have identical tissue antigens, and transplants between identical twins are not rejected.

A transplant will be rejected whenever the transplanted tissue has one or more antigenic determinants not present on tissues of the recipient, because the immune system is organized to attack nonself antigens. Finding completely compatible tissues among nontwins —tissues whose entire complement of antigenic determinants is also present in the recipient—is an almost impossible task; the enormous diversity of tissue types means that a perfect or nearly perfect match is unattainable. Luckily, not all histocompatibility loci are of equal importance. Some provide antigens that elicit a stronger immune reaction than others. In the mouse, there is one major histocompatibility locus and a dozen or so minor ones. This most important histocompatibility locus is the **major histocompatibility complex** (MHC) mentioned above and in Table 13.3 in connection with its many immunity-related functions. The mouse MHC locus is designated *H-2*, but, since its DNA organization is unknown, it is not clear whether *H-2* represents a single conventional locus that produces one polypeptide or whether it is actually a tightly linked cluster of loci with related functions. In any case, incompatibility at the *H-2* locus leads to graft rejection (in the recipient mouse) in 8 to 12 days; incompatibility at one of the minor loci leads to rejection in 15 to 300 days, depending on the locus. If the donor and recipient are incompatible at many of the minor loci simultaneously, then a graft may be rejected in 8 to 12 days even though the donor and recipient match at the *H-2* locus.

The genetic situation regarding transplants in humans is thought to be similar to that in the mouse: one major (MHC) locus and several minor ones. The human MHC locus is designated *HLA*, and it is on chromosome 6. The minor loci are scattered throughout the genome. In humans, the critical period of graft rejection is the first four months after surgery; if a severe rejection crisis is to occur, it will usually occur during this period. The important point is that the immune barrier confronting tissue antigens produced by the major histocompatibility locus in both mice and humans is a much stronger barrier than confronts antigens from minor loci. Several methods have been developed for tissue typing in the laboratory to determine whether tissue from a potential donor is compatible at the major locus with the recipient. Obviously, near relatives are genetically more similar than distant relatives or unrelated people, so the search for a suitable donor usually begins with the recipient's immediate family, particularly with brothers and sisters. Once a suitable donor is identified, the transplant can be carried out with incompatibility at only the minor loci to be overcome.

Incompatibility at minor histocompatibility loci can be treated with **immunosuppressive drugs**—chemicals that interfere with the immune system. When transplants first began to be carried out on a wide scale in the 1950s, the only immunosuppressives available knocked out almost all immunity. This left the patient wide open to infection, and such minor infections as a sore throat could explode into a lethal disease. But drugs and other procedures in use at the present time are somewhat more specific in their effects; they suppress only part of the immune system. The patient is still in danger of succumbing to severe infections, but the risk is less than it would have been in times past.

In spite of tissue typing and immunosuppressive drugs, transplants of major organs are still not easy. A suitable donor must first be found. Organ banks containing organs from cadavers are of limited usefulness, because major organs cannot be maintained for long outside the human body without undergoing irreversible damage. In addition, the transplant surgery can be extremely delicate. The major organ that can be transplanted most readily is a kidney. Kidney transplants, now carried out almost routinely with more than 75 percent success, are relatively easy surgically; one major artery, one major vein, and the ureter have to be hooked up, and the organ functions well without a nerve supply. Kidney transplants are also made practical by the fact that everyone has two kidneys, and many people are willing to donate one to an otherwise doomed relative. Moreover, patients can be kept alive for years with an artificial kidney machine (if one is available and the expense can be borne) while a search for a suitable donor goes on. The advantages of kidney transplants do not apply to such organs as the liver, lungs, and heart. Many transplants of these organs have been performed, some successfully, but countless practical problems stand in the way of their widespread use.

Ethical concerns about organ transplants seem to have been somewhat resolved. It is now taken for granted that a man or woman is free to bequeath any body parts for immediate transplantation after death or for storage in an organ bank. (A few tissues such as eye corneas can be stored without damage.) It is also generally agreed that "death" means the complete cessation of brain function rather than heart failure. This eliminates the possibility of such favorite fictional fantasies as brain transplants (which the intricacies of surgery would have made impossible anyway). It is also generally agreed that adults can freely donate such nonvital organs as one kidney for transplantation, provided only that they be fully apprised of the risks and agree to accept them.

HLA and Disease Associations

The *HLA* locus is exceedingly complex, but it can be divided into four **regions** (called *HLA*-A, *HLA*-B, *HLA*-C, *HLA*-D), which are ordered on the chromosome as D-B-C-A. Each region is involved in regulating different aspects of the immune system, and each region has its own series of alternative "alleles" that can be identified by the appropriate laboratory tests. Each chromosome has one "allele" for each region, and the collection of such "alleles" constitutes the chromosome's **haplotype**.

For example, a chromosome carrying *HLA*-Dw2, *HLA*-B7, *HLA*-Cw4, and *HLA*-A2 has the haplotype Dw2, B7, Cw4, and A2. Since every normal individual has two copies of chromosome 6 (the *HLA*-bearing chromosome), every normal individual carries two *HLA* haplotypes. The number of possible haplotypes is enormous because of the large number of D, B, C, and A "alleles" that exist. Among European Caucasians, for example, there are at least 12 forms of the D region, at least 32 of B, at least 7 of C, and at least 20 of A—and more are being discovered rapidly. Theoretically, any of these regions can be found in association with any of the others, so there are at least $12 \times 32 \times 7 \times 20 = 53,760$ possible haplotypes. However, certain haplotypes are much more common than others, and some are exceedingly rare or nonexistent. Nevertheless, there is a great deal of haplotype variation in the human population.

As might be expected of a locus that plays a fundamental role in regulating the immune response, there is an association between certain *HLA* haplotypes and certain diseases. Several examples are listed in Table 13.4, where it can be seen that the strongest associa-tion is between **ankylosing spondylitis** (a disease of the joints) and B27; individuals who have a haplotype carrying B27 have an 88-fold increased risk of developing this condition as compared with non-B27 individuals. The other associations in Table 13.4 are significant but weaker than the ankylosing spondylitis–B27 association.

The ABO Blood Groups

The most widespread kind of tissue transplant is an ordinary blood transfusion. Transfusions are not usually thought of as transplants, perhaps partly because they are so commonplace. But in addition, the immune barrier to transfusions does not involve the major histocompatibility locus (*HLA*), since histocompatibility antigens are not found on red blood cells. Other antigens predominate on these cells, and these antigens define the familiar ABO and Rh blood groups as well as dozens of other blood groups that are less widely known. The ABO and Rh blood groups were discussed briefly in Chapter 3 as examples of simple Mendelian inheritance in humans. Here a somewhat more detailed discussion is in order.

TABLE 13.4 SOME DISORDERS ASSOCIATED WITH PARTICULAR HLA HAPLOTYPES

Disease	*HLA* haplotype	Relative risk (approximate)
Ankylosing spondylitis (joint disease)	B27	88
Rheumatoid arthritis (joint disease)	Dw4; Cw3	16
Juvenile diabetes (sugar metabolism)	DR3; DR4	4
Addison disease (adrenal glands)	B8; Dw3	4–11
Chronic active hepatitis (liver)	A1; B8	2–3
Ulcerative colitis (gastrointestinal tract)	B5	9
Psoriasis (skin)	B13; Bw17; Bw37	4–5
Anterior uveitis (eye disease)	B27	15
Hodgkin disease (malignancy)	A1, B5, B8, Bw18	1–2

Source: Data from F. Vogel and A. G. Motulsky, 1979, Human Genetics: Problems and Approaches, Springer-Verlag, New York.

The **ABO blood groups** are the most important ones in transfusions. The antigens are determined by one locus on chromosome 9 at which there are three alleles: I^A, I^B, and I^O. (Actually there are several kinds of I^A alleles, but this is a detail we need not cover here.) Individuals who are genetically I^A/I^A have on their red blood cells a particular antigen called the A antigen. (Remember that an antigen represents a collection of antigenic determinants.) People who are genetically I^B/I^B have on their red blood cells an antigen called B. Heterozygotes of genotype I^A/I^B have both the A and B antigens on their red blood cells. The I^O allele does not produce a specific antigen, and the red blood cells of I^O/I^O individuals carry neither the A nor the B antigen. A person's ABO blood type is determined by which antigens are present on the red blood cells (Table 13.5). I^A/I^A and I^A/I^O individuals have **blood type A** because only A antigens are present on their red blood cells; I^B/I^B and I^B/I^O individuals have **blood type B**; I^A/I^B individuals have **blood type AB**; and I^O/I^O individuals are said to have **blood type O**, which means the absence of both A and B antigens. The frequencies of the ABO blood types vary from population to population, as illustrated in the maps in Figure 13.11. Among Caucasians (Londoners), the frequencies are O (48%), A (42%), B (8%), and AB (1%). Among Chinese, the frequencies are O (34%), A (31%), B (28%), and AB (7%). As with *HLA*, certain disease associations involve the ABO blood groups. Examples are stomach cancer (20 percent greater risk in type A than in type O), duodenal ulcers (30 percent greater risk in type O than in other blood groups), and obstructive blood clots (thromboembolic disease—60 percent greater risk in types A, B, and AB than in type O).

From a chemical point of view, the A and B antigens correspond to allele-specific modifications of a complex carbohydrate found on the surface of all red blood cells (Figure 13.12). The I^A and I^B alleles code for alternative forms of an enzyme that attaches one or another sugar to the end of this molecule; the I^A form of the enzyme attaches one sugar, and the I^B form of the enzyme attaches the alternative

TABLE 13.5 IMPORTANT PROPERTIES OF ABO, Rh, AND MN BLOOD GROUPS

Blood group and genotype	Antigens on red blood cells	Antibodies present in blood	Antibodies present in saliva and other body fluids?
ABO blood group:			
I^A/I^A or I^A/I^O	A	Anti-B	Yes, if secretor (*Se/Se* or *Se/se*)
I^B/I^B or I^B/I^O	B	Anti-A	Yes, if secretor (*Se/Se* or *Se/se*)
I^A/I^B	A and B	None	No
I^O/I^O	None	Anti-A and anti-B	Yes, if secretor (*Se/Se* or *Se/se*)
Rh blood group:			
DD	Rh$^+$	None	
Dd	Rh$^+$	None	
dd	None	Anti-Rh$^+$ if exposed to Rh$^+$ antigen	
MN blood group:			
MM	M	None	
MN	M and N	None	
NN	N	None	

sugar. The I^O form of the enzyme causes no extra sugar to be attached to the basic carbohydrate, and the unmodified form is often called the **H substance**. Under certain genetic conditions, the A and B antigens not only are present on red blood cells but are also secreted into the saliva, sweat, semen, and other body fluids. The secretion of these antigens is controlled by another locus—the **secretor locus**. The secretor locus has two alleles, *Se* and *se*, and the allele leading the secretion, *Se*, is dominant. *Se/Se* and *Se/se* genotypes are therefore secretors, whereas *se/se* genotypes are not. Thus, in secretors, the ABO phenotype of a person can be determined from the body fluids alone (see Table 13.5). Among Caucasians, about 77 percent of all individuals are secretors (i.e., *Se/Se* or *Se/se*).

Another locus involved in the ABO blood groups should be mentioned. Certain rare individuals are homozygous for a recessive mutation associated with the absence of α-L-fucose from the carbohydrate precursor of the ABO substances (see Figure 13.12). With such a precursor, the I^A and I^B encoded enzymes cannot modify the precursor, so such individuals cannot express the ABO antigens on their red blood cells, whatever their ABO genotype may be; they are said to have the **Bombay phenotype**.

The A and B antigens are highly antigenic; the antibodies against them are very potent. Unlike most antigens, an individual need not have been exposed to A or B antigens on red blood cells to produce antibodies. That is, a person not carrying the A antigen (and therefore of blood type B or O) will have so-called naturally occurring antibodies against A even though he or she was never exposed to red blood cells carrying the A antigen. How this happens is not clear, but it may result from the "A-like" or "B-like" antigenicity exhibited by several classes of microorganisms. Therefore, although a person may never have been ex-posed to type A blood, repeated exposure to substances that have antigens similar to A will induce the production of anti-A antibodies. The same goes for the B antigen. Remember, however, that individuals do not make antibodies against their own antigens because of immunological tolerance. The result of all this is that a person will have circulating in the blood antibodies against the A or B antigens not present on his or her own red blood cells. Therefore individuals with type A red blood cells will possess anti-B antibody; people with type B blood will possess anti-A antibodies; people with type O blood will have both anti-A and anti-B antibodies; and those with blood type AB will have neither anti-A nor anti-B antibodies. (These naturally occurring antibodies are of the IgM type).

If a blood transfusion is performed using red blood cells against which the recipient has antibodies, the antibodies in the recipient will immediately attack the introduced cells and cause them to clump together (**agglutinate**) in various parts of the body. These clumps block the tiny capillary vessels, and parts of the brain, heart, and other vital organs are deprived of their oxygen supply. The patient goes into shock and may die, so blood to be transfused must not carry any major antigens that the recipient does not also possess. Conversely, antibodies in the donor blood may clump the cells of the recipient. Therefore best medical practice requires a two-way crossmatch of donor and recipient. However, one can transfuse limited amounts of blood that carries antibodies against the recipient's antigens; for example, people with type AB blood have A and B antigens but they may receive blood from type O people even though it contains antibodies against both A and B antigens. The reason is that the small volume of transfused antibody is diluted so rapidly in the large volume of the recipient's blood that it does no harm. The main transfusion rule that must be

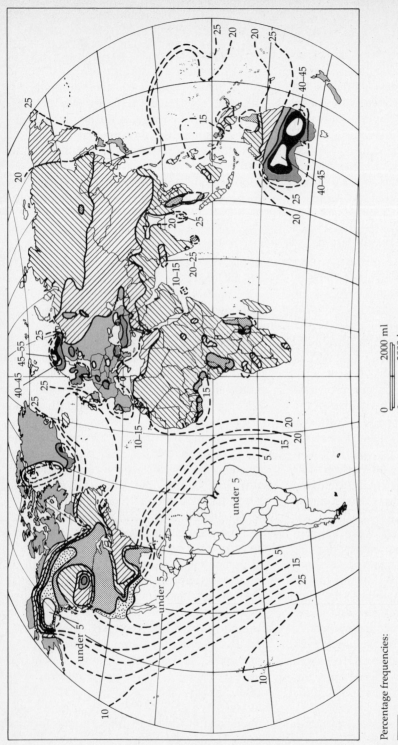

Percentage frequencies:

5–10
10–15
15–20
20–25
25–30
30–35
35–40

2000 ml

0

3000 km

0

(a)

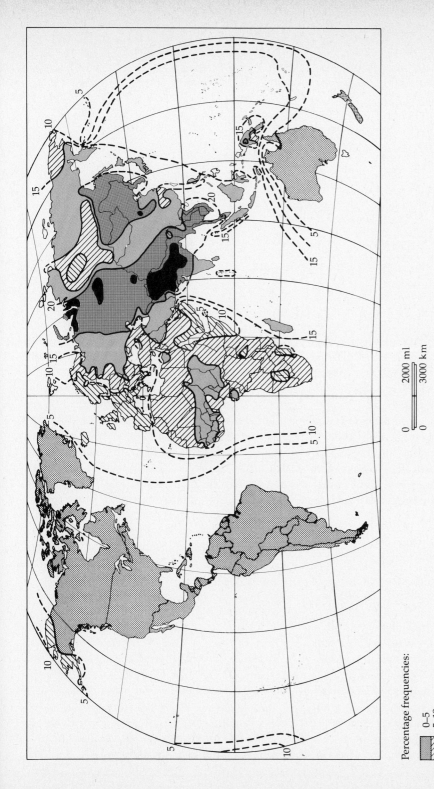

Percentage frequencies:

- 0–5
- 5–10
- 10–15
- 15–20
- 20–25
- 25–30

0   2000 ml

0   3000 km

(b)

Figure 13.11 Maps showing the frequencies of *(a)* the I^A allele and *(b)* the I^B allele among aboriginal populations of the world. The frequencies of both I^A and I^B are low among the aboriginal Indians of southern North America and South America.

Figure 13.12 Function of I^A and I^B alleles in producing ABO blood group antigens.

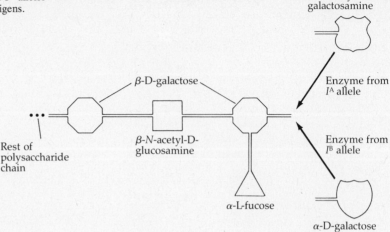

observed, then, is that transfused cells can have no antigen unless the recipient has the same antigen. Therefore, individuals of blood type AB can receive blood of type A, B, AB, or O; type A people can receive A or O blood; type B people can receive B or O blood; and type O individuals can receive only type O blood. (Type AB people were formerly called **universal recipients** because they can receive blood of any type; type O individuals were known as **universal donors** because they can donate blood to anyone.)

The Rh Blood Groups

The one other blood group system that is of significant medical importance is the **Rh system**. The genetic basis of the Rh system is rather complex, and it is not clear whether it involves a single locus on chromosome 1 with many alleles or a tightly linked cluster of loci. For our purposes, we may treat it as a single locus with two alleles, usually called D and d, although the actual situation regarding the number of loci and the number of alleles is still unknown. The D allele produces on the surface of red blood cells an antigen, chemically unre-

lated to the A and B antigens, called the Rh$^+$ antigen; the d allele does not produce a distinct antigen. Therefore, the red blood cells of both D/D and D/d genotypes carry the Rh$^+$ antigen; they are said to be Rh$^+$ or **Rh positive**. The red blood cells of d/d genotypes do not carry the Rh$^+$ antigen; they are said to be Rh$^-$ or **Rh negative** (see Table 13.5). Individuals who are Rh$^-$ are capable of producing antibodies against Rh$^+$ antigen. Unlike the ABO system, however, Rh$^-$ individuals will not produce anti-Rh$^+$ antibodies unless they have previously been exposed to Rh$^+$ cells.

The medical significance of the Rh system arises in those cases in which an Rh$^-$ mother who is producing anti-Rh$^+$ antibody due to previous exposure to Rh$^+$ antigen becomes pregnant with an Rh$^+$ fetus. (This can occur when the mother is d/d and the father is D/D or D/d.) In such a case, some of the anti-Rh$^+$ antibodies from the mother can traverse the placenta, since these antibodies are of the IgG class. The antibodies that invade the fetus's blood begin to attack the fetus's red blood cells, leading to a serious blood condition called **erythroblastosis fetalis**, sometimes called **hemolytic disease of the newborn**. In the most severe cases, the fetus cannot replace its red cells quickly enough and dies. Less severely affected infants may be born in apparent good health but soon become jaundiced due to toxic by-products from damaged red cells. If the infant lives, brain damage and mental retardation may result. Emergency treatment of the child may require an exchange transfusion—removing virtually all the child's blood and replacing it with Rh$^-$ blood. Gradually the Rh$^-$ red blood cells break down and are replaced with new Rh$^+$ cells made by the child, but the child does not produce anti-Rh$^+$ antibodies and therefore no subsequent problems arise. The time factor in these transfusions is critical because the untransfused child suffers from a shortage of oxygen. In the absence of treatment, three-quarters of the affected infants would die.

Actually, an Rh$^-$ woman will not produce anti-Rh$^+$ antibodies unless she has been exposed to Rh$^+$ antigens. This exposure usually requires one or more pregnancies during which she carries Rh$^+$ embryos. With each pregnancy, some of the Rh$^+$ cells from the fetus seep across the placenta into the mother's bloodsteam, especially at the time of birth. The mother produces antibodies against these cells. Because of the need for prior exposure, the first Rh$^+$ child is often not at risk, but it serves to immunize the mother. Often the mother does not produce sufficient antibody in response to the first exposure to cause problems with even the second Rh$^+$ child. But third and subsequent Rh$^+$ children incur a substantial risk of erythroblastosis fetalis. (Only 5 percent of infants with Rh hemolytic disease are first-born, roughly 45 percent are second-born, and about half are the third or later births. At one time the frequency of the condition in the United States was 1 in 150 births, but today's smaller families have reduced the incidence.) In matings of d/d mothers with D/d fathers, half the children will be d/d; these d/d children

do not have the Rh$^+$ antigen and they are never at risk, nor do they sensitize the mother.

The incidence of erythroblastosis fetalis in a population depends on the frequency of marriages of Rh$^-$ women with Rh$^+$ men. Among U.S. whites, the frequency of Rh$^-$ blood is about 15 percent and that of Rh$^+$ blood is 85 percent. Thus the type of mating in question occurs in about 13 percent of white marriages (i.e., $0.15 \times 0.85 = 0.13$). Rh hemolytic disease is virtually nonexistent in Japan, where more than 99 percent of the people are Rh$^+$. Among the Basques, a genetically semi-isolated group that lives in the Pyrenees Mountains between France and Spain, the incidence of Rh$^-$ is exceptionally high—about 43 percent. (A similar situation exists among Basque descendants who live in northern Nevada.) When Rh$^-$ women from these semi-isolates marry outside their own cultural group, in the majority of cases they marry Rh$^+$ men and therefore they run the risk of erythroblastosis fetalis among their children.

For Rh$^-$ women who are presently coming into their reproductive years and have never been transfused with Rh$^+$ blood, the risk of Rh hemolytic disease has all but been eliminated by a preventive treatment. The treatment involves injecting anti-Rh$^+$ antibodies into the Rh$^-$ mother soon after the birth of her first Rh$^+$ child (Figure 13.13). Any Rh$^+$ cells from the child that make it into the mother's blood are destroyed by these antibodies before they can immunize the mother. By the time the next pregnancy occurs, the injected antibodies have disappeared. (All proteins, including antibodies, break down or are destroyed in time.) Neither can a second Rh$^+$ child immunize the mother if, as before, she is injected with anti-Rh$^+$ antibody. This treatment is very effective provided that an Rh$^-$ woman is not already producing anti-Rh$^+$ antibody. The injected anti-Rh$^+$, by destroying any Rh$^+$ cells that might be present, prevents the mother's im-

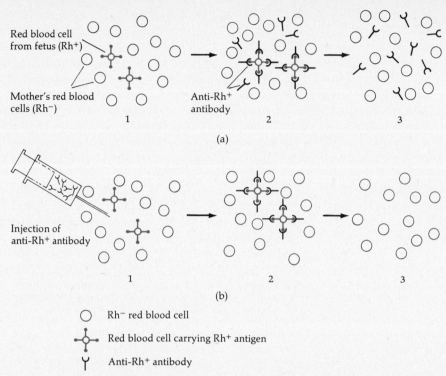

Figure 13.13 Preventive treatment of Rh hemolytic disease. *(a)* A portion of the Rh⁻ mother's bloodstream in the absence of treatment. Rh⁺ red blood cells from an Rh⁺ fetus seep into the mother's bloodstream (1). These cells stimulate production of anti-Rh⁺ antibody and are destroyed (2). However, the mother continues to produce antibody (3), and these antibodies may cross the placenta in a subsequent pregnancy. If the fetus in the subsequent pregnancy is Rh⁺, the antibodies will cause Rh hemolytic disease. *(b)* The preventive treatment: Anti-Rh⁺ antibody is injected into the mother (1), and this injected antibody destroys the Rh⁺ cells (2). The Rh⁺ cells are destroyed before they can stimulate production of anti-Rh⁺ antibody by the mother. Thus, after the injected antibody has broken down, the mother's blood remains free of anti-Rh⁺ (3).

mune system from being stimulated to produce its own anti-Rh⁺ antibody. Strangely enough, a similar kind of prevention sometimes occurs naturally. If red blood cells from the fetus carry an A or B antigen not present in the mother in addition to the Rh⁺ antigen, then the mother's anti-A or anti-B antibodies will destroy the cells before they can stimulate the production of anti-Rh⁺ antibody. In this way ABO incompatibility prevents Rh disease.

One may wonder why hemolytic disease of the newborn can come about from Rh incompatibility but not from ABO incompati-

bility. For example, why is it that hemolytic disease of the newborn does not ordinarily occur in type A or AB children whose mothers produce anti-A antibody? One part of the answer is that a type of ABO hemolytic disease *does* occur, but rarely—affecting perhaps 0.1 percent of all newborns. One reason the condition is so rare is that the antibodies responsible are not the same ones involved in ABO transfusion incompatibility. The usual anti-A and anti-B antibodies important in transfusion are of the IgM class, which are unable to cross the placenta. The antibodies responsible for ABO hemolytic disease are of the IgG class, which can traverse the placental barrier. Under ordinary circumstances, this IgG class of ABO antibody is not produced. Nevertheless, there is a smaller than expected frequency of I^A/I^O children arising from matings of I^O/I^O mothers with I^A/I^O fathers. The amount of the deficiency varies from study to study. (In one Japanese study the deficiency was more than 10 percent of the number of I^A/I^O children expected.) This deficiency has been attributed to very early spontaneous abortion brought about by ABO incompatibility between mother and fetus.

Other Blood Groups

Blood groups have application extending beyond genetics and medicine. Besides the ABO system and the Rh system, there are literally dozens of blood group systems. (A **blood group system** refers to the several antigens specified by the different alleles at a single locus.) Other than the ABO and Rh systems, most of these antigens are not recognized as being very antigenic by human lymphocytes; they therefore do not usually play a significant role in blood transfusions or pregnancy. They are, however, antigenic in other animals such as rabbits. Red blood cells from one person can be injected into an animal and at some later time antibodies can be isolated. These antibodies are then tested against the red blood cells from other people, and in some instances there will be an antigen-antibody reaction (the red blood cells will clump) but not in other instances. The people whose blood cells are recognized by the antibody and clump must have the same antigen as the person whose cells were originally used to immunize the animal, but those whose red cells are not recognized by the antibody must lack this antigen. Studies of this kind are then combined with family studies to determine the mode of inheritance of the antigenic difference.

One blood group that was discovered in almost the way just described is the **MN system**. The MN blood group antigens are due to two alleles at a single locus on the long arm of chromosome number 2. (The true genetic situation is known to be slightly more complicated but is not important here.) By injecting red blood cells of the appropriate kind into rabbits, anti-M and anti-N antibodies can be obtained. Anti-M antibody reacts only with cells carrying the M antigen; anti-N antibody reacts only with cells carrying the N antigen. The alleles that produce the M and N antigens are also called *M* and *N*. Three genotypes at the *MN* locus can be found (see Table 13.5). Some people are genetically *M/M* (their red blood cells react only with anti-M), others are *N/N* (their red blood cells react only with anti-N), and still others are *M/N* (their red blood cells react with both anti-M and anti-N). In the MN system, the antigens carried on the red blood cells can be used to identify the genotype of an individual at this locus. Because of this fact, the Mendelian law of genetic segregation can be observed directly. Matings of *MM* × *MM* produce only *MM* children; *MN* × *MM* matings produce half *MN* and half *MM* progeny; *MN* × *MN* matings produce one-fourth *MM*, one-half *MN*, and one-fourth *NN* offspring. One can easily work out the offspring distribution of matings involving *MM* × *NN*, *MN* × *NN*, and *NN* × *NN*. It should be realized that these numbers are averages, that they summarize what is found in a large number of matings between the given genotypes. In any one family, because the number of children is small, chance fluctuations will often occur, and these ratios will not always be obeyed exactly.

Applications of Blood Groups

The dozens of different blood group systems used collectively can identify a person almost as uniquely as fingerprints. Although many people have the same ABO blood type in common, or the same Rh blood type in common, or the same MN blood type in common, when these blood groups are considered separately, many fewer people have the same ABO, Rh, and MN blood types simultaneously. This is simply to say that there are fewer people with the precise blood type O;Rh⁻;MN than there are people with blood type O with no regard to their Rh and MN types. Now if one considers, say, 30 blood groups instead of only three, then the odds would be exceedingly high that any two people would have a different genotype with respect to one or more of the 30 groups. Every human being is genetically unique, different from all who have come before and all who will follow; the only exceptions are identical twins, triplets, and quadruplets.

The property of blood groups to characterize individuals and groups of people has been used widely by anthropologists to study the origin and migrations of various human populations. In these kinds of studies, blood group alleles that are common in one population but less common or rare in others are very important. For example, the B type antigen is virtually absent in American Indians (with a few exceptions, such as the Blackfeet). Another example: Among the many alleles of the Rh locus (the previous discussion divided these alleles into only two classes, *D* and *d*), one allele is highly prevalent in African Negroes but rare elsewhere; a different allele is frequent in European Caucasians but rare elsewhere. Provided sufficient supporting information can be gained from, for example, similarities in language, alleles of this sort can sometimes be used as markers of ancestry. In a similar vein, comparisons of humans and other primate species have been pursued using blood group similarities to measure genetic relationship.

The many blood group systems have found legal application in such litigations as paternity suits. The principle behind these applications is that a person falsely accused of fathering a child can sometimes be exonerated on the grounds that, based on genetic evidence, he could not possibly have fathered the child. For example, a man of blood type AB cannot have a type O child, except in the extremely unlikely event of mutation. As another example, a man of blood type O;M cannot have been the father of an A;MN child whose mother is AB;M, because although the ABO system works out satisfactorily (the child could be genotypically I^A/I^O, having received the I^O gene from the father), the MN system does not work out. Such a child would have to receive the gene for the N antigen from its father because the mother does not have it. But the accused father does not have this gene either, and therefore the real father must be somebody else. The same kind of reasoning applies to matching infants to their natural mothers in the unlikely event of a mixup in the hospital nursery. An O;Rh⁺;M woman could not, for instance, be the mother of an A;Rh⁻;N child.

The ABO antigens are important in certain kinds of police work. These antigens are extremely stable and persistent, so minute quantities of dried blood (or body fluids, in the case of secretors) can be typed. Indeed, the ABO blood types of some Egyptian mummies can even be determined! A minute amount of an assailant's blood under a victim's fingernail can sometimes be decisive in excluding innocent suspects; knowledge that the assailant's blood type is AB will exclude 99 percent of innocent white suspects in London, for example.

It seems appropriate to end this section with an account of one of the earliest criminal cases involving the use of blood typing for evidence:

It was a Japanese who took the lead. In 1928 K. Fujiwara, Director of the Institute of Forensic Medicine in Niigata, reported a crime which had been solved by means of blood group determination from seminal stains. A sixteen-year-old girl, Yoshiko Hirai, who went from village to village selling fortune-telling slips, had been found strangled. Witnesses had seen the girl at 7:30 P.M. on her way from the village to the railroad station. She must have been raped and killed shortly afterward. Fujiwara found whitish spots which contained well-preserved semen on the girl's body. Her blood group was O. Fujiwara tested the semen spots by the absorption method, adding anti-A and anti-B serums, and thus determined that the semen had the group characteristic A. Meanwhile the police had arrested two suspects. One of them was a twenty-four-year-old

mentally retarded beggar named Mochitsura Tagami, who straightway confessed that he had committed the murder. The second suspect, Iba Hoshi, denied all guilt. The police were on the point of accepting Tagami's confession and releasing Hoshi when Fujiwara asked for time to determine the blood groups of the prisoners. The results proved that Tagami, who had made the confession, could not possibly be the criminal; his blood group was O. Hoshi, on the other hand, had group A. Confronted with the data, Hoshi confessed; that Fujiwara knew his blood and his semen had something in common impressed [him as] witchcraft.[1]

[1]From J. Thorwald, 1966, Crime and Science, Harcourt, Brace & World, New York, p. 90.

SUMMARY

1. The **immune response** is the body's defense against invading substances. The response is mediated by two types of **lymphocytes** (white blood cells): **thymus-dependent** or **T cells**, which undergo differentiation in the thymus gland and include **cytolytic** (killer) T cells and other agents of **cellular immunity**; and **bone-marrow-derived** or **B cells,** which differentiate primarily in the bone marrow (or, in birds, in an organ called the **bursa of Fabricius**) and which produce circulating **antibodies** and are thus agents of **humoral immunity**.

2. An **antigen** is any substance that is able to elicit an immune response by virtue of certain three-dimensional configurations of atoms on its surface; these surface landmarks are known as **antigenic determinants** or **epitopes**, and a typical antigen carries many different antigenic determinants. An antigenic determinant elicits an immune response by combining with a complementary **receptor** molecule on the surface of a B cell or a T cell. Any B cell or T cell has only one type of receptor, which can combine with only one type of antigenic determinant. The antigenic determinant-receptor complex is essential for B-cell or T-cell stimulation, but proper stimulation also requires **cell cooperation** involving **helper** T cells and a class of white blood cells called **antigen-presenting macrophages**. Many small molecules (**haptens**) are not themselves antigenic but can stimulate an immune response when attached to a larger antigen (the **carrier**); the role of the antigen-presenting macrophage in this context is to combine with the carrier and "present" the hapten to the appropriate lymphocyte.

3. Stimulated lymphocytes undergo successive mitotic divisions to give rise to a **clone** of cells; each is capable of recognizing the antigenic determinant that caused the stimulation. Since stimulation is selective in the sense that only those lymphocytes that have the appropriate receptor molecules can respond to an antigenic determinant, the mitosis-stimulating process is called **clonal selection**. In the case of B cells, certain members of the clone differentiate into **plasma cells**, which are the actual antibody-producing cells. Other members of the clone do not immediately differentiate into plasma cells but are held in readiness to mount an antibody response against reinvasion by the same antigenic determinant. This readiness to mount a secondary response is referred to as **immunity**.

4. Five principal **classes** of antibody can

be distinguished: IgG, IgM, IgA, IgD, and IgE, where the symbol Ig stands for **immunoglobulin**. Each class consists of a number of **light** polypeptide chains, which may be of either the κ type or the λ type but not both, and a number of **heavy** polypeptide chains, which may be of five types: γ (found in IgG), μ (IgM), α (IgA), δ (IgD), or ε (IgE); the type of heavy chain thus determines the class of antibody. However, there are alternative genes for some of the heavy chains—four for the γ chain, for example. The particular heavy-chain gene used in an antibody determines the antibody's **subclass**. In addition, some heavy-chain genes have normal allelic variants that are inherited in simple Mendelian fashion; these allelic variants are known as **allotypes**.

5. IgG is the prevalent type of antibody in the blood serum, and it is the only class able to cross the placenta. IgM is a large and powerful antibody which, in addition to other functions, constitutes the B-cell receptor. (Although the B-cell receptor is a complete antibody, the T-cell receptor consists of only heavy chains). IgA is the class of antibody found in the gut, respiratory tract, urogenital tract, and other mucous membranes. IgE seems to be involved in allergic reactions. The functions of IgD are not well understood, but it may also serve as a receptor molecule.

6. All antibodies consist of one or more Y-shaped units. IgG, IgD, and IgE each have one such unit, IgM has five, and IgA has a variable number. Each Y-shaped unit is actually a double Y, like this: Ⴤ . The short lines in the tines of the Y represent the light chains, the long lines the heavy chains. The entire structure is held together by disulfide (-S-S-) bridges, and each class of antibody has its own type of carbohydrate attached to the heavy chains in the stalk of the Y. The tips of the tines of the Y are the **antigen-binding sites**—sites able to combine with an antigenic determinant because of their complementary surface land-

scape. This surface landscape of the antigen-binding sites establishes the antibody's **idiotype**. The surface landscapes of the antigen-binding sites are determined by the folding and interaction of the light and heavy chains in the upper half of the tines of the Y. The amino acid sequences in this portion of the antibody are said to be **variable** because they differ in each particular type of antibody. The lower half of the light chains and the lower three-quarters of the heavy chains have a **constant** amino acid sequence; that is, their amino acid sequence is determined by the particular light-chain gene and the particular heavy-chain gene used in coding for the antibody.

7. Any individual organism is thought to be capable of producing at least 1 million (10^6) different antibodies. Two broad hypotheses have been proposed regarding the origin of so much variability. According to the **germ-line theory**, a large number of genes code for different light chains and a large number of genes code for different heavy chains; the theory requires at least 1000 (10^3) genes of each type. According to the **somatic theory**, there are relatively few antibody-coding genes; the great diversity of antibody types is generated anew in each individual by mutation and recombination. We now know that neither hypothesis is completely right and neither is completely wrong. Relative to each type of light chain, there is a moderate (i.e., in the hundreds) number of coding sequences, each coding for most of the variable (V) part of one light chain; the rest of the variable part of a light chain is coded by a much smaller number of **joining** (J) sequences, which are separated from the corresponding constant (C) coding region by an intervening sequence. During lymphocyte differentiation, the DNA undergoes a process of **V-J joining**, during which one V sequence is spliced onto one J sequence, thus creating an intact V-J-C gene for a light chain. The origin of heavy-chain variability is analogous, but there is an additional small set of **diversity** (D) regions between heavy-chain V and heavy-chain J sequences; the DNA splicing in the case of heavy chains thus involves **V-D-J joining**. Such **combinatorial joining**, as exemplified by V-J and V-D-J joining, is a principal source of antibody diversity. In addition, there is a propensity for nucleotide substitutions near the splicing junctions and for heteroduplex formation among the V genes followed by mismatch repair; these processes are an additional, somatic source of antibody variability. Although V-J and V-D-J joining appears to occur on both homologous chromosomes in any cell, only one light chain and one heavy chain are actually expressed by the cell (called **allelic exclusion**).

8. DNA splicing is also involved in **class switching**, in which a B cell can undergo a transition from producing one type of antibody (usually IgM) to another (usually IgG). The class switch is mediated by a splice that moves the intact V-D-J region from proximity with one constant region (e.g., a μ gene) to proximity with another constant region (e.g., a γ gene).

9. The homogeneous antibody produced by a single clone of B cells is called a **monoclonal antibody**. Recent methods for large-scale production of monoclonal antibody involve fusing an antibody-producing B cell with a **myeloma** tumor cell. The hybrid cells, called **hybridomas**, can proliferate indefinitely in cell culture and continue to produce antibody.

10. Breakdowns of immunity include **immunodeficiency** diseases, in which patients lack B-cell function or T-cell function or both, and **autoimmune** diseases, in which the immune system attacks one's own antigens. **Rheumatoid arthritis** is an example of an autoimmune disease, as is the heart damage that sometimes accompanies **rheumatic fever**. Autoimmunity results from a breakdown of **immunological tolerance**, the process whereby the body comes to recognize and to tolerate "self" antigens while retaining the ability to attack "nonself" antigens. Nonself antigens include abnormal antigens on the surface of malignant cells, which provide the basis of **immune surveillance** against malignancies. On the other hand, certain kinds of self-recognition are essential in the immune response. For example, cell cooperation requires that the cooperating cells have identical antigens coded by a locus (or tightly linked loci) called the **major histocompatibility complex** (MHC). Self-recognition is also important in the formation of **anti-idiotype** antibodies, which are important in the **network hypothesis** of regulation of the immune system.

11. Rejection of transplanted tissues is primarily a T-cell function directed against antigens produced by more than a dozen loci called **histocompatibility loci**. The general transplantation rule is that a recipient individual will reject any tissue that carries histocompatibility antigens that he or she does not possess. However, some antigens are much more potent than others, and the locus responsible for the most potent antigens is the MHC mentioned above. In the mouse, the MHC is called the *H-2* locus. The MHC locus in humans is called *HLA*, and it is on chromosome 6. The success of transplants is greatly increased if donor and recipient are compatible for *HLA* (i.e., if the donor has no *HLA* antigens not also present in the recipient). With *HLA* compatibility, rejection reactions due to the minor loci can often be controlled by **immunosuppressive drugs**.

12. *HLA* plays a key role in the immune response, and the "locus" can be subdivided into four functionally distinct units (order D-B-C-A on the chromosome), with each unit having many alternative "alleles." The **haplotype** of a chromosome refers to its particular combination of "alleles" corresponding to the D, B, C, and A regions. (It is not yet clear whether *HLA* codes for one complex polypeptide or whether it is actually a cluster of four separate loci.) Certain diseases have an increased prevalence among individuals who have a particular haplotype. One example is **ankylosing spondylitis**, which tends to be associated with haplotypes containing B27.

13. Clinical problems in blood transfusions do not involve *HLA* but primarily two other loci: one responsible for the **ABO blood groups** and the other for the **Rh blood groups**. An individual's ABO blood group is determined by the presence of particular carbohydrate antigens on the surface of the red blood cells. Presence of these antigens is, in turn, controlled by a single locus on chromosome 9 at which there are three alleles: I^A, I^B, and I^O. Genotypes I^A/I^A and I^A/I^O produce the A antigen and are said to have **blood type A**; genotypes I^B/I^B and I^B/I^O produce the B antigen and are said to have **blood type B**; genotype I^A/I^B produces both A and B antigens and has **blood type AB**; and genotype I^O/I^O produces neither antigen and has **blood type O**. Because the A and B antigens are similar to certain antigens of microorganisms, virtually all individuals are stimulated to produce antibodies (usually IgM) against antigens not present on their own red cells. Thus, type A individuals will also have anti-B antibodies, type B individuals will have anti-A, type O will have both anti-A and anti-B, and type AB will have neither anti-A nor anti-B. Difficulties in transfusion arise when donor red blood cells carry antigens against which the recipient has antibodies, for then the antibodies will cause **agglutination** of the red blood cells and blockage of capillaries. In transfusions, type O is called a **universal donor** (because O cells have neither A nor B antigen) and type AB is called a **universal recipient** (because AB individuals have neither anti-A nor anti-B antibody). As with *HLA*, there is an elevated risk of particular diseases in persons with certain blood types, such as stomach cancer with type A and duodenal ulcers with type O. Secretion of the A and B antigens into the body fluids is controlled by a dominant allele at another locus called the **secretor locus**.

14. The **Rh blood groups** are determined by a red cell antigen distinct from the A and B antigens. The Rh antigen is controlled by a single locus on chromosome 1 (or perhaps by a cluster of loci) at which there are two principal alleles (*D* and *d*). Genotypes *D/D* and *D/d* produce the Rh antigen and are said to be Rh$^+$ (**positive**); genotype *d/d* does not produce the antigen and is said to be Rh$^-$ (**negative**). Unlike the situation with ABO, Rh$^-$ individuals do not usually produce anti-Rh$^+$ antibody unless stimulated with Rh$^+$ antigen. However, anti-Rh$^+$ antibodies are typically of the IgG type and can cross the placenta. Thus, antibodies from an Rh$^-$ mother can cross the placenta and attack the red cells of an Rh$^+$ fetus and cause a blood condition known as **erythroblastosis fetalis** or **hemolytic disease of the newborn**. On the other hand, Rh$^-$ mothers do not usually produce anti-Rh$^+$ antibody unless they have previously carried one or two Rh$^+$ fetuses; seepage of Rh$^+$ cells from these prior fetuses is necessary for stimulation of the immune system. Modern preventive treatment for newborn hemolytic disease involves injecting the mother with anti-Rh$^+$ antibodies near the time of birth to destroy the otherwise stimulatory Rh$^+$ cells from the fetus. This sort of protection sometimes occurs naturally; matings of Rh$^-$; O mothers with Rh$^+$; AB fathers never lead to Rh hemolytic disease, because the maternal ABO antibodies destroy the fetal Rh$^+$ cells before they can stimulate production of anti-Rh$^+$ antibodies.

15. Dozens of different blood group systems analogous to ABO and Rh are known. Most involve substances that are not antigenic in humans but are antigenic in another organism. For example, the MN system involves two red cell antigens (M and N) that are antigenic in rabbits. The M antigen is determined by a single allele (*M*); the N antigen is determined by an alternative allele (*N*). Thus, *MM* genotypes have blood type M, genotype *NN* has blood type N, and genotype *MN* has blood type MN (because both antigens are expressed). The simple Mendelian inheritance of ABO, Rh, MN, and the other blood groups makes these loci particularly useful in anthropological studies, paternity exclusion, criminology, and other applications.

WORDS TO KNOW

Immune response

Lymphocyte
T cell
B cell
Macrophage
Complement
Cell cooperation
Antigen
Hapten
Carrier
Immunological tolerance
Immune surveillance
Network hypothesis

Antibody

Antigenic determinant
  (epitope)
Receptor site
Clonal selection
Plasma cell
Immunoglobulin
Class (G, M, A, D, E)

Subclass
Allotype
Heavy chain
Light chain
Constant region
Variable region
Antigen-binding site
Idiotype
Monoclonal
Hybridoma

Syndrome

Immunodeficiency disease
Autoimmune disease
Rheumatoid arthritis
Ankylosing spondylitis
Hemolytic disease of the
  newborn
Erythroblastosis fetalis

Antibody diversity

Germ-line theory
Somatic theory

DNA splicing
Combinatorial joining
V-J joining
V-D-J joining
Class switch
Allelic exclusion

Histocompatibility

Major histocompatibility
  complex (MHC)
HLA
H-2
Haplotype

Blood groups

ABO
Agglutination
H substance
Bombay phenotype
Secretor
Rh
MN

PROBLEMS

1. For discussion: A newborn boy in Dallas, Texas, was born with a severe immunodeficiency disease and would surely have died from acute infection. His life was spared by placing him inside a germ-free plastic bubble, where he has now been for nearly 10 years. Critics argue that raising a child in such a restrictive environment is cruel and inhuman and that nature should have been permitted to follow its course. Others argue that saving the boy's life was the only reasonable route and that, other than his immunodeficiency disease, the boy seems normal and healthy and well adjusted to his situation. What is your opinion about this case? If you had such a child and had to choose one way or the other, what course would you pick?

2. What are the primary functions of T cells? Of B cells?

3. What is an antibody class? An antibody subclass? An antibody allotype?

4. What is the difference between an epitope and an idiotype?

5. What is immune surveillance and how does it relate to the T antigens of transformed tumor cells discussed in Chapter 12?

6. How many possible antibody heavy chains could be formed by combinatorial joining in a mammal that has 140 V_H regions, 6 D_H regions, and 11 J_H regions?

7. How does the network hypothesis contradict the widely held belief that the immune system does not recognize "self" antigens?

8. Most individuals are likely to be heterozygous for two different *HLA* haplotypes, and any two unrelated individuals are likely to have no haplotypes in common. In this situation, what is the probability that two siblings will have identical *HLA* genotypes?

9. Among a group of four siblings, one has blood group A, one B, one AB, and one O. What are the genotypes of the parents?

10. Red blood cells from six individuals are mixed with serum from a type A individual and with serum from a type B individual. In some cases the red cells agglutinate (+); in other cases no agglutination occurs (−). What are the blood types of individuals 1 through 6?

Individual	Serum from A	Serum from B
1	−	+
2	+	+
3	−	−
4	−	−
5	+	−
6	−	+

11. An AB secretor whose father was a B nonsecretor marries an AB nonsecretor. What types of offspring are possible and what are their expected frequencies?

12. In a case of disputed paternity, the following blood types were found among the two putative fathers, the mother, and the baby.

Male 1:	A; Rh+; MN
Male 2:	B; Rh−; N
Mother:	AB; Rh+; MN
Baby:	A; Rh−; M

Which male is the baby's father?

13. Two newborns in a maternity ward lose their identification tags, and the two sets of parents are irate. In an attempt to match each newborn with its correct parents, blood typing is carried out with the following results:

Father 1:	A; Rh+; M
Mother 1:	O; Rh+; M
Father 2:	AB; Rh−; MN
Mother 2:	B; Rh+; M
Baby A:	O; Rh−; M
Baby B:	A; Rh+; M

Could couple 1 be the parents of baby A? Of baby B? Could couple 2 be the parents of baby A? Of baby B? Which baby belongs to couple 1? To couple 2?

14. Among Basques the frequency of Rh− is about 43 percent. Among whites in the United States it is about 15 percent. If a Basque female marries an American male, what is the probability that the mating will be incompatible for the Rh system?

15. In the case of the murder of Yoshiko Hirai discussed in this chapter, was Iba Hoshi a secretor or a nonsecretor? How can you tell?

FURTHER READING AND REFERENCES

Baltimore, D. 1981. Gene conversion: Some implications for immunoglobulin genes. Cell 24:592–594. Yet another manner of creating antibody variability.

Baltimore, D. 1981. Somatic mutation gains its place among the generators of diversity. Cell 26:295–296. Although DNA splicing is elegant, a great deal of antibody diversity is generated by somatic mutation.

Bernard, O., N. Hozumi, and S. Tonegawa. 1978. Sequences of mouse immunoglobulin light genes before and after somatic changes. Cell 15:1133–1144. Direct evidence of V-J joining.

Buisseret, P. D. 1982. Allergy. Scientific American 247:86–95. Role of IgE in common allergies.

Coleclough, C., R. P. Perry, K. Karjalainen, and M. Weigert. 1981. Aberrant rearrangements contribute significantly to the allelic exclusion of immunoglobulin gene expression. Nature 290:-372–378. Gene splicing in B cells is frequently imprecise and many "mistakes" are made.

Dausset, J. 1981. The major histocompatibility complex in man. Science 213:1469–1474. Nobel prize lecture reviewing the past, present, and future concepts regarding *HLA*.

Davis, M. M., S. K. Kim, and L. E. Hood. 1980. DNA sequences mediating class switching in α-immunoglobulins. Science 209:1360–1365. Which DNA sequences are involved in class switching?

Davis, M. M., S. K. Kim, and L. Hood. 1980. Immunoglobulin class switching: Developmentally regulated DNA rearrangements during differentiation. Cell 22:1–2. A review of DNA splicing and some unanswered questions.

Early, P., H. Huang, M. Davis, K. Calame, and L. Hood. 1980. An immunoglobulin heavy chain variable region gene is generated from three segments of DNA: V_H, D and J_H. Cell 19:981–992. Details of V-D-J joining.

Golub, E. S. 1980. Idiotypes and the network hypothesis. Cell 22:641–642. A key idea is that the idiotype has a dual function, acting as both antigen and antibody.

Golub, E. S. 1980. Know thyself: Autoreactivity in the immune response. Cell 21:603–604. Some degree of reaction against "self" seems essential to mount a normal immune response.

Golub, E. 1981. Suppressor T cells and their possible role in the regulation of autoreactivity. Cell 24:595–596. On the subpopulations of T cells and their relevance to the immune response.

Golub, E. S. 1981. The Cellular Basis of the Immune Response, 2nd ed. Sinauer Associates, Sunderland, Mass. A leading textbook introducing cellular immunology. Source of data in Table 13.1.

Herberman, R. B., and J. R. Ortals. 1981. Natural killer cells: Their role in defense against disease. Science 214:24–30. Special types of T cells and their functions.

Hieter, P. A., G. F. Hollis, S. J. Korsmeyer, T. A. Waldmann, and P. Leder. 1981. Clustered arrangement of immunoglobulin λ constant region genes in man. Nature 294:536–540. Six λ-like genes are arranged in tandem on a 50-kb segment of DNA.

Hood, L., M. Steinmetz, and R. Goodenow. 1982. Genes of the major histocompatibility complex. Cell 28:685–687. Transplantation antigens and other features encoded by *MHC*.

Klein, J., A. Juretič, C. N. Baxevanis, and Z. A. Nagy. 1981. The traditional and a new version of the mouse *H-2* complex. Nature 291:455–460. The manifold functions of *H-2* and a unifying hypothesis of its structure.

Koffler, D. 1980. Systemic lupus erythematosus. Scientific American 243:52–61. Discusses a serious autoimmune disease found mainly in women.

Leder, P. 1982. The genetics of antibody diversity. Scientific American 246:102–115. Review of combinatorial joining and class switching.

Liu, C.-P., P. W. Tucker, J. F. Mushinski, and F. R. Blattner. 1980. Mapping of heavy chain genes for mouse immunoglobulins M and D. Science 209:1348–1353. V-D-J joining, and a model for the dual expression of IgM and IgD in B lymphocytes.

Marcu, K. B. 1982. Immunoglobulin heavy-chain constant-region genes. Cell 29: 719–721. Review of limited number of genes encoding the constant regions of antibodies.

Mourant, A. E., A. C. Kopéc, and K. Domaniewska-Sobczak. 1976. The Distribution of the Human Blood Groups and Other Polymorphisms, 2nd ed. Oxford University Press, New York. A compendium of information on the varying incidences of blood groups in different branches of humanity. Source of Figure 13.11.

Oriol, R., J. Danilovs, and B. R. Hawkins. 1981. A new genetic model proposing that the *Se* gene is a structural gene closely linked to the *H* gene. Am. J. Hum. Genet. 33:421–431. New information suggesting close linkage between secretor and the locus involved in the Bombay phenotype.

Potash, M. J. 1981. B lymphocyte stimulation. Cell 23:7–8. What induces B lymphocytes to secrete antibody?

Reinherz, E. L., and S. F. Schlossman. 1980. The differentiation and function of human T lymphocytes. Cell 19:821–827. A good review of a highly complex population of cells.

Rose, N. R. 1980. Autoimmune diseases. Scientific American 244:80–103. Malfunctions of the immune system and strategies of treatment.

Ryder, L. P., A. Svejgaard, and J. Dausset. 1981. Genetics of *HLA* disease association. Ann. Rev. Genet. 15:169–187. Much more data on *HLA* and disease.

Seidman, J. G., A. Leder, M. Nau, B.Norman, and P. Leder. 1978. Antibody diversity. Science 202:11–17. The role of recombination in the origin of antibody diversity.

Shreffler, D. C., and C. David. 1975. The *H-2* major histocompatibility complex and the I immune response region: Genetic variation, function, and organization. Adv. Immunol. 20:125–195. Source of Table 13.3 and a thorough review of the functional properties of the mouse MHC.

Siebenlist, U., J. V. Ravetch, S. Korsmeyer, T. Waldmann, and P. Leder. 1981. Human immunoglobulin D segments encoded in tandem multigene families. Nature 294:631–635. Organization of the human D regions.

Snell, G. D. 1981. Studies in histocompatibility. Science 213:172–178. Nobel prize lecture reviewing the mouse *H-2* system from its earliest studies.

Vogel, F., and A. G. Motulsky. 1979. Human Genetics: Problems and Approaches. Springer-Verlag, New York. Source of data in Table 13.4.

chapter 14

Population Genetics and Evolution

This and the next two chapters constitute a brief summary of **population genetics**—the study of how Mendel's laws and other principles of genetics apply to entire **populations** (interbreeding groups) of organisms. On the one hand, population genetics includes studies of contemporary populations, such as tribes of Indians or major human races, in which the focus is on the genetic relationships among groups and on the reasons underlying differences in the incidence of inherited disorders or other traits. On the other hand, population genetics attempts to clarify the major forces involved in the process of **evolution**, by which we mean cumulative change in the genetic makeup of a population through time.

The first part of this chapter considers the evidence for human evolution and some major features of the process, including points that are in dispute. Then we argue that evolutionary change uses the genetic variation that occurs among individuals in virtually every population, and methods for studying genetic variation in contemporary populations are discussed; the organization of genetic variation in populations is then considered. Chapter 15 deals with the effects of major evolutionary forces, such as mutation, migration, natural selection, and random genetic drift; it also summarizes new information about evolution that has been obtained from molecular studies. Finally, Chapter 16 concerns the inheritance of traits that are influenced by several or many genes and by the effects of environment.

The Place of Humans Among Living Things

There is a natural tendency for the human mind to lump things into categories based on their similarities and differences. This tendency was first applied systematically and scientifically to living organisms by the Swedish naturalist Carolus Linnaeus in the eighteenth century. Out of his efforts grew the present system of Latin nomenclature with its major categories of **kingdom**, **phylum**, **class**, **order**, **family**, **genus**, and **species**. (Various intermediate categories such as "superclass" and "subclass" are also officially recognized.) The fundamental unit of **classification** is the **species**; in sexually reproducing organisms, a species is a group of organisms that are capable of interbreeding among themselves and producing viable and fertile offspring, but they are incapable of successful reproduction with members of other such groups. Thus, the Yanomama Indians of Brazil and the Eskimos of Alaska are in the same species because they are potentially capable of successful interbreeding. The word **population** (or, more precisely, **local population**) is typically used to denote a group of organisms of the same species living within a sufficiently restricted geographic area that they usually find mates within their own group. Thus, the Yanomama and the Eskimos are in the same species but different populations. Categories above the species level are referred to collectively as **higher categories**. Evolutionary change occurring within a population or within a species is often called **microevolution**; evolutionary change leading to the formation of a new species or higher category is often called **macroevolution**.

Linnaean classification is **natural**, which means that it is based on the totality of characteristics of organisms and uses information from embryology, anatomy, morphology, cytology, genetics, ecology, and, more recently, molecular biology. The value of a natural system is that the classification reflects evolutionary relationships and lines of descent. Two species in the same genus are more closely related (i.e., have more genes in common) than two species in different genera, two genera in the same family are more closely related than two genera in different families, and so on up the hierarchy of higher categories.

Human beings are no more exempt from biological processes such as evolution than we are exempt from the laws of physics and chemistry. Human origins are part of the evolutionary process, and in the whole sweep of evolutionary time, only a small part. Since the first appearance of life on earth, the number of species has increased like that of branches on a tree, with each branch representing one of the higher categories. The trunk represents the first primitive living cell. This grew and at one point split into two branches; the branches grew and split into more branches, which split into yet more branches, which again and again and yet again split into ever more new branches. So on and on it went until today each living species is like a leaf at the end of a twig. The branches behind the leaves are, of course, hidden from direct view. They are part of the vast evolutionary record that can be known only by careful comparisons of living organisms and of fossils that are preserved and discovered. Most branches and twigs on the tree are dead, extinct; they have no leaves, no living representatives. The approximately 1.3 million species of plants and animals alive today are estimated to represent a mere 1 percent of all the species that ever existed.

The place of humans on the evolutionary tree can be established by comparisons of humans with other living organisms—that is, by placing us correctly in the system of classification (Table 14.1). We are, of course, animals, which from an evolutionary point of view allies us more closely with sponges and insects

TABLE 14.1   BIOLOGICAL CLASSIFICATION OF MODERN HUMANS

Classification	Characteristics
Kingdom: Animalia	Cells lack cellulose walls; no photosynthesis
Phylum: Chordata	Embryos have elastic tubular structure along the back
Subphylum: Vertebrata	Adults have vertebral column made of bone
Superclass: Tetrapoda	Four limbs present
Class: Mammalia	Young are suckled
Subclass: Theria	Young are live-born
Infraclass: Eutheria	Females lack pouches (like those in kangaroos)
Order: Primates	Many special traits, among them flat nails
Suborder: Anthropoidea	Tailless
Family: Hominidae	Erect posture; relatively large brain
Genus: *Homo*	Modern brain structure; skillful hands
Species: *Homo sapiens*	Large brain; flat face with prominent nose and jutting chin; small teeth
Subspecies: *Homo sapiens sapiens*	Modern human beings; first appeared about 40,000 years ago

than with corn and mushrooms. During embryonic development we have gill slits and, along the back, an elastic tubular structure known as the notochord; these features ally us more closely with sea squirts and lancelets than with sponges and insects. We have a backbone and a spinal cord, and are thereby more closely related to fish, frogs, and birds than to sea squirts and lancelets. We are warm blooded, have hair, and feed our young with milk from mammary glands, which puts us closer to cats, mice, and pigs than to fish, frogs, and birds. We have flexible limbs and grasping hands, nails instead of claws, and stereoscopic vision; these ally us with tree shrews, lemurs, and monkeys more than with cats, mice, and pigs. We have broad, flattened chests and shortened lower spinal columns; this making us more like gibbons and orangutans than tarsiers, lemurs, and monkeys. The shape of our hands, feet, and teeth, and the position and shape of our muscles and internal organs ally us more closely with gorillas and chimpanzees than with gibbons and orangutans. Our fully erect posture, long legs, large brain, flat face with prominent nose and jutting chin, protruding lips, small teeth, and

relative hairlessness distinguish us from gorillas and chimpanzees and make us one unique species, *Homo sapiens*.

We humans are most closely related to such organisms as the chimpanzee and gorilla, which means that we have certain genes in common because of our shared ancestry. Some people take umbrage at this relationship, particularly, it is said, Victorian matrons. But humans also share genes with pigs, though fewer than with chimpanzees, and with centipedes, though fewer than with pigs; we even share genes with corn and cauliflower, though fewer than with centipedes. At one time these relationships could only be inferred from the positions of the various organisms in the scheme of natural classification. Nowadays, the similarity of shared genes can be studied directly by nucleotide sequence determinations of DNA. Perhaps the Victorians took umbrage because they misunderstood the concept of common ancestry. While it is true that modern humans and modern apes have a common ancestry, it is not correct to say that "humans descended from apes." Indeed, the common ancestor of present apes and humans was neither ape nor human but a more primitive creature having unique characteristics of its own, and sharing some but lacking other characteristic features of both modern apes and modern humans.

Outline of Human Evolution

Studies of organisms alive today provide no information about species that branched off an evolutionary line of descent and became extinct. Neither do they tell us how a species that exists today became transformed into its present form over millions of years of evolutionary time, nor do they tell us what the intermediate forms looked like. For these stories the fossil record will tell the tale—when the record is sufficiently complete. Unfortunately, the fossil record of human evolution is still only fragmentary, although discoveries of major importance have been made during the past several decades. However, fossils from the earliest stages of human evolution are not clearly distinct from the ancestors of today's chimpanzees and gorillas. Moreover, whether a fossil represents a transitional stage on the main line of evolution or a side branch that eventually became extinct is seldom known for certain. As a consequence, the existing fossil record of human evolution is still like a jigsaw puzzle with many pieces missing. Where available pieces may fit into the overall picture is uncertain. Agreement, when reached at all among physical anthropologists, is often only tentative.

The earliest ancestral apes were very likely among a diverse group of species known as the **dryopithecines**, a group represented by a large number of fossils dating back approximately 10 to 25 million years and found throughout Europe, Asia, and Africa. The dryopithecine fossils represent species ranging in size from small chimpanzees to large gorillas; although their teeth are much like those of modern apes, the shape and proportions of their limbs are more like those of modern monkeys than modern apes. Dryopithecine locomotion was not of the advanced hominoid types (branch swinging, knuckle walking, or upright walking on two legs); these types of locomotion evolved only later.

One dryopithecine-like form, known primarily from fossil jaws and teeth dating back 14 million years and found in India and Africa, is known as *Ramapithecus*. This creature was probably small (about the size of a modern 5- or 6-year-old child) and had a smaller snout than other dryopithecines; the correspondingly smaller front teeth suggest that the arms and hands were used instead of teeth for threatening displays and defense. The use of arms and hands for such purposes suggests that *Ramapithecus* may have walked upright, an inference leading some anthropologists to classify it along with the human family, the **hominids**, rather than that of the apes. The geology of certain *Ramapithecus*-bearing deposits and the fossil animals and plants found along with remains of *Ramapithecus* imply a habitat marked by broad, sluggish rivers bordered by tropical forests and wooded grasslands. Just possibly, *Ramapithecus* may have been in the main line of human descent, and some anthropologists place it as one of the earliest members of the human family.

On the other hand, the classification of *Ramapithecus* as a hominid (in the human family) is contested; some authorities place it with the pongids (the ape family) and regard it as a dryopithecine having no special affinity with later hominids. The argument will probably continue until more complete specimens of *Ramapithecus* are found. The classification of *Ramapithecus* is further confused because, if *Ramapithecus* is to be regarded as a hominid, then the separation of apes and humans would have had to occur 14 to 20 million years ago; this conflicts with the belief based on other evidence that the separation actually occurred more recently—5 to 10 million years ago.

Answers to questions about the most primitive hominids are frustrated by a great gap in the fossil record extending from approximately 10 million years ago to roughly 5 million years ago. This gap comes about because no major hominid-bearing deposits of this time period have been found. The gap ends with an extensive collection of fossil hominids found in South and East Africa, dating back 1 to about 4 million years (abbreviated 4 My) and known as the **australopithecines**. By 2 to 3 million years ago, the australopithecines surely walked erect because bones of their feet, legs, hips, and backs show strong similarities with those of modern humans. The australopithecines had relatively small brains and a low forehead (in one species there was essentially no forehead). They had large teeth, especially molars and premolars, thus giving them larger, more rugged looking faces than modern humans. Nevertheless, the australopithecines had an erect posture and used crude stone tools, which places them among the hominids as near or direct relatives of humans.

Two distinct australopithecine forms are known: a smaller form about 4 ft tall and weighing about 60 lb, and a larger form about 4.5 ft tall and weighing about 100 lb. The smaller form is known as the "gracile" australopithecine, the larger as the "robust" australopithecine. Their place in human evolution is illustrated in Figure 14.1, but there are a number of uncertainties. The gracile australopithecines consist of two species: *Australopithecus africanus* and an earlier one called *A. afarensis*. (Some authorities regard *A. afarensis* as just an earlier form of *A. africanus* and recognize only the one species.) The robust australopithecines consist of two contemporaneous species *(A. robustus* and *A. boisei)*, and, as illustrated in Figure 14.1, the gracile australopithecines preceded the robust form in time.

There is considerable disagreement about how the two australopithecine forms relate to each other. Four possibilities exist: *A. africanus* could be (1) in the main line of human descent (illustrated in Figure 14.1), (2) in the line of robust australopithecine descent, (3) in both lines of descent (in which case *A. africanus* would have split into two species), or (4) in neither line of descent (in which case *A. africanus* would be off on a branch by itself). Each of these four possibilities has its adherents. In any case, the gracile australopithecines are clearly distinct from the robust australopithecines. Patterns of tooth wear suggest that *A. africanus* included meat in its diet but *A. robustus* and *A. boisei* did not. There is also some evidence—though not conclusive—that the two forms preferred somewhat different habitats: The gracile form probably preferred dry, grassy savannas, whereas the robust form probably preferred a wetter area with heavy vegetation. Whatever the place of the australopithecines in human evolution, it is clear that the robust form became extinct about 1 million years ago.

The period 2.2 to 1.6 My has produced a number of fossils generally agreed to represent our own genus (*Homo*) and to have been in the main line of human descent. These fossils are designated as the species *Homo habilis* (see Figure 14.1). *H. habilis* was larger than *A. africanus* (averaging about 90 lb as compared with 60 lb), and the human line had already begun to develop its distinctive and large brain (*H. habilis'* cranial capacity ranged from 500 to 750 cm^3, as compared with 400 to 500 cm^3 for *A. africanus*).

H. habilis was succeeded by *H. erectus*, which lived in the period 1.6 to 0.6 My (see Figure 14.1). Still beetle-browed, the skull of *H. erectus* was becoming more humanlike; the forehead became higher and the face and teeth became smaller. *H. erectus* was larger than *H. habilis*, averaging about 110 lb, and its cranial

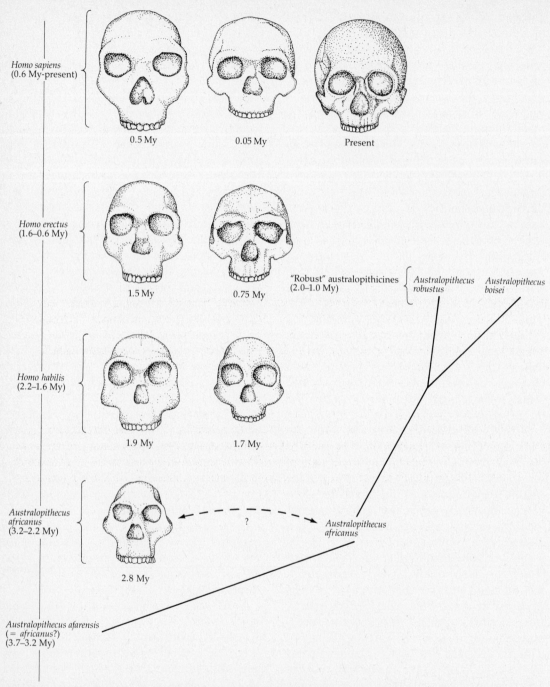

Figure 14.1   Possible schemes of human evolution. Whether *Australopithecus africanus* belongs in the direct line of human descent or in the branch leading to the robust australopithecines, or whether it is a common ancestor of both, is a matter of controversy. The position of *A. afarensis* is also uncertain.

capacity ranged from 750 to 1200 cm^3 (as compared with 1000 to 2000 cm^3 for modern *Homo sapiens*). Studies of the bones of the lower limbs have indicated that *H. erectus* had an improved ability to walk erect, and studies of the bones of the upper limbs have shown that its hands were capable of more precise manipulation of objects.

H. erectus differed from its ancestors not only in morphology but also in adaptations. Advances in tool-making ability, ability to hunt large game animals, and the use of fire (and perhaps clothing as well) allowed *H. erectus* to spread into the temperate areas of Europe and Asia. *H. erectus* was therefore a most successful species that came to occupy tropical and temperate areas of Eurasia and Africa. As it encountered different environmental conditions, the species became differentiated into a series of geographical types called **geographical subspecies**. (Major finds of *H. erectus* include specimens from Africa, Java, China, and Europe.) Somewhere along the way, *H. erectus* gave rise to an early form of *Homo sapiens*, although the place, the time, and the transitional stages through which *H. erectus* evolved into *H. sapiens* are in doubt.

By about 600,000 years ago, however, the transition from *H. erectus* to *H. sapiens* had already taken place. Most authorities place the first members of our own species in an earlier subspecies technically known as ***Homo sapiens neanderthalensis***. Specimens of *H. s. neanderthalensis* have been recovered from many places throughout the Old World, although the largest samples of the best-preserved specimens come from Europe and the Near East. (The subspecies name is derived from one of the first specimens ever discovered of a fossil hominid—part of a skull found in 1856 in the Neander Valley in Germany.)

The most important differences between *H. erectus* and *H. s. neanderthalensis* involve the shape of the skull and the ways the species adapted to their environments. The skull of *H. s. neanderthalensis* is marked by large jaws, large eyes sockets, and massive brow ridges that extend continuously across the brow, quite unlike the brow ridges of *H. erectus*. Although the forehead of *H. s. neanderthalensis* remains low, the brain case is much larger than that of *H. erectus*. Having an average cranial capacity of about 1600 cm^3, the brain of *H. s. neanderthalensis* was some 60 percent larger than that of *H. erectus* and even somewhat larger than the average of modern humans.

Neandertal stone tools were more complex than those of *H. erectus*, which probably indicates that this subspecies was making specialized tools for particular jobs. Neandertal culture is the first that gives strong evidence of symbolic behavior among hominids. Certain European caves contain the skeletons of cave bears arranged in specially built stone boxes along with Neandertal tools. Moreover, Neandertals seem to have buried their own dead, at least some of them, and in some cases they left offerings of flowers and meat in the graves. This behavior has been interpreted as indicating a religious concern for death among their own kind and for animals that the Neandertals hunted. Indeed, some anthropologists argue that this type of symbolic behavior would have required the presence of a complex language among the Neandertals, a language that would presumably have begun to develop in *H. erectus*. The idea is disputed, however, and others maintain that the Neandertals were anatomically incapable of speech.

In any case, *H. s. neanderthalensis* was not only an inhabitant of tropical and temperate areas of the Old World but was able, for the first time in hominid evolution, to adapt to essentially arctic habitats. The Neandertals lived in Europe during the time of the **ice ages**—advances and retreats of glaciers that affected the climate of the entire world. As the glaciers advanced southward they produced arctic environments in what is now temperate Europe. In these areas many Neandertal sites have been found that contain the bones of animals and the pollen of plants that are characteristic of arctic climates.

Between 35,000 and 40,000 years ago, the geographical race of *H. s. neanderthalensis* that inhabited Europe (often referred to as "Classic Neandertals") was rather abruptly replaced by the **Cro-Magnon** people, named after a fossil find in Cro-Magnon, France. The Cro-Magnons represent an early race of the modern subspecies *Homo sapiens sapiens*, the subspecies to which today's human beings belong. In the Near East, on the other hand, the transition from *H. s. neanderthalensis* to *H. s. sapiens* seems to have proceeded more slowly and continuously than it did in Europe. This apparent difference in the rate of transition has led some anthropologists to propose that *H. s. sapiens* evolved first in the Near East and spread throughout the Old World, perhaps along such migration routes as those suggested in Figure 14.2. It is thus proposed that, as populations of *H. s. sapiens* spread, they caused the extinction of populations of *H. s. neanderthalensis*. Other authorities contend that the differences in transition rates are only apparent and are caused by the lack of data from critical time periods and critical places. From mainly archeological evidence, they argue that the transition from *H. s. neanderthalensis* to *H. s. sapiens* took place throughout the entire range of the former subspecies. Whatever the exact mode of transition between these two subspecies may have been, *H. s. sapiens* seems to have been well established throughout the Old World by about 35,000 years ago.

The most important differences between *H. s. neanderthalensis* and *H. s. sapiens* are found in their skulls, their cultures, and their geographic distributions. The jaws and teeth of *H. s. sapiens* are smaller than those of the Neandertals, and the brow ridges are reduced to very small bumps. The forehead of modern humans is much higher than that of the Neandertals, and the whole brain case is much more spherical. Finally, the modern subspecies has a chin on the lower jaw. The culture of *H. s. sapiens* became technically and socially more complex than it had been in the Neandertals,

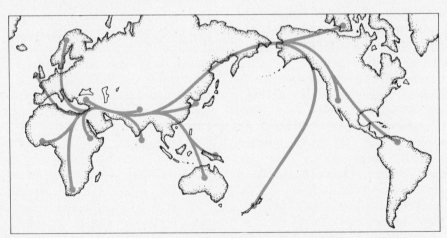

Figure 14.2 Map of the world showing possible migration routes of the modern human subspecies (*Homo sapiens sapiens*), which originated some 40,000 years ago, perhaps in the Middle East.

and thus cultural advance provided new ways of adapting to the environment. *H. s. sapiens* was the first hominid to radiate into the New World; its movement there is thought to have taken place 20,000 to 30,000 years ago.

Thus today's human beings—their skeletons essentially unchanged for the last 40,000 years—are biologically quite different from their earliest ancestors in stature, in erect posture and gait, in dexterity of the hands, in brain size, and in many other ways. The signal event in human evolution was without doubt the creation of culture as a primary means of contending with the environment. Culture relieved mankind of its utter dependence on the long, slow process of evolution to alter the **gene pool** (the aggregate of genes in a species) and thereby accommodate the demands of the environment; culture provided the means whereby men and women could adapt to the environment rapidly and directly or could even alter the environment to suit themselves. In early cultural times, naked bodies were protected in shelters warmed by fire, the arm became a lethal weapon when the hand held a club; later on, seeds were collected and sown and the crops harvested. Of supreme importance, the transmission of culture from generation to generation by means of language has led to an ever-accelerating rate of change, like the force of a hurricane, which seems to feed on its own furious winds. The wood fire in the hearth became fossil fuel and then nuclear energy. The club became a spear, a bow with arrows, a rifle, a bazooka, a bomb, a nuclear warhead, a guided missile. Sowing and reaping became modern agribusiness, its yields boosted by artificial breeding of crops and by fertilizers and insecticides. In our own century, the power of mankind to control the environment has become so vast that, carelessly or thoughtlessly applied, it causes worldwide and often irreversible degradation and pollution.

Gradual Versus Punctuated Evolution

Although most of the principal stages of hominid evolution have probably been discovered, the fossil record is sadly incomplete. (It has been estimated that the probability of a hominid becoming fossilized and discovered at a favorable site is approximately 10^{-6}.) Thus, the transitional forms between *A. africanus*, *H. habilis*, *H. erectus*, and *H. sapiens* are much in doubt. The main line of human evolution has been marked by an increase in cranial capacity and stature (as well as by changes in many other traits), but the pattern of phenotypic change in these traits is in dispute. Two extreme views are known as **gradual evolution** and **punctuated evolution**. Gradual evolution proposes that the transitional forms, if discovered, would exhibit a steady, gradual pattern of change over time; punctuated evolution proposes that the pattern of phenotypic change is highly unsteady and that it is marked by long periods of virtually no change at all punctuated by brief periods of relatively abrupt change. The abrupt changes envisaged in punctuated evolution are not proposed to occur in the space of just a few generations; the proposal is that major phenotypic change occurs in a relatively short span of geological time, but this span may involve tens or hundreds of thousands of generations.

Gradual and punctuated views of the evolution of cranial capacity are illustrated in Figure 14.3. The shaded bars represent the average (dot) and the range (smallest to largest) in cranial capacity found among the species indicated. *N* refers to the number of specimens of each type, and the bar for *H. sapiens* includes both *H. s. neanderthalensis* and *H. s. sapiens*. The gradual hypothesis is depicted with the broken line; there is a steady increase in cranial capacity from species to species. The punctuated view is illustrated by the solid line; each species is presumed to maintain a relatively constant average cranial

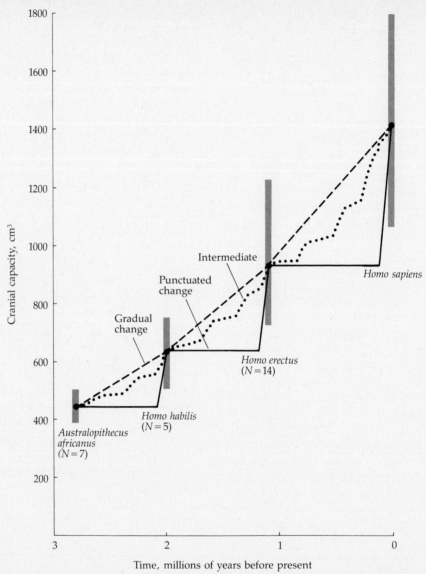

Figure 14.3 Average cranial capacity (dots) and range (bars) for various species in human evolution. The symbol N refers to the number of skulls, and the data for *H. sapiens* include the Neandertals. Broken line: hypothetical changes for gradual evolution, in which changes occur slowly over time. Solid line: hypothetical changes for punctuated evolution, in which phenotype changes little for long periods and then relatively rapidly. Dotted line: an intermediate view.

capacity for a long period, followed by a rather abrupt change to the succeeding species. Of course, both gradual and punctuated evolution are extremes. The situation may actually be

something like that illustrated with the dotted line, where change is neither gradual nor punctuated but a sort of mixture of periods of relatively slow or no change accompanied by other periods of relatively rapid change. Which pattern of change best depicts the actual evolution of cranial capacity is unfortunately not known owing to the paucity of fossils, the uncertainties of dating, and other problems.

Natural Selection and Genetic Variation

Evolution—cumulative change in the genetic characteristics of a population—requires the occurrence of inherited genetic variation; if all individuals in a population were genotypically identical, evolution would not be possible. This requirement for inherited genetic variation was first brought forward by Charles Darwin in his monumental book *The Origin of Species,* first published in 1859, in which he argued that the genetic characteristics of populations are molded by external forces of the environment acting on preexisting hereditary variants that occur among members of the population. (Similar ideas were proposed by Alfred Russel Wallace at about the same time.)

Darwin called his theory of the mechanism of evolutionary change **natural selection**, and the idea rests on three observations:

1. All populations produce more young than can possibly survive and reproduce.

2. Since not all individuals are equally likely to survive and reproduce, there will be phenotypic variation in this ability from individual to individual.

3. At least some of the phenotypic variation in ability to survive and reproduce is inherited; that is, it is due to differences in genotype among individuals.

Now, the argument continues, since there will certainly be variability in the ability of individuals to survive and reproduce, those most able to survive and reproduce in a given environment will do so, thereby contributing their genes disproportionately to individuals of the next generation. Since part of the parents' ability to survive and reproduce is heritable, their offspring will receive the beneficial genes. Thus, genes that enhance the ability of an organism to survive and reproduce in its environment will increase in frequency in the population as time goes on. The population thereby becomes progressively more adapted to its environment because of the action of natural selection. A key element in natural selection is the environment, and perhaps it would be more precise to say that a population becomes adapted to its environment because the environment selects the adaptive traits. The arguments leading to the theory of natural selection seem rather obvious to biologists today, even self-evident. But Darwin perceived them in the middle of a century when most people disputed the very occurrence of evolution, let alone worried about its mechanism. (A major part of *The Origin of Species* is devoted to detailing the evidence for the occurrence of evolution itself rather than to proving the occurrence of natural selection.)

Populations of humans and most other organisms have a prodigious amount of genetic variation, and it occurs at several levels. Some genetic variation involves differences in the chromosomal complements of individuals. One obvious example is sex: Human males are normally XY, females normally XX. Genetic diversity in humans sometimes involves other chromosomal differences—trisomies, translocations, inversions, and variations in the length of such chromosomes as the Y. Notwithstanding the importance of sex and of chromosomal abnormalities in human genetics, chromosomal differences constitute only a small fraction of the total genetic variation in humans. With few exceptions, all normal individuals have the same chromosomal complement with the same number of genes coding for molecules with the same cellular functions.

Additional genetic diversity among humans occurs at the level of DNA. One type of diversity involves variation in the number and location of transposable DNA sequences (see Chapter 11). Another type involves differences among alleles carried at individual loci. Some of this allelic diversity is accounted for by carriers of dominant mutations; examples include achondroplasia and Huntington disease. Such mutant alleles are rare; the amount of genetic diversity due to harmful dominant mutations is small. Another sort of allelic diversity is apparent in children who suffer from syndromes caused by autosomal recessive genes— genes of the sort that result in cystic fibrosis, sickle cell anemia, Tay-Sachs disease, and albinism. (Of course, not all recessive genes are harmful and not all recessive genes are rare; the I^O allele in the ABO blood groups and the d allele in the Rh system are examples of recessive alleles that are neither harmful nor rare.) Harmful recessive alleles are often present in heterozygous parents, and these alleles segregate in meiosis and may come together in fertilization to produce a homozygous recessive child. However, as emphasized in connection with Figure 4.3 in Chapter 4, only a minority of the total number of harmful recessive genes in a population are actually present in homozygous genotypes. Most harmful recessive genes are hidden by virtue of being heterozygous in normal, healthy individuals. These hidden harmful recessive alleles are part of the total genetic diversity, but only a part. Most human genetic variation is due to alleles that are normal (not abnormal or harmful) and common (not rare).

Anyone can see enormous phenotypic diversity in such traits as facial characteristics: in noses (whether long or short), ears (large or small), the shape of the hairline, the set of the eyes, hair color, hair texture (curly, waved,

straight), eye color, skin color (pink, black, brown, yellow, red), height (lanky or short), weight, growth rate, athletic skills, even certain mental abilities. What these traits have in common is that they are all influenced by several or many genes acting together, and such traits as height, weight, athletic skills, and mental ability are profoundly influenced by such environmental factors as diet or training. This interaction between genes and environment makes it difficult to determine how much of the differences among individuals is due to heredity and how much is due to varying environments. Special statistical procedures have been developed to help attack this question, and it is known that at least part of the variation is genetic. Although a wealth of genetic diversity is involved in determining such traits as these, it is impossible to measure genetic diversity precisely except in special well-studied cases. Nevertheless, the amount of genetic diversity for traits determined by multiple loci and influenced by environmental factors is very great.

Genetic diversity is easiest to study with traits that are not strongly influenced by environment. Some of these were discussed in Chapter 3, including the inherited ability to taste the chemical phenylthiocarbamide (PTC), for which nontasters are homozygous recessives. Substantial genetic variation is also evident for *HLA* haplotypes and the ABO, Rh, and MN blood groups (see Chapter 12), and genetic variation also occurs in many other blood group systems with names such as Lewis, Lutheran, Kell, Duffy, Kidd, Diego, Dombrock, Auberger, and Stoltzfus. In all these cases, genetic diversity among individuals is evident. Bear in mind that all this diversity is common and usual in the population; it is normal. At most of these loci there is not a single "best" or "most normal" genotype. Every genotype within the normal range seems just as good as any other.

Measuring Genetic Diversity

A particularly useful technique for revealing hidden genetic variation is **protein electrophoresis**, a separation procedure for proteins discussed in Chapter 11. (**Hidden** genetic variation is variation that is not apparent from casual inspection of phenotype.) Recall that proteins placed near the edge of a moist gel and subjected to an electric field will move across the gel in response to the field. The speed at which a protein moves is determined largely by the net electric charge of the molecule—in other words, on the balance between positively charged amino acids (lysine, arginine, and histidine) and negatively charged ones (aspartic acid and glutamic acid). After electrophoresis, the position of an enzyme in the gel can be visualized by treating the gel with a staining solution that changes color and so creates a band of color wherever the enzyme-catalyzed reaction occurs. (See Figure 14.4 for an example involving the starch-degrading enzyme **amylase** in *Drosophila melanogaster*.) As noted in Chapter 11, any alleles leading to amino acid substitutions that alter the electrophoretic characteristics of an enzyme will be detectable by electrophoresis. (Of course, genetic studies must always be carried out to confirm that observed electrophoretic variation is due to genetic variation at the locus coding for the enzyme in question.) In Figure 14.4, two alleles (*F* and *S*) and three genotypes (*F/F*, *F/S*, and *S/S*) are represented among the six individuals. The *F* allele codes for a fast-migrating enzyme, and *F/F* homozygotes exhibit only the fast-moving enzyme band; and *S* allele codes for a slow-migrating

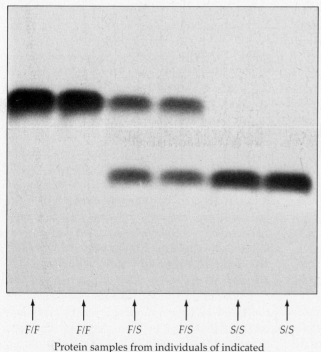

Direction of current flow and of migration of proteins

F/F   F/F   F/S   F/S   S/S   S/S

Protein samples from individuals of indicated genotypes placed in slots on this edge of gel

Figure 14.4   Gel slab used in electrophoresis showing dark bands that mark the position to which a particular enzyme has migrated in response to the electric field. The enzyme here is amylase in *Drosophila melanogaster*, and the electrophoretic pattern of three genotypes (*F/F*, *F/S*, and *S/S*) is shown. (Negative print of gel photo courtesy of J. Coyne and D. Hickey.)

enzyme, and *S/S* homozygotes exhibit only the slow-moving enzyme band. *F/S* heterozygotes (in the middle in Figure 14.4) exhibit both the fast-enzyme form (corresponding to the *F* allele) and the slow-enzyme form (corresponding to the *S* allele), so *F* and *S* are **codominant** (i.e., both expressed in heterozygotes). Enzymes that differ in electrophoretic mobility because of genetic differences at the corresponding locus are called **allozymes,** and codominance is the usual situation regarding allozyme-associateed alleles. In any case, by the simple and rapid procedure of electrophoresis, many different enzymes from hundreds of individuals can be examined, and at least some of the allelic variation in coding regions of DNA can be detected.

The utility of electrophoresis in population genetics is that loci coding for enzymes can be examined without regard to whether or not the loci are variable. (By contrast, blood group studies have a built-in bias because new blood groups can be discovered only when antigenic differences preexist in the population.) Consequently, electrophoresis allows the level of genetic variation to be estimated quantitively. This estimate of the amount of genetic variation in a population is usually expressed in terms of two numbers. One of these, the **proportion of polymorphic loci**, conveys what proportion of the studied loci have two or more common alleles; what one means by *common* is a matter of judgment, but, in population genetics, **polymorphic loci** are conventionally defined as loci at which the most common homozygote has a frequency of less than 90 percent. The other summary number is the **average heterozygosity**, which refers to the average frequency of heterozygotes among the loci in question.

The concepts of proportion of polymorphic loci and average heterozygosity are illustrated for 14 allozyme loci in Europeans in Table 14.2. The loci are arranged in decreasing order of the most common homozygote, and the observed proportion of heterozygotes for each locus is indicated. Since a frequency of 90 percent for the most common homozygote is the cutoff in the definition of polymorphism, the eight loci from *ADA* through *GPT* are polymorphic, and the observed proportion of polymorphic loci in this group of loci is therefore $\frac{8}{14}$, or 57 percent. (Loci that are not polymorphic—*PEP-D* through *AK* in this example—are often said to be **monomorphic**.) The average heterozygosity among these loci is just the average of the numbers in the last column, and in this case it turns out to be 23 percent. (If the number of individuals examined varies greatly between loci, then this should be taken into account in the averaging; such statistical procedures are beyond our scope, however.) An average heterozygosity of 23 percent implies that a typical individual will be heterozygous at about one-fourth of the loci in Table 14.2.

Actually, the loci in Table 14.2 are not highly representative of allozyme variation in humans. In the largest electrophoretic survey yet carried out, involving 71 loci in Europeans, the frequency of polymorphic loci was found to be about 30 percent and the average heterozygosity about 7 percent. Mammals on the whole are somewhat less variable (polymorphism 15 percent, heterozygosity 4 percent), plants come next on the scale (polymorphism 25 percent, heterozygosity 7 percent), and invertebrates tend to be highly variable (polymorphism 40 percent, heterozygosity 11 percent). Such levels of allozyme variation are not restricted to diploids; similar or even greater amounts are found in natural populations of *Escherichia coli*.

Genetic variation can also be studied at the DNA level; an important example is illustrated in Figure 14.5, which depicts the genetic organization of regions flanking the β-globin gene. When human DNA is digested with

TABLE 14.2 FREQUENCIES OF MOST COMMON HOMOZYGOTES AND HETEROZYGOTES AT VARIOUS ENZYME LOCI IN EUROPEANS

Enzyme	Locus	Frequency of most common homozygote	Frequency of heterozygotes
Peptidase D	PEP-D	0.986	0.014
Peptidase C	PEP-C	0.978	0.016
Glutamic oxaloacetate transaminase	GOT-M	0.966	0.033
Phosphogluconate dehydrogenase	PGD	0.962	0.037
Alcohol dehydrogenase-2	ADH_2	0.941	0.058
Adenylate kinase	AK	0.920	0.077
Adenosine deaminase	ADA	0.897	0.100
Esterase D	ES-D	0.815	0.175
Phosphoglucomutase-1	PGM_1	0.573	0.367
Phosphoglucomutase-3	PGM_3	0.550	0.383
Alkaline phosphatase (placental)	PL	0.394	0.502
Acid phosphatase (red cell)	ACP_1	0.358	0.519
Alcohol dehydrogenase-3	ADH_3	0.352	0.482
Glutamic pyruvic transaminase	GPT	0.255	0.495

Proportion of polymorphic loci $\frac{8}{14} = 0.57$

Average heterozygosity   0.23

Source: Data from H. Harris, D. A. Hopkinson, and F. B. Robson, 1974, The incidence of rare alleles determining electrophoretic variants: Data on 43 enzyme loci in man, Ann. Hum. Genet: Lond. 37:237–253.

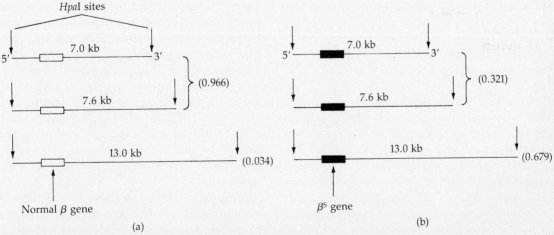

Figure 14.5 *HpaI* restriction-site polymorphism downstream from the β-globin gene. (*a*) Fragment sizes and their relative frequency among normal alleles. (*b*) Fragment sizes and their relative frequencies among β^S alleles. Note that the 13-kb fragment is common in β^S but rare in normal β. (Data from Y. W. Kan and A. M. Dozy, 1980.)

the restriction enzyme *Hpa*I, the β-globin-containing fragment can be any one of three sizes (7.0 kb, 7.6 kb, or 13.0 kb) depending on the distance between the β-globin gene and the first *Hpa*I site downstream. The numbers in Figure 14.5 are the relative frequencies of the DNA fragments found for the normal β gene [part (*a*)] and the sickle cell gene [β^S, part (b)] among American blacks. Relative to the normal gene, the vast majority are found in the 7.0-kb or 7.6-kb size classes; for β^S the situation is reversed: Most β^S genes are found in the 13.0-kb fragment. Polymorphisms like the one in Figure 14.5(*b*), which involve the positions of restriction sites, are known as **restriction-site polymorphisms**. The *Hpa*I polymorphism near the β-globin gene is especially useful, as it permits prenatal diagnosis of sickle cell anemia. For example, in a mating between genotypes β(7.6)/β^S(13.0) and β(7.0)/β^S(13.0), any offspring who has only the 13.0-kb fragment would almost certainly be β^S/β^S because there is virtually no recombination between the β-globin gene and the *Hpa*I sites. Variation involving restriction sites has also been found in mitochondrial DNA; such variation is becoming increasingly important in studies in population genetics and evolution.

Allele Frequencies and Genotype Frequencies

Polymorphisms involving allozymes, restriction sites, and blood groups establish that there is substantial genetic variation in the human population, although we do not yet know the level of variation in the genome as a whole. This section concerns the relationship between the frequency of individual alleles at a locus and the frequency of genotypes at the locus. Genetic variation in populations is, of course, organized into genotypes. **Genotype frequen-** cies (the relative proportions of the various genotypes) are determined partly by the frequencies of various kinds of matings, by the rate of mutation, by natural selection through the possibility that some genotypes may survive longer or be more fertile than others, by migration of individuals into the population, and by other factors as well. Predicting genotype frequencies might therefore seem to be virtually impossible, but it is not. In spite of these complexities, there is a remarkably simple rule, called the **Hardy-Weinberg rule**, that can often be used to calculate the genotype frequencies at a locus. The rule works when the population is large and when the effects of mutation, selection, and migration are small. It is also required that there be **random mating**, which means that mates be chosen without regard to the locus in question. In the cases of many loci in humans and other organisms, these requirements are satisfactorily met; the Hardy-Weinberg rule for finding genotype frequencies is therefore of great practical utility.

An illustration may be taken from the MN blood group system discussed in Chapter 12. In one study of 425 individuals, there were 137 *MM* genotypes, 207 *MN*, and 81 *NN* (Table 14.3). We wish to determine the relationship between these genotype frequencies and the frequencies of the individual *M* and *N* alleles. The **allele frequency** of any prescribed allele among a group of individuals is just the proportion of all alleles at the locus in question that are of the prescribed type. To be specific, the group of individuals in Table 14.3 represents a total of 850 alleles at the *MN* locus because each of the 425 individuals carries two alleles, and 425 × 2 = 850. The *MM* homozygotes represent 137 × 2 = 274 *M* alleles, the *MN* heterozygotes represent 207 *M* and 207 *N* alleles, and the *NN* homozygotes represent 81 × 2 = 162 *N* alleles (see Table 14.3). Altogether there are 274 + 207 = 481 *M* alleles and 207

TABLE 14.3 EXAMPLE OF ALLELE FREQUENCY CALCULATIONS FOR THE *MN* LOCUS

Type of Observation	Sample data			
Genotype	*MM*	*MN*	*NN*	
Number of individuals	137	207	81	(total = 425)
Number of alleles at *MN* locus	274 *M*	207 *M* + 207 *N*	162 *N*	(total = 850)
Total *M* (or *N*) alleles	481 *M*		369 *N*	

Calculation of gene frequency

Frequency of *M* allele = 481/850 = 0.5659

Frequency of *N* allele = 369/850 = 0.4341

+ 162 = 369 *N* alleles among a total of 850. Thus,

$$\text{Allele frequency of } M = \frac{481}{850} = 0.5659$$

$$\text{Allele frequency of } N = \frac{369}{850} = 0.4341$$

It may be useful to express these calculations in words:

$$\left(\begin{array}{c}\text{Allele frequency}\\\text{of } M\end{array}\right) \text{ equals}$$

$$\frac{2\text{ times}\left(\begin{array}{c}\text{number of}\\MM\text{ homozygotes}\end{array}\right)\text{plus}\left(\begin{array}{c}\text{number of}\\MN\text{ heterozygotes}\end{array}\right)}{2\text{ times (total number of individuals)}}$$

$$\left(\begin{array}{c}\text{Allele frequency}\\\text{of } N\end{array}\right) \text{ equals}$$

$$\frac{2\text{ times}\left(\begin{array}{c}\text{number of}\\NN\text{ homozygotes}\end{array}\right)\text{plus}\left(\begin{array}{c}\text{number of}\\MN\text{ heterozygotes}\end{array}\right)}{2\text{ times (total number of individuals)}}$$

It will be convenient to represent the allele frequency of *M* (0.5659) by the symbol p, as this will allow us to establish the Hardy-Weinberg rule without regard to the actual numerical value of the allele frequency. In general, p can represent any number between 0 and 1. (It cannot be less than 0 or greater than 1 because the proportion of an allele in any sample must be between 0 and 100 percent.) Although p could stand for any number between 0 and 1 in other examples, in the case at hand p is 0.5659. Note that the frequency of the *N* allele, 0.4341, equals $1 - 0.5659$, which is to say $1 - p$. It will be convenient to represent the allele frequency of *N* (0.4341) by the symbol q. Thus, $p + q = 1$.

Recall now that we have made certain assumptions: large population size; negligible effects of mutation, migration, and selection; and random mating. It is important to realize that random mating with respect to one trait—for example, the MN blood groups—does not mean indiscriminate mating with regard to other traits. Mating can be random with respect to some traits and at the same time nonrandom with respect to others. Mating in humans is random or nearly random with respect to such traits as blood groups and allozymes, but among people in the United States who have European or African ancestry, mating tends to be nonrandom with respect to such traits as height, IQ score, and skin color. In these latter cases individuals tend to mate with others more similar to themselves than would be expected by chance; tall individuals tend to choose tall mates, and so on. Such a tendency toward like-with-like mating is called **assortative mating**.

When the allele frequency of M is p, then a fraction p of all eggs and sperm will carry the M allele. Similarly, since the frequency of N is q, a fraction q of all eggs and sperm will carry the N allele. With random mating, eggs and sperm are joined at random to produce the genotypes, so the genotype frequencies can be obtained by cross multiplication in the sort of Punnett square illustrated in Figure 14.6. The expected frequency of MM genotypes (unshaded square) is p^2; that of MN genotypes (shaded squares) is $pq + pq$, which equals $2pq$; and that of NN genotypes is q^2. Substituting 0.5659 for p and 0.4341 for q leads to the following expected genotype frequencies:

$$MM: \quad p^2 \quad = (0.5659) \times (0.5659) = 0.3202$$
$$MN: \quad 2pq = 2 \times (0.5659) \times (0.4341) = 0.4913$$
$$NN: \quad q^2 \quad = (0.4341) \times (0.4341) = 0.1884$$

To obtain the expected number of individuals of each genotype among a total of 425 (i.e., the number of Table 14.3), each of the genotype frequencies must be multiplied by 425. This leads to the following comparison between the observed (obs) and expected (exp) numbers.

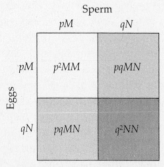

Sperm

Figure 14.6 Punnett square showing genotype frequencies expected with random mating at the MN locus. Allele frequencies of M and N are denoted p and q, respectively.

$$MM: \quad \text{obs } 137; \text{ exp } (0.3202) \times 425 = 136$$
$$MN: \quad \text{obs } 207; \text{ exp } (0.4913) \times 425 = 209$$
$$NN: \quad \text{obs } 81; \text{ exp } (0.1884) \times 425 = 80$$

So the fit between the observed number of each genotype and the number expected on the basis of Figure 14.6 is very good.

Genotype frequencies that follow the rule of p^2, $2pq$, and q^2 are known as **Hardy-Weinberg frequencies**, and any genotype frequencies that bear these relationships to one another are said to follow the Hardy-Weinberg rule.

Implications of the Hardy-Weinberg Rule

Allele Frequencies Do Not Change with Time One important implication of the Hardy-Weinberg rule is that the allele frequencies stay the same from generation to generation. To illustrate this aspect, let us calculate the expected genotype frequencies among a large number of offspring of the individuals in Table 14.3. Since their sperm and eggs join at random because of random mating, the Punnett square in Figure 14.6 is still appropriate. Thus, if a perfectly representative sample of 425 of their offspring were studied, we would find 136 MM, 209 MN, and 80 NN—the same numbers as calculated before. The allele frequencies of M and N among the offspring are thus:

$$\text{Allele freq. of } M = \frac{2 \times 136 + 209}{2 \times 425} = 0.5659$$

$$\text{Allele freq. of } N = \frac{2 \times 80 + 209}{2 \times 425} = 0.4341$$

which are the same as they were *among the parents*. Thus, allele frequencies stay the same generation after generation, provided that the assumptions of large population size and negligible effects of mutation, migration, and selection remain in force.

Case of Complete Recessive Allele
Allele frequencies were easy to calculate in the case of Table 14.3 because M and N are codominant and the heterozygote can be identified. When one allele is completely recessive, only two phenotypes (dominant or recessive) occur, yet the allele frequencies can readily be calculated when the genotypes follow the Hardy-Weinberg rule. If we let q represent the allele frequency of the recessive allele, then the genotype frequency of the homozygous recessive will be q^2. Thus, the square root of the genotype frequency of the homozygous recessive will correspond to the allele frequency of the recessive, because $\sqrt{q^2} = q$. That is to say,

$$\begin{pmatrix}\text{Allele frequency} \\ \text{of recessive}\end{pmatrix} \text{ equals } \sqrt{\begin{array}{c}\text{Genotype frequency of} \\ \text{recessive homozygote}\end{array}}$$

To take a specific example, consider the Rh blood groups among the Basques. In one study, a proportion 0.575 (i.e., 57.5 percent) was found to be Rh$^+$ (genotypes DD or Dd) and a proportion 0.425 (42.5 percent) was found to be Rh$^-$ (genotype dd). Letting q represent the allele frequency of d, and using the square root formula above, we have

$$q = \sqrt{0.425} = 0.652$$

Knowing q, we also know p because $p = 1 - q = 1 - 0.652 = 0.348$. The expected genotype frequencies are therefore

$DD:$ $p^2 = (0.348)^2 = 0.121$ $\Big\}$ 0.575
$Dd:$ $2pq = 2(0.348)(0.652) = 0.454$
$dd:$ $q^2 = (0.652)^2 = 0.425$

Note that $0.121 + 0.454 = 0.575$, which corresponds to the observed proportion of Rh$^+$ individuals.

Frequency of Carriers (Heterozygotes) of Rare Harmful Recessives In Chapter 4 we emphasized that rare recessive alleles occur much more frequently in heterozygotes than in recessive homozygotes. This is one implication of the Hardy-Weinberg rule. If q represents the allele frequency of a recessive, there will be $2pq$ heterozygotes and q^2 recessive homozygotes. If q is small (say, 0.1 or less), then $2pq$ will be substantially larger than q^2 (for $q = 0.1$, $2pq = 0.18$ but $q^2 = 0.01$). The relationship between the frequency of heterozygotes and that of homozygous recessives was illustrated in Figure 4.3; in that graph, the vertical axis corresponds to $2pq$ and the horizontal axis to q^2. In the case of phenylketonuria, for example, the incidence of homozygous recessives is about 0.0001 (i.e., 1 per 10,000). Thus, $q = \sqrt{0.0001} = 0.01$, so $p = 1 - q = 1 - 0.01 = 0.99$. The genotype frequencies are thus

Homozygous
normal: $p^2 = (0.99)^2 = 0.9801$

Heterozygous: $2pq = 2(0.99)(0.01) = 0.0198$

Homozygous
recessive: $q^2 = (0.01)^2 = 0.0001$

which means that there are 198 carriers for each affected individual.

Allele Frequency Definition of Polymorphism Earlier we defined a polymorphic locus as one at which the most common homozygote had a frequncy of less than 0.90. If the allele frequency of the most common allele is represented as p, then the most common homozygote will have a frequency of p^2. Thus, a polymorphic locus is one at which p^2 is less than 0.90, or p is less than $\sqrt{0.90} = 0.95$. Therefore, if mating is random, we could equivalently define a polymorphic locus as one at which the most common allele has a frequency less than 0.95. If mating is nonrandom, the definitions are not equivalent, and the allele frequency definition is the one usually used in practice.

Multiple Alleles   The Hardy-Weinberg rule is easily extended to loci that have more than two alleles. With multiple alleles, the expected frequency of any *homozygote* is the square of the corresponding allele frequency, and the expected frequency of any *heterozygote* is equal to twice the product of the corresponding allele frequencies. To be concrete, consider the Basques again, this time with respect to the ABO blood groups. The estimated frequency of allele I^A is 0.2661, that of I^O is 0.6928, and that of I^B is 0.0411. For the sake of remembering which allele frequency symbol goes with which allele, it will be useful to represent these allele frequencies as *a*, *o*, and *b*, respectively. Thus, $a = 0.2661$, $o = 0.6928$, and $b = 0.0411$. With random mating, the expected genotype frequencies will be:

$I^A I^A$: $a^2 = (0.2661)^2$ $= 0.071$ ⎱
$I^A I^O$: $2ao = 2(0.2661)(0.6928) = 0.369$ ⎰ 0.440 (A)

$I^B I^B$: $b^2 = (0.0411)^2$ $= 0.002$ ⎱
$I^B I^O$: $2bo = 2(0.0411)(0.6928) = 0.057$ ⎰ 0.059 (B)

$I^A I^B$: $2ab = 2(0.2661)(0.0411) = 0.022$ 0.022 (AB)
$I^O I^O$: $o^2 = (0.6928)^2$ $= 0.479$ 0.479 (O)

(The numbers at the far right are the expected frequencies of the A, B, AB, and O *phenotypes*.)

X-linked Loci   The Hardy-Weinberg rule holds for random mating involving **X-linked loci**, provided that we consider only females and that the allele frequencies are the same in both sexes. The situation is outlined in the Punnett square in Figure 14.7. Here we consider two X-linked alleles, designated *C* and *c*, with respective allele frequencies *p* and *q*. When mating is random, X-bearing sperm join randomly with X-bearing eggs, so, *among females*, the expected genotype frequencies will be:

$$CC:\ \ p^2$$
$$Cc:\ \ 2pq$$
$$cc:\ \ q^2$$

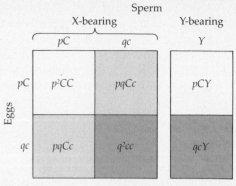

Figure 14.7   Punnett square showing genotype frequencies expected with random mating at an X-linked locus. Note that the genotype frequencies in hemizygous males correspond to the allele frequencies.

Males are hemizygous for X-linked genes because a male must receive his father's Y chromosome. Hence, *among males,* the genotype frequencies from Figure 14.7 are

$$C:\ p$$
$$c:\ q$$

Figure 14.7 can be applied to the human Xg blood groups, which are due to a pair of alleles (designated Xg^a and *Xg)* on the X chromosome. Among Britishers, the allele frequency of Xg^a (*p*) is 0.675 and that of *Xg* (*q*) is 0.325. Expected genotype frequencies are therefore

Among *females*:
Xg^a/Xg^a: $p^2 = (0.675)^2 = 0.456$
Xg^a/Xg: $2pq = 2(0.675)(0.325) = 0.439$
Xg/Xg: $q^2 = (0.325)^2 = 0.105$

Among *males*:
Xg^a: $p = 0.675$
Xg: $q = 0.325$

Rare X-linked Recessives   In Chapter 5 we emphasized that traits due to rare X-linked recessives occur much more frequently among males than among females, and this principle was illustrated graphically in Figure 5.5. The principle follows immediately from

Figure 14.7. If the allele frequency of a rare X-linked recessive is q, then the frequency of affected females will be q^2 whereas that of affected males will be q. When q is small (say, 0.1 or less), q^2 will be very much smaller than q. To consider a specific instance, recall that the frequency of the green form of X-linked color blindness among Western European males is about 0.05 (i.e., 5 percent, or 1 in 20). Thus, in this instance, $q = 0.05$, and the expected frequency of color-blind females is $q^2 = (0.05)^2 = 0.0025$ (i.e., 1 in 400). Comparisons for other values of q were illustrated in Figure 5.5, where the vertical axis corresponds to q and the horizontal axis to q^2.

Multiple Loci and Linkage Disequilibrium

The Hardy-Weinberg rule corresponds to a situation in which the alleles at a single locus are associated at random, and, with random mating, the Hardy-Weinberg proportions are attained in one or relatively few generations. Considerations involving two or more linked loci are somewhat more complex; for the sake of concreteness, we will consider two loci. Suppose one locus has alleles A and a at respective allele frequencies p_1 and q_1; and a linked locus has alleles B and b at respective allele frequencies p_2 and q_2. From the foregoing discussion of the Hardy-Weinberg rule, we know that the population will quickly attain a state in which the genotype frequencies are as follows:

$$\begin{array}{ll} AA: & p_1^2 \\ Aa: & 2p_1q_1 \\ aa: & q_1^2 \end{array}$$

and

$$\begin{array}{ll} BB: & p_2^2 \\ Bb: & 2p_2q_2 \\ bb: & q_2^2 \end{array}$$

That is to say, the A and a alleles are in random association with each other, and the B and b alleles are in random association with each other. However—and this may at first seem paradoxical—the alleles at the A locus may continue to be in nonrandom association with alleles at the B locus. To be precise about this point, we must first examine the situation in which the alleles at the loci *are* in random association. For this purpose it is convenient to consider the frequencies of the four possible types of *gametes*. With random association of alleles at the loci, the gametic frequencies will be

$$\begin{array}{ll} AB: & p_1p_2 \\ Ab: & p_1q_2 \\ aB: & q_1p_2 \\ ab: & q_1q_2 \end{array}$$

In any population that has the above gametic frequencies, the A and B loci are said to be in **linkage equilibrium**.

In any actual population the gametic frequencies may depart from the ideal random association frequencies. Whatever the actual gametic frequencies may be, we can always write them in the form

$$\begin{array}{ll} AB: & p_1p_2 + D \\ Ab: & p_1q_2 - D \\ aB: & q_1p_2 - D \\ ab: & q_1q_2 + D \end{array}$$

where D is a number that measures the degree of departure from linkage equilibrium. The state $D = 0$ represents linkage equilibrium, of course, and, in any population in which D is not 0, the A and B loci are said to be in **linkage disequilibrium**.

At this point an example is in order; we will use data involving the locus of the MN blood group and a closely linked locus of another blood group having alleles S and s. In an extensive study of an English population, the gametic frequencies were found to be:

$$MS: \quad 0.247$$
$$Ms: \quad 0.283$$
$$NS: \quad 0.080$$
$$Ns: \quad 0.390$$

The *allele* frequencies are thus:

p_1 = allele freq. of M = 0.247 + 0.283 = 0.530

q_1 = allele freq. of N = 0.080 + 0.390 = 0.470

p_2 = allele freq. of S = 0.247 + 0.080 = 0.327

q_2 = allele freq. of s = 0.283 + 0.390 = 0.673

Then, writing the frequency of the MS gamete in the form involving D (any of the gametes could be used for the calculation),

$$MS: \quad 0.247 = p_1 p_2 + D$$
$$= (0.530)(0.327) + D$$

so $\quad D = 0.247 - (0.530)(0.327) = 0.074$

A linkage disequilibrium of $D = 0.074$ may at first seem rather small, but, for the above allele frequencies, $D = 0.074$ is about 50 percent of its theoretical maximum.

To summarize the principal point: *Each of two linked loci can be in Hardy-Weinberg proportions, yet the loci can be in linkage disequilibrium.* However, D tends to go to 0 of its own accord. Indeed, with random mating in the absence of selection, migration, and so on, the value of D after n generations (call it D_n) can be shown to be

$$D_n = (1 - r)^n D_0$$

were D_0 is the value of D in the initial generation and r is the recombination fraction between the loci in question. For r less than about 0.20, the time required to reduce the value of D by $\frac{1}{2}$ (i.e., the half-time) is approximately $0.7/r$ generations.

The possibility of linkage disequilibrium has implications for the interpretation of associations between particular disorders and known genotypes, such as those discussed in Chapter 13 involving *HLA* haplotypes or ABO blood groups. When a disease association is discovered, then it cannot be determined without further information whether the association is a direct result of the locus in question or whether it is due to another locus in linkage disequilibrium with the locus in question. On the other hand, linkage disequilibrium can sometimes be helpful; the disequilibrium between the hemoglobin β^S allele and the downstream *Hpa*I restriction-site polymorphism is a preeminent example (see Figure 14.5).

Differentiation of Populations: Race

Earlier we defined a *local population* as a group of individuals living within a sufficiently restricted geographic area that matings usually occur within the group. Within-group mating preferences need not be based on geography, however. In humans, cultural similarity is important, and individuals tend to mate with others of the same or a similar cultural background. Although there is some intermating and gene exchange between all large human groups, the entire human population cannot be considered as a single, large, randomly mating entity. The exchange of genes between populations is limited, which is known as **genetic isolation**. Genetic isolation between most human populations is by no means complete, but partial isolation does occur because of the tendency for within-group matings.

Over the course of hundreds of thousands of years, and especially in the distant past when populations were small and migration was more difficult than it is at present, isolated or semi-isolated human populations came to have different allele frequencies. Some differences in allele frequency probably arose by chance, as a result of historical accident. Allele frequencies can change by chance from generation to generation, and small populations are especially susceptible to such effects. Other differences in allele frequency probably occurred as each population became adapted to its own physical environment; genotypes better adapted to a particular environment survive and reproduce more successfully, and their alleles come to be represented in greater frequency in future generations. Although all humans share essentially the same chromosomal complement, the frequency of individual alleles at many loci varies from population to population.

By way of analogy, the human gene pool is composed of a number of smaller pools—puddles—like the water puddles on a gravel street after a heavy rain, all interconnected in tiny rivulets through which material flows from one to the next. If one examines the puddles closely, one sees that they are not identical. Some are large, some small; some have more rotifers, nematodes, bacteria, or algae than others. But the differences are quantitative, not qualitative. The puddles are distinguished by how much or how little they have of each component and not by whether or not they possess the constituent. No puddle is completely isolated, so whatever one puddle has in abundance travels through the rivulets to the other ones. Because of this structure, the puddles form one large, interconnected unit, but a unit with local differences. The puddles in the analogy represent **races**—groups whose allele frequencies differ from those of other groups.

How many human races are to be distinguished is a matter of where one draws the lines. The purpose of distinguishing human races genetically is to trace the origin and history of the human population and to anticipate its evolutionary future. To draw the line at differences in allele frequency that are very small is simply not useful; the people in virtually every city or town of every country would then represent a distinct race, because allele frequencies do vary even within a single country. In practice, an arbitrary judgment must be made about how different two sets of allele

frequencies must be to consider them sufficiently distinct to define separate races. Where the line is drawn depends on the particular application, on whether large groups are to be considered or small ones. Anthropologists, using various physical and cultural criteria, have defined the number of races as anywhere between three—usually called Caucasoid, Negroid, and Mongoloid—and more than 30. A few anthropologists deny that the concept of race has any relevance at all, quite apart from the danger of misuse of the concept (such as in the nonsensical idea of "racial purity"). In any case, there is no "correct" number of human races. The number depends on how finely one makes distinctions and on the purpose, which makes the concept of "race" an arbitrary but sometimes useful tool—nothing more.

Genetic distinctions between human populations must be based on allele frequencies at many loci simultaneously. Some loci have allele frequencies that are virtually the same in all populations; others have allele frequencies that are somewhat different in all populations. Genetic distinctions must be based on the whole constellation of allele frequencies; no one locus or small group of loci gives an adequate picture. (This is why skin color is not an acceptable trait for distinguishing human populations; it puts too much emphasis on the one small group of loci that determines skin pigmentation and ignores everything else; it also gives a false impression of the amount of genetic differentiation between populations.)

The concept of race as a constellation of allele frequencies is illustrated with several blood group loci in three large human groups in Figure 14.8. Allele frequencies at some loci (e.g., MN, Lutheran) do not vary a great deal among populations, whereas at other loci (e.g., P, Rh, Lewis) the variation is substantial. A convenient quantitative measure of genetic differentiation among populations is the **fixation index**, which measures the proportionate re-

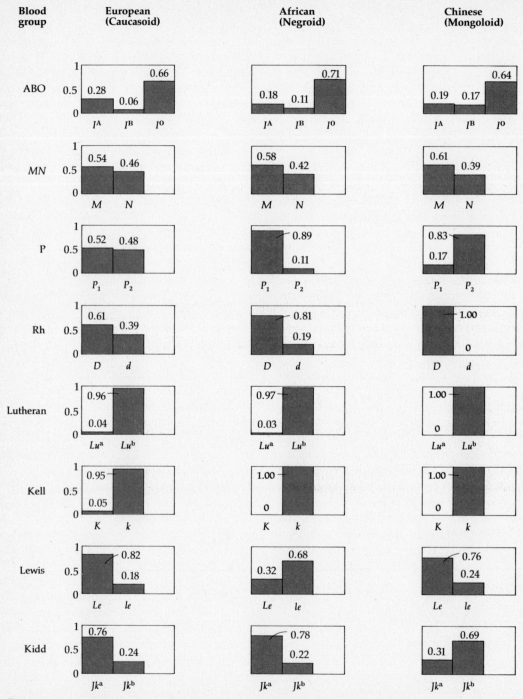

Figure 14.8   Allele frequencies at eight blood group loci in three large human populations.

duction in average heterozygosity as compared with what the heterozygosity would be if the populations were a single random-mating unit. To fix ideas, we will calculate the fixation index for the *MN* locus between Africans and Chinese. First we must calculate the heterozygosity (i.e., proportion of heterozygotes) within each population. Letting *p* represent the allele frequency of *M* and *q* that of *N*, we have

For Africans: $\quad p = 0.58$
$\quad\quad\quad\quad\quad q = 0.42$
$\quad\quad\quad\quad 2pq = 2(0.58)\,(0.42) = 0.4872$

For Chinese: $\quad p = 0.61$
$\quad\quad\quad\quad\quad q = 0.39$
$\quad\quad\quad\quad 2pq = 2(0.61)\,(0.39) = 0.4758$

The **average heterozygosity among subpopulations**, denoted H_S, is the average of these individual $2pq$ values, so

$$H_S = \frac{0.4872 + 0.4758}{2} = 0.4815$$

If the populations were to fuse into a single random-mating unit, the allele frequencies would be the average of those among the subpopulations. Representing the average allele frequency of *M* as $\bar{p}$ and that of *N* as $\bar{q}$, we have

$$\bar{p} = \frac{0.58 + 0.61}{2} = 0.595$$

$$\bar{q} = \frac{0.42 + 0.39}{2} = 0.405$$

$$2\bar{p}\bar{q} = 2(0.595)\,(0.405) = 0.4820$$

This latter heterozygosity ($2\bar{p}\bar{q}$) is called the **total heterozygosity** and designated H_T.

The fixation index measures the proportionate reduction in heterozygosity due to population subdivision; although usually represented as F_{ST}, we will denote the fixation index as simply F to avoid the subscripts. In symbols,

$$F = \frac{H_T - H_S}{H_T}$$

In the MN case at hand,

$$F = \frac{0.4820 - 0.4815}{0.4820} = 0.001$$

and, in words,

$$\binom{\text{Fixation}}{\text{index}} \text{ equals}$$

$$\frac{\begin{pmatrix}\text{heterozygosity}\\\text{calculated from}\\\text{average allele}\\\text{frequencies}\end{pmatrix} \text{minus} \begin{pmatrix}\text{average of}\\\text{actual}\\\text{hetero-}\\\text{zygosities}\end{pmatrix}}{\begin{pmatrix}\text{heterozygosity}\\\text{calculated from}\\\text{average allele}\\\text{frequencies}\end{pmatrix}}$$

In the MN case, the value of F near 0 reflects the fact that the allele frequencies in both subpopulations are very similar. With two alleles, F must always be between 0 and 1. A value of 0 indicates identical allele frequencies in the subpopulations; a value of 1 means that the populations are **fixed** (i.e., completely homozygous) for alternative alleles.

Since the allele frequencies for the P blood groups are very different in Africans and Chinese, we would expect the value of F to be much greater than in the MN case. Referring to the allele frequencies in Figure 14.8, we have

$$H_S = \frac{2(0.89)\,(0.11) + 2(0.17)\,(0.83)}{2} = 0.2390$$

$$\bar{p} = \frac{0.89 + 0.17}{2} = 0.53$$

$$\bar{q} = \frac{0.11 + 0.83}{2} = 0.47$$

$$H_T = 2\bar{p}\bar{q} = 2(0.53)\,(0.47) = 0.4982$$

$$F = \frac{H_T - H_S}{H_T} = \frac{0.4982 - 0.2390}{0.4982} = 0.520$$

The large value of F in this case (0.52) indicates very great differentiation at this locus.

The advantage of a quantity like the

fixation index is that it can be calculated for each locus and then averaged. The overall average F then summarizes the amount of genetic differentiation at many loci in terms of a single number. A surprising and important finding regarding genetic differentiation among human races is shown in Table 14.4. The upper row pertains to the three major races (Caucasoid, Negroid, and Mongoloid), and it indicates a fixation index $F = 0.069$. This number means that, of the total genetic diversity (heterozygosity) in these three races, only 6.9 percent can be attributed to genetic differentiation *between* the races; the vast majority (93.1 percent) represents diversity found *within* each race. The second row in Table 14.4 pertains to the native Yanomama Indians of Venezuela and Brazil—an exceptionally well-studied group from the standpoint of both anthropology and genetics. Here again the picture regarding differentiation is similar: Of the total genetic diversity in the Yanomama, only 7.7 percent can be ascribed to differences *between* villages; most of the diversity (92.3 percent) occurs *within* villages.

The modern concept of race is based on how the human gene pool is structured. This is quite unlike the traditional view, which was based on a few easily identifiable traits such as skin color, facial features, and body build, and which used these traits to define certain human "types"—a Caucasoid type, a Negroid type, a Mongoloid type, and so on. The traditional concept of race does not correspond to anything real in terms of genetic characteristics. There is no such thing as a "perfect specimen" of a Caucasian, a Negro, or anything else. There is tremendous genetic diversity *within* races, and to define arbitrarily a perfect "type" to which everyone in a group should conform is to imagine a genetic uniformity that does not exist. Because the traditional concept of race was mixed up with the fanciful notion of a "pure race"—one with little or no genetic variation—the concept led to fearful imaginings of the biological consequences of interracial mating, or "mongrelization" of the races. The modern viewpoint emphasizes similarities between races as much as it recognizes the differences in allele frequencies that occur. The modern view also emphasizes that genetic variation *within* races is so enormous that it all but swamps genetic differences *between* races.

Inbreeding

Inbreeding refers to matings between relatives. In Chapter 4 we used the term **consanguineous matings** to refer to matings between relatives, so the two terms—*consanguineous mating* and *inbreeding*—can be used interchangeably. The most common form of inbreeding in most human populations is mating between first cousins, designated by the letters G and H in Figure 14.9. Individual I in the figure, the offspring of the consanguineous mating between G and H, is said to be **inbred**, and a characteristic feature of all forms of inbreeding is the closed loop that occurs in the pedigree. At the top of Figure 14.9 are two individuals (A and B) who are called **common ancestors** of G and H. One of the principal consequences of inbreeding is that a particular allele in one of the common ancestors can be transmitted to both of the consanguineous partners, and their mating can thus lead to this allele becoming homozygous in the inbred offspring.

TABLE 14.4 FIXATION INDEX, *F*, AMONG VARIOUS SUBPOPULATIONS

	Number of populations	Number of of loci	F
Major races	3	35	0.069
Yanomama Indians	37	15	0.077

Source: Data from M. Nei, 1975, Molecular Population Genetics and Evolution, American Elsevier, New York.

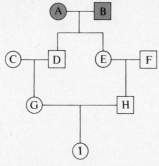

Figure 14.9 Pedigree showing a mating between first cousins (G and H). The individual I is the result of inbreeding, as A and B (shaded) are common ancestors of G and H.

Because of the increased chance of particular alleles becoming homozygous in inbred offspring, inbreeding increases the proportion of homozygous offspring at the expense of heterozygous offspring. Indeed, the amount of inbreeding can conveniently be measured in terms of the reduction in heterozygosity that occurs. To be specific, let us compare the frequency of heterozygotes in a group of inbred individuals with the frequency of heterozygotes in a group of noninbred individuals from the same local population (subpopulation). If we represent the heterozygosity among inbred individuals as H_I and the heterozygosity of noninbred individuals as H_S, then the effects of inbreeding can be measured in terms of the **inbreeding coefficient** (f) as

$$f = \frac{H_S - H_I}{H_S}$$

That is to say, the inbreeding coefficient is the proportionate reduction in heterozygosity that occurs as a result of the inbreeding. (Note: In population genetics the inbreeding coefficient is often symbolized as F_{IS} or simply F; we use f to avoid subscripts and to avoid confusion with the fixation index discussed earlier.)

This heterozygosity definition of f will now permit the genotype frequencies with inbreeding to be expressed in terms of the inbreeding coefficient. After these genotype frequencies are worked out, we can consider some practical consequences of inbreeding. In a random-mating subpopulation, the heterozygosity H_S equals $2pq$ (p and q being, of course, the allele frequencies). Since $f = (H_S - H_I)/H_S$, we have

$$f = \frac{H_S - H_I}{H_S}$$
$$H_S f = H_S - H_I$$
$$H_I = H_S - H_S f$$
$$H_I = H_S(1 - f)$$
$$H_I = 2pq(1 - f)$$

Hence, $2pq(1 - f)$ is the frequency of heterozygotes among inbred individuals.

We will not derive the frequencies of homozygotes here, but they follow from the fact that inbreeding itself does not change the allele frequencies. Thus, whatever the value of f, the genotype frequency of any homozygote plus one-half the genotype frequency of heterozygotes must equal the corresponding allele frequency. Using this principle, the homozygote frequencies can be shown to be

$$AA: \quad p^2(1 - f) + pf$$
$$aa: \quad q^2(1 - f) + qf$$

with, of course,

$$Aa: \quad 2pq(1 - f)$$

These genotype frequencies are summarized in Table 14.5 along with the special case of $f = 0$,

TABLE 14.5 GENOTYPE FREQUENCIES IN INBRED POPULATIONS

Genotype	Frequency in inbred population	Frequency with random mating ($f = 0$)
AA	$p^2(1 - f) + pf$	p^2
Aa	$2pq(1 - f)$	$2pq$
aa	$q^2(1 - f) + qf$	q^2

which corresponds to random mating. Complete inbreeding means $f = 1$, but inbreeding in humans almost never reaches such high values. The average inbreeding coefficient in human populations is often in the range $f = 0.0005$ (in France) to $f = 0.003$ (in Japan).

Consequences of Inbreeding

Values of the inbreeding coefficient of an inbred child from various degrees of consanguin-

eous mating are shown in Figure 14.10. For first-cousin matings, $f = \frac{1}{16}$; for second-cousin matings, $f = \frac{1}{64}$; and so on.

As emphasized in Chapter 4, a principal pedigree characteristic of traits due to rare autosomal recessive alleles is the frequent occurrence of consanguinity among the parents of the affected individual. We can now combine information from Figure 14.10 with the formulas in Table 14.5 to show why inbreeding is so important when dealing with rare reces-

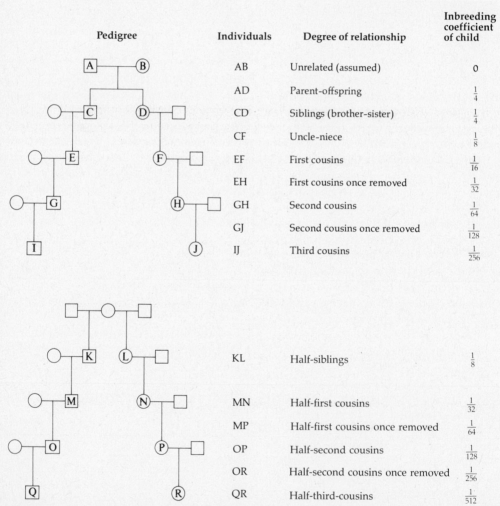

Pedigree	Individuals	Degree of relationship	Inbreeding coefficient of child
	AB	Unrelated (assumed)	0
	AD	Parent-offspring	$\frac{1}{4}$
	CD	Siblings (brother-sister)	$\frac{1}{4}$
	CF	Uncle-niece	$\frac{1}{8}$
	EF	First cousins	$\frac{1}{16}$
	EH	First cousins once removed	$\frac{1}{32}$
	GH	Second cousins	$\frac{1}{64}$
	GJ	Second cousins once removed	$\frac{1}{128}$
	IJ	Third cousins	$\frac{1}{256}$
	KL	Half-siblings	$\frac{1}{8}$
	MN	Half-first cousins	$\frac{1}{32}$
	MP	Half-first cousins once removed	$\frac{1}{64}$
	OP	Half-second cousins	$\frac{1}{128}$
	OR	Half-second cousins once removed	$\frac{1}{256}$
	QR	Half-third-cousins	$\frac{1}{512}$

Figure 14.10 Inbreeding coefficients of potential offspring of matings between relatives who have the pedigree relationships shown.

sives. Since first-cousin matings are the most important consanguineous matings in most human populations, we will focus on these matings. We will also represent the allele frequency of the recessive as q. Then, from Table 14.5, the frequency of homozygous recessives from first-cousin matings will be

$$aa \text{ (inbred): } q^2(1 - \tfrac{1}{16}) + q(\tfrac{1}{16})$$

On the other hand, the frequency of homozygous recessives among noninbred individuals is

$$aa \text{ (noninbred): } q^2$$

Now, if first-cousin matings constitute a proportion, c, of all matings (the others are assumed to be nonconsanguineous), the overall frequency of homozygous recessives in the population will be the average of the above two frequencies. Thus,

$$aa \text{ (total): } q^2(1 - c) + c[q^2(1 - \tfrac{1}{16}) + q(\tfrac{1}{16})]$$

The first term in this expression corresponds to the nonconsanguineous matings and the second to the consanguineous matings.

In the total population, the frequency of homozygous recessives attributable to first-cousin matings will be

$$aa \text{ (inbred among total): } c[q^2(1 - \tfrac{1}{16}) + q(\tfrac{1}{16})]$$

and, therefore, the overall proportion of homozygous recessive offspring who have consanguineous parents will be

$$\frac{aa \text{ (inbred among total)}}{aa \text{ (total)}}$$

This proportion can be evaluated for various values of q and c by substituting into the expressions for aa (inbred among total) and aa (total) from above.

Figure 14.11 shows the proportions for various values of q and for two values of c: $c = 0.01$ (appropriate for the United States, where

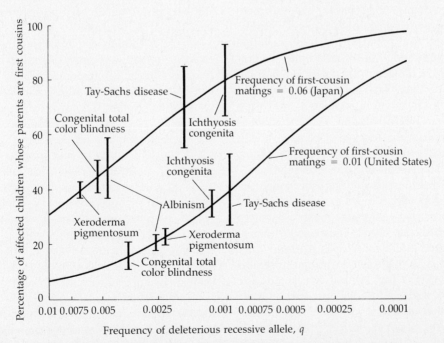

Figure 14.11 Proportion of homozygous recessive individuals who have first-cousin parents in Japan (upper curve) and in the United States (lower curve). The curves themselves are the ones expected theoretically.

approximately 1 percent of all matings are between first cousins) and $c = 0.06$ (appropriate for Japan). Also illustrated are the observed ranges of proportions for various traits due to autosomal recessives. Taking Tay-Sachs disease in the United States as an example, $q = 0.001$ (approximately) and $c = 0.01$; then

aa (inbred among total)

$$= c[q^2(1 - \tfrac{1}{16}) + q(\tfrac{1}{16})]$$
$$= (0.01)[(0.001)^2(\tfrac{15}{16}) + (0.001) (\tfrac{1}{16})]$$
$$= 6.3 \times 10^{-7}$$

and

aa total$) = q^2 (1 - c) + c[q^2(1 - \tfrac{1}{16}) + q(\tfrac{1}{16})]$
$$= (0.001)^2 (1 - 0.01) + 6.3 \times 10^{-7}$$
$$= 1.62 \times 10^{-6}$$

so the proportion is

$$\frac{aa \text{ (inbred among total)}}{aa \text{ (total)}} = \frac{6.3 \times 10^{-7}}{1.6 \times 10^{-6}}$$

$$= 0.39$$

which is the proportion plotted in Figure 14.11 corresponding to $q = 0.001$. These calculations imply that, even though first-cousin matings account for only 1 percent of all matings in the United States, they account for 39 percent of all matings that produce Tay-Sachs offspring. That is to say, first-cousin matings will be found in the pedigrees of 39 percent of all children with Tay-Sachs disease. This effect of consanguinity on the occurrence of rare homozygous recessives is one of the most important consequences of inbreeding; data pertaining to various other traits are included in Figure 14.11.

SUMMARY

1. **Population genetics** deals with the application of genetic principles to interbreeding groups (**populations**) or organisms. Population genetics thus deals with **genetic variation** in populations and with the principles governing changes in the genetic makeup of a population through time. Such cumulative changes in a population's genetic composition are often referred to as **microevolution**. Genetic changes that accompany the formation of new species or **higher categories** (e.g., genus, family, order, class, and so on) constitute **macroevolution**.

2. The oldest fossil organism thought by many anthropologists to be in the main line of human ancestry is among a group of species known collectively as the **australopithecines**. The supposed direct ancestor is *Australopithecus africanus,* which existed during the period 3.2 to 2.2 My. (The symbol My stands for millions of years before the present.) Other australopithecine species include *A. robustus* and *A. boisei,* which are larger, more "robust" offshoots of *A. africanus* that became extinct about 1.0 My. A recent australopithecine find

has been named *A. afarensis*; it existed during the period 3.7 to 3.2 My and may be a direct ancestor of *A. africanus*.

3. Although the interpretation of the earliest human ancestors is much disputed, there is substantial agreement about more recent ancestors. The earliest ancestor in our own genus (***Homo***) is known as *Homo habilis* (2.2 to 1.6 My). *H. habilis* was succeeded by yet another species (*H. erectus*—1.6 to 0.6 My), which eventually gave rise to our own species *H. sapiens* (0.6 My to the present).

4. Human evolution has been marked by progressive changes in, among other things, height, body weight, and cranial capacity. Two broad views regarding the nature of such changes have been advanced. One view, known as **gradual evolution**, proposes that the changes occur gradually over time, with each succeeding generation being slightly different for the trait in question than the generation before. The other view, called **punctuated evolution**, proposes that most of the change actually occurs in a relatively short period of time

(short relative to the duration of a species); with punctuated evolution, a species can change little over long periods and then change rapidly.

5. Whether species primarily undergo gradual or punctuated change, the underlying process by which populations become adapted to their environment is **natural selection**. Natural selection is the mechanism of evolution proposed by Charles Darwin in 1859, and it rests on three tenets: (a) all populations produce more young than can survive and reproduce, (b) individuals in a population vary in their ability to survive and reproduce and some of this variation is inherited, and therefore (c) each succeeding generation will have a disproportionate representation of those alleles that promote survival and reproduction in the environment in question.

6. The occurrence of natural selection thus requires the presence of genetic variation, which is found abundantly at several levels in most natural populations. One level of genetic variation involves chromosomal aberrations, but most genetic variation involves DNA alterations that are not visible through the light microscope. Some of these mutations are responsible for particular disorders, such as cystic fibrosis or Huntington disease, but most are not strongly associated with any phenotypic abnormality. Much genetic variation occurs in the form of alleles that have individually small effects on such multifactorial traits as height or weight, but these alleles can at present be studied only on a statistical basis. The most widely studied form of genetic variation involves alleles associated with easily identified phenotypes, such as those determining various blood groups, those influencing the rate of movement of particular enzymes in response to an electric field (**electrophoretic mobility**), or, more recently, those associated with the presence or absence of a particular restriction site in a fragment of DNA.

7. Proteins that have an altered electro-phoretic mobility as a result of mutation in the corresponding structural gene are said to be **allozymes**. Genetic variation associated with allozymes is widespread in natural populations of virtually all organisms. The amount of allozyme variation in a population is usually summarized in terms of two numbers. One is the **proportion of polymorphic loci**. A **polymorphic locus** is one at which the most common homozygote has a frequency of less than 90 percent, and the proportion of polymorphic loci is thus the proportion of all loci studied that are polymorphic. The other summary number is the **average heterozygosity**, which is simply the average frequency of heterozygous genotypes among all the loci in question. In most human populations (although Europeans have been most extensively studied), the proportion of polymorphic loci is about 0.30 and the average heterozygosity is about 0.07.

8. The frequencies of genotypes in a population are determined in part by the frequencies of the various types of matings that can occur. The simplest situation regarding types of matings is known as **random mating**, in which mating pairs occur randomly with respect to the locus in question. With random mating, and in the absence of such evolutionary forces as mutation, migration, and selection, genotype frequencies are related in a simple manner to the **allele frequencies** at the locus in question. (The allele frequency of a prescribed allele among a group of individuals is equal to the proportion of all alleles at the locus that are of the prescribed type.) For two alleles A and a at respective allele frequencies p and q (where $p + q = 1$), the genotype frequencies with random mating are given by the **Hardy-Weinberg rule**:

$$AA: \quad p^2$$
$$Aa: \quad 2pq$$
$$aa: \quad q^2$$

9. The Hardy-Weinberg rule has important implications: (a) The allele frequencies remain constant from generation to generation, provided that the assumptions of no mutation, selection, and so on, are valid; (b) in the case of recessive alleles, the allele frequency of the recessive is given by the square root of the genotype frequency of the recessive homozygote; (c) when a recessive allele is rare, there will always be many more heterozygotes than recessive homozygotes; (d) with multiple alleles and random mating, the genotype frequency of any homozygote is equal to the square of the corresponding allele frequency, and the genotype frequency of any heterozygote is equal to two times the product of the corresponding allele frequencies; (e) for X-linked loci, the genotype frequencies in females are given by the Hardy-Weinberg rule, whereas the genotype frequencies in hemizygotes males are given by the corresponding allele frequencies.

10. Two linked loci can each be in Hardy-Weinberg proportions, yet the alleles at each locus can be in nonrandom association with the alleles at the other locus. Such nonrandom associations are referred to as **linkage disequilibrium**. With random mating, and in the absence of forces that can change allele frequency, alleles at linked loci do gradually attain a state of random association (i.e., **linkage equilibrium**), and the rate of approach to linkage equilibrium depends on the recombination fraction between the loci. The possibility of linkage disequilibrium introduces ambiguity into the interpretation of associations between particular loci and particular disorders.

11. Populations among which gene exchange is limited are said to have partial **genetic isolation.** Such populations can come to have different allele frequencies as time goes on, and such differences are conveniently measured in terms of the **fixation index** (usually represented as F_{ST} or simply F), which equals the proportionate reduction in heterozygosity among subpopulations as compared with a hypothetical situation in which the subpopulations were to fuse and undergo random mating. That is to say,

$$F = \frac{H_T - H_S}{H_T}$$

where H_T represents the heterozygosity of the hypothetical fused population and H_S represents the average heterozygosity among the actual subpopulations.

12. Populations with differing constellations of allele frequencies constitute **races**. In terms of this allele frequency concept of race, the human species can be said to have anywhere between three races and hundreds of races, depending on how finely one wishes to make distinctions. In terms of measures of genetic differentiation such as F_{ST}, however, there is much more genetic variation *within* races than *between* races.

13. Inbreeding refers to matings between relatives (i.e., consanguineous matings). The consanguineous parents of an inbred individual thus have one or more **common ancestors**, and the principal effect of inbreeding is that the inbred individual can inherit the same allele from one of these common ancestors. The effects of inbreeding are conveniently measured in terms of the **inbreeding coefficient**, symbolized as F_{IS} or just f, which is the reduction in heterozygosity expected in an inbred individual as compared with a noninbred individual. That is to say,

$$f = \frac{H_S - H_I}{H_S}$$

where H_I and H_S represent the probability of heterozygosity at a locus in inbred and noninbred individuals, respectively.

14. When there is inbreeding, the genotype frequencies are no longer given by the Hardy-Weinberg rule. Genotype frequencies among a group of individuals who have inbreeding coefficient f are given by

AA: $p^2(1 - f) + pf$
Aa: $2pq(1 - f)$
aa: $q^2(1 - f) + qf$

where p and q are the allele frequencies of A and a, respectively. From these genotype frequencies, it is clear that inbreeding increases the frequency of homozygous genotypes. In human populations the average inbreeding coefficient is usually relatively small, typically in the range $f = 0.0005$ to $f = 0.003$.

15. Because of the increased homozygosity in inbred individuals, individuals who are homozygous for rare recessive alleles will have consanguineous parents more often than would be expected by chance. For a rare recessive with allele frequency q, the incidence of homozygotes among inbred individuals will be $q^2(1 - f) + qf$, whereas the incidence among noninbred individuals will be q^2. Important values of f in human populations are $f = \frac{1}{16}$ (for first-cousin matings) and $f = \frac{1}{64}$ (for second-cousin matings).

WORDS TO KNOW

Evolution	**Genetic Diversity**	**Random Mating**
Classification	Electrophoresis	Hardy-Weinberg rule
Microevolution	Allozyme	X-linked loci
Macroevolution	Polymorphic locus	Linkage equilibrium
Species	Monomorphic locus	Linkage disequilibrium
Subspecies	Heterozygosity	
Gradual	Restriction-site polymorphism	**Inbreeding**
Punctuated		Consanguineous mating
Natural selection	**Population**	Common ancestor
	Genotype frequencies	Inbreeding coefficient
Human Evolution	Allele frequencies	Genotype frequencies
Australopithecines	Genetic isolation	
Homo habilis	Differentiation	
Homo erectus	Race	
Neandertals	Total heterozygosity	
Cro-Magnon		

PROBLEMS

1. For discussion: What is your opinion of the view that all teaching of evolution in public schools must be accompanied by an equal amount of time teaching the story of creation as set forth in Genesis? Do all Christian religions interpret Genesis literally? Certain religious fundamentalists portray evolutionists as claiming that "humans evolved from apes." Considering what you have read in this chapter, is that statement consistent with current knowledge and understanding? In what ways was *Australopithecus africanus* like an "ape"? Was *Homo erectus* an "ape"? Was *Homo sapiens neanderthalensis*? Evolutionists claim that modern humans and modern worms both evolved from a remote common ancestor. Would it be correct to say: "Evolutionists claim that humans evolved from worms"? Does knowledge that humans share a remote common ancestry with other living organisms diminish the value of human life?

2. What is the process of natural selection and on what underlying premises is it based?

3. How can a single ancestral species be the progenitor of two descendant species?

4. The diagram here shows the electrophoretic pattern of an enzyme in three genotypes (*FF*, *FS*, and *SS*). How can the heterozygote have three enzyme bands? (Hint: Consider what would happen if the active enzyme molecule were composed of more than one polypeptide chain.)

F/F   F/S   S/S

5. The diagram shows the electrophoretic patterns found with an enzyme whose mobility is controlled by two alleles at a single locus. Note that the possible electrophoretic patterns are different in females and males. What hypothesis could you make about the mode of inheritance of this locus to account for these data?

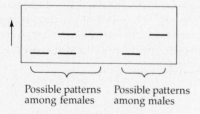

Possible patterns   Possible patterns
among females   among males

6. Among a sample of 1000 Britishers, the number of individuals with each of the MN blood group phenotypes was as follows:

$$M : 298$$
$$MN: 489$$
$$N : 213$$

What is the *M* allele frequency? What is the *N* allele frequency? What number of each of the genotypes would be expected with random mating?

7. The accompanying diagram is of a hypothetical

electrophoresis gel stained for an enzyme whose mobility is controlled by two codominant alleles (*F* and *S*) at a single locus. Which individuals have genotype *FF*? Which have genotype *FS*? Which have genotype *SS*? In this sample of individuals, what is the frequency of the *F* allele? What is the allele frequency of *S*?

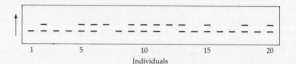

1   5   10   15   20

Individuals

8. Ten allozyme loci were studied in a large number of individuals from a particular population, and the proportion of heterozygous genotypes was tabulated. What is the average heterozygosity among this group of loci?

Locus	Proportion of heterozygotes
1	0.03
2	0
3	0.21
4	0.08
5	0
6	0
7	0.31
8	0.01
9	0.02
10	0

9. Six allozyme loci are studied in a large number of individuals from a particular population. At each locus the allele frequency of the most abundant allele is calculated, with the results shown here. Which loci are polymorphic in this population? Which are monomorphic?

Locus	Frequency of most common allele
1	0.88
2	0.96
3	0.98
4	1.00
5	0.93
6	0.91

10. A certain randomly mating population has a frequency of Rh$^-$ blood types of 16 percent. What is the frequency of the d allele? What is the frequency of the D allele? What are the expected genotype frequencies?

11. In Zurich, Switzerland, the allele frequencies of I^A, I^B, and I^O are 0.27, 0.06, and 0.67, respectively. What are the expected frequencies of blood types A, B, AB, and O?

12. The frequency of the green form of X-linked color blindness among Western European males is about 0.05. What is the frequency of carrier females?

13. As discussed in Chapter 3, the dry type of ear cerumen ("wax") is due to homozygosity for a simple Mendelian recessive. Among American Indians the frequency of dry-cerumen individuals is 66 percent. What is the frequency of the recessive allele? What is the overall frequency of heterozy-gotes? Among individuals with the wet type of cerumen, what is the frequency of heterozy-gotes?

14. Among American blacks the frequency of the dry-cerumen phenotype is 5 percent (see Problem 13). What are the allele frequencies of the dominant and recessive alleles in this population? What is the frequency of heterozygotes?

15. Considering only the locus controlling wet *versus* dry ear cerumen, use the information in Problems 13 and 14 to calculate the fixation index F between American Indians and American blacks.

16. A certain phenotype caused by homozygosity for a simple Mendelian recessive allele has an incidence among the offspring of unrelated individuals of 4×10^{-4} (i.e., 1 per 2500). Calculate the incidence of the condition among the offspring of first-cousin matings.

FURTHER READING AND REFERENCES

Cronin, J. E., N. T. Boaz, C. B. Stringer, and Y. Rak. 1981. Tempo and mode in hominid evolution. Nature 292:113–122. Source of data for Figures 14.1 and 14.3.

Crow, J. F., and M. Kimura. 1970. An Introduction to Population Genetics Theory. Harper and Row, New York. Famous textbook on mathematical aspects of population genetics.

Denaro, M., H. Blanc, M. J. Johnson, K. H. Chen, E. Wilmsen, and L. L. Cavalli-Sforza. 1981. Ethnic variation in *Hpa*I endonuclease cleavage patterns of human mitochondrial DNA. Proc. Natl. Acad. Sci. U.S.A. 78:5768–5772. More detail on mitochondrial DNA polymorphisms.

Harris, H., D. A. Hopkinson, and E. B. Robson. 1974. The incidence of rare alleles determining electrophoretic variants: Data on 43 enzyme loci in man. Ann. Hum. Genet. Lond. 37:237–253. Data for Table 14.2.

Hartl, D. L. 1980. Principles of Population Genetics. Sinauer Associates, Sunderland, Mass. A nonmathematical introduction to population genetics. Source of Figure 14.11.

Hartl, D. L. 1981. A Primer of Population Genetics. Sinauer Associates, Sunderland, Mass. A brief introduction to various aspects of the field for those wishing an overview.

Huether, C. A., and E. A. Murphy. 1980. Reduction of bias in estimating the frequency of recessive genes. Am. J. Hum. Genet. 32:212–222. Estimating the frequency of recessive alleles as the square root of the frequency of homozygous recessives can lead to errors when the sample is small; a somewhat better though more laborious procedure is suggested here.

Jacquard, A. 1978. Genetics of Human Populations, trans. D. M. Yermanos. Freeman, Cooper and Co., San Francisco. An excellent elementary introduction to theoretical population genetics as applied to humans.

Kan, Y. W., and A. M. Dozy. 1978. Polymorphism of DNA sequence adjacent to human β-globin structural gene: Relationship to sickle mutation. Proc. Natl. Acad. Sci. U.S.A. 75:5631–5635. An important report demonstrating the usefulness of recombinant DNA procedure in diagnosis.

Kan, Y. W., and A. M. Dozy. 1980. Evolution of hemoglobin S and C in world populations. Science 209:388–390. Tracking globin evolution by means of restriction-site polymorphisms.

Lewontin, R. C. 1974. The Genetic Basis of Evolutionary Change. Columbia University Press, New York. An easy-to-read discussion of the problems in studying the importance of genetic polymorphisms.

Lovejoy, C. O. 1981. The origin of man. Science 211:341–350. What makes humans unique?

Neel, J. V. 1978. The population structure of an Amerindian tribe, the Yanomama. Ann. Rev. Genet. 12:365–413. What principal features of population structure have a pronounced impact on human evolution?

Nei, M. 1975. Molecular Population Genetics and Evolution. American Elsevier, New York. Data for Table 14.4.

Oxnard, C. E. 1975. The place of the australopithecines in human evolution: Grounds for doubt? Nature 258:389–395. A minority view that the australopithecines have no direct place in human ancestry.

Panny, S. R., A. F. Scott, K. D. Smith, J. A. Phillips III, H. H. Kazazain, Jr., C. C. Talbot, Jr., and C. D. Boehm. 1981. Population heterogeneity of the *Hpa*I restriction site associated with the β-globin gene: Implications for prenatal diagnosis. Am. J. Hum. Genet. 33:25–35. The association between the sickle cell gene and a flanking *Hpa*I restriction-site polymorphism varies among black populations.

Stebbins, G. L., and F. J. Ayala. 1981. Is a new evolutionary synthesis necessary? Science 213:967–971. What is the relationship between microevolution and macroevolution?

Trinkaus, E., and W. W. Howells. 1979. The Neanderthals. Scientific American 241:118–133. Who were these people, where did they come from, where did they go?

Walker, A., and R. E. F. Leakey. 1978. The hominids of East Turkana. Scientific American 239:54–66. A view of the evolution of early humans.

Wallace, B. 1981. Basic Population Genetics. Columbia University Press, New York. A well-written introduction to population genetics.

chapter 15
The Causes of Evolution

- Mutation
- Migration
- Selection
- Heterozygote Superiority or Inferiority
- Mutation-Selection Balance
- Random Genetic Drift
- Isolate Breaking
- Founder Effects
- Molecular Evolution
- Sequence Divergence
- Gene Families
- Macromutations: The Exon Shuffle

Allele frequencies in populations are determined largely by the forces of mutation, migration, natural selection, and random genetic drift. **Mutation** refers to spontaneous changes in genes (see Chapter 11); **migration** means movement of individuals into or out of a population; **selection** refers to increased or diminished length of life or number of offspring from certain genotypes; and **random genetic drift** refers to random changes in allele frequency caused by restricted size of populations. Although the effect on allele frequencies of each of these forces acting alone is relatively easy to understand, it is not often clear which forces are the most important in determining allele frequencies when all of them are acting in concert, as they often do in real populations.

Furthermore, the actual magnitudes of mutation, migration, selection, and random genetic drift are extremely difficult to measure except in very special instances. As a consequence of these uncertainties, it is still unclear how the major evolutionary forces act together to shape the kind and amount of genetic variation found in populations of humans, other animals, plants, and microorganisms.

Mutation

Mutation is the original, ultimate source of all genetic variation: *New alleles arise only by mutation.* A population may receive a new allele by way of migration, of course, but this simply means that the original mutation occurred in another population somewhere else at an earlier time.

In spite of its importance in creating new genetic variation, mutation by itself is a rather weak force for changing allele frequencies. To appreciate how weak the force of mutation can be, consider the case of a degenerating gene—a gene that serves no useful purpose to the organism and so accumulates mutations. (Pseudogenes, discussed in Chapter 9, are thought by some to be examples of degenerating genes.) For convenience we will designate the functional form of the gene as A, and the allele frequency of A will be designated as p. Allele A undergoes mutational inactivation because of **forward** mutation, and we will suppose that the rate of forward mutation per generation is μ. (In reasonable examples, μ could be 10^{-5}.) In any one generation, the fraction of A alleles that mutate is μ and the fraction of A alleles that do not mutate is $1 - \mu$. Therefore, in any generation, the fraction of A alleles that maintain their status as functional alleles is $1 - \mu$, and the allele frequency of A thus changes according to the rule:

$$\left(\begin{array}{c} \text{Allele frequency of } A \\ \text{in any generation} \end{array} \right) \text{ equals}$$

$$\left(\begin{array}{c} \text{allele frequency} \\ \text{of } A \text{ in previous} \\ \text{generations} \end{array} \right) \text{ times } \left(\begin{array}{c} \text{fraction of alleles} \\ \text{not mutating} \end{array} \right)$$

or, if we let p_0, p_1, p_2, p_3, etc. represent the A allele frequency in the 0th, 1st, 2d, 3d, etc. generations,

$$p_1 = p_0(1 - \mu)$$
$$p_2 = p_1(1 - \mu)$$
$$p_3 = p_2(1 - \mu)$$

and so on. Now if we substitute p_1 from the first equation into the second, we obtain

$$p_2 = p_0(1 - \mu)(1 - \mu) = p_0(1 - \mu)^2$$

Substituting this into the equation for p_3, we obtain

$$p_3 = p_0(1 - \mu)^2(1 - \mu) = p_0(1 - \mu)^3$$

and the general rule thus becomes clear as

$$p_n = p_0(1 - \mu)^n$$

where p_n represents the allele frequency of the active form of the gene (A) in the nth generation.

If we assume that all the alleles in the population are functional at time $n = 0$, then $p_0 = 1$, and the equation becomes

$$p_n = (1 - \mu)^n$$

or

$$\ln p_n = n \ln(1 - \mu)$$

so

$$n = \frac{\ln p_n}{\ln(1 - \mu)}$$

which provides the number of generations (n) to reach a specified allele frequency (p_n) for a given mutation rate (μ). When $\mu = 10^{-5}$, we have

$$p = 1 \text{ at } n = 0 \text{ generation}$$
$$p = 0.5 \text{ at } n = 69{,}314 \text{ generations}$$
$$p = 0.25 \text{ at } n = 138{,}628 \text{ generations}$$
$$p = 0.125 \text{ at } n = 207{,}942 \text{ generations}$$
$$p = 0.0625 \text{ at } n = 277{,}256 \text{ generations}$$
$$p = 0.03125 \text{ at } n = 346{,}570 \text{ generations}$$

Note that the number of generations increases by the constant amount 69,314 generations for each halving of the allele frequency (i.e., $69{,}314 + 69{,}314 = 138{,}628$; $138{,}628 + 69{,}314 = 207{,}942$; and so on). That is to say, *when* $\mu = 10^{-5}$, *it requires 69,314 generations to reduce the allele frequency from any specified value to half that value.* The force of mutation is thus exceedingly weak. In humans, where there are about 20 yr per generation, each halving of the allele frequency for $\mu = 10^{-5}$ would require $69{,}314 \times 20 = 1{,}368{,}280$ yr, or about 1.4 My. Thus, if a human gene had begun to degenerate at the time of *Homo erectus* 1.4 My ago, about 50 percent of the alleles in the present-day population would still be functional.

In this discussion of degenerating genes, we have ignored the possibility of **reverse** mutation (mutation that restores gene activity) because, when an inactive gene undergoes mutation, it is much more likely to sustain a second inactivating change than to undergo a change that exactly reverses the effects of the original mutation. (As mentioned in Chapter 11, mutations caused by insertions of transposable elements are exceptional in this regard.) Reverse mutation can easily be taken into account, however. Suppose we have two forms of a gene, A and a, and forward mutation (A to a) occurs at the rate μ per generation and reverse mutation (a to A) occurs at the rate ν per generation. ($\mu = 10^{-5}$ and $\nu = 10^{-6}$ are reasonably realistic examples.) Then it is clear that A cannot disappear from the population because, when A is very rare, reverse mutation (a to A) will increase the A allele frequency. Similarly, allele a cannot disappear because, when a is very rare, forward mutation (A to a) will increase the a allele frequency. In the long run the population attains a state of balance (or **equilibrium**) in which both A and a are present at allele frequencies that remain the same generation after generation. At equilibrium, the loss in the number of A alleles due to forward mutation is exactly counterbalanced by the gain in new A alleles due to reverse mutation. Indeed, the equilibrium allele frequencies of A and a depend on the forward and reverse mutation rates as follows:

$$\frac{\text{Equilibrium frequency of } A}{\text{Equilibrium frequency of } a} = \frac{\text{reverse mutation rate}}{\text{forward mutation rate}}$$

or

$$\hat{p}/\hat{q} = \nu/\mu$$

where $\hat{p}$ and $\hat{q}$ represent the equilibrium frequencies of A and a, respectively. Since $\hat{q} = 1 - \hat{p}$,

$$\hat{p} = \frac{\nu}{\mu}(1 - \hat{p}) = \frac{\nu}{\mu} - \frac{\nu}{\mu}\hat{p}$$

$$\hat{p}\left(1 + \frac{\nu}{\mu}\right) = \frac{\nu}{\mu}$$

$$\hat{p}\left(\frac{\mu + \nu}{\mu}\right) = \frac{\nu}{\mu}$$

$$\hat{p} = \frac{\nu}{\mu + \nu}$$

In the case of $\mu = 10^{-5}$ and $\nu = 10^{-6}$,

$$\hat{p} = \frac{0.000001}{0.00001 + 0.000001} = \frac{1}{11} = 0.0909$$

Again, the times involved are very long. When $\mu = 10^{-5}$ and $\nu = 10^{-6}$, it requires 63,013 generations for any specified allele frequencies to go halfway to their equilibrium values. (The half-time derived earlier—69,314— is different because here we are allowing reverse mutation to occur.)

Migration

Migration—the movement of individuals between populations—makes populations genetically more similar than they would otherwise be; migration is a sort of "glue" that holds populations together. Although patterns of migration among human populations can be extremely complex, two rather simple situations emphasize the general effects of migration.

One-Way Migration Migration can conveniently be measured as the proportion of alleles in a population in question that derived from migrant individuals in the previous generation. In any generation, the allele frequency in the population of interest will be a sort of average of the allele frequencies among nonmigrants and migrants. That is to say,

$$\begin{pmatrix}\text{Allele frequency in} \\ \text{any generation}\end{pmatrix} \text{ equals}$$

$$\begin{pmatrix} \text{allele frequency in} \\ \text{previous generation} \end{pmatrix} \text{times} \begin{pmatrix} \text{proportion of} \\ \text{nonmigrants} \end{pmatrix}$$

plus

$$\begin{pmatrix} \text{allele frequency} \\ \text{among migrants} \end{pmatrix} \text{times} \begin{pmatrix} \text{proportion of} \\ \text{migrants} \end{pmatrix}$$

or

$$p_1 = p_0(1 - m) + p^*m$$

where p_1 and p_0 represent the allele frequency in the population of interest in the 1st and 0th generation, p^* is the allele frequency among migrants, and m is the proportion of alleles in any generation that derive from migrants. The above equation for p_1 can conveniently be rewritten by subtracting p^* from both sides; thus,

$$\begin{aligned} p_1 - p^* &= p_0(1 - m) + p^*m - p^* \\ &= p_0(1 - m) + p^*(m - 1) \\ &= p_0(1 - m) - p^*(1 - m) \\ &= (p_0 - p^*)\,(1 - m) \end{aligned}$$

The relationship between p_2 and p_1 is identical to that between p_1 and p_0, so

$$p_2 - p^* = (p_1 - p^*)(1 - m)$$
$$p_3 - p^* = (p_2 - p^*)(1 - m)$$

and so on. These generation-by-generation expressions greatly resemble those obtained earlier for mutation, and direct substitution from each equation into the next reveals the general rule

$$p_n - p^* = (p_0 - p^*)(1 - m)^n$$

where n is the number of generations during which migration occurred.

The above expression for p_n involves five quantities: the allele frequency in the population after n generations (p_n), the original allele frequency (p_0), the allele frequency among migrants (p^*), the migration rate (m), and the number of generations (n). If any four of these

quantities are known, the other one can be determined. Consider, for example, American blacks whose ancestors were brought as slaves to the United States from West Africa about 10 generations ago (i.e., $n = 10$). Because of interracial mating, genes from U.S. whites were introduced into the black population; this transfer of genes can be regarded as one-way migration from whites to blacks because individuals of mixed racial ancestry are regarded as blacks. Since we know the number of generations ($n = 10$), all we need are the appropriate allele frequencies to use the formula to estimate the amount of one-way migration (m). These allele frequencies will differ from locus to locus, of course, but for purposes of illustration we will use the Fy^a allele of the Duffy blood group system. Among present-day blacks in Claxton, Georgia, the allele frequency is 0.045 (this allele frequency corresponds to p_n or, in this case, p_{10}); among present-day whites in Claxton, the allele frequency is 0.422 (which corresponds to p^*). The allele frequency among the African ancestors of Claxton blacks must have been very close to 0, as the Fy^a allele is extremely rare in most of West Africa; thus we may put $p_0 = 0$. Altogether we have

$$\begin{aligned} p_0 &= 0 \\ p_{10} &= 0.045 \\ p^* &= 0.422 \\ n &= 10 \end{aligned}$$

which, when substituted into the one-way migration formula, yields

$$0.045 - 0.422 = (0 - 0.422)\,(1 - m)^{10}$$

$$-0.377 = -0.422(1 - m)^{10}$$

Then $\quad (1 - m)^{10} = \dfrac{-0.377}{-0.422} = 0.893$

$$10\ln(1 - m) = \ln(0.893) = -0.113$$

$$\ln(1 - m) = \frac{-0.113}{10} = -0.0113$$

$$1 - m = e^{-0.0113} = 0.9888$$

$$m = 1 - 0.9888 = 0.0112$$

The value $m = 0.0112$ implies that, in each generation, about 1 percent of all alleles among Claxton blacks were actually from the white population. Somewhat different values of m are obtained for alleles at other loci, but the overall average for many loci is close to 1 percent.

Island Migration One-way migration involves just two populations, with migrants from one entering the other. A somewhat more complex situation is illustrated in Figure 15.1. Here a number of populations (each consisting of N breeding individuals) are interconnected by migration (at the rate m per generation). This sort of population structure is often called an **island** structure because it is easiest to visualize if each local population is assumed to inhabit a little island of its own. Because of their population structure, the populations in Figure 15.1 will undergo genetic divergence (as measured by, say, the fixation index F discussed earlier). However—and this is the principal point—the amount of genetic divergence is kept at moderate levels by surprisingly small amounts of migration.

A full-fledged analysis of island migration is beyond the scope of this book, but the most important result is summarized in Figure 15.2, where the fixation index, F, is plotted against

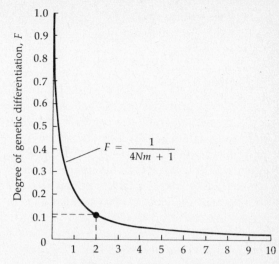

Figure 15.2 Decrease in ultimate steady-state value of fixation index, F, with increase in number of migrants per generation, Nm, in the island model. As few as two migrants per generation reduce the equilibrium value of F to 0.11.

the number of migrants into each population in each generation. (The equation of the curve is $F = 1/(4Nm + 1)$; since the population size is N and the proportion of migrants is m, the quantity Nm corresponds to the actual *number* of migrants in each generation.) When the number of migrants is 0, which corresponds to complete isolation of the populations, the fixation index ultimately goes to 1 (see Figure 15.2). However, the ultimate value of F decreases rapidly as the number of migrants increases, and for migration as little as two individuals per generation, the ultimate value of F is 0.11 (black dot on curve), which is generally considered to correspond to a moderate amount of genetic divergence. Although the population structure in Figure 15.1 is a less than perfect rendition of the actual structure of human populations, the conclusion that a small amount of migration prevents a large amount of genetic differentiation is nevertheless valid, and it goes a long way toward

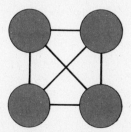

Figure 15.1 Island migration. Each circle represents a subpopulation that interchanges migrants with every other subpopulation.

explaining why the values of F among human populations are moderate (see Table 14.4 in Chapter 14).

Selection

Of the four principal forces that can change allele frequency—mutation, migration, selection, and random genetic drift—selection is the most difficult to summarize in simple generalizations. **Selection** refers to differences in survival or fertility among different genotypes, and it is because of selection that populations become progressively more adapted to their environment. The ability of a genotype to survive and reproduce is reflected in the average number of offspring born to individuals with that genotype, and this number is called the genotype's **fitness**. Suppose, to be specific, that each individual of genotype AA has, on the average, 1.100 offspring, that each individual of genotype Aa has an average of 1.100 offspring, and that each individual of genotype aa has an average of 1.045 offspring. Then we would say that the fitnesses of the genotypes are:

$$AA: \quad 1.100$$
$$Aa: \quad 1.100$$
$$aa: \quad 1.045$$

Usually it is convenient to deal in terms of **relative fitness**, which is usually obtained by dividing each of the fitnesses in question by the largest among them. In the example at hand, the relative fitnesses of three genotypes are

$$AA: \quad \frac{1.100}{1.100} = 1.00$$

$$Aa: \quad \frac{1.100}{1.100} = 1.00$$

$$aa: \quad \frac{1.045}{1.100} = 0.95$$

Thus, in this example, the aa genotype has a reduction in fitness of 0.05 (or 5 percent, obtained as $1.00 - 0.95$) relative to a value of 1

for the other genotypes. This reduction in fitness of a genotype is often called the **selection coefficient** against the genotype. (In this example the selection coefficient against aa is 0.05.) We could summarize the present example by saying that A is a favored dominant allele with a 5 percent selective advantage; alternatively, we could say that a is a disfavored recessive allele with a 5 percent selective disadvantage.

All current understanding of selection is based on the concept of fitness (usually relative fitness), but the possibilities for fitness are so numerous that selection defies easy generalities. The fitness of a genotype depends on the environment, and if a species occupies a broad geographic range encompassing many different sorts of environments, the fitness of a genotype may have one value in one part of the species' range but a different value somewhere else. Because the environment changes from season to season and year to year owing to climatic variation, the fitness of a genotype may change through time. The fitness of a genotype in a population may also depend on various attributes of the population: its **age structure** (which refers to the proportions of individuals of various ages in the population, from newborn to old age) and its **density** (which refers to the number of organisms living in an area of a certain size). Indeed, the fitness of a genotype may depend on the genotype frequencies at other loci or even on the genotype frequencies at the locus in question. Overarching these problems is the fact that fitness is extremely difficult to measure in practice; in most organisms, estimates of fitness tend to be rather imprecise except in special cases (e.g., the fitness of a lethal or sterile genotype is 0).

Despite these problems, it is worthwhile to consider a special case of selection to discuss several important attributes of selection; the special case involves the situation when fitnesses are **constant**, which is to say that they depend only on the genotype and are unchanging from generation to generation. To simplify matters further, we will assume random mating and ignore the effects of mutation, migration, and random genetic drift.

If, among the three genotypes, *AA*, *Aa*, and *aa*, genotype *AA* has the highest fitness and *aa* has the lowest, with the fitness of *Aa* not falling outside these extremes, then, in each succeeding generation, the frequency of the *A* allele will increase because *A*-bearing genotypes reproduce more than *a*-bearing genotypes. The patterns of increase in the *A* allele frequency are illustrated for three cases in Figure 15.3. In each case, the initial frequency of the favored allele (*A*) is 0.05, and the selection coefficient against the disfavored homozygote (*aa*) is 5 percent. (The formula used for the generation-by-generation calculation of allele frequency is given and explained in the legend of Figure 15.3.) The thick black line corresponds to a **favored dominant** allele, in which the fitnesses are

$$AA: 1.0$$
$$Aa : 1.0$$
$$aa : 0.95$$

The thick gray line corresponds to a **favored recessive** with fitnesses

$$AA: 1.0$$
$$Aa : 0.95$$
$$aa : 0.95$$

(This case is apt to be a little confusing because here the uppercase symbol *A* represents a recessive allele and the lowercase symbol *a* represents a dominant.)

The third example in Figure 15.3 is an intermediate situation in which the fitness of

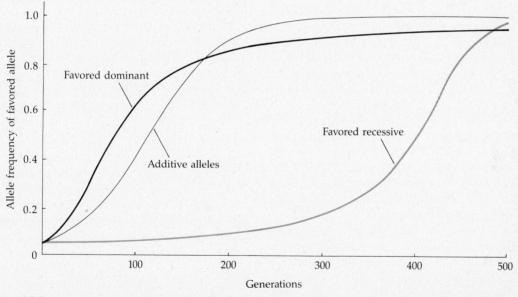

Figure 15.3 Pattern of increase in the allele frequency of a favored dominant (thick black line), a favored recessive (thick gray line), and an additive (thin black line) allele. In each case the selection coefficient against the disfavored homozygote is 5 percent and the initial frequency of the favored allele is 0.05. If p and q represent the allele frequencies of the favored and disfavored alleles, respectively, and w_1, w_2, and w_3 represent the relative fitnesses of favored homozygote, heterozygote, and disfavored homozygote, respectively, then the allele frequency in the next generation is given by $(p^2w_1 + pqw_2)/\overline{w}$, where $\overline{w} = p^2w_1 + 2pqw_2 + q^2w_3$. This formula has been used to calculate the generation-by-generation frequencies plotted in the figure.

the heterozygote is exactly the average of the homozygote fitnesses. In this situation the alleles are said to be **additive**, and the fitnesses illustrated in the figure are

$$AA: 1.0$$
$$Aa: 0.975$$
$$aa: 0.95$$

Three general characteristics of selection as exemplified in Figure 15.3 are to be noted:

1. The most rapid *changes* in allele frequency occur when the frequency of the favored allele is intermediate (i.e., between about 0.2 and 0.8).

2. Selection of a favored dominant is much more effective when the favored allele is rare than when it is common (thick black line—note how slowly the allele frequency changes once the favored allele reaches a frequency of about 0.95).

3. Selection of a favored recessive is much more effective when the favored allele is common than when it is rare (thick gray line—note how slowly the allele frequency changes when the frequency of the favored allele is around 0.05).

Points 2 and 3 are illustrated in a somewhat different form in Figure 15.4. The vertical axis gives the time required for an allele frequency to change from any initial value (horizontal axis) to a value halfway between the initial value and its ultimate destination. (Note that the scale of the vertical axis is logarithmic.) For a favored allele, the ultimate destination is an allele frequency of 1.0 (called **fixation** of the allele); for a disfavored allele, the ultimate destination is an allele frequency of 0 (called **loss** of the allele). Thus, Figure 15.4 gives the half-times (in generations) to fixation of a favored allele or loss of a disfavored allele. (Actually, what is plotted is the half-time multiplied by the selection coefficient against the disfavored homozygote. To obtain the half-

time in any particular case, the plotted value must be divided by the selection coefficient.)

For a favored dominant or disfavored recessive (thick black line), note how the half-times increase dramatically when the favored dominant is close to fixation (lower horizontal axis) or (to say the same thing) when the disfavored recessive is rare (upper horizontal axis). The situation for a favored recessive (disfavored dominant) is the reverse (thick gray line); selection is relatively inefficient when the favored recessive is rare (or the disfavored dominant common). The reason for the relative inefficiency of selection in these cases relates to the principle that rare alleles are found much more frequently in heterozygotes than in homozygotes. Selection for or against a recessive allele will thus be inefficient when the allele frequency of the recessive is small, as most recessive alleles will then occur in heterozygotes and therefore will not be exposed to selection.

Inefficient selection against rare recessives can be seen most dramatically in the case of a recessive lethal. In this case the fitnesses of *AA, Aa,* and *aa* are 1, 1, and 0, respectively, and, according to the formula in the legend of Figure 15.3, the frequency of A (p) changes as

$$p_n = \frac{1}{1 + q_{n-1}}$$

or (since $q_n = 1 - p_n$)

$$q_n = \frac{q_{n-1}}{1 + q_{n-1}}$$

which has the solution in terms of q_0:

$$q_n = \frac{q_0}{1 + nq_0}$$

Because $1 + nq_0 = 2$ when $n = 1/q_0$, this expression implies that selection requires $n = 1/q_0$ generations to reduce the allele frequency from any specified value (q_0) to half that value ($q_0/2$). When $q_0 = 0.01$, for example, it will require 100 generations of selection just to reduce the allele frequency to 0.005.

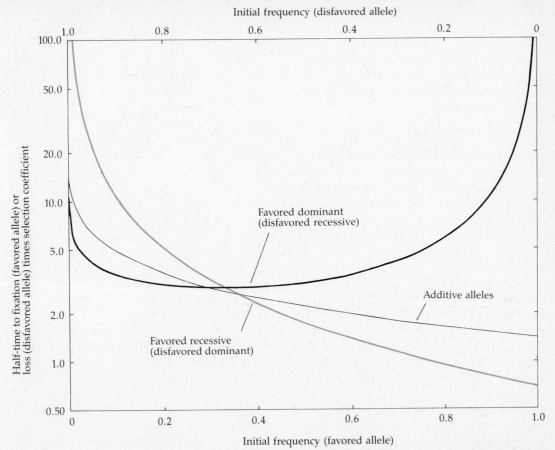

Figure 15.4 Half-times to fixation of a favored allele or loss of a disfavored allele for initial frequencies given by the horizontal axis. Note that the scale of the vertical axis is logarithmic. Plotted values are the product of the half-time and the selection coefficient against the disfavored homozygote. Actual half-times (in generations) can thus be obtained by dividing the plotted value by the selection coefficient. The curves involve an approximation that is valid so long as the selection coefficient is small.

Heterozygote
Superiority or Inferiority

So far we have considered the simplest possible case of selection when one homozygote has the highest fitness, the other homozygote has the lowest fitness, and the fitness of the heterozygote does not fall outside these extremes. In all cases, the favored allele increases in frequency and ultimately goes to fixation. Moreover, at every stage in the selection process, selection acts to increase the average fitness of the population. The **average fitness** of a population consisting of AA, Aa, and aa genotypes is calculated as

$$\begin{pmatrix} \text{Average fitness} \\ \text{of population} \end{pmatrix}$$

$$\text{equals} \begin{pmatrix} \text{frequency} \\ \text{of } AA \end{pmatrix} \text{times} \begin{pmatrix} \text{fitness} \\ \text{of } AA \end{pmatrix}$$

$$\text{plus} \begin{pmatrix} \text{frequency} \\ \text{of } Aa \end{pmatrix} \text{times} \begin{pmatrix} \text{fitness} \\ \text{of } Aa \end{pmatrix}$$

$$\text{plus}\left(\begin{array}{c}\text{frequency}\\\text{of } aa\end{array}\right)\text{times}\left(\begin{array}{c}\text{fitness}\\\text{of } aa\end{array}\right)$$

With random mating, the genotype frequencies will be p^2, $2pq$, and q^2. The average fitness of a population is usually represented by the symbol $\overline{w}$, so we can write

$$\overline{w} = p^2w_1 + 2pqw_2 + q^2w_3$$

where w_1, w_2, and w_3 represent the fitnesses of *AA, Aa,* and *aa,* respectively. If w_1 is the highest of the three fitnesses, w_3 is the lowest, and w_2 is not outside these extremes, then the average fitness of the population is greatest (**maximized**) when $p = 1$.

What happens when w_2 *is* outside the limits established by w_1 and w_3? Here we must distinguish two possibilities. First, w_2 could be greater than both w_1 and w_3; heterozygotes thus have the highest fitness, and this case is called **heterozygote superiority** or **overdominance**. Alternatively, w_2 could be less than both w_1 and w_3; this case is called **heterozygote inferiority** because heterozygotes have the lowest fitness.

With overdominance, the principle of selection acting to maximize average fitness still holds. Figure 15.5 illustrates how average fitness depends on allele frequency for a case of overdominance, and the arrows on the curve indicate the direction of allele frequency change. The illustration assumes a 5 percent selection coefficient against the *AA* homozygote and a 10 percent selection coefficient against the *aa* homozygote, so the relative fitnesses are:

$$AA: \quad w_1 = 0.95$$
$$Aa: \quad w_2 = 1.0$$
$$aa: \quad w_3 = 0.90$$

The maximum average fitness in this example is 0.967, and it occurs when the allele frequency of A is about 0.67 ($\frac{2}{3}$, to be exact). When p is less than $\frac{2}{3}$, the average fitness will increase if p increases; when p is greater than $\frac{2}{3}$, on the other

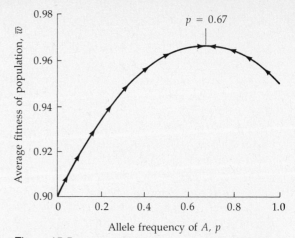

Figure 15.5   A case of overdominance showing how the allele frequency changes (arrows) to maximize average fitness. Fitnesses of *AA, Aa,* and *aa* are 0.95, 1.0, and 0.90 in this example. The value $p = 0.67$ represents the equilibrium allele frequency.

hand, the average fitness will increase if p decreases. As emphasized by the direction of the arrows, selection does indeed maximize average fitness. Whatever the initial value of p may be, selection will ultimately bring the allele frequency to $p = \frac{2}{3}$, after which the allele frequency will no longer change. This stable value of p is called a **stable equilibrium** because selection moves the allele frequency toward the equilibrium. When there is overdominance, the equilibrium *ratio* of A and a allele frequencies is given by

$$\frac{\text{Equilibrium frequency of } A}{\text{Equilibrium frequency of } a}$$

$$\text{equals} \frac{\text{selection coefficient against } aa}{\text{selection coefficient against } AA}$$

For the example in Figure 15.5, the selection coefficients are 0.05 (against AA) and 0.10 (against aa), so the equilibrium ratio is

$$\frac{p}{q} = \frac{0.10}{0.05} = 2$$

which corresponds to $p = \frac{2}{3}$ and $q = \frac{1}{3}$.

A famous example of overdominance in human populations involves the β^S sickle cell polymorphism in West Africa. Recall from Chapter 4 that homozygotes for the β^S allele are severely anemic. Until relatively recent medical advances, the homozygous condition had been essentially lethal, so the fitness of β^S/β^S homozygotes throughout most of human history has been 0 (i.e., a selection coefficient against β^S/β^S of 1). However, heterozygotes (β^S/β) are more resistant to malarial infections by the parasite *Plasmodium falciparum* than are normal homozygotes (β/β). Although the occurrence of mild malarial infections is about the same in β^S/β and β/β genotypes, the occurrence of severe malarial infections is about twice as high among β/β homozygotes as among β^S/β heterozygotes. In some African populations the overall survival of β/β genotypes is reduced by about 15 percent as compared with β^S/β heterozygotes, which corresponds to a selection coefficient against β/β of 0.15. Assuming that all three genotypes in these populations have about the same fertility, the equilibrium ratio of β and β^S allele frequencies would be expected to be

$$\frac{\text{Equilibrium frequency of } \beta}{\text{Equilibrium frequency of } \beta^S}$$

$$\text{equals } \frac{\text{selection coefficent against } \beta^S/\beta^S}{\text{selection coefficient against } \beta/\beta}$$

$$= \frac{1.0}{0.15} = 6.67$$

which corresponds to allele frequencies of about 0.87 and 0.13 for β and β^S, respectively. These frequencies are close to the averages found throughout most of West Africa, although there is considerable variation in allele frequency among local populations, with the β^S frequency ranging from about 0.05 to 0.15.

In the case of heterozygote inferiority, the principle of maximization of average fitness no longer holds. Figure 15.6 illustrates the case

$$AA: w_1 = 1.05$$
$$Aa : w_2 = 1.00$$
$$aa : w_3 = 1.10$$

so there is a selection coefficient of 5 percent in favor of AA and a selection coefficient of 10 percent in favor of aa (relative to a value of 1.00 for the fitness of Aa). With heterozygote inferiority there is still an equilibrium (i.e., a point at which allele frequencies do not change), and it is given by

$$\frac{\text{Equilibrium frequency of } A}{\text{Equilibrium frequency of } a} \text{ equals}$$

$$\frac{\text{selection coefficent in favor of } aa}{\text{selection coefficient in favor of } AA}$$

which is

$$\frac{p}{q} = \frac{0.10}{0.05} = 2$$

corresponding to $p = \frac{2}{3}$ and $q = \frac{1}{3}$ in the example in Figure 15.6. However, when there

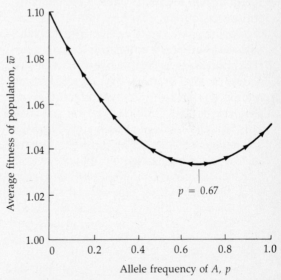

Figure 15.6 A case of heterozygote inferiority showing how allele frequency changes (arrows). The value $p = 0.67$ is an unstable equilibrium and corresponds to a minimum of average fitness. Fitnesses of AA, Aa, and aa are 1.05, 1.00, and 1.10, respectively, in this example.

is heterozygote inferiority, the equilibrium is **unstable**, which means that allele frequencies near (but not equal to) the equilibrium become progressively farther away from the equilibrium. In Figure 15.6, if p is less than $\frac{2}{3}$, p ultimately goes to 0, so in this eventuality the average fitness of the population is maximized; however, if p is greater than $\frac{2}{3}$, p ultimately goes to 1, and the average fitness attains the value 1.05, which is not its maximum.

Figure 15.6 shows that natural selection is not all-powerful; average fitness is not always maximized. It is useful to think of the fitness curve in Figure 15.6 as a sort of surface having two **peaks** (points of high fitness, corresponding here to allele frequencies of $p = 0$ and $p = 1$) and one **pit** (a point of low fitness, here $p = \frac{2}{3}$). As shown by the direction of the arrows, natural selection tends to change allele frequencies so that average fitness climbs a peak, but it climbs the nearest peak and not necessarily the highest. In this nearest-peak sense, natural selection is shortsighted. The same sort of phenomenon can occur when alleles at many loci interact in their effects on fitness. In this more complex but more realistic situation, the fitness surface (called an **adaptive topography** in this context) is an abstract surface in many dimensions, but it may have regions corresponding to peaks, pits, valleys, and even saddle shapes. Nevertheless, natural selection causes average fitness to climb nearby peaks even though they may not be the highest peaks on the whole fitness surface.

The theoretical importance of Figure 15.6 is thus in showing that natural selection can trap a population on a submaximal fitness peak. There is, however, a process that can rectify the situation. Random genetic drift, as its name implies, can change allele frequency in a random fashion. (The process will be discussed in a few pages.) Indeed, with a sufficiently small population, random genetic drift can change allele frequency in such a way

as to *decrease* average fitness. For a population stranded near the $p = 1$ fitness peak in Figure 15.6, for example, random genetic drift could in a succession of generations shift the allele frequency to the left against the arrows and bring it to a point where p is less than $\frac{2}{3}$. Then selection can take over and will of its own accord move the allele frequency toward $p = 0$ and the population toward its maximum fitness. In this manner random effects on allele frequency permit a population to explore the full range of its adaptive topography and ultimately achieve its maximum fitness.

Mutation-Selection Balance

In actual populations, all the forces that can change allele frequency act in concert, with some tending to reinforce one another and others acting in opposition. Forces that act in opposition are particularly important because, with one force acting to increase allele frequency and another acting to decrease it, the allele frequency can eventually attain an equilibrium at which the forces exactly balance each other. In this section we consider the combined effects of mutation and selection when these forces oppose each other.

We will first consider the case of a recessive lethal because the action of selection in this case has already been established. Accordingly, let A be the normal allele and a the recessive lethal, and let their allele frequencies be denoted p and q, respectively. Recall the recessive lethal formula

$$q_n = \frac{q_{n-1}}{1 + q_{n-1}}$$

which implies that the *change* in allele frequency due to selection (i.e., $q_n - q_{n-1}$) is

$$q_n - q_{n-1} = \frac{q_{n-1}}{1 + q_{n-1}} - q_{n-1}$$

$$= \frac{-q^2_{n-1}}{1 + q_{n-1}}$$

Now consider the effect of mutation. Let μ be the forward (A to a) mutation rate per generation. (It will turn out that the equilibrium allele frequency of a is sufficiently small that reverse mutation can safely be ignored.) Previously we established that, with mutation alone,

$$p_n = p_{n-1}(1 - \mu)$$

or, since $q_n = 1 - p_n$,

$$1 - q_n = (1 - q_{n-1})(1 - \mu)$$

or

$$q_n = 1 - (1 - q_{n-1})(1 - \mu)$$

$$= q_{n-1} + (1 - q_{n-1})\mu$$

so, with mutation alone, the change in allele frequency is

$$q_n - q_{n-1} = (1 - q_{n-1})\mu$$

At equilibrium the changes due to selection and mutation must exactly cancel each other. Population geneticists frequently denote equilibrium allele frequencies with circumflexes, so we will write $\hat{q}$ instead of q_{n-1}. Using the above expressions for the changes due to selection and mutation, equilibrium must occur when

$$\frac{\hat{q}^2}{1 + \hat{q}} = (1 - \hat{q})\mu$$

so that the negative force of selection will exactly balance the positive force of mutation. Thus

$$\hat{q}^2 = (1 - \hat{q})(1 + \hat{q})\mu$$

$$= (1 - \hat{q}^2)\mu$$

$$= \mu - \mu\hat{q}^2$$

Now $\hat{q}$ is going to be small and μ smaller still, so the term $\mu\hat{q}^2$ can be ignored because it will be very close to 0. Hence

$$\hat{q}^2 = \mu$$

to a good approximation, or

$$\hat{q} = \sqrt{\mu}$$

is the equilibrium frequency of the recessive lethal.

More generally, the equilibrium allele frequency of a recessive is given by

Equilibrium frequency
  of recessive allele

$$\text{equals } \sqrt{\frac{\text{forward } (A \text{ to } a) \text{ mutation rate}}{\text{selection coefficient against } aa}}$$

$\hat{q} = \sqrt{\mu}$ in the case of a recessive lethal because the selection coefficient against lethal homozygotes is 1.

As an example of how these mutation-selection considerations can be applied, we may consider Tay-Sachs disease. Recall from Chapter 4 that the incidence of the condition among non-Jews is about 1 per 400,000, which equals 2.5×10^{-6}; this represents the genotype frequency of the homozygous recessive and so corresponds to $\hat{q}^2$. Since Tay-Sachs disease is lethal, we use the formula

$$\hat{q}^2 = \mu$$

so, substituting for $\hat{q}^2$ (and assuming that the population is at equilibrium),

$$\mu = 2.5 \times 10^{-6}$$

This sort of indirect estimate of the mutation rate is an important technique in human genetics, and many of the mutation rates in Table 11.2 of Chapter 11 were derived from analogous indirect considerations.

When an allele is completely recessive, its equilibrium frequency is determined by the forward mutation rate and the selection coefficient against the recessive homozygotes. However, when an allele is not completely recessive, the selection coefficient against the recessive homozygote is almost irrelevant to its equilibrium frequency; in this case the selection occurs mainly through the heterozygotes because they are so much more numerous than the recessive homozygotes; to a very good approximation the equilibrium frequency is given by

Equilibrium frequency
of nonrecessive allele

equals $\dfrac{\text{forward } (A \text{ to } a) \text{ mutation rate}}{\text{selection coefficient against } Aa}$

(Note that there is no square root involved in this case.)

A small amount of selection against heterozygotes (called **partial dominance**) has a major effect on the equilibrium frequency with mutation-selection balance. Suppose the forward mutation rate is 10^{-5}, for example, and imagine a complete recessive with a selection coefficient against homozygotes of 0.75. The fitnesses would then be

AA: 1.0
Aa : 1.0
aa : 0.25

and the equilibrium frequency of a would be

$$\hat{q} = \sqrt{\frac{10^{-5}}{0.75}} = 0.00365$$

On the other hand, if there is a small amount of selection against heterozygotes (say, 1 percent), the fitnesses would be

AA: 1.0
Aa : 0.99
aa : 0.25

and the equilibrium frequency becomes

$$\hat{q} = \frac{10^{-5}}{0.01} = 0.00100$$

so that the small amount of dominance of the harmful allele reduces the equilibrium allele frequency by a factor of 3.65.

Considerations of mutation-selection balance can be used to estimate mutation rates in the case of incomplete recessives. Recall from Chapter 11 that Huntington disease is a late-onset degeneration of the neuromuscular system. In terms of neuromuscular degeneration, the allele that causes the condition is dominant. However, because of the late age of onset, the allele is not completely dominant in its effects on fitness. In one large study in Michigan, the selection coefficient against heterozygotes was estimated as 0.19, and in this population the frequency of the Huntington allele was 5×10^{-5}. Assuming equilibrium and using the formula for incomplete dominance, we have

$$5 \times 10^{-5} = \frac{\mu}{0.19}$$

or

$$\mu = (0.19)(5 \times 10^{-5}) = 9.5 \times 10^{-6}$$

as an estimate of the mutation rate.

Random Genetic Drift

Random genetic drift is the process whereby allele frequency changes in a random (i.e., undirected) fashion from one generation to the next. The process is most pronounced in small populations; for purposes of illustration, it will be convenient to imagine a population that consists of just five diploid breeding individuals in each generation (a very small population indeed). Suppose there are two alleles (A and a) at some locus in this population, and suppose that the effects of mutation, migration, and selection can be ignored. Since there are only five individuals, only 10 alleles at the locus are found in the population, so the allele frequency of A (call it p) must assume one of the values $\frac{0}{10}, \frac{1}{10}, \frac{2}{10}, \ldots, \frac{10}{10}$. For convenience, suppose $p = \frac{5}{10}$ (i.e., $p = \frac{1}{2}$) at the outset. Now, the five breeding individuals will produce a huge number of potential gametes—so huge that the allele frequency among the gametes will precisely reflect the allele frequency among adults—namely, $p = \frac{1}{2}$. However, only 10 of these gametes will be represented in the next generation because the next generation will again consist of five individuals. So, to create the next generation, exactly 10 gametes are to be chosen at random from the huge (i.e., infinite) pool of gametes, in which $p = \frac{1}{2}$. Random genetic drift of allele frequency occurs in this situation because the 10 lucky gametes that combine to form the next generation may not perfectly reflect the allele frequencies of the previous generation.

Indeed, since $p = \frac{1}{2}$ at the outset, any one of the lucky 10 gametes could be A or a, each with probability $\frac{1}{2}$. We may as well think of this case as a coin-tossing situation, with A representing heads and a representing tails. The various possible numbers of A alleles among the 10 lucky gametes, and their associated probabilities, correspond precisely to the number of heads to be obtained in 10 flips of a coin. Indeed, the probability of obtaining exactly i heads in 10 flips is given by

$$\frac{10!}{i!(10-i)!} \left(\frac{1}{2}\right)^i \left(\frac{1}{2}\right)^{10-i}$$

where, recall from Chapter 3, the symbol $i!$ means $1 \times 2 \times 3 \times \ldots \times i$ (with 0! defined as 1).

So, the probability that $i = 5$ in the next generation (i.e., $p = \frac{1}{2}$, like the previous generation) is

$$\Pr\{i = 5\} = \frac{10!}{5!5!} \left(\frac{1}{2}\right)^5 \left(\frac{1}{2}\right)^5$$

$$= \frac{3,638,800}{(120)(120)} \left(\frac{1}{32}\right) \left(\frac{1}{32}\right)$$

$$= 0.24609$$

On the other hand, the probability that $i = 0$ (i.e., A becomes **lost** so $p = 0$) is

$$\Pr\{i = 0\} = \frac{10!}{0!10!} \left(\frac{1}{2}\right)^0 \left(\frac{1}{2}\right)^{10}$$

$$= \frac{1}{1024}$$

$$= 0.00098$$

Likewise the probabilities that $i = 1, 2, 3$, etc. may be calculated. These values are plotted in the histogram corresponding to $n = 1$ (n is the number of generations of random genetic drift) in Figure 15.7. Here we use the symbol N to represent population size, with $2N$ being the number of alleles at any particular locus; since $N = 5$ in this example, $2N = 10$. The horizontal axis in Figure 15.7 is labeled "Proportion of populations" instead of "Probability" because it is often easiest to think in terms of a large number of completely isolated subpopulations, each starting with the same allele frequency and each of the same size. So, with this interpretation, all populations (100 percent) begin with a number of A alleles equal to 5 ($p = \frac{1}{2}$—see the histogram for $n = 0$). After $n = 1$ generation of random genetic drift, a proportion 0.24609 (see above) of the populations will have $p = \frac{1}{2}$, a proportion 0.00098 will have $p = 0$, and so on.

After the first generation of random genetic drift, things become more complicated because, in most populations, p no longer equals $\frac{1}{2}$. In general, if a population consists of N breeding individuals and has an allele frequency of A of p, the probability that exactly i copies of the A allele will be represented in the next generation is

$$\Pr\{i \text{ copies of } A\} = \frac{(2N)!}{i!(2N-i)!} p^i (1-p)^{2N-i}$$

So to calculate the probability that there are exactly i copies of allele A after two generations of random genetic drift, we have to calculate the probability for each value of $p = 0/2N, 1/2N, 2/2N, \ldots, 2N/2N$, then multiply each of these by the probability that p is actually $0/2N, 1/2N, 2/2N, \ldots, 2N/2N$ in the

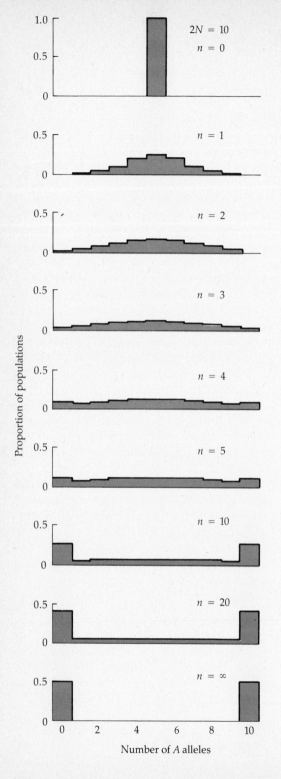

previous generation, and then add these numbers together.

The calculations just described are tedious and repetitive; they are best carried out on some sort of computer or programmable calculator. The results for various values of n (generations of random genetic drift) when N = 5 are summarized in Figure 15.7, where the principal features of random genetic drift are much in evidence but warrant some emphasis:

1. Ignoring the cases of fixation ($p = 1$) and loss ($p = 0$), the distribution of allele frequencies becomes progressively flatter so that, after a time, the allele frequency is as likely to be any one value as any other; to be more concise, the distribution of allele frequencies in **unfixed** or **segregating** populations (i.e., populations in which $p \neq 0$ and $p \neq 1$) becomes flat.

2. Ultimately, all populations become fixed for one allele or the other (see Figure 15.7 when $n = \infty$).

3. The *average* allele frequency among all populations remains what it was at the outset (in this case, the average allele frequency remains $p = \frac{1}{2}$ for any number of generations of random genetic drift).

4. Putting 2 and 3 together, if the probability of ultimate fixation of A is, say, x and the initial frequency of A is p, then we must have the average allele frequency still equaling p, so

$$x(1) + (1 - x)(0) = p$$

or $x = p$. That is to say, *the probability of ultimate fixation of an allele is equal to its frequency in the initial population* (provided, of course, that we ignore mutation, migration, and selection).

Figure 15.7 Dispersion of allele frequency caused by various numbers (*n*) of generations of random genetic drift in a large collection of ideal populations, each of size $N = 5$. Note that all populations ultimately become fixed for either *A* or *a*.

The pattern of allele frequency change in Figure 15.7 is typical of random genetic drift, but the process is accelerated in this case because the population is so small. Larger populations undergo the same sort of process, but more slowly.

A key element in determining the rate of random genetic drift in an actual population is the population's **effective population size.** This concept of effective population size is necessary because actual populations never match a theoretically perfect (or **ideal**) population. A theoretically ideal population maintains a constant size from generation to generation, all individuals in the population reproduce at the same time, there are equal numbers of males and females, and each individual has an equal chance of contributing successful gametes to the next generation. Although an actual population is almost never theoretically ideal, it does have a certain rate of random genetic

drift, so we can invent an imaginary ideal population that has the same rate of random genetic drift as occurs in the actual population; the size of this imaginary population is defined as the actual population's effective size.

An experimental example of random genetic drift among 107 isolated populations of *Drosophila melanogaster* is illustrated in Figure 15.8. Each population was maintained at a constant size of 16 by selecting at random 8 males and 8 females in each generation. The frequency of A in all populations was $p = \frac{1}{2}$ at the outset, and the figure shows how the allele frequencies among the populations changed in 19 successive generations of random genetic drift.

The overall pattern of change in Figure 15.8 is similar to that of Figure 15.7: The initially humped distribution of allele frequencies (background) becomes progressively flatter as the proportion of populations in which A

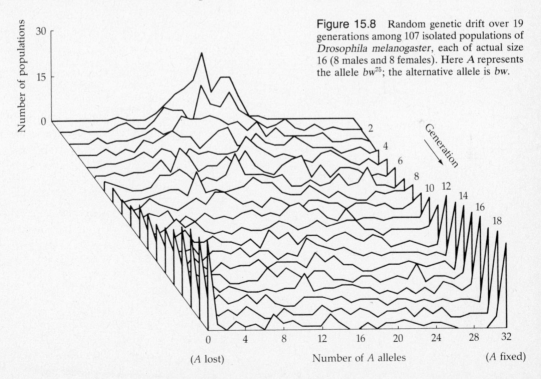

Figure 15.8 Random genetic drift over 19 generations among 107 isolated populations of *Drosophila melanogaster*, each of actual size 16 (8 males and 8 females). Here A represents the allele bw^{75}; the alternative allele is bw.

is fixed or lost steadily increases (fixation and loss correspond to the peaks at the edges of the surface). However, the detailed pattern in Figure 15.8 is not what would be expected in ideal populations of size 16 (the actual size). Rather, the pattern matches that of ideal populations of size 9, so we say that the "effective size" of the populations in Figure 15.8 is 9. Figure 15.8 illustrates another important point about effective population size: The effective size of a population is usually less, and sometimes much less, than its actual size.

It should be evident from Figures 15.7 and 15.8 that one principal effect of random genetic drift is to cause genetic differentiation among populations because their allele frequencies become progressively dispersed. Thus, a convenient measure of the effects of random genetic drift is in terms of the **fixation index** (F or F_{ST}) defined earlier as

$$F = \frac{H_T - H_S}{H_T}$$

where, recall, H_T is the total heterozygosity that would be expected if the population were fused into one large random-mating unit, and H_S is the average actual heterozygosity among subpopulations.

A complete analysis of the effects of random genetic drift on F is beyond our scope, but the principal result of such an analysis is illustrated in Figure 15.9. The symbol N in the figure represents effective population size, and the curves show the increase in F for various values of N. [The curves are based on the rule that the quantity $(1 - F)$ decreases by a fraction $1/(2N)$ in each generation.] Note that random genetic drift can produce significant levels of genetic differentiation (F) even for effective sizes of 500 or greater. Of course, genetic differentiation accumulates more slowly in larger populations.

Figure 15.9 raises a subtle but fundamentally important aspect of random genetic drift. If we think of any one of a large number of subpopulations undergoing random genetic

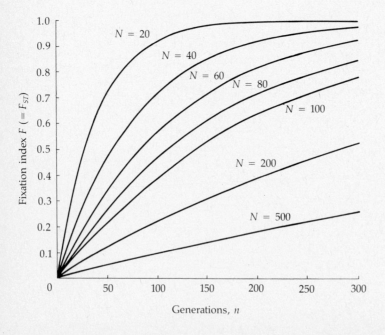

Figure 15.9 Increase in fixation index, F, among ideal populations as a function of time and effective population number, N.

drift, the genotype frequencies within this subpopulation will be given by the Hardy-Weinberg rule because random mating is assumed to occur within each subpopulation. Thus, in a subpopulation with an A allele frequency of p_i, the genotype frequencies will be

$$AA: p_i^2$$
$$Aa : 2p_iq_i$$
$$aa : q_i^2$$

However, if we consider the average genotype frequencies among all subpopulations, the situation is very different. There is an overall deficiency of heterozygotes (as measured by F), and the same reasoning as used in the context of inbreeding (Table 14.5 in Chapter 14) leads to the following genotype frequencies for the total population:

$$AA:\quad p^2(1 - F) + pF$$
$$Aa:\quad 2pq(1 - F)$$
$$aa:\quad q^2(1 - F) + qF$$

where p and q are the average allele frequencies of A and a, respectively, among all subpopulations.

Thus, random genetic drift leads to a sort of paradox: Even though mating is random within each subpopulation, there is a kind of "inbreeding" in the population considered as a whole. The paradox is resolved because the populations are not infinitely large, and random mating in a finite population will sometimes entail mating between relatives. This is a subtle point, but consider two extreme cases: Random mating in a population of size $N = 1$ amounts to repeated self-fertilization, and random mating in a bisexual population of size $N = 2$ amounts to repeated brother-sister mating. The situation with larger (but not infinite) populations differs from these extreme cases only in degree, and it accounts for the inbreeding-like effect of random genetic drift.

Isolate Breaking

The inbreeding-like phenomenon associated with random genetic drift has important implications for the overall frequency of genotypes that are homozygous for rare, harmful recessives. To take a concrete example, consider two isolated populations of the same size, one having an a allele frequency of 0.10 and the other an a allele frequency of 0.01. The two populations would then have aa frequencies of

$$aa:\ (0.10)^2 = 0.0100$$

and

$$aa:\quad (0.01)^2 = 0.0001$$

for an average frequency of

$$aa:\quad \frac{0.0100 + 0.0001}{2} = 0.0050$$

However, if the populations were to become fused (often called **isolate breaking**), the a allele frequency in the fused population would be the average of 0.10 and 0.01, or $(0.10 + 0.01)/2 = 0.055$. The frequency of aa homozygotes in the fused population would then be

$$aa:\quad (0.055)^2 = 0.0030$$

which is substantially smaller than the average aa frequency in the unfused populations (i.e., 0.0050) calculated above.

More generally, the reduction in the frequency of aa homozygotes upon population fusion is related to the fixation index among the unfused populations by the relationship

$$aa \text{ (before fusion)} - aa \text{ (after fusion)} = pqF$$

where p and q are the respective average allele frequencies of A and a among subpopulations. In this example, $p = 0.945$, $q = 0.055$, and F works out to 0.039. The quantity pqF is 0.0020, which does equal the difference in aa frequencies before and after fusion.

Founder Effects

As noted earlier, a certain amount of mating between relatives is unavoidable in small populations. Even when individuals choose their mates at random in a small population, some fraction of the time they will be mating with relatives. The inbreeding-like effect of population subdivision is most important in very small populations. In a small religious community (the Hutterites) in South Dakota and Minnesota, for example, the average inbreeding coefficient of individuals was found to be 0.04. However, almost all the inbreeding is due to random mating in a population of restricted size; almost none of it is due to actual nonrandom mating.

Another example of inbreeding due to small population size is the Kel Kummer Tuareg (Figure 15.10), a group that lives in the southern Sahara Desert. Complete genealogies of all 2420 living and dead members of the Kel Kummer tribe have been compiled (a task that took the French ethnologist Chaventré 15 years). These genealogies trace the ancestry of every member of the group back to 156 people who founded the group some 300 years ago. (Figure 15.11 shows part of the complex pedigree of the Kel Kummer.) As of 1970 the size of the Kel Kummer Tuareg was 367 people. Their ancestry was extraordinarily complex; each pair of members of the tribe had at least 15 ancestors in common. Furthermore, the 156 original founders did not contribute equally to the genetic composition of the present population. The parents of the great warrior Kari-Denna contributed more than 10 percent of the genes in the present population, and Kari-Denna's wife's parents were the origin of another 20 percent of the genes in the present population. Indeed, 40 percent of the genes in the tribe today trace back to just six individuals among the founders, and virtually all the present-day genes trace back to a mere 25 individuals. These 25 individuals were the true genetic founders of the Kel Kummer tribe, even though 156 individuals were present at the outset.

This founder aspect of the history of the Kel Kummer illustrates an important principle known as the **founder effect**. When a small group of individuals moves away from a larger group and establishes a genetically semi-isolated population, then the allele frequencies in the new population may be very different from those in the original population, simply because the founders of the new population may not be perfectly representative of the population from whence they came. (Founder effects are thus special varieties of random genetic drift.) Founder effects are especially important if the founders are few in number, because the smaller the group, the less genetically representative it is likely to be. Being nonrepresentative genetically may come about because of chance differences in allele frequency among the founders, but it may also occur because founders of a new population are sometimes groups of relatives. One striking example of the founder effect concerns the allele causing Huntington disease. In the Australian state of Tasmania there are about 350,000 people, about 120 of them affected with Huntington disease. Every single affected individual has an ancestry that traces back to a Huguenot woman who left England in 1848 to settle in the islands off New Zealand.

To summarize this section on random genetic drift, it may be well to recall the discussion of Table 3.1 in Chapter 3, where differences in the incidence of various inherited conditions among human populations were emphasized. It may also be useful to recall that allele frequencies of allozymes, blood groups, and restriction-site polymorphisms also differ from population to population. Although some of these population differences may be attributable to natural selection (e.g., the β^S sickle

Figure 15.10 Several present-day members of the Kel Kummer Tuareg.

cell allele in West Africa), other differences are likely to have resulted from random genetic drift and founder effects.

Molecular Evolution

Techniques of molecular biology such as amino acid sequencing of proteins or nucleotide sequencing of DNA or RNA provide an approach to the study of evolution that is not dependent on the fossil record. Although macromolecules such as DNA, RNA, and proteins are totally destroyed during the process of fossilization, molecular studies can be carried out in a wide variety of living organisms. Since biological information is largely contained in *sequence* information (amino acids in proteins or nucleotides in DNA and RNA), compari-

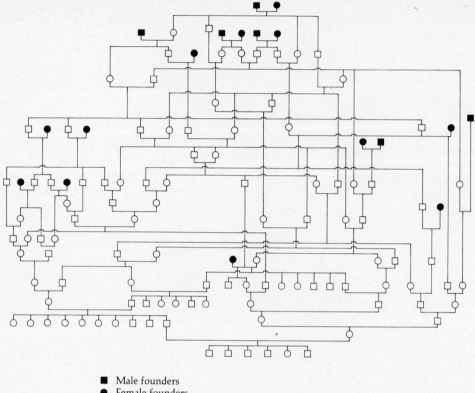

■ Male founders
● Female founders
□ Other males
○ Other females

Figure 15.11 Partial pedigree of Kel Kummer population showing the extensive interbreeding made necessary in part by the small size of the population. Marriages in this pedigree are denoted by horizontal lines somewhat below the level of the marriage partners.

sons of sequence information among living forms can be used to decipher how such sequences have become altered during the course of evolutionary time. Sequence studies of this sort, especially those involving nucleotide sequences, are still in their infancy, and we can expect significant new insights into the evolutionary process in the future. In this section we consider some of the established and emergent principles of molecular evolution that have already been achieved.

Sequence Divergence

One of the most intensively studied proteins is **cytochrome c**, an ancient, relatively small protein involved in electron transfer during respiration and found in all eukaryotes and some prokaryotes. In one extensive amino acid sequence comparison of cytochrome c, 87 species were involved. Some of the sequences were identical (like that of cytochrome c in humans and chimpanzees), others had small differences (such as one amino acid difference between humans and rhesus monkeys), and others had many differences (such as 13 amino acid differences between human and kangaroo cytochrome c). Such amino acid differences indicate that, during the course of evolutionary time, the original, prevalent form of a gene can be lost and a new, mutant form of the gene fixed. Such a replacement of one gene form by another is called **gene substitution**. At a finer level one can imagine individual nucleotides being replaced with others, which is called **nucleotide substitution**. Of course, over a long period of evolutionary time, genes can undergo successive substitutions, and thus particular genes in two distinct species become increasingly divergent.

If one now makes the reasonable assumption that species with very similar genes are more closely related than species with very different genes, then amino acid sequence comparisons can be used to reconstruct the evolutionary history of the species involved. Usually more than one possible evolutionary tree is compatible with the data, but the "best" tree is typically considered to be the one that involves the smallest number of nucleotide substitutions; such a tree is said to be the **most parsimonious tree**.

The most parsimonious tree for the cytochrome c's of 87 species is illustrated in Figure 15.12. The tree originates at the upper right and proceeds downward and to the left, terminating with the living species. Each number in the tree is related to the number of nucleotide substitutions in the corresponding branch. Note that the species fall into five rather well-defined groups, which supports one scheme of classification that divides living organisms into five kingdoms: **animalia** (animals), **plantae** (plants), **fungi** (such as yeast), **protista** (includes single-celled eukaryotes such as protozoans), and **monera** (bacteria—the letters

R.v., R.p., and so on in Figure 15.12 refer to various cytochrome c-containing bacteria).

Amino acid sequence comparisons are also of interest because the hypothetical sequence of long-extinct common ancestors can be reconstructed. Some examples for cytochrome c are illustrated in Figure 15.13, where the amino acids are represented by their single-letter abbreviations (see Table 9.1 in Chapter 9). Across the top is the hypothetical sequence of cytochrome c in the common ancestor of plants, animals, and fungi; lower down in the figure are hypothetical sequences of various more recent human ancestors, and the last row is the actual sequence in modern humans. Each vertical line indicates that the corresponding amino acid is identical in ancestor and descendant; each solid dot indicates an amino acid substitution. As can be determined from the figure, humans still retain 70 out of 104 amino acids of the hypothetical sequence in the plant-animal-fungal ancestor, 88 out of 104 of the hypothetical sequence of the vertebrate ancestor, 90 out of 104 amino acids are shared with the mammalian ancestor, and 96 out of 104 with the anthropoid ancestor.

While the results of amino acid substitutions are evident in Figures 15.12 and 15.13, the causes of such substitutions are by no means clear. Two popular but contradictory views are, first, that most substitutions result from the action of natural selection replacing one allele with a more beneficial mutant, or, alternatively, that most substitutions result from random genetic drift replacing one allele with an equivalent or nearly equivalent mutant. (Such mutations that have little or no effects on fitness are called **neutral alleles**.)

Proponents of the neutral theory derive support from two observations. First, many amino acid substitutions involve the replacement of an amino acid with a chemically similar amino acid. For example, in comparing the

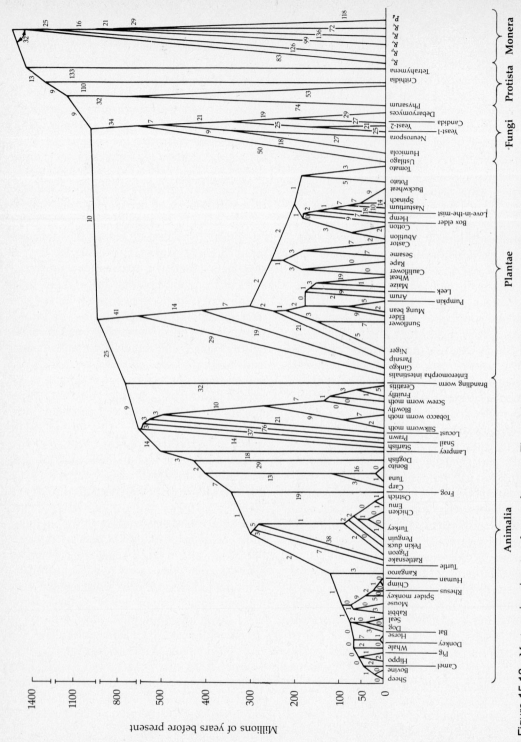

Figure 15.12 Most parsimonious tree for cytochrome c. The numbers are the inferred number of nucleotide substitutions in each segment of the tree. (Abbreviations: R.v., *Rhodomicrobium vannielii*; R.p., *Rhodopseudomonas palustris*; R.r., *Rhodospirillum rubrum*; R.s., *Rhodopseudomonas spheroides*; R.c., *Rhodopseudomonas capsulata*; and P.d., *Paracoccus dentrificans*.)

upper two sequences in Figure 15.13, an alanine (A) at position 3 is replaced with a chemically similar valine (V), and a valine at position 43 is replaced with an alanine. A second line of evidence for the neutral theory comes from the **rate** of amino acid substitution, which refers to the probability of an amino acid being substituted in an interval of time, such as 10^7 or 10^8 yr. For many proteins, the rate of amino acid substitution is roughly constant for long periods of time, although each protein has its own particular rate. For about the last 500 million yr, for example, cytochrome c has been evolving at the rate of about 0.17 amino acid substitution per amino acid site per 10^9 yr, whereas the comparable rate for hemoglobin is 0.57. This apparent constancy in rate of substitution is referred to as a **molecular clock** (each amino acid substitution is a sort of "tick"), and differences in rates among proteins may reflect selective constraints on the molecule; those with low rates have relatively few amino acid sites at which selectively neutral substitutions can occur.

Those who espouse selection counter the neutralist arguments in two principle ways. They first point out that the chemical similarity of certain amino acids is no guarantee that they will have the same functional effects in an intact protein molecule. They also argue against the molecular clock, saying that the apparent regularity of substitutions is an artifact of averaging over tens or hundreds of millions of years. When examined more carefully, claim the selectionists, proteins exhibit different rates of substitution at different periods in their evolutionary history, and even different rates of substitution at different places in the molecule. In short, the arguments involving selection versus neutrality in amino acid substitutions are complex, and the issue is by no means resolved.

Gene Families

One of the oddest species in Figure 15.12 is the rattlesnake. Although the species is still grouped with other reptiles in the most parsimonious tree, rattlesnake cytochrome c seems to have undergone an extraordinary amount of amino acid substitution. One plausible explanation of this highly divergent molecule is that it is the product of a gene duplication. **Duplication and divergence** of genes is considered to be one of the principal ways that organisms acquire new protein functions during the course of evolution. Once a gene has duplicated, one of the copies is relatively free to accumulate nucleotide substitutions, as the original function of the gene will continue to be carried out by the other copy. (There are also forces that tend to keep the copies alike, such as gene conversion or concerted evolution by means of unequal crossing-over discussed in Chapter 9.) In the case of the rattlesnake, the cytochrome c gene may have duplicated, with one copy undergoing rapid evolution, perhaps becoming specialized to a particular tissue. In time, the other copy may have been deleted or rendered unexpressed, leaving the divergent gene the

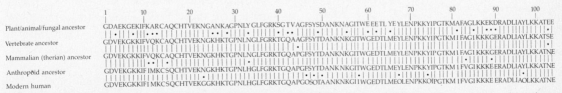

Figure 15.13 Cytochrome c amino acid sequence in modern humans as compared with inferred sequences in various distant ancestors. Vertical lines indicate amino acid identity; dots indicate nonidentity.

principal source of cytochrome c in the organism.

Repeated duplication and divergence lead to **families** of evolutionarily related genes. Such gene families are commonplace and include the hemoglobins, immunoglobulins, proteins involved in blood coagulation, certain protein-degrading enzymes, and related muscle proteins. Sequence comparisons of proteins within a family allow the evolutionary history of the family to be reconstructed. Such a reconstruction for the human hemoglobin family is illustrated in Figure 15.14. Note that the duplication leading to the α-β divergence occurred about 500 My ago and that leading to the β-γ divergence occurred about 100 My ago.

Undoubtedly the largest family of human DNA sequences is the **Alu family**, so called because all members of the family share a common restriction site for the enzyme *Alu*I. Each Alu sequence is about 300 bp long, and there are about 500,000 such sequences widely dispersed in the genome; altogether, Alu sequences constitute about 3 percent of total human DNA. The function or functions played

by the Alu family of transposable elements are unknown, but there is significant divergence among different members of the family. On the average, two members of the Alu family differ at 15 to 20 percent of their nucleotide sites. An example is shown in Figure 15.15, which gives a partial sequence of two Alu members. As in Figure 15.13, vertical lines represent sequence matches and dots represent mismatches. The open circle at position 33 represents a deletion in the clone 8 sequence relative to clone 2, and the closed triangle represents nucleotide 53, which is missing in both clones but present in some other members of the family.

Macromutations: The Exon Shuffle

Although duplication and divergence provides one important way for organisms to acquire new protein functions in the course of evolution, it is apparently not the only way. In Chapter 9 we emphasized that one of the evolutionary functions of intervening sequences may be to facilitate DNA rearrangements to create novel and perhaps beneficial combinations of exons. This evolutionary **exon shuffle** has been supported by the finding that, in at least some cases, the exons in a gene code for functionally distinct parts of the corresponding polypeptides. (Such functionally distinct parts are often called **modules** or **domains**.)

More recent and dramatic support for the exon shuffle comes from detailed studies of the three-dimensional structure of hemoglobin. The analysis involves measuring the distance between pairs of amino acids in the folded molecule; these distances are then represented in the form of a triangle (Figure 15.16). In the figure, both the horizontal and vertical axes represent the β-globin molecule, with the numbers corresponding to the amino acids. The gray regions represent pairs of amino acids whose distances are smaller than 9 Å, light

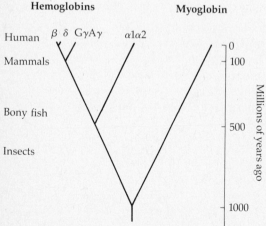

Figure 15.14 Evolution of the human globin gene family by repeated duplication and divergence.

Clone 2 5′—GAATCACTTCAACCCTAGAGGCAGAGGTTGCAGTGAGCCGAGATCTGTGCTACTGTATCCCGCCTGCTAACAGAGTGAGATTCCATCTCAAAAAAAA—3′

Clone 8 5′—GAATCCCTT–AAACCAAGAGGTGGAGGTTGCAGTGAGCCGAGATCGCACGGCTGCACTCCAGCCTGGTGACAGAGCGGAGACTCCATCTCAAAAAAAA—3′

Figure 15.15 Portion of sequence of human *Alu*-family DNA from two clones. Vertical lines indicate identical bases, dots indicate mismatches, open circle denotes a single-nucleotide deletion in clone 8 (or addition in clone 2), and closed triangle indicates a nucleotide found in some other members of the family but missing in these two.

regions correspond to distances of 9 to 27 Å, and dark regions correspond to distances greater than 27 Å. The dark regions thus delineate parts of the molecule that are distant from one another in the folded protein. Now, the principal observation is that the positions of the intervening sequences (solid horizontal and vertical lines) neatly separate these distant regions from one another. However, the functional parts of the molecule as determined by distances (F1 through F4) do not correspond exactly to the exons (E1 through E3). Indeed, exon E2 corresponds to two functional parts (F2 and F3), as if there might at one time have

been an intervening sequence separating two distinct exons for F2 and F3 (broken lines), a hypothetical intron that might have been deleted in the course of evolutionary time. Remarkably, a hemoglobin-like protein of soybeans called **leghemoglobin** has now been found to possess a third intron, and its position corresponds precisely to that of the broken line in Figure 15.16. This correspondence between exons and functionally or structurally distinct parts of proteins provides strong support for the role of exon shuffling in the evolutionary origin of new proteins.

Before leaving the subject of molecular

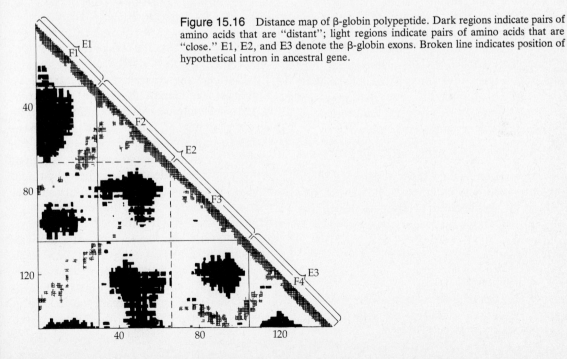

Figure 15.16 Distance map of β-globin polypeptide. Dark regions indicate pairs of amino acids that are "distant"; light regions indicate pairs of amino acids that are "close." E1, E2, and E3 denote the β-globin exons. Broken line indicates position of hypothetical intron in ancestral gene.

evolution, it should be emphasized that modern molecular techniques involving recombinant DNA have been developed relatively recently, and they have not yet been applied systematically on a wide scale for the study of

molecular evolution. As they are so applied, we can confidently expect the discovery of dramatic new principles underlying the change of macromolecules and the evolution of organisms.

SUMMARY

1. The principal forces that can change allele frequency are (a) **mutation**, which refers to spontaneous changes in genes, (b) **migration**, which refers to movement of individuals into or out of populations, (c) **selection**, which refers to genetically caused differences in survival or reproduction among individuals, and (d) **random genetic drift**, which refers to chance fluctuations in allele frequency in small populations.

2. Mutation in its widest sense includes all changes in the genetic material, from polyploidy and gross chromosomal aberrations at one extreme, to changes in the locations of transposable elements, to DNA rearrangements, to simple nucleotide substitutions. Thus, mutation is the ultimate source of all evolutionary novelty. However, from the standpoint of changing allele frequencies, mutation is a very weak force. If we let the allele frequency of A in the nth generation be p_n and represent the forward (A to a) mutation rate by μ, then, ignoring reverse mutation,

$$p_n = p_0(1 - \mu)^n$$

If, in addition, we consider reverse (a to A) mutation at the rate v, then A and a eventually attain an **equilibrium** (a state at which the allele frequencies no longer change), which is given by

$$\hat{p} = \frac{v}{\mu + v}$$

However, for realistic values of the mutation rates, the time required to achieve near-

equilibrium allele frequencies is measured in tens or hundreds of thousands of generations.

3. Migration is a cohesive force tending to make populations more similar than they would be in the absence of migration. The amount of migration is conveniently measured by means of a quantity usually denoted by the symbol m, which is the proportion of alleles in any generation that derive from migrants. We consider two rather extreme situations. In **one-way migration**, migration occurs in only one direction, from a source population into a recipient population. If p_n represents the frequency of an allele in the recipient population in generation n, and p^* represents the frequency of the same allele in the source population, then

$$p_n - p^* = (p_0 - p^*)(1 - m)^n$$

This formula has application in calculating the degree of population admixture. The other extreme of migration is **island migration**, in which a group of semi-isolated populations exchange migrants in all possible ways. In the context of island migration, the degree of genetic differentiation among the subpopulations is conveniently expressed in terms of the fixation index (F); it is noteworthy that a relatively small amount of migration (two or more migrant individuals per generation) is sufficient to keep F at a level usually considered to represent moderate or small amounts of genetic differentiation.

4. Selection is conveniently expressed in terms of genotypes' **fitness**, which refers to the ability of the genotypes to survive and reproduce in the environment in question. Population geneticists usually deal in terms of **relative fitness**, in which the fitnesses of all genotypes are expressed relative to a value of 1.00 for one of them. Although the fitnesses of genotypes can vary from generation to generation depending on many factors, for purposes of exploring the consequences of selection it is convenient to consider the case in which fitnesses are **constant** (i.e., unchanging from generation to generation). The **selection coefficient** against a genotype is the reduction in fitness of the genotype relative to a value of 1.00 for the most fit genotype.

5. Even when fitnesses are constant, selection can operate in several distinct ways. Perhaps the simplest case involves two alleles at one locus, with one homozygote having the highest fitness, the other homozygote having the lowest fitness, and the fitness of the heterozygote not falling outside the range established by the homozygotes. In such a case, selection will always increase the frequency of the allele associated with the favored homozygote, ultimately resulting in **fixation** (i.e., allele frequency 1.0) of the favored allele and **loss** of the disfavored allele. However, the rate at which selection changes allele frequency depends markedly on the dominance or recessiveness of the alleles. For a favored allele, selection is most efficient when a favored dominant is rare or when a favored recessive is common; for a disfavored allele, selection is most efficient (at decreasing the allele frequency) when a disfavored dominant is rare or when a disfavored recessive is common. In most cases, allele frequency changes due to selection are greatest when the alleles are at an intermediate frequency (i.e., in the range 0.2 to 0.8).

6. Inefficient selection against rare recessive alleles is illustrated in the case of a recessive lethal. If q_n represents the frequency of a recessive lethal after n generations of selection, then

$$q_n = \frac{q_0}{1 + nq_0}$$

and it requires $n = 1/q_0$ generations to reduce the allele frequency by half.

7. Overdominance or **heterozygote superiority** occurs when the fitness of the heterozygote is greater than that of both homozygotes. In such a case, selection against the homozygotes is in opposition, and the allele frequencies eventually attain an equilibrium. If s is the selection coefficient against AA and t is the selection coefficient against aa (relative to a fitness of 1.00 for Aa genotypes), then, at equilibrium, the ratio of $A{:}a$ allele frequencies is $\hat{p}/\hat{q} = t/s$, or

$$\hat{p} = \frac{t}{s + t}$$

The equilibrium is **stable**, which means that allele frequencies in the neighborhood of the equilibrium automatically move toward the equilibrium under the influence of selection. Moreover, the equilibrium represents the point at which the **average fitness** of the population is a maximum.

8. Heterozygote inferiority is the opposite of overdominance; with heterozygote inferiority, the fitness of both homozygotes is greater than that of the heterozygote. If s is the selection coefficient *in favor* of AA and t is the selection coefficient *in favor* of aa (relative to a fitness of 1.00 for Aa), there is again an equilibrium at

$$\hat{p} = \frac{t}{s + t}$$

but, in this case, the equilibrium is **unstable**, which means that allele frequencies in the neighborhood of the equilibrium move progressively farther away from the equilibrium. Thus, if p is less than its equilibrium value, p

eventually goes to 0, whereas if p is greater than its equilibrium value, p eventually goes to 1. In this case, the average fitness of the population is a minimum at the point of equilibrium, and selection, acting to increase fitness, moves the population "up the hill" of average fitness. (This analogy between a fitness surface and a landscape forms the basis of the **adaptive topography** concept.) However, the average fitness ultimately attained by selection may not be the highest possible, but random genetic drift permits random changes in allele frequency, allowing a population to fully "explore" its adaptive topography.

9. Many alleles are harmful, but they cannot be completely eliminated by natural selection because of recurrent mutation to the harmful allele. There are two important cases to consider. First, if the harmful allele is completely recessive, then the equilibrium frequency of the allele is given by

$$\hat{q} = \sqrt{\frac{\mu}{s}}$$

where s is the selection coefficient of the recessive homozygote and μ is the mutation rate of the normal to the harmful allele. On the other hand, if the allele is not completely recessive, its equilibrium frequency is given by

$$\hat{q} = \frac{\mu}{s'}$$

where s' represents the selection coefficient against the *heterozygote*. These formulas have had important applications in the estimation of human mutation rates.

10. Random genetic drift is a process whereby allele frequencies can change in random fashion from one generation to the next. Random genetic drift comes about because populations are not infinite in size; thus, the allele frequencies among successful gametes that give rise to each new generation may not be perfectly representative of allele frequencies in the previous generation. These random

effects tend to be larger in small populations, so random genetic drift is particularly important in populations that are small. The aspect of a population that determines the importance of random genetic drift is not the population's actual size, however. The determining factor is the population's **effective size**, which is the size of a theoretically ideal population that has the same rate of random genetic drift as the actual population; usually, though not always, the effective size of a population is less than its actual size.

11. It is often convenient to consider the effect of random genetic drift in the context of a large number of completely isolated subpopulations, each having the same effective size. The principal effect of random genetic drift is to cause a progressive dispersion of allele frequencies among the subpopulations. Although the allele frequency in any population can change randomly from generation to generation, the average allele frequency among subpopulations remains constant. Ultimately, each population becomes fixed for one allele or the other, and the probability that a population becomes fixed for any particular allele is the same as the frequency of that allele in the initial generation.

12. The dispersion of allele frequencies among subpopulations is conveniently measured in terms of the fixation index F_{ST} (or, as we have been calling it, simply F). The fixation index increases as a result of random genetic drift, and it increases in such a way that the quantity $(1 - F)$ decreases by a fraction $1/(2N)$ in each generation, with N being the effective size of each of the subpopulations. Although each subpopulation has genotype frequencies given by the Hardy-Weinberg rule in terms of its own allele frequencies, there is a deficiency of heterozygotes when the subpopulations are considered as a whole. Indeed, the average genotype frequencies among subpopulations are given by

$$AA: \quad p^2(1 - F) + pF$$
$$Aa: \quad 2pq(1 - F)$$
$$aa: \quad q^2(1 - F) + qF$$

where p and q respectively represent the average A and a allele frequencies among subpopulations. These genotype frequencies illustrate the inbreeding-like effect of random genetic drift.

13. Since random genetic drift creates inbreeding-like effects, the average frequency of recessive homozygotes among subpopulations will be greater than what the frequency would be if the subpopulations were to fuse and undergo random mating. (Such population fusion is known as **isolate breaking**.) The excess homozygosity for a recessive allele among isolated subpopulations is equal to pqF.

14. Founder effects refer to the random genetic drift that occurs when a relatively small group of individuals founds a new population, as the allele frequencies among the founders may not be perfectly representative of those in the population of origin. Founder effects may be particularly important in humans because founders of new populations are often groups of relatives. Founder effects and random genetic drift are responsible for an unknown fraction of the allele frequency variation among human populations.

15. Molecular evolution refers to changes in macromolecules that accompany the evolutionary process. Sequence comparisons of a protein or fragment of DNA among a variety of living organisms can be used to reconstruct the evolutionary history of the organisms. During the course of evolution, genes can undergo **gene substitutions**, during which one allelic form of a gene is replaced with a different allelic form. The simplest form of gene substitution is a **nucleotide substitution**, in which the alleles involved in the gene substitution differ by one nucleotide. Given information on the nucleotide sequence of a gene in a wide variety of species (or given the amino acid sequence of the corresponding protein), an evolutionary tree of the species can be constructed because the number of substitutions increases with the length of time two species have been separated. (Indeed, some authors interpret existing data as evidence of a **molecular clock**—a relatively constant rate of gene substitution per period of evolutionary time.) Usually more than one possible tree is consistent with sequence comparisons, but the best tree is considered to be the one that involves the smallest assumed number of substitutions; such a tree is called the **most parsimonious** tree.

16. A principal mechanism whereby species acquire new genes in the course of evolution is **duplication and divergence**. In this process, one of the gene copies is free to undergo genetic divergence, as the function of the original gene is carried out by the other copy. Repeated occurrence of duplication and divergence leads to a **family** of related genes, of which the hemoglobin family is an example. Among humans, one of the largest families of DNA sequences is the **Alu family**, consisting of about 500,000 copies of a 300-bp transposable sequence scattered throughout the genome.

17. Eukaryotes may also evolve new genes by means of an **exon shuffle**—a process whereby exons from different genes are brought together in novel and perhaps useful combinations. Many exons code for functionally or structurally distinct parts of the corresponding protein; these parts are known as **domains** or **modules**. The exon shuffle thus produces novel combinations of domains. In the case of hemoglobin, the original gene may have been split into four exons; in the course of evolutionary time, the central intervening sequence seems to have been deleted, leading to the present-day gene with three exons and two introns.

WORDS TO KNOW

Allele	**Migration**	**Random Genetic Drift**
Fixation	One-way	Ideal population
Loss	Island	Effective population number
Neutral		Fixation index
Additive	**Selection**	Isolate breaking
Partial dominance	Relative fitness	Founder effect
	Selection coefficient	
Mutation	Average fitness	**Molecular Evolution**
Forward	Overdominance	Sequence divergence
Reverse	Stable equilibrium	Gene substitution
Equilibrium	Heterozygote inferiority	Molecular clock
	Unstable equilibrium	Most parsimonious tree
	Sickle cell anemia	Duplication and divergence
	Adaptive topography	Exon shuffle
		Gene family

PROBLEMS

1. For discussion: What are some reasons why the allele frequency of a particular allele may vary from population to population? Would these same processes apply to variation in the allele frequency of harmful recessive alleles? Given that there is variation in the frequency of harmful recessive alleles among populations, do you believe that screening for heterozygotes should focus on high-risk populations, such as Ashkenazi Jews for Tay-Sachs disease or blacks for sickle cell anemia? Some commentators have argued against such focusing because, when a group is already the target of social discrimination, calling attention to their particular genetic diseases further isolates and stigmatizes the group. What is your reaction to this argument?

2. What is meant by the term *sequence divergence* as applied to the two copies of a duplicated gene? What processes might promote sequence divergence? What processes might impede sequence divergence?

3. A certain allele has a forward mutation rate of 3×10^{-5} and its initial allele frequency is 0.85. What is the allele frequency in the next generation? What is its frequency after 10 generations? How many generations are required to reduce the allele frequency to 0.75? What will be the allele's ultimate frequency assuming that the rate of reverse mutation is 0?

4. An allele A mutates to a at the rate 3×10^{-5}, and a reverse mutates to A at the rate 4×10^{-6}. What is the ultimate equilibrium allele frequency of A? Of a?

5. Among present-day whites in Claxton, Georgia, the allele frequency of M in the MN blood group system is 0.507. Among present-day Claxton blacks the allele frequency is 0.484, and among blacks from West Africa the allele frequency is 0.474. Assuming one-way migration for 10 generations, what migration rate is required to account for these allele frequencies?

6. Consider the genotypes AA, Aa, and aa. Genotpye AA has, on the average, 1.05 offspring in the course of a reproductive lifetime; genotype Aa has, on the average, 1.04 offspring; and genotype aa dies before reaching reproductive age. What are the relative fitnesses of AA, Aa, and aa?

7. Genotypes *AA, Aa,* and *aa* have relative fitnesses 1.00, 0.98, and 0.40, respectively. What is the selection coefficient against *Aa*? Against *aa*?

8. A recessive lethal has an allele frequency of 0.05. What would the allele frequency be in the next generation? What would it be after 10 generations?

9. Allele *A* mutates to allele *a* at the rate 4×10^{-6}, and *a* is a recessive lethal. What is the equilibrium allele frequency of *a*? At equilibrium, what is the frequency of homozygous recessive zygotes?

10. A recessive lethal has an equilibrium allele frequency of 0.005. What is the mutation rate of the normal to the recessive allele?

11. Allele *A* mutates to allele *a* at the rate 4×10^{-6}, and *a* is a recessive lethal. In addition, there is a 2 percent selection coefficient against heterozygotes. What is the equilibrium allele frequency of *a*?

12. Genotypes *AA, Aa,* and *aa* have relative fitnesses of 1.00, 0.99, and 0.97, respectively, and the allele frequency of *A* is 0.9. Assuming random mating, what is the average fitness of the population?

13. Genotypes *AA, Aa,* and *aa* have relative fitnesses of 0.90, 1.0, and 0.95, respectively. What is the equilibrium allele frequency of *A*? Of *a*? What is the average fitness of the population at equilibrium? Are there any allele frequencies that would produce a greater average fitness?

14. For a locus with two neutral alleles, *A* and *a*, among a group of isolated subpopulations, the average allele frequencies among subpopulations are 0.6 and 0.4, respectively. The average heterozygosity among subpopulations is 0.456. What is the fixation index among the subpopulations?

15. Consider a recessive allele in two equal-sized populations. In population 1 the allele frequency of the recessive is 0.2, in population 2 it is 0.02. What is the frequency of homozygous recessives in population 1? In population 2? What is the average frequency of homozygous recessives? If the two populations were to fuse and undergo random mating, what would be the frequency of homozygous recessives in the fused population?

FURTHER READING AND REFERENCES

Baba, M. L., L. L. Darga, M. Goodman, and J. Czelusniak. 1981. Evolution of cytochrome c investigated by the maximum parsimony method. J. Mol. Evol. 17:197–213. Source of Figures 15.12 and 15.13.

Barker, W. C., and M. O. Dayhoff. 1980. Evolutionary and functional relationships of homologous physiological mechanisms. BioScience 30:593–600. On the evolution of related families of proteins by duplication and divergence. Source of Figure 15.14.

Barker, W. C., L. K. Ketcham, and M. O. Dayhoff. 1978. A comprehensive examination of protein sequences for evidence of internal gene duplication. J. Mol. Evol. 10:265–281. Among 116 "superfamilies" of related proteins, 20 give evidence of internal gene duplication.

Bodmer, W. F., and L. L. Cavalli-Sforza. 1976. Genetics, Evolution, and Man. Freeman, San Francisco. Chapters 6 through 13 are relevant to human population genetics.

Dickerson, R. E. 1980. Cytochrome c and the evolution of energy metabolism. Scientific American 242:136–153. Traces the evolution of this important protein from bacteria to humans.

Dobzhansky, Th., F. J. Ayala, G. L. Stebbins, and J. W. Valentine. 1977. Evolution. Freeman, San Francisco. An introduction to the subject.

Doolittle, R. F. 1981. Similar amino acid sequences: Chance or common ancestry? Science 214:149–159. How can one tell whether similar sequences derive from a common ancestral gene?

Eiferman, F. A., P. R. Young, R. W. Scott, and S. M. Tilghman. 1981. Intragenic amplification and divergence in the mouse α-fetoprotein gene. Nature 294:713–722. An example of the creative power of the exon shuffle when combined with duplication and divergence.

Ewens, W. J., R. S. Spielman, and H. Harris. 1981. Estimation of genetic variation at the DNA level from restriction endonuclease data. Proc. Natl. Acad. Sci. U.S.A. 78:3748–3750. Gives the reasoning behind the use of restriction enzymes to estimate genetic variability; rather mathematical.

Ferris, S. D., A. C. Wilson, and W. M. Brown. 1981. Evolutionary tree for apes and humans based on cleavage maps of mitochondrial DNA. Proc. Natl. Acad. Sci. U.S.A. 78:2432–2436. Another approach to molecular relationships between species.

Friedman, M. J., and W. Trager. 1980. The biochemistry of resistance to malaria. Scientific American 244:154–164. The biochemical basis of resistance conferred to sickle cell and thalassemia heterozygotes.

Gō, Mitiko. 1981. Correlation of DNA exonic regions with protein structural units in haemoglobin. Nature 291:90–92. Source of Figure 15.16.

Hartl, D. L. 1980. Principles of Population Genetics. Sinauer Associates, Sunderland, Mass. Pages 141–266 give a more detailed account of the processes that change allele frequency.

Jacquard, A. 1974. The Genetic Structure of Populations. Springer-Verlag, New York. Source of Figure 15.11.

Jagadeeswaran, P., B. G. Forget, and S. M. Weissman. 1981. Short interspersed repetitive DNA elements in eucaryotes: Transposable DNA elements generated by reverse transcription of RNA pol III transcripts? Cell 26:141–142. How do members of the Alu family transpose, and why are there so many copies?

Miyata, T., and H. Hayashida. 1981. Extraordinarily high evolutionary rate of pseudogenes: Evidence for the presence of selective pressure against changes between synonymous codons. Proc. Natl. Acad. Sci. U.S.A. 78:5739–5743. Pseudogenes evolve faster than any other known genes.

Rubin, C. M., C. M. Houck, P. L. Deininger, T. Friedmann, and C. W. Schmid. 1980. Partial nucleotide sequence of the 300-nucleotide interspersed repeated human DNA sequences. Nature 284:372–374. Source of data in Figure 15.15.

Schimke, R. T. 1980. Gene amplification and drug resistance. Scientific American 243:60–69. Experimental evolution of gene duplications.

Schmid, C. W., and W. R. Jelinek. 1982. The Alu family of dispersed repetitive sequences. Science 216:1065–1070. A summary of what is known about the most prominent short dispersed repeat family of DNA in primates.

Scientific American, September 1978, volume 239, number 3. An entire issue on the subject of evolution.

Stansfield, W. D. 1977. Science of Evolution. Macmillan, New York. An introductory textbook of evolution.

chapter 16
Quantitative and Behavior Genetics

Much human genetic diversity involves genes which, taken individually, have small effects but which operate in concert and thereby exert an aggregate impact on the traits they influence. Traits that are influenced simultaneously by many genes having small effects are also almost always influenced to one extent or another by an individual's environment—by diet, training, experience, and other nongenetic factors. Traits in this category include those that can be measured numerically in single individuals, traits such as height, weight, girth, IQ score, intensity of skin pigmentation, and blood pressure. Such **quantitative traits**, as these are called, are supremely important in animal and plant breeding. The growth rates of

beef cattle and pigs, milk production in cattle, egg production in poultry, and yields of corn, wheat, rice, and soybeans are all important quantitative traits.

There is a second catetgory of traits strongly influenced by both heredity and environment; these are known as **threshold traits**. Important examples of threshold traits in humans are diabetes (technically diabetes mellitus, the most common form of diabetes in Caucasians) and the mental illness schizophrenia. Like quantitative traits, threshold traits are influenced by the environment and by the aggregate action of several or many loci. Unlike quantitative traits, however, the phenotype cannot be measured directly. In the case of threshold traits, the genotype and environment act together to determine a risk or **liability**—a predisposition—toward a particular condition. This liability cannot be measured directly, as can such quantitative traits as height and weight. All that can be observed about a person's liability is whether the individual has the condition in question or not. Threshold traits are either present or absent; for example, a person is either diabetic or not diabetic. If a person expresses the trait, however, then that person's liability toward the trait must be high (above a certain "threshold"); if a person does not express the trait, the person's liability must be low (below the threshold). This information can be used to make inferences about the inheritance of liability.

Determining the genetic basis of quantitative or threshold traits is no easy matter. In the first place, one cannot investigate them by compiling elaborate pedigrees and making inferences about how individual genes have trickled down through the generations. This pedigree method works fine for traits determined by single genes, of course, because the presence of a trait in a person is indicative of genotype. But in the case of quantitative and threshold traits, each individual gene may have such a small effect that its hereditary transmission cannot be discerned. Moreover, quantitative and threshold traits are influenced by genes at several loci, in many cases tens of loci or even hundreds. So even if the segregation and transmission of a single gene were discernible, the effects caused by the simultaneous segregation and transmission of all the other genes affecting the same trait would hopelessly complicate the picture.

In actual fact, all traits are potentially influenced by many genes. Even the expression of traits that are inherited in a simple Mendelian manner, such as cystic fibrosis, Tay-Sachs disease, and sickle cell anemia, are influenced

by one or more genes called **modifiers** that are polymorphic in human populations. Such modifiers explain in part why there is variation in the severity of disease among individuals affected with, say, cystic fibrosis. However, conditions that have a simple Mendelian pattern of inheritance are relatively easy to understand genetically because a mutation at a single locus has a much larger influence on the trait than the aggregate effects of all the modifiers. Although modifiers certainly do influence the severity of cystic fibrosis, one recessive allele determines whether or not the disease will be present.

In the case of quantitative and threshold traits, on the other hand, all the genes that influence the trait may be so nearly identical in the magnitude of their effect that no one of them stands out. For this reason quantitative and threshold traits are often called **multifactorial** or **polygenic** traits, and the underlying genes of individually small effects are known as **polygenes** or **minor genes**. Simple Mendelian traits such as cystic fibrosis and sickle cell anemia are known as **monogenic** traits, indicating that one locus has a major effect that is clearly distinguishable from the effects that other modifiers may have; accordingly, the major locus responsible for a simple Mendelian trait is often called a **major gene**.

A complicating factor in the study of quantitative and threshold traits is that they are influenced by environment. The strength of the environmental influence on the expression of a trait, as compared with the genetic effects, varies from trait to trait. In some cases, the environmental effects are rather small; in others the environmental and genetic factors may be about equally important; in still other cases environmental factors may be so important as to render insignificant whatever genetic contributions there may be. As before, it should be kept in mind that all traits are influenced to some extent by environment. If this were not the case, then no inherited disorders could be successfully treated by surgery, special diets, drugs, or any other medical practices. But it should also be kept in mind that the influence of the usual environments to which people are exposed on such traits as cystic fibrosis and sickle cell anemia are small compared with the effects of the single mutant gene. Indeed, the usual environmental effects on these traits are relatively minor, like the effects of inherited genetic modifiers. In the case of quantitative and threshold traits, however, the environment may be equally important or even more important than hereditary factors.

Major and Minor Genes

The distinction between major and minor genes is not entirely clear-cut. A major gene in one context can be a minor gene in another context. Such a situation is illustrated for a hypothetical quantitative trait in Figure 16.1. The vertical axis represents the measured phenotype of individuals, the horizontal axis represents an environmental gradient (such as temperature), and the black and gray lines depict the relationship between phenotype and environment for genotypes AA and aa, respectively. (For the sake of simplicity the heterozygote has been ignored.) In any range of environment, such as E_1 or E_2 in the figure, the range in phenotype of the genotypes is indicated by the vertical bars along the vertical axis, with the black bar representing AA and the gray bar representing aa.

If the trait in Figure 16.1 is examined in the range of environments denoted E_1, then alleles A and a are typical minor genes. The AA and aa genotypes have different effects on phenotype, but there is so much variation in phenotype due to alleles at other loci and due to the environment that AA and aa genotypes cannot be distinguished by examination of phenotype. Indeed, in environment E_1, the range

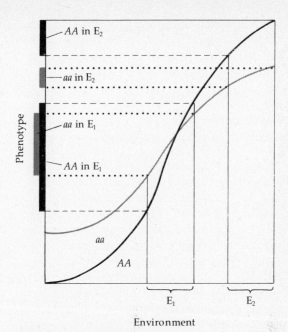

Figure 16.1 Hypothetical relationship between environment and phenotype for two genotypes. In environments denoted E_1, the locus in question is a minor gene; in environments denoted E_2, the locus is a major gene. At the far left, genotype aa is better than AA; at the far right, AA is better than aa. The relationship illustrated is an example of genotype-environment interaction.

of aa phenotypes is entirely included within that of AA phenotypes.

On the other hand, if the trait in Figure 16.1 is examined in the range of environments denoted E_2, then A and a can be considered major genes. In E_2 environments, AA and aa genotypes have different phenotypes and the range of phenotypes does not overlap. Thus, each genotype is associated with a distinct range of phenotypes, and in E_2 environments the trait in Figure 16.1 is a simple Mendelian trait.

Figure 16.1 illustrates another important point about quantitative traits: There may not be any genotype that is the "best" across a broad range of environments. In Figure 16.1, genotype aa is best at the far left of the range of environments, whereas genotype AA is best at the far right. It is thus important to remember that, for quantitative traits, phenotype depends on genotype *and* environment.

Procedures Used to Study Quantitative Traits

In spite of the complexities involved in the study of quantitative and threshold traits, progress has been made in assessing the heredity contribution to many such traits. Two principal issues are of interest: How many genes contribute to the determination of the trait? How much of the variation in the trait observed in a population can be attributed to genetic differences between individuals? There is no general answer to the first question—how many genes are involved—because the number varies from trait to trait. At one extreme are traits such as seed color in wheat, which ranges from dark red to white and is primarily influenced by the alleles at only three loci. At the other extreme are traits such as growth rate in mice, which are influenced by hundreds of loci.

   If a trait is influenced by a relatively small number of genes and if homozygous genotypes can be identified, then special methods can be applied to determine how many genes are involved in a quantitative trait. These procedures work well only for traits that are not strongly influenced by environment. To illustrate one method, suppose that the alleles at a single locus determine a trait. Suppose further that homozygous genotypes can be obtained from, for example, two different inbred lines, one line being genetically AA and the other aa. Then all offspring of a mating of $AA \times aa$ will be genetically Aa. The phenotype of these Aa hybrids indicates the direction and degree of dominance of the alleles. If the hybrids' phenotype is nearer to that of AA than that of aa, then the A allele is partially dominant; if the

hybrids' phenotype is intermediate between those of the parents, then neither A nor a is dominant. Information about the number of genes influencing the trait is obtained by crossing the hybrids among themselves. The $Aa \times Aa$ matings produce $\frac{1}{4}$ AA, $\frac{1}{2}$ Aa, and $\frac{1}{4}$ aa progeny. If there is no dominance, then only the AA and aa offspring will have phenotypes as extreme with respect to this trait as those of the original parents; in this case $\frac{1}{4} + \frac{1}{4} = \frac{1}{2}$ of the offspring will have extreme phenotypes; the rest will be intermediate. When two loci are involved in the trait, however, then the genotypes of hybrids from crosses of $AABB$ and $aabb$ inbred lines will be $AaBb$; if the A and B loci are unlinked, then the cross of $AaBb \times AaBb$ will produce $\frac{1}{16}$ $AABB$ and $\frac{1}{16}$ $aabb$ offspring. In this case only $\frac{1}{16} + \frac{1}{16} = \frac{1}{8}$ of the offspring will have phenotypes as extreme as those of the original parental inbred lines. When three unlinked loci are involved in a trait, then $\frac{1}{32}$ of the offspring of hybrid matings will be as extreme as the original homozygous strains. Thus, by determining the proportion of offspring that have phenotypes like their homozygous grandparents, a rough idea of the number of genes involved in the trait can be obtained.

Number of Loci Influencing Human Skin Pigmentation

Studies of inbred lines are far more relevant to experimental animals and plants than to human populations. Moreover, the method outlined above is useful only in cases in which the number of loci influencing a quantitative trait is relatively small. Nevertheless, the principles involved have been applied with reasonable success to the study of the number of loci contributing to skin color differences between blacks and whites. The basic information available is the distribution of skin pigmentation among the children of matings of Negro × Caucasian, hybrid × hybrid, hybrid × Negro, and hybrid × Caucasian. The first finding in such matings is that the loci involved in skin pigmentation have little or no dominance; heterozygous offspring tend to be very nearly intermediate in skin color between their homozygous parents. The second finding is that the number of loci involved in the major differences in skin pigmentation between Negroes and Caucasions is remarkably small; the patterns of segregation suggest that approximately three or four loci are involved, although the nature of the evidence is by no means conclusive.

Figure 16.2 compares the distribution of black skin pigmentation among Negro Americans (heavy line) with the theoretical distributions expected if skin pigmentation were due to a pair of alleles at each of three loci (thin line) or four loci (gray line). The fits are relatively good except at the extremes of the color range, but the theoretical curves are based on several questionable assumptions, including (1) complete lack of dominance of pigment-determining alleles, (2) absence of linkage, (3) random mating, and (4) 20 percent Caucasian-derived alleles in Negro Americans. This last assumption is very uncertain because the proportion of genes in American Negroes that derive from Caucasian ancestors is exceedingly difficult to measure. Estimates often involve one-way migration formulas of the type illustrated in Chapter 15; estimates range all the way from 6 to 30 percent and depend on the population studied—Charlotte versus Seattle, for example—and on the particular loci examined—ABO, Rh, MN, etc. (Equally difficult to assess is the proportion of genes in American Caucasians that derive from Negro ancestors). Despite the assumptions, Figure 16.2 illustrates that the major features of skin pigment distribution in Negro Americans can be accounted for by the segregation of only a handful of loci.

That only three of four loci out of tens of thousands are involved in black versus white skin pigmentation is all the more astonishing when seen in light of the prejudice, the social and economic injustice, the ignorance and racism that have been based on just these differences. It so happens that northern European Caucasians are nearly homozygous for the genes determining light pigmentation and that

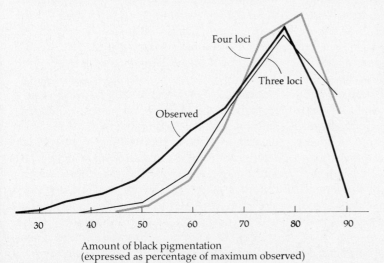

Figure 16.2 Distribution of black skin pigmentation in Negro Americans (heavy line) compared with the distributions expected if skin pigmentation were due to a pair of alleles at each of three loci (thin line) or four loci (gray line). The fits are relatively good except at the extremes of the color range.

Four loci

Three loci

Observed

30   40   50   60   70   80   90

Amount of black pigmentation
(expressed as percentage of maximum observed)

African Negroes are nearly homozygous for the genes determining dark pigmentation. Why this is true is not clear. It possibly relates to natural selection for skin color in the long-forgotten past, but whether the selection had to do with protection from the ultraviolet rays of the sun or something entirely different no one really knows. (One suggestion is that dark pigmentation prevents overproduction of vitamin D, a substance whose synthesis is facilitated by the sun's ultraviolet rays, although recent evidence makes this hypothesis doubtful.) The few loci involved in human skin color are exceptional in that one race has a high preponderance of one type of allele and another race another. How different recent human history might have been, especially in the United States, if these differences in gene frequency were in the alleles at three or four loci determining blood groups, and therefore not externally visible, rather than in the genes affecting skin color.

Although the alleles for light skin pigmentation are preponderant in Caucasians and those for dark skin pigmentation are preponderant in Negroes, the alleles are not completely fixed; there are alleles for darker pigmentation in most Caucasian populations and alleles for lighter pigmentation in most Negro populations, but the frequency of these alleles is low. The chance that a newborn will be homozygous for several of these rare alleles at once is therefore quite small, but occasionally it happens that two rather light-skinned parents will have a dark-skinned child or two rather dark-skinned parents will have a light-skinned child. This is especially true in countries like the United States where partial genetic mixing among races has elevated the frequency of dark pigmentation alleles among light-skinned people and the frequency of light pigmentation alleles among dark-skinned people.

Success in determining the number and action of genes affecting skin color has been achieved partly because the number of loci involved is small and because the allele frequencies at these loci are markedly different in two populations. When more loci are involved, or when the allele frequencies are not strikingly different in different populations, or when environmental influences on a trait are strong, then matters are much more difficult. Most quantitative and threshold traits are of this sort. Special methods are therefore required to sort out genetic and environmental influences on these traits. Of special importance in the study of threshold traits are certain clones (genetically identical individuals) that occur quite routinely and normally in human populations—identical twins.

Twins

Two types of twins can occur: One type consists of **monozygotic** twins (also called **MZ** twins or **identical** twins); the other type is **dizygotic** twins (also called **DZ** twins or **fraternal** twins). As the name implies, MZ twins originate from a single fertilized egg (zygote) that gives rise to two separate embryos, as would happen if an embryo in an early stage of development were to fall apart into two distinct clumps of cells. Because they derive from one fertilized egg, MZ twins are often called **one-egg** twins. The falling apart of embryonic cells can occur at almost any early stage of embryonic development. If it occurs early enough, each twin may have its own amnion, chorion, and placenta. [See Figure 16.3 for diagrams of the several possible intrauterine relationships of twins; those that have a separate amnion, chorion, and placenta are illustrated in part (a).] If the separation occurs later, the twins may have separate amnions but will be inside the same chorion [see Figure 16.3(c)]. If still later, the twins will be within the same amnion [see Figure 16.3(d)].

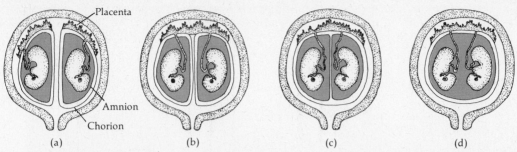

Figure 16.3   The several possible intrauterine relationships of twins. *(a)* The twins (either MZ or DZ) have separate amnions, chorions, and placentas. *(b)* The twins (again either MZ or DZ) have separate amnions and chorions but a fused placenta. *(c)* MZ twins may have separate amnions but the same chorion and placenta. *(d)* MZ twins may share the same amnion, chorion, and placenta.

Sometimes it happens that the separation of the two embryos is not complete or that parts of the embryos fuse together. Such fused embryos give rise to **Siamese** twins—twins who are joined together. Siamese twins can often be separated surgically if, for example, they are joined by the arms or legs. In other cases the twins are so fused that they share one vital organ, like the heart; the life of one twin must then be sacrificed for the sake of the other. Fortunately, this dilemma arises only infrequently because Siamese twins are extremely rare.

Dizygotic twins arise from a double ovulation, with each egg being fertilized by a different sperm. DZ twins almost always have their own amnion and chorion, and they very often have their own placentas [see Figure 16.3(*a*) and (*b*)]. DZ twins arise from two eggs and are therefore often called **two-egg** twins. DZ twins are genetically related as siblings—as two brothers, or two sisters, or brother and sister. MZ twins, because they originate from a single egg and a single sperm, are genetically identical.

The frequency of twinning varies from population to population. Among whites in the United States, about 1 in 85 births is twins; among Japanese, about 1 in 145 births is twins. This rather large difference is due almost en-

tirely to differences in the rate of DZ twinning. The rates of DZ twinning in the two populations are about 1 in 135 births and 1 in 370 births, respectively; the rates of MZ twinning are 1 in 256 and 1 in 238—almost the same. There does seem to be an inherited tendency toward two-egg twinning, but the tendency is not strong and the mode of inheritance is complex.

The importance of twins in the study of threshold traits derives from the genetic identity of one-egg twins. Whereas differences between two-egg twins with regard to any trait may arise because of environmental or genetic factors—or combinations of the two—differences between one-egg twins must be due only to environmental factors because the twins are gentically identical. Consequently, if the occurrence or nonoccurrence of a trait has a genetic component, then identical twins will more often be alike with regard to the trait in question than will fraternal twins. Conversely, if the genetic component is small and unimportant, then one-egg twins will be no more similar with respect to the trait than two-egg twins. The importance of genetic factors in influencing such traits as facial features is illustrated in the sometimes striking resemblance between MZ twins (Figure 16.4).

One difficulty that arises in twin studies—

Figure 16.4 Monozygotic twins.

and it is a serious one—is that the environment of one-egg twins may be more similar than that of two-egg twins. Within the uterus, one-egg twins frequently share the same embryonic membranes and placenta; two-egg twins rarely do. After their birth, the striking resemblances in facial features, body conformation, and so on between identical twins probably lead their parents, teachers, and friends to treat them more similarly than if they were two-egg twins. To the extent that these factors are important relative to the trait in question, resemblances between identical twins due to similarities in their environments will mistakenly be attributed to their genetic identity. This difficulty can be avoided in part by studying identical twins who, owing to a family breakup, were separated from each other shortly after birth and raised in different environments; such cases are not numerous, though. Even when reared separately, however, MZ twins may have resemblances resulting, not from their identical genes, but from their special and intimate intrauterine association. The overall conclusion to keep in mind is that the results of twin studies are to be interpreted with caution.

MZ and DZ Twins in the Study of Threshold Traits

The extent to which twins are alike with respect to any threshold trait is expressed as a number called the **concordance**. A concordance is simply the percentage of cases in which both twins of a pair have the trait when it is known that at least one has it. Suppose, for example, that 70 sets of identical twins were found in which only one member of the pair was affected with some disorder and that 30 sets were found in which both were affected; then the concordance would be $^{30}/_{(30\ +\ 70)} = 30$ percent.

The concordance varies according to the degree of genetic determination of a trait and according to the mode of inheritance. A trait that is completely determined by heredity will have a concordance in one-egg twins of 100 percent. The concordance of the same trait in two-egg twins will depend on how many genes are involved and on how these interact. To take a simple case, the concordance in fraternal twins of a trait determined by a single dominant gene with complete penetrance will be 50 percent (if only one parent is affected); if the trait is determined by a single recessive gene then the concordance in fraternal twins will be 25 percent (if neither parent is affected). If penetrance is incomplete or if several genes are involved in the inheritance, then the concordance cannot easily be calculated. It will depend on the number of loci and where they are on the chromosomes, on the allele frequencies in the population, on whether any or all of the loci are dominant, and on whether the loci interact in a complicated fashion. However, the difference in the concordance rates of one-egg and two-egg twins will reveal, in a general way, the importance of genetic factors in the causation of a condition.

With respect to traits known to be almost totally genetic in origin, the concordance rates are as expected. Concordance in one-egg twins with respect to blood groups is 100 percent. Concordance in fraternal twins depends on the particular blood group under consideration, but it is as anticipated. (Because of the 100 percent concordance in one-egg twins, the various blood groups are an easy and reliable way to determine whether newborn twins of the same sex are identical or fraternal.) The con-

cordance in one-egg twins with respect to whether eye color is blue or brown is again 100 percent; among two-egg twins it is 55 percent. A finer classification of eye color by shade reveals slight differences in some one-egg twins; these differences must be environmental in origin. With respect to light versus dark hair color, the concordance in one-egg and two-egg twins is 100 percent and 30 percent, respectively, although again a finer classification of color by shade reveals environmentally caused differences in some one-egg twins.

Table 16.1 lists the concordance rates in one-egg twins and two-egg twins for 14 threshold traits. The two traits set off at the bottom of the table—cancer at any site and death from acute infection—do not have significantly higher concordances in one-egg than in two-egg twins. Thus there is not a strong genetic component in these traits, at least insofar as can be ascertained from twins. All the other traits in the table do have a significant genetic component as judged by the concordance rates. These include cancers occurring at the same site, hypertension, bronchial asthma, and diabetes; the infectious diseases tuberculosis and rheumatic fever; the autoimmune disease rheumatoid arthritis; the nervous disorders classified as mental deficiency, manic-depressive psychosis, epilepsy, and schizophrenia; and the behavioral trait of tobacco smoking. Thus, even though the genetic diversity concerned with such traits as these cannot be directly observed, the studies of twins indicate that the widespread genetic diversity in blood groups and in enzymes as detected by electrophoresis does have a counterpart in the loci that influence threshold traits.

Tuberculosis and rheumatic fever are both caused by bacteria. The diseases themselves are therefore caused by an environmental condition—by exposure to the infectious agents—and in these cases the genetic contribution to the traits must represent susceptibili-

TABLE 16.1 CONCORDANCES OF ONE-EGG AND TWO-EGG TWINS FOR SEVERAL TRAITS

Trait	Concordance in one-egg twins, percent*	Concordance in two-egg twins, percent*
Schizophrenia	34 (203)	12 (222)
Tuberculosis	37 (135)	15 (513)
Manic-depressive psychosis	67 (15)	5 (40)
Hypertension	25 (80)	7 (212)
Mental deficiency	67 (18)	0 (49)
Rheumatic fever	20 (148)	6 (428)
Rheumatoid arthritis	34 (47)	7 (141)
Bronchial asthma	47 (64)	24 (192)
Epilepsy	37 (27)	10 (100)
Diabetes mellitus	47 (76)	10 (238)
Smoking habits (females only)	83 (53)	50 (18)
Cancer (at same site)	7 (207)	3 (767)
Cancer (at any site)	16 (207)	13 (212)
Death from acute infection	8 (127)	9 (454)

*Numbers in parentheses denote number of twin pairs studied.
Source: Based on data in L. L. Cavalli-Sforza and W. F. Bodmer, 1971, The Genetics of Human Populations, Freeman, San Francisco.
Note: The *concordance* of a trait in a group of twins is the percentage of cases in which, when one twin is affected, the other twin is also affected. The difference in concordance rates between one-egg and two-egg twins provides a very crude and often difficult to interpret measure of the importance of hereditary factors in the determination of a trait. In the two examples set off at the bottom of the table, the concordance rates between one-egg and two-egg twins are not significantly different. The concordance rates for all the other traits are significantly different, thus providing evidence of some degree of genetic involvement in susceptibility to the various conditions.

ty to infection. Several other traits in Table 16.1 are worth special mention because they are so widespread. Diabetes affects about 1 in 250 people and so is very common. (The incidence of most of these traits varies with age. In the case of diabetes, for example, about 1 in 1000 people under age 25 is affected whereas about 1 in 30 people aged 60 to 70 are affected. The incidences given for diabetes and the other traits are averages.) Epilepsy and manic-depressive psychosis each affect about 1 in 250 people. Mental deficiency (which can have any of a multitude of causes) affects about 1 in 130 people. Schizophrenia affects about 1 in 1000 people (and about half of all persons committed to mental hospitals are schizophrenics). Altogether, some 2 percent of all people in the United States—1 in 50—suffer from one of these five traits. And all these traits have a significant genetic component, although each is also influenced by an individual's environment. The concordance in identical twins for these traits averages around 50 percent. This means that when a person has a genotype with a strong liability or predisposition toward one of these threshold conditions, the chance is still not 100 percent that the person will actually become affected; indeed, the chance is roughly half of what one would expect if the occurrence or nonoccurrence of the trait were entirely determined by genes.

Quantitative Traits: The Normal Distribution

Threshold traits are somewhat difficult to deal with because there are only two classes of individuals: affected and unaffected. Quantitative traits are somewhat easier because the magnitude of the trait can be measured directly in each individual. Moreover, there are many classes of individuals, as many as the precision of a particular measuring device will allow. Take height, for example. In principle, every person has a slightly different height from every other one. But to detect this difference in two individuals of nearly the same height might require measuring height in angstrom units. In practice, therefore, height is measured to the nearest inch or centimeter or some other convenient interval. This means that some people will register the same height; differences between them cannot be detected in terms of the units of measurement. There will still be many height categories—36, for example, if height is measured to the nearest inch and the range of heights is from 4 ft to 7 ft. Each of these categories will contain a different proportion of people, but 36 heights and 36 proportions

present too much information to be comprehended easily. A comprehension of the relationships among a range of heights is made somewhat easier by drawing a picture, a graph in which the horizontal axis represents height and the vertical axis represents the proportion of people in each height category. When the points on the graph are connected by a smooth, curving line, then for height and many other quantitative traits a bell-shaped curve like the one shown in Figure 16.5 results. This curve is called the **normal curve** or the **normal distribution**; it is extremely important in statistics and biology because any trait whose value is the sum of a great many independent factors, each of small individual effect, will have approximately a normal distribution when graphed for a large number of individuals. Sometimes the scale of measurement must be changed to obtain a normal distribution; what may have to be graphed, for example, is the square root or the logarithm of the measurement rather than the measurement itself. In any case, many quantitative traits are approximately normally distributed when measured on an appropriate scale.

A distribution curve like the one in Figure 16.5 does not contain any less information than the original data, of course; it merely presents the data in another kind of organization. A graph is useful for some purposes—it allows one to see the distribution of a trait

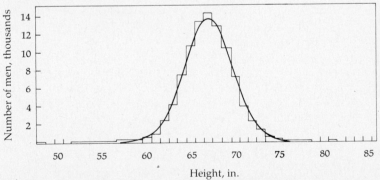

Figure 16.5 Distribution of height observed among 91,163 English draftees called for military service in 1939. The smooth curve (heavy line) is the normal curve that best fits the observed data. It has a mean of 67.5 in and a standard deviation of 2.62 in. The equation of a normal curve is

$$y = \frac{1}{\sqrt{2\pi}\sigma}e^{-(x-\mu)^2/2\sigma^2}$$

where μ is the mean and σ^2 the variance.

pictorially—but it is helpful to summarize the information in a more condensed form. This can easily be done. One of the most convenient attributes of the normal distribution is that its entire shape is completely determined by two numbers. One of these has to do with the point on the horizontal axis over which the curve has its peak, and the other has to do with the breadth of the curve. The peak of a normal, bell-shaped curve is the average, or **mean**. The mean is obtained simply by averaging all the measurements. Imagine a somewhat unusual population consisting of 10 percent 4-footers, 70 percent 5-footers, and 20 percent 6-footers. The average height or mean would be $(0.10 \times 4) + (0.70 \times 5) + (0.20 \times 6) = 5.1$ ft. The broadness of the distribution can be measured in any number of ways, but the most useful measure is known as the **variance**. This is calculated by subtracting from each measurement the value of the mean, squaring this difference, and averaging. The variance in height in the above example would be $0.10 \times (4 - 5.1)^2 + 0.70 \times (5 - 5.1)^2 + 0.20 \times (6 - 5.1)^2$, which equals 0.29. The variance is numerically equal to the mean of the squares minus the square of the mean. In this case the mean of the squares is $(0.10 \times 4^2) + (0.70 \times 5^2) + (0.20 \times 6^2) = 26.30$, and the square of the mean is $5.1^2 = 26.01$, so the variance is $26.30 - 26.01 = 0.29$—in agreement with the alternative method of calculation above.

Figure 16.5 is of interest because it is based on an enormous amount of data—the heights of 91,163 English men called for military service in 1939. The mean of the height distribution is 67.5 in and its variance is 6.86 in². (Since calculation of the variance involves averaging the squared deviations from the mean, the units of a variance will always be the square of the original unit of measurement.) As noted earlier, the variance is a measure of the spread of a distribution, but it is often convenient to measure the spread in terms of the square root of the variance to preserve the original units. The square root of the variance is called the **standard deviation** and is usually symbolized by the Greek letter sigma (σ). In the case of Figure 16.5,

$$\sigma = \sqrt{6.86 \text{ in}^2} = 2.62 \text{ in}$$

One significance of the normal distribution is that it permits a simple summary of a vast amount of data. One aspect of such a summary is illustrated in Figure 16.6, which depicts a normal distribution having some mean (symbolized here as μ) and a standard deviation σ. As indicated by the braces at the

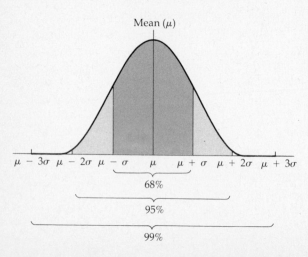

Mean (μ)

$\mu - 3\sigma$ $\mu - 2\sigma$ $\mu - \sigma$ μ $\mu + \sigma$ $\mu + 2\sigma$ $\mu + 3\sigma$

68%

95%

99%

Figure 16.6 With a normal distribution, shown here, approximately 68 percent of the population lies within 1 standard deviation (σ) of the mean (μ), approximately 95 percent lies within 2 standard deviations, and approximately 99 percent within 3 standard deviations.

bottom of the figure, 68 percent of the entire population is expected to be within 1 standard deviation of the mean (heavy shading), 95 percent is expected to be within 2 standard deviations (light shading), and 99 percent within 3 standard deviations. These relationships apply to any normal distribution, and in particular to the data in Figure 16.5, where the mean is 67.5 and the standard deviation is 2.62. According to Figure 16.6, 68 percent of the heights will therefore be in the range

$$\mu - \sigma \quad \text{to} \quad \mu + \sigma$$

which is

$$67.5 - 2.62 \quad \text{to} \quad 67.5 + 2.62$$

or

$$64.88 \quad \text{to} \quad 70.12$$

Similarly, 95 percent of the heights will be in the range

$$\mu - 2\sigma \quad \text{to} \quad \mu + 2\sigma$$

$$67.5 - 2(2.62) \quad \text{to} \quad 67.5 + 2(2.62)$$

$$62.26 \quad \text{to} \quad 72.74$$

Thus, knowing the mean and variance of a normal distribution provides a simple and accurate depiction of the distribution of a quantitative trait in a population. It should, however, be remembered that the distribution of any quantitative trait is due to the joint effects of genotype and environment. Knowledge that a trait is normally distributed tells us nothing at all about its genetic basis. Indeed, traits that are determined exclusively by environmental factors can nevertheless have a normal distribution.

Quantitative Traits and Selection

Allele frequencies at loci that influence quantitative or threshold traits are subject to change by the same forces that alter allele frequencies at other types of loci—mutation, migration, selection, and random genetic drift. For our purposes, the most important of these forces is selection because consideration of selection with quantitative traits involves concepts not encountered in the single-locus selection situations in Chapter 15. First we should note that selection can be of two broad types. **Natural selection** is the type discussed in Chapter 15, which follows as an inevitable consequence of intrinsic differences in survival or fertility among different genotypes. The other type of selection is known as **artificial selection**, and it is particularly important relative to quantitative traits. Artificial selection is the type of selection used by animal or plant breeders whenever they choose a phenotypically select group of individuals to be the parents of the next generation. Of course, artificial selection and natural selection of a trait can occur simultaneously. Sometimes natural selection may act to reinforce the artificial selection imposed by a breeder; more usually, natural selection acts against the breeder's aims.

Whether selection is natural or artificial, it can act in any of three fundamentally different ways. These different modes of selection are illustrated for a quantitative trait at the left side of Figure 16.7. In each case, the shaded part of the distribution represents those selected individuals that are to become the parents of the next generation. Figure 16.7(a) represents **directional** selection, in which selected individuals are near one of the phenotypic extremes. Directional selection is the type most often practiced by animal or plant breeders, as, for example, in the selection of greater egg laying in poultry or higher yields in corn. The long-term effect of directional selection is shown at the right of Figure 16.7(a); the effect is to shift the mean of the phenotypic distribution in the direction of selection. (Note that the mean of the left-hand distribution is 20, whereas that of the right-hand distribution is 30.) It should, however, be pointed out that directional selection (or any other type of

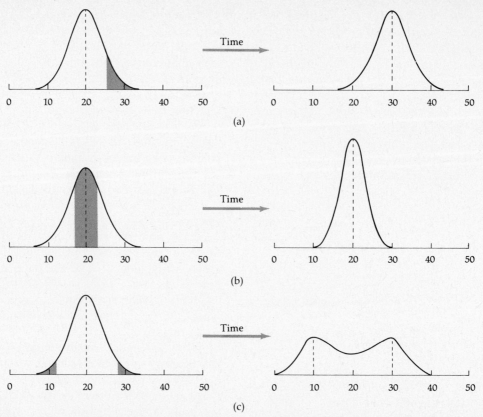

Figure 16.7 Three principal modes of selection of quantitative traits (left) and their expected long-term consequences (right). In each case, the shaded region represents those individuals who become parents of the next generation. *(a)* Directional selection. *(b)* Stabilizing selection. *(c)* Diversifying selection.

selection) can alter the phenotypic distribution of a trait only if there is sufficient genetic variation underlying the trait in the first place.

Figure 16.7(*b*) illustrates **stabilizing** (or **normalizing**) selection, in which selected individuals are those nearest the mean of the entire population. Stabilizing selection is thus a conservative evolutionary force tending to maintain the phenotypic distribution at its present mean. Indeed, stabilizing selection may be the most common type of natural selection because, for many quantitative traits, phenotypic extremes tend to be disfavored. An example of stabilizing selection in humans involves birth weight and is illustrated in Figure 16.8, which

shows the relationship between the birth weight of newborn males and females (horizontal axis) and the mortality rate for each weight (vertical axis; note that the scale is logarithmic). The stablizing selection on birth weight is evidenced by the shape of the mortality curve; mortality is higher among very light and very heavy babies than it is in babies of intermediate birth weight. The expected effects of long-continued stabilizing selection are shown at the right of Figure 16.7(*b*). The mean phenotype of the population has not changed, but the variance of the distribution has become smaller.

The third principal mode of selection is

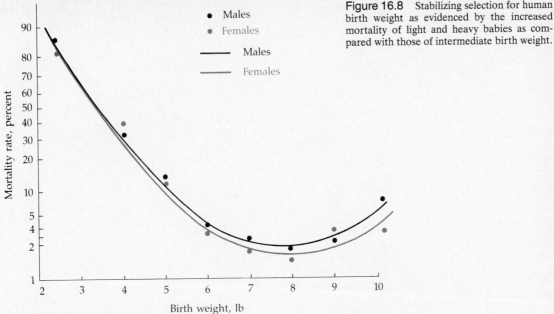

diversifying selection, illustrated in Figure 16.7(*c*). With diversifying selection, individuals with phenotypes at *both* extremes of the range are selected. The expected long-term effects of diversifying selection are to spread the distribution out (i.e., to increase its variance); sometimes diversifying selection leads to a distribution that has two peaks, as shown at the right in Figure 16.7(*c*).

When we considered selection in Chapter 15, we were principally concerned with changes in allele frequency at the major locus under consideration. Selection of a quantitative trait involves simultaneous changes in allele frequency at several or many loci. Although formulas for the theoretical change in such allele frequencies can be derived, the information usually of prime interest is how selection will alter the overall distribution of phenotypes, particularly the mean. The issue as it typically arises in practice is illustrated in Figure 16.9. Part (*a*) shows the distribution of a hypothetical quantitative trait in a population, and its

mean (denoted μ) is indicated. The phenotypes selected as parents of the next generation are indicated by the shading, and the mean phenotype among these selected parents is denoted μ_s. When these selected individuals are mated randomly with one another, they give rise to the next generation, the phenotypic distribution of which is shown in Figure 16.9(*b*); the mean of the progeny generation is denoted μ'. Note that the selection in Figure 16.9 need not be actual selection as practiced by an animal or plant breeder. It could as well be a sort of "paper selection." For example, if one had records of a large number of individuals in the parental generation and records of their offspring, then one could determine (on paper) the distribution of offspring whose parents were in the shaded region of Figure 16.9(*a*). This procedure represents a sort of directional "selection," but one that in no way interferes with the normal course of human mating and reproduction.

In the situation of Figure 16.9, μ' will

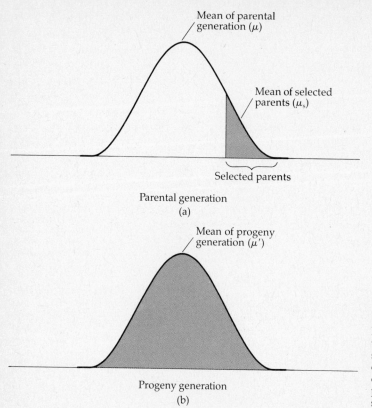

Parental generation
(a)

Progeny generation
(b)

Figure 16.9   Directional selection. *(a)* Distribution of trait in parental generation, with selected parents represented by shading. *(b)* Distribution of trait among offspring of selected parents. Note that offspring mean is greater than mean of parental generation but less than mean of selected parents.

almost always turn out to be greater than μ but less than μ_S. That is to say, the mean of the progeny generation will almost always be greater than the mean of the entire parental generation, but it will be smaller than the mean of the selected parents. This relationship is true for two reasons: First, μ' will be greater than μ because some of the phenotypic variation is due to underlying genotypic variation; second, μ' will be smaller than μ_S because some of the phenotypic variation is due to variation among environments, and favorable environments are not transmitted through genes to the next generation. Furthermore, it is possible to specify more quantitatively the relationship among μ', μ, and μ_S. Indeed, given knowledge of μ and μ_S (and one other

number), it is possible to derive a **prediction equation**, which predicts the value of μ' for the given μ and μ_S. The "one other number" referred to is the heritability of the trait in the population in question, and it is to this concept of heritability that we now turn.

Heritability

The distribution of a quantitative trait such as height in a population is determined by both genetic and environmental factors. An important element in understanding quantitative traits is to evaluate the relative importance of genetic and environmental factors in contributing to phenotypic variation. The key to evaluating these contributions is to realize that there are many different genotypes in a large population, each having its own miniature distribution. If the genotypes could be identified, the distribution of the trait in the population would be much as depicted in Figure 16.10. In this picture, the overall distribution of the trait is represented by the heavy line. The overall distribution is actually composed of five separate distributions, with each of these smaller ones representing the distribution of the trait in individuals who have an identical genotype with respect to the relevant genes. (In most actual cases, of course, there would be many more than five genotypes.) The vertical lines mark the means of each of the five distributions, numbered 1 through 5; if these means were known, then the variance among them would be the **genotypic variance** because it derives solely from differences in genotype. The variance of the heavy curve, which is the variance of the trait in the whole population, is composed of two parts: the genotypic variance (symbolized V_g) and the **environmental variance** (V_e). The usefulness of variance as a measure of breadth of the distribution is that these two components often add together; thus, the **total variance** (V_t) is given by

(Total variance) equals (genotypic variance) plus (environmental variance) or, in symbols,

$$V_t = V_g + V_e$$

This partitioning of the total variance into genetic and environmental components is possible, strictly speaking, only when the quantitative trait satisfies two requirements. One is that the best genotype in one environment be the best in all, the second best in one environment be the second best in all, the third best in one environment be the third best in all, and so on; when genotypes follow this pattern there is said to be an absence of **genotype-environment interaction**. (It should be noted that the genotypes illustrated in Figure 16.1 *do* exhibit genotype-environment interaction.) The second requirement is that the genetic and environmental effects be independent, that certain genotypes not be associated with certain environments more often than would be expected by chance; when this requirement is met, there is said to be an absence of **genotype-environment association**. Often—but by no means always—these two requirements are satisfied to a good approximation, in which case V_t will equal V_g plus V_e.

The genotypic variance can be subdivided still further. One can calculate what the genotypic variance in a hypothetical population would be if the hypothetical population had the same mean as the actual population but if there were random mating, no dominance, and no interaction between loci. The genotypic variance of this hypothetical population, which is called the **additive genetic variance**, may be represented as V_a. One can then calculate how much of the genetic variation in the actual

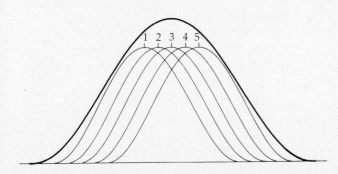

Figure 16.10 The heavy line depicts the overall distribution of phenotypes of a quantitative trait in a population. The smaller, lighter curves represent the phenotypic distributions of each of five distinct genotypes in the population. Each individual genotype has its own mean (numbers). The variance among these genotypic means is the genotypic variance.

population can be attributed to dominance over and above the additive variance; the variance due to dominance, called the **dominance variance**, may be represented as V_d. A similar sort of calculation reveals how much variance is due to interaction between loci over and above the additive and dominance effects. The variance due to interaction among loci is called the **interaction variance** or the **epistatic variance** and is symbolized V_i. With random mating, the genotypic variance can be shown to be equal to the sum of these components:

$$V_g = V_a + V_d + V_i$$

provided that the joint effects of, for example, dominance and interaction are so small that they can safely be neglected. The total variance in the population can then be written as

$$V_t = V_g + V_e$$

$$= V_a + V_d + V_i + V_e$$

Note that the subdivision of the variance has nothing whatever to do with what individual genes are actually contributing to the trait; these variances are cumulative, aggregate, abstract statistical attributes of the genes taken collectively.

Subdividing the variance as above is fine in theory. How the variance components are to be estimated in actual practice is not easy, however, especially in humans. The principle behind the estimates is somewhat like this: If one studied many sets of MZ twins, then all the variance in a trait between the members of each pair would have to be environmental in origin, because each pair of twins is genotypically identical. If one then compared the variance between identical twins with the total variance (V_t) in the general population, one would have in the first instance involving MZ twins an estimate of V_e and in the second instance an estimate of $V_g + V_e$ (because $V_t =$

$V_g + V_e$). Subtracting V_e from V_t provides an estimate of V_g because

$$V_t - V_e = (V_g + V_e) - V_e = V_g$$

(All of this, of course, assumes that V_e is really the same for both the twins and the entire population and that the genetic and environmental effects are independent, so their interactions can safely be ignored.) In most studies of quantitative traits in humans, comparisons of the trait between parent and offspring, siblings, uncle and nephew, and so on are used rather than comparisons of MZ twins. It happens that V_a, V_d, and V_i make different contributions to the amount of variation to be found between parent and offspring, siblings, and so on, and by making the appropriate comparisons V_a, V_d, and V_i can be estimated, although it is beyond the scope of this book to elaborate on the details.

The estimate of **additive genetic variance,** V_a, is especially important to the animal or plant breeder who wishes to improve a strain or herd by breeding superior individuals as outlined in Figure 16.9. Selecting superior individuals to be the parents of the next generation will indeed result in genetic improvement of the population, provided at least some of the variation in the population is genetic in origin —but by how much will it be improved? Here is where the additive genetic variance becomes important. The ratio of the additive variance to the total variance—V_a/V_t—determines how much genetic improvement can be achieved. The ratio V_a/V_t is given a special name—the **heritability**—and it is customarily represented by the symbol h^2. (The symbol is written with a square—h^2—because it is a ratio of variances, which themselves involve squared terms.) Thus,

$$h^2 = V_a/V_t$$

For reasons that will become clear in a few

pages, the heritability defined as V_a/V_t is often called the **narrow-sense heritability**.

To the animal or plant breeder, the concept of narrow-sense heritability (the ratio of the additive genetic variance to the total variance) is useful because it allows the prediction of how rapidly a breed can be improved by directional selection. As an example of the application of narrow-sense heritability in animal breeding, suppose that the mean of some trait in a population is 100 and that the breeder selects as parents individuals with a mean of 120. What will be the mean of the trait in their offspring? The offspring mean depends on h^2, and the prediction equation for the type of selection shown in Figure 16.9 is

$$\begin{pmatrix} \text{mean of} \\ \text{progeny} \\ \text{generation} \end{pmatrix} \text{minus} \begin{pmatrix} \text{mean of} \\ \text{parental} \\ \text{generation} \end{pmatrix} \text{equals}$$

$$\begin{pmatrix} \text{narrow-sense} \\ \text{heritability} \end{pmatrix} \text{times} \left[\begin{pmatrix} \text{mean of} \\ \text{parents} \end{pmatrix} \text{minus} \begin{pmatrix} \text{mean of} \\ \text{parental} \\ \text{generation} \end{pmatrix} \right]$$

or, in the symbols of Figure 16.9,

$$\mu' - \mu = h^2(\mu_S - \mu)$$

or just

$$\mu' = \mu + h^2(\mu_S - \mu)$$

In the hypothetical example above, we had μ = 100 and μ_S = 120. Suppose that the heritability is 25 percent. (Heritabilities of important traits in farm animals and plants are usually less than 50 percent; see Table 16.2 for some examples). With $h^2 = 0.25$, the mean of the progeny will be

$$\mu' = 100 + (0.25)(120 - 100)$$
$$= 100 + (0.25)(20)$$
$$= 105 \text{ in.}$$

As noted earlier, μ' is larger than the mean of the previous generation but smaller than the mean of the parents—a relationship that is

TABLE 16.2 RANGES OF NARROW-SENSE HERITABILITIES OBSERVED FOR VARIOUS TRAITS IN FARM ANIMALS

Animal	Trait	Heritability (narrow sense)
Cattle	Milk yield	0.2–0.4
	Percent fat in milk	0.4–0.8
	Percent protein in milk	0.4–0.7
	Feed efficiency	0.3–0.4
Sheep	Fleece weight	0.3–0.6
	Fiber diameter	0.2–0.5
	Body weight	0.2–0.4
Poultry	Eggs per hen	0.05–0.15
	Rate of egg production	0.15–0.30
	Body weight	0.3–0.7
	Egg weight	0.4–0.7
Swine	Daily gain in weight	0.1–0.5
	Feed efficiency	0.15–0.60
	Back-fat thickness	0.50–0.70
	Litter size	0.10–0.20

Source: Based on data in F. Pirchner, 1969, Population Genetics in Animal Breeding, Freeman, San Francisco.
Note: Narrow-sense heritability is the ratio of additive genetic variance in a trait to the total variance in the trait. Its value differs from population to population. Nevertheless, the narrow-sense heritability is a useful concept to animal breeders because it provides a measure of the speed with which a particular population can be improved by selective breeding.

virtually always found with selective breeding practices.

The above prediction equation can be written in a somewhat simplified form because the quantities $(\mu' - \mu)$ and $(\mu_S - \mu)$ are given special names. The difference $(\mu' - \mu)$ (symbolized R) is called the **response** to selection, and the difference $(\mu_S - \mu)$ (symbolized S) is called the **selection differential**. Thus,

$$R = \mu' - \mu$$
$$S = \mu_S - \mu$$

and so

$$R = h^2 S$$

As an application to humans, we may consider the height distribution of men in Figure 16.5, where the mean was 67.5. For simplicity, we will ignore the fact that females tend to be somewhat shorter than males, and we will suppose that the narrow-sense heritability of height in this population is 50 percent. What would be the expected mean height of individuals whose parents' heights averaged 72 in? Applying the prediction equation, we have

$$\mu' = \mu + h^2(\mu_S - \mu)$$
$$= 67.5 + (0.50)(72.0 - 67.5)$$
$$= 67.5 + (0.50)(4.5)$$
$$= 69.75 \text{ in.}$$

Note that the heritability of a trait depends on variation in the environment because

$$h^2 = \frac{V_a}{V_t}$$
$$= \frac{V_a}{V_g + V_e}$$

Increasing the variation in the environment (V_e) therefore decreases h^2; decreasing V_e increases h^2. Therefore, the heritability of a trait has meaning only with respect to a particular population in a particular range of environments. Moreover, a high heritability does not mean that the performance of a breed cannot be improved by improving the environment. The meaning of $h^2 = 1.0$ is that $V_e = 0$. This says only that there is no *variation* in the environment; it does not say that the environment is a good one. A group of half-starved cattle could conceivably have a heritability of growth rate of 100 percent, but the best and quickest way to improve the herd would be to feed it better. To carry this example one step further, a farmer underfeeding his cattle may interpret the 100 percent heritability as indicating that the breed is genetically inferior. This interpretation would be wrong, of course— an absolute, complete, total, unmitigated mistake; the problem is the food.

Although animal and plant breeders are often interested in narrow-sense heritability (the ratio of additive genetic variance to total phenotypic variance), human geneticists are frequently interested in another variance ratio —the ratio of genotypic variance to total phenotypic variance—because this ratio conveys the fraction of phenotypic variation that is attributable to differences in genotype among individuals. There is great potential for confusion here because this latter variance ratio is also called "heritability." More precisely it should be called the **broad-sense heritability**; we will use the symbol H^2 for broad-sense heritability to distinguish it from the narrow-sense heritability h^2 that is important for purposes of prediction. Thus,

$$H^2 = \frac{V_g}{V_t}$$

Table 16.3 illustrates broad-sense heritabilities for various quantitative traits in Caucasians, and the great variation in H^2 among traits should be noted. On the whole, the numbers in Table 16.3 are likely to be overestimates of the true values because, in some

TABLE 16.3 BROAD-SENSE HERITABILITIES OF SOME TRAITS IN CAUCASIANS

Trait	Heritability*
Fertility	10–20
Birth weight	16
Longevity	(29)
Handedness	31
Serum lipid levels	(44)
Diastolic blood pressure	44
Systolic blood pressure	57
Body weight	63
Amino acid excretion	(72)
Stature	85
Total fingerprint ridge count	80–95

Source: Data from C. Smith, 1975, Quantitative inheritance, in G. Fraser and O. Mayo (eds.), *Textbook of Human Genetics,* Blackwell Scientific, Oxford.
*Estimates in parentheses are considered tentative.

cases, assortative mating for the traits in question has not been taken into account (see Chapter 14 for a discussion of assortative mating), nor have nongenetic similarities in environment among members of a family been taken into account. Nevertheless, it is clear from Table 16.3 that some of the phenotypic variation of quantitative traits in humans is due to genotypic differences among individuals, but the precise magnitude of the genotypic contribution is variable among traits and uncertain in amount.

Heritabilities of Threshold Traits

Threshold traits are those that are either present or absent, and in this sense they are unlike quantitative traits. Threshold traits are like quantitative traits, however, in that the **liability** or risk of becoming affected with the trait is controlled by many genes acting together with strong environmental influences. Examples of threshold traits in humans are diabetes and schizophrenia, as mentioned earlier, but they also include such traits as hypertension (high blood pressure) and bronchial asthma as well as such frequent birth defects as anencephaly (smallness or absence of the brain), spina bifida (exposed spinal cord), congenital heart malformation, harelip and cleft palate, clubfoot, pyloric stenosis (obstructed digestive tract), and congenital dislocation of the hip.

Although the importance of genetic factors in the causation of threshold traits can be evaluated in studies of twins (see Table 16.1), the results of such studies are sometimes questionable because, among other objections, the environments of identical twins may be more similar than the environments of fraternal twins. A more recent approach to the study of threshold traits involves the use of conceptual tools developed for dealing with quantitative traits.

These applications make special use of the idea that every person has a certain liability toward a threshold disease, and they assume that the liabilities are distributed in the population approximately according to the normal curve. Any person whose liability is above a certain value—the **threshold**—will actually develop the trait; those whose liabilities are below this value will not. Recall that the liability of a person cannot be observed directly; the only information available is the incidence of the condition among related people. However, this information can be used to make inferences about the underlying distribution of liability, and by special statistical procedures the heritability of liability can be estimated.

Using these methods, the heritability of liability to schizophrenia has been estimated as about 80 percent, for example, and the heritability of liability to diabetes is 30 to 40 percent. (These are broad-sense heritabilities.) The heritability of liability toward schizophrenia being about 80 percent means that, within the population studied (in this case Caucasians in the United States), about 80 percent of the variation in liability to schizophrenia is genetic in origin. This number does *not* mean that schizophrenia is genetically determined and that environment is unimportant; it means only that genotypic differences among individuals can account for about 80 percent of the variation in liability among individuals.

Genetic Counseling

Genetic counseling—advising patients of the genetic risks faced by them or their relatives—is an appropriate subject to consider in the context of threshold traits because the most common birth defects are threshold traits. Genetic counseling makes available to everyone information about human heredity that would otherwise be known only to geneticists. Such knowledge is of great benefit because it prevents legitimate anxieties about one's own inheritance from deteriorating into fears based on rumor or superstition. Genetic counselors assist people in assessing the risk of hereditary illness befalling themselves, their relatives, or their children. Counselors do not make decisions for their patients. They help them evaluate risks. Final decisions are left to the patient and the patient's family. About 85 percent of the people who request assistance at clinics that specialize in genetic counseling request counseling because they have a child affected with some disorder and wish to know the risk to a subsequent child. (This risk is known as the **recurrence risk**). The balance of the patients are divided about equally between those who have medical problems themselves that might be transmitted to their children, and those who have an affected relative and wonder whether this poses special dangers to themselves or their children. Perhaps 1 person in 10, or 1 in 20, in the general population is in serious need of genetic counseling for one of these reasons. But almost everyone could benefit from genetic counseling at one time or another because most of us have relatives with an ailment that is determined in part by heredity—ailments such as hypertension, diabetes, schizophrenia, epilepsy, or the most common kinds of birth defects. For most of us, counseling would offer comfort and reassurance.

Fortunately, about two-thirds of the people who request counseling because they have had an abnormal child and wish to know the risk of recurrence usually receive rather good news—their recurrence risk will be relatively low. Recurrence risks range from essentially 0 to 50 percent, depending on the trait. High-risk traits are generally those with simple, single-gene inheritance like phenylketonuria or cystic fibrosis; risk of recurrence in such cases is 25 percent if the gene is autosomal recessive, 50 percent if the gene is autosomal dominant. (In the present discussion, a high-risk trait will be

considered as a trait with a recurrence risk of 20 percent or more; a low-risk trait will be considered as a trait with a recurrence risk of 10 percent or less). Low-risk traits are generally those with complex polygenic inheritance such as diabetes or schizophrenia; these traits are variable in expression, heterogeneous in cause, and multifactorial in their underlying genetic basis. As a group, high-risk traits are much less common than low-risk traits. This is shown graphically in Figure 16.11, which includes many of the conditions ordinarily encountered in Caucasians. The vertical axis on the graph represents the recurrence risk in brothers and sisters of affected individuals. Low-risk traits are denoted on the graph by dots, high-risk traits by squares. The open circles on the graph indicate traits for which the recurrence risks are intermediate (10 to 20 percent). The heavier lines represent the recurrence risks of traits inherited as simple Mendelian recessives or simple Mendelian dominants, or as polygenic traits with the broad-sense heritabilities as noted. Of the 36 traits in the graph, 21 are low risk, 10 are high risk, and 5 are intermediate. The 36 traits have a combined incidence of 4.8 percent. This number is made up of 3.2 percent low-risk traits, 0.4 percent high-risk traits, and 1.2 percent intermediates. Thus, to repeat, about two-thirds of the families being appraised of their recurrence risk will be reassured by rather good news—their recurrence risk will be low.

Figure 16.11 makes a counselor's job look easy. The counselor merely identifies which trait is involved, finds where it is on the graph, and presto—there is the recurrence risk. Counseling is not so straightforward, however, and a graph like Figure 16.11 would not actually be used. First, the numbers graphed in this figure are mere averages. Proper counseling will take into account the racial or ethnic background of the parents (Japanese have higher frequency of harelip and cleft

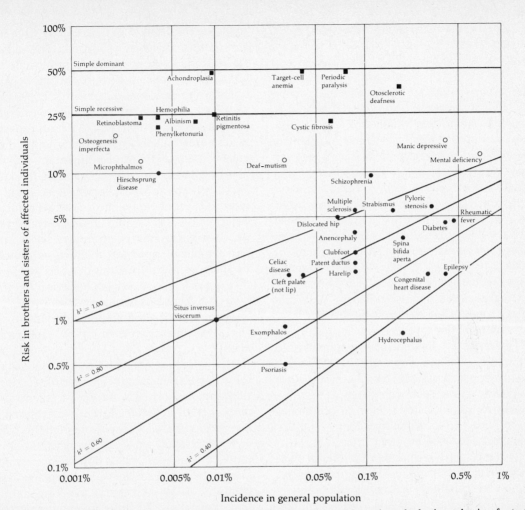

Figure 16.11 Graph of recurrence risks of common abnormalities. The numbers along the horizontal axis refer to the frequency of each trait in the population as a whole; the numbers along the vertical axis refer to the risk of each trait among brothers and sisters of affected individuals (the recurrence risk). The heavier lines on the graph denote the recurrence risks expected for traits inherited as simple dominants, simple recessives, or threshold traits with the broad-sense heritabilities as noted. The incidences and recurrence risks plotted here are those observed in Caucasians.

palate), where they live (the incidence of spina bifida is lower in the western part of the United States than in the eastern part), their social class (anencephaly is more frequent among the children of unskilled workers, at least in Ireland), how many normal children the parents have had (congenital dislocation of the hip is more frequent in firstborn babies), and the sex of the affected child (pyloric stenosis is five times more common in boys than girls; congenital hip dislocation is six times more common in girls than boys). Many other factors such as family history must also be considered.

Adding to the difficulties encountered by

genetic counselors is the problem of proper diagnosis. For example, the most common variety of cleft palate is polygenically inherited and has a low recurrence risk, but there is a rare form, recognized by tiny indentations on the inner surface of the lip, that is inherited as a simple Mendelian dominant. Also, many traits that are inherited in a simple manner can be mimicked by nongenetic disorders, and the counselor has to sort out which is which. Genetic counseling is always a delicate business, best left to those especially trained for it.

At one time, the best a counselor could do was advise parents of their risks. This has all changed. Counselors now have an arsenal of medical technology at their disposal. In many cases of recessive inheritance, they can identify heterozygous carriers; counselors need no longer wait for an affected child to be born, they can alert the parents in advance. In high-risk pregnancies, amniocentesis can be used to determine whether the embryo is normal. When abnormal children are born, they are recipients of medical technology—corrective surgery for harelip, cleft palate, pyloric stenosis, and spina bifida; antibiotics for cystic fibrosis; insulin for diabetes; other drugs for schizophrenia; special diets for phenylketonuria—the list could be extended indefinitely. The arsenal of medical technology has multiplied by many times the power of counselors to advise and the power of physicians to treat.

Genetics of Human Behavior

There is perhaps no more controversial an application of genetics than its relation to human behavior. Mistrust seems to derive from two sources: one psychological and the other scientific. On a psychological level, many people seem to believe that the finding of any significant genetic influence on behavior would be tantamount to a sort of genetic predestination, a kind of enslavement to our genes, that puts our own behavior beyond our own control. This biology-is-destiny interpretation is unjustified. From discussions throughout this chapter and, indeed, throughout the book, it ought to be clear that a genetic influence on a trait does not at all imply that the environment is unimportant. For quantitative traits especially, the environment can exert profound influences on a trait and even reverse any possible genetic predispositions, as illustrated in Figure 16.1. Perhaps it would be most accurate to interpret genetic influences on behavior as determining a set of potentialities, which may be very broad; the environment then determines which of these potentialities will actually be achieved. Seen from this perspective, genes do not predetermine human behavior, but they may establish limits in much the way genes put limits on our physical abilities; no human can run a two-minute mile or fly by flapping his arms.

The second objection to applying genetics to human behavior, a scientific objection, is substantive. Normal variation in behavior, if influenced by genes at all, must be analyzed as a quantitative trait. Animal and plant breeders have successfully analyzed many quantitative traits, but in these fields special precautions can be taken to minimize genotype-environment interaction and special experimental designs can be used to eliminate genotype-environment association. Such experimental control is not available to the human geneticist. For many behavioral traits, assortative mating may be important, and environmentally relevant factors may tend to be shared among members of a family. These complications and others tend to make estimates of genotypic variation greater than the true value may actually be. Moreover, it may be argued that statistical techniques that are appropriate for animal and plant breeding because they permit prediction of response to selection are quite inappropriate in human genetics. Components of variation, if they can be estimated accurately at all, tell us nothing about how many loci may influence the trait in question, nor about the relative importance of these loci, nor about what the loci do. For these reasons, some human geneticists have adopted a more direct approach by studying genetic variation in hormones, enzymes, or other molecules thought to be involved in brain function.

In some instances, genes have an overriding influence on behavior. Examples include genetically caused abnormalities in the development of the nervous system such as Tay-Sachs disease (see Chapter 4) and phenylketonuria (see Chapter 9), genetically caused degeneration of the nervous system such as Huntington disease (see Chapter 11), and the bizarre behavioral effects of Lesch-Nyhan syndrome (see Chapter 5). At one time there was thought to be a strong association between the XYY chromosome complement and violent criminality, but this association has not been confirmed in later studies; the XYY situation, discussed in detail in Chapter 6, illustrates some of the serious methodological difficulties involved in the study of genetic influences on human behavior.

Cases like Tay-Sachs disease, phenylketonuria, and Lesch-Nyhan syndrome, while demonstrating the importance of genes in the development and maintenance of the nervous system, are individually rare disorders; thus, they leave open the question of whether genotypic variation accounts for a significant fraction of variation in behavior among normal individuals. That there exists a potential for such genetic influences is illustrated in the animal experiment in Figure 16.12. Involved in these experiments are two strains of rats: a strain that learns to run mazes well (unshaded bars) and another strain that learns poorly (shaded bars). As indicated by the large difference in learning ability in the average environment, the strains clearly differ from each other

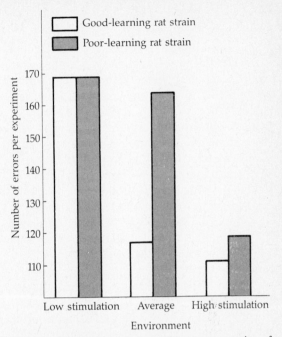

Figure 16.12 Maze-running ability among strains of rats selected for good learning and poor learning in an average environment. Strong genotype-environment interaction is indicated because the difference between strains virtually disappears in high-stimulation or low-stimulation environments.

and the difference is due to genotype. (Indeed, the strains had both been derived by means of directional selection from a genetically heterogeneous base population; the good-learning strain was obtained by repeated "up" selection for good learners, the poor-learning strain was obtained by repeated "down" selection for poor learners).

Although Figure 16.12 illustrates a genetically based difference in learning ability in the average environment, it also illustrates a pronounced genotype-environment interaction. When rats are reared in a high-stimulation environment enriched with ramps, mirrors, balls, tunnels, and so on, the difference in learning ability is very much reduced; and when rats are reared in a low-stimulation environent, the difference disappears entirely.

Figure 16.12 is not intended to imply that maze learning in rats is in any way comparable to learning in humans. Human learning is undoubtedly much more complex. However, since Figure 16.12 illustrates a pronounced genotype-environment interaction for the simple maze-learning situation in rats, all the more surely we can expect genotype-environment interaction to be an important factor in human learning. With regard to human behavior, about all we have to go on are twin studies and measures of resemblance between relatives, and, as emphasized earlier, these approaches are fraught with difficulties when applied to humans. Indeed, some would argue that the likelihood of genotype-environment interaction and the possibility of genotype-environment association invalidate the assumptions of the entire quantitative genetic approach, and thus they argue that estimates from such analysis are completely meaningless.

In the rest of this chapter, we consider two examples in which a proposed genetic influence on human behavior is highly controversial. One case involves behavioral differences between the sexes and the other involves IQ.

Male and Female Behavior

In Chapter 5 we considered the developmental basis of sex and discussed the importance of various hormones, particularly testosterone, in switching sexual differentiation from a preprogrammed female direction to a male direction. These early gonadal hormones are also known to influence differentiation of the central nervous system, and, in experimental animals (primarily rodents), such early hormones can be demonstrated to influence nonreproductive behaviors such as overall activity, aggression, play, taste preferences, and maze learning. In these experimental organisms, there do seem

to be innate behavioral differences between the sexes.

On the other hand, it may be argued that experimental studies of behavior in rodents have little or no relevance to human behavior. The human brain is vastly more complex than the rodent brain, and human family and social structures are much more sophisticated than those of rats. So, the question remains: Are there indeed inborn behavioral differences between the sexes in humans, and, if so, how important are they?

It should be emphasized at the outset that the answers to such questions in humans are unknown. Not only are they unknown, but they might be unknowable from direct observation because innate behavioral traits become hopelessly intertwined with external environmental influences from the moment of a child's birth and sometimes prior to birth. The only way to sort out the effects of "nature and nurture" would be to isolate newborns from *all* human contact and care for them with machines of some unholy design. Such an experiment would be grotesque and immoral, unthinkable in any society.

Such experiments have, however, been carried out with monkeys; while certainly not humans, monkeys are much nearer relatives than rodents. Extensive studies of behavior in rhesus monkeys have been carried out by Harry F. Harlow and collaborators at the University of Wisconsin. The observations are made possible because newborn monkeys can be raised in complete isolation from adult monkeys. They do need body contact, but this is supplied by a simple terrycloth-covered wire dummy, an inanimate "surrogate" mother (Figure 16.13). Infants raised individually with surrogate mothers are socially and psychologically abnormal. Harlow puts it this way:

The first 47 baby monkeys were raised during the first year of life in wire cages so arranged that the

Figure 16.13   A baby monkey with its terrycloth surrogate mother. Monkey infants raised in complete isolation from other monkeys become severely disturbed both socially and psychologically.

infants could see and hear and call to other infants but not contact them. Now they are five to seven years old and sexually mature. As month after month and year after year have passed, these monkeys have appeared to be less and less normal. We have seen them sitting in their cages strangely mute, staring fixedly into space, relatively indifferent to people and other monkeys. Some clutch their heads in both hands and rock back and forth. . . . Others, when approached or even left alone, go into violent frenzies of rage, grasping and tearing at their legs with such fury that they sometimes require medical care. Eventually we realized that we had a laboratory full of neurotic monkeys.[1]

[1]H. F. Harlow, 1962 The heterosexual affectional system in monkeys, The American Psychologist 17:1–9.

These monkeys are sexually incompetent. The males are unable to mate even with sexually experienced females, who react to the males' repeated failures with contempt and frustration. The motherless females, when mated with experienced and patient males, turn out to be cruel and heartless mothers, quite devoid of maternal behavior toward their own young.

As it happens, differences in behavior between normally raised male and female monkeys can be identified in the first several months of age. For example, threat behavior, marked by a distinct facial grimace, is observed about five times as often in males as females. Passivity, observed as a turning away from an approaching animal or as a relaxed body posture, is three to four times as frequent in females as males. Rigidity, which is some four times as frequent in females as males, is a response evoked by an approaching monkey and identified by a rigid body posture with the limbs extended, often with the tail erect and the head looking back over the shoulder at the approaching animal. There is great overlap in the behavior of male and female monkeys. Any particular animal can and does sometimes exhibit behavior found most frequently in the opposite sex, but the statistical differences between the sexes are quite marked. Behavioral patterns in the sexes not only differ on the average, but the differences become greater as the monkeys age, probably because certain kinds of behavior are reinforced by other animals. But a notable point is that at least some underlying differences in behavior seem to be innate. According to Harlow, rhesus monkeys raised in isolation exhibit behavior appropriate to their own sex with little or no alteration.

The innate behavioral patterns in monkeys can overpower even strong social influences. As mentioned before, monkeys raised in complete isolation are hopelessly neurotic. They can, however, be "rehabilitated" by "therapy," the "therapist" of a 6-month-old "patient" being a 3-or-4-month-old, socially normal monkey, a monkey too young to be aggressive but old enough to cling to the patient and interact with it. Upon first seeing the therapist, the patient withdraws and huddles in a remote corner of the cage; the therapist follows along and clings to the patient. In a few days, the patient is clinging back. In weeks, the patient and therapist are playing enthusiastically with each other. After 6 months of therapy the patient's abnormal behavior has almost completely disappeared. In one series of cases, quite by chance, all the patients were males and all the therapists were females. The male

patients therefore had only female therapists to pattern their behavior after. Yet, when identifible sexual behavioral patterns in the patients began to emerge, they were typically male patterns. Thus some sort of innate tendency toward male behavior is not swamped out by the presence of behavioral models of the opposite sex.

The innate sex difference in behavior in rhesus monkeys is evidently due to prenatal hormone influences, as females artificially treated with male hormone during embryonic development and childhood behave in a way resembling their male counterparts. On the other hand, such conclusions must not be applied to humans without great care and thoughtfulness—and much more evidence. The biological determinants of behavior could be much stronger in rhesus monkeys than in other species of monkeys or in humans. Possibly, whatever biological differences in human behavior patterns may exist may be too small to measure; if innate differences occur, they may easily be swamped out or altered beyond recognition by conscious or unconscious learning. On the other hand, perhaps there are innate differences in behavior between boys and girls. Among girls with the adrenogenital syndrome, a defect in hormone metabolism causing a buildup of testosterone-like substances (see Chapter 5), there is an increased incidence of "tomboyish" behavior, although this syndrome is surely a rare and perhaps unrepresentative situation.

Whatever the case regarding innate behavioral differences may be, it is clear, even from studies of rhesus monkeys, that behavior is variable, that the sexes overlap in their behavior, and that behavioral differences between the sexes can only be stated as statistical averages. From such uncertainties alone, it seems to follow that human society should not stereotype sex roles and should not squeeze people into them; it should rather extend to each person an equal measure of individual freedom and the same room to develop his or her own potentials.

Race and IQ

Another application of genetics that has been highly controversial in recent years is the use of quantitative genetic techniques in the investigation of IQ scores, particularly in regard to racial differences in average IQ. The controversy arises in attempting to interpret the sort of data illustrated in Figure 16.14, which shows the distribution of IQ scores among a "standard" sample of white school children in the United States (black curve) and a sample of 1800 black school children from the southern United States (gray curve); the mean of the white children is about 100, that of black children is about 85, and the issue is how to account for this 15-point difference. It is important to note at the outset that there is a large overlap in the distributions (shaded area). Because of this overlap, 11 percent of blacks score above the average of whites, and 18 percent of whites score below the average of blacks. Thus, the distributions in Figure 16.14 have little or no relevance when applied to particular individuals. The IQ controversy deals only with overall averages. (Incidentally, the average IQ among contemporary Japanese is about 111, which is the highest in the world.)

One possible response to Figure 16.14 is to say that the data are completely meaningless on grounds that the IQ test is invalid. The IQ tests used today evolved from those developed by Alfred Binet and his colleagues in Paris in

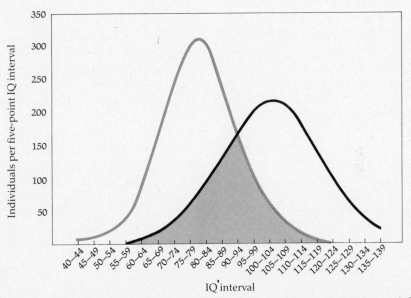

Figure 16.14   Distributions of IQ score among school children in the United States. The black curve is for a "standard" sample of white school children; the gray curve is for 1800 black school children from the southern United States. The mean of the IQ distribution of whites is about 100, that of blacks about 85, but note the extensive overlap of the distributions (shaded area). The source of the difference in the distributions is unknown, but a majority of geneticists and psychologists believe that the difference is environmental in origin and is related to the cultural bias of the IQ test itself.

the early 1900s. Binet's goal was to develop a test that would distinguish mildly subnormal children from normal ones with a view to providing the mildly subnormal children with special attention in schools. Later on, the tests were revised and extended and used widely to classify *normal* individuals with regard to their IQ. Modern tests involve items that examine verbal and geometrical skills, mathematical and other abstract tasks, and memory. Relative to middle-class U.S. whites, IQ scores are fairly reliable predictors of performance in schools. In fact, they were designed to be predictors of school performance. By trial and error, questions that were poor predictors were eliminated, and questions that were better predictors were added. Thus, it is important to realize that an IQ score is not a measure of "innate" intelligence. The test is intentionally culture-bound in the sense that it measures how well one has acquired the symbols, motivations, and values that lead to success in the classroom. The word *intelligence* as usually used is not one, single attribute; it is a complex of many different skills and abilities, and reasonable people may disagree about the importance of these. It is doubtful whether there could be a single test of any sort that could accurately access all these abilities and assign to each an acceptable measure of importance. The term *intelligence quotient* (IQ) is thus a grave misnomer.

Additional evidence for the culture-bound nature of IQ tests arises from the following observations: (1) When comparing southern whites with northern blacks, the average IQ of northern blacks is higher than that of southern whites; (2) when tests analogous to IQ tests are developed and standardized for blacks raised in ghettos in large cities, the blacks score higher than the whites; (3) black children do better on IQ tests when administered by black examiners than when administered by white examiners; and (4) there is a

relatively strong association between IQ score and social class.

In spite of these reservations about what the IQ test actually measures, there have been considerable discussion and disagreement about the interpretation of the data in Figure 16.14. It is in this context that the concepts of quantitative genetics have been brought in, and, unfortunately, they have been brought in incorrectly. From twin studies and the resemblance between relatives, the broad-sense heritability of IQ scores among whites has been variously estimated as between 40 and 80 percent. As noted earlier, such studies are on shaky ground where human behavior is concerned because of the likelihood of genotype-environment interaction and the questionable nature of certain other assumptions; some geneticists have claimed that the estimates of heritability are completely meaningless. Nevertheless, if we take the heritability estimate seriously, it still does not speak to the data in Figure 16.14. Recall that the broad-sense heritability of a trait in a population tells how much of the total phenotypic variation in the trait is due to all genetic effects combined; these genetic effects include additive gene effects, dominance, gene interaction, and, in the case of IQ score, assortative mating. A broad-sense heritability in IQ of, say, 40 percent only means that, *within* the population of U.S. whites, 40 percent of the variation in IQ among individuals is due to the sort of genetic effects listed above. If a reliable estimate of broad-sense heritability were available for blacks, then a similar statement could be made about variation in IQ *within* the black population.

Unfortunately, the meaning of heritability as applied to IQ score has been completely misinterpreted. A broad-sense heritability of 40 percent, even if it is accepted as real, does *not* imply that 40 percent of the average IQ difference between blacks and whites is due to genetic differences between the races, nor does it imply that *any* genetic differences necessarily exist. What heritability has to do with the difference in average IQ between blacks and whites is nothing, absolutely nothing. The reason heritability is irrelevant in this context is that, in the absence of any detailed knowledge of the genetic similarity between blacks and whites regarding genes influencing IQ, all or none of the difference in average IQ score could be environmental in origin. Indeed, the broad-sense heritability of IQ in each population could be 100 percent, yet the difference in average IQ could still be entirely environmental in origin.

It would not be impossible, in theory, to find out the source of the difference. If the detailed genetic relationship between blacks and whites with regard to genes influencing IQ were known, then this could be done. If these genes were at the same frequencies in blacks and whites, then all the difference would have to be environmental in origin. But, of course, no one really knows whether blacks and whites differ with regard to these genes; that is the subject of the argument. One may speculate, however, based on the genetic diversity with regard to blood groups and enzyme differences as detected by electrophoresis; for these traits the general rule is that genetic variation *within* racial groups is usually much larger than that *between* racial groups. From this point of view one would expect whatever genetic differences might exist to be rather small. Here is another point to remember: If the environments of blacks and whites were to become equal, the direction of any remaining difference in average IQ score would not be predictable. If blacks and whites were to be granted the same environmental advantages, the average IQ of blacks could be, for example, 100 or 98.6 or 102.4. Inasmuch as genetic differences between races with regard to genes influencing IQ are likely to be rather small, it hardly matters whether the difference is in favor of blacks or whites. Given comparable environments, the IQ distributions would probably overlap almost completely anyway. At the present time, the use of heritability estimates to justify educational or social policies is a misinterpretation of the facts and a misuse of genetics.

SUMMARY

1. Multifactorial (or **polygenic**) traits are traits that are determined by the combined action of alleles at several or many loci and usually also by the effects of environment. Thus, such traits are not inherited in simple Mendelian fashion. Two important kinds of multifactorial traits can be distinguished. **Quantitative traits** are traits such as height or weight that can be measured in single individuals. **Threshold traits** are traits such as diabetes that are either present or absent in an individual; threshold traits can be interpreted as quantitative traits if there is assumed to be an underlying **liability** (or risk) toward the trait, with all individuals who have a liability greater than a certain level actually developing the trait. The genes involved in multifactorial traits are often called **minor genes** to distinguish them from genes of large effect (**major genes**) that are involved in simple Mendelian inheritance.

2. The distinction between major and minor genes can sometimes depend on the environment. If all the genotypes at a locus produce overlapping phenotypes within a particular range of environments, then the locus in question is a minor gene because the genotypes cannot be distinguished by examination of phenotype. However, in a different range of environments, these same genotypes may be associated with distinct phenotypes and thus, in this range of environments, the locus in question would be considered a major gene.

3. For quantitative traits that are influenced by relatively few loci, the approximate number of loci involved can be determined by comparisons of the distribution of phenotypes among offspring of matings between homozygous genotypes, back-crosses of hybrids to either homozygous genotype, and matings among the hybrids. The genetic basis of black versus white skin pigmentation in humans is amenable to such studies, and the results suggest that (a) there is little dominance among the pigment-determining alleles, and (b) most of the genetic variation in black versus white skin pigmentation can be accounted for by alleles at only a few loci, approximately four to six.

4. Quantitative and threshold traits are often studied by comparisons of twins. **Monozygotic (MZ)** twins arise from a single zygote that separates into two parts at an early stage of embryonic development and goes on to give rise to two genetically identical embryos. MZ twins are often called **one-egg** or **identical** twins. **Dizygotic (DZ)** twins arise from two zygotes and are thus genetically related in the same way as ordinary siblings. DZ twins are often called **two-egg** or **fraternal** twins. The frequency of twinning varies among populations, being about 1 per 88 births in the United States. Most of the interpopulational variation in twinning rates is due to variation in the rate of DZ twinning. In theory, differences between MZ twins are due entirely to environment, whereas differences between DZ twins are due to genetic and environmental factors. However, MZ twins more often share certain embryonic membranes than do DZ twins, so intrauterine environmental effects may affect MZ and DZ twins differently. In addition, MZ twins are often treated more similarly by parents, teachers, and peers than are DZ twins, so some of the similarities between MZ twins may

actually trace to common environmental factors.

5. For threshold traits, twin comparisons are often expressed in terms of **concordance**. The concordance is the proportion of cases in which both twins have a trait when it is known that at least one has it. If the concordance among MZ twins is significantly greater than that among DZ twins, then this suggests there is a genetic component involved in the trait in question. Higher MZ twin concordance is observed for such traits as hypertension, bronchial asthma, and diabetes, but not for such traits as cancer at any site and death from acute infection.

6. Many quantitative traits in populations are distributed approximately according to a smooth, bell-shaped curve known as the **normal distribution**. This distribution is completely characterized by two quantities: the **mean** (average) and **variance** (average of squared deviations from the mean). It is often convenient to deal in terms of the square root of the variance, which is called the **standard deviation**. In a normal distribution, approximately 68 percent of the observations will be within 1 standard deviation from the mean, 95 percent will lie within 2 standard deviations, and 99 percent within 3 standard deviations.

7. Selection on a trait can be of two types: **natural** (that which occurs because of intrinsic differences in fitness among the various phenotypes) and **artificial** (that imposed by an animal or plant breeder). Whether selection is natural or artificial, it can occur in one of three fundamentally different ways. **Directional** selection occurs when individuals at one extreme of the phenotypic range are favored; **stabilizing** (or **normalizing**) selection occurs when individuals of intermediate phenotype are favored; and **diversifying** selection occurs when individuals at *both* extremes of the phenotypic range are favored.

8. Predicting the response to selection by means of a **prediction equation** involves the concept of **heritability** (more precisely, **narrow-sense heritability**). Under certain assumptions (discussed below), the **total phenotypic variance** of a trait (V_t) can be expressed as the sum of **genotypic variance** (V_g—the variance due to differences in genotype) and **environmental variance** (V_e—the variance due to environmental factors). The genotypic variance can be further subdivided into the **additive genetic variance** (V_a—that part of the genotypic variance that can be accounted for by the additive effects of genes), the **dominance variance** (V_d—due to dominance effects), and the **interaction** (or **epistatic**) **variance** (V_i—due to interactions among loci). Two types of heritability are important in quantitative genetics. The **narrow-sense heritability** (h^2), given by $h^2 = V_a/V_t$, is the fraction of the total variance due to additive effects. The **broad-sense heritability** (here symbolized H^2), given by $H^2 = V_g/V_t$, is the fraction of the total variance due to all genetic effects combined. For variance subdivision to be valid, it is necessary that there be no **genotype-environment interaction**; that is, the relative rankings of the genotypes must be the same in all relevant environments. In addition, there must be no **genotype-environment association** (i.e., no systematic tendency for certain genotypes to occur in certain environments).

9. For predicting the response to directional selection, narrow-sense heritability (h^2) is the important quantity. The prediction equation for directional selection reads

$$\mu' = \mu + h^2(\mu_S - \mu)$$

where μ' is the mean in the progeny generation, μ is the mean in the parental generation, and μ_S is the mean among selected parents. The quantity ($\mu' - \mu$) is often called the **response** to selection (R), and ($\mu_S - \mu$) is often called the **selection differential** (S). Thus, the prediction equation can also be written

$$R = h^2 S$$

10. Heritabilities of threshold traits can be calculated as they are for other quantitative traits, from the resemblance between relatives. However, in the case of threshold traits, the variances and heritabilities refer to the underlying distribution of liability.

11. Threshold traits are particularly important in **genetic counseling** because many common birth defects (e.g., congenital heart disease, pyloric stenosis, anencephaly, cleft palate) are threshold traits. When a couple has had an affected child, the risk of a subsequent child having the same defect is called the **recurrence risk**. In theory, recurrence risks can be calculated from the population incidence of the trait and the broad-sense heritability. In practice, many other factors must be taken into account, including the racial background of the parents and the sex of the affected child.

12. Although a large number of rare, simple Mendelian traits are known that affect intellectual performance or behavior (e.g., Tay-Sachs disease, Huntington disease, phenylketonuria, and Lesch-Nyhan syndrome), behavior within the normal range of variation must be considered as a quantitative trait influenced by multiple loci and by environment. The genetic contribution to normal behavioral variation is unknown, possibly small, and doubtless different for different behavioral traits. Attempts to apply conventional methods of quantitative genetics (such as variance subdivision) to behavioral traits are of dubious validity because of the likelihood of genotype-environment interaction and genotype-environment association.

13. Investigations in experimental animals such as rodents and monkeys support the view that there are innate sex differences in nonreproductive behavior that result from the prenatal action of certain sex-related hormones, particularly testosterone. The corresponding situation in humans is by no means certain, but such differences may occur in humans as well. However, the sex differences observed are averages, and there is substantial individual variability in behavior and overlap between the sexes.

    14. IQ tests were originally developed to identify mildly subnormal children in need of special help in school, but they have since been widely used to classify normal individuals. IQ tests are not a valid measure of "intelligence." They are reasonably accurate predictors of success in school, but this is what they were designed for. On standard tests, the average of U.S. white school children is about 15 points higher than the average of southern black school children. (However, there is a substantial overlap in the distributions, and northern blacks score higher than southern whites.) Cer-

tain nongeneticists have claimed that this IQ difference is largely genetic in origin because the broad-sense heritability of IQ is "high" (40 to 80 percent, depending on the study). Apart from the questionable validity of such estimates because of genotype-environment interaction and genotype-environment association, these commentators seem unaware that heritability has no relevance to comparisons between populations. Heritability deals only with variation *within* populations (i.e., within the white population and within the black population). All existing data are consistent with the view that the IQ difference between blacks and whites is entirely environmental in origin.

WORDS TO KNOW

Trait	**Twins**	**Selection**	**Variance**
Multifactorial (polygenic)	MZ	Natural	Total (phenotypic)
Quantitative	One-egg	Artificial	Genotypic
Threshold	Identical	Directional	Environmental
Major gene	DZ	Stabilizing	Additive
Minor gene	Two-egg	Diversifying	Dominance
Genotype-environment interaction	Fraternal	Prediction equation	
Genotype-environment association	Concordance	Selection differential	
Genetic counseling	**Distribution**	Response	
Recurrence risk	Mean	**Heritability**	
	Variance	Broad-sense	
	Standard deviation	Narrow-sense	
	Normal	Liability	

PROBLEMS

    1. For discussion: Some people feel that research on innate sex differences or on genetic differences in intellectual capability should not be carried out. They correctly point out that such research is extremely difficult to design properly, and they fear that poor research may be worse than no research at all. Even if such research could be well designed and involved no hidden invalid assumptions, they argue,

the information is likely to be misinterpreted and misused for malevolent social purposes by sexists and racists. Others feel that it is the obligation of scientists to learn the truth as best it can be learned, and the possible misuse of scientific information is not a valid reason for halting inquiry. What are your views on these issues?

    2. Why are pedigree methods inadequate for the

study of quantitative traits so that more elaborate statistical studies of populations must be used instead?

3. What is a major gene? A minor gene? Are major genes and minor genes really two different types of genes?

4. What is genotype-environment interaction and how does it differ from genotype-environment association?

5. What is the distinction between broad-sense heritability and narrow-sense heritability? Under what unusual conditions will these two types of heritability be equal?

6. What are some of the nongenetic reasons why MZ twins may be more alike than same-sex DZ twins?

7. Suppose the secondary sex ratio (i.e., the proportion of males and females at birth) is exactly 50 percent. What fraction of newborn DZ twins will be male-male pairs? Male-female pairs? Female-female pairs?

8. Suppose the secondary sex ratio among DZ twins is exactly $\frac{1}{2}$ (see Problem 7). In a collection of 1000 twin pairs, 650 are of the same sex. What proportion of the twin pairs are MZ?

9. For a rare recessive allele causing a disorder such as cystic fibrosis, what is the concordance of the disorder among DZ twins? Among MZ twins?

10. For a rare dominant allele that causes a disorder such as Huntington disease, what is the concordance among DZ twins. Among MZ twins?

11. Two inbred lines of plants fixed for alternative alleles at four unlinked loci influencing a quantitative trait are crossed to produce an F_1 generation, and the F_1's are intercrossed to produce an F_2. What proportion of the F_2 generation will have genotypes as extreme as their inbred grandparents for the four relevent loci?

12. A certain flower has a variable number of petals, and the distribution of petal number in a hypothetical population is shown in the table. Calculate the mean, variance, and standard deviation of petal number.

Proportion of flowers	Petal number
$\frac{1}{9}$	4
$\frac{2}{9}$	5
$\frac{3}{9}$	6
$\frac{2}{9}$	7
$\frac{1}{9}$	8

13. A population of *Drosophila melanogaster* has a normal distribution of the number of bristles on certain body segments. The mean number of bristles is 40 and the variance in bristle number is 4. What range of bristle number will include 68 percent of the population? What range will include 95 percent of the population?

14. In the *Drosophila* population described in Problem 13, directional selection is carried out by selecting individuals with a mean bristle number of 44 to be parents of the next generation. What is the selection differential? The narrow-sense heritability of bristle number in this population is 50 percent. What is the average bristle number in the progeny generation?

15. A population of flour beetles has a normal distribution of pupa weight with a mean of 2200 μg. Parents averaging 2400 μg are selected and mated at random, and their progeny have an average pupa weight of 2240 μg. What is the heritability of pupa weight in this population? Is this broad-sense or narrow-sense heritability?

FURTHER READING AND REFERENCES

Anderson, A. M. 1982. The great Japanese IQ increase. Nature 297:180–181. How can the remarkable increase be explained?

Battistuzzi, G., G. J. F. Esan, F. A. Fasuan, G. Modiano, and L. Luzzatto. 1977. Comparison of *Gd*[a] and *Gd*[b] activities in Nigerians. A study of the variation of the G6PD activity. Am. J. Hum. Genet. 29:31–36. How quantitative variation can occur even within simple Mendelian classes.

Blank, R. H. 1981. The Political Implications of Human Genetic Technology. Westview Press,

Boulder, Colo. Virtually all aspects of research and intervention are dispassionately discussed.

Bulmer, M. G. 1980. The Mathematical Theory of Quantitative Genetics. Oxford University Press, New York. An advanced treatise of quantitative genetics.

Cavalli-Sforza, L. L., and W. F. Bodmer. 1971. The Genetics of Human Populations. Freeman, San Francisco. One of the best available books on the subject. Source of data in Table 16.1.

Cooper, R. M., and J. P. Zubek. 1958. Effects of enriched and restricted early environments on the learning ability of bright and dull rats. Can. J. Psychol. 12:159–164. Discussed in connection with Figure 16.12.

Ehrhardt, A. A., and H. F. L. Meyer-Bahlburg. 1981. Effects of prenatal sex hormones on gender-related behavior. Science 211:1312–1318. To what degree do prenatal sex hormones influence psychosexual development?

Falconer, D. S. 1981. Introduction to Quantitative Genetics, 2nd ed. Longman, New York. A classic textbook well worth the effort to study.

Feldman, M. W., and R. C. Lewontin. 1975. The heritability hang-up. Science 190:1163–1168. Takes the point of view that heritability studies are of no real value in human genetics.

Fraser, F. C. 1980. The William Allan memorial award address: Evolution of a palatable multifactorial threshold model. Am. J. Hum. Genet. 32:796–813. A down-to-earth nonmathematical examination of counseling with regard to threshold traits.

Gottesman, I. I., and J. Shields. 1982. Schizophrenia: The Epigenetic Puzzle. Cambridge University Press, Cambridge. A thorough and critical review of the importance of genetic and environmental factors in the causation of this famous form of "madness."

Goy, R. W. 1978. Development of play and mounting behavior in female rhesus virilized prenatally with esters of testosterone or dihydrotestosterone. Recent Adv. Primatol. 1:449–462. Evidence of the profound effects of male hormone in monkeys.

Harrison, G. A., J. S. Wiener, J. M. Tanner, and N. A. Barnicot. 1964. Human Biology: An Introduction to Human Evolution, Variation and Growth. Oxford University Press, London. Source of data in Figure 16.5.

Kempthorne, O. 1978. Logical, epistemological and statistical aspects of nature-nurture data interpretation. Biometrics 34:1–23. A view that traditional methods of quantitative genetics are relevant to human genetics when properly used.

Kennedy, W. A., V. van deRiet, and J. C. White, Jr. 1963. Monographs of the Society for Research in Child Development 28:ser. 90. Source of data for Figure 16.14.

Loehlin, J. C., G. Lindzey, and J. N. Spuhler. 1975. Race Differences in Intelligence. Freeman, San Francisco. A cautious, critical, and comprehensive assessment of the race-IQ controversy.

Lynn, R. 1982. IQ in Japan and the United States shows a growing disparity. Nature 297: 222–223. Over the course of the last generation Japanese IQ has increased by about 7 points.

McClearn, G. E., and J. C. DeFries. 1973. Introduction to Behavioral Genetics. Freeman, San Francisco. A short but pungent introduction.

Newcombe, H. B. 1964. In M. Fishbein (ed.). Papers and Discussions of the Second International Conference on Congenital Malformations. International Medical Congress, New York. Source of data in Figure 16.11.

Notkins, A. L. 1979. The causes of diabetes. Scientific American 241:62–73. On the types of diabetes and their genotype-environmental causation.

Pirchner, F. 1969. Population Genetics in Animal Breeding. Freeman, San Francisco. A good textbook on the assessment and use of heritability in animal breeding. Source of data in Table 16.2.

Plomin, R., J. C. DeFries, and G. E. McClearn. 1980. Behavioral Genetics: A Primer. Freeman, San Francisco. A useful basic introduction to the study of behavioral genetics.

Rubin, R. T., J. M. Reinisch, and R. F. Haskett. 1981. Postnatal gonadal steroid effects on human behavior. Science 211:1318–1324. Summary of research on testosterone and aggression in men, mood and the menstrual cycle in women, and pubertal sex-role reversal in pseudohermaphrodites.

Smith, C. 1975. Quantitative inheritance. In G. R. Fraser and O. Mayo (eds.). Textbook of Human Genetics. Blackwell, London, pp. 382–441. A good review. Source of data in Table 16.3.

Stern, C. 1970. Model estimates of the number of gene pairs involved in pigmentation variability of the Negro-American. Human Heredity 20:165–168. Source of Figure 16.2.

Tsung, M. X. T., and R. Vandermey. 1980. Genes and the Mind: Inheritance of Mental Illness. Oxford University Press, New York. A brief and readable introduction to the genetics of mental disorders.

Vogel, F., and A. G. Motulsky. 1979. Human Genetics: Problems and Approaches. Springer-Verlag, New York. Contains good discussion of human behavioral genetics.

Wills, C. 1981. Genetic Variability. Clarendon, Oxford. Source of Figure 16.8.

Glossary

A: (1) Conventional symbol for the deoxyribonucleotide deoxyadenosine phosphate or for the ribonucleotide adenosine phosphate; also used to represent the base adenine. (2) A group of human chromosomes comprising numbers 1 through 3. (3) A blood type in the ABO blood group system. (4) A class of immunoglobulin containing alpha heavy chains.

AB: A blood type in the ABO blood group system.

ABO blood groups: Refers to various antigens on the surface of red blood cells coded by alleles at a locus denoted *I*.

Abortive infection: An unproductive, failed viral infection.

Acentric: A chromosome with no centromere.

Achondroplasia: A dominantly inherited form of dwarfism.

Acidic proteins: Proteins rich in aspartic and glutamic acids thought to be important in eukaryotic gene regulation.

Acridine: One of a group of substances that stack between the bases in DNA and so cause small additions or deletions.

Acrocentric: A chromosome that has its centromere nearly at one of its ends.

Activation: Of enzymes, rendering active or functional.

Acute dose: An amount of radiation delivered all at once.

Adaptive topography: Theoretical surface depicting the relationship between allele frequency or gametic frequency and average fitness.

Additive alleles: Alleles that interact in such a way that the phenotype of the heterozygote is exactly the average of the phenotypes of the two corresponding homozygotes.

Additive variance: That part of the genotypic variance that can be attributed to the additive effects of genes.

Adenine: Purine base in DNA and RNA.

Adjacent-1 segregation: Segregation of a heterozygous translocation in which each part of the reciprocal translocation proceeds to the same anaphase I pole as the normal chromosome bearing the nonhomologous centromere.

Adjacent-2 segregation: Segregation of a heterozygous translocation in which each part of the reciprocal translocation proceeds to the same anaphase I pole as the normal chromosome bearing the homologous centromere.

Adrenogenital syndrome: Hormone imbalance leading to a buildup of testosterone-like substance and therefore causing masculinization of affected females.

Age of onset: The age at which a trait first appears. *Variable age of onset* refers to differing ages of onset of the same trait in different individuals.

Agglutination: Clumping or aggregation caused by an antigen-antibody complex.

Albinism: Absence of pigment usually due to homozygosity for an autosomal recessive allele.

Alcaptonuria: Recessively inherited defect in the breakdown of tyrosine leading to excretion of alcapton in the urine, which oxidizes and turns black.

Alkylating agent: One of a group of highly reactive chemicals that becomes attached to DNA and thereby causes mutations.

Allele: Alternative forms of a gene that can be present at any particular locus.

Allele frequency: Relative proportion of all alleles at a locus that are of a designated type.

Allelic: The state of being alleles.

Allelic exclusion: Production of antibody corresponding to the antibody genes present on only one of a pair of homologous chromosomes.

Allosteric regulation: Activation or inhibition of an enzyme by a substance chemically unrelated to the substrate of the enzyme.

Allotypes: Alternative forms of the same immunoglobulin subclass coded by alternative alleles at one locus.

Allozymes: Enzymes having different electrophoretic mobilities and coded by alternative alleles at the same locus.

Alpha hemoglobin: One of the polypeptide chains in adult and fetal hemoglobin.

Alternate segregation: Segregation of a heterozygous translocation in which both parts of the reciprocal translocation go to one anaphase I pole and both normal chromosomes go to the other.

Amber: $5'-UAG-3'$ termination codon.

Ames test: A bacterial test for mutagens.

Amino acid: Fundamental chemical subunit of proteins.

Amino acid substitution: Replacement of one amino acid for a different amino acid in a polypeptide.

Aminoacyl-tRNA synthetase: One of 20 or more enzymes that attaches an amino acid to its corresponding tRNA.

Amino end: The end of a polypeptide that has a free amino $(-NH_2)$ group; this is the end of the polypeptide at which synthesis begins.

Amniocentesis: Procedure in which fetal cells are obtained from the amniotic fluid for diagnosis.

Amnion: The innermost membrane surrounding the embryo or fetus.

Amplification: Preferential replication of certain DNA sequences.

Anaphase: A stage of cell division preceding telophase in which chromosomes begin to move toward opposite poles. In metaphase of mitosis or metaphase II of meiosis, this separation involves splitting of the centromere. In metaphase I of meiosis, it does not involve centromere splitting but rather separation of homologous centromeres.

Aneuploid: Having an abnormal relative dosage of genes or chromosomes as compared with a diploid; a trisomic is an example.

Ankylosing spondylitis: Joint disease strongly associated with B27 haplotype.

Antibody: Serum protein produced in response to an antigen and capable of binding to the antigen.

Antibody diversity: The variety of antibodies produced or capable of being produced by an organism.

Anticodon: Three-base sequence in tRNA that undergoes base pairing with corresponding codon in mRNA.

Antigen: Any substance able to elicit an immune response.

Antigen-binding site: That part on an antibody molecule that adheres to the corresponding antigen.

Antigenic determinant: That part of an antigen that stimulates an immune response against itself.

Antigenic drift: Minor alterations in influenza antigens.

Antigenic shift: Major alterations in influenza antigens.

Antiparallel: Orientation of strands in duplex DNA, one running 3′ to 5′ as the other runs 5′ to 3′.

Antisense strand: The DNA strand that is the complement of the sense strand.

Artificial selection: Selection imposed by a breeder in which individuals of only certain phenotypes are permitted to breed.

Ascertainment bias: Incorrect ratio of phenotypes due to overrepresentation of some types of individuals because of the way they were discovered in the population or study.

Asexual inheritance: See Inheritance.

Assortative mating: Mating based on phenotype. In positive assortative mating, like phenotypes mate more frequently than would be expected by chance; in negative assortative mating, the reverse occurs.

Attachment site: The site at which bacteriophage DNA is integrated into the host DNA.

Attenuation: Regulatory process occurring in the transcription of amino acid-synthesizing operons in prokaryotes in which translation of an included leader polypeptide terminates subsequent transcription at a site called the attenuator.

Attenuator: See Attenuation.

Australopithecines: Group of several species existing 3.7 to 1.0 My, at least one of which may have been in the main line of human descent.

Autoimmune disease: Disorder marked by partial breakdown of immunological tolerance.

Autoradiography: Procedure whereby cellular structures such as chromosomes or DNA photograph themselves by means of incorporated radioactive atoms.

Autosome: Any chromosome that is not a sex chromosome.

Average fitness: The arithmetic mean of the fitnesses of all individuals in a population at a given time.

Average heterozygosity among subpopulations: Among a group of subpopulations, the arithmetic mean or average of the heterozygosities.

Avian sarcoma virus: A retrovirus causing connective tissue tumors in birds.

B: A group of human chromosomes comprising numbers 4 and 5. Also, a blood type in the ABO blood type system.

Bacteriophage: A virus that infects bacterial cells.

Balanced: A chromosomal rearrangement in which no genes are lost or present in excess.

Balanced translocation: Translocation in which all genes are represented the normal number of times.

Baldness: See Common baldness.

Barr body: See Sex-chromatin body.

Base: A chemical constituent of nucleotides in DNA and RNA. Five principal bases are of significance in genetics: adenine, guanine, thymine (found only in DNA), cytosine, and uracil (found only in RNA). Pyrimidine bases contain a single carbon-nitrogen ring; purine bases have a fused double-ring structure.

Base analogue: A substance chemically similar to one of the normal bases and incorporated into DNA.

Base substitution: Replacement of one nucleotide with a different nucleotide in DNA.

B cell: Lymphocyte responsible for synthesis of antibody.

B DNA: DNA in its right-handed double-helix configuration.

Beta hemoglobin: One of the polypeptide chains in the predominant form of adult hemoglobin.

Binding: Adherence.

Bivalent: Chromosomal structure occurring only in meiosis I and consisting of two replicated and paired homologous chromosomes held together by chiasmata.

Blastycyst: Hollow ball of cells containing the inner cell mass from which the embryo develops.

Blood group: A collection of alternative red blood cell antigens inherited in simple Mendelian fashion, like ABO, Rh, MN, etc.

Bloom syndrome: A condition characterized by enhanced sister-chromatid exchange, elevated mutation rate, and frequently cancer.

Bombay phenotype: Inability to express A and B antigens of the ABO blood groups on red blood cells owing to absence of the H substance.

bp: base pairs; a double-stranded DNA molecule containing 100 nucleotides is 100 bp long.

Brachydactyly: A dominantly inherited form of short fingers.

Broad-sense heritability: Ratio of genotypic variance to total phenotypic variance.

Bubble: In DNA replication, an oval structure of DNA with a replication fork at each end.

BUdR: Bromodeoxyuridine, an analogue of thymidine.

C: Conventional symbol for the deoxyribonucleotide deoxycytidine phosphate or for the ribonucleotide cytidine phosphate or for the base cytosine. Also, a group of human chromosomes comprising 6 through 12 plus the X.

cAMP: Cyclic adenosine monophosphate; important regulatory molecule.

CAP protein: Catabolite activator protein; involved in the positive regulation of the lactose operon and other operons in *Escherichia coli.*

Capsid: Part of a virus that encloses the genetic material.

Cap site: Position at which is added an unusual mucleotide in an unusual orientation to the 5′ end of eukaryotic mRNA's forming a cap.

Carboxyl end: The end of a polypeptide that has a free carboxyl (−COOH) group.

Carcinogen: An agent that causes cancer.

Carrier: (1) Heterozygous, as for a chromosome abnormality or mutant gene. (2) A molecule to which is attached another molecule for purposes of eliciting an immune response.

Casettes: Location of certain potentially expressible DNA sequences that require transposition for their expression.

Cell: The smallest unit of any organism that exhibits the properties of living matter.

Cell cooperation: Interaction between cells essential in a normal immune response.

Cell fusion: The joining together of somatic cells from two individuals or species to form a new type of hybrid cell with the chromosomes of both.

Centromere: Part of a chromosome connected to spindle fibers and involved in normal chromosome movement.

Chiasma: Cross-shaped structure connecting non-sister chromatids of homologous chromosomes.

Chimera: See Mosaic.

Chi structure: Cross-shaped configuration involving two DNA molecules formed during recombination.

Chromatid: One of the two strands produced by chromosome replication and sharing a common centromere.

Chromatin: Fundamental structural fiber of the chromosome; visible in interphase as a diffuse granular-appearing substance.

Chromosomal mosaic: An individual composed of two or more chromosomally different types of cells.

Chromosome: Structure containing the genetic material in a highly condensed chemical form and visible with appropriate staining through the light microscope.

Chromosome breakage: Interruption of the physical continuity of a chromosome.

Chromosome fusion: Physical joining of chromosomes or parts of chromosomes.

Chronic dose: An amount of radiation delivered as a series of smaller doses.

Chronic myelogenous leukemia: A type of leukemia frequently found in association with a particular translocation involving chromosomes 9 and 22.

Cistron: DNA sequence coding for a polypeptide.

Class: See Immunoglobulin class.

Classification: Categorization; *natural classification* refers to the systemic study and categorization of species with respect to their evolutionary relationships.

Class switch: Change in production of one class of antibody to production of a different class.

Clonal selection: Process in which an antigenic determinant stimulates mitosis of appropriate B-cell precursor leading to a clone.

Clone: A group of genetically identical cells.

Cloning: See DNA cloning.

Coding sequence: Sequence in DNA or RNA that codes for amino acids.

Codominant: Alleles that are both expressed in heterozygotes.

Codon: Three-base sequence in mRNA specifying an amino acid.

Cohesive ends: Complementary single-stranded ends of lambda DNA.

Colchicine: A drug that inhibits formation of the spindle.

Color Blindness: Inability to discriminate certain colors; the common forms are the red and green types, both due to X-linked recessives.

Combinatorial joining: Theory that antibody genes are formed by DNA splicing of sequences corresponding to parts of the complete antibody gene.

Common ancestor: An ancestor of both one's mother and one's father.

Common baldness: Thought to be due to an allele that is dominant in males but recessive in females.

Complement: Series of serum proteins that interact with antibody-bound cells to cause cell lysis.

Complementary pairing: Pairing of bases facilitated by hydrogen bonds; in genetics the most important examples of complementary pairing are A-T (or A-U) and G-C.

Complementation: Genetic situation in which an individual carrying two recessive alleles expresses a normal phenotype owing to the alleles being at different loci and therefore doubly heterozygous.

Complexity: A measure of the amount of sequence information in DNA.

Concordance: Among a group of pairs of individuals, the proportion of pairs in which both members have a particular trait when at least one member of each pair has the trait.

Conditional mutation: A mutation whose phenotypic effects are expressed under some environmental conditions but not under others, like a temperature-sensitive mutation.

Congenital: Present from birth.

Consanguineous mating: A mating between relatives.

Constant region: The portion of a heavy chain or light chain that varies little in amino acid sequence among antibodies within the same subclass.

Conversion: See Gene conversion.

Covalent bond: Strong chemical bond formed by two atoms that share a common electron.

Cro-Magnon: An early race of *Homo sapiens sapiens* first appearing about 40,000 years ago.

Crossing-over: Physical exchange of parts between nonsister chromatids of homologous chromosomes.

Cultural inheritance: See Inheritance.

Curie: Unit of radiation equal to 3.7×10^{10} disintegrations per second.

Cystic fibrosis: A recessively inherited glandular disorder.

Cytoplasm: All cellular substances except the nucleus.

Cytoplasmic inheritance: See Inheritance.

Cytosine: Pyrimidine base in DNA and RNA.

D: (1) a group of human chromosomes comprising 13 through 15. (2) A class of immunoglobulins containing delta heavy chains. (3) Diversity region; a DNA sequence that joins the variable region with the J region in an antibody heavy chain.

Deamination: Removal of an $-NH_2$ group from a molecule.

Deficiency: Having a gene or part of a chromosome missing.

Deletion: See Deficiency.

Delta hemoglobin: The betalike chain in the minority form of adult hemoglobin.

Denaturation: Unfolding of a protein molecule to render it nonfunctional; also separation of complementary strands of duplex DNA.

Deoxyribonucleotide: A nucleotide in which the sugar is deoxyribose.

Development: Process by which a new organism is formed from the zygote.

Deviation: Difference.

Dicentric: A chromosome with two centromeres.

Differentiation: Occurrence of differing allele frequencies among different populations; see also Development.

Diminution: Preferential elimination of certain DNA sequences.

Diploid: Possessing two complete sets of homologous chromosomes.

Directional selection: Selection favoring either phenotypic extreme.

Dispersed repeated gene family: Group of identical or similar transposable elements found at scattered sites in the chromosomes.

Distribution: Statistical description of the phenotypes that occur in a population and their relative proportions.

Diversifying selection: Selection simultaneously favoring both phenotypic extremes.

DNA: Deoxyribonucleic acid; composed of two intertwined strands, with each strand being a long sequence of nucleotides.

DNA cloning: Production of many identical copies of a defined fragment of DNA.

DNA ligase: An enzyme that covalently connects

the 3′ end of one DNA strand with the 5′ end of another.

DNA polymerase: An enzyme that catalyzes the production of a DNA strand that is complementary in base sequence to a template strand.

DNA splicing: Excision of certain DNA sequences in the production of an intact antibody gene.

DNA-uracil glycosidase: An enzyme that removes uracil-containing nucleotides from DNA.

Dominance variance: That part of the genotypic variance that can be attributed to the dominance effects of genes.

Dominant: An allele that conceals the presence of another allele—the recessive—in heterozygotes.

Dosage: Refers to the relative number of copies of genes or chromosomes.

Dosage compensation: Regulation of X-linked genes in such a way that such genes are equally active in males and females.

Dosage-sensitive locus: A locus that causes death or severe abnormality unless present exactly twice in a diploid.

Dose-response relationship: Description of how the amount of genetic or biological damage from radiation depends on the dose.

Double helix: A duplex DNA molecule that has a coiled or helical configuration.

Doubling dose: The amount of radiation necessary to increase mutation or biological damage to twice the background level.

Downstream: In RNA or the antisense strand of DNA, toward the 3′ end; in the sense strand of DNA, toward the 5′ end.

Down syndrome: Trisomy 21—i.e., 47,+ 21.

Down-syndrome translocation: Robertsonian translocation involving chromosome 21.

Duplex DNA: DNA molecule that has two complementary strands.

Duplication: Occurrence of one excess copy of a gene or region of chromosome. Also used in a more general sense to indicate any number of excess copies of a gene or region of chromosome.

Duplication and divergence: Process in which new genes are created by the initial duplication of an ancestral gene followed by the sequence divergence of the two copies.

DZ twins: Nonidentical twins; twins arising from two fertilized eggs.

E: (1) A group of human chromosomes comprising 16 through 18. (2) A class of immunoglobulins containing epsilon heavy chains.

Ear cerumen: Waxy substance produced in the ear canal.

Edwards syndrome: Trisomy 18—i.e., 47,+18.

Effective population number: The size of an ideal population that has the same rate of random genetic drift as occurs in an actual population.

Electrophoresis: Separation of molecules by means of an electric field applied across a jellylike slab.

Embryo: In humans, the developing organism from the second to the seventh week of development.

Emphysema: A lung disorder characterized by distension of air sacs in the lungs. See Familial emphysema.

Endogenous virus: Chromosomal DNA sequence or sequences obviously similar in structure and characteristics to known integrated proviruses.

Endonuclease: An enzyme that attacks one or more interior positions in a DNA molecule.

Enrichment: In mutation studies, a procedure designed to increase the relative proportion of a desired mutant type so that the type can more frequently be found in later screening.

Envelope: Part of a virus that encloses the capsid.

Environmental variance: That part of the phenotypic variance that can be attributed to differences in environment among individuals.

Enzyme: A protein able to accelerate certain chemical reactions without itself being altered in the process.

Epidemic: A local outbreak of disease.

Epitope: An antigenic determinant.

Epsilon hemoglobin: One of the polypeptide chains of embryonic hemoglobin.

Equilibrium: A state of stasis; absence of change.

Erythroblastosis fetalis: See Hemolytic disease of the newborn.

Euchromatin: Chromatin so organized that it does not stain during interphase.

Eukaryote: An organism whose cells normally possess a nucleus surrounded by a nuclear envelope and in which true mitosis and/or meiosis occurs.

Euploid: Having the same relative dosage of genes and chromosomes as occurs in a normal diploid, like, for example, a triploid.

Evolution: Cumulative change in the genetic characteristics of a species through time.

Excision repair: Process in which abnormal or improper nucleotides in a strand of DNA are cut out of the molecule and the excised portion is resynthesized by enzymes that use the remaining strand as a template.

Exon: A sequence of DNA that codes for amino acids. See Coding sequence.

Exonuclease: An enzyme that digests a DNA molecule from its ends.

Exon shuffle: Hypothesis that new genes can be created by combining exons from preexisting genes.

Expressivity: The degree or severity with which a trait is expressed; *variable expressivity* refers to the varying severity of a trait among different individuals.

F: A group of human chromosomes comprising 19 and 20. Also, symbol for the fixation index as a measure of genetic divergence and equaling the proportionate reduction in heterozygosity compared with the heterozygosity expected with random mating.

Familial: Tending to occur in groups of relatives.

Familial emphysema: An inherited lung disease associated with a deficiency of the protein α_1-antitrypsin.

Familial hypercholesterolemia: Inherited high cholesterol levels due to a dominant gene.

Favism: See G6PD deficiency.

Feedback inhibition: The rendering inactive of the first enzyme in a metabolic pathway by the end product of the pathway.

Fetus: In humans, the developing organism from the seventh week to term.

First-division nondisjunction: Nondisjunction occurring in the first meiotic division.

Fitness: Measure of ability to survive and reproduce.

Five-prime (5′) end: The end of a DNA or RNA strand that has a free 5′ phosphate.

Fixation: State at which the allele frequency of a designated allele is 1.0.

Fixation index: Measure of genetic differentiation defined as the proportionate reduction in average heterozygosity as compared with the total heterozygosity that would be expected with random mating.

Fork: Y-shaped structure defining the position of unwinding during DNA replication.

Forward mutation: Change of a normal allele into a mutant form.

Founder effect: Genetic differences between an original population and an offshoot population caused by the alleles in the founders of the offshoot population being nonrepresentative of the original population; a special variety of random genetic drift.

Fragmentation mapping: Production of chromosome fragments by x-rays in somatic cells to enable the localization of genes of interest.

Frameshift mutation: Insertion or deletion of a number of nucleotides in a coding region to alter the translational reading frame downstream of the mutation.

Fraternal twins: See DZ twins.

Frequency: Proportion.

G: (1) Conventional symbol for the deoxyribonucleotide deoxyguanosine phosphate or for the ribonucleotide guanosine phosphate or for the base guanine. (2) A group of human chromosomes comprising numbers 21, 22, and the Y. (3) A class of immunoglobulins containing gamma heavy chains.

Gamete: Reproductive cell, such as sperm or egg in animals.

Gamma hemoglobin: The betalike chain in fetal homoglobin.

G bands: Chromosome bands produced by Giemsa stain.

Gene: The fundamental unit of inheritance; used in several different senses such as unit of function, of mutation, and of recombination.

Gene conversion: The change of one allele to another by means of enzymatic comparison and alteration.

Gene family: Group of genes that have evolved from a common ancestral gene.

Gene frequency: See Allele frequency.

Generalized transduction: Transduction that can involve virtually any genes of the host.

Gene substitution: Replacement of one allele by another in the course of evolution.

Genetic code: Correspondence between three-base codons in mRNA and amino acids incorporated into a polypeptide.

Genetic counseling: Imparting advice and knowledge to individuals with regard to their own genetic constitution or with regard to their offspring or other relatives.

Genetic diversity: Occurrence of alternative genotypes at many loci in a population.

Genetic map: Diagram showing relative positions of genes along a chromosome.

Genome: Refers collectively to all genes carried by a cell.

Genotype: An individual's genetic constitution.

Genotype-environment association: Nonrandom occurrence of genotype-environment combinations.

Genotype-environment interaction: Inability to predict the relative performance of genotypes in one environment from knowledge of their relative performance in a different environment.

Genotype frequency: Relative proportion of individuals who have a designated genotype.

Genotypic variance: That part of the phenotypic variance that can be attributed to differences in genotype among individuals.

Germinal mutation: A mutation occurring in a germ cell.

Germ-line theory: Theory that antibody diversity has its origin in a large number of distinct genes in the germ line.

Goldberg-Hogness box: Eukaryotic binding site for RNA polymerase; usually involves the sequence 5'–TATA–3'.

Gonadal dose: The amount of radiation reaching the gonads.

Gonads: The sexual organs in which gametes are formed.

Gout: Severe painful swelling of certain joints associated with excess uric acid.

G6PD: Glucose 6-phosphate dehydrogenase enzyme, or the gene coding for the enzyme.

G6PD deficiency: Disease associated with lack or low levels of enzyme glucose 6-phosphate dehydrogenase; symptoms include anemia induced by certain drugs or by fava beans.

Gradual evolution: Evolutionary change occurring in a smooth, cumulative, nonabrupt manner.

Guanine: Purine base in DNA and RNA.

H-2: Designation of mouse MHC.

Haploid: Possessing only one set of chromosomes.

Haplotype: The group of particular alleles carried by loci of the MHC on one chromosome.

Hapten: A small molecule able to elicit an immune response when attached to a carrier molecule.

Hardy-Weinberg rule: Genotype frequencies expected with random mating.

HAT medium: Growth medium containing hypoxanthine, aminopterin, and thymine, which selects for $HGPRT^+, TK^+$ cells.

Heavy chain: The longest type of polypeptide in an antibody.

Hemizygous: Refers to X-linked genes in males to which the terms *heterozygous* and *homozygous* do not apply.

Hemolytic disease of the newborn: Blood disorder caused by breakdown of fetal Rh^+ cells by mother's anti-Rh^+ antibody crossing the placenta.

Hemophilia: Disorder characterized by abnormally slow or absent blood clotting; the best-known form is an X-linked condition associated with Queen Victoria's descendants and known as Royal hemophilia.

Heritability: In quantitative genetics, a ratio of two variances.

Hermaphroditism: Presence of functional male and female sexual structures in the same individual.

Herpesvirus: Type of virus associated with characteristic runny cold sores.

Heterochromatin: Chromatin so organized that it tends to stain during interphase.

Heteroduplex: A region of duplex DNA formed by the pairing of complementary strands from two other molecules of DNA.

Heterogeneous nuclear RNA: Refers to large transcripts found in eukaryotic nuclei.

Heterozygosity: Proportion of heterozygotes at a locus, or average proportion of heterozygotes among a group of loci.

Heterozygote inferiority: A situation in which the fitness of a heterozygote is less than that of both homozygotes.

Heterozygous: A genetic situation in which homologous loci in an individual carry different alleles.

HGPRT: The enzyme hypoxanthine guanine phosphoribosyl transferase, or the locus that codes for the enzyme.

HGPRT deficiency: See Lesch-Nyhan syndrome.

Highly repetitive DNA: Highly redundant simple-sequence DNA tending to be associated with centromeric regions.

Histocompatibility: Acceptance by a recipient of transplanted tissue from a donor.

Histone: Type of protein rich in lysine and arginine found in association with DNA.

HLA: Designation of human MHC.

hnRNA: See Heterogeneous nuclear RNA.

Homo erectus: Thought to be a direct predecessor of *H. sapiens;* lived 1.6 to 0.6 My.

Homo habilis: An early representative of the genus *Homo;* lived 2.2 to 1.6 My.

Homologous: Genetically matched in some way, as two chromosomes matched in size and structure and carrying alleles of the same genes, or loci at corresponding positions on homologous chromosomes.

Homo sapiens neanderthalensis: See Neandertals.

Homo sapiens sapiens: Modern humans.

Homozygous: A genetic situation in which homologous loci in an individual carry the same allele.

Host specific: A virus or parasite able to live only within one or a restricted range of hosts.

Hotspot: Nucleotide or part of a gene that has a much higher than normal mutation rate.

H substance: The precursor of A and B antigens in the ABO blood groups.

Huntington disease: Autosomal dominant degeneration of the neuromuscular system having an age of onset approximately in midlife. Also called Huntington chorea.

Hybrid cells: Cells formed by cell fusion that have the chromosomes of both parental cells.

Hybridization: In nucleic acid chemistry, the coming together of nucleic acids that have sufficiently complementary base sequences. In genetics in general, formation of offspring (hybrids) by mating of two species or varieties.

Hybridoma: Hybrid between a plasma cell and a myeloma tumor cell that produces a monoclonal antibody.

Hydrogen bond: Weak chemical bond formed by nearby atoms such as oxygen and/or nitrogen exerting an attraction on a common hydrogen.

Hypercholesterolemia: See Familial hypercholesterolemia.

Hyperploid: Having certain genes or chromosomes represented more times than in a diploid.

Hypoploid: Having certain genes or chromosomes represented fewer times than in a diploid.

Ideal population: Imaginary population that has theoretically perfect characteristics.

Identical twins: See MZ twins.

Idiotype: An antigenic determinant in the antigen-binding site of an antibody.

Immune response: The body's reaction to viruses, foreign cells, or tissues.

Immune surveillance: Theory that one of the important normal functions of the immune system is to screen the body for and destroy cells that have undergone malignant transformation.

Immune tolerance: See Tolerance.

Immunodeficiency disease: Any disease in which part of the immune response is absent or weak.

Immunoglobulin: Antibody molecule.

Immunoglobulin class: Type of immunoglobulin defined by the nature of its heavy polypeptide chain.

Inborn errors of metabolism: Genetic defects leading to nonfunctional enzymes that upset body chemistry.

Inbreeding: Mating between relatives.

Inbreeding coefficient: Measure of the genetic effects of inbreeding in terms of the proportionate reduction in heterozygosity in an inbred individual as compared with the heterozygosity expected with random mating.

Incomplete penetrance: See Penetrance.

Independent assortment: Genetic situation in which the segregation of one pair of alleles has no influence on the segregation of a pair of alleles at a different locus; Mendel's second law.

Inducer: A substance that permits genes to be expressed. In prokaryotes, inducers bind with their corresponding repressors, thus permitting transcription and translation.

Influenza: Respiratory illness caused by a virus.

Inheritance: Passage of genetic elements from one generation to the next. *Sexual* inheritance involves the process of meiosis. *Asexual* inheritance involves mitosis or, in prokaryotes, fission. *Cytoplasmic* inheritance involves synthesis and distribution of cytoplasmic organelles such as mitochondria. *Cultural* inheritance involves learned behavior transmitted by social or cultural means.

Inhibition: Of enzymes, rendering inactive or nonfunctional.

Initiation codon: First codon translated, typically 5′−AUG−3′ in cytoplasmic mRNA's.

Insertion sequence: Certain prokaryotic DNA sequences capable of transposition.

Interphase: A stage of cell division preceding prophase in which chromosomes are not condensed and characterized by the occurrence of DNA synthesis; conventionally separated into three parts—G_1 (the first part, prior to DNA synthesis), S (the second part, in which DNA synthesis occurs), and G_2 (the third part, after DNA synthesis has ceased). Although DNA synthesis occurs in interphase of mitosis and interphase I of meiosis, interphase II of meiosis is exceptional in that DNA synthesis does not occur.

Intervening sequence: Sequence of DNA that interrupts coding sequences.

Intron: See Intervening sequence.

Inversion: A chromosome with part of its genes in reverse of the normal order.

Inversion loop: Circular structure formed by homologous pairing between the inverted region of a chromosome and its normal homologue.

Island migration: Migration occurring uniformly among a group of populations.

Isochromosome: A chromosome that has two genetically identical arms.

Isolate breaking: Reduction in average frequency of rare homozygotes that occurs with population fusion.

J: DNA sequence that joins the variable region with the constant region in an antibody light chain, or that joins the V-D region with the constant region in an antibody heavy chain.

Karyotype: Chromosomal constitution.

kb: Kilobase pairs; a unit of nucleic acid length consisting of 1000 nucleotides.

Kindred: A group of relatives.

Klinefelter syndrome: The 47,XXY sex-chromosome abnormality; affected individuals are male.

LacI gene: Codes for the repressor of the lactose operon in *E. coli.*

Lactose: Milk sugar, composed of linked pair of galactose and glucose sugars.

LacY gene: Codes for β-galactoside permease in *E. coli.*

LacZ gene: Codes for β-galactosidase in *E. coli.*

Lambda: A bacteriophage that infects *E. coli.*

Leader sequence: Part of RNA molecule upstream (i.e., toward the 5′ end) of the first coding sequence.

Left-handed DNA: See Z DNA.

Lesch-Nyhan syndrome: Severe X-linked nervous disorder due to lack of enzyme HGPRT.

Lethal: A condition that is incompatible with life.

Liability: Risk or predisposition.

Ligase: See DNA ligase.

Light chain: The shortest type of polypeptide in an antibody.

Linkage: Genetic situation in which alleles at two loci do not exhibit independent assortment owing to their being sufficiently close together on the same chromosome.

Linkage disequilibrium: Absence of linkage equilibrium.

Linkage equilibrium: Condition in which the proportion of gametes carrying a particular combination of alleles is equal to the product of the corresponding allele frequencies.

Locus: Position on a chromosome.

Loss: State at which the allele frequency of a designated allele is 0.

Lymphocyte: White blood cell.

Lyon hypothesis: Dosage compensation in mammals by the inactivation of one X chromosome in each somatic cell of females.

Lysis: The bursting open of virus-infected or bacteriophage-infected cells.

Lysogeny: Alternative life cycle of temperate bacteriophage in which bacteriophage DNA becomes incorporated into that of the host.

Lytic cycle: Viral or bacteriophage life cycle in which progeny viruses are formed and the host cell is killed as these are liberated.

M: A class of immunoglobulins containing μ heavy chains.

Macroevolution: Evolutionary changes leading to the formation of new species or higher categories.

Macrophage: Accessory cell important in the immune response.

Major gene: Major locus.

Major histocompatibility complex: Group of tightly linked loci involved with important cellular antigens and other aspects of the immune response.

Major locus: A locus whose alleles are primarily responsible for the presence or absence of a trait.

Map distance: A measure of the separation between two syntenic loci; obtained by summation of recombination fractions between intervening loci. For loci sufficiently close together, the map distance equals the recombination fraction.

Map unit: The unit of distance on a genetic map; each map unit corresponds to 1 percent recombination if the loci involved are sufficiently close together.

Masked mRNA: Messenger RNA's present in the cytoplasm but unexpressed.

Mean: Arithmetic average.

Meiosis: Process of cell division resulting in production of gametes.

Messenger RNA: See mRNA.

Metabolism: Body chemistry.

Metacentric: A chromosome with its centromere approximately at its center.

Metaphase: A stage of cell division preceding anaphase in which chromosomes align on the metaphase plate. In metaphase of mitosis or metaphase II of meiosis, each chromosome consists of two chromatids as it aligns. In metaphase I of meiosis, the bivalents—paired and replicated homologous chromosomes—align.

5-methyl cytosine: A chemically modified cytosine frequently found in DNA.

7-methyl guanosine: Unusual ribose-base unit attached to the 5' end of certain mRNA's and exemplifying a 5' cap.

MHC: Major histocompatibility complex.

Microevolution: Evolutionary changes within a species.

Migration: Movement of individuals or gametes among populations.

Minor gene: A locus having sufficiently small effects that its segregation in pedigrees cannot be discerned.

Mismatch repair: Correction of mispaired nucleotides in heteroduplex regions of DNA.

Missense mutation: A mutation that creates an amino acid substitution.

Mitosis: A mode of cell division in eukaryotes in which a dividing cell produces two genetically identical daughter cells.

Mitotic nondisjunction: Nondisjunction occurring in mitosis.

MN: A blood group system in humans controlled by two codominant alleles.

Moderately repetitive DNA: DNA sequences found tens to hundreds of times per haploid genome and tending to be dispersed throughout the chromosomes.

Modification: Chemical alteration of certain bases in DNA or RNA.

Modification enzyme: An enzyme that modifies certain bases in or near a restriction site and renders the site insensitive to cutting by the corresponding restriction enzyme.

Modifier gene: A gene or allele that influences the expression of a trait associated with a different locus.

Molecular clock: Hypothesis that gene substitutions have a certain clocklike regularity in their occurrence in the course of evolution.

Molecular evolution: Cumulative change in macromolecules during the course of evolution.

Monoclonal: Antibodies produced by a single clone of cells.

Monomorphic locus: Conventionally, a locus at which the most common allele has an allele frequency greater than 0.95.

Monosomic: Having one chromosome missing in an otherwise diploid individual.

Morula: The small clump of cells formed by successive divisions of the zygote.

Mosaic: An individual composed of two or more genetically different types of cells.

Most parsimonious tree: The evolutionary tree that requires the fewest assumptions.

mRNA: Fully processed transcript exported to the cytoplasm for translation.

Müllerian ducts: Embryonic ducts associated with internal sexual structures of females.

Multifactorial: A trait whose occurrence or expression is determined by the combined effects of alleles at several or many loci.

Multiple alleles: The occurrence of more than two alleles at a locus in a population of organisms.

Mutagen: An agent that causes mutations.

Mutagenesis: Creation of mutations.

Mutation: A heritable chemical alteration of the genetic material.

Mutation rate: Probability of occurrence of a new mutation in a gene in one generation.

My: Millions of years before the present time.

MZ twins: Monozygotic twins; twins arising from a single fertilized egg.

Narrow-sense heritability: Ratio of additive genetic variance to total phenotypic variance.

Natural selection: Process in which individuals who are best able to survive and reproduce in a particular environment leave a disproportionate share of the offspring and thus gradually increase the overall ability of the population to survive and reproduce in the environment.

Neandertals: An early subspecies of *Homo sapiens* designated *H.s. neanderthalensis*; first appeared about 600,000 years ago.

Negative control: Manner of regulation in which a gene is transcribed unless it is complexed with a particular protein.

Negative selection: In mutation studies, selection carried out in such a way that only those cells or individuals with a mutant gene at a specified locus can survive.

Network hypothesis: Theory that the immune system regulates itself by means of an interacting network of idiotype and anti-idiotype interactions.

Neutral allele: An allele that has no effects on its carrier's ability to survive or reproduce.

Nick: A break in the sugar-phosphate backbone of a strand of DNA or RNA.

Nitrous acid: A potent deaminating agent and mutagen.

Noncomplementation: Genetic situation in which an individual carrying two recessive alleles expresses a mutant phenotype owing to the alleles being at the same locus and therefore homozygous.

Nondisjunction: Cell division occurring in such a way that daughter cells have extra or missing chromosomes.

Nonsense mutation: A mutation that creates a chain-terminating codon.

Nonsense suppressor: A mutation in a tRNA gene that can read through a termination codon and so suppress nonsense mutations.

Nonsister chromatids: Chromatids that originate by replication of different (though possibly homologous) chromosomes.

Normal distribution: The frequently encountered smooth, bell-shaped distribution curve.

Nucleosome: Tiny spherical particle containing DNA in association with histone molecules.

Nucleosome fiber: Chromosome fiber formed by adjacent nucleosomes.

Nucleotide: A unit of nucleic acid structure consisting of a five-carbon sugar to which is attached a phosphate group (on the 5′ carbon of the sugar) and a base (on the 1′ carbon of the sugar). In DNA the sugar is deoxyribose; in RNA it is ribose.

Nucleotide substitution: Replacement of one nucleotide with a different nucleotide in DNA.

Nucleus: The cellular structure containing chromatin.

O: A blood type in the ABO blood group system. Also used to designate the absence of a chromosome, as in the designation XO for 45,X Turner syndrome.

Ochre: 5′−UAA−3′ termination codon.

Okazaki fragments: Relatively short molecules of DNA formed by the discontinuous nature of replication of one of the parental DNA strands.

Oncogenic: Refers to cancer-causing virus.

One-egg twins: See MZ twins.

One gene—one enzyme hypothesis: See One gene—one polypeptide hypothesis.

One gene—one polypeptide hypothesis: Hypothesis that one gene codes for one and only one polypeptide; usually correct in eukaryotes and prokaryotes, but violated in some viruses with overlapping genes.

One-way migration: Migration occurring in one direction, from one population to another.

Oocyte: Cell destined to undergo meiosis to produce an egg.

Opal: 5′−UGA−3′ termination codon; also called umber.

Operator: A binding site for repressor in prokaryotic operons.

Operon: In prokaryotes, a group of cistrons transcribed together and thereby simultaneously expressed.

Overdominance: A situation in which the fitness of a heterozygote is greater than the fitness of both homozygotes.

Overlapping gene: A gene that produces two polypeptides by virtue of the corresponding mRNA being translated in two reading frames.

Pandemic: A worldwide outbreak of disease.

Paracentric inversion: A chromosome with an inverted region that does not include the centromere.

Partial dominance: Any situation in which the phenotype of the heterozygote is between the phenotypes of the corresponding homozygotes, but excluding cases in which the phenotype of the heterozygote is the same as the phenotype of one of the homozygotes.

Patau syndrome: Trisomy 13—i.e., 47,+13.

Pedigree: A diagram expressing the genetic relationships among individuals.

Penetrance: The proportion of cases in which individuals who have a relevant genotype express the corresponding phenotype. *Complete penetrance* refers to a penetrance of 100 percent. *Incomplete penetrance* refers to any degree of penetrance less than 100 percent.

Peptide bond: Chemical linkage between the carboxyl ($-COOH$) group of one amino acid and the amino ($-NH_2$) group of another; characteristic linkage in polypeptides.

Pericentric inversion: A chromosome with an inverted region that includes the centromere.

Permissive temperature: See Temperature-sensitive mutation.

Phenocopy: An environmentally caused trait resembling one known to be inherited.

Phenotype: An individual's appearance with respect to a trait, either physical, mental, or biochemical.

Phenotypic variance: The variance in phenotype among individuals in a population.

Phenylketonuria: Recessively inherited defect in phenylalanine metabolism leading to severe mental retardation if untreated by a low-phenylalanine diet.

Phosphate group: Chemical unit with the general molecular formula $-O-PO_3H_2$.

Placenta: Organ responsible for the exchange of gases and nutrients between a mother's bloodstream and that of her embryo.

Plasma cell: A B-cell derivative that secretes antibody.

Plasmid: In prokaryotes, a self-replicating molecule of DNA.

Polar body: Tiny product of meiosis I or meiosis II in females. The polar body produced in meiosis I sometimes undergoes meiosis II to produce two even tinier polar bodies.

Poly-A tail: Run of consecutive A's added to the 3' end of many eukaryotic transcripts.

Polycistronic mRNA: An mRNA coding for two or more polypeptides in tandem.

Polydactyly: Extra fingers or toes: inherited as a dominant in some kindreds, multifactorial in others.

Polygenic: See Multifactorial.

Polymorphic locus: Conventionally, a locus at which the allele frequency of the most common allele is less than 0.95.

Polyoma: A rodent DNA-bearing tumor virus.

Polypeptide: Linear string of amino acids linked by peptide bonds.

Polyploid: Having more than two haploid sets of chromosomes.

Polyprotein: A polypeptide chain that is enzymatically cleaved into two or more shorter, functional polypeptides.

Polysomic: A diploid individual who has one or more extra copies of a particular chromosome.

Polytene: Giant chromosomes found in salivary glands and other tissues of certain dipterans formed by several consecutive replications with no intervening cell division.

Population: Group of members of the same species that usually find mates within their own group.

Positive control: Manner of regulation in which a gene is not transcribed unless the appropriate region is complexed with a particular protein.

Positive selection: In mutation studies, selection carried out in such a way that only those cells or individuals with a nonmutant gene at a specified locus can survive.

Prediction equation: An expression relating the response to selection and the selection differential.

Pribnow box: Prokaryotic binding site for RNA

polymerase; usually involves the sequence 5′−TATA−3′.

Primary nondisjunction: Nondisjunction involving a normal pair of chromosomes.

Primary sex ratio: The sex ratio at the time of fertilization.

Primary structure: The amino acid sequence of a polypeptide.

Primer: In DNA replication, a relatively short molecule of RNA that is elongated with deoxyribonucleotides by DNA polymerase.

Primordial germ cells: Cells set aside in early development that are destined to give rise to spermatocytes or oocytes.

Probability: Mathematical expression of the degree of confidence that certain events will or will not occur.

Probe: A radioactive fragment of DNA used to identify the presence or location of complementary sequences.

Processing: See RNA processing.

Prokaryote: An organism whose cells lack a well-defined nucleus and in which cells divide by means of fission.

Promoter: Recognition sequence for binding of RNA polymerase.

Proofreading: Checking a DNA strand for correct base pairing with its partner strand.

Prophage: Bacteriophage DNA that has been integrated into the host DNA.

Prophase: A stage of cell division preceding metaphase in which chromosomes first become visible in the light microscope. In prophase of mitosis the chromosomes are visibly double as soon as they appear; each chromosome consists of two sister chromatids sharing a common centromere. In prophase I of meiosis the chromosomes appear as single strands when first visible; prophase I is a relatively long stage characterized by (1) the first appearance of the chromosomes, (2) pairing of homologous chromosomes, (3) paired chromosomes becoming visibly double producing a four-stranded bivalent, (4) increased thickening of chromosomes leading to the clear appearance of chiasmata, and (5) apparent repulsion of homologous centromeres. Prophase II of meiosis is a relatively brief stage in which chromosomes recondense after a slight relaxation in interphase II.

Protein: A molecule composed of one or more amino acid-containing polypeptide chains.

Protein kinase gene: A gene coding for an enzyme that adds phosphate groups to other proteins or enzymes and thereby influences their ability to function.

Provirus: Circular duplex DNA complementary in sequence to a retrovirus RNA.

Pseudogene: A DNA sequence that has obvious homology with a known gene but is neither transcribed nor translated.

Pseudohermaphroditism: Presence of nonfunctional male and female sexual structures in the same individual.

Punctuated evolution: Evolution occurring in relatively rapid and abrupt bursts intermittent with long periods of relatively little change.

Punnett square: A cross-multiplication square devised to aid in calculation of types of offspring from a mating and their expected frequencies.

Purine: Base formed from a fused carbon-nitrogen double ring.

Pyrimidine: Base formed from a single carbon-nitrogen ring.

Q bands: Fluorescent chromosome bands produced by such substances as quinacrine.

Quadrivalent: Four-armed structure formed by meiotic pairing involving a reciprocal translocation and its two normal homologues.

Quantitative trait: A trait that can be measured in terms of a relatively continuous scale, such as height or weight.

Race: A population or group of populations that has undergone sufficient genetic differentiation to be regarded as distinctive in some arbitrary sense but not so much genetic differentiation as to be regarded as a subspecies.

Rad: Unit of absorbed radiation dose in terms of the energy imparted per mass of absorbing material; 1 rad equals 100 erg per gram of absorbing material.

Radiation: Emission of energy or subatomic particles from certain substances.

Random genetic drift: Fluctuation in allele frequency from generation to generation due to restricted population size.

Random mating: Mating without regard to genotype or phenotype.

Rate of mutation: See Mutation rate.

Ratio: Relative proportions, like a 1:2:1 ratio of genotypes in a mating of two heterozygotes, or a 3:1 ratio of phenotypes from the same mating when one of the alleles is dominant.

Reassociation: Coming together of complementary DNA strands that have previously been separated.

Receptor site: Site on a B cell or T cell that combines with the stimulatory antigenic determinant.

Recessive: An allele that must be homozygous for the corresponding trait to be expressed.

Reciprocal translocation: Interchange of parts between nonhomologous chromosomes.

Recombinant: A chromosome or DNA molecule that carries genetic information from each of two parental chromosomes or DNA molecules.

Recombinant DNA: See DNA cloning.

Recombination: Production of DNA molecule or chromosome carrying hereditary determinants from each of two preexisting DNA molecules or chromosomes.

Recombinational switch: A DNA sequence having terminal inverted repeats which, when present in a DNA molecule in one orientation, promotes transcription in one direction, but which can undergo recombination within the inverted repeats and subsequently promote transcription in the other direction.

Recombination fraction: Proportion of gametes produced by an individual that have undergone recombination.

Recurrence risk: Risk that an inherited disorder will occur again in a family.

Relative fitness: Measure of one genotype's fitness as a proportion of another genotype's fitness.

Rem: Unit of radiation dose equivalence used for comparing different types of radiation for purposes of radiation protection. The dose in rems is obtained by multiplying dose in rads by a quantity related to the type of radiation and by another quantity related to the distribution of the radiation to different parts of the body. For x-rays or electrons uniformly received, 1 rem equals 1 rad.

Replication: Synthesis of two daughter DNA molecules from a single parental molecule.

Repression: Rendering a gene unexpressible.

Repressor: A molecule, usually a protein, that binds to DNA and so prevents transcription.

Response: Change in population mean in one generation of selection.

Restriction enzyme: An enzyme that cuts duplex DNA at sites that have a particular nucleotide sequence.

Restriction fragment: A DNA molecule produced by cutting a larger molecule with a restriction enzyme.

Restriction map: Diagram of a DNA molecule showing the positions of one or more restriction sites.

Restriction site: The nucleotide sequence recognized by a restriction enzyme.

Restriction-site polymorphism: Refers to genetic variation detected by the presence or absence of a restriction site at a particular location in the DNA.

Restrictive temperature: See Temperature-sensitive mutation.

Retinoblastoma: Autosomal dominant disorder characterized by retinal malignancies.

Retrovirus: RNA viruses that produce complementary DNA strands upon infecting cells.

Reverse mutation: Change of a mutant allele back into a normal form.

Reverse transcriptase: An enzyme able to catalyze the synthesis of a complementary DNA strand from an RNA template.

Rh blood groups: Various antigens on the surface of red blood cells coded by alleles at a cluster of tightly linked loci or multiple alleles at a single locus.

Rheumatoid arthritis: A painful joint condition caused by partial breakdown of immunological tolerance.

Ribonucleic acid: See RNA.

Ribonucleotide: Chemical subunit of ribonucleic acid consisting of a ribose sugar to which is attached a phosphate (at its 5' carbon) and a base (at its 1' carbon).

Ribose: The five-carbon sugar found in ribonucleic acid.

Ribosomal RNA: See rRNA.

Ribosome: Cellular organelle that is the structure on which protein synthesis occurs.

Ring chromosome: A chromosome joined to itself to form a complete circle.

RNA: Ribonucleic acid, composed of a linear string of chemical subunits called ribonucleotides; serves several functions in the eukaryotic cell, such as mRNA, rRNA, and tRNA; serves as the genetic material in some viruses.

RNA polymerase: An enzyme capable of catalyzing transcription.

RNA processing: Alteration of transcripts that occurs in the nuclei of eukaryotic cells; includes splicing, capping, and polyadenylation.

RNA splicing: Excision of intervening sequences from eukaryotic transcripts.

RNA transcript: The RNA molecule produced in transcription that is complementary in base sequence to a sense strand of DNA.

Robertsonian translocation: Translocation involving the long arms of two acrocentric chromosomes.

Roentgen: Unit of radiation exposure in terms of the number of ions produced per mass of air; 1 r equals 2.58×10^{-4} coulombs per kilogram of air.

rRNA: The RNA molecules that form part of ribosomes.

Rubella: German measles virus; causes birth defects in embryos.

Satellite: Inconsistently staining blob of chromosomal material attached to the tip of certain chromosomes.

Screening: A survey of phenotypes of a large number of cells or individuals. In genetic counseling, usually refers to widespread tests for individuals with a genetic disorder or for carriers.

Secondary nondisjunction: Nondisjunction involving a trisomic chromosome.

Secondary sex ratio: The sex ratio at the time of birth.

Second-division nondisjunction: Nondisjunction occurring in the second meiotic division.

Secretor: Autosomal dominant trait associated with ability to secrete A and B antigens of ABO blood groups in body fluids.

Segregation: Separation of homologous chromosomes or of alternative alleles into distinct gametes during meiosis; Mendel's first law.

Selection: In mutation studies, a procedure designed in such a way that only a desired type of cell can survive, as in selection for resistance to an antibiotic. In evolutionary studies, intrinsic differences in the ability of genotypes to survive and reproduce. In animal and plant breeding, choosing individuals with certain phenotypes to be parents of the next generation.

Selection coefficient: The amount by which relative fitness is reduced or increased.

Selection differential: Deviation of the mean of selected parents from the population mean.

Selfish DNA: DNA sequences that increase in number and spread throughout a population by virtue of their ability to undergo transposition.

Semiconservative: Mode of DNA replication in which each daughter molecule contains one of the parental DNA strands intact.

Sense strand: The DNA strand that is transcribed.

Sequence divergence: Progressive reduction of similarity in amino acid or nucleotide sequences.

Serum: The fluid part of the blood.

Sex-chromatin body: Heterochromatic element in interphase nuclei corresponding to an inactivated X chromosome.

Sex chromosome: An X chromosome or a Y chromosome.

Sex influenced: A trait whose expression or severity differs between the sexes.

Sex limited: A trait that is expressed in only one sex.

Sex linked: A trait determined by genes on a sex chromosome, usually the X.

Sex ratio: Relative proportions of males and females in a population.

Sexual inheritance: See Inheritance.

Shine-Dalgarno sequence: Prokaryotic ribosome binding site, usually includes the sequence $5'-AGGA-3'$.

Sickle cell anemia: A recessively inherited severe anemia due to amino acid substitution in the β-globin chain; heterozygotes tend to be more resistant to falciparum malaria than are normal homozygotes.

Silent mutation: A mutation that creates a synonymous codon and so does not alter the amino acid sequence.

Simple Mendelian: A trait whose occurrence is determined by a major gene such as a dominant or recessive and which thereby occurs in pedigrees according to Mendel's laws.

Single active X principle: The inactivation of all X chromosomes except one in cells of mammals.

Sister chromatid: The partner strand of a chromatid attached to a common centromere.

Sister-chromatid exchange: Exchange of parts of sister chromatids; usually used with reference to mitosis.

Solenoid: Chromosome fiber formed by coiling of nucleosome fiber.

Somatic cell genetics: The study of the inherited characteristics of somatic cells grown under laboratory conditions.

Somatic mutation: A mutation occurring in a somatic cell.

Somatic theory: Theory that antibody diversity has its origin in somatic events such as mutation and somatic recombination or gene conversion.

Southern blot: Transfer of DNA fragments from an electrophoresis gel to a special type of paper for the purpose of identifying a particular fragment by means of hybridization with a radioactive complementary fragment (a probe).

Specialized transduction: Transduction that can involve at most a few genes of the host.

Species: Group of actually or potentially interbreeding organisms that is reproductively isolated from other such groups.

Spermatocyte: Cell destined to undergo meiosis to produce sperm.

Spikes: Projections from surface of influenza virus carrying important antigenic determinants.

Spindle: Football-shaped network of filaments arching through the cell during cell division and serving as the framework for chromosomal and cytoplasmic division.

Splicing: See RNA splicing and DNA splicing.

Split gene: A gene containing one or more intervening sequences.

Spontaneous abortion: Natural premature termination of pregnancy often associated with fetal abnormalities.

Spontaneous mutation: A mutation occurring in the absence of any known mutagen.

Stabilizing selection: Selection favoring intermediate phenotypes.

Stable equilibrium: Of allele frequency, an equilibrium that is automatically and gradually reestablished after a perturbation of allele frequencies.

Standard deviation: The square root of the variance.

Sterile mutation: A mutation that causes infertility.

Subclass: Refers to immunoglobulins within the same class whose heavy chains derive from different heavy-chain genes within the class.

Submetacentric: A chromosome that has its centromere neither approximately at its center nor near one of its ends.

Subspecies: Relatively isolated population or group of populations in a species that is sufficiently distinctive to be designated with a special name in addition to the species name.

Substrate: Substance acted upon by an enzyme.

Supersolenoid: Chromosome fiber formed by coiling of the solenoid fiber.

SV40: Simian virus 40; related to polyoma, but infects primates.

Synapsis: Pairing of homologous chromosomes; in humans, occurs only in meiosis.

Syndrome: A group of symptoms occurring together with sufficient regularity that it can be recognized as a distinct clinical entity.

Synteny: The state of two loci being on the same chromosome.

T: Conventional symbol for the deoxyribonucleotide thymidine phosphate or for the base thymine.

TATA box: Binding site for RNA polymerase; usually involves the sequence 5′–TATA–3′.

Tautomeric forms: Alternative configurations of atoms in a base.

Tay-Sachs disease: A recessively inherited enzyme defect causing degeneration of the central nervous system; particularly common in certain Jewish populations.

T cell: Lymphocyte responsible for cellular immunity.

Telomere: Special DNA sequence found at tips of chromosomes.

Telophase: A final stage of cell division characterized by division of the cytoplasmic contents as the cell pinches into two daughter cells. In telophase of mitosis or telophase I or telophase II of meiosis in males, division of the cytoplasm is approximately equal. In telophase I and telophase II of meiosis in females, cytoplasmic division is very unequal and produces one tiny product called a polar body.

Temperate: A virus or bacteriophage that can infect host cells without killing them.

Temperature-sensitive mutation: A mutation whose

phenotypic effects are expressed at one temperature (the restrictive temperature) but not at another (the permissive temperature).

Template: A pattern.

Teratogen: An agent that causes birth defects owing to its effects during development.

Termination codons: Codons signaling the end of synthesis of a polypeptide chain; in cytoplasmic mRNA's, corresponding to 5′−UAA−3′, 5′−UAG−3′, or 5′−UGA−3′.

Terminator codon: See Termination codons.

Tertiary structure: The three-dimensional configuration of a polypeptide.

Testicular-feminization syndrome: X-linked condition characterized by absence of development of male sexual structures in XY individuals.

Testosterone: Male sex hormone.

Tetraploid: Having four haploid sets of chromosomes.

Tetrasomic: An otherwise diploid individual who has two extra copies of a particular chromosome.

Thalidomide: A tranquilizer that is a teratogen because it inhibits normal development of the long bones in the arms and legs.

Three-prime (3′) end: The end of a DNA or RNA strand having a free 3′ hydroxyl.

Threshold trait: A multifactorial trait that is either present or absent in an individual.

Thymine: Pyrimidine base found in DNA.

Thymine dimer: Covalent linkage between adjacent thymines in a strand of DNA.

TK: The enzyme thymidine kinase, or the locus that codes for the enzyme.

Tolerance: Ability to avoid mounting an immune response against one's own antigens.

Total heterozygosity: Among a group of populations, the heterozygosity that would result from population fusion and random mating.

Total variance: Phenotypic variance.

Trait: A characteristic.

Transcript: See RNA transcript.

Transcription: The process in which an RNA strand complementary in base sequence to a DNA strand is produced.

Transduction: Transfer of genes from one cell to another by means of a virus.

Transfer RNA: See tRNA.

Transformation: (1) Treatment of cells with purified DNA for purposes of altering their genetic material. (2) Production of transformed, tumorlike cells by a viral infection.

Transition: Base substitution involving two pyrimidines or two purines.

Translation: Process of polypeptide synthesis.

Translocation: Attachment of part of a chromosome to a nonhomologous chromosome.

Transposable element: DNA sequence capable of altering its position within a DNA molecule or moving to a different DNA molecule.

Transposition: Physical movement of a DNA sequence from one place to another in the same or a different DNA molecule.

Transposon: Prokaryotic transposable element consisting of antibiotic resistance or another gene flanked by repeated sequences.

Transversion: Base substitution involving one pyrimidine and one purine.

Triplication: A gene or region of chromosome represented three times instead of once.

Triploid: Having three haploid sets of chromosomes.

Trisomic: Having one chromosome extra in an otherwise diploid individual.

Trisomy 13: 47,+13 karyotype; Patau syndrome.

Trisomy 18: 47,+18 karyotype; Edwards syndrome.

Trisomy 21: 47,+21 karyotype, Down syndrome.

tRNA: Small RNA molecules including an anticodon (which recognizes a corresponding codon in mRNA) and an amino acid that function in translation.

Turner syndrome: 45,X sex chromosome abnormality; affected individuals are female.

Two-egg twins: See DZ twins.

U: Conventional symbol for the ribonucleotide uridine phosphate or for the base uracil.

Ultraviolet light: Short-wavelength light absorbed by DNA.

Umber: See Opal.

Unbalanced: A chromosome rearrangement in which certain genes are lost or present in excess.

Unbalanced translocation: Translocation in which some genes are missing or present in excess.

Unequal crossing-over: Occurrence of crossing-over in unsymmetrically aligned duplications.

Unique-sequence DNA: A DNA sequence found once per haploid genome.

Unstable equilibrium: Of allele frequency, an equilibrium that does not become reestablished after a perturbation of allele frequencies.

Unstable mutation: A mutation that undergoes further mutation at a high rate.

Unwinding: Separation of strands of duplex DNA during replication.

Upstream: In RNA or antisense strand of DNA, toward the 5' end; in sense strand of DNA, toward the 3' end.

Uracil: Base found in ribonucleic acid that serves the role of thymine in deoxyribonucleic acid.

V: Variable region of an antibody polypeptide, or the corresponding DNA sequence.

Variable age of onset: See Age of onset.

Variable expressivity: See Expressivity.

Variable region: The portion of a heavy chain or light chain that varies greatly in amino acid sequence among antibodies in the same subclass.

Variance: Measure of spread of a distribution defined as the mean of the squared deviations from the mean.

Variation among populations: Refers to allele frequencies or traits that differ in different populations.

V-D-J joining: Type of DNA splicing involved in the creation of a heavy-chain gene.

Virulent: A virus or bacteriophage that kills the host cells.

Virus: A tiny internal parasite of cells capable of infectious transmission between cells but incapable of reproduction on its own.

V-J joining: Type of DNA splicing involved in the creation of a light-chain gene.

Wolffian ducts: Embryonic ducts associated with internal sexual structures of males.

X chromosome: One of the sex chromosomes; normal human females have two X's, males have only one.

Xeroderma pigmentosum: Autosomal recessive disorder characterized by development of multiple malignant skin tumors; some forms are associated with defective DNA repair processes.

X linked: A gene on the X chromosome.

XO: 45,X karyotype; Turner syndrome.

X-ray: Energy emitted as electromagnetic radiation; very penetrating and very mutagenic.

XXX: 47,XXX karyotype.

XXY: 47,XXY karyotype; Klinefelter syndrome.

XYY: 47,XYY karyotype.

Y chromosome: One of the sex chromosomes; normal human males have one Y chromosome, females have none.

Y linked: A gene on the Y chromosome.

Z DNA: DNA in its left-handed configuration.

Zeta hemoglobin: One of the polypeptide chains of embryonic hemoglobin.

Zygote: A fertilized egg.

Answers to Problems

CHAPTER 1

2. Similarity of mitochondrial and chloroplast DNA and organelles with their prokaryotic counterparts

3. 22

4. 6

5. 69; 92

6. Interferes with formation of spindle; metaphase

7. 92

8. Look for an elevated incidence of sister-chromatid exchange in fetal cells.

9. Normal DNA consists of two intertwined strands, and wherever one strand carries an A, the other carries a T; wherever one strand carries a G, the other carries a C.

10. ATGGTGCACCTGACT

11. GBABGAABCCBC

12. The long arm of the chromosome underwent *two* sister-chromatid exchanges at different positions.

13. In many cases, the effects of a mutant gene in a chromosome can be covered up by the presence of the normal gene in the homologous chromosome.

14. A female with a color-blindness mutation usually has a normal copy of the gene in her other X chromosome; a male, having only one X, will express the color-blindness trait. Thus, it requires two mutant genes to produce a color-blind female but only one to produce a color-blind male.

CHAPTER 2

2. D; Giemsa staining

3. X

4. They carry the same genes and thus have nearly identical DNA at every position.

5. It suggests that the genes on the heterochromatic X are inactive, which they are (see Chapter 5).

6. Father; mother

7. Mother; father

8. The number of chromosomes is determined by the number of centromeres. The first meiotic division *reduces* the number of centromeres in each resulting nucleus, whereas the second meiotic division leaves them *equal*.

9. First

10. One sperm would have chromosomes 1 through 22 and no sex chromosomes; the other would have chromosomes 1 through 22 plus two Y chromosomes. Fertilization of a normal egg by the first sperm would produce a zygote with a normal complement of autosomes and one X (designated 45,X or XO); fertilization of a normal egg by the second sperm would produce a zygote with a normal complement of autosomes, one X, and two Y's (designated 47,XYY or XYY).

11. Chromatid 1 carries *AB*; 2 carries *Ab*; 3 carries *aB*; 4 carries *ab*.

12. Chromatid 1 carries *AB*; 2 carries *AB*; 3 carries *ab*; 4 carries *ab*.

13. Father; crossing-over; chromosome 15

14. Number the bands on the normal chromosome 1 (top) through 12 (bottom). Breaks between 1 and 2 and between 6 and 7 with repair occurring in such a way that the middle segment is inverted will produce the abnormal chromosome shown. The band sequence of the abnormal chromosome is 1-6-centromere-5-4-3-2-7-8-9-10-11-12. The individual would be expected to be normal because all genes are present in normal number.

15. Long arm derives from chromosome 14, short arm from chromosome 21. Evidently, one chromosome was broken in the long arm very near the centromere, the other in the short arm very near the centromere, and the pieces were interchanged before rejoining. The missing chromosomal material consists of the short arms of both chromosomes.

16. 50 percent XX, 50 percent XY; 50 percent female, 50 percent male

17. $2^4 = 16$; $2^{23} = 8,388,608$, or about 8.4×10^6

CHAPTER 3

2. Genotype; genotype; phenotype; phenotype; genotype; phenotype; phenotype; genotype; genotype; phenotype; phenotype; phenotype

3. Genes—not traits—are inherited, so a trait cannot, strictly speaking, be "autosomal" or, for that matter, "recessive." Nevertheless, the casual terminology is widely used and even useful in cases where no misinterpretation is possible.

4. The genes involved have incomplete penetrance and the corresponding trait may or may not be expressed depending on the particular environmental conditions occurring to each twin during development.

5. Both parents could be *Dd* (i.e., Rh⁺) and have a *dd* (i.e., Rh⁻) child; the reverse situation is not possible unless a new mutation occurs.

6. All offspring will be *Tt* and therefore tasters.

7. The woman in question must be *Dd* and her husband *dd*. Expect $\frac{1}{2}$ *Dd* (i.e., Rh⁺) and $\frac{1}{2}$ *dd* (i.e., Rh⁻) offspring. Punnett square is similar to that in Figure 3.8.

8. Mother *tt*; child *tt*; thus father must be *Tt*.

9. The father has to contribute a *T* allele to the child, so he must be either *TT* or *Tt*.

10. The child had to receive the *I*ᴬ and the *d* allele from its father; thus, the father's genotype for

ABO must be $I^A I^A$, $I^A I^B$, or $I^A I^O$, and for Rh must be Dd or dd. Thus, the father's phenotype must be either A or AB and either Rh$^+$ or Rh$^-$.

11. The son must have received the B allele from his mother, so the genotype of mother is Bb, father bb, and son Bb.

12. To produce an AB child, one parent must carry I^A and the other I^B; to produce an O child, both parents must carry I^O. So the parents are $I^A I^O$ (blood type A) and $I^B I^O$ (blood type B).

13. The woman must have contributed I^O to the first child and I^A to the second, so she is $I^A I^O$ and has blood type A.

14. The probability that the oldest child is affected is simply $\frac{1}{2}$. The probability that exactly two out of four are affected is:

$$\frac{4!}{2!2!}\left(\frac{1}{2}\right)^2\left(\frac{1}{2}\right)^2 = \frac{3}{8} = 0.375$$

15. Probability of exactly two affected is $[3!/(2!1!)] \times (\frac{1}{2})^2(\frac{1}{2})^1 = 3/8$.
Probability of exactly three affected is $[3!/(3!0!)] \times (\frac{1}{2})^3(\frac{1}{2})^0 = \frac{1}{8}$.
Probability of two *or* three affected is $\frac{3}{8} + \frac{1}{8} = \frac{1}{2}$.

CHAPTER 4

2. If the recessive trait is rare, most of the abnormal alleles will occur in heterozygotes, so sterilization of homozygotes does little to reduce the number of abnormal alleles.

3. If a recessive allele is rare, it is much more likely to be heterozygous in each of two relatives than in each of two unrelated individuals.

4. Both heterozygous

5. The parents must be heterozygous, and the woman is not affected, so the probability is $\frac{2}{3}$.

6. $\{3!/[(1!)(2!)]\}(\frac{1}{4})^1(\frac{3}{4})^2 = 27/64 = 0.422$

7. Cross homozygous animals. If offspring are blind, the recessives are alleles (i.e., noncomplementing); otherwise, they are nonalleles (i.e., complementing).

8. Since the two forms of albinism are equally frequent in Caucasians, the Caucasian parent will have the Hopi form half the time and the non-Hopi form the other half. Thus, the probability of affected offspring (i.e., noncomplementation) is $\frac{1}{2}$.

9. Aminopterin is required to knock out the major pathway of DNA synthesis.

10. The X; the X

11. a, d, g, and h

12. Chromosome 7

13. $Pi^A/Pi^A;D/d$ and $Pi^A/Pi^Z;D/d$ and $Pi^A/Pi^A;d/d$ and $Pi^A/Pi^Z;d/d$ in ratio 1:1:1:1 or proportions $\frac{1}{4}:\frac{1}{4}:\frac{1}{4}:\frac{1}{4}$; illustrates independent assortment

14. Ab; AB; ab; and aB. First and last listed are nonrecombinant; middle two are recombinant.

15. $Lu\ Se/lu\ se$ gametes and frequencies:

$Lu\ Se$ frequency $(1 - 0.13)/2 = 0.435$
$Lu\ se$ frequency $\qquad 0.13/2 = 0.065$
$lu\ Se$ frequency $\qquad 0.13/2 = 0.065$
$lu\ se$ frequency $(1 - 0.13)/2 = 0.435$

$Lu\ se/lu\ Se$ gametes and frequencies:

$Lu\ se$ frequency $(1 - 0.13)/2 = 0.435$
$Lu\ Se$ frequency $\qquad 0.13/2 = 0.065$
$lu\ se$ frequency $\qquad 0.13/2 = 0.065$
$lu\ Se$ frequency $(1 - 0.13)/2 = 0.435$

CHAPTER 5

2. $(\frac{1}{2})^5 = \frac{1}{32} = 0.031$; $[5!/(4!)(1!)](\frac{1}{2})^4(\frac{1}{2})^1 = 5/32 = 0.156$

3. A sex-limited trait expressed only in males (see Chapter 3)

4. His paternal grandfather

5. XXY

6. XXX

7. None

8. 1 (she *must* be heterozygous)

9. The woman's father would *not* have had hemophilia, so the chance that the woman herself is a carrier is 0; the woman's mother *must* be a carrier, so the chances that she herself is a carrier is $\frac{1}{2}$.

10. Mother *G6PDA/G6PDB*; father *G6PDA*

11. $\frac{1}{2}$

12. $\frac{1}{2}$

13. Males: all have the green form of color blindness; females: all have normal color vision because of complementation.

14. Male; female

15. The mother's genotype must be *gcb G6PDA/ gcb⁺ G6PDB*. The son carries *G6PDA*, so the probability that the boy is color blind is the same as the probability that the *G6PDA*-bearing chromosome did *not* undergo recombination in the *gcb–G6PD* region; from Figure 5.8, this probability is $1 - 0.05 = 0.95$ (i.e., there is a 95 percent chance that the boy is color blind).

CHAPTER 6

2. A tetraploid

3. Because the Y carries very few genes and the X is subject to dosage compensation

4. 45,Y

5. 46 (i.e., the karyotype is 46,X,+21)

6. Meiotic nondisjunction occurs in meiosis, mitotic nondisjunction in mitosis. Primary nondisjunction occurs in chromosomally normal individuals and often results from such phenomena as anaphase lag; secondary nondisjunction occurs in trisomics as a normal consequence of meiosis. First-division nondisjunction occurs in the first meiotic division; second-division nondisjunction occurs in the second meiotic division.

7. The mosaic individual would have 45,X in some cells and 47,XXX in others, designated 45,X/47,XXX.

8. Second

9. XX; O; Y; Y

10. Because these karyotypes have no outstanding, characteristic, or consistent symptoms

11. The mother's genotype could be *rcb/rcb⁺* (i.e., heterozygous for the color-blindness allele). Second-division nondisjunction of the *rcb*-bearing X produces an *rcb/rcb* egg, which, when fertilized by a Y-bearing sperm, results in a *rcb/rcb/Y* zygote.

12. Theoretically XX, XY, XXX, and XXY. Virtually all offspring are either XX or XY.

13. $\frac{2}{3}$

14. It had to have occurred in the father; otherwise, the child would be color blind.

15. Because many zygotes are aborted before the pregnancy is recognized

CHAPTER 7

2. Use the gene symbols as in the illustration with a raised dot to represent the centromere. Possibilities are *a b c · d e f* (normal), *a d · c b e f* (pericentric inversion), and *a e f* (acentric) plus a ring carrying the centromere and genes *b, c,* and *d*.

3. Use the gene symbols as in the illustration with a raised dot to represent the solid centromere and a slash to represent the open centromere. Possibilities are *a b · c d* with *y / z* (normal), *a b*

· *y* with *z / c d* (reciprocal translocation), and *a b · / z* with *y c d* (a dicentric and an acentric).

4. The Giemsa banding patterns of 2p and 2q are recognizably the same as those found on two nonhomologous acrocentrics in related primates.

5.

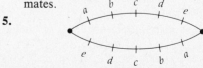

6. Its arms must be identical.

7. $a\,b\,c \cdot c\,b\,a$ and $e\,d \cdot d\,e$

8. She carries a Robertsonian translocation (see Figure 7.15).

9. An unbalanced translocation refers to a rearranged chromosome or chromosomes with some regions absent or present in excess.

10. $\cdot a\,b\,c\,a\,b\,c\,a\,d\,e$ and $\cdot a\,d\,e$

11. Depending on the mode of synapsis, complementary gametes could carry the following number of copies of the region: 4 *versus* 3; 5 *versus* 2; or 6 *versus* 1.

12. Four-strand double produces the following gametes:

$\cdot a\,b\,c'\,d'\,a' \cdot$ (dicentric)
$\cdot a\,b\,c\,d'\,a' \cdot$ (dicentric)
$e'\,b'\,c\,d\,e$ (acentric)
$e'\,b'\,c'\,d\,e$ (acentric)

Two-strand double produces the following gametes

$\cdot a\,b\,c\,d\,e$ (not involved in crossovers)
$\cdot a'\,d'\,c'\,b'\,e'$ (not involved in crossovers)
$\cdot a\,b\,c'\,d\,e$ (monocentric because of two-strand double)

$\cdot a'\,d'\,c\,b'\,e'$ (monocentric because of two-strand double)

13. First breaks in a–b and e–f; second breaks in e–d and b–f.

14. In adjacent-1 segregation (see Figure 7.13), homologous centromeres go to opposite anaphase I poles. Thus, one class of gametes from adjacent-1 segregation will carry a duplication of 21p and a deficiency of part of the short arm of 5 (i.e., −5pter), and the resulting zygote will give rise to a *cri-du-chat* infant; the complementary class of gametes will carry a deficiency of 21p and +5pter and the resulting zygote will thus have the *cri-du-chat* segment duplicated. Both types of offspring can survive. In adjacent-2 segregation (see Figure 7.13), homologous centromeres go to the same pole, so one class of gametes will carry a duplication of 21q and a deficiency of most of 5 (forms lethal zygotes), and the other class of gametes will carry a deficiency of 21q and a duplication of most of 5 (also forms lethal zygotes).

15. $\widehat{XX}Y$ crossed with XY produces the following types of zygotes: $\widehat{XX}X$, $\widehat{XX}Y$, XY, and YY. Surviving offspring are expected to be $\widehat{XX}Y$ females and XY males in equal proportions; the *sons* receive the father's X and the *daughters* the father's Y.

CHAPTER 8

2. $4^3 = 64$

3. The hydrogen bonds that hold the strands together are relatively weak and can be broken by heat; the covalent bonds within each strand are too strong to be broken by moderate temperatures.

4. 5′-AGCGTTCACC-3′ and 5′-AGCGTTCACC-3′
3′-TCGCAAGTGG-5′ and 3′-TCGCAAGTGG-5′
At the left, the bottom strand is new; at the right, the top strand is new; this is called semiconservative replication.

5. 9.826×10^9 Å $= 98.26$ cm $= 38.71$ in

6. The base pairs must match as follows: A in one strand corresponds to T in the other; T in one strand corresponds to A in the other; G in one strand corresponds to C in the other; and C in one strand corresponds to G in the other.

7. Excise the middle nucleotide of the *top* strands in part *(b)*.

8. Probability of a site occurring is $(\frac{1}{4})^4$ for *Alu*I, and the reciprocal of this is the average distance between sites; thus, for *Alu*I, the answer is $4^4 = 256$. For *Asu*I, the average distance is $4^5 = 1024$. For *Eco*RI, the average distance is 4096. For *Acy*I, the probability of a site occurring is $(\frac{1}{4})(\frac{1}{2})(\frac{1}{4})(\frac{1}{4})(\frac{1}{2})(\frac{1}{4}) = 1/1024$, so the average distance between sites is 1024.

9. The sequence is a perfect palindrome except

for the $_C^G$ in the center, so the hairpin structure will resemble that in Figure 8.19 with no "tails" at the ends.

10. *Alu*I:6213; *Bam*HI:6240, 6264; *Eco*RI:6231, 6273; *Sal*I:6246, 6258; *Pst*I:6252; *Xho*II:6240, 6264; *Hind*III: site does not occur

11. 5'-GAATTCGACCCTTTCAAGAATTC-3'
    3'-CTTAAGCTGGGAAAGTTCTTAAG-5'

12. Introducing pure DNA into a living cell; proves DNA is genetic material because introducing *only* DNA changes the phenotype of the cell from antibiotic sensitive to antibiotic resistant

13. 3'-CTACACAG-5'

14. Replication terminates because the dideoxynucleotide lacks the 3' hydroxyl required for continued synthesis; dideoxynucleotides must be added as triphosphates so they can be incorporated into the growing strand.

15. Because DNA polymerase cannot initiate synthesis; it requires a primer.

CHAPTER 9

2. The presence of a cap (see Figure 9.22)

3. Palindromic nucleotide sequences in single-stranded RNA

4. Gene 2: will grow with C or A but not with O or G; gene 1: will grow with O or C or A but not with G

5. C $\xrightarrow{\text{Gene 4}}$ B $\xrightarrow{\text{Gene 3}}$ A $\xrightarrow{\text{Gene 1}}$ E$^{\text{Gene 2}}$ → D

6. Amino terminal end at left, carboxyl end at right. Peptide bond is the C-N bond near the center of the dipeptide. Amino end is gly.

7. mRNA reads:

   5'-ACACGAGCUAAUUCAUUGGCCACCGAAGAUUAA-3'

   Polypeptide: TRANSLATED

8. 5'-AGUGCGAUUCCAAAGC-3'

9. 5'-ATTCGTACTTTGCCAAT-3'
   (antisense)
   3'-TAAGCATGAAACGGTTA-5'   (sense)

10. 3'-AGC-5' would pair with 5'-UCG-3' (ser codon)
    3'-GUU-5' would pair with 5'-CAA-3' (gln codon)
    3'-UGU-5' would pair with 5'-ACA-3' (thr codon)

5'-ACG-3' would pair with 5'-CGU-3' (arg codon)
5'-GUA-3' would pair with 5'-UAC-3' (tyr codon)
5'-CAU-3' would pair with 5'-AUG-3' (met codon)

11. R-R-R-R-R-R; illustrates synonymous codons

12. S-R-I-L-D-Q-S; R-V-Y-W-T-N-Q; A-Y-T-G-P-I-S

13. Amino acids alternating ile and tyr

14. Intron is
    5'-GTCGACCAAATTTCGCAGGAG-3'.

15. The bottom strand is thus like the mRNA in sequence (although, of course, U replaces T in RNA), and it must be translated from the 5' to the 3' end (i.e., right to left as depicted). Sequence of amino acids is thus S-T-S-R-R-S-R-Q-L. After *Eco* RII digestion and ligation, sequence in plasmid is

    5'-AAGCTGCCTGGACGTCGA-3'
    3'-TTCGACGGACCTGCAGCT-5'

    Amino acid sequence would now be S-T-S-R-Q-L.

CHAPTER 10

2. In eukaryotes, transcription and translation are automatically uncoupled, with transcription occurring in the nucleus and translation in the cytoplasm.

3. CAP in *E. coli* is catabolite activator protein; the mRNA cap is an unusual nucleotide in an unusual orientation at the 5' end of a eukaryotic mRNA.

4. A polycistronic mRNA (prokaryotes) produces several polypeptides during translation, each initiated by an AUG and terminated by a "stop." A polyprotein (eukaryotes) is produced as one long polypeptide that is cleaved into segments after translation. Can serve same regulatory role by simultaneous regulation of two or more polypeptide-coding genes.

5. Combinatorial joining involves DNA joining; the normal sort of splicing to excise introns involves RNA.

6. 3'-TAAGCATTGAAGCCCTGGG-5'; right to left

7. A Southern blot is a pattern of DNA bands that have been blotted onto a special type of filter paper, hybridized with a radioactive DNA probe, and autoradiographed. A probe is a particular DNA sequence that can be used in hybridization to locate the occurrence or position on a gel of homologous sequences. The larger the DNA fragment that carries a sequence homologous to the probe, the shorter is the distance it will have migrated in the original gel prior to the Southern blot.

8. *Hpa*II: 5'-A$\overset{*}{C}$ -3' and 5'-CGGTC -3' and
   3'-TGGC-5'   3'- $\overset{*}{C}$AGGC-5'

   5'-CGGGAC$\overset{*}{C}$GGATC$\overset{*}{C}$GG-3'
   3'-   CCTGG$\overset{*}{C}$CTAGG$\overset{*}{C}$C-5'

   *Msp*I: 5'-A$\overset{*}{C}$ -3' and 5'-CGGTC -3' and
   3'-TGGC-5'   3'- $\overset{*}{C}$AGGC-5'

   5'-CGGGAC -3' and 5'-$\overset{*}{C}$GGAT$\overset{*}{C}$ -3' and
   3'-   CCTGG$\overset{*}{C}$-5'   3'- CTAGG$\overset{*}{C}$-5'

   5'-$\overset{*}{C}$GG-3'
   3'- $\overset{*}{C}$-5'

9. The recombination event would cause the de-

letion of a segment of DNA carrying *hin*, the promoter, and one of the direct repeats.

10. The *hin* gene codes for a recombination enzyme that catalyzes the site-specific recombination. A *hin*⁻ mutation would be unable to switch flagellar type. In such a temperature-sensitive mutation, switching could occur at 30°C (because the *hin* gene product could function) but not at 37°C (because the *hin* gene product would denature and be nonfunctional).

11. *lac* operon would remain in the state depicted in Figure 10.10*(a)*, and normal induction by lactose would not occur.

12. *lac* operon would remain in the state depicted in Figure 10.11*(a)*, and normal induction by lactose would not occur.

13. In both cases the operon remains in the state depicted in Figure 10.11*(b)* even in the absence of lactose. For *lacI* mutations this is because the repressor cannot bind to *lacO*; for *lacO* mutations it is because the normal repressor cannot complex with the mutant *lacO*. In both cases the enzymes are produced constitutively.

14. The *trpL* peptide could be translated even when the levels of tryptophan were very low. Transcription would thus always terminate. The operon could not be normally induced by low tryptophan levels.

15. Inhibition of the last enzyme would still permit accumulation of useless (or perhaps even harmful) intermediates. Inhibition of the first enzyme also eliminates production of these intermediates.

CHAPTER 11

2. The phenotype is expressed with a normal diet but not with a low-phenylalanine diet. Indicates that genes and environment are both important, even with simple Mendelian traits.

3. Forward is mutation from normal to mutant, and reverse is from mutant to normal; many base changes can mutate a normal gene, but only a few can reverse a mutation that has already occurred.

4. Transitions, because a purine still pairs with a pyrimidine, although with the wrong one

5. Transitions, which would cause U/C or A/G interchanges

6. mRNA is

   5'-GGGGAGAACGAGACAAUCUGCUCG-3'.

   Polypeptide: GENETICS

7. DNA: 5'-GGGGAGAACGAGAGAATCTGCTCG-3'
         3'-CCCCTCTTGCTCTCTTAGACGAGC-5'

   mRNA: 5'-GGGGAGAACGAGAGAAUCUGCUCG-3'

Polypeptide: GENERICS

Mutation is a transversion and a missense.

8. DNA: 5'-GGGGAGAATGAGACAATCTGCTCG-3'
3'-CCCCTCTTACTCTGTTAGACGAGC-5'

mRNA: 5'-GGGGAGAAUGAGACAAUCUGCUCG-3'

Polypeptide: GENETICS

Mutation is a transition and is silent.

9. DNA: 5'-GGTGGAGAACGAGACAATCTGCTCG-3'
3'-CCACCTCTTGCTCTGTTAGACGAGC-5'

mRNA: 5'-GGUGGAGAACGAGACAAUCUGCUCG-3'

Polypeptide: GGERDNLL

Mutation is a frameshift mutation.

10. DNA: 5'-GGGAGAACGAGACAATCTGCTCG-3'
3'-CCCTCTTGCTCTGTTAGACGAGC-5'

mRNA: 5'-GGGAGAACGAGACAAUCUGCUCG-3'

Polypeptide: GRTRQSA

Mutation is a frameshift mutation.

11. DNA: 5'-GGGGAGAACTAGACAATCTGCTCG-3'
3'-CCCCTCTTGATCTGTTAGACGAGC-5'

mRNA: 5'-GGGGAGAACUAGACAAUCUGCUCG-3'

Polypeptide: GEN

Mutation is a transversion and (amber) nonsense.

12. With amber suppressor, polypeptides would be GEN (if no readthrough) or GENLTICS (if readthrough); with ochre suppressor only GEN would be produced.

13. Two (positions 9–10 and 23–24). Methylation and deamination would change the ^C_G pairs into ^T_A pairs. Mutation at 9 would produce GENETICS and so would be silent. Mutation at 23 would produce GENETICL and so would be a missense. Mutations at 10 and 24 would produce GENKTICS (missense) and GENETICS (silent), respectively.

14. If the element were inserted in the proper orientation, transcription of the gene would occur. In the reverse orientation, transcription of the opposite DNA strand would occur, and in the opposite direction.

15. Sequence (a) PWLL; (b) PLGFYD

Recombinants: 5'-CCCTGGGCTTCTATGATTGAT-3'
3'-GGGACCCGAAGATACTAACTA-5'
codes for PWASMID (double mutant) and

5'-CCCCTGGCTTCTATGATTGAT-3'
3'-GGGGACCGAAGATACTAACTA-5'
codes for PLASMID (normal)

CHAPTER 12

2. The capsid (mainly if not exclusively protein) surrounds the viral genetic material; the envelope (lipid, protein, and carbohydrate) surrounds the capsid.

3. Since only the DNA gets into the infected cells, it follows that the genetic information for making progeny viruses must reside in the DNA.

4. To enzymatically disrupt the cell walls

5. A temperate phage has two alternative life cycles—the lytic cycle and lysogeny—whereas a virulent phage has only the lytic cycle. Thus, temperate phage can undergo lysogeny whereas virulent phage cannot.

6. Endogenous viruses are viruslike DNA sequences found in multiple copies integrated into the host genome. With copialike elements, they share the features of integration into the host genome, multiple copies, and, very likely, transposability.

7. Antigenic drift involves a relatively minor change in viral antigens caused by conventional mutation. Antigenic shift involves a major antigenic change caused by viruses that have picked up genes from other nonhuman influenza viruses. An epidemic is a local outbreak of disease; a pandemic is a worldwide epidemic.

8. Volume of polyoma $= \frac{4}{3}(3.14)(25)^3 = 6.54 \times 10^4$ Å^3.
Volume of average cell $= \frac{4}{3}(3.14)(0.5 \times 10^5)^3 = 5.23 \times 10^{14}$ Å^3.
Ratio of volumes is about 8×10^9, or 8 billion.

9. Lambda can insert itself into the bacterial chromosome and carries a gene that codes for the enzyme that catalyzes this process.

10. Three: Gene 1 is represented by mutations 1, 4, and 5; gene 2 by mutations 2 and 6; and gene 3 by mutation 3.

11. Infect a bacterial strain known to carry several mutations at scattered locations on the chromosome. Then determine by selection whether transduction of the normal allele can occur at

any or all of these loci. If transduction occurs at many of the loci, the phage is a generalized transducing phage; otherwise, it is a specialized transducing phage or perhaps a nontransducing phage.

12. *uxaA–tolC–metC*

13. *uxaA–ebgA–tolC–metC*

14. There is consanguinity among the parents of the affected individual; the parents are first cousins.

15. $\frac{2}{3}$

CHAPTER 13

2. T cells are agents of cellular immunity and have killer functions, helper functions, and so on; B cells are agents of humoral immunity and produce antibody.

3. An antibody class is determined by the general type of heavy chain, such as γ chains in IgG. Subclasses are determined by the particular type of heavy chain within a class, such as γ^1. Allotypes are genetic variants of genes coding for a particular subclass of heavy chains.

4. An epitope is an antigenic determinant on an antigen; an idiotype is an antigenic determinant within an antigen-binding site on an antibody.

5. Immune surveillance refers to the ability of the immune system to recognize and attack certain antigens present on the surface of cancer cells. The T antigens involved in transformation are suspectible to such attack.

6. $140 \times 6 \times 11 = 9240$

7. The network hypothesis entails the production of anti-idiotype antibodies (i.e., antibodies against other antibodies); however, one's own antibodies are certainly part of one's "self."

8. $\frac{1}{4}$

9. I^A/I^O and I^B/I^O

10. Serum from A contains anti-B; serum from B contains anti-A. Blood types are A (individuals 1 and 6), B (individual 5), AB (individual 2), and O (individuals 3 and 4).

11. One parent's genotype is I^A/I^B;*Se/se*, the other's is I^A/I^B;*se/se*. Possible offspring are:

$\frac{1}{8}$ I^A/I^A;*Se/se* (A, secretor)
$\frac{1}{8}$ I^A/I^A;*se/se* (A, nonsecretor)
$\frac{1}{4}$ I^A/I^B;*Se/se* (AB, secretor)
$\frac{1}{4}$ I^A/I^B;*se/se* (AB, nonsecretor)
$\frac{1}{8}$ I^B/I^B;*Se/se* (B, secretor)
$\frac{1}{8}$ I^B/I^B;*se/se* (B, nonsecretor)

12. Male 1 (because the baby must have received an *M* allele from its father)

13. Couple 1 could be the parents of baby A or baby B. Couple 2 could be the parents of baby B but not of baby A (because of ABO blood types). So, baby A belongs to couple 1 and baby B to couple 2.

14. $0.43 \times 0.85 = 0.37$, or 37 percent

15. A secretor, as one of the body fluids (in this case the semen) carried the ABO antigens (in this case A)

CHAPTER 14

2. Natural selection is the process whereby genes that promote survival and reproduction in a particular environment increase in frequency in the population. Premises: (a) more young are born than can survive and reproduce, (b) there is phenotypic variation in the chance of survival and reproduction, and (c) some of this phenotypic variation is associated with genetic variation.

3. Two (or more) isolated populations can undergo genetic divergence and ultimately achieve the status of species.

4. If the enzyme were a dimer (i.e., composed of two aggregated polypeptide chains), heterozygotes would have three classes of dimer—namely, F + F, F + S, and S + S. The third (intermediate mobility) band is the F + S dimer.

5. The enzyme locus is X linked.

6. Frequency of $M = (596 + 489)/2000 = 0.542$; of $N = 0.458$. Expected numbers: $M = 293.8$; $MN = 496.5$; $N = 209.8$. (The numbers do not add to exactly 1000 because of round-off error.)

7. Genotype F/F: 7, 12 (i.e., 2 of them); F/S: 2, 5, 6, 9, 10, 11, 13, 15, 18, 20 (i.e., 10); S/S: 1, 3, 4, 8, 14, 16, 17, 19 (i.e., 8). Allele frequency of $F = (4 + 10)/40 = 0.35$. Allele frequency of $S = (16 + 10)/40 = 0.65$.

8. $H = 0.066$

9. Polymorphism implies that the frequency of the most common allele is greater than 0.95. So, monomorphic populations are 2, 3, and 4; polymorphic populations are 1, 5, and 6.

10. Frequency of $d = 0.4$; of $D = 0.6$. Expected genotype frequencies: $DD = 0.36$; $Dd = 0.48$; $dd = 0.16$.

11. Expected frequencies: A = 0.0729 + 0.3618 = 0.4347; B = 0.0036 + 0.0804 = 0.0840; AB = 0.0324; O = 0.4489.

12. Frequency of heterozygous females = 0.095, or nearly 10 percent

13. Frequency of recessive allele = 0.8124. Frequency of heterozygotes = 0.3048. Frequency of dominant homozygotes = 0.0352. Frequency of heterozygotes *among* those with wet cerumen = 0.3048/(0.3048 + 0.0352) = 0.896, or about 90 percent.

14. Frequency of recessive allele = 0.2236; frequency of dominant allele = 0.7764; frequency of heterozygotes = 0.3472, or about 35 percent.

15. Average heterozygosity among subpopulations = (0.3048 + 0.3472)/2 = 0.3260. Allele frequency of recessive in hypothetical fused population is (0.8124 + 0.2236)/2 = 0.5180, of dominant allele = 1 − 0.5180 = 0.4820, so total heterozygosity = 2(0.5180)(0.4820) = 0.4994. Thus, $F = (0.4994 − 0.3260)/0.4994 = 0.35$; this is a very large amount of genetic differentiation at this locus.

16. Allele frequency of recessive = 0.02. For first cousins, $f = \frac{1}{16} = 0.0625$. Among offspring of first cousins the incidence is:

$$(0.02)^2(1 − 0.0625) + 0.02(0.0625) = 0.0016 = 16 \times 10^{-4}$$

CHAPTER 15

2. Sequence divergence refers to differences in nucleotide sequence that accumulate during the course of evolution. Principal causes include mutation, random genetic drift, and perhaps selection. Divergence is impeded by gene conversion or concerted evolution (see Chapter 8).

3. Frequency in next generation is $0.85(1 − 3 \times 10^{-5}) = 0.85(0.99997) = 0.8499745$. Frequency after 10 generations is $0.85(0.99997)^{10} = 0.849745$. Equation relevant to allele frequency of 0.75 is $0.75 = (0.85)(0.99997)^n$, so $n = [\ln(0.75) − \ln(0.85)]/\ln(0.99997) = 4172$ generations. Allele frequency ultimately goes to 0.

4. Ultimate equilibrium allele frequency of A is $(4 \times 10^{-6})/(4 \times 10^{-6} + 3 \times 10^{-5}) = 0.1176$; ultimate frequency of a is $1 − 0.1176 = 0.8824$.

5. Equation to be solved is
$$0.484 − 0.507 = (0.474 − 0.507)(1 − m)^{10}$$
$$(1 − m)^{10} = 0.69697$$
$$\ln(1 − m) = −0.036$$
$$m = 0.035$$

6. Relative fitnesses of AA, Aa, and aa are respectively 1.00, 0.99048, and 0.

7. Selection coefficient against $Aa = 0.02$; against $aa = 0.60$.

8. Allele frequency in next generation is $0.05/(1 + 0.05) = 0.04762$. After 10 generations allele frequency will be $0.05/(1 + 10 \times 0.05) = 0.03333$.

9. Equilibrium allele frequency of $a = \sqrt{4 \times 10^{-6}} = 0.002$.
Frequency of aa zygotes $= (0.002)^2 = 4 \times 10^{-6}$.

10. $q = \sqrt{\mu}$ in this case, so $\mu = (0.005)^2 = 2.5 \times 10^{-5}$.

11. From the formula for partial dominance in the text, $q = (4 \times 10^{-6})/0.02 = 0.0002$.

12. Average fitness is $(0.81)(1.00) + (0.18)(0.99) + (0.01)(0.97) = 0.9979$.

13. Ratio of allele frequencies is $0.05/0.10 = 0.5$, so allele frequency of $A = 0.5/(1 + 0.5) = 1/3$ and a is $2/3$. Average fitness at equilibrium is $(1/3)^2(0.90) + 2(1/3)(2/3)(1.0) + (2/3)^2(0.95) =$ 0.967. No other allele frequencies can produce a greater average fitness with random mating.

14. Total heterozygosity in hypothetical fused population is $2(0.6)(0.4) = 0.48$; $F = (0.48 - 0.456)/(0.48) = 0.05$.

15. Population 1 frequency of $aa = 0.04$; population 2 frequency $= 0.0004$; average $= 0.0202$. Fused population would have allele frequency of $(0.2 + 0.02)/2 = 0.11$, and frequency of homozygous recessives would be 0.0121.

CHAPTER 16

2. Quantitative traits involve the segregation of minor genes, which, by definition, cannot be followed individually in pedigrees as can the major genes involved in simple Mendelian traits.

3. A major gene is one whose phenotypic effects can be detected separately from the effects of other genes or environmental factors; a minor gene is one whose phenotypic effects are obscured by other genetic and environmental factors. They are not separate classes of genes but refer to the manner in which phenotypes are examined and the range of environments involved.

4. Genotype-environment interaction refers to a situation in which a genotype's phenotype in one environment cannot be predicted from knowledge of the phenotype in a different environment. Genotype-environment association refers to genotype-environment combinations occurring nonrandomly.

5. Broad-sense heritability involves the genotypic variance; narrow-sense involves the additive genetic variance. These will be equal if all alleles affecting the trait are additive and if there is random mating.

6. MZ twins may share a more intimate association of embryonic membranes and may be treated more similarly by parents, sibs, peers, teachers, and others.

7. $\frac{1}{4}, \frac{1}{2}, \frac{1}{4}$

8. Let m = proportion of MZ twins. Overall proportion of like-sex twins will be
$$m(1) + (1 - m)(\tfrac{1}{2}) = \frac{1 + m}{2}$$
In this case $(1 + m)/2 = 0.65$, so $m = 0.30$; i.e., 30 percent of the twins are MZ.

9. $\frac{1}{4}$; 1

10. $\frac{1}{2}$; 1

11. For one locus, $\frac{1}{4}$ of the F_2 will have a genotype as extreme as one grandparental inbred and $\frac{1}{4}$ will have a genotype as extreme as the other grandparental inbred. For four unlinked loci, the analogous quantities are $\left(\frac{1}{4}\right)^4 = \frac{1}{256}$ and $\left(\frac{1}{4}\right)^4 = \frac{1}{256}$, for a total of $\frac{1}{128}$.

12. Mean $= (\tfrac{1}{9})(4) + (\tfrac{2}{9})(5) + (\tfrac{3}{9})(6) + (\tfrac{2}{9})(7) + (\tfrac{1}{9})(8) = 6$.
Variance $= (\tfrac{1}{9})(-2)^2 + (\tfrac{2}{9})(-1)^2 + (\tfrac{3}{9})(0) + (\tfrac{2}{9})(1)^2 + (\tfrac{1}{9})(2)^2 = \frac{12}{9} = 1.33333$.
Standard deviation $= 1.1547$.

13. Standard deviation is 2. About 68 percent of the population will have bristles in the range $40 \pm 2 = 38$ to 42; about 95 percent will be in the range $40 \pm 2(2) = 36$ to 44.

14. Selection differential is $44 - 40 = 4$. Predicted response $= 0.5(4) = 2$; predicted mean of progeny is thus $40 + 2 = 42$.

15. Selection differential $= 200$; response $= 40$; heritability $= \frac{40}{200} = 0.20$. This corresponds to narrow-sense heritability, but it is often called the *realized heritability* when obtained by this sort of experiment.

Credits

Chapter 1

1-1 The American Museum of Natural History, New York City.

1-3 (a) M. E. Bayer, The Institute for Cancer Research, Philadelphia.
(b) M. E. Doohan, University of Wisconsin, Madison.
(c) K. R. Porter, University of Colorado.

1-4 (a) K. R. Porter, University of Colorado.
(b) R. A. Dilley, Purdue University.

1-5 Redrawn from W. DeWitt, Biology of the Cell (Philadelphia: Saunders, 1977).

1-8 B. John, Australian National University, Canberra.

1-9 Cytogenetics Laboratory, University of California, San Francisco.

1-10 B. E. Schuh, Monmouth Medical Center, Long Branch, New Jersey.

1-12 G. Albrecht-Buehler, Cold Spring Harbor Laboratory.

1-14 A. P. Craig-Holmes, M.D., Dept. of Medicine, Baylor College of Medicine, Houston.

1-15 J. M. Opitz, M.D., Shodair Children's Hospital, Helena, Montana.

1-16 R. J. Green, Center for Disease Control, Dept. of Health and Human Services, Atlanta, Georgia.

Chapter 2

2-2 Cytogenetics Laboratory, University of California, San Francisco.

2-3 M. W. Shaw, M.D. Anderson Hospital, The University of Texas at Houston.

2-4 A. P. Craig-Holmes, M.D., Dept. of Medicine, Baylor College of Medicine, Houston.

2-5 C. C. Lin, M.D., University of Calgary.

2-6 A. P. Craig-Holmes, M.D., Dept. of Medicine, Baylor College of Medicine, Houston.

2-8 After J. J. Yunis, Science, 1976, 191:1268–1270. Copyright © 1976 by the American Association for the Advancement of Science.

2-9 (a, c, d) L. F. Meisner, Cytogenetics Unit, State Laboratory of Hygiene, University of Wisconsin, Madison.
(b) H. A. Lubs, Dept. of Pediatrics, University of Colorado Medical Center, Denver.

2-10 B. John, Australian National University, Canberra.

2-13 B. John, Australian National University, Canberra.

2-14 A. T. L. Chen, Ph.D., Center for Disease Control, Dept. of Health and Human Services, Atlanta, Georgia.

2-15 B. John, Australian National University, Canberra.

Chapter 3

3-1 V. Orel, Mendelianum of the Moravian Museum, Brno.

3-6 From O. L. Mohr, The Journal of Heredity, 1932, 23:345–352.

3-7 After O. L. Mohr, The Journal of Heredity, 1932, 23:345–352.

3-10 (a, b) U.P.I.
(c) K. Taysi, M.D., Dept. of Pediatrics, St. Louis Children's Hospital, St. Louis, Missouri.
(d) A. M. Winchester.
(e) Rephotographed from P. R. Dodge, page 492, fig. XVI.9. R. D. Adams, L. M. Eaton, and G. M.

Shy (eds) Congenital Neuromuscular Disorders, 1960 38:479–527, (Baltimore: Williams & Wilkins, 1960).
(f) J. Miller, M.D., Dept. of Ophthalmology, St. Louis Children's Hospital, St. Louis, Missouri.
(g) J. G. Hall, M.D., Dept. of Medical Genetics, The University of British Columbia, Vancouver.
(h) J. de Framond, Dept. of Genetics, Washington University School of Medicine, St. Louis, Missouri.

3-11 R. S. Lees, M.D., Arteriosclerosis Center E17–421, Massachusetts Institute of Technology, Cambridge.

3-12 D. L. Rimoin, M.D., Ph.D., Harbor–UCLA Medical Center.

3-13 B. O. Hall, University of California, San Francisco.

Chapter 4

4-2 After E. W. Sinnott, L. C. Dunn, and T. Dobzhansky, Principles of Genetics (New York: McGraw-Hill, 1950).

4-4 A. Cerami, Ph.D., The Rockefeller University.

4-5 Adapted from L. L. Cavalli-Sforza, "The Genetics of Human Populations," in Scientific American September 1974, 231:80–89, Copyright © 1974 by Scientific American, Inc. All rights reserved.

4-7 The Field Museum of Natural History, Chicago.

4-8 After C. M. Woolf and F. C. Dukepoo, Science, 1969, 164:30–37. Copyright © 1969 by the American Association for the Advancement of Science.

4-9 After C. Stern, Principles of Human Genetics, W. H. Freeman, San Francisco, 1973.

4-17 R. S. Kucherlapati, Princeton University.

Chapter 5

5-2 Copyright © by Camera M.D. Studios, Inc., New York.

5-4 Pedigree after W. Bulloch and P. G. Fildes, The Treasury of Human Inheritance, 1911, 1, No. 490.

5-7 Culver Pictures.

5-10 (a, b) B. E. Schuh, Monmouth Medical Center, Long Branch, New Jersey.
(c, d) The National Foundation - March of Dimes.

5-11 Dr. Pat Calarco, San Francisco Medical Center, University of California.

5-12 (a) Dr. Pat Calarco, San Francisco Medical Center, University of California.
(b) Dr. A. T. Hertig and the Carnegie Institution of Washington, Dept. of Embryology, Davis Division.

5-13 (a) Carnegie Institution of Washington, Dept. of Embryology, Davis Division.
(b, c) Photos by M. Jacobson, D. Yeager, Copyright © Camera M.D. Studios, Inc., New York.

5-14 Wide World Photos.

5-17 Photograph: J. M. Opitz, M.D., Shodair Children's Hospital, Helena, Montana.

Karyotype: L. F. Meisner, Cytogenetics Unit, State Laboratory of Hygiene, University of Wisconsin, Madison.

Chapter 6

6-4 Photograph: J. M. Opitz, M.D., Shodair Children's Hospital, Helena, Montana.
Karyotype: C. M. Moore, M.D., Dept. of Pediatrics, University of Texas Medical School, Houston.

6-5 Photograph: The National Foundation - March of Dimes.
Karyotype: B. E. Schuh, Monmouth Medical Center, Long Branch, New Jersey.

6-6 M. W. Shaw, M.D., Anderson Hospital, The University of Texas at Houston.

6-7 (a, b, c) Child Development and Mental Retardation Center, University of Washington.
(d) Rainier School, Buckley, Washington.

6-8 Data from E. B. Hook and A. Lindsjö, American Journal of Human Genetics, 1978, 30:19–27 (Chicago: The University of Chicago Press, 1978).

6-9 Photograph: K. Taysi, M.D., St. Louis Children's Hospital, St. Louis, Missouri.
Karyotype: M. W. Shaw, M.D., Anderson Hospital, The University of Texas at Houston.

6-10 Photograph: K. Taysi, M.D., St. Louis Children's Hospital, St. Louis, Missouri.
Karyotype: M. W. Shaw, M.D., Anderson Hospital, The University of Texas at Houston.

6-11 The National Foundation - March of Dimes.

6-12 (a) B. E. Schuh, Monmouth Medical Center, Long Branch, New Jersey.
(b) M. W. Shaw, M.D., Anderson Hospital, The University of Texas at Houston.

Chapter 7

7-2 P. Roberts, Ph.D., Oregon State University.

7-3 Photograph: K. H. Rothfels, Ph.D., University of Toronto. Micrograph: E. Patau, University of Wisconsin, Madison.

7-4 (a, b) K. Taysi, M.D., St. Louis Children's Hospital, St. Louis, Missouri.

7-8 K. H. Rothfels, Ph.D., University of Toronto.

7-15 M. W. Shaw, M.D., Anderson Hospital, The University of Texas at Houston.

Chapter 8

8-10 J. A. Huberman, Roswell Park Memorial Institute, Buffalo, New York.

8-12 Concerted strand exchange and formation of Holliday structures by E. Coli RecA protein by C. DasGupta, A. M. Wu, R. Kahn, R. P. Cunningham, and C. M. Radding, in Cell (1981) 25:507–516. Copyright © 1981, MIT Press, Cambridge.

8-15 A. L. Olins, Ph.D., Oak Ridge Graduate School of Biomedical Sciences, University of Tennessee.

8-17 T. E. Rucinsky, Ph.D., Cancer Electron Microscope Facility, Washington University School of Medicine, St. Louis, Missouri.

8-21 Data from Y. Nishioka and P. Leder, Cell, 1979, 18:875–882. Copyright © 1979, MIT Press, Cambridge.

Chapter 9

9-4 W. R. Centerwall, M.D., Director of the Medical Genetics Service, School of Medicine and Medical Center, University of California, Davis.

9-9 O. L. Miller, Jr., and B. R. Beatty, Journal of Cellular Physiology, 1969, 74, Supplement 1:225–282 (New York: Alan R. Liss, Inc., 1969).

9-10 (a) Data from R. C. Dickson, J. Abelson, W. M. Barnes, and W. S. Reznikoff, Science, 1975, 187:27–35. Copyright © 1975 by the American Association for the Advancement of Science.
(b) Data from Y. Nishioka and P. Leder, Cell, 1979, 18:875–882; numbering of nucleotides is in original reports. Copyright © 1979, MIT Press, Cambridge.

9-11 Data from Y. Nishioka and P. Leder, Cell, 1979, 18:875–882. Copyright © 1979, MIT Press, Cambridge.

9-12 Partially after J. L. Sussman and S. H. Kim, Science, 1976, 192:853–858. Copyright © 1976 by the American Association for the Advancement of Science.

9-13 A. Rich, Massachusetts Institute of Technology, Cambridge.

9-16 Data from B. G. Barrell, G. M. Air, and C. A. Hutchison III, Nature, 1976, 264:34. Copyright © 1976, Macmillan Journals Ltd., London.

9-17 Data from Y. Nishioka and P. Leder, Cell, 1979, 18:875–882. Copyright © 1979, MIT Press, Cambridge.

9-19 T. E. Rucinsky, Ph.D., Cancer Electron Microscope Facility, Washington University School of Medicine, St. Louis, Missouri.

9-21 After E. A. Zimmer et al., Proceedings of the National Academy of Sciences, U.S.A., 1980, 77:2158.

Chapter 10

10-1 Based in part on E. R. Huehns, N. Dance, G. H. Beaven, F. Hecht, and A. G. Motulsky, Cold Spring Harbor Symposium on Quantitative Biology, 1964, 29:327–331.

10-2 N. J. Proudfoot et al., Science, 1980, 209:1329–1336. Copyright © 1980 by the American Association for the Advancement of Science. Data from L. H. T. Van de Ploeg and R. A. Flavell, Cell, 1980, 19:947–958. Copyright © 1980, MIT Press, Cambridge.

10-4 Data from M. R. Lauth et al., Cell, 1976, 7:67–74. Copyright © 1976, MIT Press, Cambridge.

10-5 J. Hicks et al., Nature, 1979, 282:478–483. Copyright © 1979, Macmillan Journals Ltd., London.

10-8 After K. Illmensee and P. C. Hoppe, Cell, 1981, 23:9–18. Copyright © 1981, MIT Press, Cambridge.

10-9 B. Mintz, The Institute for Cancer Research, Philadelphia.

10-13 P. Gruss et al., Proceedings of the National Academy of Sciences, U.S.A., 1979, 76:4317–4321. V. B. Reddy et al., Science, 1978, 200:494–502. Copyright © 1978 by the American Association for the Advancement of Science.

10-14 Data from C. Yanofsky, Nature, 1981, 289:751–758. Copyright © 1981, Macmillan Journals Ltd., London.

10-15 S. Horinouchi and B. Weisblum, Proceedings of the National Academy of Sciences, U.S.A., 1980, 77:7079–7083.

Chapter 11

11-1 M. W. Strickberger, Genetics, 2d ed. (New York: Macmillan, 1976); after J. V. Neel and W. J. Schull, Human Heredity (Chicago: The University of Chicago Press, 1954).

11-9 Data from C. A. Marotta et al., Journal of Biological Chemistry, 1977, 252:5040–5053.

11-13 E. Strobel, Ph.D., Purdue University.

11-15 Data from S. J. Rothstein et al., Cell, 1980, 19:795–805. Copyright © 1980, Macmillan Journals Ltd., London. And from S. J. Rothstein and W. S. Reznikoff, Cell, 1981, 23:191–199. Copyright © 1981, Macmillan Journals Ltd., London.

11-16 Cytogenetics Laboratory, University of California, San Francisco.

11-17 Based on J. Schultz, in B. M. Duggar (ed.), Biological Effects of Radiation (New York: McGraw-Hill, 1936).

11-19 L. Bergman and Associates.

11-21 B. N. Ames, Science, 1974, 204:587–593. Copyright © 1974 by the American Association for the Advancement of Science.

Chapter 12

12-1 B. Roizman, The University of Chicago.

12-2 R. C. Williams, Virus Laboratory, University of California, Berkeley.

12-3 J. D. Griffith, Stanford University.

12-4 From H. Dorn, Acta Geneticae Medicae Gemellologiae, 1959, 8:395–408 (New York: Alan R. Liss, Inc., 1959).

12-9 Reprinted from Natural History, 1973, Vol. 82, No. 1. Copyright © 1973, The American Museum of Natural History, New York.

12-11 R. D. Goldman, Carnegie-Mellon University.

12-12 R. C. Williams, Virus Laboratory, University of California, Berkeley.

12-14 From B. E. Griffin et al., in John Tooze (ed.), DNA Tumor Viruses (New York: Cold Spring Harbor Laboratory, 1980).

12-15 Data from J. M. Coffin, in J. R. Stephenson (ed.),

Molecular Biology of DNA Tumor Viruses (New York: Academic Press, 1980).

Chapter 13

13-1 Reprinted with the permission of E. Golub, Purdue University.

13-11 A. E. Mourant et al., The Distribution of the Human Blood Groups and Other Polymorphisms, 2d ed. (London: Oxford University Press, 1976).

13-12 From D. L. Hartl, 1980. Principles of Population Genetics, Fig. 20, p. 36 (Sunderland, Massachusetts: Sinauer Associates, 1980).

Chapter 14

14-1 Skulls redrawn from J. E. Cronin et al., Nature, 1981, 292:113–122. Copyright © 1981, Macmillan Journals Ltd., London.

14-2 Based on Cavalli-Sforza and W. F. Bodmer, The Genetics of Human Populations (San Francisco: W. H. Freeman, 1971).

14-3 Data from J. E. Cronin et al., Nature, 1981, 292:113–122. Copyright © 1981, Macmillan Journals Ltd., London.

14-4 Negative print of gel photo: J. Coyne, Ph.D., University of California, Davis, and D. Hickey, Harvard University.

14-5 Data from Y. W. Kan and A. M. Dozy, Science, 1980, 209:388–391. Copyright © 1980 by the American Association for the Advancement of Science.

14-8 Data from Cavalli-Sforza and W. F. Bodmer, The Genetics of Human Populations (San Francisco: W. H. Freeman, 1971).

14-11 From D. L. Hartl, Principles of Population Genetics (Sunderland, Massachusetts: Sinauer Associates, 1980).

Chapter 15

15-8 Data from Buri, Evolution, 1956, 10:367–402.

15-9 After D. L. Hartl, Principles of Population Genetics (Sunderland, Massachusetts: Sinauer Associates, 1980).

15-10 A. Jacquard, Institut National D'Etudes Demographiques, Paris.

15-11 From A. Jacquard, The Genetic Structure of Populations (New York: Springer-Verlag, 1974).

15-12 M. G. Baba et al., Journal of Molecular Evolution, 1981, 17:197–213 (New York: Springer-Verlag, 1981).

15-13 Human data from R. V. Eck and M. O. Dayhoff, Atlas of Protein Sequence and Structure, National Biomedical Research Foundation, Silver Springs, Maryland, 1966. Others from M. G. Baba et al., Journal of Molecular Evolution, 1981, 17:197–213 (New York: Springer-Verlag, 1981).

15-14 From W. C. Barker and M. O. Dayhoff. Reprinted from BioScience, 1980, 30:597. Copyright © 1980 by the American Institute of Biological Sciences. Reprinted with permission of the copyright holder.

15-15 Data from C. M. Rubin et al., Nature, 1980, 284:372–374. Copyright © 1980, Macmillan Journals Ltd., London.

15-16 M. Gō, Nature, 1981, 291:90-92. Copyright © 1981, Macmillan Journals Ltd., London.

Chapter 16

16-2 C. Stern, Human Heredity, 1970, 20:165–168. Copyright © 1970 by S. Karger AG, Basel.

16-4 H. Stein.

16-5 Data from G. A. Harrison, J. S. Wiener, J. M. Tanner, and N. A. Barnicot, Human Biology: An Introduction to Human Evolution, Variation and Growth (London: Oxford University Press, 1964).

16-8 Modified from C. Wills, Genetic Variability (Oxford: Clarendon Press, 1981); after data of M. N. Karn and L. S. Penrose, Annals of Human Genetics, 15:147–164, 1951.

16-11 Data from H. B. Newcombe, in M. Fishbein (ed.), Papers and Discussions of the Second International Conference on Congenital Malformations (New York: The International Medical Congress, Ltd., 1964).

16-12 Adapted from F. Vogel and A. G. Motulsky, Human Genetics: Problems and Approaches (New York: Springer-Verlag, 1979), based on data of R. M. Cooper and J. P. Zubeck, Canadian Journal of Psychiatry, 1958, 12:159–164.

16-13 H. F. Harlow, University of Wisconsin Primate Laboratory.

16-14 W. A. Kennedy et al., Monographs of the Society for Research in Child Development, 1963, 28, ser. 90. © The Society for Research in Child Development, Inc.

Index